1895	**Roentgen** discovers X rays.
1896	**Bequerel** discovers radioactivity.
1897	**Thomson** identifies cathode rays as negative corpuscles (electrons).
1900	**Planck** introduces the quantum idea.
1905	**Einstein** introduces the light corpuscle (photon) concept.
1905	**Einstein** advances the special theory of relativity.
1911	**Rutherford** reveals the nuclear atom.
1913	**Bohr** gives a quantum theory of the hydrogen atom.
1915	**Einstein** advances the general theory of relativity.
1923	**Compton's** experiments confirm the existence of the photon.
1924	**de Broglie** advances the wave theory of matter.
1925	**Goudsmit** and **Uhlenbeck** establish the spin of the electron.
1925	**Pauli** states the exclusion principle.
1925	**Davisson** and **Germer** and **Thomson** verify the wave nature of electrons.
1926	**Schrödinger** develops the wave theory of quantum mechanics.
1927	**Heisenberg** proposes the uncertainty principle.
1928	**Dirac** blends relativity and quantum mechanics in a theory of the electron.
1929	**Hubble** discovers the expanding universe.
1932	**Anderson** discovers antimatter in the form of the positron.
1932	**Chadwick** discovers the neutron.
1932	**Heisenberg** gives the neutron-proton explanation of nuclear structure.
1934	**Fermi** proposes a theory of the annihilation and creation of matter.
1938	**Meitner** and **Frisch** interpret results of **Hahn** and **Strassmann** as nuclear fission.
1939	**Bohr** and **Wheeler** give a detailed theory of nuclear fission.
1942	**Fermi** builds and operates the first nuclear reactor.
1945	**Oppenheimer's** Los Alamos team creates a nuclear explosion.
1947	**Bardeen** and **Brattain** develop the transistor.
1956	**Reines** and **Cowan** identify the antineutrino.
1957	**Feynman** and **Gell-Mann** account for all weak interactions with a "left-handed" neutrino.
1960	**Maiman** invents the laser.
1965	**Penzias** and **Wilson** discover background radiation in the universe left over from the Big Bang.
1967	**Bell** and **Hewish** discover pulsars, which are neutron stars.
1968	**Wheeler** names black holes.
1969	**Gell-Mann** suggests quarks as the building blocks of nucleons.
1969	First manned lunar landing.
1980	**Rubin** postulates that the universe contains much "dark matter."
1994	**Princeton Plasma Physics Laboratory** achieves more than 10 MW of fusion power.
1996	Discovery of evidence for possible life in Martian meteorite.

CONCEPTUAL
Physics

EIGHTH EDITION

CONCEPTUAL
Physics
EIGHTH EDITION

written and illustrated by

City College of San Francisco

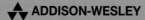 **ADDISON-WESLEY**

An imprint of Addison Wesley Longman, Inc.

Reading, Massachusetts • Menlo Park, California • New York • Harlow, England
Don Mills, Ontario • Sydney • Mexico City • Madrid • Amsterdam

Sponsoring Editor: Julia Berrisford
Developmental Editor: Amy J. Berk
Project Editor: Ann-Marie Sargent
Art Development Editor: Vita Jay/Heather Cooper
Design Administrator: Jess Schaal
Text and Cover Design: Kay Fulton
Cover Lettering: Ernie Brown
Cover Photo: Pekka Parviainen, Polar Image, Finland
Photo Research: Michelle E. Ryan
Production Administrator: Randee Wire
Marketing Manager: Brenda L. Bravener
Compositor: Interactive Composition Corporation
Printer and Binder: R. R. Donnelley & Sons Company
Cover Printer: The Lehigh Press, Inc.

For permission to use copyrighted material, grateful acknowledgment is made to the copyright holders on pp. 727–728 which are hereby made part of this copyright page.

Periodic Table Insert: John and Tracy Suchocki

Conceptual Physics, Eighth Edition

Copyright © 1998 Paul G. Hewitt

Library of Congress Cataloging-in-Publication Data

Hewitt, Paul G.
    Conceptual physics/written & illustrated by Paul G. Hewitt.--
8th ed.
        p.   cm.
    Includes bibliographical references and index.
    ISBN 0-321-00971-1
    1. Physics.   I. Title.
QC23.H56   1997
530--dc21   96-47678
                                                    CIP

1 2 3 4 5 6 7 8 9 10—DOW—00 99 98 97

To Kenneth W. Ford

eminent physicist,

great human being

Contents in Brief

Contents in Detail   viii
Conceptual Physics Photo Album   xiv
To the Student   xv
To the Instructor   xvi
Acknowledgments   xix

About Science   2

part 1   MECHANICS   17

2   Linear Motion   18
3   Nonlinear Motion   36
4   Newton's Laws of Motion   56
5   Momentum   82
6   Energy   100
7   Rotational Motion   118
8   Gravity   142
9   Satellite Motion   166

part 2   PROPERTIES OF MATTER   181

10   The Atomic Nature of Matter   182
11   Solids   197
12   Liquids   215
13   Gases and Plasmas   234

part 3   HEAT   255

14   Temperature, Heat, and Expansion   256
15   Heat Transfer   270
16   Change of Phase   288
17   Thermodynamics   304

part 4   SOUND   323

18   Vibrations and Waves   324

H - DO @ HOME, ASSIGNMENT

C - DO IN CLASS, LECT

L - DO MOSTLY IN LAB

X - SKIP

- COMBINE

CL - LAB, LECT CLASS ½ + ½

(C) IN CLASS BUT ONLY PARTS

L **19** Sound 342

20 Musical Sounds 359

part 5 ELECTRICITY AND MAGNETISM 371

CL **21** Electrostatics 372
CL **22** Electric Current 398
CL **23** Magnetism 418
CL **24** Electromagnetic Induction 436

part 6 LIGHT 451

C **25** Properties of Light 452
26 Color 470
LC **27** Reflection and Refraction 486
CL **28** Light Waves 516
L **29** Light Emission 541
C **30** Light Quanta 559

part 7 ATOMIC AND NUCLEAR PHYSICS 577

C **31** The Atom and the Quantum 578
CL **32** The Atomic Nucleus and Radioactivity 590
C **33** Nuclear Fission and Fusion 612

part 8 RELATIVITY 633

C **34** Special Theory of Relativity 634
35 General Theory of Relativity 668

Epilogue 682
Appendix A   Systems of Measurement 685
Appendix B   More About Motion 689
Appendix C   Graphing 693
Appendix D   More About Vectors 697
Appendix E   Exponential Growth and Doubling Time 703
Glossary 711
Photo Credits 727
Index 729

Contents in Detail

Contents in Detail   viii
Conceptual Physics Photo Album   xiv
To the Student   xv
To the Instructor   xvi
Acknowledgments   xix

1 **About Science**   **2**
Scientific Measurements   3
Size of the Earth   3
Size of the Moon   4
Distance to the Moon   6
The Distance to the Sun   6
The Size of the Sun   7

Mathematics—The Language of Science   8
The Scientific Method   9
The Scientific Attitude   9
Science, Art, and Religion   12
Science and Technology   13
Physics—The Basic Science   13
In Perspective   14

part 1 **MECHANICS**   **17**

2 **Linear Motion**   **18**
Aristotle on Motion   18
Copernicus and the Moving Earth   20
Galileo and the Leaning Tower   20
Galileo's Inclined Planes   21
Description of Motion   23
Speed   23
Velocity   25
Acceleration   26
Acceleration on Galileo's Inclined Planes   27
Free Fall   28
How Fast   28
How Far   29
Free Fall—How Quickly How Fast Changes   30

3 **Nonlinear Motion**   **36**
Motion Is Relative   36
Velocity—A Vector Quantity   37
Projectile Motion   40
Fast-Moving Projectiles—Satellites   46
Circular Motion   49

4 **Newton's Laws of Motion 56**
Newton's First Law of Motion   56
Mass   57
Newton's Second Law of Motion   60
When Acceleration Is Zero—Equilibrium   62

When Acceleration Is g—Free Fall   66
When Acceleration Is Less Than g—Nonfree Fall   67
Newton's Third Law of Motion   70
Summary of Newton's Three Laws   76

5 **Momentum**   **82**
Momentum   82
Impulse   83
Impulse Changes Momentum   84
Case 1: Increasing Momentum   85
Case 2: Decreasing Momentum over a Long Time   85
Case 3: Decreasing Momentum over a Short Time   86
Bouncing   87
Conservation of Momentum   88
Collisions   90
More Complicated Collisions   94

6 **Energy**   **100**
Work   101
Power   101
Mechanical Energy   102
Potential Energy   103
Kinetic Energy   105
Work-Energy Theorem   105

Conservation of Energy   106
Machines   108
Efficiency   109
Comparison of Kinetic Energy
  and Momentum   110
Energy for Life   113

7   Rotational Motion   118
Rotational Inertia   118
Torque   121
Center of Mass and Center of Gravity   123
Locating the Center of Gravity   124
Stability   125
Centripetal Force   128
Centrifugal Force   129
Centrifugal Force in a Rotating Frame   130
Simulated Gravity   131
Angular Momentum   133
Conservation of Angular Momentum   134

8   Gravity   142
Copernicus, Brahe, and Kepler   142
Kepler's Laws   143

Newton's Law of Universal Gravitation   144
The Universal Gravitational Constant, G   145
*Gravity and Distance: The Inverse-Square
  Law   147*
Weight and Weightlessness   149
Ocean Tides   150
Tides in the Earth and Atmosphere   153
Tides on the Moon   154
Gravitational Fields   154
The Gravitational Field Inside a Planet   155
Einstein's Theory of Gravitation   157
Black Holes   157
Universal Gravitation   159

9   Satellite Motion   166
The Falling Apple   166
The Falling Moon   167
Satellite Motion   168
Circular Orbits   168
Elliptical Orbits   170
Energy Conservation and Satellite Motion   172
Escape Speed   176

part 2   PROPERTIES OF MATTER   181

10   The Atomic Nature of Matter   182
Atoms   183
Molecules   186
Molecular and Atomic Masses   187
Elements, Compounds, and Mixtures   188
Atomic Structure   189
Antimatter   192
Phases of Matter   193

11   Solids   197
Müller's Micrograph   197
Crystal Structure   198
Density   200
Elasticity   201
Tension and Compression   203
Arches   205
Scaling   206

12   Liquids   215
Pressure   215
Pressure in a Liquid   216

Buoyancy   219
Archimedes' Principle   219
What Makes an Object Sink or Float?   222
Flotation   223
Pascal's Principle   225
Surface Tension   226
Capillarity   228

13   Gases and Plasmas   234
The Atmosphere   234
Atmospheric Pressure   236
Barometers   238
Boyle's Law   240
Buoyancy of Air   242
Bernoulli's Principle   243
Applications of Bernoulli's Principle   244
Plasma   248
Plasma in the Everyday World   248
Plasma Power   249

part 3   **HEAT**   **255**

14 Temperature, Heat, and Expansion 256
Temperature 256
Heat 258
Measurement of Heat 259
Specific Heat Capacity 260
Expansion 262
Expansion of Water 263

15 Heat Transfer 270
Conduction 270
Convection 272
Why Fast-Moving Molecules Rise in Air 273
Why Rising Air Cools 274
Radiation 275
Absorption and Reflection of Radiant Energy 276
Emission of Radiant Energy 277
Cooling at Night by Radiation 278
Newton's Law of Cooling 279
The Greenhouse Effect and Global Warming 280
Solar Power 281
The Thermos Bottle 283

16 Change of Phase 288
Evaporation 288
Condensation 290
Condensation in the Atmosphere 291
Fog and Clouds 292
Boiling 293
Geysers 293
Boiling Is a Cooling Process 294
Boiling and Freezing at the Same Time 294
Melting and Freezing 295
Regelation 295
Energy and Changes of Phase 296

17 Thermodynamics 304
Absolute Zero 304
Internal Energy 306
First Law of Themodynamics 306
Adiabatic Processes 308
Meteorology and the First Law 309
Second Law of Thermodynamics 312
Heat Engines 312
Order Tends to Disorder 316
Entropy 318

part 4   **SOUND**   **323**

18 Vibrations and Waves 324
Vibration of a Pendulum 325
Wave Description 325
Wave Motion 328
Wave Speed 328
Transverse Waves 329
Longitudinal Waves 330
Interference 331
Standing Waves 332
Doppler Effect 333
Bow Waves 335
Shock Waves 336

19 Sound 342
Origin of Sound 342
Nature of Sound in Air 343
Media That Transmit Sound 344
Speed of Sound in Air 345

Reflection of Sound 345
Refraction of Sound 346
Energy in Sound Waves 348
Forced Vibrations 348
Natural Frequency 348
Resonance 349
Interference 350
Beats 353

20 Musical Sounds 359
Noise Versus Music 359
Pitch 360
Sound Intensity and Loudness 360
Quality 362
Musical Instruments 363
Fourier Analysis 364
Compact Discs 367

| part 5 | ELECTRICITY AND MAGNETISM | 371 |

21 Electrostatics 372
Electrical Forces 373
Electric Charges 373
Conservation of Charge 374
Coulomb's Law 376
Conductors and Insulators 377
Semiconductors 378
Superconductors 378
Charging 379
Charging by Friction and Contact 379
Charging by Induction 379
Charge Polarization 381
Electric Field 383
Electric Shielding 386
Electric Potential 388
Electric Energy Storage 390
Van de Graaff Generator 392

22 Electric Current 398
Flow of Charge 398
Electric Current 399
Voltage Sources 400
Electrical Resistance 401
Ohm's Law 401
Ohm's Law and Electric Shock 402
Direct Current and Alternating Current 404
Converting ac to dc 405
Speed and Source of Electrons
   in a Circuit 405
Electric Power 407
Electric Curcuits 408
Series Circuits 408

Parallel Circuits 410
Parallel Circuits and Overloading 411
Safety Fuses 412

23 Magnetism 418
Magnetic Forces 419
Magnetic Poles 419
Magnetic Fields 420
Magnetic Domains 421
Electric Currents and Magnetic Fields 423
Electromagnets 424
Magnetic Force on Moving Charged
   Particles 424
Magnetic Force on Current-Carrying Wires 425
Electric Meters 426
Electric Motors 427
Earth's Magnetic Field 428
Biomagnetism 431

24 Electromagnetic Induction 436
Electromagnetic Induction 436
Faraday's Law 438
Generators and Alternating Current 439
Power Production 441
Turbogenerator Power 441
MHD Power 441
Transformers 442
Self-Induction 445
Power Transmission 446
Field Induction 447
In Perspective 447

| part 6 | LIGHT | 451 |

25 Properties of Light 452
Electromagnetic Waves 452
Electromagnetic Wave Velocity 453
The Electromagnetic Spectrum 454
Transparent Materials 456
Opaque Materials 458
Shadows 459
Seeing Light—The Eye 462

26 Color 470
Selective Reflection 471
Selective Transmission 472

Mixing Colored Light 473
Mixing Colored Pigments 475
Complementary Colors 477
Why the Sky Is Blue 478
Why Sunsets Are Red 480
Why Clouds Are White 481
Why Water Is Greenish Blue 482

27 Reflection and Refraction 486
Reflection 486
Principle of Least Time 487
Law of Reflection 487

Plane Mirrors 489
Diffuse Reflection 490
Refraction 492
Cause of Refraction 496
Dispersion 497
Rainbows 498
Total Internal Reflection 501
Lenses 505
Image Formation by a Lens 506
Lens Defects 509

28   Light Waves 516
Huygens' Principle 516
Diffraction 519
Interference 521
Single-Color Thin Film Interference 526
Interference Colors by Reflection from Thin Films 527
Polarization 529
Three-Dimensional Viewing 532
Holography 535

29   Light Emission 541
Excitation 542
Emission Spectra 544
Incandescence 546
Absorption Spectra 547
Fluorescence 548
Fluorescent Lamps 550
Phosphorescence 550
Lasers 551

30   Light Quanta 559
Birth of the Quantum Theory 560
Quantization and Planck's Constant 560
Photoelectric Effect 562
Wave-Particle Duality 564
Double-Slit Experiment 565
Particles as Waves: Electron Diffraction 567
Uncertainty Principle 569
Complementarity 572

part 7   ATOMIC AND NUCLEAR PHYSICS **577**

31   The Atom and the Quantum 578
Discovery of the Atomic Nucleus 578
Atomic Spectra: Clues to Atomic Structure 579
Bohr Model of the Atom 580
Relative Sizes of Atoms 582
Explanation of Quantized Energy Levels:
  Electron Waves 583
Quantum Mechanics 584
Correspondence Principle 587

32   The Atomic Nucleus and Radioactivity 590
X Rays and Radioactivity 590
Alpha, Beta, and Gamma Rays 591
The Nucleus 593
Isotopes 594
Why Atoms Are Radioactive 595
Half-Life 596

Radiation Detectors 597
Natural Transmutation of Elements 600
Artificial Transmutation of Elements 602
Radioactive Isotopes 603
Carbon Dating 604
Uranium Dating 606
Effects of Radiation on Humans 606

33   Nuclear Fission and Fusion 612
Nuclear Fission 613
Nuclear Fission Reactors 615
Plutonium 617
Breeder Reactors 618
Mass-Energy Equivalence 620
Nuclear Fusion 623
Controlling Fusion 625

| part 8 | RELATIVITY | 633 |

34 Special Theory of Relativity 634
Motion Is Relative 635
Michelson-Morley Experiment 635
Postulates of the Special Theory of
 Relativity 637
Simultaneity 638
Spacetime 639
Time Dilation 641
The Twin Trip 645
Addition of Velocities 651
Space Travel 652
Length Contraction 654

Relativistic Momentum 657
Mass, Energy, and $E = mc^2$ 658
The Correspondence Principle 661

35 General Theory of Relativity 668
Principle of Equivalence 669
Bending of Light by Gravity 670
Gravity and Time: Gravitational Red Shift 673
Gravity and Space: Motion of Mercury 675
Gravity, Space, and a New Geometry 676
Gravitational Waves 678
Newtonian and Einsteinian Gravitation 679

Epilogue 682
Appendix A   Systems of Measurement 685
United States Customary System 685
Système International 685
Meter 686
Kilogram 687
Second 687
Newton 687
Joule 687
Ampere 687
Kelvin 687
Area 688
Volume 688
Scientific Notation 688
Appendix B   More About Motion 689
Computing Velocity and Distance Traveled on an
 Inclined Plane 689
Computing Distance When Acceleration Is
 Constant 691

Appendix C   Graphing 693
Graphs—A Way to Express Quantitative
 Relationships 693
Cartesian Graphs 693
Slope and Area Under the Curve 695
Graphing with Conceptual Physics 695
Appendix D   More About Vectors 697
Vectors and Scalars 697
Adding Vectors 697
Finding Components of Vectors 698
Examples 698
Sailboats 700
Appendix E   Expontial Growth and Doubling Time 703

Glossary 711
Photo Credits 727
Index 729

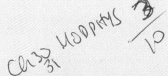

The Conceptual Physics Photo Album

Conceptual Physics is a very personalized book, which is reflected in the many photographs of family and friends that grace its pages. Foremost is the photo of glider pilot Ken Ford on page 352, to whom this edition is dedicated. Formerly the CEO of The American Institute of Physics, Ken currently teaches physics to high school students at Germantown Academy in Pennsylvania.

The opening photo of this book, page 1, is of another remarkable human being, Charlie Spiegel, with his great granddaughter Sarah Stafford, motivated by the chickie, ponder a positive future for humankind. Charlie is shown again on page 458. Part openers with the cartoon style blurbs about physics are of family and close friends. Part 1 on page 17 features dear little Andrea Wu; Part 2 on page 181 is of my grandson Alexander; Part 3 on page 255 is Terrence Jones, son of my niece Corine Jones; Part 4 on page 323 is my grandson Alexander again; Part 5 on page 371 is Sarah Yee, daughter of CCSF friend and colleague David Yee; Part 6 on page 451 is James Hewitt, my nephew Davey's son; Part 7 on page 577 is of Kerrine Lee, my neighbor in Hilo, Hawaii; and Part 8 on page 633 is W. J. Akil Marshall, son of artist and UH Hilo colleague, Mike Marshall.

To celebrate this eighth edition, we've included the photographs that comprised the covers of the previous seven editions. The first three editions all featured the micrograph that serves as the chapter opener on page 197. The micrograph was violet-magenta in the First Edition, teal green in the Second Edition, and warm brown in the Third Edition. The cover of the Fourth Edition was the photo of interference colors from gasoline on a wet street that now opens Chapter 28 on page 516. The Fifth Edition featured the photo of overlapping waves and reflections of the Exploratorium, which is now on page 324. The Don Briggs photo of water with reflections of Grand Canyon walls, now on page 215, graced the cover of the Sixth Edition, with particle tracks now on page 598 added for a physics-in-nature theme. The dramatic lightning bolt photo by Bill Bickel that made up the cover of the Seventh Edition now opens Chapter 21 on page 372.

Friends who teach physics begin on page 92, where CCSF colleagues Annette Rappleyea and Will Maynez demonstrate the air track designed and built by Will. To the left of Annette are my former teaching assistants, Glenda Gin and Ralph Diaz. To the right of Will are his former assistants, Ray Choi and Kumiko Furukawa. On page 111, demonstrating momentum conservation, are dear friends David Vasquez and Helen Yan. David teaches computer graphics in San Francisco and created the CD ROM materials that accompany the high-school version of *Conceptual Physics*. Helen appears again on page 277, posing with the box she first posed with for the Fifth Edition, when she was my teaching assistant. Now she is a physics instructor, is co-author of the *Next-Time Questions,* and did the hand lettering that adorns this book. On page 112 is friend and dedicated Sacramento physics teacher, Pablo Robinson, who risks his body for science on the bed of nails. Pablo is the author of the lab manual that accompanies this book. An old black and white photo of Pablo's children, David and Kristin, is shown on page 140. On page 230 we see dear friend Marshall Ellenstein, one of Chicago's finest physics teachers, walking barefoot on broken glass. Marshall, a long-time contributor to *Conceptual Physics,* edited the *Conceptual Physics Alive!* video series. University of Pennsylvania at Johnstown physics prof David Willey, a contributor to the test bank, walks barefoot on hot coals on page 299. In the photo on entropy on page 318, at the far left, is dedicated Washington DC physics teacher Howie Brand, a contributor to

the *Conceptual Physics Practice Book.* Howie's friendship goes back to undergraduate school days when he and I edited the college newspaper. CCSF friend and colleague Dave Wall is shown at the right in the photo of the laser on page 554. Dave's daughter, Ellender, who was shown as a little girl working with electronics in the Fourth Edition, now co-authors a physics text with her dad (that her "Uncle Paul" illustrated).

Family photos begin with the touching photo on page 75 of my brother Steve with daughter Gretchen on their coffee farm in Costa Rica. The lovely girl on page 184 is my daughter Leslie, now a geologist/geographer, teacher, and earth-science co-author with me on *Conceptual Physical Science*, Addison Wesley Longman, 1994. This photo of Leslie, sort of a trademark of *Conceptual Physics*, has been in all my books since the Third Edition, when black and white photos were standard fare. A more recent photo of Leslie, with husband Bob Abrams, is on page 554. My son Paul, U.S. Navy, is shown on pages 308, 309, and in more youthful days on page 116. His lovely wife, Ludmila, holds crossed Polaroids on page 532, and theirs and Alexander's dog Hanz pants on page 289. My late son James is shown on pages 82, 133, 361, and 508. He left me my first grandson, Manuel, who is shown on pages 205, 349, and 657. Manuel's grandmother, Millie Hewitt, bravely holds her hand above the active pressure cooker on page 274. Brother Dave (no, not a twin) and his wife Barbara demonstrate atmospheric pressure on page 239. Sister Marjorie Hewitt Suchocki, an author and theologian at a graduate school in the Claremont Colleges, illustrates reflection on page 489. Marjorie's son and my nephew John Suchocki (pronounced Su-hock'-ee, with silent c), my chemistry co-author of *Conceptual Physical Science*, walks fearlessly across hot coals on page 270. My songsters niece, Nancy Hewitt, appropriately opens the chapter on musical sounds on page 359 by singing a piece from her CD, *3 Wishes 4 Eve.* Marques Jones, the youngest son of niece Corine, is shown on page 286. The group listening to music on page 366 are part of the wedding party of nephew John and Tracy Suchocki; from left to right, Butch Orr, my niece Cathy Orr, bride and groom, niece Joan Lucas, sister Marjorie, Tracy's parents Sharon and David Hopwood, teacher friends Kellie and Mark Werkmeister, and myself.

Personal friends feature the Hu family, of Honolulu. Foremost is Meidor on pages 279, 352, and 393, who in addition to supplying photos of her family, took many others for this and the previous edition. Her mom Ping is on page 121, her dad Wai Inn on page 391, brother Tin Hoy on page 421, sister Mei Tuck on page 344, and uncle Chiu Man Wu, Andrea's dad, on page 289. Meidor and I share a common love for Beezes the bunny, on page 472. Close and dear friend Tim Gardner demonstrates Bernoulli's principle on page 244. Dear friend Tenny Lim draws her bow on page 103, as she has been doing since the Sixth Edition. New friend and former student Alexei Cogan demonstrates center of gravity on page 127. Another student friend, Cliff Braun, is at the left of the photo on page 133, with nephew Robert Baruffaldi at the right. Lifelong friend Paul Ryan drags his finger through molten lead on page 299. Another lifelong friend, Ernie Brown, the designer of the physics logo on this and previous editions, is shown to the right in the entropy photo on page 318, and in the same photo, standing on the porch (thanks to Adobe Photoshop!), is my wildlife biologist-photographer buddy from graduate school days, Jim Morgan.

The inclusion of these people who are so dear to me makes *Conceptual Physics* all the more my labor of love.

To the Student

You know you can't enjoy a game unless you know its rules; whether it's a ball game, a computer game, or simply a party game. Likewise, you can't fully appreciate your surroundings until you understand the rules of nature. Physics is about the rules of nature--so beautifully elegant that it can be neatly described mathematically. That's why many physics courses are treated as applied mathematics. But introductory physics that emphasizes computation misses something essential-- *comprehension* -- a gut feeling for the concepts. This book emphasizes comprehension before computation. We treat physics *conceptually*-- in down-to-earth *English*. You'll see the mathematical structure of physics in frequent equations, but more than being recipes for computation, you'll see the equations as *guides to thinking*.

I enjoy physics, and you will too--because you'll understand it. If you get hooked and take a follow-up course, *then* you can focus on mathematical problems. Go for comprehension of concepts now, and if computation follows, it will be with understanding.

Enjoy your physics!

Paul G. Hewitt

To the Instructor

Because physics is the basic science—the foundation of chemistry, biology, and all disciplines of science—it should be part of the educational mainstream for both science and nonscience students. Unfortunately, mathematical prerequisites often deter average nonscience students from an encounter with physics. For these students, the math requirement is an unfortunate roadblock, for, when the ideas of physics are presented conceptually and when equations are seen as guides to thinking rather than recipes for algebraic manipulation, our discipline is accessible to all students—with or without a math foundation. And, for students who will continue in the study of physics, recent compelling evidence indicates that physics should be first understood conceptually, before being used as a base for applied mathematics.

This book seeks to build that conceptual base. For the nonscience student, it is a base from which to view nature more perceptively—to see that a surprisingly few relationships make up its rules. For the science student, it is this and a springboard to a greater involvement in physics. A first-semester overview of Newtonian and modern physics will help to correct a missing essential in the science major's physics education: The practice of conceptualizing before calculating. For nonscience and science students alike, a conceptual way of looking at physics shapes analytical thinking.

This eighth edition of *Conceptual Physics* continues and builds upon the traditions established for it in earlier editions, with the goal of providing a comprehensive, accurate, and always interesting and enjoyable introduction to the concepts of physics.

Organization and Content

Following a sequence from classical mechanics to modern physics, the overall organization remains much the same as in previous editions. Based on extensive input from reviewers, however, every chapter within this organization has been substantially polished to ensure that the material is as clear as possible, accurate, and up-to-date. Key changes in content include the following.

- Chapter 1 kicks off with an analytical and interest-generating treatment of measurements of the earth, moon, and sun by the early Greeks.
- Circular motion is extended in Chapter 3 to include a fascinating application of $v = r\omega$—the tapered shape of railroad train wheels and how the shape keeps a train on a track.
- Static equilibrium is embellished in Chapter 4 for further vector development in the accompanying *Conceptual Physics Practice Book*.
- Chapter 8 treats the inverse-cube nature of tidal forces for the first time, which is also further developed in the *Practice Book*.
- There is new emphasis on the similarity of solar and terrestrial radiation in Chapter 15.
- To keep abreast of common usage, the term *phase* replaces *state* in this edition, and Chapter 16 is now titled "Change of Phase."
- Chapter 17 now includes information on OTEC, a contender for future power production.

- The topic of chaos is now treated in Chapter 31.
- Chapter 34 now treats relativistic momentum rather than relativistic mass.

Features

Key features of the Eighth Edition include the following.

- A conversational writing style aimed at engaging student interest.
- In-text questions that pose a question, with answers provided on the same page. These questions primarily reinforce ideas presented before the student moves further through the topics.
- Brief boxed biographies of prominent physicists.
- Brief boxes of physics sidelights in the high-interest topics of sports and technology (*new to this edition*).
- Extensive full-color line art and photo program, including Hewitt cartoons.

Other pedagogical features to assist the student are:

- *Summary of Terms* to help the student learn the language of physics.
- *Review Questions* at the end of each chapter. As the name implies, these are a simple review of the chapter material. Their purpose is to provide a structured way for the student to review the essentials of the chapter content. They are not meant to be challenging, as, in most cases, the answers can simply be looked up within the chapters. As in previous editions, the number of questions average about 30 to 40 per chapter.
- *Projects* that can be done in or out of class on topics likely to appeal to the student. Several new projects have been added in selected chapters.
- *Exercises,* unlike the Review Questions, stress thinking rather than mere recollection of information, and call for an understanding of the definitions, principles, and relationships of the chapter material. The purpose of the exercises is mainly to help the student apply the ideas of physics to familiar situations. *Exercises in this edition have undergone a major overhaul.* Whereas previous editions usually had fewer than 30 exercises per chapter, this edition has an average of 40 per chapter, with a cap of 50 in the meatier chapters. More new exercises are in the Instructor's Manual. *The net result is a 35% increase in the overall number of exercises in this edition.*
- *Problems,* which were first introduced in the seventh edition of *Conceptual Physics,* highlight concepts more clearly understood with numerical values and straightforward calculations. The calculations are to enhance learning of concepts, and not the other way around. The primary goal of the book is still to help students develop a gut feel for the concepts of physics in their everyday language, not to develop avid numerical problem solvers—to put comprehension before computation. Whereas there were relatively few problems in 22 chapters of the previous edition, there are *now nearly twice as many problems* in 31 of the 35 chapters. Only the most qualitative chapters, four in all, have no problem sets. To help ensure that the problem sets don't supplant the more important exercises (in this author's opinion), there are a maximum of ten problems in any particular chapter.
- *Fold-Out Periodic Table, which is new to this edition and included for quick and easy reference.*
- *Glossary* that contains definitions of many physics and science terms most likely to be unfamiliar to students.

- *Appendices* provide more information about measurement, motion, graphing, vectors, and exponential growth. *The appendix on graphing is new to this edition.*
- As in previous editions, units of measurement are not emphasized. When used, they are almost exclusively expressed in SI (exceptions include such units as calories, grams per centimeter cubed, and light-years). Mathematical derivations are avoided in the main body of the text and appear in footnotes or in the appendices.

Support Package

The overall *Conceptual Physics* "instructional package" provides complete supporting materials for the book for both students and faculty.

For the Student

- *Practicing Physics,* by Hewitt, which he considers his most creative work, guides students to a sometimes computational way of doing physics. It is a most unusual workbook, with a user-friendly tone that makes wide use of analogies and intriguing situations. It can be used in-class to develop concepts, or as an out-of-class tutorial. Learning is greatly increased when students do the step-by-step concept development in this book.
- *Conceptual Physics Laboratory Manual,* by Paul Robinson, contains interesting lab experiments, as well as a wide range of activities similar to the Projects in *Conceptual Physics.* These activities guide students to experience phenomena before they quantify the same phenomena in a follow-up laboratory experiment. An accompanying *Instructor's Manual* provides teaching tips and suggestions for how activities can effectively precede experiments.

For the Instructor

- *Instructor's Manual,* by Hewitt, is a very different instructor's manual containing a variety of sample syllabi, many lecture ideas and topics not treated in the book, as well as teaching tips and suggested step-by-step lectures and demonstrations. Answers to the Exercises and Problems are available to instructors in a format suitable for photocopying and posting for students to review. It also contains "Conceptual Physics Illustrations," which can help make your chalkboard presentation more interesting.
- *Next-Time Questions,* by Paul G. Hewitt and Helen Yan, provides cartooned questions and answers to delight and stimulate your students. Photocopy and post selected questions weekly in an area where students can ponder them, or make overhead transparencies of them to supplement your lectures.
- *Acetates*—A set of four-color acetates of figures and tables from the text, with teaching guide.
- *Printed and Computerized Test Bank,* with questions written by Hewitt, has been expanded, and is available for the IBM-PC or Macintosh computer.

Acknowledgments

I am enormously grateful to Ken Ford for checking every paragraph of this edition for accuracy, for supplying many timely suggestions, and for helping me revamp the end-of-chapter material with fresh exercises and problems. Many years ago I admired Ken's own books, one of which, *Basic Physics,* inspired me to first write *Conceptual Physics.* Today, I am honored that Ken has given so much of his time and energy in helping to make this edition my best ever. Errors invariably crop up after manuscript is submitted, so I take full responsibility for any errors that survive his scrutiny. This edition is appropriately dedicated to Ken Ford.

I am also grateful to Bosnian physicist, Josip Slisko, presently teaching at the Faculty of Physical and Mathematical Sciences at the University of Puebla in Mexico, for his insightful pedagogical critique of the book. A stickler for accurate wording, Josip's many suggestions have produced a clearer-reading text.

I am thankful to Charlie Spiegel, who as in previous editions, made many helpful suggestions. Charlie passed away before he could see the results of his efforts in this edition. He is shown in the opening photo with his great granddaughter on page 1. We always referred to his ideas being formulated as "fish in the frying pan," and in this spirit, Charlie poses with a frying pan for Figure 25.9 on page 458. I am thankful to another friend, now departed, Lester Hirsch, of California State University in Los Angeles. It was Lester who convinced me to add problem sets to the previous edition, and encouraged their expansion in this edition. Both Charlie and Lester led long, full, and productive lives. Both are much missed.

For valued suggestions, I thank my friends Howie Brand, Marshall Ellenstein, John Hubisz, and Pablo Robinson. For input from my City College of San Francisco friends and colleagues, I am grateful to Jim Court, Jerry Hosken, Dack Lee, Chelcie Liu, Will Maynez, Annette Rappleyea, Dave Wall, Norman Whitlatch, and David Yee. Also in San Francisco, I am grateful to my Exploratorium friends and colleagues, Judith Brand, Paul Doherty, Gabe Espinda, and Ron Hipschman. For input from my University of Hawaii at Hilo friends and colleagues, I thank Richard Crowe, George Curtis, Marlene Hapai, Suk Hwang, and Walter Steiger. For information on OTEC, I am indebted to Tom Daniel and Adrianne Greenlees of the Natural Energy Laboratory of Hawaii Authority. Special thanks to nephew, John Suchocki for all his work on the periodic table.

For critiquing the previous edition and suggesting many valuable improvements, I thank Chuck Hatchett, my former student at City College. Students at the University of Hawaii at Hilo who also made suggestions include Ana Churchill, Jocelyne Comstock, John F. Gornick, Gretchen Grove, Cara Hafer, Susan Rezentes, Nancy Snedecor, and Carol Tauriainen-Ernst. I'm grateful to Steve Moser for contributing several creative exercises to this edition. For other creative ideas, I thank Bob Becker, Ken Ganezer, and Henry A. Garon. Thanks also to Donald P. Manuel, Jr. and Bill Sones for bringing errors in the previous edition to my attention.

I am grateful to my sister, Marjorie Hewitt Suchocki, for valued philosophical suggestions. I thank my pen pal, Marilyn Hromatko, for her insightful ideas. Thanks to dear friend Huey Dan Johnson for global environmental input. Appreciation goes to

Cedric Linder and Greg Hillhouse for their contributions to the *Practice Book,* and for extensively testing the effectiveness of *Conceptual Physics* for mainstream and service physics courses in South African universities (*The Physics Teacher,* 34(6): 332–338, 1996).

For photos, I thank my brother Dave, my son Paul, Milo Patterson, and David Willey. For photo-computer rendering, I'm grateful to Marissa Kunz. For proofreading, I thank Danny Franks. And for help with the test bank, I thank Lonnie Grimes and David Willey.

I remain grateful to the authors of books that initially served as influences and references many years ago: Theodore Ashford, *From Atoms to Stars;* Albert Baez, *The New College Physics—A Spiral Approach;* John N. Cooper and Alpheus Smith, *Elements of Physics;* Richard P. Feynman, *The Feynman Lectures on Physics;* Ken Ford, *Basic Physics;* Eric Rogers, *Physics for the Inquiring Mind;* Alexander Taffel, *Physics: Its Methods and Meanings;* UNESCO, *700 Science Experiments for Everyone;* and Harvey E. White, *Descriptive College Physics.*

I am particularly grateful to Meidor Hu for her assistance in all phases of book and ancillary preparation, and for her extensive photography. I thank Jay Pasachoff for directing me to Pekka Parviainen in Turku, Finland, who supplied the cover photographs of the aurora borealis. Thanks go to my life-long friend, Ernie Brown, for designing the physics logo. I thank dear friend, Helen Yan, for her hand lettering that adorns the book and the *Next-Time Questions,* which she co-authored.

I extend my sincere thanks for their suggestions on improvements for this edition to the following colleagues.

Royal Albridge, Vanderbilt University

A. F. Barghouty, Roanoke College

Elizabeth Behrman, Wichita State University

Terry M. Carlton, Stephen F. Austin University

David E. Clark, University of Maine

Jefferson Collier, Bismark State College

James R. Crawford, Southwest Texas State University

Yuan Ha, Temple University

Teresa Hein, American University

Robert Hudson, Roanoke College

P. Jena, Virginia Commonwealth University

Ronald H. Kaitchuck, Ball State University

R. L. Kernell, Old Dominion University

Jim Knowles, North Lake College

Allen Miller, Syracuse University

Fred Otto, Winona State University

Robert G. Packard, Baylor University

Joseph Pizzo, Lamar University

David Raffaelle, Glendale Community College

Phyllis Salmons, Embry-Riddle Aeronautical University

Mikolaj Sawicki, John A. Logan College

Julius Sigler, Lynchburg College

Barbara Z. Thomas, Brevard Community College

David Wright, Tidewater Community College

Thanks also go to Robert Hudson (Roanoke College), Mark Buesing (Libertyville High School), and Norman M. Gelfand (Fermilab), who did a final review of the material once the book was in page proofs to ensure accuracy.

Finally, I am grateful to the staff at the Glenview, Illinois, office of Addison Wesley Longman for their dedication to this edition. A special note of appreciation is due my sponsoring editor, Julia Berrisford, for her professional concern and assistance. Special thanks to development editor, Amy J. Berk, whose many suggestions and uncanny knack for catching glitches still amazes me. I also deeply appreciate the excellent work of my project editor, Ann-Marie Sargent. I am thankful to Vita Jay and Heather Cooper, who managed the complex art program; to Jess Schaal, who managed the design of the book and its cover; and to Michelle E. Ryan, who researched the new photographs. For high-quality copyediting, I'm particularly grateful to Suzanne Lyons, whose talent for clear expression is enhanced by her background in physics (U.C. Berkeley) and graduate education (Stanford). I've been blessed with a first-rate book team!

Paul G. Hewitt
Hilo, Hawaii

1

· · · · · · · · · · · · ·

About Science

The earth's rotation makes the stars appear to move across the sky in this time-exposure photograph.

First of all, science is the body of knowledge about nature that represents the collective efforts, findings, insights, and wisdom of the human race. Second, science is a human activity, with the functions of discovering the orderliness of nature and the causes that govern this order. Science had its beginnings before recorded history, when people first discovered regularities and relationships in nature, such as star patterns in the night sky and weather patterns—when the rainy season started or the days grew longer. From these regularities, people learned to make predictions that gave them some control over their surroundings.

Science made great headway in Greece in the third and fourth centuries BC. Subsequently, it spread throughout the Mediterranean world. Scientific advances came to a halt in Europe when the Roman Empire fell in the fifth century AD. Barbarian hordes destroyed almost everything in their path as they overran Europe and ushered in the Dark Ages. During this time the Chinese were charting the stars and the planets and Arab nations were developing new mathematics. Greek science was reintroduced to Europe by Islamic influences that penetrated into Spain during the tenth, eleventh, and twelfth centuries. Universities emerged in Europe in the thirteenth century, when scholars developed the philosophy that reason is compatible with faith. The fifteenth century saw art and science beautifully blended by Leonardo da Vinci. Scientific thought was furthered in the sixteenth century with the advent of the printing press and increased communication.

The sixteenth-century Polish astronomer Copernicus caused great controversy when he published a book proposing that the sun is stationary and that the earth revolves around it. These ideas conflicted with the popular view that the earth was the center of the universe. They also conflicted with Church teachings and were banned for 200 years. The Italian physicist Galileo was arrested for popularizing the Copernican theory and for his other contributions to scientific thought. Yet a century later, Copernican advocates were respected.

This kind of cycle happens age after age. In the early 1800s geologists met with violent condemnation because they differed with the Genesis account of creation. Later in the same century geology was accepted, but theories of evolution were condemned and the teaching of them forbidden. Every age has its groups of intellectual rebels who are persecuted, condemned, or suppressed at the time but who later seem harmless and often essential to the elevation of human conditions. "At every crossway on the road that leads to the future, each progressive spirit is opposed by a thousand men appointed to guard the past."*

Scientific Measurements

Measurements are a hallmark of good science. How much you know about something is often related to how well you can measure it. This was well put by the famous physicist Lord Kelvin in the last century: "I often say that when you can measure something and express it in numbers, you know something about it. When you cannot measure it, when you cannot express it in numbers, your knowledge is of a meager and unsatisfactory kind. It may be the beginning of knowledge, but you have scarcely in your thoughts advanced to the stage of science, whatever it may be." Scientific measurements are not something new but go back to ancient times. In the third century BC, for example, fairly accurate measurements were made of the sizes of the earth, moon, sun, and the distances between them.

Size of the Earth

The size of the earth was first measured in Egypt by the geographer and mathematician Eratosthenes about 235 BC.[†] Eratosthenes calculated the circumference of the earth in the following way. He knew that the sun is highest in the sky at noon on June 22, the summer solstice. At this time a vertical stick casts its shortest shadow. If the sun is directly overhead, a vertical stick casts no shadow at all. This occurs in Syene, a city south of Alexandria (where the Aswan Dam stands today). Eratosthenes learned that the sun was directly overhead at Syene from library information, which reported that at this

---

* From Count Maurice Maeterlinck's "Our Social Duty."

[†] Eratosthenes was second chief librarian at the University of Alexandria in Egypt, founded by Alexander the Great. Eratosthenes was one of the foremost scholars of his time and wrote on philosophy and scientific and literary matters. As a mathematician he invented a method for finding prime numbers. His reputation among his contemporaries was immense—Archimedes dedicated a book to him. As a geographer he wrote *Geography,* the first book to give geography a mathematical basis and to treat the earth as a globe divided into Frigid, Temperate and Torrid zones. It long remained a standard work, and was used a century later by Julius Caesar. Eratosthenes spent most of his life in Alexandria and died there in 195 BC.

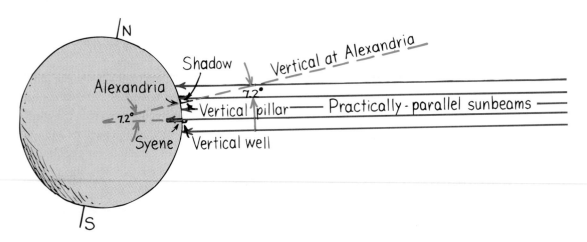

FIGURE 1.1 When the sun is directly overhead at Syene, it is not directly overhead in Alexandria, 800 km north. When the sun's rays shine directly down a vertical well in Syene, they cast a shadow of a vertical pillar in Alexandria. The verticals at both locations extend to the center of the earth, and make the same angle that the sun's rays make with the pillar at Alexandria. Eratosthenes measured this angle to be $\frac{1}{50}$ of a complete circle. Therefore the distance between Alexandria and Syene is $\frac{1}{50}$ the earth's circumference. (Equivalently, the shadow cast by the pillar is $\frac{1}{8}$ the height of the pillar, which means the distance between the locations is $\frac{1}{8}$ the earth's radius.)

unique time sunlight shines directly down a deep well in Syene and is reflected back up again. Eratosthenes reasoned that if the sun's rays were extended into the earth at this point, they would pass through the center. Likewise, a vertical line extended into the earth at Alexandria (or anywhere else) would also pass through the earth's center. In Alexandria at this time, a vertical stick casts a shadow—which is very easy to measure.

At noon on June 22, Eratosthenes measured the shadow cast by a vertical pillar in Alexandria and found it to be $\frac{1}{8}$ the height of the pillar (Figure 1.1). This corresponds to a 7.2° angle between the sun's rays and the vertical pillar. Since 7.2° is 7.2/360, or 1/50 of a circle, Eratosthenes reasoned the distance between Alexandria and Syene must be $\frac{1}{50}$ the circumference of the earth. Thus the circumference of the earth becomes 50 times the distance between these two cities. This distance, quite flat and frequently traveled, was measured by surveyors to be 5000 stadia (800 kilometers). So Eratosthenes calculated the earth's circumference to be 50 × 5000 stadia = 250,000 stadia. This is within 5% of the currently accepted value of the earth's circumference.

We get the same result by bypassing degrees altogether and comparing the length of the shadow cast by the pillar to the height of the pillar. Geometrical reasoning shows to a close approximation, the ratio of *shadow length/pillar height* is the same ratio as the *distance between Alexandria and Syene/earth's radius*. So just as the pillar is 8 times greater than its shadow, the radius of the earth must be 8 times the distance between Alexandria and Syene.

Since the circumference of a circle is 2π times its radius ($C = 2\pi r$), the earth's radius is simply its circumference divided by 2π. In modern units the earth's radius is 6370 kilometers, and its circumference is 40,000 km.

Size of the Moon

Aristarchus was perhaps the first to suggest that the earth spins on an axis daily, which accounts for the daily motion of the stars. He also hypothesized that the earth moves

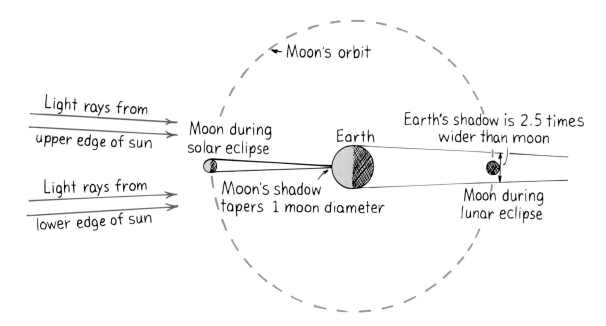

FIGURE 1.2   During a lunar eclipse the earth's shadow is observed to be 2.5 times as wide as the moon's diameter. Because of the sun's large size, the earth's shadow must taper. The amount of taper is evident during a solar eclipse, where the moon's shadow tapers a whole moon diameter from moon to earth. So the earth's shadow tapers the same amount in the same distance. Therefore the earth diameter must be 3.5 moon diameters. (Not to scale.)

around the sun in a yearly orbit, and that the other planets do likewise.* He correctly measured the moon's diameter and its distance from the earth. He did all this in about 240 BC, seventeen centuries before his findings became fully accepted.

Aristarchus compared the size of the moon with the size of the earth by watching an eclipse of the moon. The earth, like any body in sunlight, casts a shadow. An eclipse of the moon is simply the event wherein the moon passes into this shadow. Aristarchus carefully studied this event and found that the width of the earth's shadow out at the moon was 2.5 moon diameters. This would seem to indicate that the moon's diameter is 2.5 times smaller than the earth's. But the earth's shadow tapers, as evidenced during a solar eclipse (Figure 1.2 shows this in exaggerated scale). At that time the earth intercepts the moon's shadow—but just barely. The moon's shadow tapers almost to a point at the earth's surface, evidence that the taper of the moon's shadow at this distance is one moon diameter. So during a lunar eclipse the earth's shadow, covering the same distance, must also taper one moon diameter. Taking the tapering of the sun's rays into account, the earth's diameter must be $(2.5 + 1)$ times the moon's diameter. In this way Aristarchus showed that the moon's diameter is $\frac{1}{3.5}$ that of the earth's. The presently accepted diameter of the moon is 3640 km, within 5% of the value calculated by Aristarchus.

FIGURE 1.3   Correct scale of solar and lunar eclipses, which shows why eclipses are rare. (They are even rarer because the moon's orbit is tilted about 5° from the plane of the earth's orbit about the sun.)

* Aristarchus was unsure of his heliocentric hypothesis, likely because the earth's seasons are not equal in length, and this doesn't support the idea that the earth circles the sun. More important, it was noted that the moon's distance from earth varies, clear evidence that it does not perfectly circle the earth. If the moon does not follow a circular path about the earth, it was hard for Aristarchus to argue that the earth follows a circular path about the sun. The explanation, the elliptical paths of planets, was not discovered until centuries later by Kepler. In the meantime, the epicycles proposed by other astronomers accounted for these discrepancies. It is interesting to speculate about the course of astronomy if the moon didn't exist. Its irregular orbit would not have contributed to the early discrediting of the heliocentric theory, which may have taken hold centuries earlier!

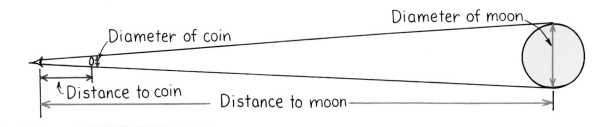

$$\frac{\text{Coin diameter}}{\text{Coin distance}} = \frac{\text{Moon diameter}}{\text{Moon distance}} = \frac{1}{110}$$

FIGURE 1.4   An exercise in ratios: When the coin barely "eclipses" the moon, then the diameter of the coin compared to the distance between you and the coin is equal to the diameter of the moon compared to the distance between you and the moon (not to scale here). Measurements give a ratio of $\frac{1}{110}$ for both.

Distance to the Moon

Tape a small coin like a dime to a window and view it with one eye so that it just blocks out the full moon. This occurs when your eye is about 110 coin diameters away. Then the ratio of *coin diameter/coin distance* is about $\frac{1}{110}$. Geometrical reasoning of similar triangles shows this is also the ratio of *moon diameter/moon distance* (Figure 1.4). So the distance to the moon is 110 times the moon's diameter. The early Greeks knew this. Aristarchus' measurement of the moon's diameter was all that was needed to calculate the earth-moon distance. So the early Greeks knew both the size of the moon and its distance from earth.

With this information, Aristarchus made a measurement of the earth-sun distance.

The Distance to the Sun

Repeat the coin-on-the-window-and-moon exercise for the sun and guess what: The ratio of *sun diameter/sun distance* is also $\frac{1}{110}$. This is because the sun and moon both have the same size to the eye. They both taper the same angle (about 0.5°). So although the ratio of diameter to distance was known to the early Greeks, diameter alone or distance alone would have to be determined by some other means. Aristarchus found a means of doing this and made a rough estimate. Here's what he did.

Aristarchus watched for the phase of the moon when it was *exactly* half full, with the sun still visible in the sky. Then the sunlight must be falling on the moon at right angles to his line of sight. The lines between earth and the moon, between earth and the sun, and between the moon and the sun form a right triangle (Figure 1.5).

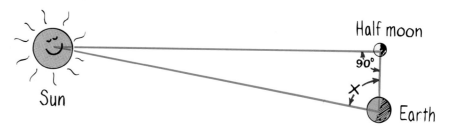

FIGURE 1.5 When the moon appears exactly half full, the sun, moon, and earth form a right triangle (not to scale). The hypotenuse is the earth-sun distance. By simple trigonometry, the hypotenuse of a right triangle can be found if you know the value of either nonright angle and the length of one side. The earth-moon distance is a known side. Measure angle X and you can calculate the earth-sun distance.

A rule of trigonometry states that if you know all the angles in a right triangle plus the length of any one of its sides, you can calculate the length of any other side. Aristarchus knew the distance from earth to moon. At the time of the half-moon he also knew one of the angles, 90°. All he had to do was measure the second angle between the line of sight to the moon and the line of sight to the sun. Then the third angle, a very small one, is 180° minus the sum of the first two angles (the sum of the angles in any triangle = 180°).

Measuring the angle between the lines of the sight to the moon and sun is difficult to do without a modern transit. For one thing, both the sun and moon are not points, but are relatively big. Aristarchus had to sight on their centers (or either edge) and measure the angle between—quite large, almost a right angle itself! By modern-day measure, his measurement was very crude. He measured 87°, while the true value is 89.8°. He figured the sun to be about 20 times the moon's distance, when in fact it is about 400 times as distant. So although his method was ingenious, his measurements were not. Perhaps Aristarchus found it difficult to believe the sun is so far away, and he erred on the nearer side. We don't know.

Today we know the sun to be an average of 150,000,000 km away. It is somewhat closer in December (147,000,000 km), and somewhat farther in June (152,000,000 km).

The Size of the Sun

Once the distance to the sun is known, the $\frac{1}{110}$ ratio of *diameter/distance* enables a measure of the sun's diameter. Another way to measure the $\frac{1}{110}$ ratio, besides the method of Figure 1.4, is to measure the diameter of the sun's image cast through a pinhole opening. You should try this. Poke a hole in a sheet of opaque cardboard and let sunlight shine on it. The round image that is cast on a surface below is actually an image of the sun. You'll see that the size of the image does not depend on the size of the pinhole, but on how far away the pinhole is from the image. Bigger holes make brighter images, not bigger ones. Of course if the hole is very big, no image is formed. Careful measurement will show the ratio of image size to pinhole distance is $\frac{1}{110}$—the same as the ratio of *sun diameter/sun-earth distance* (Figure 1.6).

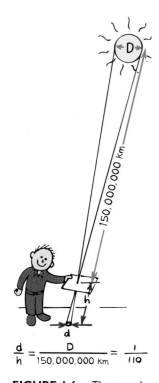

$$\frac{d}{h} = \frac{D}{150,000,000 \text{ km}} = \frac{1}{110}$$

FIGURE 1.6 The round spot of light cast by the pinhole is an image of the sun. Its *diameter/distance* ratio is the same as the *sun diameter/sun's distance* ratio: $\frac{1}{110}$. The sun's diameter is $\frac{1}{110}$ its distance from earth.

FIGURE 1.7 Renoir accurately painted the spots of sunlight on her dress—images of the sun cast by relatively small openings in the leaves above.

FIGURE 1.8 The crescent-shaped spots of sunlight are images of the sun when it is partially eclipsed.

Interestingly, at the time of a partial solar eclipse, the image cast by the pinhole will be a crescent shape—the same as that of the partially covered sun! This provides an interesting way to view a partial eclipse without actually looking at the sun.

Have you noticed that the spots of sunlight you see on the ground beneath trees are perfectly round when the sun is overhead, and spread into ellipses when the sun is low in the sky? These are pinhole images of the sun, where light shines through openings in the leaves that are small compared to the distance to the ground below. A round spot 10 cm in diameter is cast by an opening that is 110×10 cm above ground. Tall trees make large images; short trees make small images. And at the time of a partial solar eclipse, the images are crescents (Figure 1.8).

Mathematics— The Language of Science

Science and human conditions advanced dramatically after the discovery almost four centuries ago that nature could be analyzed and described mathematically. When the ideas of science are expressed in mathematical terms, they are unambiguous. They don't have the double meanings that so often confuse the discussion of ideas expressed in common language. When findings in nature are expressed mathematically, they are easier to verify or disprove by experiment. The abstract math developed by mathematicians is often years later found to be the exact language by which nature can be described. The methods of mathematics and experimentation led to enormous success in science.*

---

* The mathematical structure of physics is evident in the equations you will encounter throughout this book. You will see that equations are simply shortcut expressions of relationships that can be expressed in words. The focus of this book, however, is on comprehending concepts—in familiar English. Equations are compact statements that help guide thinking, and are not used in this book as recipes for mathematical problem solving. A premature effort at mathematical problem solving too often obscures the physics, so emphasis on solving math problems is best postponed until after you understand the concepts—perhaps in a follow-up course. Note the small number of problems at the ends of the chapters in this book, compared to the number of exercises. Conceptual physics puts comprehension comfortably before computation.

The Scientific Method

The Italian physicist Galileo Galilei and the English philosopher Francis Bacon are usually credited as being the principal founders of the **scientific method**—a method that is extremely effective in gaining, organizing, and applying new knowledge. This method, introduced in the sixteenth century, is essentially as follows:

1. Recognize a problem.
2. Make an educated guess—a *hypothesis*.
3. Predict the consequences of the hypothesis.
4. Perform experiments to test predictions.
5. Formulate the simplest general rule that organizes the three main ingredients—hypothesis, prediction, experimental outcome—into a *theory*.

Although this cookbook method has a certain appeal, it has not always been the key to the discoveries and advances in science. Many cases of trial and error, experimentation without guessing, and just plain accidental discovery have accounted for much of the progress in science. The success of science has more to do with an attitude common to scientists than with a particular method. This attitude is one of inquiry, observation, experimentation, and humility.

The Scientific Attitude

It is common to think of a fact as something that is unchanging and absolute. But in science, a **fact** is generally a close agreement by competent observers of a series of observations of the same phenomenon. For example, though it was once considered a fact that the universe is unchanging and permanent, today it is a scientific fact that the universe is expanding and evolving. A scientific **hypothesis**, on the other hand, is an educated guess that is only considered factual after it has been demonstrated by experiments. When a hypothesis has been tested over and over again and has not been contradicted, it may become known as a **law** or *principle*.

If a scientist finds evidence that contradicts a hypothesis, law, or principle, then, in the scientific spirit it must be changed or abandoned. Revision is called for regardless of the reputation or authority of the persons advocating the belief (unless the new evidence turns out to be wrong—which sometimes happens). For example, the greatly respected Greek philosopher Aristotle (384–322 BC) claimed that an object falls at a speed proportional to its weight. This false idea was held to be true for nearly 2000 years because of Aristotle's compelling authority. In the scientific spirit, however, a single verifiable experiment to the contrary outweighs any authority, regardless of reputation or the number of followers or advocates. In modern science, argument by appeal to authority has little value.

Scientists must accept their experimental findings even when they would like them to be different. They must strive to distinguish between what they see and what they wish to see, for scientists, like most people, have a vast capacity for fooling themselves.* People have always tended to adopt general rules, beliefs, creeds, ideas, and hypotheses without thoroughly questioning their validity. Further, ideas are often retained long after they have been shown to be meaningless, false, or at least questionable. The most widespread assumptions are often the least questioned. Most often, when an idea is adopted, particular attention is given to cases that seem to support it, while cases that seem to refute it are distorted, belittled, or ignored.

---

* In your education it is not enough to be aware that other people may try to fool you; it is more important to be aware of your own tendency to fool yourself.

Scientists use the word *theory* in a way that differs from its usage in everyday speech. In everyday speech a theory is no different from a hypothesis—a supposition that has not been verified. A scientific **theory**, on the other hand, is a synthesis of a large body of information that encompasses well-tested and verified hypotheses about certain aspects of the natural world. Physicists, for example, speak of the theory of the atom, chemists speak of the molecular theory, and biologists speak of the cell theory.

The theories of science are not fixed, but undergo change. Scientific theories evolve as they go through stages of redefinition and refinement. During the past hundred years, for example, the theory of the atom has been repeatedly refined as new evidence of atomic behavior has been gathered. Similarly, chemists have refined their view of the way molecules bond together, and biologists have refined the cell theory. The refinement of theories is a strength of science, not a weakness. Many people feel that it is a sign of weakness to change their minds. Competent scientists must be experts at changing their minds. They change their minds, however, only when confronted with solid experimental evidence to the contrary or when a conceptually simpler hypothesis forces them to a new point of view. More important than defending beliefs is improving them. Better hypotheses are made by those who are honest in the face of experimental evidence.

Facts are revisable data about the world.

Theories interpret facts.

Away from their profession, scientists are inherently no more honest or ethical than most other people. But in their profession they work in an arena that puts a high premium on honesty. The cardinal rule in science is that all hypotheses must be testable—they must be susceptible, at least in principle, to being shown *wrong*. In science, it is more important that there be a means of proving an idea wrong than that there be a means of proving it right. This is a major factor that distinguishes science from nonscience. At first this may seem strange, for when we wonder about most things, we concern ourselves with ways of finding out whether they are true. Scientific hypotheses are different. In fact, if you want to distinguish whether a hypothesis is scientific or not, look to see if there is a test for proving it wrong. If there is no test for its possible wrongness, then the hypothesis is not scientific. Albert Einstein put it well when he stated, "No number of experiments can prove me right; a single experiment can prove me wrong."

Consider biologist Charles Darwin's hypothesis that organisms evolve from simpler to more complex life forms. This could be proved wrong if paleontologists found that more complex forms of life appeared before their simpler counterparts. Einstein hypothesized that light is bent by gravity. This might be proved wrong if starlight that grazed the sun and could be seen during a solar eclipse were not deflected from its normal path. As it turns out, less complex life forms are found to precede their more complex counterparts and starlight is found to bend as it passes close to the sun, which support the claims. If and when a hypothesis or scientific claim is confirmed, it is regarded as useful and a stepping stone to additional knowledge.

Consider the hypothesis "The alignment of planets in the sky determines the best time for making decisions." Many people believe it, but this hypothesis is not scientific. It cannot be proven wrong, nor can it be proven right. It is *speculation*. Likewise, the hypothesis "Intelligent life exists on other planets somewhere in the universe" is not scientific. Although it can be proven correct by the verification of a single instance of intelligent life existing elsewhere in the universe, there is no way to prove it wrong if no life is ever found. If we searched the far reaches of the universe for eons and found no life, we would not prove that life doesn't exist "around the next corner." A hypothesis that is capable of being proved right but not capable of being proved wrong is not a scientific hypothesis. Many such statements are quite reasonable and useful, but they lie outside the domain of science.

None of us has the time, energy, or resources to test every idea, so most of the time we take somebody's word. How do we know whose word to take? To reduce the likelihood of error, scientists accept only the word of those whose ideas, theories, and findings are testable—if not in practice, at least in principle. Speculations that cannot be tested are regarded as "unscientific." This has the long-run effect of compelling honesty—findings widely publicized among fellow scientists are generally subjected to further testing. Sooner or later, mistakes (and deception) are found out; wishful thinking is exposed. A discredited scientist does not get a second chance in the community of scientists. Honesty, so important to the progress of science, thus becomes a matter of self-interest to scientists. There is relatively little bluffing in a game where all bets are called. In fields of study where right and wrong are not so easily established, the pressure to be honest is considerably less.

Question Which of these is a scientific hypothesis?

(a) Atoms are the smallest particles of matter that exist.

(b) Space is permeated with an essence that is undetectable.

(c) Albert Einstein is the greatest physicist of the twentieth century.

The ideas and concepts most important to our everyday life are often unscientific; their correctness or incorrectness cannot be determined in the laboratory. Interestingly enough, it seems that people honestly believe their own ideas about things to be correct, and almost everyone is acquainted with people who hold completely opposite views—so the ideas of some (or all) must be incorrect. How do you know whether or not *you* are one of those holding erroneous beliefs? There is a test. Before you can be reasonably convinced that you are right about a particular idea, you should be sure that you understand the objections and the positions of your most articulate antagonists. You should find out whether your views are supported by sound knowledge of opposing ideas or by your *misconceptions* of opposing ideas. You make this distinction by seeing whether or not you can state the objections and positions of your opposition to *their* satisfaction. Even if you can successfully do this, you cannot be absolutely certain of being right about your own ideas, but the probability of being right is considerably higher if you pass this test.

Answer Only *a* is scientific, because there is a test for falseness. The statement is not only *capable* of being proved wrong, but it in fact *has* been proved wrong. Statement *b* has no test for possible wrongness and is therefore unscientific. Likewise for any principle or concept for which there is no means, procedure, or test whereby it can be shown to be wrong (if it is wrong). Some pseudoscientists and other pretenders of knowledge will not even consider a test for the possible wrongness of their statements. Statement *c* is an assertion that has no test for possible wrongness. If Einstein was not the greatest physicist, how could we know? It is important to note that because the name Einstein is generally held in high esteem, it is a favorite of pseudoscientists. So we should not be surprised that the name of Einstein, like that of Jesus and other highly respected sources, is cited often by charlatans who wish to bring respect to themselves and their points of view.

Question Suppose in a disagreement between two people, A and B, you note that person A states and restates only one point of view, whereas person B clearly states both her own position and that of person A. Who is more likely to be correct?

Although the notion of being familiar with counter points of view seems reasonable to most thinking people, just the opposite—shielding ourselves and others from opposing ideas—has been more widely practiced. We have been taught to discredit unpopular ideas without understanding them in proper context. With the $20/20$ vision of hindsight, we can see that many of the "deep truths" that were the cornerstones of whole civilizations were shallow reflections of the prevailing ignorance of the time. Many of the problems that plagued societies stemmed from this ignorance and the resulting misconceptions; much of what was held to be true simply wasn't true. This is not confined to the past. Every scientific advance is by necessity incomplete and partly inaccurate, for the discoverer sees, more or less, with the blinders of the day.

Science, Art, and Religion

The search for order and meaning in the world around us has taken different forms: one is science, another is art, and another is religion. Although the roots of all three go back thousands of years, the traditions of science are relatively recent. More important, the domains of science, art, and religion are different, although they often overlap. Science is principally engaged with discovering and recording natural phenomena, the arts are concerned with the value of human interactions as they pertain to the senses, and religion addresses the source, purpose, and meaning of it all.

The principal values of science and the arts are comparable. In art we find what is possible in human experience. We can learn about emotions ranging from anguish to love, even if we haven't yet experienced them. The arts do not necessarily give us those experiences, but describe them to us and suggest what may be in store for us. A knowledge of science similarly tells us what is possible in nature. Scientific knowledge helps us predict possibilities in nature even before these possibilities have been experienced. It provides us with a way of connecting things, of seeing relationships between and among them, and of making sense of the myriad natural events we find around us. Science broadens our perspective of the natural environment of which we are a part. A knowledge of both the arts and the sciences makes for a wholeness in the way we view the world and the decisions we make about it and ourselves. A truly educated person is knowledgeable in both the arts and the sciences.

Science and religion are different from each other. Science is both a body of knowledge and a method of probing nature's secrets. Religious beliefs and practices normally have to do with the faith and worship of God and the creation of human com-

Answer Who knows for sure? Person B may have the cleverness of a lawyer who can state various points of view and still be incorrect. We can't be sure about the "other guy." The test for correctness or incorrectness suggested here is not a test of others, but of and for *you.* It can aid your personal development. As you attempt to articulate the ideas of your antagonists, be prepared, like scientists who are prepared to change their minds, to discover evidence contrary to your own ideas—evidence that may alter your views. Intellectual growth often comes in this way.

munity, not with the experimental practices of science. In this respect, science and religion are as different as apples and oranges and do not contradict each other. While science is concerned with the workings of cosmic processes, religion addresses itself to the purpose of the cosmos. The two complement rather than contradict each other.

When we study the nature of light later in this book, we will treat light first as a wave and then as a particle. Waves and particles are contradictory; to the person who knows a little bit about physics, light can be only one or the other, and we have to choose between them. But to the enlightened physicist, waves and particles complement each other and provide a deeper understanding of light. In a similar way, it is mainly people who are either uninformed or misinformed about the deeper natures of both science and religion who feel that they must choose between believing in religion and believing in science. Unless one has a shallow understanding of either or both, there is no contradiction in being both religious and scientific in one's thinking.

Science and Technology

Science and technology are also different from each other. Whereas science has to do with discovering evidence that explains natural phenomena, technology has to do with tools, techniques, and procedures for putting the findings of science to use.

Science and technology are human enterprises, but in different ways. In deciding what problems to work on, scientists are guided by their own tastes—what they find interesting and exciting—and also, sometimes, by a desire to help other people or to serve their nation. Most often scientists are primarily driven by curiosity, the simple urge to know. They pursue knowledge, insofar as possible, free of current fashions, beliefs, and value judgments. What scientists discover may shock or anger some people—as did Darwin's theory of evolution. But science, by itself, does not intrude on human life. Technology does. It can hardly be ignored once developed. We do not have the option of refusing to hear the sonic boom produced by a supersonic aircraft flying overhead, we do not have the option of refusing to breathe polluted air, and we do not have the option of living in a nonnuclear age. Unlike science, advances in technology must be measured in terms of the human factor, with the understanding that technology is our slave and not the reverse. The legitimate purpose of technology is to serve people—people in general, not just some people, and future generations, not just those who currently wish to gain advantage for themselves. Technology must be humanistic if it is to lead to a better world.

We are all familiar with the abuses of technology. Many people blame technology itself for widespread pollution, resource depletion, and even social decay in general— so much so that the promise of technology is obscured. That promise is a cleaner and healthier world. It is much wiser to combat the misuse of technology with knowledge than with ignorance. Wise applications of science and technology *can* lead to a better world.

Physics—The Basic Science

Science is the present-day equivalent of what used to be called *natural philosophy.* Natural philosophy was the study of unanswered questions about nature. As the answers were found, they became part of what is now called *science.* The study of science today branches into the study of living things and nonliving things: the life sciences and the physical sciences. The life sciences branch into such areas as biology, zoology, and botany. The physical sciences branch into such areas as geology, astronomy, chemistry, and physics.

Physics is more than a part of physical sciences. It is the *basic* science. It's about the nature of basic things such as motion, forces, energy, matter, heat, sound, light, and the insides of atoms. Chemistry is about how matter is put together, how atoms combine to form molecules, and how the molecules combine to make up the many kinds of matter around us. Biology is more complex and involves matter that is alive. So underneath biology is chemistry, and underneath chemistry is physics. The concepts of physics reach up to these more complicated sciences. That's why physics is the most basic science.

An understanding of science begins with an understanding of physics. The following chapters present physics conceptually so that you can enjoy understanding it.

Question Which of the following activities involves the utmost human expression of passion, talent, and intelligence?

(a) painting and sculpture   (b) literature   (c) music   (d) religion   (e) science

---

In Perspective

Only a few centuries ago the most talented and most skilled artists, architects, and artisans of the world directed their genius and effort to the construction of the great cathedrals, synagogues, temples, and mosques. Some of these architectural structures took centuries to build, which means that nobody witnessed both the beginning and the end of construction. Even the architects and early builders who lived to a ripe old age never saw the finished results of their labors. Entire lifetimes were spent in the shadows of construction that must have seemed without beginning or end. This enormous focus of human energy was inspired by a vision that went beyond worldly concerns—a vision of the cosmos. To the people of that time, the structures they erected were their "spaceships of faith," firmly anchored but pointing to the cosmos.

Today the efforts of many of our most skilled scientists, engineers, artists, and artisans are directed to building the spaceships that already orbit the earth and others that will voyage beyond. The time required to build these spaceships is extremely brief compared to the time spent building the stone and marble structures of the past. Many people working on today's spaceships were alive before the first jetliner carried passengers. Where will younger lives lead in a comparable time?

Answer All of them! The human value of science, however, is the least understood by most individuals in our society. The reasons are varied, ranging from the common notion that science is incomprehensible to people of average ability to the extreme view that science is a dehumanizing force in our society. Most of the misconceptions about science probably stem from the confusion between the *abuses* of science and science itself.

Some people who view science as cold and impersonal seek psychic comfort in mysticism and other counterscientific concepts. Still more embrace a combination of science and mysticism, such as astronomy and astrology, the normal and the paranormal. In straddling science and superstition, they stand with one foot in the twentieth century and the other in the thirteenth. It is unfortunate that they do not see basic science as an enchanting human activity shared by a wide variety of people who, with present-day tools and know-how, are reaching further and finding out more about themselves and their environment than people in the past were ever able to do. The more you know about science, the more passionate you feel toward your surroundings. There is physics in everything you see, hear, smell, taste, and touch!

We seem to be at the dawn of a major change in human growth, not unlike the stage of a chicken embryo before it fully matures. When the chicken embryo exhausts the last of its inner-egg resources and before it pokes its way out of its shell, it may seem to be at its last moments. But what seems like an end is really only a beginning. Are we like the hatching chicks ready to poke through to a whole new range of possibilities? Are our space-faring efforts the early signs of a new human era?

The earth is our cradle and has served us well. But cradles, however comfortable, are one day outgrown. So with the inspiration that in many ways is similar to the inspiration of those who built the early cathedrals, synagogues, temples, and mosques, we aim for the cosmos.

We live in an exciting time!

Summary of Terms

Fact   A close agreement by competent observers of a series of observations of the same phenomenon.

Hypothesis   An educated guess; a reasonable explanation of an observation or experimental result that is not fully accepted as factual until tested over and over again by experiment.

Law   A general hypothesis or statement about the relationship of natural quantities that has been tested over and over again and has not been contradicted. Also known as a *principle*.

Scientific method   An orderly method for gaining, organizing, and applying new knowledge.

Theory   A synthesis of a large body of information that encompasses well-tested and verified hypotheses about certain aspects of the natural world.

Suggested Reading

Bronowski, Jacob, *Science and Human Values.* New York: Harper & Row, 1965.

Cole, K. C. *Sympathetic Vibrations: Reflections of Physics as a Way of Life.* New York: Morrow, 1984.

Feynman, Richard P. *Surely You're Joking, Mr. Feynman.* New York: Norton, 1986.

Gleick, James, *Genius—The Life and Science of Richard Feynman.* New York, Pantheon Books, 1992.

Pagels, H. R. *The Cosmic Code: Quantum Physics as the Language of Nature.* New York: Simon & Schuster, 1982.

Sagan, Carl. *The Demon-Haunted World.* Random House, New York, 1995.

Wolpert, Lewis. *The Unnatural Nature of Science.* Cambridge, MA, Harvard University Press, 1993.

Review Questions

1. Briefly, what is science?
2. Throughout the ages, what has been the general reaction to new ideas about established "truths"?

Scientific Measurements

3. When the sun was directly overhead in Syene, why was it not directly overhead in Alexandria?
4. The earth, like everything else illuminated by the sun, casts a shadow. Why does this shadow taper?
5. Why did Aristarchus make his measurements of the sun's distance at the time of a half-moon?

Mathematics—The Language of Science

6. What was the effect of the discovery three centuries ago that a mathematical structure underlies nature?

The Scientific Method

7. Outline the steps of the scientific method.

The Scientific Attitude

8. Distinguish among a scientific fact, a hypothesis, a law, and a theory.
9. In daily life people are often praised for maintaining some particular point of view, for the "courage of their convictions." A change of mind is seen as a sign of weakness. How is this different in science?
10. In daily life we see many cases of people who are caught misrepresenting things and who soon thereafter are excused and accepted by their contemporaries. How is this different in science?
11. What test can you perform to increase the chance in your own mind that you are right about a particular idea?

Science, Art, and Religion

12. Why are students of the arts encouraged to learn about science and science students encouraged to learn about the arts?

13. Why do many people believe they must choose between science and religion?

Science and Technology

14. Clearly distinguish between science and technology.

Physics—The Basic Science

15. Of physics, chemistry, and biology, which science is the least complex? The most complex? (At your school, which of these is the "easiest" and which is the "hardest" as a science course?)

Project

Poke a hole in a piece of cardboard and hold the cardboard in the sunlight. Note the image of the sun that is cast below. To convince yourself that the round spot of light is an image of the round sun, try holes of different shapes. A square or triangular hole will still cast a round image when the distance to the image is large compared to the size of the hole. When the sun's rays and the image surface are perpendicular, the image is a circle; when the sun's rays make an angle with the image surface, the image is a "stretched-out" circle, an ellipse. Let the solar image fall upon a coin, say a dime. Position the cardboard so the image just covers the coin. This is a convenient way to measure the diameter of the image—the same as the diameter of the easy-to-measure coin. Then measure the distance between the cardboard and the coin. Your ratio of image size to image distance should be about $\frac{1}{110}$. This is the ratio of solar diameter to solar distance to the earth. Using the information that the sun is 150,000,000 kilometers distant, calculate the diameter of the sun.

Exercises

1. Which of the following are scientific hypotheses? (a) Chlorophyll makes grass green. (b) The earth rotates about its axis because living things need an alternation of light and darkness. (c) Tides are caused by the moon.

2. In answer to the question, "When a plant grows, where does the material come from?" Aristotle hypothesized by logic that all material came from the soil. Do you consider his hypothesis to be correct, incorrect, or partially correct? What experiments do you propose to support your choice?

3. The great philosopher and mathematician Bertrand Russell (1872–1970) wrote about ideas in the early part of his life that he rejected in the latter part of his life. Do you see this as a sign of weakness or a sign of strength in Bertrand Russell? (Do you speculate that your present ideas about the world around you will change as you learn and experience more, or do you speculate that further knowledge and experience will solidify your present understanding?)

4. Bertrand Russell wrote, "I think we must retain the belief that scientific knowledge is one of the glories of man. I will not maintain that knowledge can never do harm. I think such general propositions can almost always be refuted by well-chosen examples. What I will maintain—and maintain vigorously—is that knowledge is very much more often useful than harmful and that fear of knowledge is very much more often harmful than useful." Think of examples to support this statement.

5. When you step from the shade into the sunlight the sun's heat is as evident as the heat from hot coals in a fireplace in an otherwise cold room. You feel the sun's heat not because of its high temperature (higher temperatures can be found in some welder's torches), but because the sun is big. Which do you estimate is larger, the sun's radius or the distance between the moon and the earth? Check your answer in the data on the inside cover. Do you find your answer surprising?

6. The shadow cast by a vertical pillar in Alexandria at noon during the summer solstice is found to be $\frac{1}{8}$ the height of the pillar. The distance between Alexandria and Syene is $\frac{1}{8}$ the earth's radius. Is there a geometric connection between these two 1-to-8 ratios?

7. What is probably being misunderstood by a person who says, "But that's only a scientific theory"?

8. Scientists call a theory that unites many ideas in a simple way "beautiful." Are unity and simplicity among the criteria of beauty outside of science? Give two examples to support your answer.

Problem

Tape a coin to your window when the full moon is visible in the sky. Position the coin and yourself so that the coin is directly in your line of sight to the moon. If you're at just the right distance from the coin, actually 110 coin diameters, the coin and the moon appear the same size and the coin barely eclipses the moon. Aha! Knowing that the average earth-moon distance is 3.84×10^5 km, calculate the diameter of the moon.

Intriguing! The number of balls released into the array of balls is always the same number emerging from the other side. But why? There's gotta be a reason -- mechanical rules of some kind. I'll know why the balls behave so predictably after I learn the rules of mechanics in the following nine chapters. Best of all, learning these rules will provide a keener intuition for understanding the world around me!

2

· · · · · · · · · · · · ·

Linear Motion

⋮ **S**he has high speed but no acceleration unless her motion changes.

More than 2000 years ago, the ancient Greek scientists were familiar with some of the ideas in physics that we study today. They had a very good understanding of some of the properties of light, but they were confused about motion. Probably the first to study motion seriously was Aristotle, the most outstanding philosopher-scientist in ancient Greece. Aristotle attempted to clarify motion by classification.

Aristotle on Motion

Aristotle divided motion into two main classes: *natural motion* and *violent motion*. We shall briefly consider each, not as study material, but only as a background to present-day ideas about motion.

Natural motion was thought to proceed from the "nature" of objects. In Aristotle's view, every object in the universe had a proper place, determined by this "nature"; any object not in its proper place would "strive" to get there. Being of the earth, an unsupported lump of clay properly fell to the ground; being of the air, an unimpeded puff of smoke properly rose; being a mixture of earth and air but predominantly earth, a feather properly fell to the ground but not as rapidly as a lump of clay. Heavier objects were expected to strive harder. Hence, objects were thought to fall at speeds proportional to their weights: the heavier the object, the faster it was thought to fall.

Natural motion could be either straight up or straight down, as in the case of all things on earth, or it could be circular, as in the case of celestial objects. Unlike up-and-

Aristotle (384–322 BC)

Greek philosopher, scientist, and educator Aristotle was the son of a physician who personally served the king of Macedonia. At 17 he entered the Academy of Plato, where he worked and studied for 20 years until Plato's death. He then became the tutor of young Alexander the Great. Eight years later he formed his own school. Aristotle's aim was to systematize existing knowledge, just as Euclid had systematized geometry. Aristotle made critical observations, collected specimens, and gathered together, summarized, and classified almost all existing knowledge of the physical world. His systematic approach became the method from which Western science later arose. After his death, his voluminous notebooks were preserved in caves near his home and were later sold to the library at Alexandria. Scholarly activity ceased in most of Europe through the Dark Ages, and the works of Aristotle were forgotten and lost. Some of the early Greeks' ideas, however, became incorporated in the scholarship that continued in the Byzantine and Islamic empires. Various texts were reintroduced to Europe during the eleventh and twelfth centuries and translated into Latin. The Church, the dominant political and cultural force in Western Europe, first prohibited the works of Aristotle and then accepted and incorporated them into Christian doctrine. Any attack on Aristotle was an attack on the Church itself.

down motion, circular motion was seen as being without beginning or end, repeating itself without deviation. Aristotle believed that different rules applied in the heavens, and he asserted that celestial bodies were perfect spheres made of a perfect and unchanging substance, which he called *ether.* (The only celestial object with any detectable change or imperfection was the moon, which, being nearest the earth, was thought by medieval Christians to be somewhat contaminated by the corrupted earth.)

Violent motion, Aristotle's other class of motion, resulted from pushing or pulling forces. Violent motion was imposed motion. A person pushing a cart or lifting a heavy weight imposed motion, as did someone hurling a stone or winning a tug-of-war. The wind imposed motion on ships. Floodwaters imposed it on boulders and tree trunks. The essential thing about violent motion was that it was externally caused and was imparted to objects; they moved not of themselves, but were pushed or pulled.

The concept of violent motion had its difficulties, for the pushes and pulls responsible for it were not always evident. For example, a bowstring moved an arrow until the arrow left the bow; after that, further explanation of the arrow's motion seemed to require some other pushing agent. Aristotle imagined, therefore, that a parting of the air by the moving arrow resulted in a squeezing effect on the rear of the arrow as the air rushed back to prevent a vacuum from forming. The arrow was propelled through the air as a bar of soap is propelled in the bathtub when you squeeze one end of it.

To sum up, Aristotle taught that all motions resulted either from the nature of the moving object or from a sustained push or pull. Provided that an object was in its proper place, it would not move unless subjected to a force. Except for celestial objects, the normal state was one of rest.

Aristotle's statements about motion were a beginning in scientific thought, and although he did not consider them to be the final words on the subject, his followers for nearly 2000 years regarded his views as beyond question. Implicit in the thinking of ancient, medieval, and early Renaissance times was the notion that the normal state of objects was one of rest. Since it was evident to most thinkers until the sixteenth century that the earth must be in its proper place, and since a force capable of moving the earth was inconceivable, it seemed quite clear that the earth did not move.

Copernicus and the Moving Earth

It was in this climate that the astronomer Copernicus formulated his theory of the moving earth. Copernicus reasoned from his astronomical observations that the earth traveled around the sun. For years he worked without making his thoughts public—for two reasons. The first was that he feared persecution; a theory so completely different from common opinion would surely be taken as an attack on established order. The second reason was that he had grave doubts about it himself; he could not reconcile the idea of a moving earth with the prevailing ideas of motion—the concept of intertia was unknown to him. Finally, in the last days of his life, at the urging of close friends he sent his *De Revolutionibus* to the printer. The first copy of his famous exposition reached him on the day he died—May 24, 1543.

Most of us know about the reaction of the medieval Church to the idea that the earth traveled around the sun. Because Aristotle's views had become so formidably a part of Church doctrine, to contradict them was to question the Church itself. For many Church leaders, the idea of a moving earth threatened not only their authority but the very foundations of faith and civilization as well. Their fears were well founded. For better or for worse, this new idea was to overturn their conception of the cosmos.

Galileo and the Leaning Tower

It was Galileo, the foremost scientist of the early seventeenth century, who gave credence to the Copernican view of a moving earth. He accomplished this by discrediting the Aristotelian ideas about motion. Although not the first to point out difficulties in Aristotle's views, Galileo was the first to provide conclusive refutation through observation and experiment.

Aristotle's falling-body hypothesis was easily demolished by Galileo. He is said to have dropped objects of various weights from the top of the Leaning Tower of Pisa and compared their falls. Contrary to Aristotle's assertion, he found that a stone twice as heavy as another did not fall twice as fast. Except for the small effect of air resistance, Galileo found that objects of various weights, when released at the same time, fell together and hit the ground at the same time. On one occasion, Galileo allegedly attracted a large crowd to witness the dropping of a light object and a heavy object from the top of the tower. Legend has it that many observers of this demonstration who saw the objects hit the ground together scoffed at the young Galileo and continued to hold fast to their Aristotelian teachings.

FIGURE 2.1 Galileo's famous demonstration.

Galileo Galilei (1564–1642)

Galileo was born in Pisa in the same year that Shakespeare was born and Michelangelo died. He studied medicine at the University of Pisa and then changed to mathematics. He developed an early interest in the mechanics of motion and was soon at odds with his contemporaries, who held to Aristotelian ideas on falling bodies. He left Pisa to teach at the University of Padua and became an advocate of the new Copernican theory of the solar system. He was one of the first to build a telescope, and he was the first to direct it to the nighttime sky and discover mountains on the moon and the moons of Jupiter. Because he published his findings in Italian instead of in the Latin expected of so reputable a scholar, and because of the recent invention of the printing press, his ideas reached a wide readership. He soon ran afoul of the Church and was warned not to teach and not to hold to Copernican views. He restrained himself publicly for nearly 15 years and then defiantly published his observations and conclusions, which were counter to Church doctrine. The outcome was a trial in which he was found guilty, and he was forced to renounce his discoveries. By then an old man broken in health and spirit, he was sentenced to perpetual house arrest. Nevertheless, he completed his studies on motion and his writings were smuggled from Italy and published in Holland. Earlier he damaged his eyes looking at the sun through a telescope, which led to blindness at the age of 74. He died 4 years later.

Galileo's Inclined Planes

Slope downward—
Speed increases

Slope upward—
Speed decreases

No slope—
Does speed change?

FIGURE 2.2 Motion of a ball on various planes.

Aristotle was an astute observer of nature, and he dealt with problems around him rather than with abstract cases that did not occur in his environment. Motion always involved a resistive medium such as air or water. He believed a vacuum to be impossible and therefore did not give serious consideration to motion in the absence of an interacting medium. That's why it was basic to Aristotle that an object requires a push or pull to keep it moving. And it was this basic principle that Galileo denied when he stated that if there is no interference with a moving object, it will keep moving in a straight line forever; no push, pull, or force of any kind is necessary.

Galileo tested this hypothesis by experimenting with the motion of various objects on inclined planes. He noted that balls rolling on downward-sloping planes picked up speed, while balls rolling on upward-sloping planes lost speed. From this he reasoned that balls rolling along a horizontal plane would neither speed up nor slow down. The ball would finally come to rest not because of its "nature" but because of friction. This idea was supported by Galileo's observation of motion along smoother surfaces: when there was less friction, the motion of objects persisted for a longer time; the less the friction, the more the motion approached constant speed. He reasoned that in the absence of friction or other opposing forces, a horizontally moving object would continue moving forever.

This assertion was supported by a different experiment and another line of reasoning. Galileo placed two of his inclined planes facing each other. He observed that a ball released from a position of rest at the top of a downward-sloping plane rolled down and then up the slope of the upward-sloping plane until it almost reached its initial height. He reasoned that only friction prevented it from rising to exactly the same height, for the smoother the planes, the more nearly the ball rose to the same height. Then he reduced the angle of the upward-sloping plane. Again the ball rose to the same height, but it had to go farther. Additional reductions of the angle yielded similar results; to reach the same height, the ball had to go farther each time. He then asked the question, "If I have a long horizontal plane, how far must the ball go to reach the same height?" The obvious answer is "Forever—it will never reach its initial height."*

* From Galileo's *Dialogues Concerning the Two New Sciences.*

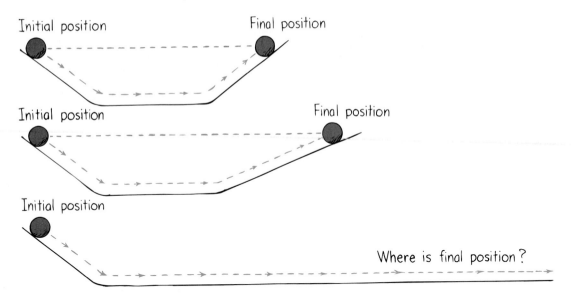

Initial position   Final position

Initial position   Final position

Initial position

Where is final position?

FIGURE 2.3 A ball rolling down an incline on the left tends to roll up to its initial height on the right. The ball must roll a greater distance as the angle of the incline on the right is reduced.

Galileo analyzed this in still another way. Because the downward motion of the ball from the first plane is the same for all cases, the speed of the ball when it begins moving up the second plane is the same for all cases. If it moves up a steep slope, it loses its speed rapidly. On a lesser slope, it loses its speed more slowly and rolls for a longer time. The less the upward slope, the more slowly it loses its speed. In the extreme case where there is no slope at all—that is, when the plane is horizontal—the ball should not lose any speed. In the absence of retarding forces, the tendency of the ball is to move forever without slowing down. This property of a moving object to continue moving he called **inertia**.

Galileo's concept of inertia discredited the Aristotelian theory of motion. Aristotle did not recognize the idea of inertia because he failed to imagine what motion would be like without friction. In his experience, all motion was subject to resistance, and he made this fact central to his theory of motion. Aristotle's failure to recognize friction for what it is—namely, a force like any other—impeded the progress of physics for 2000 years, until the time of Galileo. An application of Galileo's concept of inertia would show that no force was required to keep the earth in motion. The way was open for Isaac Newton to synthesize a new vision of the universe. We'll return to Newton in later chapters, and now acquaint ourselves with some of the terms that Galileo introduced to describe motion.

Question Would it be correct to say that inertia *causes* a moving object to continue in motion?

Answer In a strict sense, no. We don't know *why* objects have this property. Nevertheless, we call the property to behave in this predictable way *inertia*. We understand many things and have labels and names for these things. There are many things we do not understand, and we have labels and names for these things also. Education consists not so much in acquiring new names and labels but in learning which we understand and which we don't.

Description of Motion

FIGURE 2.4 A speedometer gives readings in both miles per hour and kilometers per hour.

In Aristotle's description of motion, the *distance* of an object from its proper place was fundamentally important. Galileo broke with this traditional concept and realized that time was the important missing ingredient in describing motion. Galileo described motion in terms of *time rates of change*. A time *rate* of change of a quantity is the change in that quantity divided by the time. It tells how fast something happens, or how much something changes in a certain amount of time. The rates that describe motion are *speed, velocity,* and *acceleration.**

Speed

Things in motion travel certain distances in given times. An automobile, for example, travels so many kilometers in an hour. **Speed** is a measure of how fast something is moving. It is the rate at which distance is covered, and is always measured in terms of a unit of distance divided by a unit of time. In general,

$$\text{Speed} = \frac{\text{distance}}{\text{time}}$$

Any combination of distance and time units is legitimate for measuring speed; for motor vehicles (or long distances) the units kilometers per hour (km/h) or miles per hour (mi/h or mph) are commonly used. For shorter distances, meters per second (m/s) are often useful units. The slash symbol (/) is read as *per* and means "divided by." Throughout this book we'll primarily use meters per second (m/s). Table 2.1 shows some comparative speeds in different units.[†]

AVERAGE SPEED   A car does not always move at the same speed. On any trip its speed usually varies somewhat, so we consider its *average speed.*

$$\text{Average speed} = \frac{\text{total distance covered}}{\text{time interval}}$$

Average speed can be calculated rather easily. For example, if we drive a distance of 80 kilometers in a time of 1 hour, we say our average speed is 80 kilometers per hour. Likewise, if we travel 320 kilometers in 4 hours,

$$\text{Average speed} = \frac{\text{total distance covered}}{\text{time interval}} = \frac{320\,\text{km}}{4\,\text{h}} = 80\,\text{km/h}$$

We see that when a distance in kilometers (km) is divided by a time in hours (h), the answer is in kilometers per hour (km/h).

Since average speed is the whole distance covered divided by the total time of travel, it doesn't indicate the different speeds and variations that may have taken place during shorter time intervals. Usually we experience a variety of speeds on most trips, so the average speed is often quite different from the speed at any instant.

If we know average speed and time of travel, distance traveled is easy to find. A simple rearrangement of the definition above gives

$$\text{Total distance covered} = \text{average speed} \times \text{time}$$

If your average speed is 80 kilometers per hour on a 4-hour trip, for example, you cover a total distance of 320 kilometers.

Table 2.1 Approximate Speeds in Different Units

20 km/h =	12 mi/h =	6 m/s
40 km/h =	25 mi/h =	11 m/s
60 km/h =	37 mi/h =	17 m/s
65 km/h =	40 mi/h =	18 m/s
80 km/h =	50 mi/h =	22 m/s
88 km/h =	55 mi/h =	24 m/s
100 km/h =	62 mi/h =	28 m/s
120 km/h =	75 mi/h =	33 m/s

* It would be nice if this chapter helps you to master these concepts, but it will be enough for you to become familiar with them and to be able to distinguish among them. The next few chapters will sharpen your understanding of these concepts.

[†] Conversion is based on 1 h = 3600 s, 1 mi = 1609.344 m.

INSTANTANEOUS SPEED   The speed that something has at any one instant is called **instantaneous speed**. It is the speed registered by the speedometer of a car. When we say that the speed of a car at some particular instant is 60 kilometers per hour, we are specifying its instantaneous speed, and we mean that if the car continued moving as fast for an hour, it would travel 60 kilometers. If it continued at that speed for half an hour, it would cover only half the distance: 30 kilometers. If it continued for only 1 minute, it would cover only 1 kilometer.

Instantaneous speed and average speed are often quite different, particularly when the time interval is large. For extremely short time intervals, on the other hand, average speed and instantaneous speed are the same. Whether we talk about average speed or instantaneous speed, we are talking about the rates at which distance is traveled.

Questions

1. What is the average speed of a cheetah that sprints 100 m in 4 seconds? How about if it sprints 50 m in 2 s?

2. If a car moves with an average speed of 60 km/h for an hour, it will travel a distance of 60 km.

   (a) How far would it travel if it moved at this rate for 4 h?

   (b) For 10 h?

   (c) Would it be possible for the car that starts from rest and ends at rest to attain an average speed of 60 km/h and never exceed a reading of 60 km/h on the speedometer?

Answers
(Are you reading this before you have formulated a reasoned answer in your thinking? If so, do you also exercise your body by watching others do push-ups? Exercise your thinking: when you encounter the many questions as above throughout this book, think before you read the footnoted answer!)

1. In both cases the answer is 25 m/s:

$$\text{Average speed} = \frac{\text{distance covered}}{\text{time interval}} = \frac{100 \text{ meters}}{4 \text{ seconds}}$$

$$= \frac{50 \text{ meters}}{2 \text{ seconds}} = 25 \text{ m/s}$$

2. The distance traveled is the average speed × time of travel, so

   (a) Distance = 60 km/h × 4 h = 240 km

   (b) Distance = 60 km/h × 10 h = 600 km

   (c) No, if the trip started from rest and ended at rest, then there are intervals with an instantaneous speed less than 60 km/h. Unless there is compensation during periods of speed greater than 60 km/h, it would not be possible to yield an average of 60 km/h. In practice, average speeds are usually appreciably less than peak instantaneous speeds.

Velocity

FIGURE 2.5 The car on the circular track may have a constant speed, but its velocity is changing every instant. Why?

Loosely speaking, we can use the words *speed* and *velocity* interchangeably. Strictly speaking, however, there is a distinction between the two. When we say that something travels at 60 kilometers per hour, we are specifying its speed. But if we say that something travels at 60 kilometers per hour to the north, we are specifying its velocity. A race-car driver is concerned primarily with his speed—how fast he is moving; an airplane pilot is concerned with her velocity—how fast and in what direction she is moving. When we describe speed and the *direction* of motion, we are specifying **velocity**.*

We distinguish between average velocity and instantaneous velocity as we do for speed. By custom, the word *velocity* alone is assumed to mean instantaneous velocity. Likewise for the word *speed* alone. If something moves at an unchanging or constant velocity, then, its average and instantaneous velocities will have the same value. The same is true for speed. When something moves at constant velocity or constant speed, then *equal distances* are covered in equal intervals of time. Constant velocity and constant speed, however, can be very different. Constant velocity means constant speed with no change in direction. A car that rounds a curve at a constant speed does not have a constant velocity—its velocity changes as its direction changes.

Questions

1. "She moves at a constant speed in a constant direction." Say the same sentence in fewer words.
2. The speedometer of a car moving to the east reads 100 km/h. It passes another car that moves to the west at 100 km/h. Do both cars have the same speed? Do they have the same velocity?
3. During a certain period of time, the speedometer of a car reads a constant 60 km/h. Does this indicate a constant speed? A constant velocity?

Answers

1. "She moves at constant velocity."
2. Both cars have the same speed, but they have opposite velocities because they are moving in opposite directions.
3. The constant speedometer reading indicates a constant speed but not a constant velocity, because the car may not be moving along a straight-line path, in which case it is accelerating. (We'll see in Chapter 4 that whenever something accelerates, a force must be acting on it.)

---

* A quantity described by both magnitude (how much) and direction (which way) is called a *vector quantity.* A quantity described only by magnitude, such as speed, is a *scalar quantity.* We will treat the vector nature of velocity in the next chapter.

FIGURE 2.6 We say that a body undergoes acceleration when there is a *change* in its state of motion.

Acceleration

We can change the velocity of something by changing its speed, by changing its direction, or by changing both its speed *and* its direction. We define the rate of change of velocity as **acceleration**:

$$\text{Acceleration} = \frac{\text{change of velocity}}{\text{time interval}}$$

We are all familiar with acceleration in an automobile. In driving, we call it "pickup" or "getaway"; we experience it when we tend to lurch toward the rear of the car. The key idea that defines acceleration is *change*. Suppose we are driving and in 1 second we steadily increase our velocity from 30 kilometers per hour to 35 kilometers per hour, and then to 40 kilometers per hour in the next second, to 45 in the next second, and so on. We change our velocity by 5 kilometers per hour each second. This change of velocity is what we mean by acceleration.

$$\text{Acceleration} = \frac{\text{change of velocity}}{\text{time interval}} = \frac{5 \text{ km/h}}{1 \text{ s}} = 5 \text{ km/h} \cdot \text{s}$$

In this case the acceleration is 5 kilometers per hour second (abbreviated as 5 km/h·s). Note that a unit for time enters twice: once for the unit of velocity and again for the interval of time in which the velocity is changing. Also note that acceleration is not just the total change in velocity; it is the *time rate of change, or change per second,* of velocity.

Questions

1. A particular car can go from rest to 90 km/h in 10 s. What is its acceleration?
2. In 2.5 s a car increases its speed from 60 km/h to 65 km/h while a bicycle goes from rest to 5 km/h. Which undergoes the greater acceleration? What is the acceleration of each vehicle?

The term *acceleration* applies to decreases as well as to increases in velocity. We say the brakes of a car, for example, produce large retarding accelerations; that is, there is a large decrease per second in the velocity of the car. We often call this *deceleration*. We experience deceleration when we tend to lurch toward the front of the car.

Answers

1. Its acceleration is 9 km/h·s. Strictly speaking, this would be its average acceleration, for there may have been some variation in its rate of picking up speed.
2. The accelerations of both the car and the bicycle are the same; 2 km/h·s.

$$\text{Acceleration}_{car} = \frac{\text{change of velocity}}{\text{time interval}} = \frac{65 \text{ km/h} - 60 \text{ km/h}}{2.5 \text{ s}} = \frac{5 \text{ km/h}}{2.5 \text{ s}} = 2 \text{ km/h} \cdot \text{s}$$

$$\text{Acceleration}_{bike} = \frac{\text{change of velocity}}{\text{time interval}} = \frac{5 \text{ km/h} - 0 \text{ km/h}}{2.5 \text{ s}} = \frac{5 \text{ km/h}}{2.5 \text{ s}} = 2 \text{ km/h} \cdot \text{s}$$

Although the velocities involved are quite different, the rates of change of velocity are the same. Hence the accelerations are equal.

We accelerate whenever we move in a curved path, even if we are moving at constant speed, because our direction and hence our velocity is changing. We experience this acceleration as we tend to lurch toward the outer part of the curve. We distinguish speed and velocity for this reason and define *acceleration* as the rate at which velocity changes, thereby encompassing changes both in speed and in direction.

Anyone who has stood in a crowded bus has experienced the difference between velocity and acceleration. Except for the effects of a bumpy road, you can stand with no extra effort inside a bus that moves at constant velocity, no matter how fast it is going. You can flip a coin and catch it exactly as if the bus were at rest. It is only when the bus accelerates—speeds up, slows down, or turns—that you experience difficulty.

In much of this book we will be concerned only with motion along a straight line. When straight-line motion is being considered, it is common to use *speed* and *velocity* interchangeably. When the direction is not changing, acceleration may be expressed as the rate at which *speed* changes.

$$\text{Acceleration (along a straight line)} = \frac{\text{change in speed}}{\text{time interval}}$$

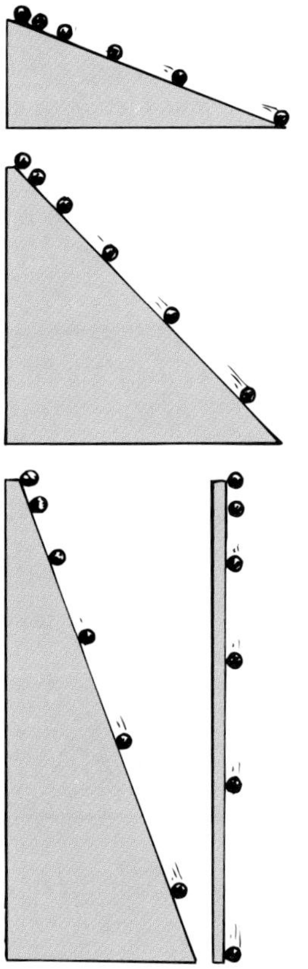

Acceleration on Galileo's Inclined Planes

Galileo developed the concept of acceleration in his experiments on inclined planes. His main interest was falling objects, and because he lacked suitable timing devices, he used inclined planes to effectively slow down accelerated motion and investigate it more carefully.

Galileo found that a ball rolling down an inclined plane will pick up the same amount of speed in successive seconds; that is, the ball will roll with unchanging acceleration. For example, a ball rolling down a plane inclined at a certain angle might be found to pick up a speed of 2 meters per second for each second it rolls. This gain per second is its acceleration. Its instantaneous velocity at 1-second intervals, at this acceleration, is then 0, 2, 4, 6, 8, 10, and so forth meters per second. We can see that the instantaneous speed or velocity of the ball at any given time after being released from rest is simply equal to its acceleration multiplied by the time:*

$$\text{Velocity acquired} = \text{acceleration} \times \text{time}$$

If we substitute the acceleration of the ball in this relationship, we can see that at the end of 1 second, the ball is traveling at 2 meters per second; at the end of 2 seconds, it is traveling at 4 meters per second; at the end of 10 seconds, it is traveling at 20 meters per second; and so on. The instantaneous speed or velocity at any time is simply equal to the acceleration multiplied by the number of seconds it has been accelerating.

Galileo found greater accelerations for steeper inclines. The ball attains its maximum acceleration when the incline is tipped vertically. Then the acceleration is the same as that of a falling object (Figure 2.7). Regardless of the weight or size, Galileo discovered that when air resistance is small enough to be neglected, all objects fall with the same unchanging acceleration.

FIGURE 2.7 The greater the slope of the incline, the greater the acceleration of the ball. What is its acceleration if the incline is vertical?

---

* Note that this relationship follows from the definition of acceleration. From $a = v/t$, simple rearrangement (multiplying both sides of the equation by t) gives $v = at$.

Free Fall

FIGURE 2.8 Pretend that a falling rock is equipped with a speedometer. In each succeeding second of fall, we would find that the rock's speed increases by the same amount: 10 m/s. (Table 2.2 shows the speeds we would read at various seconds of fall.)

Table 2.2 Free-Fall from Rest

Time of Fall (seconds)	Velocity Acquired (meters/second)
0	0
1	10
2	20
3	30
4	40
5	50
.	.
.	.
.	.
t	10t

How Fast

Things fall because of the force of gravity. When a falling object is free of all restraints—no friction, air or otherwise, and falls under the influence of gravity alone, the object is in a state of **free fall**. (We'll consider the effects of air resistance on falling in Chapter 4). Table 2.2 shows the instantaneous speed of a freely falling object at 1-second intervals. The important thing to note in these numbers is the way the speed changes. *During each second of fall, the object gains a speed of 10 meters per second.* This gain per second is the acceleration. Free-fall acceleration is approximately equal to 10 meters per second each second, or, in shorthand notation, 10 m/s^2 (read as 10 meters per second squared). Note that the unit of time, the second, enters twice—once for the unit of speed and again for the time interval during which the speed changes.

In the case of freely falling objects, it is customary to use the letter g to represent the acceleration (because the acceleration is due to *gravity*). Although the value of g varies slightly in different parts of the world, its average value is equal to 9.8 meters per second each second, or, in shorter notation, 9.8 m/s^2. We round this off to 10 m/s^2 in our present discussion and in Table 2.2 to establish the ideas involved more clearly; multiples of 10 are more obvious than multiples of 9.8. Where accuracy is important, the value of 9.8 m/s^2 should be used.

Note in Table 2.2 that the instantaneous speed or velocity of an object falling from rest is consistent with the equation that Galileo deduced with his inclined planes:

$$\text{Velocity acquired} = \text{acceleration} \times \text{time}$$

The instantaneous velocity v of an object falling from rest* after a time t can be expressed in shorthand notation as

$$v = gt$$

To see that this equation makes good sense, take a moment to check it with Table 2.2. Note that the instantaneous velocity or speed in meters per second is simply the acceleration $g = 10$ m/s^2 multiplied by the time t in seconds.

Question What would the speedometer reading on the falling rock shown in Figure 2.8 be 3.5 s after it drops from rest? How about 6 s after it is dropped? 100 s?

So far we have been considering objects moving straight downward in the direction of gravity. How about an object thrown straight upward? Once released, it continues to move upward for a while and then comes back down. At the highest point, when it is changing its direction of motion from upward to downward, its instantaneous speed is zero. Then it starts downward just as if it had been dropped from rest at that height.

Answer The speedometer readings would be 35 m/s, 60 m/s, and 1000 m/s respectively. You can reason this from Table 2.2 or use the equation $v = gt$, where g is replaced by 10 m/s^2.

---

* If instead of being dropped from rest the object is thrown downward at speed v_0, the speed v after any elapsed time t is $v = v_0 + gt$. We will not be concerned with this added complication here, and will instead learn as much as we can from the most simple cases. That will be a lot!

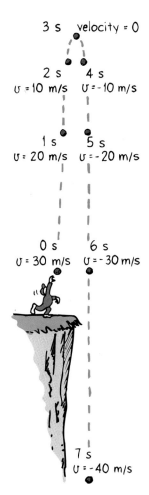

FIGURE 2.9 The rate at which the velocity changes each second is the same.

Table 2.3 Distance Fallen in Free Fall

Time of Fall (seconds)	Distance Fallen (meters)
0	0
1	5
2	20
3	45
4	80
5	125
.	.
.	.
.	.
t	$\frac{1}{2}(10t^2)$

During the upward part of this motion, the object slows from its initial upward velocity to zero velocity. Its speed decreases—evidence of acceleration (or deceleration in the upward direction). How much does its speed decrease each second? It should come as no surprise that it decreases at the rate of 10 meters per second each second—the same acceleration it experiences on the way down. So interestingly enough, as Figure 2.9 shows, the instantaneous speed at points of equal elevation in the path is the same whether the object is moving upward or downward. The velocities are opposite, of course, because they are in opposite directions. Note that the downward velocities have a negative sign, indicating the downward direction (it is customary to call *up* positive, and *down* negative). Whether motion is upward or downward, during each second the speed or the velocity changes by 10 meters per second. The acceleration is a downward 10 meters per second squared the whole time, whether the object is moving upward or downward.

How Far

How *far* an object falls is altogether different from how *fast* it falls. With his inclined planes Galileo found that the distance a uniformly accelerating object travels is proportional to the *square of the time*. The details of this relationship are in Appendix B. We will state here only the results. The distance traveled by a uniformly accelerating object starting from rest is

Distance traveled = $\frac{1}{2}$ (acceleration × time × time)

This relationship applies to the distance something falls. We can express it for the case of a freely falling object in shorthand notation as

$$d = \frac{1}{2} gt^2$$

where d is the distance something falls when the time of fall in seconds is substituted for t and squared.* If we use 10 m/s² for the value of g, the distance fallen for various times will be as shown in Table 2.3.

Note that an object falls a distance of only 5 meters during the first second of fall, although its speed at 1 second is 10 meters per second. This may be confusing, for we may think that the object should fall a distance of 10 meters. But for it to fall 10 meters in its first second of fall, it would have to fall at an *average* speed of 10 meters per second for the entire second. It starts its fall at 0 meters per second, and its speed is 10 meters per second only in the last instant of the 1-second interval. Its average speed during this interval is the average of its initial and final speeds, 0 and 10 meters per second. To find the average value of these or any two numbers, we simply add the two numbers and divide by 2. This equals 5 meters per second, which over a time interval of 1 second gives a distance of 5 meters. As the object continues to fall in succeeding seconds, it will fall through ever-increasing distances because its speed is continuously increasing.

---

* d = average velocity × time

$d = \dfrac{\text{initial velocity} + \text{final velocity}}{2} \times \text{time}$

$d = \dfrac{0 + gt}{2} \times t = \frac{1}{2} gt^2$ (See Appendix B for further explanation.)

FIGURE 2.10 Pretend that a falling rock is equipped with an odometer. The readings of distance fallen increase with time by $\frac{1}{2} gt^2$ and are shown in Table 2.3.

FIGURE 2.11 A feather and a coin fall at equal accelerations in a vacuum.

Question A cat steps off a ledge and drops to the ground in 1/2 second.
(a) What is its speed on striking the ground?
(b) What is the average speed during the 1/2 second?
(c) How high is the ledge from the ground?

It is a common observation that all objects do not fall with equal accelerations. A leaf, a feather, or a sheet of paper may flutter to the ground slowly. The fact that air resistance is responsible for those different accelerations can be shown very nicely with a closed glass tube containing light and heavy objects—a feather and a coin, for example. In the presence of air, the feather and coin fall with quite different accelerations. But if the air in the tube is removed by a vacuum pump and the tube is quickly inverted, the feather and coin fall with the same acceleration (Figure 2.11). Although air resistance appreciably alters the motion of things like falling feathers, the motion of heavier objects like stones and baseballs at ordinary low speeds is not appreciably affected by the air. The relationships $v = gt$ and $d = \frac{1}{2} gt^2$ can be used to a very good approximation for most objects falling in air from an initial position of rest.

Free Fall—How Quickly How Fast Changes

Much of the confusion that arises in analyzing the motion of falling objects comes about because it is easy to get "how fast" mixed up with "how far." When we wish to specify how fast something is falling, we are talking about speed or velocity, which is expressed as $v = gt$. When we wish to specify how far something falls, we are talking about distance, which is expressed as $d = \frac{1}{2} gt^2$. Speed or velocity (how fast) and distance (how far) are entirely different from each other.

A most confusing concept, and probably the most difficult encountered in this book, is "how quickly does how fast change"—acceleration. What makes acceleration so complex is that it is a *rate of a rate*. It is often confused with velocity, which is itself a rate (the rate of change of position). Acceleration is not velocity, nor is it even a change in velocity. Acceleration is the rate at which velocity itself changes.

Please remember that it took people nearly 2000 years from the time of Aristotle to reach a clear understanding of motion, so be patient with yourself if you find that you require a few hours to achieve as much!

Answer If we round g off to 10 m/s^2, we find
(a) Speed: $v = gt = 10 \text{ m/s}^2 \times \frac{1}{2} \text{ s} = 5 \text{ m/s}$

(b) Average speed: $\bar{v} = \dfrac{\text{initial } v + \text{final } v}{2} = \dfrac{0 \text{ m/s} + 5 \text{ m/s}}{2} = 2.5 \text{ m/s}$

We put a bar over the symbol to denote average speed: $\bar{v}$
(c) Distance: $d = \bar{v}t = 2.5 \text{ m/s} \times \frac{1}{2} \text{ s} = 1.25 \text{ m}$
Or equivalently,

$$d = \tfrac{1}{2} gt^2 = \tfrac{1}{2} \times 10 \text{ m/s}^2 \times (\tfrac{1}{2} \text{ s})^2 = \tfrac{1}{2} \times 10 \text{ m/s}^2 \times \tfrac{1}{4} \text{ s}^2 = 1.25 \text{ m}$$

Notice that we can find the distance by either of these equivalent relationships.

$$\text{Speed} = \frac{\text{distance}}{\text{time}}$$

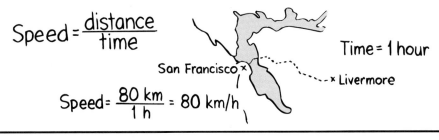

Time = 1 hour

San Francisco ×

× Livermore

$$\text{Speed} = \frac{80 \text{ km}}{1 \text{ h}} = 80 \text{ km/h}$$

$$\text{Velocity} = \left\{ \begin{array}{l} \text{speed } and \\ \text{direction} \end{array} \right\}$$

San Francisco ×

E

Velocity = 300 km/h, east

$$\text{Acceleration} = \left\{ \begin{array}{l} \text{Rate of} \\ \text{change in} \\ \text{velocity} \end{array} \right\} \text{ due to } \left\{ \begin{array}{l} \text{change in speed} \\ \text{and/or direction} \end{array} \right\}$$

40 km/h   80 km/h   0 km/h

Change in speed
but *not* direction

40 km/h

40 km/h

Change in direction
but *not* speed

40 km/h

Change in speed
and direction

$$\text{Acceleration} = \frac{\text{change in velocity}}{\text{time}}$$

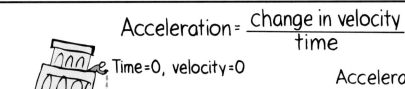

Time = 0, velocity = 0

Time = 1 s, velocity = 10 m/s

$$\text{Acceleration} = \frac{20}{2 \text{s}} \text{ m/s}$$

$$a = 10 \frac{\text{m/s}}{\text{s}}$$

$$a = 10 \text{ m/ss}$$

$$a = 10 \text{ m/s}^2$$

Time = 2 s, velocity = 20 m/s

FIGURE 2.12   Motion analysis.

Hang-Time

Basketball players, ballet dancers, and others who have developed jumping ability when leaping straight up seem to hang in the air in defiance of gravity. We say they have a long "hang-time"—the amount of time airborne with feet off the ground. A common misconception is that the greatest jumpers have hang-times of more than 2 seconds. Surprisingly, the hang-time of the greatest jumpers is almost always less than 1 second! Our perception of a longer hang-time is one of many illusions we have about nature.

Jumping ability is best measured by a standing vertical jump. Stand facing a wall, and with feet flat on the floor and arms extended upward, make a mark on the wall at the top of your reach. Then make your jump and at the peak make another mark. The distance between these two marks measures your vertical leap. If it's more than 2 feet (0.6 meter), you're exceptional.

Here's the physics. When you leap upward, jumping force is applied only as long as your feet are still in contact with the ground. The greater the force, the greater your launch speed and the higher you'll jump. Interestingly, as soon as your feet leave the ground, whatever upward speed you attain immediately decreases at the steady rate of g, 10 m/s^2. Maximum height is reached when your upward speed decreases to zero. You then begin falling, gaining speed at exactly the same rate, g. If you land as you took off, upright with legs extended, then time rising equals time falling. Hang-time is the sum of rising and falling times. While airborne, no amount of leg or arm pumping or other bodily motion will change your hang-time.

The relationship between rising or falling time and vertical height is given by

$$d = \tfrac{1}{2} g t^2$$

If we know the vertical height, we can rearrange this expression to read

$$t = \sqrt{\frac{2d}{g}}$$

Basketball player Spud Webb's record jumping height is 1.25 m (4 ft). Setting d equal to 1.25 m*, and using the more precise 9.8 m/s^2 for g, we find that t, half the hang-time, is

$$t = \sqrt{\frac{2d}{g}} = \sqrt{\frac{2(1.25 \text{ m})}{9.8 \text{ m/s}^2}} = 0.50 \text{ s}$$

Doubling this, we see Spud's record hang-time is 1 second!

We're talking here of vertical motion. How about running jumps? We'll see in Chapter 3 that hang-time depends only on the jumper's vertical speed at launch. While airborne, the jumper's horizontal speed remains constant while the vertical speed undergoes acceleration. Interesting physics!

* The value of 1.2 m for d represents the maximum height of the jumper's center of gravity, not the height of the bar. The height gained by the jumper's center of gravity is what's important in determining jumping ability. We'll learn about center of gravity in Chapter 7.

Summary of Terms

Inertia The sluggishness or apparent resistance an object offers to changes in its state of motion. Mass is the measure of inertia.

Speed How fast something moves. The distance traveled per unit of time.

Instantaneous speed The speed at any instant.

Velocity The speed of an object and specification of its direction of motion.

Acceleration The rate at which velocity changes with time; the change in velocity may be in magnitude or direction or both.

Free fall Motion under the influence of gravity only.

Summary of Equations

$$\text{Speed} = \frac{\text{distance}}{\text{time}}$$

$$\text{Average speed} = \frac{\text{total distance covered}}{\text{time interval}}$$

$$\text{Acceleration} = \frac{\text{change of velocity}}{\text{time interval}}$$

$$\text{Acceleration (along a straight line)} = \frac{\text{change in speed}}{\text{time interval}}$$

Velocity acquired in free fall, from rest: $v = gt$

Distance fallen in free fall, from rest: $d = \frac{1}{2}gt^2$

Review Questions

*Each chapter in this book concludes with a set of review questions and exercises, and for some chapters, problems. The **Review Questions** are designed to help you fix ideas and catch the essentials of the chapter material. You'll notice that answers to the questions can be found within the chapters. The **Exercises** stress thinking rather than mere recall of information and call for an understanding of the definitions, principles, and relationships of the chapter material. In many cases the intention of particular exercises is to help you to apply the ideas of physics to familiar situations. Unless you cover only a few chapters in your course, you will likely be expected to tackle only a few exercises for each chapter. Answers should be in complete sentences, with an explanation or sketches when applicable. The large number of exercises is to allow your instructor a wide choice of assignments. **Problems** feature concepts that are more clearly understood with numerical values and straightforward calculations. Be sure to include units of measurement in your answers. The problems are relatively few in number so as to avoid an undue emphasis on problem solving that could obscure the primary goal of* Conceptual Physics—*to develop a gut feel for the concepts of physics in your everyday language. Calculations are to enhance the learning of concepts, and not the other way around.*

Aristotle on Motion

1. Contrast Aristotle's ideas of natural motion and violent motion.
2. Did Aristotle think that a force acts on the moon as it revolves around the earth? On earth, did he think a force keeps a ball rolling along a smooth, level surface?

Copernicus and the Moving Earth

3. Contrast the relationship between the earth and sun as seen by Aristotle and by Copernicus.

Galileo and the Leaning Tower

4. What did Galileo discover in his legendary experiment on the Leaning Tower?

Galileo's Inclined Planes

5. What did Galileo discover in his experiments with inclined planes?
6. What does it mean to say an object has inertia? Give an example.

Description of Motion

7. What is a "time rate of change"? Give three examples.

Speed

8. What two units of measurement are necessary for describing speed?
9. Distinguish speed in general from instantaneous speed. Give an example.
10. What is the average speed of a horse that gallops a distance of 15 km in a time of 30 min?
11. How far does a horse travel if it gallops at an average speed of 25 km/h for 30 min?

Velocity

12. Distinguish between speed and velocity.
13. If a car moves with a constant velocity, does it also move with a constant speed? Explain.
14. If a car moves with a constant speed, can you say that it also moves with a constant velocity? Give an example to support your answer.

Acceleration

15. Distinguish between velocity and acceleration.
16. What is the acceleration of a car that increases its velocity from 0 to 100 km/h in 10 s?
17. What is the acceleration of a car that maintains a constant velocity of 100 km/h for 10 s? (Why do many students who correctly answer the last question get this question wrong?)
18. When are you most aware of motion in a moving vehicle—when it is moving steadily in a straight line or when it is accelerating? If a car moved with absolutely constant velocity (no bumps at all), would you be aware of motion?
19. Acceleration is generally defined as the time rate of change of velocity. When can it be defined as the time rate of change of speed?

Acceleration on Galileo's Inclined Planes

20. What was the relationship among velocity, acceleration, and time that Galileo discovered on his inclined planes?
21. How did Galileo tip his inclined planes to better understand the role of gravity in making things move?

Free Fall

How Fast

22. What exactly is meant by a "freely falling" object?

23. What is the gain in speed per second for a freely falling object?

24. What is the velocity acquired by a freely falling object 5 s after being dropped from a rest position? What is it 6 s after?

25. The acceleration of free fall is about 10 m/s². Why does the seconds unit appear twice?

How Far

26. What relationship between distance traveled and time of travel did Galileo discover with his inclined planes?

27. What is the distance fallen for a freely falling object 5 s after being dropped from a rest position? What is it 6 s after?

Free Fall—How Quickly How Fast Changes

28. Consider these measurements: 10 m, 10 m/s, and 10 m/s². Which is a measure of distance, which of speed, and which of acceleration?

Hang-Time

29. When using the formula $d = \frac{1}{2} gt^2$ to calculate hang-time, is time t the time it takes to jump up, the time to come down, or the total up-and-down time? Why?

30. If a friend claims that in a standing jump he can remain off the ground for 1 second, then how high a jump does this correspond to? For 2 seconds?

Project

Stand flatfooted next to a wall and make a mark at the highest point you can reach. Then jump vertically and mark your highest point. The distance between the marks is your vertical jumping distance. Use this to calculate your hang-time.

Exercises

1. A ball is rolling across the top of a billiard table and slowly rolls to a stop. How would Aristotle interpret this observation? How would Galileo interpret it?

2. Why did Galileo use inclined planes to investigate free fall?

3. Is a fine for speeding based on one's average speed or instantaneous speed? Explain.

4. One airplane travels due north at 300 km/h while another travels due south at 300 km/h. Are their speeds the same? Are their velocities the same? Explain.

5. Can an automobile with a velocity toward the north have an acceleration toward the south? Explain.

6. Can an object reverse its direction of travel while maintaining a constant acceleration? If so, give an example. If not, explain why.

7. You are driving north on a highway. Then, without changing speed, you round a curve and drive east. (a) Does your velocity change? (b) Do you accelerate? Explain.

8. Correct your friend who says, "The dragster rounded the curve at a constant velocity of 100 km/h."

9. Harry says acceleration is how fast you go. Carol says acceleration is how fast you get fast. They look to you for confirmation. Who's correct?

10. Starting from rest, one car accelerates to a speed of 50 km/h, and another car accelerates to a speed of 60 km/h. Can you say which car underwent the greater acceleration? Why or why not?

11. Cite an example of something that undergoes acceleration while moving at constant speed. Can you also give an example of something that accelerates while traveling at constant velocity? Explain.

12. (a) Can an object be moving when its acceleration is zero? If so, give an example. (b) Can an object be accelerating when its speed is zero? If so, give an example.

13. Can you give an example wherein the acceleration of a body is opposite in direction to its velocity? Do so if you can.

14. On which of these hills does the ball roll down with increasing speed and decreasing acceleration along the path? (Use this example if you wish to explain to someone the difference between speed and acceleration.)

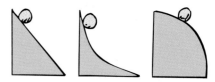

15. Suppose that the three balls shown in Exercise 14 start simultaneously from the tops of the hills. Which one reaches the bottom first? Explain.

16. What is the acceleration of a car that moves at a steady velocity of 100 km/h for 100 seconds? Explain your answer.

17. Correct your friend who says, "In free fall, air resistance is more effective in slowing a feather than a coin."

18. Suppose that a freely falling object were somehow equipped with a speedometer. By how much would its speed reading increase with each second of fall?

19. Suppose that the freely falling object in the preceding exercise were also equipped with an odometer. Would the readings of distance fallen each second indicate equal or different falling distances for successive seconds?

20. For a freely falling object dropped from rest, what is the acceleration at the end of the 5th second of fall? The 10th second? Defend your answer.

21. When a ball player throws a ball straight up, by how much does the speed of the ball decrease each second while ascending? In the absence of air resistance, by how much does it increase each second while descending? How much time is required for rising compared to falling?

22. Someone standing at the edge of a cliff (as in Figure 2.9) throws a ball nearly straight up at a certain speed and another ball nearly straight down with the same initial speed. If air resistance is negligible, which ball will have the greater speed when it strikes the ground below?

23. Answer the previous question for the case where air resistance is *not* negligible—where air drag affects motion.

24. If you drop an object, its acceleration toward the ground is 10 m/s^2. If you throw it down instead, would its acceleration after throwing be greater than 10 m/s^2? Why or why not?

25. In the preceding exercise can you think of a reason why the acceleration of the object thrown downward through the air would be appreciably less than 10 m/s^2?

26. If it were not for air resistance, why would it be dangerous to go outdoors on rainy days?

27. Extend Tables 2.2 and 2.3 to include times of fall of 6 to 10 seconds, assuming no air resistance.

28. Two balls are released simultaneously from rest at the left end of the tracks A and B as shown. Which ball reaches the end of its track first?

29. In this chapter, we studied idealized cases of balls rolling down smooth planes and objects falling with no air resistance. Suppose a classmate complains that all this attention focused on idealized cases is valueless because idealized cases simply don't occur in the everyday world. How would you respond to this complaint? How do you suppose the author of this book would respond?

30. Make up two multiple-choice questions that would check a classmate's understanding of the distinction between velocity and acceleration.

Problems

1. The ocean's level is currently rising at about 1.5 mm per year. At this rate, in how many years will sea level be 3 meters higher than now?

2. What is the acceleration of a vehicle that changes its velocity from 100 km/h to a dead stop in 10 s?

3. A racing car moving at 10 m/s north increases its velocity to 16 m/s north in a time interval of 3 s. What is its average acceleration during this time interval? What is the direction of this acceleration?

4. A ball is thrown straight up with an initial speed of 30 m/s. How high does it go, and how long is it in the air (neglecting air resistance)?

5. A ball is thrown with enough speed straight up so that it is in the air several seconds. (a) What is the velocity of the ball when it gets to its highest point? (b) What is its velocity 1 s before it reaches its highest point? (c) What is the change in its velocity during this 1-s interval? (d) What is its velocity 1 s after it reaches its highest point? (e) What is the change in velocity during this 1-s interval? (f) What is the change in velocity during the 2-s interval? (Careful!) (g) What is the acceleration of the ball during any of these time intervals and at the moment the ball has zero velocity?

6. What is the instantaneous velocity of a freely falling object 10 s after it is released from a position of rest? What is its average velocity during this 10-s interval? How far will it fall during this time?

7. A climber near the summit of a vertical cliff accidentally knocks loose a large rock. She sees it shatter at the bottom of the cliff 8 s later. What was the speed of impact? How far did the rock fall?

8. A car takes 10 s to go from $v = 0$ to $v = 50$ m/s at approximately constant acceleration. If you wish to find the distance traveled using the equation $d = \frac{1}{2}at^2$, what value should you use for a?

9. A reconnaissance plane flies 600 km away from its base at 200 km/h, then flies back to its base at 300 km/h. What is its average speed?

10. Surprisingly, very few athletes can jump more than 2 feet (0.6 m) straight up. Use $d = \frac{1}{2}gt^2$ and solve for the time one spends moving upward in a 2-foot vertical jump. Then double it for the "hang-time"—the time one's feet are off the ground.

3

Nonlinear Motion

: **T**he curved path is a composite of horizontal and vertical components of
: motion.

In the previous chapter we distinguished between two kinds of *linear motion*—constant-velocity motion and accelerated motion. This chapter extends these ideas to *nonlinear motion*—motion along a curved path. We will see that the curved motion of a tossed baseball, or any projectile, is a combination of two "components" of motion—horizontal motion (without acceleration) and vertical motion (under the acceleration of gravity). These two components of motion are completely independent of each other. The vertical part doesn't depend on the horizontal part, and vice versa—interesting stuff.

Motion Is Relative

Everything moves. Even things that appear to be at rest are moving. They move with respect to, or relative to, the sun and stars. A book at rest relative to the table it lies on, is moving at about 30 kilometers per second relative to the sun. And it moves even faster relative to the center of our galaxy. When we discuss the motion of something, we describe its motion relative to something else. When we say that a space shuttle moves at 8 kilometers per second, we mean relative to the earth below. When we say an express train travels at 200 kilometers per hour, of course we mean relative to the track. Unless stated otherwise, an object's speed is taken to be relative to the surface of the earth.

Consider a slow-moving airplane. If there is no wind blowing, an airplane that travels at 100 kilometers per hour relative to the air also travels at 100 kilometers per hour relative to the ground below. But if there is wind, the speed of the airplane relative to the air and its speed relative to the ground below are different. If the plane flies in the same direction as the wind, its ground speed is greater; opposite to the wind, its ground speed is less. If it flies in a crosswind direction, its speed relative to the ground is different still. These different speeds can be shown by a useful technique that involves *vectors*.

Velocity— A Vector Quantity

FIGURE 3.1 This vector, scaled so that 1 cm equals 20 km/h, represents a velocity of 60 km/h to the right.

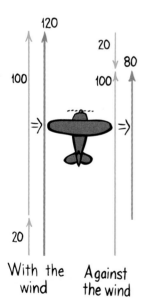

With the wind   Against the wind

FIGURE 3.2 The velocity of an airplane relative to the ground depends on its velocity relative to still air and to the wind.

Recall from Chapter 2 the difference between speed and velocity—speed is a measure of "how fast"; velocity is a measure of both how fast and "which way." If the speedometer in a car reads 100 kilometers per hour, for example, you know your *speed*. If there is also a compass on the dashboard, indicating the car is moving due north, you know your *velocity*—100 kilometers per hour north. To know your velocity is to know your speed *and* your direction.

Any quantity that requires both magnitude (the amount of, or how much) and direction (which way) for a complete description is a **vector quantity**. Thus velocity is a vector quantity. Other examples are force and acceleration. By contrast, a quantity that can be described by magnitude only, not involving direction, is called a **scalar quantity**. Speed is a scalar quantity, as are mass and volume. So speed is a scalar quantity and velocity is a vector quantity. In this chapter we will consider the vector nature of velocity.

Pictures are often more descriptive than words. It is useful to represent a vector quantity by an arrow. Whenever the length of the arrow represents the magnitude of the quantity and the direction of the arrow represents the direction of the quantity, we call the arrow a **vector**. The vector in Figure 3.1 is scaled so that 1 centimeter represents 20 kilometers per hour; it is 3 centimeters long and points to the right, and therefore it represents a velocity of 60 kilometers per hour to the right.

Consider an airplane flying due north at 100 kilometers per hour relative to the surrounding air. We can represent this by an arrow drawn to scale—the length of the arrow corresponding to 100 kilometers per hour and its direction corresponding to the direction north. There is a tailwind (wind from behind) that also moves due north at a velocity of 20 kilometers per hour. This example is represented with vectors in Figure 3.2 left. Here the velocity vectors are scaled so that 1 centimeter represents 20 kilometers per hour. Thus, the 100-kilometer-per-hour velocity of the airplane is shown by the 5-centimeter vector and the 20-kilometer-per-hour tailwind is shown by the 1-centimeter vector. You can see (with or without the vectors) that the resulting velocity is going to be 120 kilometers per hour. Without the tailwind, the airplane would travel 100 kilometers in 1 hour relative to the ground below. With the tailwind, it would travel 120 kilometers in one hour.

Suppose the aircraft makes a U turn, so that the wind is head-on (a headwind). Now the velocity vectors are in opposite directions (Figure 3.2 right). The result is 100 kilometers per hour minus 20 kilometers per hour, which equals 80 kilometers per hour. Flying against a 20-kilometer-per-hour headwind, the airplane would travel only 80 kilometers relative to the ground in 1 hour.

Questions

1. Consider a motorboat that normally travels 10 km/h in still water. If the boat travels in a river that flows also at a rate of 10 km/h, what will be its velocity relative to the shore when it heads directly upstream? When it heads directly downstream?

2. Suppose the boat is oriented at right angles to the water flow. With its motor still running as before, will its speed be greater, less, or the same as 10 km/h?

Adding vectors that act along parallel directions is simple enough: if they are in the same direction, they add; if they are in opposite directions, they subtract. The sum of two or more vectors is called their **resultant**. To find the resultant of two vectors that are at angles to each other, we use the *parallelogram rule.** Construct a parallelogram wherein the two vectors are adjacent sides—the diagonal of the parallelogram shows the resultant. This is shown in Figure 3.3, where the parallelogram is a rectangle. (In this chapter we'll treat only vectors that are either parallel or that make rectangles. A more general treatment of vectors is in Appendix D on page 697.)

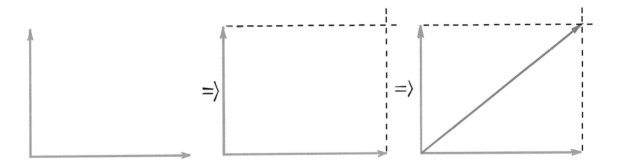

FIGURE 3.3   The pair of vectors at right angles to each other make two sides of a rectangle, the diagonal of which is their resultant.

With the vector technique we can correct for the effect of a crosswind on the velocity of an airplane. Consider a slow-moving airplane that flies north at 80 km/h and is caught in a strong crosswind of 60 km/h blowing east. Figure 3.4 shows vectors for the airplane velocity and wind velocity. The scale here is 1 cm:20 km/h. The diagonal of the constructed parallelogram (rectangle in this case) measures 5 cm, which represents 100

Answers

1. When the boat heads upstream its velocity is zero relative to the land (a velocity of +10 added to a velocity of −10 equals zero). When the boat heads directly downstream its velocity is 20 km/h in the direction of river flow (a velocity of +10 added to a velocity of +10 equals +20 in the same direction).

2. Greater than 10 km/h. How much greater? Read on!

---

* A parallelogram is a four-sided plane figure having the opposite sides parallel and equal.

FIGURE 3.4 An 80-km/h airplane flying in a 60-km/h crosswind has a resultant speed of 100 km/h relative to the ground.

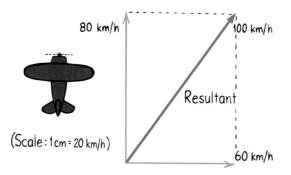

km/h. So the airplane moves at 100 km/h relative to the ground, in a direction between north and northeast.*

There is a special case of the parallelogram that often occurs. When two vectors that are equal in magnitude and at right angles to one another are to be added, the parallelogram becomes a square. Since for any square the length of a diagonal is $\sqrt{2}$, or 1.414, times one of the sides, the resultant is $\sqrt{2}$ times one of the vectors. For example, the resultant of two equal vectors of magnitude 100 acting at right angles to each other is 141.4.

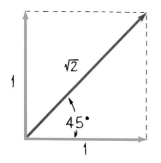

FIGURE 3.5 The diagonal of a square is $\sqrt{2}$ the length of one of its sides.

Question Again consider the motorboat that normally travels 10 km/h in still water. If the boat heads directly across the river, which also flows at a rate of 10 km/h, what will be its velocity relative to the shore?

We can look at the vector configuration of Figure 3.5 from another point of view. Just as a vertical and a horizontal vector combine to produce a single resultant vector at some angle between them, a single vector at any angle can be "resolved" (separated) into a pair of vectors that are at right angles to each other. This pair of vectors are the *components* of the single vector.

Answer When the boat heads cross-stream (at right angles to the river flow) its velocity is 14.14 km/h, 45° downstream (in accord with the diagram in Figure 3.5).

* Whenever the vectors are at right angles to each other, their resultant can be found by the Pythagorean theorem, a well-known tool of geometry. It states that the square of the hypotenuse of a right-angle triangle is equal to the sum of the squares of the other two sides. Note that two right triangles are present in the parallelogram (rectangle in this case) in Figure 3.4. From either one of these triangles we get:

$$\text{resultant}^2 = (60 \text{ km/h})^2 + (80 \text{ km/h})^2$$

$$= 3600 \text{ (km/h)}^2 + 6400 \text{ (km/h)}^2$$

$$= 10,000 \text{ (km/h)}^2$$

The square root of 10,000 (km/h)^2 is 100 km/h, as expected.

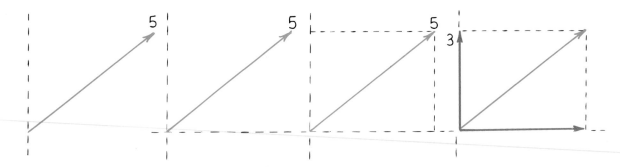

FIGURE 3.6 Construction of the vertical and horizontal components of a vector.

This is illustrated in Figure 3.6 for a vector that has magnitude 5 and is 37° from the horizontal. We can resolve it into horizontal and vertical components. What would be the magnitude of each component? For this special case of 37°, the horizontal is 4 and the vertical is 3.

Any vector may be represented by horizontal and vertical components. Figure 3.7, as an example, shows the horizontal and vertical velocity components of a stone that is thrown into the air. Its initial path is greater than 45° to the horizontal. Notice that the vertical component of velocity is greater than the horizontal component. Because of gravity, the vertical component of velocity will decrease as the stone rises and increase when the stone heads back toward ground level. So the path of the stone is curved.

Projectile Motion

A **projectile** is any object that is projected by some means and continues in motion under the influence of gravity (and air drag). The path of a projectile is called a *trajectory*. A stone thrown into the air, a cannonball shot from a cannon, and a ball that rolls off the edge of a table are all projectiles that follow curved trajectories, which at first glance seem rather complicated. However, these trajectories will seem surprisingly simple when we look at the horizontal and vertical components of motion separately.

The horizontal component of motion for a projectile is no more complex than the horizontal motion of a bowling ball rolling freely along a level bowling alley. If the retarding effect of friction can be ignored, the bowling ball moves at constant velocity. Since there is no force acting horizontally on the ball, it rolls of its own inertia and

FIGURE 3.7 Vertical and horizontal components of the stone's velocity.

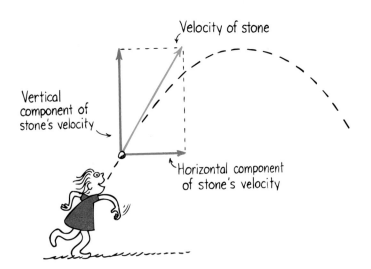

FIGURE 3.8 (top) Roll a ball along a level surface, and its velocity is constant because no component of gravitational force acts horizontally. (left) Drop it, and it accelerates downward and covers a greater vertical distance each second.

covers equal distances in equal intervals of time. It rolls without accelerating. The *horizontal* part of a projectile's motion is just like the bowling ball's motion along the alley (Figure 3.8 top).

The *vertical* component of motion for a projectile following a curved trajectory is just like the motion described in Chapter 2 for a freely falling object (Figure 3.8 left). Like a ball dropped in midair, the projectile, drawn by gravity, accelerates downward. We will see in the next chapter that force produces acceleration, and both act in the same direction. It is because of the vertical force of gravity that projectiles accelerate in the vertical direction.

Interestingly enough, the horizontal component of motion for a projectile is completely independent of the vertical component of motion. Unless air drag or some other horizontal force acts, the constant horizontal velocity component is not affected by the vertical force of gravity. Gravity affects only the vertical component of motion. It is important to understand that *each component acts independently of the other.* Their combined effects produce the trajectories that projectiles follow.

These ideas are nicely illustrated in the simulated multiple-flash exposure in Figure 3.9, which shows equally timed successive positions for a ball rolled off a horizontal table. Investigate the "photo" carefully, for there's a lot of good physics there. The trajectory of the ball is best analyzed by considering the horizontal and vertical components of motion separately. There are two important things to notice. The first is that the ball's horizontal component of motion doesn't change. The ball travels the same horizontal distance in the equal times between each flash. That's because there is no component of force acting horizontally. Gravity acts only downward, so the only acceleration of the ball is downward. The second thing to note from the figure is that the vertical positions become farther apart with time—the same as if the ball were simply dropped. It is interesting to note that the downward motion of the ball is the same as that which occurs in free fall. Figure 3.10 is an actual strobe-light photograph of two balls that start to move from the same height at the same time.

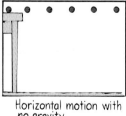
Horizontal motion with no gravity

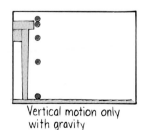
Vertical motion only with gravity

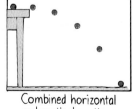
Combined horizontal and vertical motion

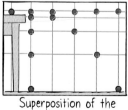
Superposition of the preceding cases

FIGURE 3.9 Simulated photographs of a moving ball illuminated with a strobe light.

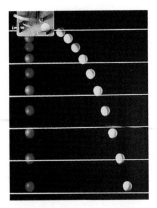

FIGURE 3.10 A strobe-light photograph of two golf balls released simultaneously from a mechanism that allows one ball to drop freely while the other is projected horizontally.

The trajectory of a projectile that accelerates only in the vertical direction while moving at a constant horizontal velocity is a **parabola**. A trajectory will be parabolic when air resistance can be neglected. This usually occurs for slow-moving projectiles or for ones that are very heavy compared to the forces of air resistance.

Question At the instant a horizontally held rifle is fired over a level range, a bullet held at the side of the rifle is released and drops to the ground. Which bullet, the one fired downrange or the one dropped from rest, strikes the ground first?

Consider a cannonball shot at an upward angle (Figure 3.11). Pretend for a moment that there is no gravity; according to the law of inertia, the cannonball would follow the straight-line path shown by the dashed line. But there is gravity, so this doesn't happen. What really happens is that the cannonball continuously falls beneath the imaginary line until it finally strikes the ground. Get this: The vertical distance it falls beneath any point on the dashed line is the same vertical distance it would fall if it were dropped from rest and had been falling for the same amount of time. This distance, as introduced in Chapter 2, is given by $d = \frac{1}{2}gt^2$, where t is the elapsed time.

FIGURE 3.11 With no gravity, the projectile would follow a straight-line path (dashed line). But because of gravity, it falls beneath this line the same vertical distance it would fall if released from rest. Compare the distances fallen with those given in Table 2.3 in Chapter 2. (With $g = 9.8$ m/s², these distances are more accurately 4.9 m, 19.6 m, and 44.1 m.)

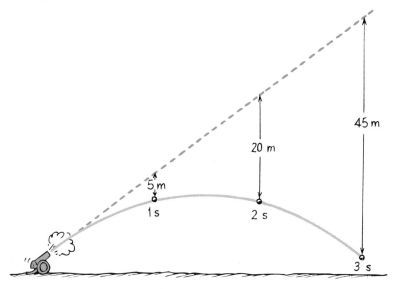

5 m
1 s
20 m
2 s
45 m
3 s

Answer Both bullets fall the same vertical distance with the same acceleration g due to gravity and therefore strike the ground at the same time. Can you see that this is consistent with our analysis of Figure 3.9 and 3.10? We can reason this another way by asking which bullet would strike the ground first if the rifle were pointed at an upward angle. In this case, the bullet that is simply dropped would hit the ground first. Now consider the case where the rifle is pointed downward. The fired bullet hits first. So upward, the dropped bullet hits first; downward, the fired bullet hits first. There must be some angle at which there is a dead heat—where both hit at the same time. Can you see it would be when the rifle is neither pointing upward nor downward—when it is horizontal?

We can put it another way: Shoot a projectile skyward at some angle and pretend there is no gravity. After so many seconds t, it should be at a certain point along a straight-line path. But because of gravity, it isn't. Where is it? The answer is that it's directly below this point. How far below? The answer in meters is $5t^2$ (or, more precisely, $4.9t^2$). How about that!

Note another thing from Figure 3.11. The cannonball moves equal horizontal distances in equal time intervals. That's because no acceleration takes place horizontally. The only acceleration is vertical, in the direction of earth's gravity. The vertical distance it falls below the imaginary straight-line path during equal time intervals continuously increases with time.

Questions

1. Suppose the cannonball in Figure 3.11 were fired faster. How many meters below the dashed line would it be at the end of the 5 s?

2. If the horizontal component of the cannonball's velocity were 20 m/s, how far downrange would the cannonball be at the end of 5 s?

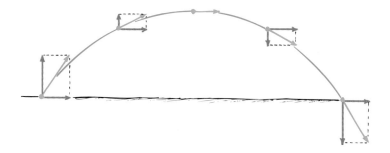

FIGURE 3.12 The velocity of a projectile at various points along its path. Note that the vertical component changes and the horizontal component is the same everywhere.

In Figure 3.12 we see vectors representing both horizontal and vertical components of velocity for a projectile following a parabolic trajectory. Notice that the horizontal component is everywhere the same, and only the vertical component changes.

Answers

1. Using $g = 10$ m/s^2, the vertical distance beneath the dashed line at the end of 5 s is 125 m $[d = 5t^2 = 5(5)^2 = 5(25) = 125]$. (If we use $g = 9.8$ m/s^2, then this distance is 122.5 m/s.) Interestingly enough, this distance doesn't depend on the angle of the cannon. If air resistance is neglected, any projectile will fall $5t^2$ meters below where it would have reached if there were no gravity.

2. In the absence of air resistance, the cannonball will travel a horizontal distance of 100 m $[d = \bar{v}t = (20)(5) = 100]$. Note that since gravity acts only vertically and there is no acceleration in the horizontal direction, the cannonball travels equal horizontal distances in equal times. This distance is simply its horizontal component of velocity multiplied by the time (and not $5t^2$, which applies only to vertical motion under the acceleration of gravity).

Note also that the actual velocity is represented by the vector that forms the diagonal of the rectangle formed by the vector components. At the top of the trajectory the vertical component is zero, so the actual velocity there is just equal to the horizontal component of velocity. Everywhere else the magnitude of velocity is greater (just as the diagonal of a rectangle is greater than either of its sides).

Figure 3.13 shows the trajectory traced by a projectile launched with the same speed at a steeper angle. Notice the initial velocity vector has a greater vertical component than when the angle of launch is smaller. This greater component results in a trajectory that reaches a greater height. But the horizontal component is less, and the range is less.

FIGURE 3.13 Path for a steeper projection angle.

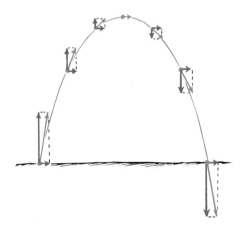

Figure 3.15 shows the paths of several projectiles, all with the same initial speed but different launching angles. The figure neglects the effects of air resistance, so the trajectories are all parabolas. Notice that these projectiles reach different *altitudes,* or heights above the ground. They also have different *horizontal ranges,* or distances traveled horizontally. The remarkable thing to note from Figure 3.15 is that the same range is obtained from two different launching angles—angles that add up to 90°! An object thrown into the air at an angle of 60°, for example, will have the same range as if it were thrown at the same speed at an angle of 30°. For the smaller angle, of course, the object remains in the air for a shorter time. The greatest range occurs when the launching angle is 45°.

FIGURE 3.14 Maximum range is attained when a ball is batted at an angle of nearly 45°. For common cases in which the weight of the projectile is significant in comparison with the force that launches it, some of the launching force lifts instead of accelerating the projectile, resulting in smaller speeds for greater projection angles. (Just as when throwing a heavy boulder, the greater the launching angle, the smaller the speed it leaves your hand.) So when the weight of the projectile is a factor, maximum range occurs for angles less than 45°.

FIGURE 3.15 Ranges of a projectile shot at the same speed at different projection angles.

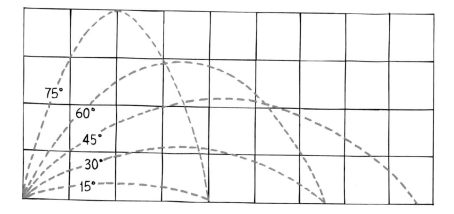

We have emphasized the special case of projectile motion without air resistance. When there is air resistance, the range of the projectile is diminished, the altitude of the projectile less, and the trajectory is not a true parabola (Figure 3.16).

FIGURE 3.16 In the presence of air resistance, the trajectory of a high-speed projectile falls short of a parabolic path.

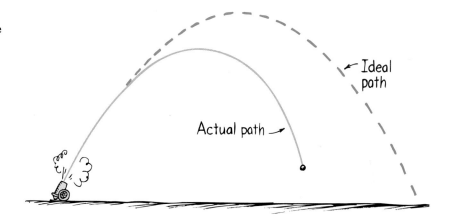

Questions

1. A projectile is shot at an angle into the air. If air resistance is negligible, what is the acceleration of its vertical component of motion? Of its horizontal component of motion?

2. At what part of its trajectory does a projectile have minimum speed?

Answers

1. The acceleration in the vertical direction is g because the force of gravity is along the vertical direction. (We will see in the following chapter that acceleration is always in the direction of the force that acts on an object.) The acceleration is zero in the horizontal direction because no horizontal force acts on the projectile.

2. The speed of a projectile is minimum at the top of its trajectory. If it is launched vertically, its speed at the top is zero. If it is projected at an angle, the vertical component of speed is zero at the top, leaving only the horizontal component. So the speed at the top is equal to the horizontal component of the projectile's velocity at any point. Isn't that nice?

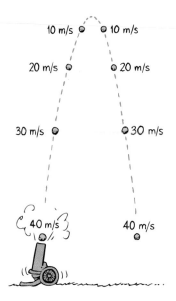

FIGURE 3.17 Without air resistance, speed lost while going up equals speed gained while coming down; time going up equals time coming down.

When air resistance is small enough to be negligible, a projectile will rise to its maximum height in the same time it takes to fall from that height to the ground (Figure 3.17). This is because its deceleration by gravity while going up is the same as its acceleration by gravity while coming down. The speed it loses while going up is therefore the same as the speed it gains while coming down. So the projectile arrives at the ground with the same speed it had when it was projected from the ground.

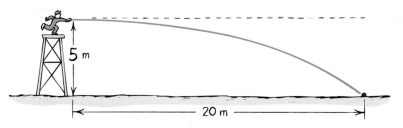

FIGURE 3.18 How fast is the ball thrown?

Question The boy on the tower throws a ball 20 m downrange as shown in Figure 3.18. What is his pitching speed?

Baseball games normally take place on level ground. For the short-range projectile motion on the playing field, the earth can be considered to be flat because the flight of the baseball is not affected by the earth's curvature. For very long-range projectiles, however, the curvature of the earth's surface must be taken into account. We'll now see that if an object is projected fast enough, it will fall all the way around the earth and be an earth satellite.

Fast-Moving Projectiles— Satellites

Consider the baseball pitcher on the tower in Figure 3.18. If gravity did not act on the ball, the ball would follow a straight-line path shown by the dashed line. But there is gravity, so the ball falls below this straight-line path. In fact, 1 second after the ball leaves the pitcher's hand it will have fallen a vertical distance of 5 meters below the dashed line—whatever the pitching speed. It is important to understand this, for it is the crux of satellite motion.

Answer The ball is thrown horizontally, so the pitching speed is horizontal distance divided by time. A horizontal distance of 20 m is given, but the time is not stated. However, you can find the time because you know the vertical distance the ball drops—5 m, which takes 1 s! This means it travels horizontally 20 m in 1 s. So its horizontal component of velocity must be 20 m/s. From the equation for constant speed (which applies to horizontal motion) $v = d/t = (20 \text{ m})/(1\text{s}) = 20$ m/s. It is interesting to note that consideration of the equation for constant speed, $v = d/t$ guides thinking about the crucial factor in this problem—the time.

Hang-Time Revisited

In Chapter 2 we stated that airborne time during a jump is independent of horizontal speed. Now we see why this is so—horizontal and vertical components of motion are independent of each other. The rules of projectile motion apply to jumping. Once the feet are off the ground, only the force of gravity acts on the jumper (neglecting air resistance). Hang-time depends only on the vertical component of lift-off velocity. It turns out that jumping lift-off force can be somewhat increased by the action of running, so hang-time for a running jump usually exceeds hang-time for a standing jump. But once the feet are off the ground, only the vertical component of lift-off velocity determines hang-time.

FIGURE 3.19 Throw a stone at any speed and one second later it will have fallen 5 m below where it would have been without gravity.

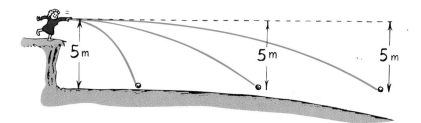

An earth satellite is simply a projectile that falls *around* the earth rather than *into* it. The speed of the satellite must be great enough to ensure that its falling distance matches the earth's curvature.* A geometrical fact about the curvature of our earth is that its surface drops a vertical distance of 5 meters for every 8000 meters tangent to the surface (Figure 3.20).† This means that if you were floating in a calm ocean, you would be able to see only the top of a 5-meter mast on a ship 8 kilometers away. So if a baseball could be thrown fast enough to travel a horizontal distance of 8 kilometers during the time (1 second) it takes to fall 5 meters, then it would follow the curvature of the earth.

FIGURE 3.20 Earth's curvature—not to scale!

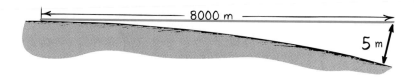

* The conventional definition of *to fall* is to "to get closer to the earth"; satellites such as the moon do not do this. In science we will find many cases where the technical definition differs from the conventional. For example, we say "the sun rises" and "the moon sets," but technically they do not.

† A tangent to a circle or to the earth's surface is a straight line that touches the circle or surface at one place, so its direction matches the direction of the circle at the point of contact.

FIGURE 3.21 If the speed of the stone and the curvature of its trajectory are great enough, the stone may become a satellite.

A little thought will show that this speed is 8 kilometers per second. If this doesn't seem fast, convert it to kilometers per hour and you get an impressive 29,000 kilometers per hour (or 18,000 miles per hour)!

At this speed, atmospheric friction would burn the baseball or even a piece of iron to a crisp. This is the fate of bits of rock and other meteorites that enter the earth's atmosphere and burn up, appearing as "falling stars." That is why satellites such as the space shuttles are launched to altitudes of 150 kilometers or more. A common misconception is that satellites orbiting at high altitudes are free from gravity. Nothing could be further from the truth. We will see later that the force of gravity on a satellite 200 kilometers above the earth's surface is almost as strong as at the surface. The high altitude is to put the satellite beyond the earth's atmosphere, where air resistance is almost totally absent, but not beyond earth's gravity. We'll return to satellite motion and gravity in Chapters 8 and 9.

FIGURE 3.22 The space shuttle is a projectile in a constant state of free fall. Because of its tangential velocity, it falls around the earth rather than vertically into it.

Circular Motion

Which moves faster on a merry-go-round—a horse near the outside rail or a horse near the inside rail? Ask different people this question, and you'll get different answers. That's because it's easy to get linear speed confused with rotational speed.

Linear speed is what we have been calling simply *speed*—the distance in meters or kilometers moved per unit of time. A point on the outside of a merry-go-round or turntable moves a greater distance in one complete rotation than a point on the inside. The linear speed is greater on the outside of a rotating object than inside and closer to the axis. The speed of something moving along a circular path can be called **tangential speed** because the direction of motion is always tangent to the circle. For circular motion we can use the terms *linear speed* and *tangential speed* interchangeably.

FIGURE 3.23 When a phonograph record turns, a ladybug farther from the center travels a longer path in the same time.

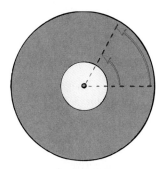

Rotational speed (sometimes called angular speed) refers to the number of rotations or revolutions per unit of time. All parts of the rigid merry-go-round and turntable turn about the axis of rotation *in the same amount of time.* All parts share the same rate of rotation, or *number of rotations or revolutions per unit of time.* It is common to express rotational rates in revolutions per minute (RPM).* Phonograph records that were common not long ago, for example, rotate at $33\frac{1}{3}$ RPM. A ladybug sitting anywhere on the surface of the record revolves at $33\frac{1}{3}$ RPM.

FIGURE 3.24 The entire phonograph record rotates at the same rotational speed, but ladybugs at different distances from the center travel at different linear speeds. A ladybug sitting twice as far from the center moves twice as fast.

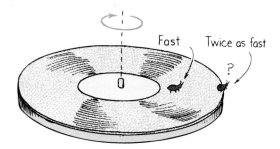

Tangential speed and rotational speed are related. Have you ever ridden on a giant, rotating round platform in an amusement park? The faster it turns, the faster is your tangential speed. This makes sense; the greater the RPMs, the faster your speed in meters per second. More exactly, if you are at a given distance from the center, there is a direct proportion between tangential speed and rotational speed. If, for example, you double the RPMs, you double your tangential speed; triple the RPMs and you triple your tangential speed. We say that tangential speed is *directly proportional* to rotational speed (at a fixed radial distance).

* Physics types usually describe rotational speed in terms of the angle turned in a unit of time. Then the symbol for rotational speed is ω (the Greek letter *omega*).

FIGURE 3.25 The tangential speed of each person equals the rotational speed of the platform multiplied by the distance from the central axis.

Tangential speed, unlike rotational speed, depends on the distance from the axis (Figure 3.25). At the very center of the rotating platform, you have no speed at all; you merely rotate. But as you approach the edge of the platform you find yourself moving faster and faster. Tangential speed is directly proportional to distance from the axis (for a given rotational speed). Move out twice as far from the rotational axis at the center and you move twice as fast. Move out three times as far and you have three times as much tangential speed. If you find yourself in any rotating system whatever, your tangential speed depends on how far you are from the axis of rotation. When a row of people locked arm in arm at the skating rink make a turn, the motion of "tail-end Charlie" is evidence of this greater speed.

So tangential speed is directly proportional to both rotational speed and radial distance.*

Question On a rotating platform like that shown in Figure 3.25, if you sit halfway between the rotating axis and the outer edge and have a rotational speed of 20 RPM and a tangential speed of 2 m/s, what will be the rotational and tangential speeds of your friend who sits at the outer edge?

Answer Since the rotating platform is rigid, all parts have the same rotational speed, so your friend also rotates at 20 RPM. Tangential speed is a different story; since she is twice as far from the axis of rotation, she moves twice as fast—4 m/s.

---

* If you take a follow-up physics course you will learn that when the proper units are used for tangential speed v, rotational speed ω, and radial distance r, the direct proportion of v to both r and ω becomes the exact equation $v = r\omega$. So in a rotating system wherein all parts simultaneously have the same ω, like a wheel, disk, or rigid wand, the tangential speed varies with r. Planets orbit the sun at both different distances and different rotational speeds, so their tangential speeds vary with different values of both ω and r. (We will learn in Chapter 9 that the innermost planets in a planetary system have both the greatest rotational speed and greatest tangential speed.)

Railroad Train Wheels

FIGURE 3.26 Because the wide part of the cup rolls faster than the narrow part, the cup rolls in a curve.

Why does a moving railroad train stay on the tracks? Most people assume the wheel flanges keep the wheels from rolling off. But if you look at these flanges you'll note they may be rusty. They seldom touch the track, except when they follow slots that switch the train from one set of tracks to another. So how do the wheels of a train stay on the tracks? They stay on the track because their rims are slightly tapered.

If you roll a tapered cup across a surface, it makes a curved path (Figure 3.26). The larger-diameter end rolls a greater distance per revolution and has a greater linear speed than the smaller end. If you fasten a pair of cups together at their wide ends (simply taping them together) and roll the pair along a pair of parallel tracks (Figure 3.27), the cups will remain on the track and center themselves whenever they roll off center. This occurs because when the pair rolls to the left of center, say, the wider part of the left cup rides on the left track while the narrow part of the right cup rides on the right track. This steers the pair toward the center. If it "overshoots" toward the right, the process repeats, this time toward the left, as the wheels tend to center themselves. Likewise for a railroad train, where passengers feel the train swaying as these corrective actions occur.

This tapered shape is essential on the curves of railroad tracks. On any curve, the distance along the outer part is longer than the distance along the inner part (as we saw in Figure 3.23). So whenever a vehicle follows a curve, its outer wheels travel faster than its inner wheels. For an automobile this is no problem because the wheels are freewheeling, and roll independently of each other. For a train, however, like the pair of fastened cups, pairs of wheels are firmly connected so that they rotate together.

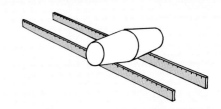

FIGURE 3.27 A pair of fastened cups stay on the tracks as they roll, for when they roll off center, the different speeds due to the taper cause them to self-correct toward the center of the track.

FIGURE 3.28 Wheels of a railroad train are slightly tapered (shown exaggerated here).

Opposite wheels have the same RPM at any time. But due to the slightly tapered rim of the wheel, its speed along the track depends on whether it rides on the narrow part of the rim or the wide part. On the wide part it travels faster. So when a train rounds a curve, wheels on the outer track ride on the wider part of the tapered rims while opposite wheels ride on their narrow parts. In this way, the wheels have different linear speeds for the same rotational speed. Can you see that if the wheels were not tapered, scraping would occur and the wheels would squeal when a train rounds a curve?

FIGURE 3.29 (left) Along a track that curves to the left, the right wheel rides on the wide part and goes faster while the left wheel rides on the narrow part and goes slower. (right) Vice versa when the track curves to the right.

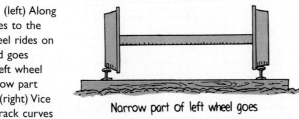

Narrow part of left wheel goes slower. so wheels curve to left

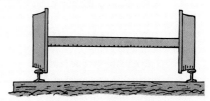

Wide part of left wheel goes faster, so wheels curve to right

Summary of Terms

Vector quantity   A quantity that has both magnitude and direction. Examples are force, velocity, and acceleration.

Scalar quantity   A quantity that has magnitude, but not direction. Examples are mass, volume, and speed.

Vector   An arrow drawn to scale used to represent a vector quantity.

Resultant   The net result of a combination of two or more vectors.

Projectile   Any object that moves through the air or through space under the influence of gravity.

Parabola   The curved path followed by a projectile under the influence of gravity only.

Linear speed   The distance moved per unit of time. Also called simply *speed*.

Tangential speed   The linear speed along a curved path.

Rotational speed   The number of rotations or revolutions per unit of time; often measured in rotations or revolutions per second or per minute.

Review Questions

1. Distinguish between linear motion and nonlinear motion.

Motion Is Relative

2. What is meant by saying that motion is relative?

3. If you walk at 1 m/s down the aisle of a bus that is moving at 10 m/s along the road, how fast are you moving relative to the road when you walk toward the front of the bus? Toward the rear of the bus?

Velocity—A Vector Quantity

4. What is a vector quantity? Give three examples.

5. What is a scalar quantity? Give three examples.

6. An airplane travels 200 km/h through the air. If it heads directly against a 40-km/h wind, what is its speed relative to the ground below?

7. Is a rectangle a parallelogram? Is a square a parallelogram?

8. If an airplane that travels 100 km/h through the air is caught in a 100-km/h crosswind, what will be its velocity relative to the ground?

9. A vector making an angle of 45° to the horizontal (like the one in Figure 3.5) has a magnitude of 141. What are the magnitudes of its horizontal and vertical components?

10. A rock is thrown upward at an angle. What happens to the vertical component of its velocity as it rises? As it falls?

Projectile Motion

11. What exactly is a projectile? Give three examples.

12. A rock is thrown upward at an angle. What happens to the horizontal component of its velocity as it rises? As it falls?

13. True or false: When air resistance does not affect the motion of a projectile, it covers equal horizontal distances in equal time intervals.

14. True or false: When air resistance does not affect the motion of a projectile, its horizontal and vertical components of velocity remain constant.

15. True or false: Changes in the vertical component of velocity for a projectile are proportional to changes in the horizontal component of velocity.

16. A projectile falls beneath the straight-line path it would follow if there were no gravity. How many meters does it fall below this line if it has been traveling for 1 s? For 2 s?

17. Does your answer to the last question depend on the angle at which the projectile is launched?

18. Why does the vertical component of velocity for a projectile change with time, whereas the horizontal component of velocity doesn't?

19. A projectile is launched upward at an angle of 75° from the horizontal and strikes the ground a certain distance down range. For what other angle of launch at the same speed would this projectile land just as far away?

20. A projectile is launched vertically at 100 m/s. If air resistance can be neglected, at what speed will it return to its initial level?

Fast-Moving Projectiles—Satellites

21. How can a projectile "fall around the earth"?

22. Why will a projectile that moves horizontally at 8 km/s follow a curve that matches the curvature of the earth?

23. Why is it important that the projectile in the last question be above the earth's atmosphere?

Circular Motion

24. Distinguish between tangential speed and rotational speed.

25. If the rotational speed of a platform is doubled, how does the tangential speed anywhere on the platform change?

26. If the rotational speed on a platform is unchanged but your distance from the center is doubled, what happens to your tangential speed?

Projects

1. Drop a stone from a bridge at the same time your friend tosses a stone horizontally, and notice that both stones strike the water at the same time.

2. Fasten a pair of foam cups together at their wide ends and roll them along a pair of meter sticks that simulate railroad tracks. See Figure 3.27. Note how they self correct whenever their path departs from the center. Question: If you taped the cups together at their narrow ends, so they tapered oppositely, would the pair of cups self-correct, or self-destruct, when rolling slightly off center?

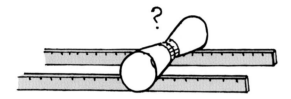

Exercises

1. What is the impact speed when a car moving at 100 km/h bumps into the rear of another car traveling in the same direction at 98 km/h?

2. What is the impact speed of a 10-km/h wind against the sail of a sailboat sailing in the same direction at 4 km/h? What would the wind impact speed be if the boat moved at 10 km/h in the same direction as the wind?

3. Which of the following are scalar quantities, which are vector quantities, and which are neither? (a) velocity; (b) age; (c) color; (d) speed; (e) acceleration; (f) shape; (g) temperature.

4. When two vectors sum to zero, how must they be related?

5. Why does vertically falling rain make slanted streaks on the side windows of a moving automobile? If the streaks make an angle of 45°, what does this tell you about the relative speed of the car and the falling rain?

6. If you are standing in a bus that moves at constant velocity and drop a ball from your outstretched hand, you'll see its path as a vertical straight line. How will the path appear to a friend standing at the side of the road?

7. A heavy crate accidentally falls from a high-flying airplane just as it flies directly above a shiny red Camaro smartly parked in a car lot. Relative to the Camaro, where will the crate crash?

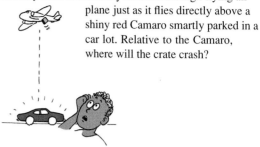

8. Suppose you drop an object from an airplane traveling at constant velocity, and further suppose that air resistance doesn't affect the falling object. What will be its falling path as observed by someone at rest on the ground, not directly below but off to the side where a clear view can be seen? What will be the falling path as observed by you looking downward from the airplane? Where will the object strike the ground, relative to you in the airplane? Where will it strike in the more realistic case where air resistance does affect the fall?

9. How does the downward component of the motion of a projectile compare to the motion of vertical free fall?

10. When you throw a rock off a cliff, what does or doesn't happen to the horizontal component of velocity along the rock's trajectory?

11. At what point in its trajectory does a batted baseball have its minimum speed? How does this compare with the horizontal component of its velocity at other points? (Ignore air drag.)

12. How far below an initial straight-line path will a projectile fall in 1 s? Does your answer depend on the angle of launch? Does it depend on the initial speed of the projectile? Defend your answer.

13. Suppose you are on a ledge in the dark and wish to estimate the height of the ledge above the ground. So you drop a stone over the edge and hear it strike the ground 1 second later. What is the height of the ledge? How could you use the stone to estimate the height if you weren't close enough to the edge to simply drop the stone? Explain.

14. A friend claims that bullets fired by some high-powered rifles travel for many meters in a straight-line path before they start to fall. Another friend disputes this claim and states that all bullets from any rifle drop beneath a straight-line path a vertical distance given by $\frac{1}{2}gt^2$ and that the curved path is apparent at low velocities and less apparent at high velocities. Now it's your turn: Will all bullets drop the same vertical distance in equal times? Explain.

15. At what angle should you hold a garden hose so that the stream of water will go farthest if air resistance is unimportant?

16. When a rifle is being fired at a distant target, why isn't the barrel lined up so that it points exactly at the target?

17. A park ranger shoots a monkey hanging from a branch of a tree with a tranquilizing dart. The ranger aims directly at the monkey, not realizing that the dart will follow a parabolic path and thus fall below the monkey. The monkey, however, sees the dart leave the gun and lets go of the branch to avoid being hit. Will the monkey be hit anyway? Does the velocity of the dart affect your answer, assuming it is great enough to travel the horizontal distance to the tree before hitting the ground? Defend your answer.

18. A projectile is fired straight upward at 141 m/s. How fast is it moving upward at the instant it reaches the top of its trajectory? Suppose instead it were fired upward at 45°. What would be its speed at the top of its trajectory?

19. When you jump upward, your hang-time is the time your feet are off the ground. Does hang-time depend on your vertical component of velocity when you jump, your horizontal component of velocity, or both? Defend your answer.

20. The hang-time of a basketball player who jumps a vertical distance of 2 feet (0.6 m) is about $\frac{2}{3}$ second. What will be the hang-time if the player reaches the same height while jumping 4 feet (1.2 m) horizontally?

21. When you take the curvature of the earth into account, will the range of a projectile be greater than, less than, or the same as the range figured for a flat earth?

22. Assuming no air resistance, why does an 8-km/s horizontally fired projectile not strike the earth's surface?

23. Since the moon is gravitationally attracted to the earth, why doesn't it simply crash into the earth?

24. When the space shuttle coasts in a circular orbit at constant speed about the earth, is it accelerating? If so, in what direction? If not, why not?

25. A ladybug sits halfway between the axis and the edge of a phonograph record. What will happen to its tangential speed if the RPM rate is doubled? At this doubled rate, what will happen to its tangential speed if it has crawled out to the edge?

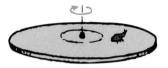

26. A large wheel is coupled to a wheel with half the diameter as shown. How does the rotational speed of the smaller wheel compare with that of the larger wheel? How do the tangential speeds at the rims compare (assuming the belt doesn't slip)?

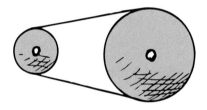

27. If you replace the regular tires on your car with tires of larger diameter, how will your speedometer reading differ?

28. Harry and Sue cycle at the same speed. The tires on Harry's bike are larger diameter than those on Sue's bike. Which, if either, has the greatest rotational speed? The greatest tangential speed?

29. Wheels on an automobile can rotate independently. Pairs of wheels on railroad cars, however, are rigidly connected so that opposite wheels rotate at identical rates. Why?

30. Make up a multiple-choice question suitable for an exam that distinguishes between rotational speed and tangential speed.

Please let your primary goal in learning physics be acquainting yourself with the information and insights within the chapters, and *not* primarily in doing the exercises, which are intended as some mental pushups to try *after* you have studied the chapter material!

Problems

1. Consider an airplane that normally has an airspeed of 100 km/h in a 100-km/h crosswind blowing from west to east. Calculate its ground velocity when its nose is pointed north in the crosswind.

2. A canoe is paddled at 4 km/h directly across a river that flows at 3 km/h, as shown in the figure. (a) What is the resultant speed of the canoe? (b) How fast and in what direction can the canoe be paddled to reach a destination directly across the river?

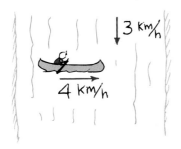

3. An airplane is flying horizontally with speed 1000 km/h (280 m/s) when an engine falls off. Neglecting air resistance, if it takes 30 s for the engine to hit the ground, (a) how high is the airplane, and (b) how far horizontally does the engine travel while it falls? (c) If the airplane somehow continues to fly as if nothing had happened, where is the engine relative to the airplane at the moment the engine hits the ground?

4. A cannonball shot with an initial velocity of 141 m/s at an angle of 45° follows a parabolic path and hits a balloon at the top of its trajectory. Neglecting air resistance, how fast is the cannonball going when it hits the balloon?

5. Students in a lab measure the speed of a steel ball launched horizontally from a table top to be 4.0 m/s. If the table top is 1.5 m above the floor, where should they place a 20-cm tall tin coffee can to catch the ball when it lands?

6. John and Tracy look from their 80-m high-rise balcony to a swimming pool below—not exactly below, but rather 20 m out from the bottom of their building. They wonder how fast they would have to jump horizontally to succeed in reaching the pool. What is the answer?

7. Ignoring air drag, what is the maximum speed possible for a horizontally moving tennis ball as it clears the net 1.0 m high and strikes within the court's border, 12.0 m distant?

8. Calculate a person's hang-time if he moves horizontally 3 m during a 1.25-m high jump. What is his hang-time if he moves 6 m horizontally during this jump?

9. Consider a bicycle that has wheels with a circumference of 2 m. What is the linear speed of the bicycle when the wheels rotate at 1 revolution per second?

10. What is the tangential speed of a passenger on a Ferris wheel that has a radius of 10 m and rotates once in 30 s?

4

· · · · · · · · · · · ·

Newton's Laws of Motion

: **T**erminal velocity occurs when air drag equals weight.

In 1642, several months after Galileo died, Isaac Newton was born. By the time Newton was 23, he developed his famous laws of motion, which completed the overthrow of the Aristotelian ideas that had dominated the thinking of the best minds for 2000 years. In this chapter we will consider these laws in order. The first is a restatement of the concept of inertia as proposed earlier by Galileo. The second law relates acceleration to the cause that produces it, force. The third is the well-known law of action and reaction. These three important laws first appeared in one of the most important books of all time, Newton's famous *Principia*.

Newton's First Law of Motion

Aristotle's idea that a moving object must be propelled by a steady force was completely turned around by Galileo, who stated that in the *absence* of a force, a moving object will continue moving. The tendency of things to resist changes in motion was what Galileo called *inertia*. Newton refined Galileo's idea and made it his first law, appropriately called the **law of inertia**. From Newton's *Principia* (translated from the original Latin):

> **Law 1: Every object continues in its state of rest, or of uniform motion in a straight line, unless it is compelled to change that state by forces impressed upon it.**

The key word in this law is *continues:* an object *continues* to do whatever it happens to be doing unless a force is exerted upon it. If it is at rest, it *continues* in a state of rest. If it is moving, it *continues* to move without turning or changing its speed. In short, the law says that an object does not accelerate of itself; acceleration must be imposed against the tendency of an object to retain its state of motion. Things at rest tend to stay at rest—things moving tend to continue moving. This tendency of things to resist changes in motion is **inertia**.

FIGURE 4.1 Examples of inertia.

Why will the coin drop into the glass when a force accelerates the card?

Why does the downward motion and sudden stop of the hammer tighten the hammerhead?

Why is it that a slow continuous increase in the downward force breaks the string above the massive ball, but a sudden increase breaks the lower string?

Mass

Every material object possesses inertia; how much depends on the amount of matter in the substance of the object—the more matter, the more inertia. In speaking of how much matter something has, we use the term *mass.* The greater the mass of an object, the greater its inertia. **Mass** is a measure of the inertia of a material object.

Mass corresponds to our intuitive notion of **weight**. We casually say that something has a lot of mass if it weighs a lot. But there is a difference between mass and weight. We can define each as follows:

> **Mass: The quantity of matter in an object. It is also the measure of the inertia or sluggishness that an object exhibits in response to any effort made to start it, stop it, or change its state of motion in any way.**

> **Weight: The force upon an object due to gravity.**

Mass and weight are directly proportional to each other. If the mass of an object is doubled, its weight is also doubled; if the mass is halved, the weight is halved. Because of this, mass and weight are often interchanged. Also, mass and weight are sometimes confused because it is customary to measure the quantity of matter in things (mass) by their gravitational attraction to the earth (weight). But mass is more fundamental than weight; it is a fundamental quantity that completely escapes the notice of most people.

There are times when weight corresponds to our unconscious notion of inertia. For example, if you are trying to determine which of two small objects is the heavier, you might shake them back and forth in your hands or move them in some way instead of lifting them. In doing so, you are judging which of the two is more difficult to get

Isaac Newton (1642-1727)

Isaac Newton was born prematurely and barely survived on Christmas Day, 1642, the same year that Galileo died. Newton's birthplace was his mother's farmhouse in Woolsthorpe, England. His father died several months before his birth, and he grew up under the care of his mother and grandmother. As a child he showed no particular signs of brightness, and at the age of $14\frac{1}{2}$ he was taken out of school to work on his mother's farm. As a farmer he was a failure, preferring to read books he borrowed from a neighboring druggist. An uncle sensed the scholarly potential in young Isaac and prompted him to study at the University of Cambridge, which he did for 5 years, graduating without particular distinction.

A plague swept through London, and Newton retreated to his mother's farm—this time to continue his studies. At the farm, at the age of 23, he laid the foundations for the work that was to make him immortal. Seeing an apple fall to the ground led him to consider the force of gravity extending to the moon and beyond, and he formulated the law of universal gravitation (which he later proved); he invented the calculus, an indispensable mathematical tool in science; he extended Galileo's work and formulated the three fundamental laws of motion; and he formulated a theory of the nature of light and showed with prisms that white light is composed of all colors of the rainbow. It was his experiments with prisms that first made him famous.

When the plague subsided, Newton returned to Cambridge and soon established a reputation for himself as a first-rate mathematician. His mathematics teacher resigned in his favor and Newton was appointed the Lucasian professor of mathematics. He held this post for 28 years. In 1672 he was elected to the Royal Society, where he exhibited the world's first reflector telescope. It can still be seen, preserved at the library of the Royal Society in London with the inscription: "The first reflecting telescope, invented by Sir Isaac Newton, and made with his own hands."

It wasn't until Newton was 42 that he began to write what is generally acknowledged as the greatest scientific book ever written, the *Principia Mathematica Philosophiae Natu-*

ralis. He wrote the work in Latin and completed it in 18 months. It appeared in print in 1687 and wasn't printed in English until 1729, 2 years after his death. When asked how he was able to make so many discoveries, Newton replied that he found his solutions to problems were not by sudden insight but by continually thinking very long and hard about them until he worked them out.

At the age of 46, his energies turned somewhat from science when he was elected a member of Parliament. He attended the sessions in Parliament for 2 years and never gave a speech. One day he rose and the House fell silent to hear the great man. Newton's "speech" was very brief; he simply requested that a window be closed because of a draft.

A further turn from his work in science was his appointment as warden and then as master of the mint. Newton resigned his professorship and directed his efforts toward greatly bettering the workings of the mint, to the dismay of counterfeiters who flourished at that time. He maintained his membership in the Royal Society and was elected president, then was re-elected each year for the rest of his life. At the age of 62, he wrote *Opticks,* which summarized his work on light. Nine years later he wrote a second edition to his *Principia.*

Although Newton's hair turned gray at 30, it remained full, long, and wavy all his life, and unlike others in his time he did not wear a wig. He was a modest man, very sensitive to criticism, and never married. He remained healthy in body and mind into old age. At 80, he still had all his teeth, his eyesight and hearing were sharp, and his mind was alert. In his lifetime he was regarded by his countrymen as the greatest scientist who ever lived. In 1705 he was knighted by Queen Anne. Newton died at the age of 85 and was buried in Westminister Abbey along with England's kings and heroes.

Newton showed that the universe ran according to natural laws that were neither capricious nor malevolent—a knowledge that provided hope and inspiration to scientists, writers, artists, philosophers, and people of all walks of life and that ushered in the Age of Reason. The ideas and insights of Isaac Newton truly changed the world and elevated the human condition.

FIGURE 4.2 An anvil in outer space, between the earth and moon for example, may be weightless, but it is not massless.

FIGURE 4.3 The astronaut in space finds it is just as difficult to shake the "weightless" anvil as it would be on earth. If the anvil is more massive than the astronaut, which shakes more—the anvil or the astronaut?

moving, seeing which of the two is most resistant to a change in motion. You are really comparing the inertias of the objects.

In the United States, the quantity of matter in an object has commonly been described by the gravitational pull between it and the earth, or its weight. This has usually been expressed in *pounds*. In most of the world, however, the measure of matter is expressed more correctly in a mass unit, the **kilogram**. At the surface of the earth, a brick with a mass of 1-kilogram weighs 2.2 pounds. In the metric system of units, the unit of force is the **newton**, which is equal to a little less than a quarter pound (like the weight of a quarter-pounder hamburger *after* it is cooked). A 1-kilogram brick weighs about 10 newtons (more precisely, 9.8 N).* Away from the earth's surface, where the influence of gravity is less, a 1-kilogram brick weighs less. It would also weigh less on the surface of planets with less gravity than the earth. On the moon, for example, where the gravitational force on things is only $\frac{1}{6}$ as strong as on earth, a 1-kilogram object weighs about 1.6 newtons (or 0.36 pounds). On planets with stronger gravity it would weigh more. But the mass of the brick is the same everywhere. The brick offers the same resistance to speeding up or slowing down regardless of whether the earth, moon, or anything at all is attracting it. In a drifting space ship where a scale with a brick on it reads zero, the brick still has mass. Even though it doesn't press down on the scale, the brick has the same resistance to a change in motion as it has on the earth. Just as much force would have to be exerted by an astronaut in the space ship to shake the brick back and forth as would be required to shake it back and forth while on earth. It would take the same push to accelerate a Cadillac limousine to a given speed on a level surface on the moon as on earth. The difficulty of lifting it against gravity (weight) is something else. Mass and weight are different from each other (Figures 4.2 and 4.3).

It is also easy to confuse mass and size—or, more specifically, **volume**. Perhaps this is because when we think of a massive object, we think of a big object—that is, a voluminous object that occupies much space. But an object can be massive—an automobile storage battery, for example—without occupying much space at all. This confusion might also occur because mass and volume are often proportional to each other, at least for the same material. For example, 2 kilograms of sugar will fill a bag of twice the volume as 1 kilogram of sugar. But this does not mean that mass *is* volume. Because 2 kilograms of sugar have twice the sweetening power of 1 kilogram, we don't say that mass *is* sweetening power. Two loaves of bread may have the same mass but quite unequal volumes. You can always squeeze a loaf of bread and change its volume, but the mass doesn't change; it contains the same amount of matter. So you see that mass is neither weight nor volume.

Although Galileo introduced the idea of inertia, Newton grasped its significance. The law of inertia defines natural motion and tells us what kinds of motion are the result of applied forces. Whereas Aristotle maintained that the forward motion of an arrow through the air required a steady force, Newton's law of inertia instead tells us that the behavior of the arrow is natural; if there is no friction or air resistance, constant speed along a straight line (or, simply, constant velocity) requires no force. And whereas Aristotle and his followers held that the circular motions of heavenly bodies were natural, requiring no applied forces, the law of inertia clearly states that in the absence of forces of some kind the planets would not move in the divine circles of ancient and medieval

* So 2.2 lb equal 9.8 N, or 1 N is approximately equal to 0.22 lb—about the weight of an apple. In the metric system it is customary to specify quantities of matter in units of mass (in grams or kilograms) and rarely in units of weight (in newtons). In the United States and places that use the British system of units, however, quantities of matter are customarily specified in units of weight (in pounds). (The British unit of mass, the *slug*, is not well known.) See Appendix A for more about systems of measurement.

astronomy but would move instead in straight-line paths off into space. Newton maintained that the curved motion of the planets was evidence of some kind of force. We shall see in later chapters that his search for this force led to the law of gravitation.

Questions

1. A hockey puck sliding across the ice finally comes to rest. How would Aristotle interpret this behavior? How would Galileo and Newton interpret it? How would you interpret it?

2. Does a 2-kg iron brick have twice as much *inertia* as a 1-kg iron brick? Twice as much *mass*? Twice as much *volume*? Twice as much *weight*?

3. Would it be easier to lift a Cadillac limousine on the earth or to lift it on the moon?

Newton's Second Law of Motion

FIGURE 4.4   The greater the mass, the greater the force must be for a given acceleration.

Every day we see things that do not continue in a constant state of motion: objects initially at rest later may move; moving objects may follow paths that are not straight lines; things in motion may stop. Most of the motion we observe undergoes changes and is the result of one or more applied forces. The overall net force, whether it be from a single source or a combination of sources, produces acceleration. The relationship of acceleration to force and inertia is given in Newton's second law.

> **Law 2: The acceleration of an object is directly proportional to the net force acting on the object, is in the direction of the net force, and is inversely proportional to the mass of the object.**

In summarized form, this is

$$\text{Acceleration} \sim \frac{\text{net force}}{\text{mass}}$$

In symbol notation, this is simply

$$a \sim \frac{F}{m}$$

Answers

1. Aristotle would probably say that the puck slides to a stop because it seeks its proper and natural state, one of rest. Galileo and Newton would probably say that once in motion the puck would continue in motion, and that what prevents continued motion is not its nature or its proper rest state, but the friction between the puck and the ice. This friction is small compared to the friction between the puck and a wooden floor, which is why the puck slides so much farther on ice. Only you can answer the last question.

2. The answers to all parts are yes. A 2-kg iron brick has twice as many iron atoms and therefore twice the amount of matter and mass. In the same location, it also has twice the weight. And since both bricks have the same density (the same mass/volume), the 2-kg brick also has twice the volume.

3. A Cadillac limousine would be easier to lift on the moon because the gravitational force is less on the moon. When you *lift* an object, you are contending with the force of gravity (its weight). Although its mass is the same on the earth, the moon, or anywhere, its weight is only $\frac{1}{6}$ as much on the moon, so only $\frac{1}{6}$ as much effort is required to lift it there. To move it horizontally, however, you are not pushing against gravity. When mass is the only factor, equal forces will produce equal accelerations whether the object is on the earth or the moon.

Force of hand
accelerates
the brick

Twice as much force
produces twice as
much acceleration

Twice the force on
twice the mass gives
the same acceleration

FIGURE 4.5   Acceleration is directly proportional to force.

We shall use the wiggly line ~ as a symbol meaning "is proportional to." We say that acceleration *a* is directly proportional to the overall net force *F* and inversely proportional to the mass *m*. By this we mean that if *F* increases, *a* increases by the same factor (if *F* doubles, *a* doubles); but if *m* increases, *a* decreases by the same factor (if *m* doubles, *a* is cut in half). With appropriate units of *F*, *m*, and *a*, the proportionality may be expressed as an exact equation:

$$a = \frac{F}{m}$$

An object is accelerated in the direction of the force acting on it. Applied in the direction of the object's motion, a force will increase the object's speed. Applied in the opposite direction, it will decrease the speed of the object. Applied at right angles, it will deflect the object. Any other direction of application will result in a combination of speed change and deflection. *The acceleration of an object is always in the direction of the net force.*

A **force**, in the simplest sense, is a push or a pull. Its source may be gravitational, electrical, magnetic, or simply muscular effort. In the second law, Newton gives a more precise idea of force by relating it to the acceleration it produces. He says in effect that *force is anything that can accelerate an object.* Furthermore, he says that the larger the force, the more acceleration it produces. For a given object, twice the force results in twice the acceleration; three times the force, three times the acceleration; and so forth. Acceleration is directly proportional to net force (Figure 4.5).

We say *net* force because oftentimes more than a single force acts on an object. Let's examine more carefully what we mean by *net*. If you and a friend pull in the same direction with equal forces on an object, the forces add to produce a net force twice as great as your single force. The combination of forces produces twice the acceleration than you pulling alone. If, however, you each pull with equal forces but in opposite directions, the object will not accelerate. Because they are oppositely directed, the forces on the object cancel one another. One of the forces can be considered to be the negative of the other, and they add algebraically to zero. The net force is zero.

Suppose you pull on an object with a force of 20 newtons and your friend pulls in the opposite direction with a force of 15 newtons. Then the force and resulting acceleration are the same as if you pulled alone with a force of 5 newtons. The resulting 5 newtons is the net force. If the forces are in the same direction they are added; if in opposite directions they are subtracted (Figure 4.6). It is the net force that accelerates the object.

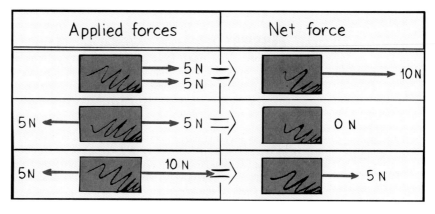

FIGURE 4.6   Net force.

Force of hand accelerates the brick

The same force accelerates 2 bricks ½ as much

3 bricks, ⅓ as much acceleration

FIGURE 4.7 Acceleration is inversely proportional to mass.

We see that force involves both magnitude (amount) and direction. Force is a *vector quantity*. When two or more forces are exerted on an object, the forces combine by vector rules. When the forces are parallel, in equal or opposite directions as just discussed, they simply add algebraically. But when two or more forces are exerted at angles to one another so they are neither in the same nor opposite directions, they combine via the parallelogram rule as shown in Chapter 3. To simplify our study of physics, we will not treat forces at angles in this chapter, and leave such to Appendix D and the *Conceptual Physics Practice Book*.

Force and mass have opposite effects on acceleration (Figure 4.7). The more massive the object, the less its acceleration. For the same force, twice the mass results in half the acceleration; three times the mass, one-third the acceleration. Increasing the mass decreases the acceleration. For example, if we put identical Ford engines in a Cadillac limousine and a Honda Civic, we would expect quite different accelerations even though the driving force in each car is the same. The limousine with its greater mass has a greater resistance to a change in velocity than does the smaller Civic. Consequently, the Cadillac limousine requires a more powerful engine to achieve the same acceleration. To achieve the same acceleration, a larger mass requires a correspondingly larger force. We say that acceleration is inversely proportional to mass.*

The acceleration of an object, then, depends both on the net force exerted on the object and on the mass of the object.

Question In Chapter 2 acceleration was defined to be the time rate of change of velocity; that is, a = (change in v)/time. Are we in this chapter saying that acceleration is instead the ratio of force to mass; that is, $a = F/m$? Which is it?

When Acceleration Is Zero—Equilibrium

When the acceleration of an object is zero, we say the object is in **mechanical equilibrium**. Whatever forces may act on it balance out. The net force on an object in equilibrium is zero. A book lying motionless on a table is in equilibrium because it is not accelerating. And the same book sliding at constant velocity across a smooth surface is also in equilibrium, because it also is not accelerating. In both cases the net force on the book is zero. Let's consider the case without motion first. We call this *static* equilibrium.

Answer Acceleration is *defined* as the time rate of change of velocity and is *produced* by a force. How much force/mass (the cause) determines the rate change in v/time (the effect). So whereas we defined acceleration in Chapter 2, in this chapter we define the terms that produce acceleration.

* Mass can be operationally defined as the proportionality constant between force and acceleration in Newton's second law, rearranged to read $m = F/a$. A 1-unit mass is that which requires 1 unit of force to produce 1 unit of acceleration. So 1 kg is the mass of a body, which under 1 N of force will accelerate 1 m/s². (In British units, 1 slug is the mass of a body that 1 pound of force will accelerate 1 ft/s².) We shall see later that mass is simply a form of concentrated energy.

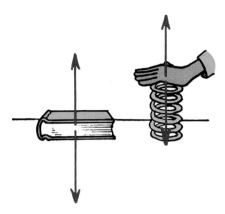

FIGURE 4.8 (Left) The table pushes up on the book with as much force as the downward force of gravity on the book. (Right) The spring pushes up on your hand with as much force as you exert to push down on the spring.

Consider a book lying motionless on a table. Its acceleration is zero. According to Newton's second law, the net force must also be zero. Weight acts downward, so another force must act upward on the book. The other force is the *support force* of the table (also called the *normal force*).* The table exerts an upward support force on the book that is equal in magnitude to the downward force of gravity on the book—its weight (Figure 4.8 left).

To better understand that the table pushes up on the book, compare the case of compressing a spring (Figure 4.8 right). Push the spring down and you can feel the spring pushing up on your hand. Similarly, the book lying on the table compresses atoms in the table, which behave like microscopic springs, producing the support force. Since the book is in equilibrium, the support force on the book is equal and opposite to the weight of the book. An ant trapped between the book and the table would feel itself being squashed from both sides, down from the weight of the book, and up due to the support force of the table.

When you step on a bathroom scale, the downward pull of gravity and the upward support force of the floor compress a spring that is calibrated to give your weight. In effect, the scale shows the support force. Since you are not accelerating, the net force on you is zero, which means the support force and your weight are equal in magnitude. For any object that is not accelerating, either no force is acting on it or there is a combination of forces that cancel to zero. This idea that the net force is zero on things in equilibrium is often quite useful. For example, if you see that a force is exerted on something that doesn't accelerate, then you know there is another force acting—one that is equal in magnitude and opposite in direction. The net force on an object in equilibrium is always zero.

When you hang from a rope, the atoms in the rope are *not* compressed, but are stretched apart. A *tension force* is produced in the rope. A rope under tension "twangs" if you pluck it. If you hang at rest from a vertical rope, then the tension equals your weight. The rope pulls up and the force of gravity pulls down. Hang from two vertical ropes (Figure 4.9), and the sum of the upward tensions equals your weight. You are in static equilibrium. Cases involving non-vertical ropes are left to the *Conceptual Physics Practice Book*.

FIGURE 4.9 The sum of the upward pulls by the rings must equal her weight.

* This force acts at right angles to the surface. When we say "normal to" we are saying "at right angles to," which is why this force is called a normal force.

Questions

1. Suppose you stand on two bathroom scales with your weight evenly divided between the two scales. What will each scale read? How about if you stand with more of your weight on one foot than the other?

2. Pull on a rope tied to a wall with a force of 100 N. What is the net force on the rope? What is the tension force in the rope? Why are your answers different?

FIGURE 4.10 The hockey puck has practically zero acceleration after it has been struck and slides across the ice.

Dynamic equilibrium occurs when an object moves without accelerating. Consider a hockey puck that slides across the ice at constant velocity. It slides without accelerating because the net force on it is zero. A bowling ball rolling at constant velocity along a bowling alley is in dynamic equilibrium—until it interacts with the pins. When you push a crate at constant velocity across a factory floor, the crate is in dynamic equilibrium. In this case the force you exert is balanced by the force of friction between the crate and the floor. The net force is zero, so the crate slides without accelerating. Any object that moves without accelerating is said to be in dynamic equilibrium.

Zero acceleration does not mean zero velocity. Zero acceleration means that the object will maintain the velocity it happens to have, neither speeding up nor slowing down nor changing direction. Most things that are moved in our environment must be pushed or pulled to overcome friction.

FRICTION   When surfaces slide or tend to slide over one another, a force of **friction** acts. Friction is caused by the irregularities in the surfaces in mutual contact, and depends on the kinds of material and how much they are pressed together. Even surfaces that appear to be very smooth have microscopic irregularities that obstruct motion. Atoms cling together at many points of contact. When one object slides against another, it must either rise over the irregular bumps or else scrape atoms off. Either way requires force.

The direction of the friction force is always in a direction opposing motion. An object sliding down an incline experiences friction directed up the incline; an object that slides to the right experiences friction toward the left. Thus, if an object is to move at

Answers

1. The readings on both scales add up to your weight. This is because the sum of the scale readings, which equals the support force by the floor, must counteract your weight so the net force on you will be zero. If you stand equally on each scale, each will read half your weight. If you lean more on one scale than the other, more than half your weight will be read on that scale but less on the other, so they will still add up to your weight. For example, if one scale reads two-thirds your weight, the other scale will read one-third your weight. Get it?

2. The net force on the rope is zero, as evidenced by its state of rest. (We'll see shortly that the rope is pulled at both ends—by your hand on one end and the *wall* on the other, producing zero net force on the rope.) The tension force acts in opposite directions within the rope and has a magnitude of 100 N, for it is being stretched by equal and opposite pulls. Our answers are different because the questions are different; the first asks for the net force *on* the rope, the result of external forces acting on the rope; the second asks for the tension force *within* the rope, which is produced by these external forces.

FIGURE 4.11 The crate slides to the right because of an applied force of 75 N. A friction force of 75 N opposes motion and results in a zero net force on the crate, so the crate slides at constant velocity (zero acceleration).

FIGURE 4.12 Friction between the tire and the ground is nearly the same whether the tire is wide or narrow. The purpose of the greater contact area is to reduce heating and wear.

constant velocity, a force equal to the opposing force of friction must be applied so that the two forces exactly cancel each other. The zero net force then results in zero acceleration.

No friction exists on a crate that sits at rest on a level floor. But disturb the contact surfaces by pushing horizontally on the crate and friction is produced. How much? If the crate is still at rest, then the friction that opposes motion is just enough to cancel your push. If you push horizontally with, say, 70 newtons, the friction is 70 newtons. You push harder, say 100 newtons. The crate is on the verge of sliding—the friction between the crate and floor opposes your push with 100 newtons. If 100 newtons is the most the surfaces can muster, then when you push a bit harder the clinging gives way and the crate slides.

Interestingly enough, the friction of sliding is somewhat less than the friction that builds up before sliding takes place. Physicists and engineers distinguish between static friction and sliding friction. To avoid information overload we won't pursue this distinction further, except to cite an important example—braking a car in an emergency stop. It is very important that you not jam on the brakes so as to make the tires lock in place. When tires lock, they slide, providing less friction than if they are made to roll to a stop. While the tire is rolling, its surface does not slide along the road surface, and friction is static friction—and therefore greater than sliding friction. The difference between static and sliding friction is also apparent when your car takes a corner too fast. Once the tires start to slide, the frictional force is reduced and off you go! A skilled driver (or an anti-lock brake system) keeps the tires below the threshold of breaking loose into a slide.

It's also interesting that the force of friction does not depend on speed. A car skidding at low speed has approximately the same friction as at high speed. If the friction force of a crate that slides against a floor is 90 newtons at low speed, to a close approximation it is 90 newtons at a greater speed. It may be more when the crate is at rest and on the verge of sliding; but once sliding, the friction force remains approximately the same.

More interesting still, friction does not depend on the area of contact. Slide the crate on its smallest surface and all you do is concentrate the same weight on a smaller area with the result that the friction is the same. So those extra wide tires you see on some cars provide no more friction than narrower tires. The wider tire simply spreads the weight of the car over more surface area to reduce heating and wear. Similarly, the friction between a truck and the ground is the same whether the truck has 4 tires or 18! More tires spread the load over more ground area and reduce the pressure per tire. Interestingly, stopping distance when brakes are applied is not affected by the number of tires—but the wear that tires experience, very much depends on the number of tires.

Friction is not restricted to solids sliding over one another. Friction occurs also in liquids and gases, collectively called **fluids** (because they flow). Just as the friction between solid surfaces depends on the nature of the surfaces, fluid friction depends on the nature of the fluid: for example, it is greater in water than it is in air. But unlike the friction between solid surfaces, such as the crate sliding across the floor, fluid friction does depend on speed and area of contact. This makes sense, for the amount of fluid pushed aside by a boat or airplane depends on the size and the shape of the craft. A slow-moving boat or airplane encounters less friction than faster boats or airplanes. And wide boats and airplanes must push aside more fluid than narrow crafts. For slow motion through water, the friction force is approximately proportional to the speed of the object. In air, the friction at most speeds is proportional to the square of the speed. So if an airplane doubles its speed it encounters four times as much frictional force (or "air drag").

At very high speed, however, the simple rules break down when the fluid flow becomes erratic and such things as vortices and shock waves develop.*

Question A jumbo jet cruises at constant velocity of 1000 km/h when the thrusting force of its engines is a constant 100,000 N. What is the acceleration of the jet? What is the force of air resistance on the jet?

$$\frac{F}{m} = g \qquad \frac{2F}{2m} = g$$

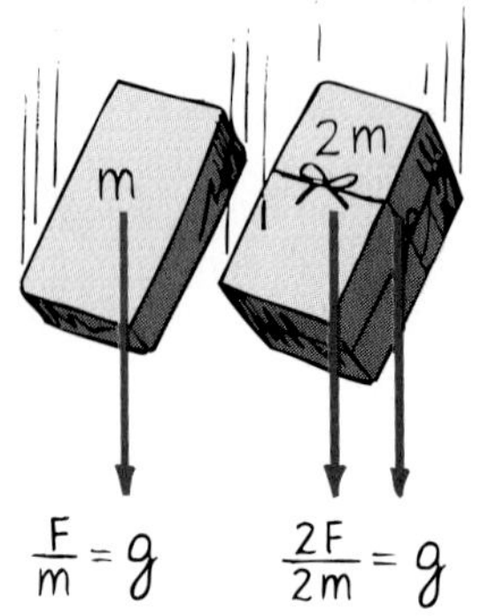

FIGURE 4.13 The ratio of weight (*F*) to mass (*m*) is the same for all objects in the same locality; hence, their accelerations are the same in the absence of air resistance.

When Acceleration Is *g*—Free Fall

Recall from Chapter 2 that Galileo was the first to measure acceleration and relate it to freely falling objects. Interestingly enough, although Galileo founded both the concepts of acceleration and inertia, he failed to notice their relationship to force. Hence he could not explain why objects of various masses fall with equal accelerations. This relationship is Newton's second law, and we'll see that it provides the explanation.

We know that a falling object accelerates toward the earth because of the gravitational force of attraction between the object and the earth. We call the force of gravity that acts on an object the *weight* of the object.[†] When this is the only force that acts on an object—that is, when air resistance and the like are negligible—we say that the object is in a state of **free fall**.

A heavy object is attracted to the earth with more force than a light object. The double brick in Figure 4.13, for example, is attracted with twice as much gravitational force as the single brick. Why then, as Aristotle supposed, doesn't the double brick fall twice as fast? The answer is that the acceleration of an object depends not only on the force—in this case, the weight—but on the mass as well. Whereas force tends to accelerate things, mass tends to resist the acceleration. So twice the force exerted on twice the inertia produces the same acceleration as half the force exerted on half the inertia. Both accelerate equally. The acceleration due to gravity is *g*.

The constant ratio of weight to mass for freely falling objects is similar to the constant ratio of circumference to diameter for circles, which is π. The ratio of weight to mass is the same for both heavy and light objects, just as the ratio of circumference to diameter is the same for both large and small circles (Figure 4.14). The acceleration of

Answer The acceleration is zero because the velocity is constant. Since the acceleration is zero, it follows from Newton's second law that the net force is zero, which means that the force of air resistance must just equal the thrusting force of 100,000 N and act in the opposite direction. So the air resistance on the jet is 100,000 N. (Note that we don't need to know the velocity of the jet to answer this question. We need only know that it is constant, our clue that acceleration and therefore net force is zero.)

---

* Even though it may not seem so yet, most of the concepts in physics are not really complicated. But friction is different. Unlike most concepts in physics, it is a very complicated phenomenon, and the findings are empirical (gained from a wide range of experiments) and the predictions approximate (also based on experiment).

[†] Weight and mass are directly proportional to each other, and the constant of proportionality is *g*. We see that weight = *mg*, so 9.8 N = (1 kg)(9.8 m/s^2).

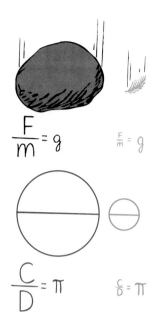

$$\frac{F}{m} = g$$

$$\frac{C}{D} = \pi$$

FIGURE 4.14 The ratio of weight (F) to mass (m) is the same for the large rock and the small feather; similarly, the ratio of circumference (C) to diameter (D) is the same for the large and the small circle.

free fall is g. We use the symbol g, rather than a, to denote that acceleration is due to gravity alone.

So we see the acceleration of free fall is independent of weight. A boulder 100 times more massive than a pebble falls at the same acceleration as the pebble. Although the force on the boulder (its weight) is 100 times greater than the force (or weight) on the pebble, its resistance to a change in motion (mass) is 100 times that of the pebble. The greater force offsets the equally greater mass.

Question In a vacuum, a coin and a feather fall equally, side by side. Would it be correct to say that equal forces of gravity act on both the coin and the feather when in a vacuum?

When Acceleration Is Less Than g—Nonfree Fall

Objects falling in a vacuum are one thing, but what of the practical cases of objects falling in air? Although a feather and a coin will fall equally fast in a vacuum, they fall quite differently in air. How do Newton's laws apply to objects falling in air? The answer is that Newton's laws apply for *all* objects, freely falling or falling in the presence of resistive forces. The accelerations, however, are quite different for the two cases. The important thing to keep in mind is the idea of *net force*. In a vacuum or in cases where air resistance can be neglected, the net force is the weight because it is the only force acting on a falling object. In the presence of air resistance, however, the net force is the difference between the weight and the force of air resistance.*

The force of air resistance acting on a falling object depends primarily on two things. First, it depends on the size of the falling object—that is, on the amount of air the object must plow through in falling. Second, it depends on the speed of the falling object; the greater the speed, the greater the number of air molecules an object encounters per second and the greater the force of molecular impact. Air resistance depends on the size and the speed of a falling object.

Answer No, no, no, a thousand times no! These objects accelerate equally not because the forces of gravity on them are equal, but because the *ratios* of their weights to masses are equal. Although air resistance is not present in a vacuum, gravity is (you'd know this if you stuck your hand into a vacuum chamber and a Mack truck rolled over it). If you answered yes to this question, let this be an alarm to be more careful when you think physics!

---

* In mathematical notation,

$$a = \frac{F_{net}}{m} = \frac{mg - R}{m}$$

where mg is the weight and R is the air resistance. Note that when $R = mg$, $a = 0$; then, with no acceleration, the object falls at constant velocity. With elementary algebra we can go another step and get

$$a = \frac{F_{net}}{m} = \frac{mg - R}{m} = g - \frac{R}{m}$$

We see that the acceleration a will always be less than g if air resistance R impedes falling. Only when $R = 0$ does $a = g$.

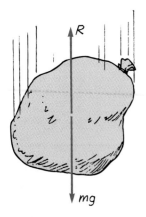

FIGURE 4.15 When weight *mg* is greater than air resistance *R*, the falling sack accelerates. At higher speeds, *R* increases. When *R* = *mg*, acceleration reaches zero, and the sack reaches its terminal velocity.

A feather dropped in the air accelerates very briefly and then "floats" to the ground at constant speed. This happens because the air resistance acting against the feather quickly increases until it is equal and opposite to the weight. The net force on the feather quickly reaches zero, and no further acceleration takes place. The feather is very light and has a relatively large surface area, so it doesn't have to fall very fast for air resistance to equal its weight. The same idea applies to all objects falling in air. As a falling object gains speed, the force of air resistance finally builds up until it equals the weight of the falling object. When this happens, the *net* force becomes zero and the object no longer accelerates; it falls at a constant speed. When acceleration terminates, we say the object has reached its **terminal speed**. If we are concerned with direction, down for falling objects, we say the object has reached its **terminal velocity**. For a feather, terminal speed is a few centimeters per second, whereas for a skydiver it is about 200 kilometers per hour. A skydiver varies terminal speed by varying position. Head or feet first is a way of encountering less air resistance and attaining maximum terminal speed. Minimum terminal speed is attained by spreading oneself out like a flying squirrel.

Consider a man and woman parachuting together from the same altitude (Figure 4.16). Suppose that the man is twice as heavy as the woman and that their same-sized chutes are open. The same size chute means that at equal speeds the air resistance is the same on each (air resistance on the two bodies is negligible compared to the air resistance on the parachutes). Who gets to the ground first—the heavy man or the lighter woman? The answer is that the person with the greater terminal speed gets to the ground first. At first we might think that because the chutes are the same, the terminal speeds for each would be the same, and therefore both would reach the ground together. This doesn't happen because air resistance also depends on speed. The woman will reach her terminal speed when air resistance against her chute equals her weight. When this happens, the air resistance against the chute of the man will not yet equal his weight. Force equilibrium occurs first for her, while he continues to accelerate to greater speeds. He must fall faster than she does for air resistance to match his greater weight.* Terminal

FIGURE 4.16 The heavier parachutist must fall faster than the lighter parachutist for air resistance to cancel his greater weight.

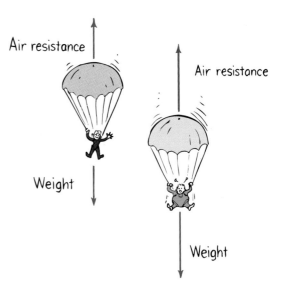

---

* Terminal speed for the twice-as-heavy man will be about 41 percent greater than the woman's terminal speed, because the retarding force of air resistance is proportional to speed squared. ($v^2_{man}/v^2_{woman} = 1.41^2 = 2.$)

speed is greater for the heavier person, with the result that the heavier person reaches the ground first.

> **Question** A skydiver jumps from a high-flying helicopter. As she falls faster and faster through the air, does her acceleration increase, decrease, or remain the same?

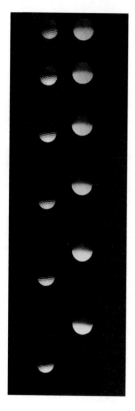

Consider a pair of tennis balls, one a hollow regular ball and the other filled with iron pellets. Although they are the same size, the iron-filled ball is considerably heavier than the regular ball. If you hold them above your head and drop them simultaneously you'll see that they appear to strike the ground at the same time. But if you drop them from a greater height, say from the top of a building, you'll note the heavier ball strikes the ground first. Why? In the first case the balls do not gain much speed in their short fall. Air resistance encountered is small compared to their weights, even for the regular ball. The tiny difference in their arrival time is not noticed. But when dropped from a greater height, the greater speeds of fall are met with greater air resistance. At any given speed each ball encounters the same air resistance because each has the same size. This air resistance may be a lot compared to the weight of the lighter ball, but only a little compared to the weight of the heavier ball (like the parachutists in Figure 4.16). So even with equal air resistances, the accelerations of each are different. There is a moral to be learned here. Whenever you consider the acceleration of something, use the equation of Newton's second law to guide your thinking: the acceleration is equal to the ratio of *net* force to mass. In this case, acceleration decreases as net force decreases. And if the air resistance builds up to equal the weight of the falling object, then the net force becomes zero and acceleration terminates.

FIGURE 4.17   A stroboscopic study of a golf ball (left) and a Styrofoam ball (right) falling in air. The air resistance is negligible for the heavier golf ball, and its acceleration is nearly equal to g. Air resistance is not negligible for the lighter Styrofoam ball, which reaches its terminal velocity sooner.

Answer Acceleration decreases because the net force on her decreases. Net force is equal to her weight minus her air resistance, and since air resistance increases with increasing speed, net force and hence acceleration decrease. By Newton's second law,

$$a = \frac{F_{net}}{m} = \frac{mg - R}{m}$$

where *mg* is her weight and *R* is the air resistance she encounters. As *R* increases, *a* decreases. Note that if she falls fast enough so that $R = mg$, $a = 0$, then with no acceleration she falls at constant speed.

Newton's Third Law of Motion

FIGURE 4.18 In the interaction between the hammer and the stake, each exerts the same amount of force on the other.

In the simplest sense, a force is a push or a pull. Looking closer, however, a force is not a thing in itself but is due to an *interaction* between one thing and another. Consider for example the interaction between a hammer and a stake. A hammer exerts a force on a stake and drives it into the ground. But this force is only half the story, for there must be a force to halt the hammer in the process. What exerts this force? The stake! Newton reasoned that while the hammer exerts a force on the stake, the stake exerts a force on the hammer. So in the interaction between the hammer and the stake there is a *pair* of forces—one acting on the stake and the other acting on the hammer. Such observations led Newton to his third law—the law of action and reaction.

Law 3: Whenever one object exerts a force on a second object, the second object exerts an equal and opposite force on the first.

One force is called the **action force**, and the other the **reaction force**. It doesn't matter which force we call *action* and which we call *reaction*. The important thing is that they are co-parts of a single interaction and that neither force exists without the other. They are equal in strength and opposite in direction. Newton's third law is often stated thus: "For every action there is an equal and opposite reaction."

In every interaction, the forces always occur in pairs. The action and reaction pair of forces make up the interaction between two things. When you walk across the floor, for example, you push against the floor, and the floor pushes against you. Both you and the floor push against each other simultaneously—there is an interaction between you and the floor. Likewise, the tires of a car push against the road while the road pushes back on the tires—the tires and road push against each other. In swimming you push the water backward, and the water pushes you forward—you and the water push against each other. In each case there is a pair of forces, one action and the other reaction, that make up the interaction. The reaction forces are what account for our motion in these cases. These reaction forces are created by friction; a person or car on ice, for example, may not be able to exert the action force to produce the needed reaction force.

We know that forces can cancel when they are equal and act in opposite directions on the same object. Even though action and reaction forces are equal and oppositely directed, they don't cancel each other when they are acting on different objects. Consider

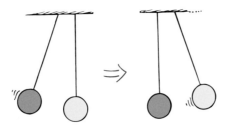

FIGURE 4.19 The impact forces between the blue and yellow balls move the yellow ball and stop the blue ball.

FIGURE 4.20 Action and reaction forces. Note that when action is "A exerts force on B," the reaction is then simply "B exerts force on A."

Action: tire pushes on road   Reaction: road pushes on tire

Action: rocket pushes on gas   Reaction: gas pushes on rocket

Action: man pulls on spring   Reaction: spring pulls on man

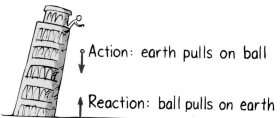

Action: earth pulls on ball

Reaction: ball pulls on earth

FIGURE 4.21 When an apple pulls on an orange, the orange accelerates—period! The fact that the orange pulls back on the apple affects the apple—not the orange. You can't cancel a force on the orange with a force on the apple.

the force pair between the apple and the orange in Figure 4.21. Suppose we ignore the apple and consider only the orange. We draw an imaginary circle around the orange and call what is within the circle the *system*. The pull from the apple supplies a force on the system, and the system accelerates. Here the interaction is between the system (the orange) and something external (the apple), so the action and reaction forces don't cancel in the system. The fact that the orange simultaneously exerts a force on the apple, which is external to the system, affects the apple but not the orange.

Now if we consider the system to be both the orange and apple, the force pair is internal to the system. In such a case, when the action and reaction forces are taken to be inside the same system, the forces do cancel each other. If there were no friction or other forces acting on the system, the apple and orange would move closer but, due to the cancellation of internal forces, the system's "center of mass" (Chapter 7) would be in the same place before and after the pulling. The system itself would not accelerate though its internal parts would. Similarly, the many force pairs between the atoms in a baseball may hold the ball together, but they don't produce acceleration of the ball. An external force is needed to accelerate the ball. Likewise, a force external to both the orange and apple is needed to produce acceleration of the orange-apple system. (Friction between the floor and the apple's feet would do the trick.) If this is confusing, it may be well to note that Newton had difficulties with the third law himself.

Questions

1. Does a speeding missile possess force?
2. We know that the earth pulls on the moon. Does this mean that the moon also pulls on the earth?
3. Since action and reaction are exactly equal and oppositely directed, how can there ever be a net force on an object?
4. Can you identify the action and reaction forces in the case of an object falling in a vaccum?
5. Why does a book sitting on a table never accelerate "spontaneously" in response to the trillions of interatomic forces acting within it?

FIGURE 4.22 The earth is pulled up by the boulder with just as much force as the boulder is pulled downward by the earth.

Strange as it may seem, a falling object pulls upward on the earth as much as the earth pulls downward on it. The downward pull on the object seems normal because the acceleration of 10 meters per second each second is quite noticeable. The same amount of force acts upward on the huge mass of the earth. However, it produces an acceleration so small it cannot be noticed or measured. A less extreme example might make this

Answers

1. No, a force is not something an object *has,* like mass, but is half of an interaction between one object and another. A speeding missile may possess the capability of exerting a force on another object when interaction occurs, but it does not possess force as a thing in itself. We will see in the two chapters to follow that a speeding missile possesses both momentum and kinetic energy.
2. Yes, both pulls make up an action-reaction pair of forces associated with the gravitational interaction between the earth and the moon. We can say that (1) the earth pulls on the moon, and (2) the moon likewise pulls on the earth; but it is more insightful to think of this as a single interaction—the earth and moon simultaneously pull on each other.
3. The net force on any object is a combination of forces that act *on* the object—say, the action forces. The reaction forces *by* the object act on other things. For example, when you kick a can, the can kicks back on you. But your foot provides the only force exerted on the can. The can flies through the air because it experiences only the action force; the reaction force is exerted *by* the can, not *on* the can. The reaction force does decelerate your foot. However, the can experiences only what happens to it, not what it does to your foot. You can't cancel the force exerted on an object with the force it simultaneously exerts on something else.
4. To identify a pair of action-reaction forces in any situation, first identify the pair of interacting objects involved. Something is interacting with something. In this case the world is interacting (gravitationally) with the falling object. So the world pulls downward on the object (call it action), while the object pulls upward on the world (reaction).
5. Every one of these interatomic forces is part of an action-reaction pair within the book. These forces add up to zero, no matter how many of them there are. This is what makes Newton's *first* law apply to the book. The book has zero acceleration unless an *external* force acts on it.

FIGURE 4.23 The force exerted against the recoiling rifle is just as great as the force that drives the bullet. Why then, does the bullet accelerate more than the rifle?

clearer. When a rifle is fired, the force exerted on the bullet is as great as the reaction force exerted on the rifle; hence, the rifle kicks. Since the forces are equal in magnitude, why doesn't the rifle recoil with the same speed as the bullet? In analyzing changes in motion, Newton's second law reminds us that we must also consider the masses involved. Suppose we let F represent both the action and reaction force, m the mass of the bullet, and m the mass of the more massive rifle. The accelerations of the bullet and the rifle are then found by taking the ratio of force to mass. The acceleration of the bullet is given by

$$\frac{F}{m} = a$$

while the acceleration of the recoiling rifle is

$$\frac{F}{m} = a$$

We see why the change in motion of the bullet is so huge compared to the change of motion of the rifle. A given force divided by a small mass produces a large acceleration, while the same force divided by a large mass produces a small acceleration. We have used different-sized symbols to indicate the differences in masses and resulting accelerations. Going back to the example of the falling object, if we similarly exaggerated symbols to represent the acceleration of the earth reacting to a falling object, the symbol m for the earth's mass would be astronomical in size. The force F, the weight of the falling object, divided by this large mass would result in a microscopic a to represent the acceleration of the earth toward the falling object.

We can see that the earth accelerates slightly in response to a falling object in still another way by considering the exaggerated examples of two planetary bodies, a through e in Figure 4.24. The forces between bodies A and B are equal in magnitude and oppositely directed in each case. If acceleration of planet A is unnoticeable in a, then it is more noticeable in b where the difference between the masses is less extreme. In c, where both bodies have equal mass, acceleration of object A is as evident as it is for B. Continuing, we see the acceleration of A becomes even more evident in d and even more so in e. So strictly speaking, when you step off the curb, the street comes up ever so slightly to meet you.

If we extend the idea of a rifle recoiling or "kicking" from the bullet it fires, we can understand rocket propulsion. Consider a machine gun recoiling each time a bullet is fired. If the machine gun is fastened so it is free to slide on a vertical wire

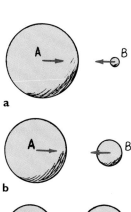

a

b

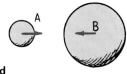

c

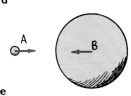

d

e

FIGURE 4.24 Which falls toward the other, A or B? Do the accelerations of each relate to their relative masses?

FIGURE 4.25 (left) The machine gun recoils from the bullets it fires and climbs upward.

FIGURE 4.26 (right) The rocket recoils from the "molecular bullets" it fires and climbs upward.

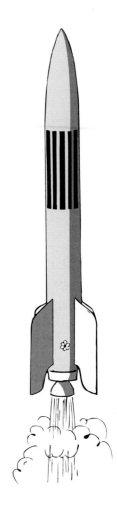

(Figure 4.25), it accelerates upward as bullets are fired downward. A rocket accelerates the same way as it continually "recoils" from the ejected exhaust gas. Each molecule of exhaust gas is like a tiny bullet shot from the rocket (Figure 4.26).

A common misconception is that a rocket is propelled by the impact of exhaust gases against the atmosphere. In fact, in the early 1900s before the advent of rockets, many people thought that sending a rocket to the moon was impossible because of the absence of an atmosphere for the rocket to push against. But this is like saying a gun wouldn't kick unless the bullet had air to push against. Not true! Both the rocket and recoiling gun accelerate not because of any pushes on the air, but because of the reaction forces by the "bullets" they fire—air or no air. A rocket works better, in fact, above the atmosphere where there is no air resistance to oppose its speed.

Using Newton's third law, we can understand how a helicopter gets its lifting force. The whirling blades are shaped to force air particles down (action), and the air forces the blades up (reaction). This upward reaction force is called *lift*. When lift equals the weight of the craft, the helicopter hovers in midair. When lift is greater, the helicopter climbs upward.

This is true for birds and airplanes. Birds fly by pushing air downward. The air simultaneously pushes the bird upward. When the bird is soaring, the wing must be shaped so that moving air particles are deflected downward. Slightly tilted wings that deflect oncoming air downward produce the lift on an airplane. Air must be pushed downward continuously to maintain lift. This supply of air is obtained by the forward

FIGURE 4.27 You cannot touch without being touched —Newton's third law.

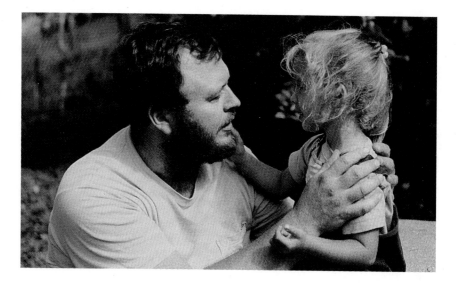

motion of the aircraft. On propeller-driven airplanes, the propellers push air backward, the air simultaneously pushes the propellers forward. For jet aircraft, the jet pushes exhaust gases backward and the exhaust gases push the jet forward. We will learn in Chapter 13 that the curved surface of a wing is an airfoil, which enhances the lifting force.

Questions

1. A car accelerates along a road. What is the force that moves the car?
2. A high-speed bus and an innocent bug have a head-on collision. The force of impact splatters the poor bug over the windshield. Is the corresponding force that the bug exerts against the windshield greater, less, or the same? Is the resulting deceleration of the bus greater than, less than, or the same as that of the bug?

We see Newton's third law at work everywhere. A fish pushes the water backward with its fins, and the water pushes the fish forward. The wind pushes against the branches of a tree, and the branches push back on the wind and we have whistling

Answers

1. It is the road that pushes the car along. Really! Except for air resistance, only the road provides a horizontal force on the car. How does it do this? The rotating tires of the car push back on the road (action). The road simultaneously pushes forward on the tires (reaction). How about that!
2. The magnitudes of both forces are the same, for they constitute an action-reaction force pair that makes up the interaction between the bus and the bug. The accelerations, however, are very different because the masses involved are different! The bug undergoes an enormous and lethal deceleration, while the bus undergoes a very tiny deceleration—so tiny that the very slight slowing of the bus is unnoticed by its passengers. But if the bug were more massive—as massive as another bus, for example— the slowing down would be quite evident!

sounds. Forces are interactions between different things. Every contact requires at least a "twoness"; there is no way that an object can exert a force on nothing. Forces, whether large shoves or slight nudges, always occur in pairs, each of which is opposite to the other. Thus, we cannot touch without being touched.

Summary of Newton's Three Laws

An object at rest tends to remain at rest; an object in motion tends to remain in motion at constant speed along a straight-line path. This tendency of objects to resist change in motion is called *inertia.* Mass is a measure of inertia. Objects will undergo changes in motion only in the presence of a net force.

When a net force acts on an object, the object will accelerate. The acceleration is directly proportional to the net force and inversely proportional to the mass. Symbolically, $a \sim F/m$. Acceleration is always in the direction of the net force. When objects fall in a vacuum, the net force is simply the weight, and the acceleration is g (the symbol g denotes that acceleration is due to gravity alone.) When objects fall in air, the net force is equal to the weight minus the force of air resistance, and the acceleration is less than g. If and when the force of air resistance equals the weight of a falling object, acceleration terminates, and the object falls at constant speed (called *terminal speed*).

Whenever one object exerts a force on a second object, the second object exerts an equal and opposite force on the first. Forces come in pairs, one action and the other reaction, both of which comprise the interaction between one thing and the other. Neither force exists without the other. Whenever they act on different objects, action and reaction don't cancel out.

Summary of Terms

Inertia   The property of things to resist changes in motion.

Mass   The quantity of matter in an object. More specifically, it is the measure of the inertia or sluggishness that an object exhibits in response to any effort made to start it, stop it, deflect it, or change in any way its state of motion.

Weight   The force due to gravity on an object.

Kilogram   The fundamental SI unit of mass. One kilogram (symbol kg) is the mass of 1 liter (l) of water at 4°C.

Newton   The SI unit of force. One newton (symbol N) is the force that will give an object of mass 1 kg an acceleration of 1 m/s^2.

Volume   The quantity of space an object occupies.

Force   Any influence that can cause an object to be accelerated, measured in newtons (or pounds in the British system).

Mechanical equilibrium   The state of an object or system of objects for which any impressed forces cancel to zero and no acceleration occurs.

Friction   The resistive force that opposes the motion or attempted motion of an object past another with which it is in contact, or through a fluid.

Free fall   Motion under the influence of gravitational pull only.

Terminal speed   The speed at which the acceleration of a falling object terminates because air resistance balances its weight.

Review Questions

Newton's First Law of Motion

1. Is inertia the *reason* for objects to maintain their states of motion, or the *name* given to this property of matter?

Mass

2. Clearly distinguish among *mass, weight,* and *volume.*

3. Which is more fundamental, *mass* or *weight*? Does an object with mass have weight? Does an object with weight have mass?

4. Could an object have mass without having weight? Could an object have weight without having mass?

5. Does a 2-kg iron brick have twice as much *inertia* as a 1-kg block of wood? Twice as much *volume*? (Why are your answers different?)

6. What kind of path would the planets follow if suddenly no force acted on them?

Newton's Second Law of Motion

7. If we say that one quantity is *proportional* to another quantity, does this mean they are *equal* to each other? Explain briefly, using mass and weight as an example.

8. A cart is pulled to the left with a force of 100 N, and to the right with a force of 30 N. What is the net force on the cart?

9. Why do we say that force is a vector quantity?

10. If the net force acting on a sliding block is somehow tripled, by how much does the acceleration increase?

11. If the mass of a sliding block is tripled while a constant net force is applied, by how much does the acceleration decrease?

12. If the mass of a sliding block is somehow tripled at the same time the net force on it is tripled, how does the resulting acceleration compare to the original acceleration?

When Acceleration Is Zero—Equilibrium

13. What is the net force on something that is in mechanical equilibrium?

14. Consider a book that weighs 15 N at rest on a flat table. How many newtons of support force does the table provide? What is the net force on the book in this case?

15. Consider a woman weighing 500 N who stands with her weight evenly divided on a pair of bathroom scales. What is the reading on each scale? If she shifts her weight so one of the scales reads 300 N, what will the other scale read?

16. What is the acceleration of an object that moves at constant velocity? What is the net force on the object in this case?

17. Why is it more difficult to slide a crate from a position of rest across the floor than it is to keep it in motion once it is sliding?

18. If you push horizontally with a force of 50 N on a crate and make it slide at constant velocity, how much friction acts on the crate? If you increase your force, will the crate accelerate? Explain.

19. What effect does the speed of a sliding object have on the friction that acts on it? What effect does the area of contact have on friction?

20. As the speed of an object moving through a fluid increases, what happens to the friction acting on it? As the area of the object increases, what happens to the fluid friction?

When Acceleration Is g—Free Fall

21. What is meant by *free fall*?

22. What is the net force that acts on a 10-N freely falling object?

23. Why doesn't a heavy object accelerate more than a light object when both are freely falling?

When Acceleration Is Less Than g—Nonfree Fall

24. What is the net force that acts on a 10-N falling object when it encounters 4 N of air resistance? 10 N of air resistance?

25. What two principal factors affect the force of air resistance on a falling object?

26. What is the acceleration of a falling object that has reached its terminal velocity?

27. Why does a heavy parachutist fall faster than a lighter parachutist who wears the same size parachute?

Newton's Third Law of Motion

28. Consider hitting a baseball with a bat. If we call the force on the bat against the ball the *action* force, identify the *reaction* force.

29. When do action and reaction pairs of forces not cancel one another?

30. If the forces that act on a bullet and the recoiling gun from which it is fired are equal in magnitude, why do the bullet and gun have very different accelerations?

31. How does a helicopter get its lifting force?

Projects

1. Ask a friend to drive a small nail into a piece of wood placed on top of a pile of books on your head. Why doesn't this hurt you?

2. Drop a sheet of paper and a coin at the same time. Which reaches the ground first? Why? Now crumple the paper into a small, tight wad and again drop it with the coin. Explain the difference observed. Will they fall together if dropped from a second-, third-, or fourth-story window? Try it and explain your observations.

3. Drop a book and a sheet of paper and note that the book has a greater acceleration—g. Place the paper beneath the book and it is forced against the book as both fall, so both fall at g. How do the accelerations compare if you place the paper on top of the raised book and then drop both? You may be surprised, so try it and see. Then explain your observation.

4. Drop two balls of different weights from the same height, and at small speeds they practically fall together. Will they roll together down the same inclined plane? If each is suspended from an equal length of string, making a pair of pendulums, and displaced through the same angle, will they swing back and forth in unison? Try it and see; then explain using Newton's laws.

5. The net force acting on an object and the resulting acceleration are always in the same direction. You can demonstrate this with a spool. If the spool is pulled horizontally to the right, in which direction will it roll?

6. Hold your hand like a flat wing outside the window of a moving automobile. Then slightly tilt the front edge upward and notice the lifting effect. Can you see Newton's laws at work here?

Exercises

Please do not be intimidated by the large number of exercises in this and other meaty chapters. If your course work is to cover many chapters, your instructor will likely assign only a few exercises from each.

1. In the orbiting space shuttle you are handed two identical boxes, one filled with sand and the other filled with feathers. How can you tell which is which without opening the boxes?

2. Your empty hand is not hurt when it bangs lightly against a wall. Why is it hurt if it does so while carrying a heavy load? Which of Newton's laws is most applicable here?

3. Why is a massive cleaver more effective for chopping vegetables than an equally sharp knife?

4. In tearing a paper towel or plastic bag from a roll, why is a sharp jerk more effective than a slow pull?

5. Each bone in the chain of bones forming your spine is separated from its neighbors by disks of elastic tissue. What happens, then, when you jump heavily on your feet from an elevated position? (*Hint:* Think about the hammerhead in Figure 4.1.) Can you think of a reason why you are a little taller in the morning than the night before?

6. Before the time of Galileo and Newton, some learned scholars thought that a stone dropped from the top of a tall mast of a moving ship would fall vertically and hit the deck behind the mast by a distance equal to how far the ship had moved forward while the stone was falling. In light of your understanding of Newton's laws, what do you think about this?

7. Because the earth rotates once per 24 hours, the western wall in your room moves in a direction toward you at a linear speed that is probably more than 1000 km/h per

hour (the exact speed depends on your latitude). When you stand facing the wall you are carried along at the same speed, so you don't notice it. But when you jump upward, with your feet no longer in contact with the floor, why doesn't the high-speed wall slam into you?

8. The chimney of a stationary toy train consists of a vertical spring gun that shoots a steel ball a meter or so straight into the air—so straight that the ball always falls back into the chimney. Suppose the train moves at constant speed along the straight track. Do you think the ball will still return to the chimney if shot from the moving train? How about if the train accelerates along the straight track? How about if it moves at a constant speed on a circular track? Why are your answers different?

9. When a junked car is crushed into a compact cube, does its mass change? Its weight? Explain.

10. Gravity on the moon is only $\frac{1}{6}$ as strong as gravity on the earth. What is the weight of a 10-kg object on the moon and on the earth? What is its mass on the moon and on the earth?

11. What is your own mass in kilograms? Your weight in newtons?

12. A rocket becomes progressively easier to accelerate as it travels through space. Why is this so? (*Hint:* About 90% of the mass of a newly launched rocket is fuel.)

13. Aristotle claimed the speed of a falling object depends on its weight. We now know that objects in free fall, whatever their weights, undergo the same gain in speed. Why does weight not affect acceleration?

14. If an object has no acceleration, can you conclude that no forces are exerted on it? Explain.

15. Can an object round a curve without any force acting on it?

16. To pull a wagon across a lawn with constant velocity, you have to exert a steady force. Reconcile this fact with Newton's first law, which says that motion with constant velocity requires no force.

17. Free fall is motion in which gravity is the only force acting. (a) Is a sky diver who has reached terminal speed in free fall? (b) Is a satellite circling the earth above the atmosphere in free fall?

18. As you stand on a floor, does the floor exert an upward force against your feet? How much force does it exert? Why are you not moved upward by this force?

19. As you are leaping into the air, how does the force that you exert on the ground compare with your weight?

20. A common saying goes, "It's not the fall that hurts you; it's the sudden stop." Translate this into Newton's laws of motion.

21. Nellie Newton hangs at rest from the ends of the rope as shown. How does the reading on the scale compare to her weight?

22. Harry the painter swings year after year from his bosun's chair. His weight is 500 N and the rope, unknown to him, has a breaking point of 300 N. Why doesn't the rope break when he is supported as shown to the left below? One day Harry is painting near a flagpole, and, for a change, he ties the free end of the rope to the flagpole instead of to his chair as shown to the right. Why did Harry end up taking his vacation early?

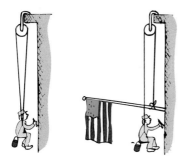

23. Consider the two forces acting on the person who stands still, namely, the downward pull of gravity and the upward support of the floor. Are these forces equal and opposite? Do they form an action-reaction pair? Why or why not?

24. For each of the following forces, what is the equal and opposite force required by Newton's third law? (a) The force of a hammer on a nail. (b) The force of gravity pulling down on a book. (c) The force of a helicopter blade pushing down on the air. (d) The force of air resistance acting on a thrown baseball.

25. You hold an apple over your head. (a) Identify all the forces acting on the apple and their reaction forces. (b) When you drop the apple, identify all the forces acting on it as it falls and the corresponding reaction forces.

26. Within a book on a table there are billions of forces pushing and pulling on all the molecules. Why is it that these forces never by chance add up to a net force in one direction, causing the book to accelerate "spontaneously" across the table?

27. Two 100-N weights are attached to a spring scale as shown. Does the scale read 0, 100, or 200 N, or give some other reading? (*Hint:* Would it read any differently if one of the ropes were tied to the wall instead of to the hanging 100-N weight?)

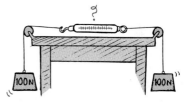

28. If we find an object that is not moving even though we know it to be acted on by a force, what inference can we draw?

29. When your car moves along the highway at constant velocity, the net force on it is zero. Why, then, do you have to keep running your engine?

30. A "shooting star" is usually a grain of sand from outer space that burns up and gives off light as it enters the atmosphere. What exactly causes this burning?

31. What is the net force on an apple that weighs 1 N when you hold it at rest above your head? What is the net force on it after you release it?

32. Does a stick of dynamite contain force?

33. When the athlete holds the barbell overhead, the reaction force is the weight of the barbell on his hand. How does this force vary for the case where the barbell is accelerated upward? Downward?

34. Why can you exert greater force on the pedals of a bicycle if you pull up on the handlebars?

35. The strong man will push the two initially stationary freight cars of equal mass apart before he himself drops straight to the ground. Is it possible for him to give either of the cars a greater speed than the other? Why or why not?

36. Suppose two carts, one twice as massive as the other, fly apart when the compressed spring that joins them is released. How fast does the heavier cart roll compared with the lighter cart?

37. If you exert a horizontal force of 200 N to slide a crate across a factory floor at constant velocity, how much friction is exerted by the floor on the crate? Is the force of friction equal and oppositely directed to your 200-N push? If the force of friction is not the reaction force to your push, what is?

38. If a Mack truck and Honda Civic have a head-on collision, upon which vehicle is the impact force greater? Which vehicle experiences the greater acceleration? Explain your answers.

39. Ken and Joanne are astronauts floating some distance apart in space. They are joined by a safety cord whose ends are tied around their waists. If Ken starts pulling on the cord, will he pull Joanne toward him, or will he pull himself toward Joanne, or will both astronauts move? Explain.

40. Which team wins in a tug of war; the team that pulls harder on the rope, or the team that pushes harder against the ground? Explain.

41. Two people of equal mass attempt a tug-of-war with a 12-m rope while standing on frictionless ice. When they pull on the rope, they each slide toward each other. How do their accelerations compare, and how far does each person slide before they meet?

42. A horse pulls a heavy wagon with a certain force. The wagon, in turn, pulls back with an opposite but equal force on the horse. Doesn't this mean the forces cancel one another, making acceleration impossible? Why or why not? (*Hint:* Consider the role of the ground.)

43. How does the weight of a falling body compare with the air resistance it encounters before it reaches terminal velocity? After?

44. Why is it that a cat that falls from the top of a 50-story building will hit the ground no faster than if it fell from the 20th story?

45. If and when Galileo dropped two balls from the top of the Leaning Tower of Pisa, air resistance was not really negligible. Assuming both balls were the same size, one made of wood and one of metal, which ball struck the ground first? Why?

46. If you drop a pair of tennis balls simultaneously from the top of a building, they will strike the ground at the same time. If you fill one of the balls with lead pellets and then drop them together, which one will hit the ground first? Which one will experience greater air resistance? Defend your answers.

47. What is the acceleration of a rock at the top of its trajectory when thrown straight upward? (Is your answer consistent with Newton's second law?)

48. In the absence of air resistance, if a ball is thrown vertically upward with a certain initial speed, on returning to its original level it will have the same speed. When air resistance is a factor, will the ball be moving faster, the same, or slower than its throwing speed when it gets back to the same level? Why? (Physicists often use a "principle of exaggeration" to help them analyze a problem. Consider the exaggerated case of a feather, not a ball, because the effect of air resistance on the feather is more pronounced and therefore easier to visualize.)

49. If a ball is thrown vertically into the air in the presence of air resistance, would you expect the time during which it rises to be longer or shorter than the time during which it falls? (Again use the "principle of exaggeration.")

50. Make up three multiple-choice questions, one for each of Newton's laws, that would check a classmate's understanding of these laws.

Problems

1. A 400-kg bear grasping a vertical tree slides down at constant velocity. What is the friction force that acts on the bear?

2. If a mass of 1 kg is accelerated 1 m/s^2 by a force of 1 N, what would be the acceleration of 2 kg acted on by a force of 2 N?

3. How much acceleration does a 747 jumbo jet of mass 30,000 kg experience in takeoff when the thrust for each of four engines is 30,000 N?

4. If you stand next to a wall on a frictionless skateboard and push the wall with a force of 30 N, how hard does the wall push on you? If your mass is 60 kg, what's your acceleration?

5. A firefighter of mass 80 kg slides down a vertical pole with an acceleration of 4 m/s^2. What is the friction force that acts on the firefighter?

6. What will be the acceleration of a skydiver when air resistance builds up to be equal to half her weight?

7. Sprinting near the end of a race, a runner with a mass of 60 kg accelerates from a speed of 6 m/s to a speed of 7 m/s in 2 s. (a) What is the runner's average acceleration during this time? (b) To gain speed the runner produces a backward force on the ground, so that the ground pushes the runner forward providing the force necessary for acceleration. Calculate this average force.

8. Before going into orbit, an astronaut has a mass of 55 kg. While in orbit, a measurement determines that a force of 100 N causes her to move with an acceleration of 1.90 m/s^2. To regain her original weight, should she go on a diet or start eating more candy?

Remember, review questions provide you with a self check of whether or not you grasp the central ideas of the chapter. The exercises and problems are extra "pushups" for you to try after you have at least a fair understanding of the chapter and can handle the review questions.

5

· · · · · · · · · · · ·

Momentum

⦂ Inertia in motion.

In Chapter 2 we introduced Galileo's idea of inertia, and in Chapter 4 we showed how this idea was incorporated into Newton's laws of motion. We discussed inertia in terms of objects at rest and objects in motion. In this chapter we will concern ourselves only with the inertia of moving objects. When we combine the ideas of inertia and motion, we are dealing with momentum. *Momentum* refers to moving things.

Momentum

We all know that a heavy truck is harder to stop than a small car moving at the same speed. We state this fact by saying that the truck has more momentum than the car. By **momentum** we mean inertia in motion. More specifically momentum is defined as the product of the mass of an object and its velocity; that is,

$$\text{Momentum} = \text{mass} \times \text{velocity}$$

Or, in shorthand notation,

$$\text{Momentum} = mv$$

When direction is not an important factor, we can say:

$$\text{Momentum} = \text{mass} \times \text{speed}$$

which we still abbreviate mv.

FIGURE 5.1 Why are the engines of a supertanker normally cut off 25 km from port?

We can see from the definition that a moving object can have a large momentum if either its mass or its velocity is large, or if both its mass and its velocity are large. The truck has more momentum than the car moving at the same speed because it has a greater mass. We can see that a huge ship moving at a small velocity can have a large momentum, as can a small bullet moving at a high velocity. And, of course, a huge object moving at a high velocity, such as a massive truck rolling down a steep hill with no brakes, has a huge momentum, whereas the same truck at rest has no momentum at all.

Impulse

Changes in momentum may occur when there is either a change in the mass of an object, a change in velocity, or both. If momentum changes while the mass remains unchanged, as is most often the case, then the velocity changes. Acceleration occurs. And what produces an acceleration? The answer is a force. The greater the force that acts on an object, the greater will be the change in velocity and, hence, the change in momentum.

But something else is important in changing momentum: time—how long a time the force acts. Apply a force briefly to a stalled automobile, and you produce a small change in its momentum. Apply the same force over an extended period of time, and a greater change in momentum results. A long sustained force produces more change in momentum than the same force applied briefly. So for changing the momentum of an object, both force and the time during which the force acts are important. We name the product of force and this time interval—**impulse**.

$$\text{Impulse} = \text{force} \times \text{time interval}$$

Or, in shorthand notation,

$$\text{Impulse} = Ft$$

Whenever you exert a net force on something, you also exert an impulse. The resulting acceleration depends on the net force; the resulting change in momentum depends on both the net force and the time during which that force acts.

Impulse Changes Momentum

Impulse changes momentum in much the same way that force changes velocity. The relationship of impulse to change of momentum comes from Newton's second law ($a = F/m$). The time interval of impulse is "buried" in the term for acceleration (change in velocity/time interval).

Rearrangement of Newton's second law gives*

$$\text{Force} \times \text{time interval} = \text{change in (mass} \times \text{velocity)}$$

or, equivalently,

$$\text{Impulse} = \text{change in momentum}$$

We can express all terms in this relationship in shorthand notation and introduce the delta symbol Δ (a letter in the Greek alphabet used to mean "change in" or "difference in"):

$$Ft = \Delta(mv)$$

which reads, "force multiplied by the time-during-which-it-acts equals change in momentum."

The impulse-momentum relationship helps us to analyze many examples in which forces act and motion changes. Sometimes the impulse can be considered to be the cause of a change of momentum. Sometimes a change of momentum can be considered to be the cause of an impulse. It doesn't matter which way you think about it. The important thing is that impulse and change of momentum are always linked. Here we will consider some ordinary examples in which impulse is related to (1) increasing momentum, (2) decreasing momentum over a long time, and (3) decreasing momentum over a short time.

*In Newton's second law ($F/m = a$) we can insert the definition of acceleration ($a = $ (change in v)/t), to get $F/m = $ (change in v)/t. Then we can rearrange this equation by multiplying both sides by mt. This gives $Ft = $ change in (mv), or, in delta notation, $Ft = \Delta$ (mv).

FIGURE 5.2 Impact force against a golf ball.

Case 1: Increasing Momentum

If you wish to increase the momentum of something as much as possible, you not only apply the greatest force you can, you also extend the time of application as much as possible. Hence the different results in pushing briefly on a stalled automobile and giving it a sustained push.

Long-range cannons have long barrels. The longer the barrel, the greater the velocity of the emerging cannonball or shell. Why? The force of exploding gunpowder in a long barrel acts on the cannonball for a longer time. This increased impulse produces a greater momentum. Of course the force that acts on the cannonball is not steady—it is strong at first and weaker as the gases expand. In this and many other cases the forces involved in impulses vary over time. The force that acts on the golf ball in Figure 5.2, for example, increases rapidly as the ball is distorted and then diminishes as the ball comes up to speed and returns to its original shape. When we speak of impact forces in this chapter, we mean the *average* force of impact.

Case 2: Decreasing Momentum over a Long Time

Imagine you are in a car out of control, and you have a choice of slamming into either a concrete wall or a haystack. You needn't know much physics to make the better decision, but knowing some physics helps you to understand why hitting something soft is entirely different from hitting something hard. In the case of hitting either the wall or the haystack, your momentum will be decreased by the *same* amount, and this means that the impulse needed to stop you is the same. The same impulse means the same product of force and time—not the same force or the same time. You have a choice. By hitting

FIGURE 5.3 A large change in momentum over a long time requires a small force.

the haystack instead of the wall, you extend the time of impact—you extend the time during which your momentum is brought to zero. The longer time is compensated by a lesser force. If you extend the time of impact 100 times, impact force is reduced 100-fold. So whenever you wish the force of impact to be small, extend the time of impact.

FIGURE 5.4 A large change in momentum over a short time requires a large force.

FIGURE 5.5 A large change in momentum over a long time requires a safely small average force.

A wrestler thrown to the floor tries to extend his time of arrival on the floor by relaxing his muscles and spreading the crash into a series of impacts as foot, knee, hip, ribs, and shoulder fold onto the floor in turn. The increased time of impact reduces the force of impact.

A person jumping from an elevated position to a floor below bends his knees upon making contact, thereby extending the time during which his momentum is being reduced by 10 to 20 times that of a stiff-legged, abrupt landing. Such knee bending reduces the forces experienced by the bones by 10 to 20 times. Of course, falling on a mat is preferable to falling on a solid floor, for this also increases the time of impact.

Bungee jumping puts the impulse-momentum relationship to a thrilling test (Figure 5.5). The momentum gained during the fall must be decreased to zero by an impulse of equal magnitude. The long stretching time of the cord ensures a small average force to bring the jumper to a safe halt before hitting the ground. Bungee cords typically stretch to about twice their original length during the fall.

Ballet dancers prefer a wooden floor with "give" to a hard floor with little or no "give." The wooden floor allows a longer time of impact whenever the dancer lands, thus reducing the force of impact and reducing the chance of injury. A safety net used by acrobats provides an obvious example of small impact force over a long time to provide the required impulse to reduce the momentum of fall.

If you're about to catch a fast baseball with your bare hand, you extend your hand forward so you'll have plenty of room to let your hand move backward after you make contact with the ball. You extend the time of impact and thereby reduce the force of impact. Similarly, a boxer rides or rolls with the punch to reduce the force of impact (Figure 5.6).

FIGURE 5.6 In both cases the boxer's jaw provides an impulse that reduces the momentum of the punch. (a) The boxer is moving away when the glove hits, thereby extending the time of contact. This means the force is less than if the boxer had not moved. (b) The boxer is moving into the glove, thereby lessening the time of contact. This means that the force is greater than if the boxer had not moved.

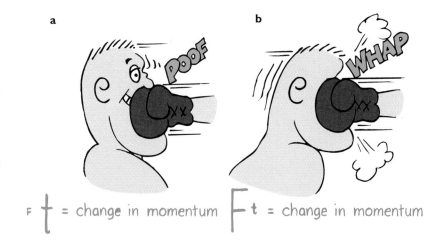

a

b

$F\ t$ = change in momentum $F\ t$ = change in momentum

Case 3: Decreasing Momentum over a Short Time

When boxing, move into a punch instead of away, and you're in trouble. Likewise if you catch a high-speed baseball while your hand moves toward the ball instead of away upon contact. Or when out of control in a car, drive it into a concrete wall instead of a haystack and you're really in trouble. In these cases of short impact times, the impact forces are large. Remember that for an object brought to rest, the impulse is the same, no matter how it is stopped. But if the time is short, the force will be large.

FIGURE 5.7 He imparts a large impulse to the bricks in a short time and produces a considerable force.

The idea of short time of contact explains how a karate expert can sever a stack of bricks with the blow of his bare hand (Figure 5.7). He brings his arm and hand swiftly against the bricks with considerable momentum. This momentum is quickly reduced when he delivers an impulse to the bricks. The impulse is the force of his hand against the bricks multiplied by the time his hand makes contact with the bricks. By swift execution he makes the time of contact very brief and correspondingly makes the force of impact huge. If his hand is made to bounce upon impact, the force is even greater.

Questions

1. If the boxer in Figure 5.6 is able to make the duration of impact three times as long by riding with the punch, by how much will the force of impact be reduced?
2. If the boxer instead moves into the punch such as to decrease the duration of impact by half, by how much will the force of impact be increased?
3. A boxer being hit with a punch contrives to extend time for best results, whereas a karate expert delivers a force in a short time for best results. Isn't there a contradiction here?
4. When does impulse equal momentum?

Bouncing

You know that if a flowerpot falls from a shelf onto your head, you may be in trouble. And whether you know it or not, if it bounces from your head, you may be in more serious trouble. Impulses are greater when bouncing takes place. This is because the impulse required to bring something to a stop and then, in effect, "throw it back again" is greater than the impulse required merely to bring something to a stop. Suppose, for example, that you catch the falling pot with your hands. Then you provide an impulse to catch it and reduce its momentum to zero. If you were to then throw the pot upward, you would have to provide additional impulse. So it would take more impulse to catch it and throw it back up than merely to catch it. The same greater impulse is supplied by your head if the pot bounces from it.

An interesting application of the greater impulse that occurs when bouncing takes place was employed with great success in California during the gold rush days. The water wheels used in gold-mining operations were ineffective. A man named Lester A.

Answers

1. The force of impact will be three times less than if he didn't pull back.
2. The force of impact will be two times greater than if he were hit at rest. This force is further increased because of the additional impulse produced when his momentum of approach is stopped. The increased impulse and short time of impact result in forces that account for many knockouts.
3. There is no contradiction because the best results for each are quite different. The best result for the boxer is reduced force, accomplished by maximizing time, and the best result for the karate expert is increased force delivered in minimum time.
4. Generally, impulse equals a *change* in momentum. If the initial momentum of an object is zero when the impulse is applied, then impulse = momentum.

FIGURE 5.8 The Pelton wheel. The curved blades cause water to bounce and make a U-turn, which produces a greater impulse to turn the wheel.

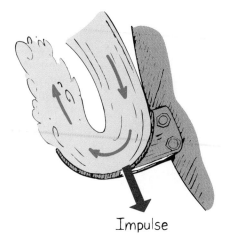

Impulse

Pelton saw that the problem had to do with their flat paddles. He designed curve-shaped paddles that would cause the incident water to make a U-turn upon impact—to "bounce." In this way the impulse exerted on the water wheels was greatly increased. Pelton patented his idea and made more money from his invention, the Pelton wheel, than any of the gold miners made from gold.

Questions

1. In reference to Figure 5.7, how does the force that the man exerts on the bricks compare to the force exerted on his hand?
2. How will the impulse resulting from the impact differ if his hand bounces back upon striking the bricks?

Conservation of Momentum

Recall from Chapter 4 what Newton's second law tells us: If we want to accelerate an object, we must apply a force on it. We say much the same thing in this chapter when we say that to change the momentum of an object we must apply an impulse on it. In any case, the force or impulse must be exerted on the object by *something external* to the object. Internal forces don't count. For example, the molecular forces within a baseball

Answers

1. In accord with Newton's third law, the forces will be equal. Only the resilience of the human hand and the training he has undergone to toughen his hand allow this feat to be performed without broken bones.
2. The impulse will be greater if his hand bounces from the bricks upon impact. If the time of impact is not correspondingly increased, a greater force is then exerted on the bricks (and his hand!).

have no effect on the momentum of the baseball, just as a person sitting inside an automobile pushing against the dashboard, with the dashboard pushing back, has no effect in changing the momentum of the automobile. This is because these forces are internal forces, that is, forces that act and react within the objects themselves. An outside, or external, force acting on the baseball or automobile is required for a change in momentum. If no external force is present, then no change in momentum is possible.

When a bullet is fired from a rifle, the forces present are internal forces. The total momentum of the system comprising the bullet and rifle therefore undergoes no net change (Figure 5.9). By Newton's third law of action and reaction, the force exerted on the bullet is equal and opposite to the force exerted on the rifle. The forces acting on the bullet and rifle act at the same time, resulting in equal but oppositely directed impulses and therefore equal and oppositely directed momenta (the plural form of *momentum*). The recoiling rifle has just as much momentum as the speeding bullet.* Although both the bullet and rifle by themselves have gained considerable momentum, the bullet and rifle together as a system experience no net change in momentum. Before firing, the momentum was zero; after firing, the net momentum is still zero. No momentum is gained and no momentum is lost.

FIGURE 5.9 The momentum before firing is zero. After firing, the net momentum is still zero, because the momentum of the rifle is equal and opposite to the momentum of the bullet.

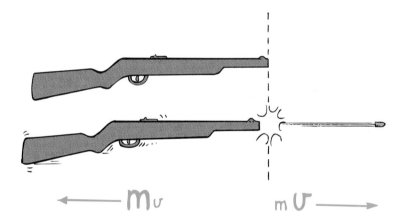

Two important ideas are to be learned from the rifle-and-bullet example. The first is that momentum, like velocity, is a quantity that is described by both magnitude and direction; we measure both "how much" and "which way." Momentum is a *vector quantity*. Therefore, when momenta act in the same direction, they are simply added; when they act in opposite directions, they are subtracted.

The second important idea to be taken from the rifle-and-bullet example is the idea of *conservation*. The momentum before and after firing is the same. For the system of rifle and bullet, no momentum was gained; none was lost. When a physical quantity remains unchanged during a process, that quantity is said to be *conserved*. We say momentum is conserved.

*Here we neglect the momentum of ejected gases from the exploding gunpowder, which can be considerable. Firing a gun with blanks at close range can be lethal because of the momentum of ejected gases. In 1984, actor Jon-Erik Hexum accidentally killed himself when he held a gun loaded with blanks to his head and pulled the trigger, fracturing his skull.

Collisions

Momentum is conserved in collisions—that is, the total momentum of a system of colliding objects is unchanged before, during, and after the collision. This is because the forces that act during the collision are internal forces—forces acting and reacting within the system itself. There is only a redistribution or sharing of whatever momentum exists before the collision.

In any collision, we can say

Total momentum before collision = total momentum after collision.

This is true no matter how the objects might be moving before they collide.

FIGURE 5.10 Elastic collisions of equally massive balls. (a) A green ball strikes a yellow ball at rest. (b) A head-on collision. (c) A collision of balls moving in the same direction. In each case, momentum is transferred from one ball to the other.

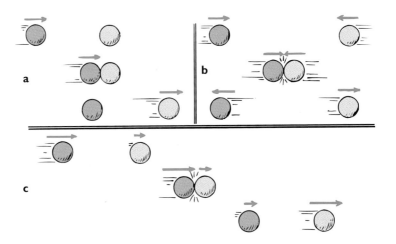

When a moving billiard ball makes a head-on collision with another billiard ball at rest, the moving ball comes to rest and the other ball moves with the speed of the colliding ball. We call this an **elastic collision**; ideally the colliding objects rebound without lasting deformation or the generation of heat (Figure 5.10). But momentum is conserved even when the colliding objects become entangled during the collision. This is an **inelastic collision**, characterized by deformation or the generation of heat or both. In a perfectly inelastic collision, both objects stick together. Consider, for example, the case of a freight car moving along a track and colliding with another freight car at rest (Figure 5.11). If the freight cars are of equal mass and are coupled by the collision, can we predict the velocity of the coupled cars after impact?

FIGURE 5.11   Inelastic collision. The momentum of the freight car on the left is shared with the freight car on the right after collision.

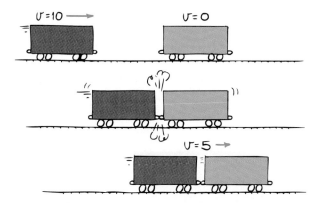

Suppose the single car is moving at 10 meters per second, and we consider the mass of each car to be m. Then, from the conservation of momentum,

$$(\text{total } mv)_{\text{before}} = (\text{total } mv)_{\text{after}}$$

$$(m \times 10 \text{ m/s})_{\text{before}} = (2m \times v)_{\text{after}}$$

By simple algebra, $v_{\text{after}} = 5$ m/s. This makes sense, for since twice as much mass is moving after the collision, the velocity must be half as much as the velocity before collision. Both sides of the equation are then equal.

Notice the importance of direction in these cases. As with any pair of vectors, momenta in the same direction are simply added. If two objects are moving toward each other, one of the momenta is considered negative, and we combine them by subtraction.

Note the inelastic collisions shown in Figure 5.12. If A and B are moving with equal momenta in opposite directions (A and B colliding head-on), then one of these is considered to be negative, and the momenta add algebraically to zero. After collision, the coupled wreck remains at the point of impact, with zero momentum.

FIGURE 5.12   Inelastic collisions. The net momentum of the trucks before and after collision is the same.

If, on the other hand, A and B are moving in the same direction (A catching up with B), the net momentum is simply the addition of their individual momenta.

If A, however, moves east with, say, 10 more units of momentum than B moving west (not shown in the figure), after collision the coupled wreck moves east with 10 units of momentum. The wreck will finally come to a rest, of course, because of the external force of friction by the ground. The time of impact is short, however, and the impact force of the collision is so much greater than the external friction force that momentum immediately before and after the collision is for practical purposes conserved. The total momentum just before the trucks collide (10 units) is equal to the combined momentum of the crumpled trucks just after impact. The same principle applies to gently docking spacecraft. Their total momentum just before docking is preserved as their total momentum just after docking.

Question Consider the air track in Figure 5.13. Suppose a gliding cart with a mass of 0.5 kg bumps into and sticks to a stationary cart that has a mass of 1.5 kg. If the speed of the gliding cart before impact is v_{before}, how fast will the coupled carts glide after collision?

FIGURE 5.13 An air track. Blasts of air from tiny holes provide a friction-free surface for the carts to glide upon.

Answer According to momentum conservation, the momentum of the first cart before the collision = the momentum of both carts stuck together afterwards.

$$(0.5 \text{ kg}) \, v_{before} = (0.5 \text{ kg} + 1.5 \text{ kg}) \, v_{after}$$

$$v_{after} = \left(\frac{0.5 \text{ kg}}{2.0 \text{ kg}}\right) v_{before} = \frac{v_{before}}{4}$$

This makes sense, because four times as much mass will be moving after the collision, so the coupled carts will glide more slowly. In keeping the momentum equal, four times the mass glides $\frac{1}{4}$ as fast.

For a numerical example of momentum conservation, consider a fish that swims toward and swallows a smaller fish at rest (Figure 5.14). If the larger fish has a mass of 5 kg and swims 1 m/s toward a 1-kg fish, what is the velocity of the larger fish immediately after lunch? We will neglect the effects of water resistance.

$$\text{Total momentum before lunch} = \text{total momentum after lunch}$$

$$(5 \text{ kg})(1 \text{ m/s}) + (1 \text{ kg})(0 \text{ m/s}) = (5 \text{ kg} + 1 \text{ kg})v_{\text{after}}$$

$$5 \text{ kg m/s} = (6 \text{ kg})v_{\text{after}}$$

$$v_{\text{after}} = 5/6 \text{ m/s}$$

Here we see that the small fish has no momentum before lunch because its velocity is zero. After lunch the combined mass of both fishes moves at velocity v_{after}, which by simple algebra is seen to be $\frac{5}{6}$ m/s. This velocity is in the same direction as that of the larger fish.

FIGURE 5.14 Two fish make up a system, which has the same momentum just before lunch and just after lunch.

Suppose the small fish in the preceding example is not at rest, but swims toward the left at a velocity of 4 m/s. It swims in a direction opposite that of the larger fish—a negative direction, if the direction of the larger fish is considered positive. In this case,

$$\text{Total momentum before lunch} = \text{total momentum after lunch}$$

$$(5 \text{ kg})(1 \text{ m/s}) + (1 \text{ kg})(-4 \text{ m/s}) = (5 \text{ kg} + 1 \text{ kg})\, v_{\text{after}}$$

$$(5 \text{ kg m/s}) - (4 \text{ kg m/s}) = (6 \text{ kg})\, v_{\text{after}}$$

$$1 \text{ kg m/s} = 6 \text{ kg}\, v_{\text{after}}$$

$$v_{\text{after}} = \tfrac{1}{6} \text{ m/s}$$

Note that the negative momentum of the smaller fish before lunch has more effect in slowing the larger fish after lunch. If the smaller fish were swimming twice as fast, then

$$\text{Total momentum before lunch} = \text{total momentum after lunch}$$

$$(5 \text{ kg})(1 \text{ m/s}) + (1 \text{ kg})(-8 \text{ m/s}) = (5 \text{ kg} + 1 \text{ kg})v_{\text{after}}$$

$$(5 \text{ kg m/s}) - (8 \text{ kg m/s}) = (6 \text{ kg})v_{\text{after}}$$

$$-3 \text{ kg m/s} = 6 \text{ kg}\, v_{\text{after}}$$

$$v_{\text{after}} = -\tfrac{1}{2} \text{ m/s}$$

Here we see the final velocity is $-\frac{1}{2}$ m/s. What is the significance of the minus sign? It means that the final velocity is *opposite* to the initial velocity of the larger fish. After lunch the two-fish system moves toward the left. We leave as a chapter-end problem the task of finding the initial velocity of the smaller fish that would halt the larger fish in its tracks.

More Complicated Collisions

The net momentum remains unchanged in any collision, regardless of the angle between the tracks of the colliding objects. Expressing the net momentum when different directions are involved can be achieved with the parallelogram rule of vector addition. We will not treat such complicated cases in great detail here but will show some simple examples to convey the concept.

In Figure 5.15 we see a collision between two cars traveling at right angles with respect to each other. Car A has a momentum directed due east, and car B's momentum is directed due north. If their individual momenta are equal in magnitude, then their combined momentum is in a northeast direction. This is the direction the coupled cars will travel after collision. We see that just as the diagonal of a square is not equal to the sum of two of the sides, the magnitude of the resulting momentum will not simply equal the arithmetic sum of the two momenta before collision. Recall the relationship between the diagonal of a square and the length of one of its sides, Figure 3.5 in Chapter 3—the diagonal is $\sqrt{2}$ times the length of the side of a square. So in this example, the magnitude of the resultant momentum will be equal to $\sqrt{2}$ times the momentum of either vehicle.*

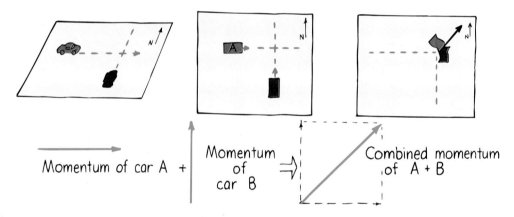

Momentum of car A + Momentum of car B ⇒ Combined momentum of A + B

FIGURE 5.15   Momentum is a vector quantity.

Figure 5.16 shows a falling Fourth-of-July firecracker exploding into two pieces. The momenta of the fragments combine by vector addition to equal the original momentum of the falling firecracker. Figure 5.17b extends this idea to the microscopic realm, where the tracks of subatomic particles are revealed in a liquid hydrogen bubble chamber.

Whatever the nature of a collision or however complicated it is, the total momentum before, during, and after remains unchanged. This extremely useful law enables us to learn much from collisions without knowing any details about the interaction forces that act in the collision. We will see in the next chapter that energy as well as momentum

*The diagonal of a square or rectangle can be calculated by the Pythagorean theorem: the square of the diagonal is equal to the sum of the squares of the two sides. Take the square root of that quantity and you have the diagonal value. In the case of a square the diagonal is $\sqrt{2}$, or about 1.41, times the length of one of the sides.

FIGURE 5.16 After the firecracker bursts, the momenta of its fragments add up (by vector addition) to the original momentum.

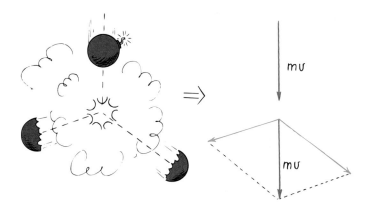

is conserved. By applying momentum and energy conservation to the collisions of subatomic particles as observed in various detection chambers, we can compute the masses of these tiny particles. We obtain this information by measuring momenta and energy before and after collisions. Remarkably, this achievement is possible without any exact knowledge of the forces that act.

Conservation of momentum and conservation of energy (next chapter) are the two most powerful tools of mechanics. Applying them yields detailed information that ranges from facts about the interactions of subatomic particles to the structure and motion of entire galaxies.

FIGURE 5.17 Momentum is conserved for colliding billiard balls and for colliding nuclear particles in a liquid hydrogen bubble chamber. In (a), billiard ball A strikes billiard ball B, which was initially at rest. In (b), proton A collides successively with protons B, C, and D. The moving protons leave tracks of tiny bubbles.

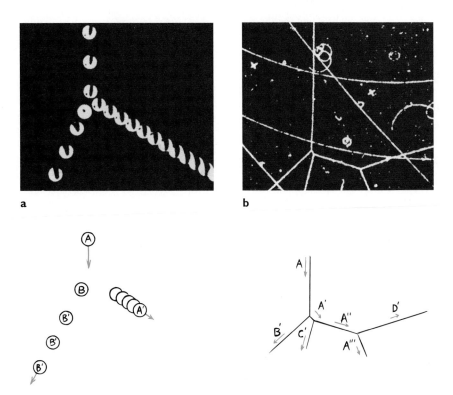

Summary of Terms

Momentum   The product of the mass of an object and its velocity; inertia in motion.

$$\text{Momentum} = mv$$

Impulse   The product of the force acting on an object and the time during which it acts. In an interaction, impulses are equal and opposite.

$$\text{Impulse} = Ft$$

Relationship of impulse and momentum   Impulse is equal to the change in the momentum of the object that the impulse acts on. In symbol notation,

$$Ft = \Delta mv$$

Conservation of momentum   When no external net force acts on an object or a system of objects, no change of momentum takes place. Hence, the momentum before an event involving only internal forces is equal to the momentum after the event:

$$mv_{\text{(before event)}} = mv_{\text{(after event)}}$$

Elastic collision   A collision in which colliding objects rebound without lasting deformation or the generation of heat.

Inelastic collision   A collision in which the colliding objects become distorted, generate heat, and possibly stick together.

Review Questions

Momentum

1. Which has a greater momentum, a heavy truck at rest or a moving skateboard?

Impulse

2. How does impulse differ from force?
3. What are the two ways to increase impulse?

Impulse Changes Momentum

4. Is the impulse-momentum relationship related to Newton's second law?
5. Why is it incorrect to say that impulse equals momentum?

Case 1: Increasing Momentum

6. To impart the greatest momentum to an object, should you exert the largest force possible, extend that force for as long a time as possible, or both? Explain.
7. For the same force, which cannon imparts the greater speed to a cannonball—a long cannon or a short one? Explain.

Case 2: Decreasing Momentum over a Long Time

8. When you are in the way of a moving object and an impact force is your fate, are you better off decreasing its momentum over a short time or over a long time? Explain.
9. Why might a wine glass survive a fall onto a carpeted floor but not onto a concrete floor?
10. Why is it a good idea to have your hand extended forward when you are getting ready to catch a fast-moving baseball with your bare hand?

Case 3: Decreasing Momentum over a Short Time

11. Why would it be a poor idea to have the back of your hand up against the outfield wall when you catch a long fly ball?
12. Which has the greater momentum when they move at the same speed—a heavy truck or a skateboard? Which requires the greater stopping force?
13. In answering the preceding question, perhaps you stated that the truck required more stopping force. Make an argument that the skateboard could require more stopping force, depending on how quickly you want to stop it.

Bouncing

14. Which undergoes the greatest change in momentum: (1) a baseball that is caught, (2) a baseball that is thrown, or (3) a baseball that is caught and then thrown back, if the baseballs have the same speed just before being caught and just after being thrown?
15. In the preceding question, in which case is the greatest impulse required?

Conservation of Momentum

16. Can you produce a net impulse on an automobile by sitting inside and pushing on the dashboard? Can the internal forces within a soccer ball produce an impulse on the soccer ball that will change its momentum?
17. Is it correct to say that if no net impulse is exerted on an object, then no change in the momentum of the object will occur?
18. What does it mean to say that a quantity is *conserved?*
19. When a bullet is fired, its momentum indeed changes! Also the momentum of the recoiling rifle changes. So momentum is not conserved for the bullet, and momentum is not conserved for the rifle. Why can we nevertheless say that when a rifle fires a bullet, momentum *is* conserved?
20. Would momentum be conserved for the *system* of rifle and bullet if momentum were not a vector quantity? Explain.

Collisions

21. Distinguish between an *elastic* collision and an *inelastic* collision. For which type of collision is momentum conserved?

22. Railroad car A rolls at a certain speed and makes a perfectly elastic collision with car B of the same mass. After the collision, car A is observed to be at rest. How does the speed of car B compare with the initial speed of car A?

23. If the equally massive cars of the previous question stick together after colliding inelastically, how does their speed after the collision compare with the initial speed of car A?

More Complicated Collisions

24. Suppose a ball of putty moving horizontally with 1 kg m/s of momentum collides and sticks to an identical ball of putty moving vertically with 1 kg m/s of momentum. Why is their combined momentum not simply the arithmetic sum, 2 kg m/s?

25. In the preceding question, what is the total momentum of the balls of putty before and after the collision?

Project

When you get a bit ahead in your studies, cut classes some afternoon and visit your local pool or billiards parlor and bone up on momentum conservation. Note that no matter how complicated the collision of balls, the momentum along the line of action of the cue ball before impact is the same as the combined momentum of all the balls along this direction after impact and that the components of momenta perpendicular to this line of action cancel to zero after impact, the same value as before impact in this direction. You'll see both the vector nature of momentum and its conservation more clearly when rotational skidding—English—is not imparted to the cue ball. When English is imparted by striking the cue ball off center, rotational momentum, which is also conserved, somewhat complicates analysis. But regardless of how the cue ball is struck, in the absence of external forces, both linear and rotational momentum are always conserved. Pool or billiards offers a first-rate exhibition of momentum conservation in action.

Exercises

1. To bring a supertanker to a stop, its engines are typically cut off about 25 km from port. Why is it so difficult to stop or turn a supertanker?

2. In terms of impulse and momentum, why do padded dashboards make automobiles safer?

3. Why are thick floor mats used by gymnasts?

4. In terms of impulse and momentum, why are nylon ropes, which stretch considerably under tension, favored by mountain climbers?

5. A person can survive a feet-first impact at a speed of about 12 m/s (27 mi/h) on concrete; 15 m/s (34 mi/h) on soil; and 34 m/s (76 mi/h) on water. Why the different values for different surfaces?

6. In terms of impulse and momentum, why is it important that an airplane wing be designed so that it deflects oncoming air downward?

7. A lunar vehicle is tested on earth at a speed of 10 km/h When it travels as fast on the moon, is its momentum more, less, or the same?

8. It is generally much more difficult to stop a heavy truck than a skateboard when they move at the same speed. State a case where the moving skateboard could require more stopping force. (Consider relative times.)

9. If you throw a raw egg against a wall you'll break it, but if you throw it with the same speed into a sagging sheet it won't break. Explain using concepts from this chapter.

10. Why is it difficult for a firefighter to hold a hose that ejects large amounts of water at a high speed?

11. Would you care to fire a gun that has a bullet ten times as massive as the gun? Explain.

12. Why are the impulses that colliding objects exert on each other equal and opposite?

13. If a ball is projected upward from the ground with 10 units of momentum, what is the momentum of recoil of the world? Why do we not feel this?

14. When an apple falls from a tree and strikes the ground without bouncing, what becomes of its momentum?

15. Why is a punch more forceful with a bare fist than with a boxing glove?

16. A boxer can punch a heavy bag for more than an hour without tiring, but will tire quickly when boxing with an opponent for a few minutes. Why? (*Hint:* When the boxer's fist is aimed at the bag, what supplies the impulse to stop the punches? When the boxer's fist is aimed at the opponent, what or who supplies the impulse to stop the punches that don't connect?)

17. Railroad cars are loosely coupled so that there is a no-ticeable time delay from the time the first car is moved and the last cars are moved from rest by the locomotive. Discuss the advisability of this loose coupling and slack between cars from the point of view of impulse and momentum.

18. If only an external force can change the velocity of a body, how can the internal force of the brakes bring a car to rest?

19. A fully dressed person is at rest in the middle of a pond on perfectly frictionless ice and must get to shore. How can this be accomplished?

20. If you throw a ball horizontally while standing on roller skates, you roll backward with a momentum that matches that of the ball. Will you roll backward if you go through the motions of throwing the ball, but instead hold on to it? Explain.

21. The examples of the two previous exercises can be ex-plained in terms of momentum conservation and in terms of Newton's third law. Assuming you've answered them in terms of momentum conservation, answer them also in terms of Newton's third law (or vice versa).

22. In Chapter 4 rocket propulsion was explained in terms of Newton's third law. That is, the force that propels a rocket is from the exhaust gases pushing against the rocket, the reaction to the force the rocket exerts on the exhaust gases. Explain rocket propulsion in terms of momentum conservation.

23. Explain how the conservation of momentum is a conse-quence of Newton's third law.

24. Go back to Exercise 36 in Chapter 4 and answer it in terms of momentum conservation.

25. Your friend says that the law of momentum conserva-tion is violated when a ball rolls down a hill and gains momentum. What do you say?

26. Bronco dives from a high flying plane and finds his mo-mentum increasing. Does this violate the conservation of momentum? Explain.

27. In a movie, the hero jumps straight down from a bridge onto a small boat that continues to move with no change in velocity. What physics is being violated here?

28. An ice sailcraft is stalled on a frozen lake on a windless day. The skipper sets up a fan as shown. If all the wind bounces backward from the sail, will the craft be set in motion? If so, in what direction?

29. Will your answer to the preceding exercise be different if the air is brought to a halt by the sail without bounc-ing?

30. Discuss the advisability of simply removing the sail in the preceding exercises.

31. When you are traveling in your car at highway speed, the momentum of a bug is suddenly changed as it splat-ters onto your windshield. Compared to the change in momentum of the bug, by how much does the momen-tum of your car change?

32. If a Mack truck and a Ford Escort have a head-on colli-sion, which vehicle will experience the greater force of impact? The greater impulse? The greater change in mo-mentum? The greater acceleration?

33. Would a head-on collision between two cars be more damaging to the occupants if the cars stuck together or if the cars rebounded upon impact?

34. Suppose there are three astronauts outside a spaceship, and they decide to play catch. All the astronauts weigh the same on earth and are equally strong. The first astro-naut throws the second one toward the third one and the game begins. Describe the motion of the astronauts as the game proceeds. How long will the game last?

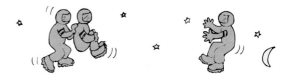

35. To throw a ball, do you exert an impulse on it? Do you exert an impulse to catch it at the same speed? About how much impulse do you exert, in comparison, if you catch it and immediately throw it back again? (Imagine yourself on a skateboard.)

36. In reference to Figure 5.7, how will the impulse at impact differ if the hand bounces back upon striking the bricks? In any case, how does the force exerted on the bricks compare to the force exerted on the hand?

37. Light consists of tiny "corpuscles" called photons that possess momentum. This can be demonstrated with a radiometer, shown in the sketch. Metal vanes painted black on one side and white on the other are free to rotate about the point of a needle mounted in a vacuum. When photons are incident on the black surface, they are absorbed; when photons are incident upon the white surface, they are reflected. Upon which surface is the impulse of incident light greater, and which way will the vanes rotate? (They rotate in the opposite direction in the more common radiometers where air is present in the glass chamber; your instructor may tell you why.)

38. A deuteron is a nuclear particle of unique mass made up of one proton and one neutron. Suppose it is accelerated up to a certain very high speed in a cyclotron and directed into an observation chamber, where it collides with and sticks to a target particle that is initially at rest and then is observed to move at exactly half the speed of the incident deuteron. Why do the observers state that the target particle is itself a deuteron?

39. When a stationary uranium nucleus undergoes fission, it breaks into two unequal chunks that fly apart. What can you conclude about the momenta of the chunks? What can you conclude about the speed of the chunks?

40. A billiard ball will stop short when it collides head-on with a ball at rest. The ball cannot stop short, however, if the collision is not exactly head-on—that is, if the second ball moves at an angle to the path of the first. Do you know why? (*Hint:* Consider momentum before and after the collision along the initial direction of the first ball and also in a direction perpendicular to this initial direction.)

Problems

1. A car with a mass of 1000 kg moves at 20 m/s. What braking force is needed to bring the car to a halt in 10 s?

2. A 1000-kg car accidentally drops from a crane and crashes at 30 m/s to the ground below and comes to an abrupt halt. Consider questions (a) and (b). Which question can be answered using the given information and which one cannot be answered? Explain. (a) What impulse acts on the car when it crashes? (b) What is the force of impact on the car?

3. A 2-kg ball of putty moving to the right has a head-on inelastic collision with a 1-kg putty ball moving to the left. If the combined blob doesn't move just after the collision, what can you conclude about the relative speeds of the balls before they collided?

4. A railroad diesel engine weighs four times as much as a freightcar. If the diesel engine coasts at 5 km/h into a freightcar that is initially at rest, how fast do the two coast after they couple together?

5. A 5-kg fish swimming 1 m/s swallows an absent minded 1-kg fish swimming toward it at a velocity that brings both fish to a halt immediately after lunch. What is the velocity v of the smaller fish before lunch?

6. Comic-strip hero Superman meets an asteroid in outer space, and hurls it at 800 m/s, as fast as a bullet. The asteroid is a thousand times more massive than Superman. In the strip, Superman is seen at rest after the throw. Taking physics into account, what would be his recoil velocity?

7. Two automobiles, each of mass 1000 kg, are moving at the same speed, 20 m/s, when they have an inelastic collision. In what direction and at what speed does the wreckage move (a) if one car was driving north and one south? (b) if one car was driving north and one east (as shown in Figure 5.15)?

8. Can you run fast enough to have the same momentum as an automobile rolling at 1 mi/h? Justify your answer.

6

· · · · · · · · · · · ·

Energy

A body with momentum also has energy of motion—kinetic energy.

Perhaps the concept most central to all of science is energy. The combination of energy and matter makes up the universe: matter is substance, and energy is the mover of substance. The idea of matter is easy to grasp. Matter is stuff that we can see, smell, and feel. It has mass and occupies space. Energy, on the other hand, is abstract. We cannot see, smell or feel most forms of energy. Surprisingly, the idea of energy was unknown to Isaac Newton, and its existence was still being debated in the 1850s. Although energy is familiar to us, it is difficult to define, because it is not only a "thing" but both a thing and a process—as if it were both a noun and a verb. Persons, places, and things have energy, but we usually observe energy only when it is happening—only when it is being transformed. It comes to us in the form of electromagnetic waves from the sun and we feel it as thermal energy; it is captured by plants and binds molecules of matter together; it is in the food we eat and we receive it by digestion. Even matter itself is condensed bottled-up energy, as set forth in Einstein's famous formula, $E = mc^2$, which we'll return to in the last part of this book. For now we'll begin our study of energy by observing a related concept: work.

Work

FIGURE 6.1 Work is done in lifting the barbell. If the weightlifter were taller, he would have to expend proportionally more energy to press the barbell over his head.

FIGURE 6.2 He may expend energy when he pushes on the wall, but if it doesn't move, no work is performed on the wall.

Power

In the previous chapter we saw that changes in an object's motion depend on both force and how long the force acts. "How long" meant time. We called the quantity "force × time" *impulse*. But "how long" need not always mean time. It can mean distance also. When we consider the quantity force × distance, we are talking about an entirely different quantity—**work**.

When we lift a load against earth's gravity, work is done. The heavier the load or the higher we lift the load, the more work is done. Two things enter the picture whenever work is done: (1) the exertion of a force and (2) the movement of something by that force. For the simplest case, where the force is constant and the motion takes place in a straight line in the direction of the force,* we define the work done on an object by an applied force as the product of the force and the distance through which the object is moved. In shorter form:

$$\text{Work} = \text{force} \times \text{distance}$$
$$W = Fd$$

If we lift two loads one story up, we do twice as much work as in lifting one load, because the *force* needed to lift twice the weight is twice as much. Similarly, if we lift a load two stories instead of one story, we do twice as much work because the *distance* is twice as much.

We see the definition of work involves both a force and a distance. A weightlifter who holds a barbell weighing 1000 newtons overhead does no work on the barbell. He may get really tired doing so, but if the barbell is not moved by the force he exerts, he does no work *on the barbell*. Work may be done on the muscles by stretching and contracting, which is force times distance on a biological scale, but this work is not done on the barbell. Lifting the barbell, however, is a different story. When the weightlifter raises the barbell from the floor, he is doing work.

The unit of measurement for work combines a unit of force (N) with a unit of distance (m); the unit of work is the newton-meter (N · m), also called the *joule* (J) (rhymes with *cool*). One joule of work is done when a force of 1 newton is exerted over a distance of 1 meter, as in lifting an apple over your head. For larger values we speak of kilojoules (kJ, thousands of joules), or megajoules (MJ, millions of joules). The weightlifter in Figure 6.1 does work in kilojoules. The energy released by 1 kilogram of gasoline is rated in megajoules.

The definition of work says nothing about how long it takes to do the work. The same amount of work is done when carrying a load up a flight of stairs, whether we walk up or run up. So why are we more tired after running upstairs in a few seconds than after walking upstairs in a few minutes? To understand this difference, we need to talk about a measure of how fast the work is done—**power**. Power is equal to the amount of work done per time it takes to do it:

$$\text{Power} = \frac{\text{work done}}{\text{time interval}}$$

* More generally, work is the product of only the component of force that acts in the direction of motion and the distance moved. For example, when a force acts at right angles to the direction of motion, with no force component in the direction of motion, no work is done. A common example is a satellite in circular orbit; the force of gravity is at right angles to its circular path and no work is done on the satellite. Hence it orbits with no change in speed.

An engine of great power can do work rapidly. An automobile engine that delivers twice the power of another does not necessarily produce twice as much work or go twice as fast as the less powerful engine. Twice the power means it can do the same amount of work in half the time or twice the work in the same time. A more powerful engine can get an automobile up to a given speed in less time than a less powerful engine can.

Here's another way to look at power: a liter (L) of fuel can do a certain amount of work, but the power produced when we burn it can be any amount, depending on how *fast* it is burned. It can power a lawnmower for a half hour or a jet engine at 3600 times the power for a half second.

The unit of power is the joule per second (J/s), also known as the watt (in honor of James Watt, the eighteenth-century developer of the steam engine). One watt (W) of power is expended when 1 joule of work is done in 1 second. One kilowatt (kW) equals 1000 watts. One megawatt (MW) equals 1 million watts. In the United States we customarily rate engines in units of horsepower and electricity in kilowatts, but either may be used. In the metric system of units, automobiles are rated in kilowatts. (One horsepower is the same as three-fourths of a kilowatt, so an engine rated at 134 horsepower is a 100-kW engine.)

FIGURE 6.3 The three main engines of a space shuttle can develop 33,000 MW of power when fuel is burned at the enormous rate of 3400 kg/s. This is like emptying an average-size swimming pool in 20 s.

Mechanical Energy

Work is done in lifting the heavy ram of a pile driver, and, as a result, the ram acquires the property of being able to do work on a piling beneath it when it falls. When work is done by an archer in drawing a bow, the bent bow has the ability to do work on the arrow. When work is done to wind a spring mechanism, the spring acquires the ability to do work on various gears to run a clock, ring a bell, or sound an alarm. In each case, something has been acquired. This "something" given to the object enables the object to do work. This "something" may be a compression of atoms in the material of an object; it may be a physical separation of attracting bodies; it may be a rearrangement of electric charges in the molecules of a substance. This "something" that enables an object to do work is **energy**.* Like work, energy is measured in joules. It appears in many forms, which will be discussed in the following chapters. For now we focus on *mechanical energy*—the form of energy due to the relative *position* of interacting bodies (potential energy) or to their *motion* (kinetic energy). Mechanical energy may be in the form of either potential energy or kinetic energy, or both.

* Strictly speaking, that which enables an object to do work is its available energy, for not all the energy in an object can be transformed to work.

FIGURE 6.4 The potential energy of Tenny's drawn bow equals the work (average force × distance) she did in drawing the arrow into position. When released, the potential energy of the drawn bow will become the kinetic energy of the arrow.

Potential Energy

An object may store energy because of its position relative to some other object. This energy is called **potential energy** (PE), because in the stored state it has the potential to do work. For example, a stretched or compressed spring has the potential for doing work. When a bow is drawn, energy is stored in the bow. A stretched rubber band has potential energy because of its position, for if it is part of a slingshot, it is capable of doing work.

The chemical energy in fuels is also potential energy, due to the relative positions of atoms in molecules—energy of position from a microscopic point of view. Such energy characterizes fossil fuels, electric batteries, and the food we eat. This energy is available when atoms are rearranged, that is, when a chemical change takes place. Any substance that can do work through chemical action possesses potential energy.

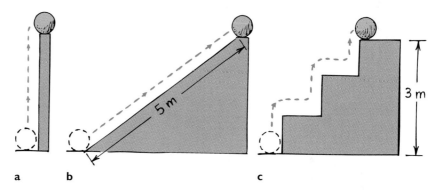

a   b   c

FIGURE 6.5 The potential energy of the 10-N ball is the same (30 J) in all three cases because the work done in elevating it 3 m is the same whether it is (a) lifted with 10 N of force, (b) pushed with 6 N of force up the 5-m incline, or (c) lifted with 10 N up each 1-m stair. No work is done in moving it horizontally (neglecting friction).

Work is required to elevate objects against earth's gravity. The potential energy of a body due to elevated positions is called *gravitational potential energy.* Water in an elevated reservoir and the ram of a pile driver have gravitational potential energy. The amount of gravitational potential energy possessed by an elevated object is equal to the work done against gravity in lifting it. The work done equals the force required to move it upward times the vertical distance it is moved ($W = Fd$). Once upward motion begins, the upward force to keep it moving at constant speed equals the weight mg of the object. So in addition to the work done in getting it started, which we'll assume for the present is negligible, the work done in lifting an object of weight mg through a height h is given by the product mgh:

$$\text{Gravitational potential energy} = \text{weight} \times \text{height}$$

$$\text{PE} = mgh$$

Note that the height h is the distance above some reference level, such as the ground or the floor of a building. The potential energy mgh is relative to that level and depends only on mg and the height h. You can see in Figure 6.5 that the potential energy of the ball at the top of the ledge depends on the height but does not depend on the path taken to get it there.

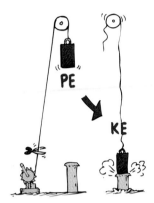

FIGURE 6.6 The potential energy of the elevated ram is converted to kinetic energy when released.

Potential energy, gravitational or otherwise, has significance only when it *changes* —when it does work or transforms to energy of some other form. For example, if the ball in Figure 6.5 falls from its elevated position and does 20 joules of work when it lands, then it has lost 20 joules of potential energy. How much *total* potential energy the ball had when it was elevated doesn't matter—in fact, it is sometimes not even possible to define total potential energy. What matters is the amount of potential energy that is converted. Only *changes* in potential energy are meaningful. One of the kinds of energy into which potential energy can change is energy of motion, or *kinetic energy.*

Questions

1. How much work is done on a 75-N bowling ball when you carry it horizontally across a 10-m-wide room?

2. How much work is done on it when you lift it 1 m? What power is expended if you lift it this distance in 1 s?

3. What is its gravitational potential energy in the lifted position?

FIGURE 6.7 The downhill "fall" of the roller coaster results in its roaring speed in the dip, and this kinetic energy sends it up the steep track to the next summit.

Answers

1. Except for the small amount of work to get the ball moving, you do no work on the ball moving it horizontally, for you apply no force (except for the tiny bit to start it) in its direction of motion. It has no more PE across the room than it had initially.

2. You do 75 J of work when you lift it 1 m ($Fd = 75 \text{ N} \times 1 \text{ m} = 75 \text{ N} \cdot \text{m} = 75 \text{ J}$). Power = 75 J/1 s = 75 W.

3. It depends. With respect to its starting position its PE is 75 J; with respect to some other reference level, it would be some other value.

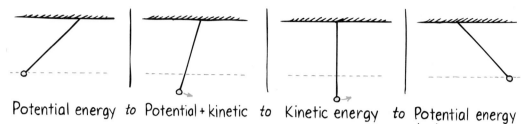

Potential energy *to* Potential + kinetic *to* Kinetic energy *to* Potential energy And so on

FIGURE 6.8 Energy transitions in a pendulum. PE is relative to the lowest point of the pendulum, when it is vertical.

Kinetic Energy

If we push on an object, we can set it in motion. More specifically, if we do work on an object, we can change the energy of motion of that object. If an object is moving, then by virtue of that motion it is capable of doing work. We call energy of motion **kinetic energy** (KE). The kinetic energy of an object depends on mass and speed. It is equal to half the mass multiplied by the *square* of the speed.

$$\text{Kinetic energy} = \tfrac{1}{2} \text{ mass} \times \text{speed}^2$$
$$\text{KE} = \tfrac{1}{2} mv^2$$

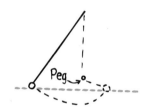

FIGURE 6.9 The pendulum bob will swing to its original height whether or not the peg is present.

The amount of kinetic energy, like the amount of speed, depends on the reference frame in which it's measured. For example, when you ride in a fast-moving car you have zero kinetic energy relative to the car, but considerable kinetic energy relative to the ground. Notice that speed is squared in the definition of KE, so if the speed of an object is doubled, its kinetic energy is quadrupled ($2^2 = 4$), and the object can do four times as much work. Accident investigators are well aware that an automobile going 100 kilometers per hour has four times the kinetic energy it would have at 50 kilometers per hour. This means a car going 100 kilometers per hour will skid four times as far when its brakes are locked as it would going 50 kilometers per hour. This is because speed is squared for kinetic energy (this squaring of speed also means that KE can only be zero or positive—never negative).

Work-Energy Theorem

When a ball is thrown, work is done on it, giving it kinetic energy. The moving ball can then hit something and push against it, doing work on what it hits. Therefore the kinetic energy of a moving object is equal to the work done in bringing it from rest to that speed or to the work a moving object can do in being brought to rest:*

$$\text{Net force} \times \text{distance} = \text{kinetic energy}$$
$$Fd = \tfrac{1}{2} mv^2$$

Put still another way, we can say that the work done is equal to the change in kinetic energy—the **work-energy theorem**:

$$\text{Work} = \Delta\text{KE}$$

---

* This can be derived as follows: If we multiply both sides of $F = ma$ (Newton's second law) by d, we get $Fd = mad$. Recall from Chapter 2 that for constant acceleration $d = \tfrac{1}{2} at^2$, so we can say $Fd = ma \left(\tfrac{1}{2} at^2\right) = \tfrac{1}{2} maat^2 = \tfrac{1}{2} m(at)^2$; and substituting $v = at$, we get $Fd = \tfrac{1}{2} mv^2$.

The work in this equation is the *net* work—that is, the work based on the net force. If, for instance, you are pushing on an object and friction is also acting on the object, the change of kinetic energy is equal to the work done by the net force, which is your push minus friction. (In this case, only part of the *total* work that you do is going into the object's kinetic energy. The rest is going into heat, which we'll discuss later.)

The change in an object's kinetic energy is equal to the work done on the object by a net force. Or the amount of work a moving object can do on another depends on its change in kinetic energy. Simply stated, the work done on or by a moving object equals the change in its kinetic energy.

Questions

1. When the brakes of a car going 90 km/h are locked, how much farther will it skid than if the brakes lock at 30 km/h?
2. Can an object have energy?
3. Can an object have work?

Kinetic energy and potential energy underlie other seemingly different forms of energy such as heat, sound, and light. Average kinetic energy of random molecular motion is related to temperature, kinetic and potential energies of vibrating air define sound intensity. Electrons in motion make electric currents. Even light energy originates from the motion of electrons within atoms. We will find there is much in common among the various forms of energy that we will investigate.

Conservation of Energy

More important than being able to state *what energy is* is understanding how it behaves—*how it transforms*. We can better understand the processes and changes that occur in nature if we analyze them in terms of transformations of energy from one form into another or of transfers from one place to another. By keeping track of energy changes we employ an extremely useful energy bookkeeping system.

As we draw back the stone in a slingshot, we do work in stretching the rubber band; we give the rubber band potential energy. When the rubber band is released, a transformation occurs and the stone gains kinetic energy equal to this potential energy.

Answers

1. Nine times farther. The car has nine times as much kinetic energy when it travels three times as fast: $\frac{1}{2} m(3v)^2 = \frac{1}{2} m9v^2 = 9(\frac{1}{2} mv^2)$. The friction force will ordinarily be the same in either case; therefore, to do nine times the work requires nine times as much sliding distance.
2. Yes, but in a relative sense. For example, an elevated object may possess PE relative to the ground below, but none relative to a point at the same elevation. Similarly, the KE that an object has is with respect to a frame of reference, usually taken to be the earth's surface. (We will see that material objects have energy of being—the congealed energy that makes up their mass.) Read on!
3. No, unlike momentum or energy, work is not something that an object *has*. Work is something that an object *does* to some other object. An object can *do* work only if it has energy.

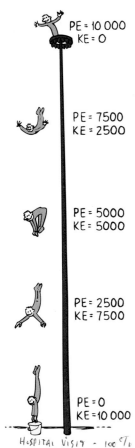

PE = 10 000
KE = 0

PE = 7500
KE = 2500

PE = 5000
KE = 5000

PE = 2500
KE = 7500

PE = 0
KE = 10 000

HOSPITAL VISIT - 100 %

FIGURE 6.10 A circus diver at the top of a pole has a potential energy of 10,000 J. As he dives, his potential energy converts to kinetic energy. Note that at successive positions one-fourth, one-half, three-fourths, and all the way down, the total energy is constant. From *Physics in Your World,* Second Edition by Karl F. Kuhn and Jerry S. Faughn. Copyright ©1980 by Saunders College Publishing. Reprinted by permission of the publisher.

The stone then transfers this energy to its target, perhaps a wooden fence post. The slight distance the post is moved multiplied by the average force of impact doesn't quite match the kinetic energy of the stone. The energy score doesn't balance. But if we investigate further, we'll find that both the stone and fence post are a bit warmer. By how much? By enough to account for the energy difference. Energy changes from one form to another. It transforms without net loss or net gain.

The study of various forms of energy and their transformations from one form into another has led to one of the greatest generalizations in physics—the law of **conservation of energy**:

> **Energy cannot be created or destroyed; it may be transformed from one form into another, but the total amount of energy never changes.**

When we consider any system in its entirety, whether it be as simple as a swinging pendulum or as complex as an exploding supernova, there is one quantity that doesn't change: energy. It may change form or it may simply be transferred from one place to another, but, as far as we can tell, the total energy score stays the same. This energy score takes into account the fact that the atoms that make up matter are themselves concentrated bundles of energy. When the nuclei (cores) of atoms rearrange themselves, enormous amounts of energy can be released. The sun shines because some of this energy is transformed into radiant energy. In nuclear reactors much of this energy is transformed into thermal energy.

Enormous compression due to gravity in the deep hot interior of the sun crushes the nuclei of hydrogen atoms together to form helium nuclei. This welding together of atomic cores is called *thermonuclear fusion.* This process releases radiant energy, some of which reaches the earth. Part of this energy falls on plants, and part of this in turn later is stored in coal. Another part supports life in the food chain that begins with plants, and part of this energy later is stored in oil. Part of the energy from the sun goes into the evaporation of water from the ocean, and part of this returns to the earth in rain that may be trapped behind a dam. By virtue of its elevated position, the water behind a dam has energy that may be used to power a generating plant below, where it will be transformed to electric energy. The energy travels through wires to homes, where it is used for lighting, heating, cooking, and operating electric gadgets. How nice that energy is transformed from one form to another!

Questions

1. Does an automobile consume more fuel when its air conditioner is turned on? When its lights are on? When its radio is on while it is sitting in the parking lot?
2. When is a body's energy evident?

Answers

1. The answer to all three questions is yes, for the energy they consume ultimately comes from the fuel. Even the energy from the battery must be given back to the battery by the alternator, which is turned by the engine, which runs from the energy of the fuel. There's no free lunch!
2. Energy is evident when it undergoes change. Likewise with the money a rich person has; it shows when it is being spent.

Machines

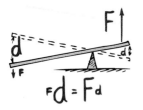

FIGURE 6.11 The lever.

A *machine* is a device for multiplying forces or simply changing the direction of forces. The principle underlying every machine is the *conservation of energy* concept. Consider one of the simplest machines, the **lever** (Figure 6.11). At the same time that we do work on one end of the lever, the other end does work on the load. We see that the direction of force is changed, for if we push down, the load is lifted up. If the heating from friction forces is small enough to neglect, the work input will be equal to the work output.

$$\text{Work input} = \text{work output}$$

Since work equals force times distance, input force $\times$ input distance = output force $\times$ output distance.

$$(\text{Force} \times \text{distance})_{\text{input}} = (\text{force} \times \text{distance})_{\text{output}}$$

If the pivot point (*fulcrum*) of the lever is relatively close to the load, then a small input force will produce a large output force. This is because the input force is exerted through a large distance and the load is moved over a correspondingly short distance. So a lever can be a force multiplier. But no machine can multiply work or multiply energy. That's a conservation of energy no-no!

A child uses the principle of the lever in jacking up the front end of an automobile. By exerting a small force through a large distance, she is able to provide a large force acting through a small distance. Consider the ideal example illustrated in Figure 6.12. Every time she pushes the jack handle down 25 centimeters, the car rises only a hundredth as far but with 100 times the force.

FIGURE 6.12 Applied force $\times$ applied distance = output force $\times$ output distance.

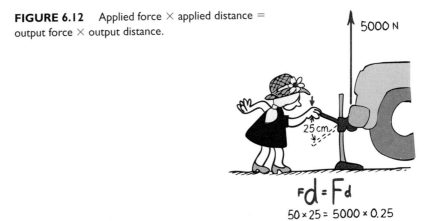

A block and tackle, or system of pulleys, is a simple machine that also multiplies force at the expense of distance. One can exert a small force through a large distance and lift a heavy load through a short distance. With the ideal pulley system shown in Figure 6.13, the man pulls 7 meters of rope with a force of 50 newtons and lifts 500 newtons through a vertical distance of 0.7 meter. The energy the man expends in pulling the rope is numerically equal to the increased potential energy of the 500-newton block. Energy is merely transformed.

Any machine that multiplies force does so at the expense of distance. Likewise, any machine that multiplies distance, such as your forearm and elbow, does so at the expense of force. No machine or device can put out more energy than is put into it. No machine can create energy; it can only transform it from one form to another.

FIGURE 6.13 Applied force $\times$ applied distance = output force $\times$ output distance.

Efficiency

The three previous examples were of *ideal machines;* 100 percent of the work input was transformed to work output. An ideal machine would operate at 100 percent efficiency. In practice, this doesn't happen and we can never expect it to happen. In any transformation some energy is dissipated to molecular kinetic energy—thermal energy. This makes the machine warmer.

Even a lever rocks about its fulcrum and converts a small fraction of the input energy into thermal energy. We may do 100 joules of work and get out 98 joules of work. The lever is then 98 percent efficient, and we degrade only 2 joules of work input into thermal energy. If the girl in Figure 6.12 puts in 100 joules of work and increases the potential energy of the car by 60 joules, the jack is 60 percent efficient; 40 joules of her input work has been applied against friction, making its appearance as thermal energy.

In a pulley system, a considerable fraction of input energy typically goes into thermal energy. If we do 100 joules of work, the forces of friction acting through the distances through which the pulleys turn and rub about their axles may dissipate 60 joules of energy as thermal energy. In that case, the work output is only 40 joules and the pulley system has an efficiency of 40 percent. The lower the efficiency of a machine, the greater the amount of energy that is degraded to thermal energy.

Inefficiency exists whenever energy in the world around us is transformed from one form to another. **Efficiency** can be expressed by the ratio

$$\text{Efficiency} = \frac{\text{useful energy output}}{\text{total energy input}}$$

An automobile engine is a machine that transforms chemical energy stored in fuel into mechanical energy. The bonds between the molecules in the petroleum fuel break up when the fuel burns. Carbon atoms in the fuel combine with oxygen in the air to form carbon dioxide, hydrogen atoms in the fuel combine with oxygen to form water, and energy is released. We'd like it if all this energy could be converted into useful mechanical energy; that is, we'd like an engine that is 100 percent efficient. This is impossible, because much of the energy is transformed into thermal energy, a little of which is used to warm the passengers in the winter but most of which is wasted. Some goes out in the hot exhaust gases, some is dissipated to the air through the cooling system or directly from hot engine parts.

Question Consider an imaginary miracle car that has a 100 percent efficient engine and burns fuel that has an energy content of 40 megajoules per liter. If the air drag and overall frictional forces on the car traveling at highway speed is 500 N, what is the upper limit in distance per liter the car could go at this speed?

Answer From the definition work = force $\times$ distance, simple rearrangement gives distance = work/force. If all 40 million J of energy in 1 L were used to do the work of overcoming the air drag and frictional forces, the distance would be:

$$\text{distance} = \frac{\text{work}}{\text{force}} = \frac{40,000,000 \text{ J/L}}{500 \text{N}} = 80,000 \text{ m/L} = 80 \text{ km/L}$$

The important point here is that even with a hypothetically perfect engine, there is an upper limit of fuel economy dictated by the conservation of energy.

FIGURE 6.14 Energy transitions. The graveyard of kinetic energy is thermal energy.

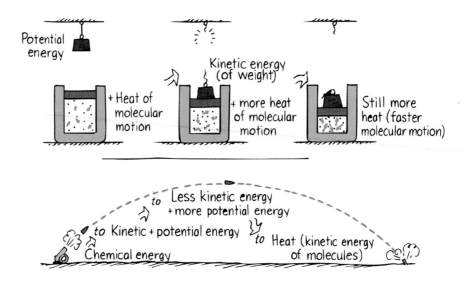

Look at the inefficiency that accompanies transformations of energy this way: In any transformation there is a dilution of available *useful energy.* The amount of usable energy decreases with each transformation until there is nothing left but thermal energy at ordinary temperature. When we study thermodynamics, we'll see that thermal energy is useless for doing work unless it can be transformed to a lower temperature. Once it reaches the lowest practical temperature, that of our environment, it cannot be used. The environment around us is the graveyard of useful energy.

Comparison of Kinetic Energy and Momentum*

Kinetic energy and momentum are both properties of motion. But they are different. Momentum, like velocity, is a vector quantity. Energy, on the other hand, like mass, is a scalar quantity. When two objects move toward each other, their momenta may partially or fully cancel. Their total momentum is less than the momentum of either one alone. But their kinetic energies cannot cancel. Since kinetic energies are always positive (or zero), the total kinetic energy of two moving objects is greater than the kinetic energy of either one alone.

For example, the momenta of two cars just before a head-on collision may add up to exactly zero, and the combined wreck after collision will have the same zero value for the momentum. But the kinetic energies add, and this energy is still there after the collision, although it's in different forms—mainly thermal energy. Or the momenta of two firecrackers approaching each other may cancel, but when they explode, there is no way their energies can cancel. Energy comes in many forms; momentum has only one form. The vector quantity momentum is different from the scalar quantity kinetic energy.

Another difference is the velocity dependence of the two. Whereas momentum is proportional to velocity (mv), kinetic energy is proportional to the square of velocity ($\frac{1}{2}mv^2$). An object that moves with twice the velocity of another object of the same mass has twice the momentum but four times the kinetic energy. It can provide twice the impulse to whatever it encounters but do four times as much work.

* This section may be skipped in a light treatment of mechanics.

Metal bullet penetrates

Rubber bullet bounces

FIGURE 6.15 Compared to the metal bullet with the same momentum, the rubber bullet is more effective in knocking the block over because it bounces on impact. The rubber bullet therefore undergoes the greater change in momentum and thereby imparts the greater impulse or wallop to the block. Which bullet does more damage?

Consider a metal bullet fired into a very large block of wood (Figure 6.15). When the bullet strikes, the bullet tips the block slightly as the momentum of the bullet is transferred to the block. If the same bullet strikes with twice the velocity, it can tip the block twice as far so it has twice the capability of knocking the block over. Call this "twice the wallop." But note that the twice-as-fast bullet has four times the kinetic energy, and will penetrate four times as far into the block. It delivers twice the wallop but four times the damage.

If our bullet is made of rubber instead of metal so that it bounces from the block instead of penetrating, it will be even more effective than a metal bullet in knocking the block over. Let's see why. If the momenta are the same for the metal and rubber bullets and they are simply brought to rest, then the change in momenta and the accompanying impulses would be the same; but only the metal bullet is brought to rest. The rubber bullet bounces, which means its change in momentum and the accompanying impulse are greater than for the metal bullet, which is only brought to rest. If the rubber bullet bounces elastically with no loss whatever in speed, its change in momentum and the impulse it delivers are twice that of the metal bullet. It hits with twice as much wallop as a penetrating bullet of the same momentum. But it does not penetrate, and, in this sense, it does little damage. This is an idealized case. In practice, such collisions are less than perfectly elastic. But now you have a better idea why rubber bullets are used by police for knockdowns with minimum injury.

This is more safely demonstrated in Figure 6.16. When the rubber-nosed dart equipped with a sharp nail makes impact against the wooden block, the collision is inelastic. The impulse is not enough to tip the block over. But when the nail is removed to expose the rubber nose, the dart swings with the same momentum and *bounces* from the block. The nearly doubled impulse knocks the block over.

a

b

c

d

FIGURE 6.16 (a) Dave releases the dart, which is brought to rest when it sticks into the wooden block (b). Dave then removes the metal point from the dart to expose its rubber nose and (c) again releases it from the same height. (d) The dart *bounces* from the block, which this time is knocked over. Dave says, "The impulse is greater for bouncing because the dart is more than simply stopped—it is thrown back out again." In terms of momentum, Helen adds, "If the dart hits with, say, positive momentum, then it bounces with negative momentum. The *change* in momentum from positive to negative in bouncing is greater than from positive to zero in sticking. The greater change in momentum results in greater impulse."

 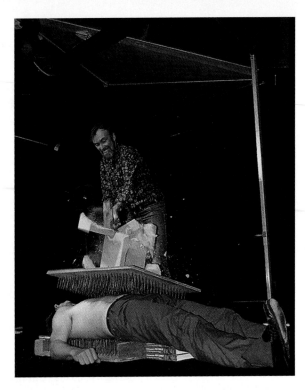

FIGURE 6.17 The author puts kinetic energy and momentum into the hammer, which strikes the block that rests on fellow physics instructor Paul Robinson, who is bravely sandwiched between beds of nails. Paul is not harmed. Why? Except for the flying cement fragments, every bit of the momentum of the hammer at impact is imparted to Paul, and subsequently to the table and the earth that supports him. But the momentum only provides the wallop; the energy does the damage. Most of the kinetic energy never gets to him, for it goes into smashing the block apart and into thermal energy. What energy remains is distributed over the more than 200 nails that make contact with his body. The driving force per nail is not enough to puncture the skin.

Question Suppose a hunter is confronted with a charging bear. Which will be more effective in knocking the bear down—a rubber bullet or a lead bullet of the same momentum?

Suppose you're carrying a football and are about to be slammed into and tackled by an approaching player whose oncoming momentum is equal but opposite to your own. The combined momentum of you and the approaching player before impact cancel to zero, and, sure enough, on impact you both stop short in your tracks. The wallop or jarring exerted on each of you is the same. This is the case whether you are tackled by a slow-moving heavy player or a fast-moving light player. If the product of his mass and his velocity matches yours, you're stopped short. Stopping power is one thing, but what about damage? Every football player knows it hurts more to be stopped by a fast-moving light player than by a slow-moving heavy player. Why? Because a light player moving with the same momentum has more kinetic energy. If he has the same momentum as a heavy player but is only half as massive, he has twice the velocity. Twice the velocity and half the mass give the lighter player twice the kinetic energy of the heavier player.* He does twice the work on you, tends to deform you twice as much, and generally does twice the damage to you. Watch out for the fast-moving little guys!

Answer A hunter will more effectively knock a bear down with a rubber bullet, but he'd better be good at running when the bear gets up.

* Note that $\frac{1}{2}(m/2)(2v)^2 = mv^2$, which is twice the value $\frac{1}{2}mv^2$ of the heavier player of mass m and speed v.

Energy for Life

Every plant, animal, and living cell is a living machine and, just like a mechanical machine, needs an energy supply. The fuels for most animal life on this planet are various hydrocarbon compounds that react with oxygen to release energy. As with the petroleum fuel that we previously discussed, the "combustion" of food in a cell (part of what is called metabolism) creates carbon dioxide and water and releases energy to sustain life.

The principal difference between the combustion of food in digestion and the combustion of fossil fuels in mechanical engines is the rate at which reactions take place. In digestion the reaction is much slower and energy is released as it is required. Like the burning of fossil fuels, the reaction is self-sustaining once started, as carbon combines with oxygen to form carbon dioxide and hydrogen with oxygen to form water. The reverse process is more difficult. Only green plants can make carbon dioxide combine with water to produce hydrocarbon compounds such as sugar. This process is photosynthesis and is energized by sunlight. Sugar is the simplest food, and all others—carbohydrates, proteins, and fats—are also synthesized compounds of carbon, hydrogen, oxygen, and other elements. How very fortunate that more energy is locked in the bonds of these hydrocarbons than in their products after combustion!

We see efficiency at work in the food chain. Larger creatures feed on smaller creatures, who in turn eat smaller creatures, and so on down the line to land plants and ocean plankton that are nourished by the sun. Moving each step up the food chain involves inefficiency. In the African bush, 10 kilograms of grass may produce 1 kilogram of gazelle. However, it will take 10 kilograms of gazelle to sustain 1 kilogram of lion. We see that each energy transformation along the food chain contributes to overall inefficiency. Interestingly enough, some of the largest creatures on the planet, the elephant and the blue whale, eat far down on the food chain. And more and more humans are considering substances such as krill or yeast as efficient sources of nourishment.

Summary of Terms

Work   The product of the force and the distance through which a body acted on by the force moves:

$$W = Fd$$

(More generally, the component of force in the direction of motion times the distance moved.)

Power   The time rate of work:

$$\text{Power} = \frac{\text{work}}{\text{time}}$$

(More generally, the rate at which energy is expended.)

Energy   The property of a system that enables it to do work.

Potential energy (PE)   The energy that a body possesses because of its position.

Kinetic energy (KE)   Energy of motion, quantified by the relationship

$$\text{Kinetic energy} = \tfrac{1}{2}mv^2$$

Conservation of energy   Energy cannot be created or destroyed; it may be transformed from one form into another, but the total amount of energy never changes.

Work-energy theorem   The work done on an object is equal to the energy gained by the object.

$$\text{Work} = \Delta KE$$

Conservation of energy for machines   The work output of any machine cannot exceed the work input. In an ideal machine, where no energy is transformed into thermal energy,

$$\text{work}_{\text{input}} = \text{work}_{\text{output}} \text{ and } (Fd)_{\text{input}} = (Fd)_{\text{output}}$$

Efficiency   The percent of the work put into a machine that is converted into useful work output.

$$\text{Efficiency} = \frac{\text{useful energy output}}{\text{useful energy input}}$$

Review Questions

1. When is energy most evident?

Work

2. A force sets an object in motion. When the force is multiplied by the time of its application, we call the quantity *impulse*, which changes the *momentum* of that object. What do we call the quantity *force × distance*, and what quantity does this change?

3. Cite an example where a force is exerted on an object without doing work on the object.

4. How many joules of work are done when a force of 1 N moves a book 2 m?

5. Which requires more work—lifting a 50-kg sack a vertical distance of 2 m or lifting a 25-kg sack a vertical distance of 4 m?

Power

6. If both sacks in the preceding question are lifted their respective distances in the same time, how does the power required for each compare? How about for the case where the lighter sack is moved its distance in half the time?

7. How many watts of power are expended when a force of 1 N moves a book 2 m in a time interval of 1 s?

Mechanical Energy

8. Exactly what is it that a body having energy is capable of doing?

Potential Energy

9. A car is lifted a certain distance in a service station and therefore has potential energy with respect to the floor. If it were lifted twice as high, how much potential energy would it have?

10. Two cars are lifted to the same elevation in a service station. If one car is twice as massive as the other, how do their potential energies compare?

11. How many joules of potential energy does a 1 kg book gain when it is elevated 4 m? When it is elevated 8 m?

Kinetic Energy

12. How many joules of kinetic energy does a 1-kg book have when it is tossed across the room at a speed of 2 m/s? How much energy is imparted to the wall it accidentally encounters?

13. Which has the greater kinetic energy—a car traveling at 30 km/h or a half-as-heavy car traveling at 60 km/h?

Work-Energy Theorem

14. A moving car has kinetic energy. If it speeds up until it is going four times as fast, how much kinetic energy does it have in comparison? Compared to its original speed, how much work must the brakes supply to stop the four-times-as-fast car?

15. What will be the kinetic energy of an arrow shot from a bow having a potential energy of 40 J?

Conservation of Energy

16. An apple hanging from a limb has potential energy because of its height. If it falls, what becomes of this energy just before it hits the ground? After it hits the ground?

17. Explain how the stored chemical energy in a green leaf is related to the energy provided by the sun.

18. Explain what the energy of electric current that operates an electric toothbrush has to do with the energy of thermonuclear fusion.

Machines

19. Can a machine multiply input force? Input distance? Input energy? (If your three answers are the same, seek help, for the last question is especially important.)

20. A force of 50 N is applied to the end of a lever, which is moved a certain distance. If the other end of the lever moves half as far, how much force can it exert?

21. If the man in Figure 6.13 pulls 1 m of rope downward with a force of 100 N, and the load rises $\frac{1}{7}$ as high (about 14 cm), what is the maximum load that can be lifted?

Efficiency

22. Referring to the previous question, if the load lifted is 500 N, what is the efficiency of the pulley system?

23. Referring to the previous question, how much work does the man do on the load? How much potential energy does the 500-N load acquire? What has become of the difference in energies?

24. Compared to work input, how much work is put out by a machine that operates at 30 percent efficiency?

25. Is a machine physically possible that has an efficiency greater than 100 percent? Discuss.

Comparison of Kinetic Energy and Momentum

26. What does it mean to say momentum is a vector quantity and energy is a scalar quantity?

27. Can momenta cancel? Can energies cancel?

28. If a moving object doubles its speed, how much more momentum does it have? How much more energy?

29. If a moving object doubles its speed, how much more impulse does it provide to whatever it bumps into (how much more wallop)? How much more work does it do as it is stopped (how much more damage)?

Energy for Life

30. The energy we require for existence comes from the chemically stored potential energy in food, which is transformed into other forms by the process of diges-

tion. What happens to a person whose work output is less than the energy he or she consumes? Whose work output is greater than the energy he or she consumes? Can an undernourished person perform extra work without extra food? Briefly discuss.

Projects

1. Fill two mixing bowls with water from the cold tap and take their temperatures. Then run an electric or hand beater in the first bowl for a few minutes. Compare the temperatures of the water in the two bowls.
2. Pour some dry sand into a tin can with a cover. Compare the temperature of the sand before and after vigorously shaking the can for a couple of minutes.

Exercises

1. Why is it easier to stop a lightly loaded truck than a heavier one that has equal speed?
2. Which requires more work to stop—a light truck or a heavy truck moving with the same momentum?
3. To determine the potential energy of Tenny's drawn bow (Figure 6.4), would it be an underestimate or an overestimate to multiply the force with which she holds the arrow in its drawn position by the distance she pulled it? Why do we say the work done is the *average* force × distance?
4. When a rifle with a longer barrel is fired, the force of expanding gases acts on the bullet for a longer distance. What effect does this have on the velocity of the emerging bullet? (Do you see why long-range cannons have such long barrels?)
5. You and a flight attendant toss a ball back and forth in an airplane in flight. Does the KE of the ball depend on the speed of the airplane? Carefully explain.
6. Can something have energy without having momentum? Explain. Can something have momentum without having energy? Defend your answer.
7. To combat wasteful habits we often speak of "conserving energy," like turning off lights not in use, turning off hot water when not being used, and keeping thermostats at a moderate level. In this chapter we also speak of "energy conservation." Distinguish between these two usages.
8. At what point in its motion is the KE of a pendulum bob a maximum? At what point is its PE a maximum? When its KE is half its maximum value, how much PE does it have?

9. A physics instructor demonstrates energy conservation by releasing a heavy pendulum bob, as shown in the sketch, allowing it to swing to and fro. What would happen if in his exuberance he gave the bob a slight shove as it left his nose? Explain.

10. Why is no work done on an object when the force is perpendicular to the velocity of the object?
11. Does the string that supports a pendulum bob do work on the bob as its swings to and fro? Does the force of gravity do any work on the bob?
12. A crate is pulled across a horizontal floor by a rope. At the same time, the crate pulls back on the rope, in accord with Newton's third law. Does the work done on the crate by the rope then equal zero? Explain.
13. Discuss the design of the roller coaster shown in the sketch in terms of the conservation of energy.

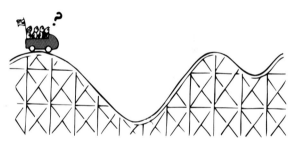

14. Suppose that you and two classmates are discussing the design of a roller coaster. One classmate says that each summit must be lower than the previous one. Your other classmate says this is nonsense, for as long as the first one is the highest, it doesn't matter what height the others are. What do you say?
15. Consider the identical balls released from rest on Tracks A and B as shown. When they reach the right ends of the tracks, which will have the greater speed? Why is this question easier to answer in this chapter than in Chapter 2?

16. Suppose an object is set sliding with a speed less than escape velocity on an infinite frictionless plane in contact with the surface of the earth, as shown. Describe its motion. (Will it slide forever at a constant velocity? Will it slide to a stop? In what way will its energy changes be similar to that of a pendulum?)

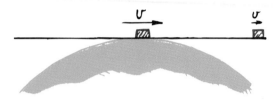

17. If a golf ball and a Ping Pong ball both move with the same kinetic energy, can you say which has the greater speed? Explain in terms of the definition of KE. Similarly, in a gaseous mixture of massive molecules and light molecules with the same average KE, can you say which have the greater speed?

18. Does a car burn more gasoline when its lights are turned on? Does the overall consumption of gasoline depend on whether or not the engine is running while the lights are on? Defend your answer.

19. An inefficient machine is said to "waste energy." Does this mean that energy is actually lost? Explain.

20. You tell your friend that no machine can possibly put out more energy than is put into it, and your friend states that a nuclear reactor puts out more energy than is put into it. What do you say?

21. This may seem like an easy question for a physics type to answer: With what force does a rock that weighs 10 N strike the ground if dropped from a rest position 10 m high? In fact, the question cannot be answered unless you know more. Why?

22. Your friend is confused about ideas discussed in Chapter 4 that seem to contradict ideas discussed in this chapter. For example, in Chapter 4 we learned that the net force is zero for a car traveling along a level road at constant velocity, and in this chapter we learned that work is done in such a case. Your friend asks, "How can work be done when the net force equals zero?" What is your explanation?

23. In the absence of air resistance, a ball thrown vertically upward with a certain initial KE will return to its original level with the same KE. When air resistance is a factor affecting the ball, will it return to its original level with the same, less, or more KE? Does your answer contradict the law of energy conservation? Defend your answer.

24. You're on a rooftop and you throw one ball downward to the ground below and another upward. The second ball, after rising, falls and also strikes the ground below.

If air resistance can be neglected and if your downward and upward initial speeds are the same, how will the speeds of the balls compare upon striking the ground? (Use the idea of energy conservation to arrive at your answer.)

25. From a rooftop, one ball is dropped from rest while another identical ball is thrown downward. Neglecting air resistance, which of the following are the same for both balls? (a) Change of KE in the first second of fall. (b) Change of PE in the first second of fall. (c) Change of KE in the first meter of fall. (d) Change of PE in the first meter of fall.

26. A falling stone gains kinetic energy as it loses potential energy so that the total of KE + PE is constant. When the stone hits the ground, it loses KE without any compensating gain of PE. How is this consistent with energy conservation?

27. A stone is dropped from a certain height and penetrates into mud. All else being equal, if it is dropped from twice the height, how much farther should it penetrate?

28. Does the KE of a car change more when it goes from 10 to 20 km/h or when it goes from 20 to 30 km/h?

29. Scissors for cutting paper have long blades and short handles, whereas metal-cutting shears have long handles and short blades. Bolt cutters have very long handles and very short blades. Why is this so?

30. Discuss the fate of the physics instructor sandwiched between the beds of nails (Figure 6.17) if the block were less massive and unbreakable and the beds contained fewer nails.

31. A pair of identical clay lumps collide and come to rest. Is momentum conserved? Is kinetic energy conserved?

32. Consider the swinging-balls apparatus. If two balls are lifted and released, momentum is conserved as two balls pop out the other side with the same speed as the released balls at impact. But momentum would also be conserved if one ball popped out at twice the speed. Can you explain why this never happens? (And why this exercise is in Chapter 6 rather than in Chapter 5?)

33. If an automobile had a 100% efficient engine, transferring all of the fuel's energy to work, would the engine be warm to your touch? Would its exhaust heat the surrounding air? Would it make any noise? Would it vibrate? Would any of its fuel go unused?

34. The energy we require to live comes from the chemically stored potential energy in food, which is transformed into other energy forms during the digestion process. What happens to a person whose combined work and heat output is less than the energy consumed? What happens when the person's work and heat output is greater than the energy consumed? Can an undernourished person perform extra work without extra food? Defend your answers.

35. Make up a multiple-choice question that would check a classmate's understanding of the distinction between work and impulse.

36. Make up a multiple-choice question that would check a classmate's understanding of the distinction between work and power.

Problems

1. This question is typical on some driver's license exams: A car moving at 50 km/h skids 15 m with locked brakes. How far will the car skid with locked brakes at 150 km/h?

2. In the hydraulic machine shown, it is observed that when the small piston is pushed down 10 cm, the large piston is raised 1 cm. If the small piston is pushed down with a force of 100 N, how much force is the large piston capable of exerting?

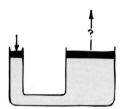

3. A 60-kg skydiver moving at terminal speed falls 50 m in 1 s. What power is the skydiver expending on the air?

4. When their car stalls on a level road, a group of students leap from the car and join in pushing it forward. The car's mass is 1000 kg, and when they push it 5 m the car gains a speed of 2.0 m/s. (a) What is its KE? (b) How much work do they do in pushing it? (c) If friction is negligible, what average force do they exert on the car?

5. Consider the inelastic collision between the two freight cars in the previous chapter (Figure 5.11). The momentum before and after the collision is the same. The KE, however, is less after the collision than before the collision. How much less, and what becomes of this energy?

6. The decrease of PE for a freely falling object equals its gain in KE, in accord with the conservation of energy. By simple algebra, find an equation for an object's speed v after falling a vertical distance h. Do this by equating KE to its change of PE.

7. Using the definitions of momentum and kinetic energy, $p = mv$ and KE $= (\frac{1}{2})mv^2$, show by algebraic manipulation that you can write KE $= p^2/2m$. This equation tells us that if two objects have the same momentum, the one of less mass has the greater kinetic energy (see Exercise 2).

8. A golf ball is hurled at and bounces from a very massive bowling ball that is initially at rest. (a) After the collision, which of the two balls has the greater momentum? (*Hint:* The golf ball has practically the same speed before and after bouncing and therefore practically the same magnitude of momentum after it rebounds. How great will be the momentum of the larger ball in order that the total momentum before and after the collision is the same?) (b) Which ball will have the greater KE after the collision?

9. How many kilometers per liter will a car obtain if its engine is 25% efficient and it encounters an average retarding force of 500 N at highway speed? Assume that the energy content of gasoline is 40 MJ/liter.

10. The power we derive from metabolism can do work and generate heat. What is the mechanical efficiency of a relatively inactive person who expends 100 W of power to produce about 1 W of power in the form of work, while generating about 99 W of heat? What is a cyclist's mechanical efficiency who, in a burst of effort, produces 100 W of mechanical power from 1000 W of metabolic power?

7

Rotational Motion

: In which direction do the riders accelerate?

In Chapter 3 we discussed circular motion without the forces that produce it. In Chapter 4 we treated Newton's laws of motion and applied the concepts of inertia, force, and acceleration to the motion of things without any rotation they might have. In this chapter we extend these ideas to rotational motion—to the rotation of an object about some axis, whether inside or outside the object. We will see how rotational inertia affects rotation, how forces applied one way produce torques that tend to rotate things, and how forces applied another way tend to move things in circular paths. Then we will extend the concept of linear momentum to rotational motion and introduce the concept of angular momentum. The concepts of this chapter are parallel to and build upon those we treated in Chapters 3 and 4—interesting stuff.

Rotational Inertia

Just as an object at rest tends to stay at rest and an object in motion tends to remain moving in a straight line, *an object rotating about an axis tends to remain rotating about the same axis unless interfered with by some external influence.* (We shall see shortly that this external influence is properly called a *torque.*) The property of an object to resist

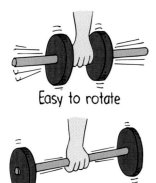

Easy to rotate

Difficult to rotate

FIGURE 7.1 Rotational inertia depends on the distribution of mass with respect to the axis of rotation.

changes in its rotational state of motion is called **rotational inertia**.* Bodies that are rotating tend to remain rotating, while nonrotating bodies tend to remain nonrotating. In the absence of outside influences, a rotating top keeps rotating, while a top at rest stays at rest.

Like the inertia for linear motion, rotational inertia relates to the mass of an object. But unlike the inertia for linear motion, rotational inertia depends on the distribution of the mass with respect to the axis of rotation. The greater the distance between the bulk of an object's mass and its axis of rotation, the greater the rotational inertia. For example, industrial flywheels are constructed so that most of their mass is concentrated along the rim. Once rotating, they have a greater tendency to remain rotating than they would if their mass were concentrated nearer the axis of rotation. The greater the rotational inertia of an object, the harder it is to change the rotational state of that object. If rotating, it is difficult to stop; if at rest, it is difficult to rotate. This fact is employed by a circus tightrope walker when he carries a long pole to aid balancing. Much of the mass of the pole is far from the axis of rotation, its midpoint. The pole therefore has considerable rotational inertia. If the tightrope walker starts to topple over, a tight grip on the pole rotates the pole. But the rotational inertia of the pole resists, giving the tightrope walker time to readjust his balance. The longer the pole, the better. And if our tightrope walker has no pole, he can at least increase the rotational inertia of his body by extending his arms full length.

The rotational inertia of the pole or any object depends on what axis it is rotated about.[†] If it rotates about an axis through its central core parallel to the length of the pole, like lead in a pencil, its mass distribution is very close to this axis and rotational inertia is very small. Rotational inertia is greater about the perpendicular midpoint axis, the axis utilized by the tightrope walker in Figure 7.2. The pole has a still greater rotational inertia when the axis is perpendicular to the pole and intersects it at one end, so the pole swings like a pendulum. The longer the pendulum, the greater its rotational inertia and the longer it takes to swing to and fro. Hence long-legged animals such as

Rotational inertia
:Sigh:

FIGURE 7.2 The tendency of the pole to resist rotation aids the acrobat.

* Often called *moment of inertia.*

[†] When the mass of an object is concentrated totally at one distance, the radius r from the axis of rotation (like a simple pendulum bob or a thin ring) rotational inertia I is equal to the mass m multiplied by the square of the radial distance. For this special case, $I = mr^2$. (For the general case, $I = \int r^2 \, dm$. This is beyond the scope of this book; it's something you may treat if you continue in your study of physics.)

FIGURE 7.3 Short legs have less rotational inertia than long legs. An animal with short legs has a quicker stride than one with long legs, just as a baseball batter can swing a short bat more quickly than a long one.

FIGURE 7.4 You bend your legs when you run to reduce rotational inertia.

giraffes, horses, and ostriches normally run with a slower gait than hippos, dachshunds and mice (Figure 7.3).

A long baseball bat held near its end has more rotational inertia than a short bat. Once moving it has a greater tendency to keep moving, but it is harder to bring it up to speed. A short bat has less rotational inertia than a long bat, and is easier to swing. Baseball players sometimes "choke up" on a bat by grasping it closer to the more massive end than normal. Choking up on the bat reduces its rotational inertia and makes it easier to bring up to speed. Similarly, when running, you bend your legs to reduce their rotational inertia so you can rotate them back and forth more quickly.

Figure 7.5 compares rotational inertias for various shapes and axes. It is not important for you to learn these values, but you can see how they vary with the shape and axis.

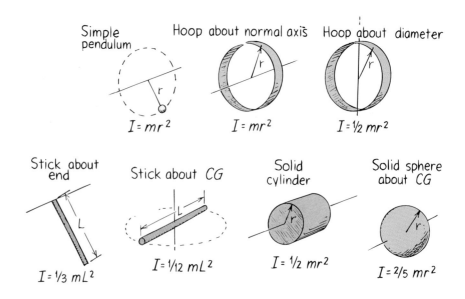

FIGURE 7.5 Rotational inertias of various objects each of mass m, about indicated axes.

Simple pendulum $I = mr^2$

Hoop about normal axis $I = mr^2$

Hoop about diameter $I = \frac{1}{2} mr^2$

Stick about end $I = \frac{1}{3} mL^2$

Stick about CG $I = \frac{1}{12} mL^2$

Solid cylinder $I = \frac{1}{2} mr^2$

Solid sphere about CG $I = \frac{2}{5} mr^2$

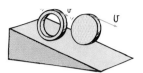

FIGURE 7.6 A solid cylinder rolls down an incline faster than a ring, whether or not they have the same mass or outer diameter. A ring has greater rotational inertia relative to its mass than a cylinder does.

Because of rotational inertia, a solid cylinder starting from rest will roll down an incline faster than a ring or hoop. The ring has its mass concentrated farthest from its axis of rotation and therefore is harder to start rolling. Any solid cylinder will beat any ring on the same incline. This doesn't seem plausible at first thought, but remember that any two objects, regardless of mass, will fall together when dropped. They will also slide together when released on an inclined plane. When rotation is introduced, the object with the larger rotational inertia *relative to its own mass* has the greater resistance to a change in its motion. Hence, any disk will roll faster than any ring on the same incline. Try it and see!

Question Consider balancing a hammer upright on the tip of your finger. If the head of the hammer is heavy and the handle long, would it be easier to balance with the end of the handle on your fingertip so that the head is at the top, or the other way around with the head at your fingertip and the end of the handle at the top?

Torque

FIGURE 7.7 Since ancient times, mass has been measured by balancing torques.

When you rotate a wrench to loosen or tighten a bolt, or make a ball spin as you throw it, or open a door, you apply a force at some distance from an axis of rotation. You apply a *torque* (rhymes with *dork*). Torque is the rotational counterpart of force. Force tends to change the motion of things; torque tends to twist or change the state of rotation of things. If you want to make a stationary object move, apply a force. If you want to make a stationary object spin, apply a torque. Torque differs from force just as rotational inertia and regular inertia differ: both torque and rotational inertia involve distance from the axis of rotation. In the case of torque, this distance, which provides leverage, is called the **lever arm**. We define **torque** as the product of this lever arm and the force that tends to produce rotation:

$$\text{Torque} = \text{lever arm} \times \text{force}$$

Torques are intuitively familiar to youngsters playing on a seesaw. Kids can balance a seesaw even when their weights are unequal. Weight alone doesn't produce rotation. Torque does, and children soon learn that the distance they sit from the pivot point

Answer Stand the hammer with the handle at your fingertip and the head at the top. Why? Because it will have more rotational inertia this way and be more resistant to a rotational change. Those acrobats you see on stage who balance their friends at the top of a long pole have an easier task when their friends are at the top of the pole. A pole empty at the top has less rotational inertia and is more difficult to balance!

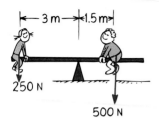

FIGURE 7.8 No rotation is produced when the torques balance each other.

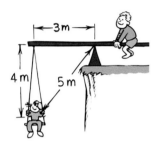

FIGURE 7.9 The lever arm is still 3 m.

is every bit as important as weight (Figure 7.8). The torque produced by the boy on the right tends to produce clockwise rotation, while torque produced by the girl on the left tends to produce counterclockwise rotation. If the torques are equal, making the net torque zero, no rotation is produced.

Suppose that the seesaw is arranged so that the half-as-heavy girl is suspended from a 4-meter rope hanging from her end of the seesaw (Figure 7.9). She is now 5 meters from the pivot point (fulcrum), and the seesaw is still balanced. We see that the lever-arm distance is still 3 meters and not 5 meters. The lever arm about any axis of rotation is the perpendicular distance from the axis to the line along which the force acts. This will always be the shortest distance between the axis of rotation and the line along which the force acts.

This is why the stubborn bolt shown in Figure 7.10 is more likely to turn when the applied force is perpendicular to the handle, rather than at an oblique angle as shown in the first figure. In the first figure the lever arm is shown by the dashed line and is less than the length of the wrench handle. In the second figure, the lever arm is equal to the length of the wrench handle. In the third figure the lever arm is extended with a pipe to provide more leverage and a greater torque.

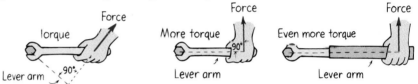

FIGURE 7.10 Although the magnitudes of the force are the same in each case, the torques are different.

Questions

1. If a pipe effectively extends a wrench handle to three times its length, by how much will the torque increase for the same applied force?
2. Consider the balanced seesaw in Figure 7.8. Suppose the girl on the left suddenly gains 50 N, such as by being handed a bag of apples. Where should she sit in order to balance, assuming the heavier boy does not move?

Answers

1. Three times more leverage for the same force gives three times more torque. (This method of increasing torque sometimes results in shearing off the bolt!)
2. She should sit $\frac{1}{2}$ m closer to the center. Then her lever arm is 2.5 m. This checks: 300 N × 2.5 m = 500 N × 1.5 m.

Center of Mass and Center of Gravity

Throw a baseball into the air, and it will follow a smooth parabolic trajectory. Throw a baseball bat spinning into the air, and its path is not smooth; its motion is wobbly, and it seems to wobble all over the place. But, in fact, it wobbles about a very special place, a point called the **center of mass**.

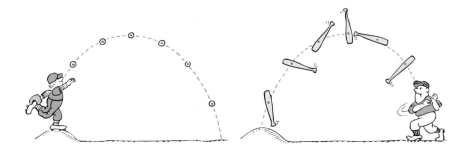

FIGURE 7.11 The center of mass of the baseball and that of the bat follow parabolic trajectories.

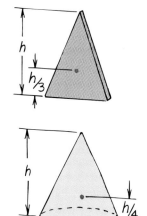

FIGURE 7.12 The center of mass for each object is shown by the colored dot.

For a given body, the center of mass is the average position of all the mass that makes up the object. For example, a symmetrical object such as a ball has its center of mass at its geometrical center; by contrast, an irregularly shaped body such as a baseball bat has more of its mass toward one end. The center of mass of a baseball bat therefore is toward the thicker end. A solid cone has its center of mass exactly one-fourth of the way up from its base.

Center of gravity is a term popularly used to express center of mass. The center of gravity is simply the average position of weight distribution. Since weight and mass are proportional, center of gravity and center of mass refer to the same point of an object.* The physicist prefers to use the term *center of mass*, for an object has a center of mass whether or not it is under the influence of gravity. However, we shall use either term to express this concept and favor the term *center of gravity* when weight is part of the picture.

The multiple-flash photograph (Figure 7.13) shows a top view of a spinning wrench sliding across a smooth horizontal surface. Note that its center of mass, indicated by the yellow dot, follows a straight-line path, while other parts of the wrench wobble as they move across the surface. Since there is no external force acting on the wrench, its center of mass moves equal distances in equal time intervals. The motion of the spinning wrench is the combination of the straight-line motion of its center of mass and the rotational motion about its center of mass.

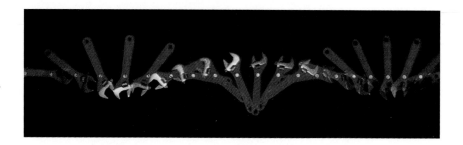

FIGURE 7.13 The center of mass of the spinning wrench follows a straight-line path.

*For almost all objects on and near the earth, these terms are interchangeable. There can be a small difference between center of gravity and center of mass when an object is large enough for gravity to vary from one part to another. For example, the center of gravity of the World Trade Center is about 1 millimeter below its center of mass. This is due to the lower stories being pulled a little more strongly by earth's gravity than the upper stories. For everyday objects (including tall buildings!) we can use the terms center of gravity and center of mass interchangeably.

FIGURE 7.14 The center of mass of the cannonball and its fragments move along the same path before and after the explosion.

If the wrench were instead tossed into the air, no matter how it rotates, its center of mass (or center of gravity) would follow a smooth parabola. The same is true for an exploding cannonball (Figure 7.14). The internal forces that act in the explosion do not change the center of gravity of the projectile. Interestingly enough, if air resistance can be neglected, the center of gravity of the dispersed fragments as they fly though the air will be in the same place that the center of gravity would be if the explosion didn't occur.

Locating the Center of Gravity

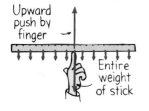

FIGURE 7.15 The weight of the entire stick behaves as if it were concentrated at its center.

The center of gravity of a uniform object such as a meter stick is at its midpoint, for the stick acts as though its entire weight were concentrated there. Supporting that single point supports the whole stick. Balancing an object provides a simple method of locating its center of gravity. In Figure 7.15 the many small arrows represent the pull of gravity all along the meter stick. All these can be combined into a resultant force acting through the center of gravity. The entire weight of the stick may be thought of as acting at this single point. Hence, we can balance the stick by applying a single upward force in a direction that passes through this point.

The center of gravity of any freely suspended object lies directly beneath (or at) the point of suspension (Figure 7.16). If a vertical line is drawn through the point of suspension, the center of gravity lies somewhere along that line. To determine exactly where it lies along the line, we have only to suspend the object from some other point and draw a second vertical line through that point of suspension. The center of gravity lies where the two lines intersect.

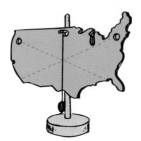

FIGURE 7.16 Finding the center of gravity for an irregularly shaped object.

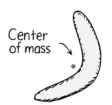

FIGURE 7.17 The center of mass can be outside the mass of a body.

FIGURE 7.18 The athlete executes a "Fosbury flop" to clear the bar while her center of gravity passes beneath the bar.

The center of mass of an object may be a point where no mass exists. For example, the center of mass of a ring or a hollow sphere is at the geometrical center where no matter exists. Similarly, the center of mass of a boomerang is outside the physical structure, not within the material making up the boomerang.

Stability

The location of the center of mass is important for stability (Figure 7.19). If we drop a line straight down from the center of mass of an object of any shape and it falls inside the base of the object, it is in stable **equilibrium**; it will balance. If it falls outside the base, it is unstable. Why doesn't the famous Leaning Tower of Pisa topple over? As we can see in Figure 7.20, a line from the center of gravity of the tower falls inside its base, so the Leaning Tower has stood for several centuries.

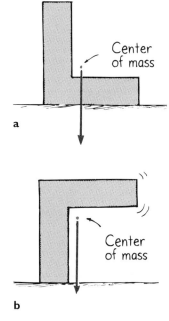

a

b

FIGURE 7.19 The center of mass of the L-shaped object is located where no mass exists. In (a), the center of mass is above a point of support, so the object is stable. In (b), it is not above a point of support, so the object is unstable and will topple over.

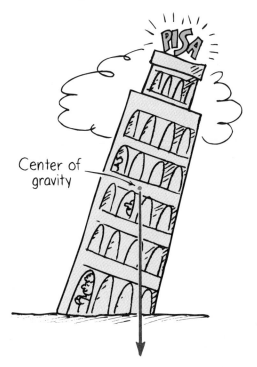

FIGURE 7.20 The center of gravity of the Leaning Tower of Pisa lies above a point of support, so the tower is in stable equilibrium.

FIGURE 7.21 The vertical distance that the center of gravity is raised in tipping determines stability. An object with a wide base and low center of gravity is more stable.

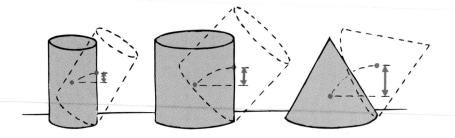

FIGURE 7.22 When you stand, your center of gravity is somewhere above the area bounded by your feet. Why do you keep your legs far apart when you have to stand in the aisle of a bumpy-riding bus?

To reduce the likelihood of tipping, it is usually advisable to design objects with a wide base and low center of gravity. The wider the base, the higher the center of gravity must be raised before the object will tip over.

When you stand erect (or lie flat), your center of gravity is within your body. Why is the center of gravity lower in an average woman than in an average man of the same height? Is your center of gravity always at the same point in your body? Is it always inside your body? What happens to it when you bend over?

If you are fairly flexible, you can bend over and touch your toes without bending your knees—providing you are not standing with your back against a wall. Ordinarily, when you bend over and touch your toes, you extend your lower extremity as shown in Figure 7.23 left, so that your center of gravity is above a point of support, your feet. If you attempt to do this when standing against a wall, however, you cannot counterbalance yourself, and your center of gravity soon protrudes beyond your feet as at the right in Figure 7.23. You are off-balance and you rotate.

FIGURE 7.23 You can lean over and touch your toes without falling over only if your center of gravity is above the area bounded by your feet.

You rotate because of an unbalanced torque. This is evident in the two L-shaped objects shown in Figure 7.24. Both are unstable and will topple unless fastened to the level surface. It is easy to see that even if both shapes have the same weight, the one on the right is less stable. This is because of the greater lever arm and, hence, greater torque.

Try balancing the pole end of a broom upright on the palm of your hand. The support base is quite small and relatively far beneath the center of gravity, so it's difficult to maintain balance for very long. After some practice you can do it if you learn to make slight movements of your hand to exactly respond to variations in balance. You learn to avoid under-responding or over-responding to the slightest variations in balance.

FIGURE 7.24 The greater torque acts on the figure in (b) for two reasons. What are they?

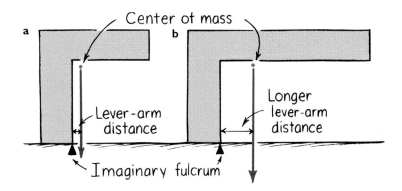

Center of mass

a   b

Lever-arm distance

Longer lever-arm distance

Imaginary fulcrum

Similarly, high-speed computers help massive rockets remain upright when they are launched. Variations in balance are quickly sensed. The computers regulate the firings at multiple nozzles to make corrective adjustments, in a way quite similar to the way your brain coordinates your adjustive action when balancing a long pole on the palm of your hand. Both feats are truly amazing.

FIGURE 7.25 Where is Alexei's center of gravity relative to his hands?

Question A uniform meter stick supported at the 25-cm mark balances when a 1-kg rock is suspended at the 0-cm end. What is the mass of the meter stick?

Answer The mass of the meter stick is 1 kg. Why? The system is in equilibrium, so any torques must be balanced: The torque that results from the weight of the rock is balanced by the equal but oppositely directed torque that results from the weight of the stick applied at its center of gravity, the 50-cm mark. The support force at the 25-cm mark is applied midway between the rock and the center of gravity of the stick, so the lever arms about the support point are equal (25 cm). This means that the weights and hence the masses of the rock and stick must also be equal. Interestingly enough, the center of gravity of the *rock and stick combination* is at the 25-cm mark—directly above the fulcrum. (Note that we don't have to go through the laborious task of considering the fractional parts of the stick's weight on either side of the fulcrum, for the center of gravity of the whole stick by itself really is at one point—the 50-cm mark!)

Centripetal Force

FIGURE 7.26 The force exerted on the whirling can is toward the center.

Any force that causes an object to follow a circular path is called a **centripetal force**.* *Centripetal* means "center-seeking," or "toward the center." The force of a rotating carnival centrifuge on an occupant inside is a center-directed force; if it ceased to act, the occupant would no longer be held in a circular path.

If we whirl a tin can on the end of a string, we find that we must keep pulling on the string—exerting a centripetal force (Figure 7.26). The string transmits the centripetal force, which pulls the can into a circular path. Gravitational and electrical forces can be transmitted across empty space to produce centripetal forces. The moon, for example, is held in an almost circular orbit by gravitational force directed toward the center of the earth. The orbiting electrons in atoms experience an electrical force toward the central nuclei.

Centripetal force is not a new kind of force but is simply the name given to any force, whether string tension, gravitational, electrical, or whatever, that is directed toward a fixed center. If the motion is circular, this force is at right angles to the path of the moving object.

When an automobile rounds a corner, the friction between the tires and the road provides the centripetal force that holds the car in a curved path (Figure 7.27). If this friction is not great enough, the car fails to make the curve and the tires slide sideways; we say the car skids.

FIGURE 7.27 (a) When a car goes around a curve, there must be a force pushing the car toward the center of the curve. (b) A car skids on a curve when the centripetal force (friction of road on tires) is not great enough.

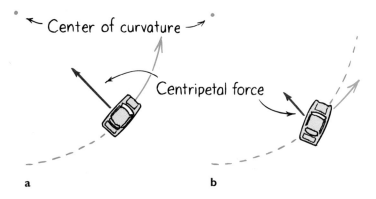

a

b

Centripetal force plays the main role in the operation of a centrifuge. A familiar example is the spinning tub in an automatic washer (Figure 7.28). In its spin cycle, the tub rotates at high speed and produces a centripetal force on the wet clothes, which are forced into a circular path against the inner wall of the tub. The tub exerts great force on the clothes, but the holes in the tub prevent the exertion of the same force on the water in the clothes and the water escapes. Strictly speaking, the clothes are forced away from the water; the water is not forced away from the clothes. Think about that.

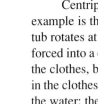

FIGURE 7.28 The clothes are forced into a circular path, but the water is not.

FIGURE 7.29 The centripetal force (adhesion of mud on the spinning tire) is not great enough to hold the mud on the tire, so it spins off in straight-line directions.

* Centripetal force depends on the mass m, tangential speed v, and radius of curvature r of the circularly moving object. If you take a follow-up physics course, you'll learn that the exact relationship is $F = mv^2/r$.

FIGURE 7.30 Large centripetal forces on the wings of an aircraft enable it to fly in circular loops. The acceleration away from the straight-line path the aircraft would follow in the absence of centripetal force is often several times greater than the acceleration due to gravity, g. For example, if the centripetal acceleration is 49 m/s^2 (five times 9.8 m/s^2), we say the aircraft is undergoing 5 g's. At the bottom of the loop, the seat presses against the pilot with an additional force five times his weight. Typical fighter aircraft are designed to withstand accelerations up to 8 or 9 g's. The pilot as well as the aircraft must withstand the centripetal acceleration. Pilots of fighter aircraft wear pressure suits to help keep the blood from flowing away from the head toward the legs, which could cause a blackout.

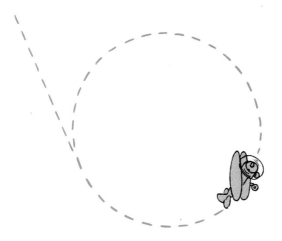

Centrifugal Force

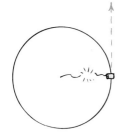

FIGURE 7.31 When the string breaks, the whirling can moves in a straight line, tangent to—not outward from—the center of its circular path.

In the preceding examples we have described the force producing circular motion as a center-directed force. Sometimes in circular motion we seem to experience an outward force. This apparent outward force is called **centrifugal force**. *Centrifugal* means "center-fleeing" or "away from the center." In the case of the whirling can, it is a common misconception to state that a centrifugal force pulls outward on the can. If the string holding the whirling can breaks (Figure 7.31), it is commonly stated that a centrifugal force pulls the can from its circular path. But the fact is that when the string breaks, the can goes off in a tangent straight-line path because *no* force acts on it. We illustrate this further with another example.

Suppose we are passengers in a car that suddenly stops short. We pitch forward against the dashboard. When this happens, we don't say that something forced us forward. In accord with the law of inertia, we pitched forward because of the absence of a force, which seat belts would have provided. Similarly, if we are in a car that rounds a sharp corner to the left, we tend to pitch outward to the right—not because of some outward or centrifugal force, but because there is no centripetal force holding us in circular motion (such as seat belts provide). The idea that a centrifugal force bangs us against the car door is a misconception. (Sure, we push against the door, but only because the door pushes on us—Newton's third law.)

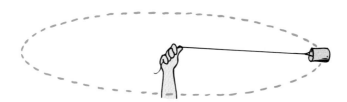

FIGURE 7.32 The only force that is exerted on the whirling can (neglecting gravity) is directed *toward* the center of circular motion. This is a centripetal force. No *outward* force acts on the can.

FIGURE 7.33 The can provides the centripetal force necessary to hold the ladybug in a circular path.

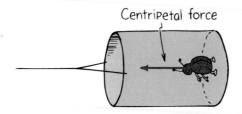

Centripetal force

Likewise when we swing a tin can in a circular path. There is no force pulling the can outward, for the only force on the can is the string pulling it inward. The outward force is on the string, not on the can. Now suppose there is a ladybug inside the whirling can (Figure 7.33). The can presses against the bug's feet and provides the centripetal force that holds it in a circular path. The ladybug in turn presses against the floor of the can, but (neglecting gravity) the only force exerted on the ladybug is the force of the can on its feet. From our outside stationary frame of reference, we see there is no centrifugal force exerted on the ladybug, just as there was no centrifugal force that banged the passenger against the car door. The centrifugal force effect is caused not by any real force but by inertia—the tendency of the moving object to follow a straight-line path. But try telling that to the ladybug!

Centrifugal Force in a Rotating Frame

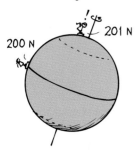

FIGURE 7.34 In the frame of reference of the spinning earth, we feel a centrifugal force that slightly decreases our weight. Like the outside horse on the merry-go-round, we have the greatest tangential speed farthest from the earth's axis, at the equator. Centrifugal force is therefore maximum for us when we are at the equator and zero at the poles where we have no tangential speed. So, strictly speaking, if you want to lose weight, walk toward the equator!

Our frame of reference* very much influences our view of nature. When we sit on the seat of a fast-moving train, we have no speed at all relative to the train but an appreciable speed relative to the reference frame of the stationary ground outside. We have just learned that in a nonrotating reference frame, the force that holds an object in circular motion is a centripetal force. Regarding the ladybug, the bottom of the can exerts a force on its feet. No other force acts on the ladybug.

But nature seen from the reference frame of the rotating system is different. In the rotating frame of the ladybug, in addition to the force of the can on the ladybug's feet, there is a centrifugal force that is exerted on the ladybug. Centrifugal force in a rotating reference frame is a force in its own right, as real as the pull of gravity. However, there is a fundamental difference. Gravitational force is an interaction between one mass and another. The gravity we experience is our interaction with the earth. But centrifugal force in the rotating frame has no such agent—it has no interaction counterpart. It feels like gravity, but with nothing pulling. Nothing produces it; it is a result of rotation. For this reason, physicists call it an "inertial" force—an apparent force—and not a real force like gravity, electromagnetic forces, and nuclear forces. Nevertheless, to observers who are in a rotating system, centrifugal force feels like, and is interpreted to be, a very real force. Just as from the surface of the earth we find gravity ever present, so within a rotating system centrifugal force seems ever present.

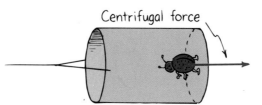

Centrifugal force

FIGURE 7.35 From the reference frame of the ladybug inside the whirling can, it is being held to the bottom of the can by a force that is directed away from the center of circular motion. The ladybug calls this outward force a *centrifugal* force, which is as real to it as gravity.

---

*A frame of reference wherein a free body exhibits no acceleration is called an *inertial* frame of reference. Newton's laws are seen to hold exactly in an inertial frame.

FIGURE 7.36 If the spinning wheel freely falls, the ladybugs inside will experience a centrifugal force that feels like gravity when the wheel spins at the appropriate rate. To the occupants, the direction "up" is toward the center of the wheel and "down" is radially outward.

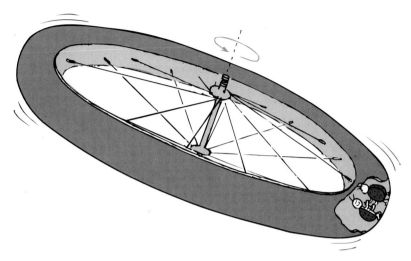

Question A heavy iron ball is attached by a spring to the rotating platform as shown in the sketch. Two observers, one in the rotating frame and one on the ground at rest, observe its motion. Which observer sees the ball being pulled outward, stretching the spring? Which sees the spring pulling the ball in a circle?

Simulated Gravity

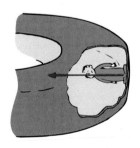

FIGURE 7.37 The interaction between the man and the floor as seen from a stationary frame of reference outside the rotating system. The floor presses against the man (action) and the man presses back on the floor (reaction). The only force exerted on the man is by the floor. It is directed toward the center and is a centripetal force.

Consider a colony of ladybugs living inside a bicycle tire—the old-fashioned balloon kind with plenty of room inside. If we toss the bicycle wheel through the air or drop it from an airplane high in the sky, the ladybugs will be in a weightless condition. They will float freely while the wheel is in free fall. Now spin the wheel. The ladybugs will feel themselves pressed to the outer part of the inner surface. If we spin the wheel not too fast and not too slowly but just right, we can provide the ladybugs with simulated gravity that will feel like the gravity to which they are accustomed. Gravity is simulated by centrifugal force. The "down" direction to the ladybugs will be what we would call radially outward, away from the center of the wheel.

FIGURE 7.38 As seen from inside the rotating system, in addition to the man-floor interaction there is a centrifugal force exerted on the man at his center of mass. It seems as real as gravity. Yet, unlike gravity, it has no reaction counterpart—there is nothing out there that he can pull back on. Centrifugal force is not part of an interaction, but results from rotation. It is therefore called a fictitious force.

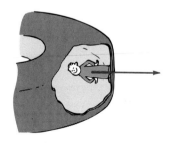

Answer The observer in the reference frame of the rotating platform states that a centrifugal force pulls radially outward on the ball, which stretches the spring. The observer in the rest frame states that a centripetal force supplied by the stretched spring pulls the ball into a circle along with the rotating platform. (The rest-frame observer can in addition state that the reaction to this centripetal force is the ball pulling outward on the spring. The rotating observer, interestingly enough, can state no reaction counterpart to the centrifugal force.)

Humans today live on the outer surface of this spherical planet and are held here by gravity. The planet has been the cradle of humankind. But we will not stay in the cradle forever. We are becoming a spacefaring people. Many people in the years ahead will likely live in huge, lazily rotating habitats in space and will be held to the inner surfaces by centrifugal force. The rotating habitats will provide a simulated gravity so that the human body can function normally.

Occupants of the space shuttle are "weightless" because they lack a support force. Over extended periods this can cause loss of muscle strength or detrimental changes in the body such as loss of calcium from the bones. Future space travelers need not be subject to weightlessness, however. A rotating space habitat for humans, like the rotating bicycle wheel for the ladybugs, can effectively supply a support force and nicely simulate gravity. Structures of small diameter would have to rotate at high rates to provide a simulated gravitational acceleration of 1 *g*. Sensitive and delicate organs in our inner ears sense rotation. Although there appears to be no difficulty at a single revolution per minute (RPM) or so, many people find it difficult to adjust to rates greater than 2 or 3 RPM (although some easily adapt to 10 or so RPM). To simulate normal earth gravity at 1 RPM requires a large structure—one 2 kilometers in diameter. This is an immense structure compared to today's space shuttle vehicles. Finances will probably dictate the size of the first inhabited structures. They are likely to be small and not rotate at all. The inhabitants will adjust as best they can to living in a weightless environment. Larger, rotating habitats will likely follow later.

FIGURE 7.39   This artist's rendering shows the interior of a space colony that would be occupied by a few thousand people.

If the structure rotates so that inhabitants on the inside of the outer edge experience 1 *g*, then halfway to the axis they would experience 0.5 *g*. At the axis itself they would experience weightlessness at 0 *g*. The variety of fractions of *g* possible from the rim of a rotating space habitat holds promise for a most different and (at this writing) as yet unexperienced environment. We would be able to perform ballet at 0.5 *g*; diving and acrobatics at 0.2 *g* and lower *g* states; three-dimensional soccer and new sports not yet conceived in very low *g* states.

Question If the earth were to spin faster about its axis, your weight would be less. If you were in a rotating space habitat that increased its spin rate, you'd "weigh" more. Explain why greater spin rates produce opposite effects in these cases.

Angular Momentum

Just as an object moving in a straight line has linear momentum, an object moving in a circular path or an object rotating about some axis has angular momentum. **Angular momentum** is a measure of the "strength" of an object's rotation about a particular axis. A space colony orbiting the sun, a rock whirling at the end of a string, and the tiny electrons whirling about atomic nuclei all have angular momentum.*

Angular momentum is defined as the product of rotational inertia and rotational velocity.

$$\text{Angular momentum} = \text{rotational inertia} \times \text{rotational velocity}$$

It is the counterpart of linear momentum:

$$\text{Linear momentum} = \text{mass} \times \text{velocity}$$

Like linear momentum, angular momentum is a vector quantity and has direction as well as magnitude. In this book, we won't treat the vector nature of angular momentum (or even of torque, which also is a vector), except to acknowledge the remarkable action of the gyroscope. The rotating bicycle wheel in Figure 7.40 shows what happens

FIGURE 7.40 Angular momentum keeps the wheel axle nearly horizontal when a torque supplied by earth's gravity acts on it. Instead of causing the wheel to topple, the torque causes the wheel's axle to turn slowly around the circle of students. This is called precession.

Answer You're on the *outside* of the spinning earth, but you'd be on the *inside* of a spinning space habitat. A greater spin rate on the outside of the earth tends to throw you *off* a weighing scale, causing it to show a decrease in weight, but *against* a weighing scale inside the space habitat to show an increase in weight.

---

*Angular momentum is a vector quantity; it has direction as well as magnitude. When a direction is assigned to rotational speed, we call it *rotational velocity* (often called *angular velocity*). Rotational velocity is a vector whose magnitude is the rotational speed. By convention, the rotational velocity vector and the angular momentum vector have the same direction and lie along the axis of rotation. If you take a follow-up course, you will learn that rotational velocity is measured in units of radians per second (rad/s), where 0.1 rad/s is approximately 1 RPM. One radian is about 57 degrees.

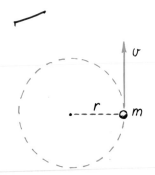

FIGURE 7.41 A small object of mass *m* whirling in a circular path of radius *r* with a speed *v* has angular momentum *mvr*.

when a torque caused by earth's gravity acts to change the direction of its angular momentum (which is along the wheel's axle). The pull of gravity that normally acts to topple the wheel over instead causes its axis to precess (move sideways) in a circular path about a vertical axis. You must do this yourself to believe it fully. You probably won't fully understand it unless you do follow-up study sometime in the future.

For the case of an object that is small compared with the radial distance to its axis of rotation, like a tin can swinging from a long string or a planet orbiting around the sun, the angular momentum can be expressed as the magnitude of its linear momentum, *mv*, multiplied by the radial distance, *r* (Figure 7.41). In shorthand notation,

$$\text{Angular momentum} = mvr$$

Just as an external net force is required to change the linear momentum of an object, an external net torque is required to change the angular momentum of an object. We can state a rotational version of Newton's first law (the law of inertia):

An object or system of objects will maintain its angular momentum unless acted upon by an unbalanced external torque.

We all know that it is easier to balance on a bicycle that is moving than on a bicycle at rest. Rotating wheels have angular momentum. To tip the wheels means a change in angular momentum, which requires a greater torque than tipping wheels at rest.

Conservation of Angular Momentum

Just as the linear momentum of any system is conserved if no net forces are acting on the system, angular momentum is conserved if no net torque acts on the system. In the absence of an unbalanced external torque the angular momentum of that system is constant. This means that the product of rotational inertia and rotational velocity at one time will be the same as at any other time.

An interesting example illustrating angular momentum conservation is shown in Figure 7.42. The man stands on a low-friction turntable with weights extended. His rotational inertia, with the help of the extended weights, is relatively large in this position. As he slowly turns, his angular momentum is the product of his rotational inertia and rotational velocity. When he pulls the weights inward, the rotational inertia of his body and the weights is considerably reduced. What is the result? His rotational speed increases! This example is best appreciated by the turning person who feels changes in rotational speed that seem to be mysterious. But it's straight physics! This procedure is used by a figure skater who starts to whirl with her arms and perhaps a leg extended and

FIGURE 7.42 Conservation of angular momentum. When the man pulls his arms and the whirling weights inward, he decreases his rotational inertia *I*, and his rotational speed *ω* correspondingly increases.

$$I\omega = I\omega$$

then draws her arms and leg in to obtain a greater rotational speed. Whenever a rotating body contracts, its rotational speed increases.

Similarly, when a gymnast is spinning freely in the absence of unbalanced torques on his body, his angular momentum does not change. However, he can change his rotational speed by simply making variations in his rotational inertia. He does this by moving some part of his body toward or away from his axis of rotation.

FIGURE 7.43 Rotational speed is controlled by variations in the body's rotational inertia as angular momentum is conserved during a forward somersault.

FIGURE 7.44 Time-lapse photo of a falling cat.

If a cat is held upside down and dropped, it is able to execute a twist and land upright even if it has no initial angular momentum. Zero-angular-momentum twists and turns are performed by turning one part of the body against the other. While falling, the cat rearranges its limbs and tail several times to change its rotational inertia. During this maneuver the total angular momentum remains zero (Figure 7.44). When it is over, the cat is falling feet downward. This maneuver rotates the body through an angle, but it does not create continuing rotation. To do so would violate angular momentum conservation.

Humans can perform similar twists without difficulty, though not as fast as a cat. Astronauts have learned to make zero-angular-momentum rotations as they orient their bodies in preferred directions when floating freely in space. The law of momentum conservation is seen in the motions of the planets and the shape of the galaxies. This law will be a fact of everyday life to inhabitants of rotating space habitats who will head for distant places.

It is fascinating to note that the conservation of momentum tells us that the moon is getting farther away from the earth. This is because the earth's daily rotation is slowly decreasing due to the friction of ocean waters on the ocean bottom, just as an automobile's wheels slow down when brakes are applied. This decrease in the earth's angular momentum is accompanied by an equal increase in the angular momentum of the moon in its orbital motion around the earth. This increase in the moon's angular momentum results in the moon's increasing distance from the earth and a decrease in its speed. This increase of distance amounts to one-quarter of a centimeter per rotation. Have you noticed that the moon is getting farther away from us lately? Well, it is; each time we see another full moon, it is one-quarter of a centimeter farther away!

Summary of Terms

Rotational inertia That property of an object that measures its resistance to any change in its state of rotation: If at rest, the body tends to remain at rest; if rotating, it tends to remain rotating and will continue to do so unless acted upon by a net external torque.

Torque The product of force and lever-arm distance, which tends to produce rotation.

$$Torque = lever\ arm \times force$$

Center of mass The average position of mass or the single point associated with an object where all its mass can be considered to be concentrated.

Center of gravity The average position of weight or the single point associated with an object where the force of gravity can be considered to act.

Equilibrium The state of an object when not acted upon by a net force or net torque. An object in equilibrium may be at rest or moving at uniform velocity; that is, not accelerating.

Centripetal force A center-seeking force that causes an object to follow a circular path.

Centrifugal force An apparent outward force due to rotation, experienced in a rotating frame of reference. It is fictitious in the sense that it is not part of an interaction, but is an apparent force in itself—a result of rotation—with no reaction-force counterpart.

Angular momentum The product of a body's rotational inertia and rotational velocity about a particular axis. For an object that is small compared with the radial distance, it is the product of mass, speed, and radial distance of rotation.

Conservation of angular momentum When no external torque acts on an object or a system of objects, no change of angular momentum takes place. Hence, the angular momentum before an event involving only internal torques is equal to the angular momentum after the event.

Suggested Reading

Brancazio, P. J. *Sport Science.* New York: Simon & Schuster, 1984.

Clarke, A. C. *Rendezvous with Rama.* New York: Harcourt Brace Jovanovich, 1973. This is the first science fiction novel to seriously consider habitation inside a spinning space facility.

NASA. *Space Settlements: A Design Study.* Washington, D.C.: U.S. Government Printing Office, 1977.

Savage, Marshall T. *The Millennial Project—Colonizing the Galaxy in Eight Easy Steps.* Boston: Little, Brown and Company, 1994.

Review Questions

1. Why is it important that Chapters 3 and 4 are studied before this chapter is attempted?

Rotational Inertia

2. Inertia depends on mass; rotational inertia depends on mass and something else. What?
3. Does the rotational inertia of an object differ for different axes of rotation?
4. Why do people with long legs generally walk with a slower stride than people with short legs?
5. Which will have the greater linear acceleration rolling down an incline, a hoop or a solid disk?

Torque

6. Compare the effects of a force exerted on an object and a torque exerted on an object.
7. What is meant by the "lever arm" of a torque?
8. In what direction should a force be applied to produce maximum torque?
9. How do clockwise and counterclockwise torques compare when a system is balanced?

Center of Mass and Center of Gravity

10. Toss a stick into the air and it appears to wobble all over the place. Specifically, what place?
11. Where is the center of mass of a baseball? Where is its center of gravity? Where are these centers for a baseball bat?

Locating the Center of Gravity

12. If you hang at rest by your hands from a vertical rope, where is your center of gravity with respect to the rope?
13. Where is the center of mass of an empty shoe box?

Stability

14. What is the relationship between center of gravity and support base for an object in stable equilibrium?
15. How far can an object be tipped before it topples over?
16. Why doesn't the Leaning Tower of Pisa topple?
17. In terms of center of gravity, support base, and torque, why can you not stand with heels and back to a wall and then bend over to touch your toes and return to your stand-up position?

Centripetal Force

18. When you whirl a can at the end of a string in a circular path, what is the direction of the force that is exerted on the can?
19. Is it an inward force or an outward force that is exerted on the clothes during the spin cycle of an automatic washer?

Centrifugal Force

20. If the string that holds a whirling can in its circular path breaks, what kind of force causes it to move in a straight-line path—centripetal, centrifugal, or no force? What law of physics supports your answer?

21. If you are not wearing a seat belt and you slide across your seat and slam against a car door when the car rounds a curve, what kind of force is responsible—centripetal, centrifugal, or no force? Support your answer.

Centrifugal Force in a Rotating Frame

22. Identify the action and reaction forces in the interaction between the ladybug and the whirling can of Figures 7.33 and 7.35.

23. Why is centrifugal force in a rotating frame called a "fictitious force"?

Simulated Gravity

24. How can gravity be simulated in an orbiting space station?

25. Why will orbiting space stations that simulate gravity likely be large structures?

26. How will the value of g vary at different distances from the hub of a rotating space station?

Angular Momentum

27. Distinguish between linear momentum and angular momentum.

28. What is the law of inertia for rotating systems in terms of angular momentum?

Conservation of Angular Momentum

29. What does it mean to say that angular momentum is conserved?

30. If a skater who is spinning pulls her arms in so as to reduce her rotational inertia to half of what it was, by how much will her angular momentum increase? By how much will her rate of spin increase? (Why are your answers different?)

Projects

1. Fasten a fork, spoon, and wooden match together as shown. The combination will balance nicely—on the edge of a glass, for example. This happens because the center of gravity actually "hangs" below the point of support.

2. Stand with your heels and back against a wall and try to bend over and touch your toes. You'll find you have to stand away from the wall to do so without toppling over. Compare the minimum distance of your heels from the wall with that of a friend of the opposite sex. Who can touch their toes with their heels nearer to the wall—males or females? On the average and in proportion to height, which sex has the lower center of gravity?

‹2 footlengths›

3. Ask a friend to stand facing a wall. With toes against the wall, ask your friend to stand on the balls of her feet without toppling backward. Your friend won't be able to do it. Now you explain why it can't be done.

4. Rest a meter stick on two extended forefingers as shown. Slowly bring your fingers together. At what part of the stick do your fingers meet? Can you explain why this always happens, no matter where you start your fingers?

5. Swing a pail of water around rapidly in a circle at arm's length and the water will not spill. Why?

6. Place the hook of a wire coat hanger over your finger. Carefully balance a coin on the straight wire on the bottom directly under the hook. You may have to flatten the wire with a hammer or fashion a tiny platform with tape. With a surprisingly small amount of practice you can swing the hanger and balanced coin back and forth and then in a circle. Centripetal force holds the coin in place.

Exercises

1. When a rolling yo-yo falls to the bottom of its cord, does it reverse rotation as it climbs back up the cord? Explain.

2. A basketball player wishes to balance a ball on his fingertip. Will he be more successful with a spinning ball or a stationary ball? What physical principle supports your answer?

3. If you walk along the top of a fence, why does holding your arms out help you to balance?

4. A popular gyroscope is the Frisbee. What is one function, besides being a place for gripping and catching, of its somewhat thicker, curved rim?

5. The front wheels located far out in front of the racing vehicle help to keep the vehicle from nosing upward when it accelerates. What physics principle is illustrated here?

6. When a car accelerates forward, why does its front end rise? Why does its front end dip when braking?

7. When a car drives off a cliff, why does it rotate forward as it falls? (Consider the torque it experiences as it rolls off the cliff.)

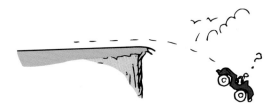

8. Which will have the greater acceleration rolling down an incline—a bowling ball or a volleyball? Defend your answer.

9. Which will roll down an incline faster, a can of water or a can of ice?

10. A friend says that a body cannot rotate when the net torque acting on it is zero. Your friend's assertion is incorrect. Correct it.

11. Is the net torque changed when a partner on a seesaw stands or hangs from her end instead of sitting? (Does the weight or the lever arm change?)

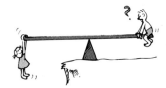

12. When you pedal a bicycle, maximum torque is produced when the pedal sprocket arms are in the horizontal position, and no torque is produced when they are in the vertical position. Explain.

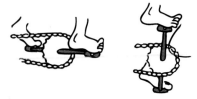

13. The spool is pulled in three ways, as shown below. There is sufficient friction for rotation. In what direction will the spool roll in each case?

14. Why is the middle seating most comfortable in a bus traveling on a bumpy road, or in a ship in a choppy sea, or in an airplane in turbulent air?

15. Explain why a long pole is more beneficial to a tightrope walker if the pole droops.

16. A carbon monoxide molecule consists of a carbon atom with a mass of 12 amu (atomic mass units) and an oxygen atom of mass 16 amu. Show in a sketch like the one below the approximate location of the center of mass of this molecule—the point about which the atoms in the molecule turn when the molecule is set into rotation. (*Note:* Practically all of the mass of an atom resides in its tiny nucleus.)

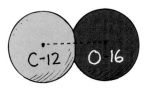

17. Why is the wobbly motion of a single star an indication that the star has a planet or system of planets?

18. Why must you bend forward when carrying a heavy load on your back?

19. Why is it easier to carry the same amount of water in two buckets, one in each hand, than in a single bucket?

20. Is it necessary that weights be placed in the middle of the pans of an equal-arm balance for accurate measurements? Why or why not?

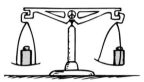

21. Nobody at the playground wants to play with the obnoxious boy, so he fashions a seesaw as shown so he can play with himself. Explain how this is done.

22. Using the ideas of torque and center of gravity, explain why a ball rolls down a hill.

23. Sometimes a kicked football sails through the air without rotating, and at other times it tumbles end over end as it travels. With respect to the center of mass of the ball, how is it kicked in both cases?

24. In playing pool or billiards, to hit the cue ball so that it rolls without skidding, should you strike it above, at, or below the center of mass of the ball?

25. How can the three bricks be stacked so that the top brick has maximum horizontal displacement from the bottom brick? For example, stacking them like the dotted lines suggest would be unstable and the bricks would topple. (*Hint:* Start with the top brick and work down. At every interface the CG of the bricks above must not extend beyond the end of the supporting brick.)

26. How does the heavy tail of a monkey enable it to reach farther when standing on a branch reaching for fruit?

27. Where is the center of mass of the earth's atmosphere?

28. Why is it dangerous to roll open the top drawers of a fully loaded file cabinet that is not secured to the floor?

29. Describe the comparative stabilities of the three objects shown in Figure 7.21, page 126, in terms of work and potential energy.

30. The centers of gravity of the three trucks parked on a hill are shown by the X's. Which truck(s) will tip over?

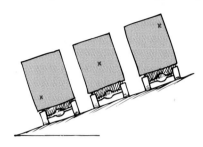

31. Why is less effort required in doing sit-ups when your arms are extended in front of you? Why is it more difficult when your arms are placed behind your head?

32. A long track balanced like a seesaw supports a golf ball and a more massive billiard ball with a compressed spring between the two. When the spring is released, the balls move away from each other. Does the track tip clockwise, tip counterclockwise, or remain in balance as the balls roll outward? What principles do you use for your explanation?

33. The value of g at the earth's surface is about 10 m/s^2. How would this value change if the earth rotated faster about its daily axis?

34. When a long-range cannonball is fired toward the equator from a northern (or southern) latitude, it lands west of its "intended" longitude. Why? (*Hint:* Consider a flea jumping from the middle to the outer edge of a moving phonograph record.)

35. If you should buy a quantity of gold at sea level in Mexico and weigh it carefully on a spring balance, would the same quantity of gold weigh more, less, or the same if weighed on the same spring balance at sea level in Alaska? Defend your answer.

36. A motorcyclist is able to ride on the vertical wall of a bowl-shaped track as shown. His weight is counteracted by the friction of the wall on the tires (vertical arrow). What supplies the centripetal force? Does this force increase or decrease if he rides faster?

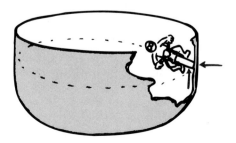

37. When a soaring eagle turns in its flight, what is the source of the centripetal force that acts on it?

38. Your friend says that a satellite in circular orbit does not accelerate, as evidenced by its constant speed. Your other friend says it does accelerate, as evidenced by the centripetal force supplied by gravity. What do you say?

39. Does the centripetal force on an object following a circular path do work on the object? How does the moon's motion about the earth supply evidence to support your answer?

40. When you are in the front passenger seat of a car turning to the left, you may find yourself pressed against the right-side door. Why do you press against the door? Why does the door press on you? Does your explanation involve a centrifugal force or Newton's laws?

41. The occupant inside a future rotating space habitat feels that she is being pulled by artificial gravity against the outer wall of the habitat (which becomes the "floor"). Explain what is going on in terms of Newton's laws and centripetal force.

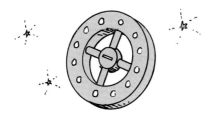

42. We humans and everything else in the world travel in a circular path as the earth rotates daily about its polar axis. Does this mean there is a centripetal force acting on us and everything else?

43. You sit at the middle of a large turntable at an amusement park as it is set spinning and then allowed to spin freely. When you crawl toward the edge of the turntable, does the rate of the rotation increase, decrease, or remain unchanged? What physics principle supports your answer?

44. A student says that when you execute a somersault and pull your arms and legs inward, your angular momentum increases. A physics instructor corrects this statement. What is the correction?

45. A sizable quantity of earth is washed down the Mississippi River and deposited in the Gulf of Mexico each year. What effect does this tend to have on the length of a day? (*Hint:* Relate this to Figure 7.42, page 134.)

46. Strictly speaking, as more and more skyscrapers are built on the surface of the earth, does the day tend to become longer or shorter? And strictly speaking, does the falling of autumn leaves tend to lengthen or shorten the 24-hour day? What physical principle supports your answers?

47. How would the length of a day be affected if all the world's inhabitants continuously walked in an easterly direction? When they stopped walking?

48. A toy train is initially at rest on a track fastened to a bicycle wheel, which is free to rotate. How does the wheel respond when the train moves clockwise? When the train backs up? Does the angular momentum of the wheel–train system change during these maneuvers? How would the resulting motions depend on the relative masses of the wheel and train?

49. Why does a typical small helicopter with a single main rotor have a second small rotor on its tail? Describe the consequence if the small rotor fails in flight.

50. We believe our galaxy was formed from a huge cloud of gas and particles. The original cloud was far larger than the present size of the galaxy, was more or less spherical, and was rotating very much more slowly than the galaxy is now. In this sketch we see the original cloud and the galaxy as it is now (seen edgewise). Explain how the law of gravitation and the conservation of angular momentum contribute to the galaxy's present shape and why it rotates faster now than when it was a larger, spherical cloud.

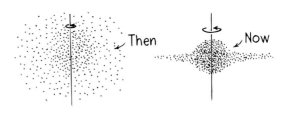

Problems

1. Hanging from one end of a light meter stick is a 1-kg weight and hanging from the other end is a 3-kg weight. Where is the center of mass of this system (the point of balance)? How does your answer relate to torque?

2. A 10,000-N vehicle is stalled one-quarter of the way across the bridge. Calculate the additional reaction forces that are supplied at the supports on both ends of the bridge.

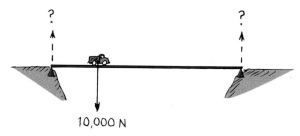

3. The rock has a mass of 1 kg. What is the mass of the measuring stick if it is balanced by a support force at the one-quarter mark?

4. To tighten a bolt, you push with a force of 80 N at the end of a wrench handle that is 0.25 m from the axis of the bolt. (a) What torque are you exerting? (b) If you move your hand inward to be only 0.10 m from the bolt, what force do you have to exert to achieve the same torque? (c) Do your answers depend on the direction of your push relative to the direction of the wrench handle?

5. Consider a too-small space habitat that consists of a rotating cylinder of radius 4 m. If a man standing inside is 2 m tall and his feet are at 1 g, what is the g force at the elevation of his head? (Do you see why projections call for large habitats?)

6. If the variation in g between one's head and feet is to be less than $\frac{1}{100} g$, then compared to one's height, what should be the minimum radius of the space habitat?

7. If a trapeze artist rotates once each second while sailing through the air, and contracts to reduce her rotational inertia to one third of what it was, how many rotations per second will result?

8. How much greater is the angular momentum of the earth orbiting about the sun than the moon orbiting about the earth? (Set up a ratio of angular momenta using data on the inside back cover.)

8

· · · · · · · · · · · ·

Gravity

If the force of gravity between the orbiting space shuttle and the earth were somehow eliminated, what would be the path of the shuttle?

For thousands of years people have looked into the night sky and wondered about the stars. Having only their naked eyes, those of long ago neither saw nor dreamed that the stars are more numerous than all the grains of sand on all the beaches of the world. The ancient Greeks envisioned the earth at the center of the universe and considered the stars to be fixed on a great revolving crystal sphere. They distinguished the planets from the stars and placed them on inner spheres with more complicated motions. They believed that the complicated motions and the positions of the planetary spheres influenced earthly events.

Copernicus, Brahe, and Kepler

In the sixteenth century the Polish astronomer Copernicus discovered that the motions of the planets could be explained much more simply if they are considered to circle the sun rather than the earth. The Copernican view was revolutionary, because it removed the earth from the center of the universe.* The Copernican idea was also highly controversial—widely considered distasteful and even insulting. Philosophical arguments against the Copernican view proliferated.

* Copernicus was not the first to have the planets revolving about the sun. The Greek scholar Aristarchus had postulated a sun-centered universe, but he did not gain many adherents.

Nicolaus Copernicus

Tycho Brahe

Johannes Kepler

The Danish astronomer Tycho Brahe, born soon after Copernicus died, was quite different from his contemporaries. Rather than carrying on with philosophical arguments, Brahe believed it would be better to accurately measure the positions of the planets in the sky and to chart their motions. This was a magnificent idea—that in formulating theories, it is better to gather facts and make careful measurements than to speculate. To implement this approach, Brahe built the world's first great observatory in his native Denmark. He did not use telescopes because they had not yet been invented. Instead he built huge brass protractor-like instruments called quadrants. With them, he cataloged the motions of the planets so accurately that his measurements are still used today. He recorded the planets' positions for 20 years to $\frac{1}{60}$ of a degree. Just before his death, Brahe entrusted a younger colleague, the German Johannes Kepler, with editing and publishing his planetary tables. From the data, Kepler discovered some very remarkable and simple laws regarding planetary motion.

Kepler's Laws

Today we know what Brahe and Kepler did not know—that the planets of the solar system are simply falling continuously around the sun (just as earth satellites continuously fall around the earth, briefly discussed in Chapter 3). Kepler, to learn how the planets moved, performed the enormous task of translating Brahe's earthbound observations to another frame of reference. He converted Brahe's measurements to values that would be obtained by a stationary observer outside the solar system. Kepler's expectation that the planets would move in perfect circles around the sun was shattered after years of effort. He found the paths to be ellipses.

Kepler also found that the planets do not go around the sun at a uniform speed but move faster when they are nearer the sun and more slowly when they are farther from the sun. They do this in such a way that an imaginary line or spoke joining the sun and the planet sweeps out equal areas of space in equal times. The triangular-shaped area swept out during a month when a planet is orbiting far from the sun (triangle ASB in Figure 8.1) is equal to the triangular area swept out during a month when the planet is orbiting closer to the sun (triangle CSD in Figure 8.1).

Ten years later Kepler discovered a third law. He had spent these years searching for a connection between the size of a planet's orbit and its period (the time required for a planet to make a complete revolution around the sun). From Brahe's data Kepler found that the square of a period is proportional to the cube of its average distance from the sun. He discovered this by noting that the fraction T^2/R^3 is the same for all the planets,

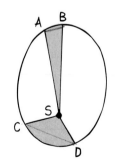

FIGURE 8.1 Equal areas are swept out in equal intervals of time.

where R is the planet's average orbit radius and T is the planet's period. Thus, **Kepler's laws of planetary motion** are:

Law 1: Each planet moves in an elliptical orbit with the sun at one focus of the ellipse.

Law 2: The line from the sun to any planet sweeps out equal areas of space in equal time intervals.

Law 3: The squares of the times of revolutions (periods) of the planets are proportional to the cubes of their average distances from the sun. ($T^2 \sim R^3$ for all planets.)

Kepler's laws apply not only to planets but also to moons or any satellite in orbit around any body. Except for Pluto (which Kepler had no knowledge of), the elliptical orbits of the planets are very nearly circular. Only the precise measurements of Brahe showed the slight differences. Kepler had no idea why the planets traced elliptical paths about the sun and no general explanation for the mathematical relationships that he had discovered. He was familiar with Galileo's ideas about inertia and accelerated motion, but he failed to apply them to his own work. Like Aristotle, he thought that the force on a moving body would be in the same direction as the body's motion. Kepler never appreciated the concept of inertia. Galileo, on the other hand, never appreciated Kepler's work and held to his conviction that the planets move in circles.* Further understanding of planetary motion required someone who could integrate the findings of these two great scientists. This task fell to Isaac Newton.

Newton's Law of Universal Gravitation

From the time of Aristotle, the arc-like motions of heavenly bodies were regarded as natural. The ancients believed that the stars, planets, and moon moved in divine circles, free from any impressed forces. As far as the ancients were concerned, this circular motion required no explanation.

Newton recognized that a force of some kind must be acting on the planets; otherwise, their paths would be straight lines. From the law of inertia, he knew this force was not directed along the path of the planets as Kepler had speculated but was, instead, in the direction in which the planets are curved toward the sun. His analysis of Kepler's second law showed that the origin of this force was the sun. From Kepler's first and third laws, he deduced that this force decreases as the square of the distance.[†] A planet twice as far from the sun is pulled toward the sun with $\frac{1}{4}$ the force; three times as far, $\frac{1}{9}$ the force; and so on. This relationship is the *inverse-square law,* which we will treat in detail shortly.

What sort of force could act from one place to another without contact? Newton was aware of such a force. When an apple falls to the ground, it is accelerated from its position on the tree. He wondered if the same force extended to the moon, holding the moon in an elliptical path around the earth, a path similar to a planet's path around the sun. This stroke of intuition—conceiving the same force to hold both the moon and

---

* It is not easy to look at familiar things through the new insights of others. We tend to see only what we have learned to see or wish to see. Galileo reported that many of his colleagues were unable or refused to see the moons of Jupiter when they peered skeptically through his telescopes. Galileo's telescopes were a boon to astronomy, but more important than a new instrument to see things better was a new way of understanding what is seen. Is this true today?

† Perhaps your instructor will show that Kepler's third law results when Newton's law of gravitation is equated to centripetal force and how T^2/R^3 equals a constant that depends only on G and M, the mass of the body about which orbiting occurs. Interesting stuff!

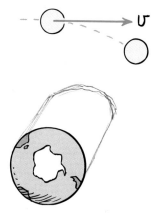

FIGURE 8.2 The tangential velocity of the moon about the earth allows it to fall around the earth rather than directly into it. If this tangential velocity were reduced to zero, what would be the fate of the moon?

apple—was a revolutionary break with the prevailing notion that there were two sets of natural laws, one for earthly events and another altogether for motions in the heavens.

To test this theory, Newton compared the fall of an apple with the fall of the moon. He realized that the moon falls in the sense that *it falls away from the straight line it would follow if there were no forces acting on it*. Because of its tangential speed, it "falls around" the round earth. Newton's test was to see if the moon's distance of fall per second was the same as the distance that an apple or anything at that distance would fall in 1 second. We'll go into the details of Newton's test in the next chapter and simply state here that his first calculations didn't check. Disappointed, but recognizing that brute fact must always win over a beautiful hypothesis, he placed his papers in a drawer where they remained for nearly 20 years. During this period he founded and developed the field of geometric optics for which he first became famous.

Newton's interest in mechanics was rekindled with the advent of a spectacular comet in 1680 and another comet two years later. He returned to the moon problem at the prodding of his astronomer friend, Edmond Halley, for whom the second comet was later named. He made corrections in the experimental data used in his earlier method and obtained excellent results. Only then did he publish what is one of the most far-reaching generalizations of the human mind: the **law of universal gravitation**.*

Everything pulls on everything else in a beautifully simple way that involves only mass and distance. According to Newton, every body attracts every other body with a force that for any two bodies is directly proportional to the product of the masses of the bodies involved and inversely proportional to the square of the distance separating them.

This statement can be expressed symbolically as

$$F \sim \frac{m_1 m_2}{d^2}$$

where m_1 and m_2 are the masses of the bodies, and d is the distance between their centers. Thus, the greater the masses m_1 and m_2, the greater the force of attraction between them—in direct proportion to their masses.[†] The greater the distance of separation d, the *weaker* the force of attraction—in inverse proportion to the square of the distance between their centers of mass.

The Universal Gravitational Constant, *G*

The proportionality form of the universal law of gravitation can be expressed as an exact equation when the constant of proportionality G, called the *universal gravitational constant*, is introduced. Then the equation is

$$F - G\frac{m_1 m_2}{d^2}$$

---

* This is a dramatic example of the painstaking effort and cross-checking that go into the formulation of a scientific theory. Contrast Newton's approach with the failure to "do one's homework," the hasty judgments, and the absence of cross-checking that so often characterize the pronouncements of people advocating less-than-scientific theories.

[†] Note the different role of mass here. Thus far we have treated mass as a measure of inertia, which is called *inertial mass*. Now we see mass as a measure of gravitational force, which in this context is called *gravitational mass*. It is experimentally established that the two are equal, and, as a matter of principle, the equivalence of inertial and gravitational mass is the foundation of Einstein's general theory of relativity.

In words, the force of gravity between two objects is found by multiplying their masses, dividing by the square of the distance between their centers, and then multiplying this result by the constant G. The magnitude of G is given by the magnitude of the force between two 1-kilogram bodies that are 1 meter apart; 0.0000000000667 newton. This is an extremely weak force. In standard units G has this same numerical value. The units of G make the force come out in newtons. In scientific notation,*

$$G = 6.67 \times 10^{-11} \text{ N} \cdot \text{m}^2/\text{kg}^2$$

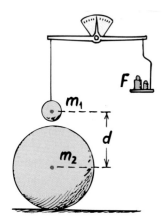

FIGURE 8.3 Jolly's method of measuring G. Balls of mass m_1 and m_2 attract each other with a force F equal to the weights needed to restore balance.

G was first measured long after the time of Newton by an English physicist, Henry Cavendish, in the eighteenth century. He accomplished this by measuring the tiny force between lead masses with an extremely sensitive torsion balance. A simpler method was later developed by Philipp von Jolly, who attached a spherical flask of mercury to one arm of a sensitive balance (Figure 8.3). After the balance was put in equilibrium, a 6-ton lead sphere was rolled beneath the mercury flask. The gravitational force between the two masses was equal to the weight that had to be placed on the opposite end of the balance to restore equilibrium. All the quantities m_1, m_2, F, and d were known, from which the ratio G was calculated:

$$\frac{F}{m_1 m_2/d^2} = 6.67 \times 10^{-11} \frac{\text{N}}{\text{kg}^2/\text{m}^2} = 6.67 \times 10^{-11} \text{ N} \cdot \text{m}^2/\text{kg}^2$$

The value of G tells us that the force of gravity is a very weak force. It is the weakest of the presently known four fundamental forces. (The other three are the electromagnetic force and two kinds of nuclear forces.) We sense gravitation only when masses like that of the earth are involved. The force of attraction between you and a battleship on which you stand is too weak for ordinary measurement. The force of attraction between you and the earth, however, can be measured. It is your weight.

Your weight depends not only on your mass, but also on your distance from the center of the earth. At the top of a mountain your mass is no different than it is anywhere else, but your weight is slightly less than at ground level because your distance from the center of the earth is greater.

Once the value of G was known, the mass of the earth was easily calculated. The force that the earth exerts on a 1-kilogram body at its surface is 9.8 newtons. The distance between the centers of mass of the 1-kilogram body and the earth is the earth's radius, 6.4×10^6 meters. Therefore, from $F = G(m_1 m_2/d^2)$, where m_1 is the mass of the earth,

$$9.8 \text{ N} = 6.67 \times 10^{-11} \text{ N} \cdot \text{m}^2/\text{kg}^2 \frac{1 \text{kg} \times m_1}{(6.4 \times 10^6 \text{ m})^2}$$

from which the mass of the earth is calculated to be $m_1 = 6 \times 10^{24}$ kilograms.

---

* The numerical value of G depends entirely on the units of measurement we choose for mass, distance, and time. The international system of choice is: for mass, the kilogram; for distance, the meter; and for time, the second. Scientific notation is discussed in Appendix A at the end of this book.

Question If there is an attractive force between all objects, why do we not feel ourselves gravitating toward massive buildings in our vicinity?

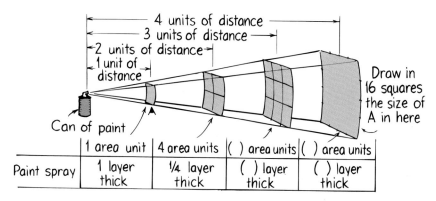

FIGURE 8.4 The inverse-square law. Paint spray travels radially away from the nozzle of the can in straight lines. Like gravity, the "strength" of the spray obeys the inverse-square law.

FIGURE 8.5 According to Newton's equation, her weight (not mass) decreases as she increases her distance from the earth's center.

Gravity and Distance: The Inverse-Square Law

We can better understand how gravity is diluted with distance by considering how paint from a paint gun spreads with increasing distance (Figure 8.4). Suppose we position a paint gun at the center of a sphere with a radius of 1 meter, and a burst of paint spray travels 1 meter to produce a square patch of paint that is 1 millimeter thick. How thick would the patch be if the experiment were done in a sphere with twice the radius? If the same amount of paint travels in straight lines for 2 meters, it will spread to a patch twice as tall and twice as wide. The paint would be spread over an area four times as big, and its thickness would be only $\frac{1}{4}$ millimeter. Can you see from the figure that for a sphere of radius 3 meters the thickness of the paint patch would be only $\frac{1}{9}$ millimeter? Can you see the thickness of the paint decreases as the square of the distance increases? This is known as the **inverse-square law**. The inverse-square law holds for gravity and for all phenomena wherein the effect from a localized source spreads uniformly throughout the surrounding space: the electric field about an isolated electron, light from a match, radiation from a piece of uranium, and sound from a cricket.

In using Newton's equation for gravity, it is important to emphasize that the distance term d is the distance between the centers of masses of the objects that are

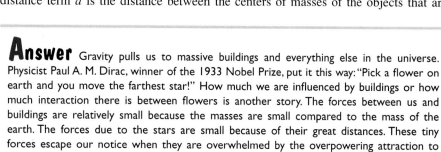

Answer Gravity pulls us to massive buildings and everything else in the universe. Physicist Paul A. M. Dirac, winner of the 1933 Nobel Prize, put it this way: "Pick a flower on earth and you move the farthest star!" How much we are influenced by buildings or how much interaction there is between flowers is another story. The forces between us and buildings are relatively small because the masses are small compared to the mass of the earth. The forces due to the stars are small because of their great distances. These tiny forces escape our notice when they are overwhelmed by the overpowering attraction to the earth.

FIGURE 8.6 If an apple weighs I N at the earth's surface, it weighs only $\frac{1}{4}$ N twice as far from the center of the earth. At three times the distance, it weighs only $\frac{1}{9}$ N. What would it weigh at four times the distance? Five times? Gravity versus distance is plotted in color.

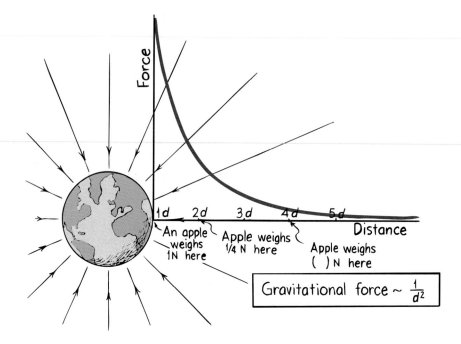

Force

An apple weighs 1N here

Apple weighs 1/4 N here

Apple weighs ( ) N here

Distance

Gravitational force ~ $\frac{1}{d^2}$

attracted to each other. Note in Figure 8.6 that the apple that normally weighs 1 newton at the earth's surface weighs only $\frac{1}{4}$ as much when it is twice the distance from the earth's center. The greater the distance from the earth's center, the less the weight of an object. A child that weighs 300 newtons at sea level will weigh only 299 newtons atop Mt. Everest. But no matter how great the distance, the earth's gravitational force approaches, but never reaches, zero. Even if you were transported to the far reaches of the universe, the gravitational influence of home would still be with you. It may be overwhelmed by the gravitational influences of nearer and/or more massive bodies, but it is there. The gravitational influence of every material object, however small or however far, is exerted through all of space. Isn't that nice?

Questions

1. By how much does the gravitational force between two objects decrease when the distance between their centers is doubled? Tripled? Increased tenfold?
2. Consider an apple at the top of a tree that is pulled by earth gravity with a force of I N. If the tree were twice as tall, would the force of gravity be only $\frac{1}{4}$ as strong? Defend your answer.

Answers

1. It decreases to one-fourth, one-ninth, and one-hundredth.
2. No, because the twice-as-tall apple tree is not twice as far from the earth's center. The taller tree would have to have a height equal to the radius of the earth (6,370 km) before the weight of the apple at its top reduces to $\frac{1}{4}$ N. Before its weight decreases by 1 percent, an apple or any object must be raised 32 km—nearly four times the height of Mt. Everest. So as a practical matter we disregard the effects of everyday changes in elevation.

Weight and Weightlessness

When you step on a spring balance such as a bathroom scale, you compress a spring inside. When the pointer stops, the elastic force of the deformed spring balances the gravitational attraction between you and the earth—nothing moves as you and the scale are in static equilibrium. The pointer is calibrated to show your weight. If you stood on a bathroom scale in a moving elevator, you'd find variations in your weight. If the elevator accelerated upward, the springs inside the bathroom scale would be more compressed and your weight reading would increase. If the elevator accelerated downward, the springs inside the scale would be less compressed and your weight reading would decrease. If the elevator cable broke and the elevator fell freely, the reading on the scale would go to zero. According to the reading, you would be weightless. Would you really be weightless? We can answer this question only if we agree on what we mean by weight.

FIGURE 8.7 Your weight equals the force with which you press against the supporting floor. If the floor accelerates up or down, your weight varies (even though the gravitational force *mg* that acts on you remains the same).

Normal weight

Greater than normal weight

Less than normal weight

zero weight

We define the weight of something as the force it exerts against the supporting floor or the weighing scale. According to this definition, you are as heavy as you feel; so in an elevator that accelerates downward, the supporting force of the floor is less and you weigh less. If the elevator is in free fall, your weight is zero (Figure 8.7). Even in this weightless condition, however, there is still a gravitational force acting on you, causing your downward acceleration. But gravity now is not felt as weight because there is no support force.

Consider an astronaut in orbit. The astronaut is weightless because he is not supported by anything (Figure 8.8). There would be no compression in the springs of a bathroom scale placed beneath his feet because the bathroom scale is falling as fast as he is. Any objects that are released fall together with him and remain in his vicinity, unlike what happens on the ground. All the local effects of gravity are eliminated. The body organs respond as though gravity forces were absent, and this gives the sensation of weightlessness. The astronaut experiences the same sensation in orbit that he would feel in a falling elevator—a state of free fall.

On the other hand, if the astronaut were in deep space far removed from any attracting objects but his spacecraft were being accelerated, he *would* have weight. Like the girl in the accelerating elevator, he would press against a scale or a supporting surface.

So we see that weight and gravity don't always have to go hand in hand. Einstein, in his general theory of relativity, explained why this is so. Floating in deep space far

FIGURE 8.8 Both are weightless.

FIGURE 8.9 The inhabitants in the Mir Space Station continuously experience weightlessness. They are in free fall around the earth. Is there a force of gravity between them and the earth?

from gravitating objects is equivalent to "floating" in free fall near a gravitating object. Weight is not directly a manifestation of gravity. It results when some force other than gravity acts (such as a floor holding you up or a rocket engine accelerating you).

The space station depicted by NASA artists in Figure 8.9 provides a weightless environment. The shuttle, station facility, and astronauts all accelerate equally toward earth, at somewhat less than 1 *g* because of their altitude. This acceleration is not sensed at all; with respect to the station, the astronauts experience zero *g*.

Question In what sense is drifting in space far away from all celestial bodies like stepping off a table?

Ocean Tides

Seafaring people have always known there was a connection between the ocean tides and the moon, but no one could offer a satisfactory theory to explain the two high tides per day. Newton showed that the ocean tides are caused by differences in the gravitational pull between the moon and the earth on opposite sides of the earth. Gravitational

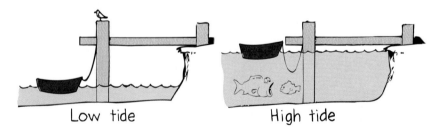

FIGURE 8.10 Ocean tides.

Low tide   High tide

Answer In both cases you'd experience weightlessness. Drifting in deep space, you would remain weightless because no force acts on you. Stepping from a table, you would be only momentarily weightless because of a momentary lapse of support force.

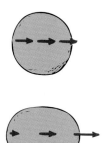

FIGURE 8.11 A ball of Jell-O remains spherical when all parts of it are pulled equally in the same direction. When one side is pulled more than the other, however, its shape is elongated.

force between the moon and earth is stronger on the side of the earth nearer to the moon, and it is weaker on the side of the earth farthest from the moon. This is simply because the gravitational force is weaker with increased distance.

To understand why the difference in gravitational pulls by the moon on opposite sides of the earth produces tides, pretend you have a big spherical ball of Jell-O. If you exert the same force on every part of the ball, it would remain spherical as it accelerates. But if you pulled harder on one side than the other, there would be a difference in accelerations and the ball would become elongated (Figure 8.11). That is what happens to this big ball we're living on. The side closer to the moon is pulled with a greater force and has a greater acceleration toward the moon than the far side—thus the earth is somewhat football shaped. But does the earth accelerate toward the moon? Yes, it must, because a force acts on it, and where there is a net force there is acceleration. It is a *centripetal* acceleration, for the earth circles the center of mass of the earth-moon system. Both the earth and the moon undergo centripetal acceleration as they circle each other about the earth-moon center of mass (a point within the earth about three-quarters of the way from the center to the surface). This makes both the earth and moon slightly elongated. Elongation of the earth is mainly in its oceans, which bulge equally on opposite sides.

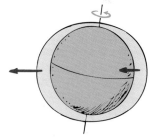

FIGURE 8.12 Two tidal bulges remain relatively fixed with respect to the moon while the earth spins daily beneath them.

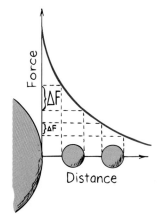

FIGURE 8.13 A plot of gravity versus distance (not to scale). The greater the distance from the sun, the smaller the force F, which varies as $1/d^2$, and the smaller difference in gravitational pulls on opposite sides of a planet, ΔF, which varies as $1/d^3$, and hence the smaller the tides.

On a world average, the ocean bulges are nearly 1 meter above the average surface level of the ocean. The earth spins once per day, so a fixed point on earth passes beneath both of these bulges each day. This produces two sets of ocean tides per day. Any part of the earth that passes beneath one of the bulges has a high tide. When the earth has made a quarter turn, 6 hours later, the water level at the same part of the ocean is nearly 1 meter below the average sea level. This is low tide. The water that "isn't there" is under the bulges that make up the high tides. A second high tidal bulge is experienced when the earth makes another quarter turn. So we have two high tides and two low tides daily. It turns out that while the earth spins, the moon moves in its orbit and appears at the same position in our sky every 24 hours and 50 minutes, so the two-high-tide cycle is actually at 24-hour-and-50-minute intervals. That is why tides do not occur at the same time every day.

The sun also contributes to ocean tides, although it is less than half as effective as the moon in raising tides—even though its pull on the earth is 180 times greater than the pull of the moon. Why doesn't the sun cause tides 180 times greater than lunar tides? The answer has to do with a key word: *difference*. Because of the great distance of the sun, the difference in its gravitational pull on opposite sides of the earth is very small (Figure 8.13). The percentage difference in the sun's pulls across the earth is only about

FIGURE 8.14   The tidal force difference due to a 1-kg body 1 m over the head of an average height person is about 60 trillionths (6×10^{-11}) N/kg. For an overhead moon, it is about 0.3 trillionth (3×10^{-13}) N/kg. So holding a melon over your head produces about 200 times as much tidal effect in your body as the moon!

0.017 percent, compared to 6.7 percent across the earth by the moon. It is only because the pull of the sun is 180 times stronger than the moon's that the sun tides are almost half as high (180×0.017 percent = 3 percent, nearly half of 6.7 percent).

Newton deduced that the difference in pulls decreases as the *cube* of the distance between the centers of the bodies—twice as far away produces $\frac{1}{8}$ the tide; three times as far, only $\frac{1}{27}$ the tide, and so on. Only relatively close distances result in appreciable tides, and so the nearby moon out-tides the enormously more massive but farther-away sun. The amount of tide also depends on the size of the body having tides. Although the moon produces a considerable tide in the earth's oceans, which are thousands of kilometers apart, it produces scarcely any in a lake. That's because no part of the lake is significantly closer to the moon than any other part, so there is no significant *difference* in moon pulls on the lake. Similarly for the fluids in your body. Any tides in the fluids of your body caused by the moon are negligible. You're not tall enough for tides. What micro-tides the moon may produce in your body are only about one two-hundredth the tides produced by a one-kilogram melon held one meter above your head (Figure 8.14)!

Question   We know that both the moon and the sun produce our ocean tides. And we know the moon plays the greater role because it is closer. Does its closeness mean it pulls with more gravitational force than the sun on the earth's oceans?

When the sun, earth, and moon are all lined up, the tides due to the sun and the moon coincide. Then we have higher-than-average high tides and lower-than-average low tides. These are called **spring tides** (Figure 8.15). (Spring tides have nothing to do

FIGURE 8.15   When the attractions of the sun and the moon are lined up with each other, spring tides occur.

Answer   No, the sun's pull is much stronger. Gravitational pull weakens as the distance squared to the body that pulls. But the *difference* in pulls across the earth's oceans weakens as the distance cubed. When the distance to the sun is squared, gravitation from the sun is still stronger than gravitation from the closer moon—because of the sun's enormous mass. But when the distance to the sun is cubed, as is the case for tidal forces, the sun's influence is less than the moon's. Distance is the key to tidal forces. If the moon were closer to earth, the tides on both the earth and the moon would increase as the closer distance cubed—which could tear the moon into pieces—catastrophic—as the planetary rings of other planets suggest.

FIGURE 8.16 When the attractions of the sun and the moon are about 90° apart (at the time of a half moon), neap tides occur.

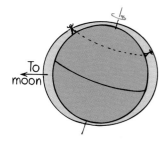

FIGURE 8.17 Inequality of the two high tides per day. Because of the earth's tilt, a person in the Northern Hemisphere may find the tide nearest the moon much lower (or higher) than the tide half a day later. Inequalities of tides vary with the positions of the moon and the sun.

with the spring season.) You can tell when the sun, earth, and moon are aligned by the full moon or by the new moon. When the moon is full, the earth is between the sun and moon. (If all three are *exactly* in line, then you have a lunar eclipse, for the full moon is in the earth's shadow.) A new moon occurs when the moon is between the sun and earth, when the non-illuminated side of the moon faces the earth. (When this alignment is perfect, the moon blocks the sun and you have a solar eclipse.) Spring tides occur at the times of a new or full moon.

All spring tides are not equally high because distances between the earth and the moon and the earth and the sun both vary; the orbital paths of the earth and the moon are not really circular, but elliptical. The moon's distance from the earth varies by about 10 percent and its effect in raising tides varies by about 30 percent. Highest spring tides occur when the moon and sun are closest to the earth.

When the moon is halfway between a new moon and a full moon, in either direction (Figure 8.16), the tides due to the sun and the moon partly cancel each other. Then, the high tides are lower than average and the low tides are not as low as average low tides. These are called **neap tides**.

Another factor that affects the tides is the tilt of the earth's axis (Figure 8.17). Even though the opposite tidal bulges are equal, the earth's tilt causes the two daily high tides experienced in most parts of the ocean to be unequal most of the time.

Our treatment of tides is quite simplified here, for tides are actually more complicated. Interfering land masses and friction with the ocean bottom, for example, complicate tidal motions. In many places the tides break up into smaller "basins of circulation," where a tidal bulge travels like a circulating wave that moves around in a small basin of water when it is properly tilted. For this reason the high tide may be hours away from an overhead moon. In mid-ocean the variation in water level—the range of the tide—is usually a meter or two. This range varies in different parts of the world; it is greatest in some Alaskan fjords and is most notable in the basin of the Bay of Fundy, between New Brunswick and Nova Scotia in southeast Canada, where tidal differences sometimes exceed 15 meters. This is largely due to the ocean floor, which funnels shoreward in a V-shape. The tide often comes in faster than a person can run. Don't dig clams near the water's edge at low tide in the Bay of Fundy!

Tides in the Earth and Atmosphere

The earth is not a rigid solid but, for the most part, is semi-molten liquid covered by a thin, solid, and pliable crust. As a result, the moon-sun tidal forces produce earth tides as well as ocean tides. Twice each day the solid crust of the earth rises and falls as much as one-quarter meter! As a result, earthquakes and volcanic eruptions have a slightly higher probability of occurring when the earth is experiencing an earth spring tide—that is, near a full or new moon.

We live at the bottom of an ocean of air that also experiences tides. Being at the bottom of the atmosphere, we don't notice these tides (just as a fish in deep water doesn't notice the ocean tides). In the upper part of the atmosphere is the ionosphere, so named because it contains many ions—electrically charged atoms that are the result of intense cosmic ray bombardment. Tidal effects in the ionosphere produce electric currents that alter the magnetic field that surrounds the earth. These are magnetic tides.

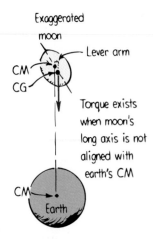

FIGURE 8.18 The earth's pull on the moon at its center of gravity produces a torque about the moon's center of mass, which tends to rotate the long axis of the moon into alignment with the earth's gravitational field (like a compass needle that aligns with a magnetic field). That's why only one side of the moon faces earth!

Gravitational Fields

FIGURE 8.19 Field lines represent the gravitational field about the earth. Where the field lines are closer together, the field is stronger. Farther away, where the field lines are farther apart, the field is weaker.

They in turn regulate the degree to which cosmic rays penetrate into the lower atmosphere. The cosmic ray penetration affects the ionic composition of our atmosphere, which in turn is evident in subtle changes in the behaviors of living things. The highs and lows of magnetic tides are greatest when the atmosphere is having its spring tides—again, near the full and new moon. Have you noticed that some of your friends seem a bit weird at the time of a full moon?

Tides on the Moon

There are two tidal bulges on the moon for the same reason there are two tidal bulges on the earth—near and far sides of each body are pulled differently. So the moon is pulled slightly away from a spherical shape into a football shape, with its long axis pointing toward the earth. But unlike earth's tides, the tidal bulges are stationary, with no "daily" rising and falling of moon tides. Since the moon takes 27.3 days to make a single revolution about its own axis (and also about the earth-moon axis), the same part of its surface faces the earth all the time. This is because the elongated moon's center of gravity is slightly displaced from its center of mass, so whenever the moon's long axis is not lined up toward earth (Figure 8.18), the earth exerts a small torque on the moon. This tends to twist the moon toward aligning with the earth's gravitational field, like the torque that aligns a compass needle with a magnetic field. So we see there is a reason why the moon always shows us its same face!

Interestingly enough, this "tidal lock" is also working on the earth. Our days are getting longer at the rate of 2 milliseconds per century. In a few billion years our day will be as long as a month and the earth will always show the same face to the moon. How about that?

The earth and the moon pull on each other. This is *action at a distance,* because the earth and the moon interact with each other even though they are not in contact. We can look at this in a different way: we can regard the moon as interacting with the **gravitational field** of the earth. The properties of the space surrounding any massive body can be considered to be altered in such a way that another massive body in this region will experience a force. This alteration of space is a gravitational field. We can think of a distant space probe being influenced by the gravitational field right where it is in space rather than by the earth and other planets or stars acting on it from a distance. The field concept plays an in-between role in our thinking about the forces between different masses.

A gravitational field is an example of a *force field,* for a body with any mass in the field space experiences a force. Another force field, perhaps more familiar, is a magnetic field. Have you ever seen iron filings lined up in patterns around a magnet. (Look ahead to Figure 23.2 on page 420, for example.) The pattern of the filings shows the strength and direction of the magnetic field at different points in the space around the magnet. Where the filings are closest together, the field is strongest. The direction of the filings shows the direction of the field at each point.

The pattern of the earth's gravitational field can be represented by field lines (Figure 8.19). Like the iron filings around a magnet, the field lines are closer together where the gravitational field is stronger. At each point on a field line, the direction of the field at that point is along the line. Arrows show the field direction. A particle, astronaut, spaceship, or any body in the vicinity of the earth will be accelerated in the direction of the field line at that location.

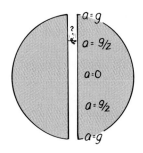

FIGURE 8.20 As you fall faster and faster in a hole bored completely through the earth, your acceleration decreases because the part of the earth's mass under you gets smaller and smaller. Less mass means less attraction, till at the center the net force is zero and acceleration is zero. Momentum carries you past the center and against a growing acceleration to the opposite end of the tunnel where acceleration is again g, directed back toward the center.

The strength of the earth's gravitational field, like the strength of its force on objects, follows the inverse-square law. It is strongest near the earth's surface and weakens with increasing distance from the earth.*

At the earth's surface the gravitational field varies slightly from location to location. Above large subterranean lead deposits, for example, the field is slightly stronger than average. Above large caverns, perhaps filled with natural gas, the field is slightly weaker. Geologists as well as mineral and oil prospectors make precise measurements of the earth's gravitational field to indicate what may lie beneath the surface.

The Gravitational Field Inside a Planet[†]

The gravitational field of the earth exists inside the earth as well as outside. Imagine a hole drilled completely through the earth from the north pole to the south pole. Forget about impracticalities such as lava and high temperatures and consider the motion you would undergo if you fell into such a hole. If you started at the north pole end, you'd fall and gain speed all the way down to the center then lose speed all the way "up" to the south pole. Without air drag, the one-way trip would take nearly 45 minutes. If you failed to grab the edge when you reach the south pole, you'd fall back toward the center, and return to the north pole in the same time.

Your acceleration, a, will be progressively less as you continue toward the center of the earth. Why? Because as you fall toward the earth's center, there is less mass pulling you toward the center. When you are at the center of the earth, the pull down is balanced by the pull up, so the net force on you is zero—$a = 0$ as you whiz with maximum speed past the center of the earth.[‡] The gravitational field of the earth at its center is zero!

The composition of the earth varies, being most dense at its core and least dense at the surface. Inside a hypothetical planet of uniform density, however, the field inside increases linearly, from zero at its center to g at the surface. We won't go into why this is so, but perhaps your instructor will provide the explanation. In any event, a plot of the gravitational field intensity inside and outside a solid planet of uniform density is shown in Figure 8.21.

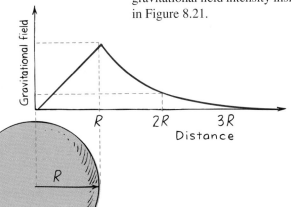

FIGURE 8.21 The gravitational field intensity inside a planet of uniform density is directly proportional to the radial distance from its center and is maximum at its surface. Outside, it is inversely proportional to the square of the distance from its center.

---

* The strength of the gravitational field g at any point is equal to the force F per unit of mass placed there. So $g = F/m$, and its units are newtons per kilogram (N/kg). The field g also equals the free-fall acceleration of gravity.

[†] This section may be skipped for a brief treatment of gravitational fields.

[‡] Interestingly enough, during the first few kilometers beneath the earth's surface you'd actually gain acceleration, because the density of surface material is much less than the condensed center. This results in a disproportionally greater pull when you are closer to the denser core material. So you'd weigh slightly more for the first few kilometers beneath the earth's surface. Farther in, your weight would decrease and would diminish to zero at the earth's center.

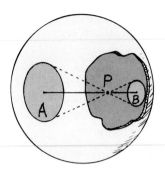

FIGURE 8.22 The gravitational field anywhere inside a spherical shell of uniform thickness and composition is zero, because the field components from all the particles of mass in the shell cancel one another. A mass at point P, for example, is attracted just as much to the larger but farther region A as it is to the smaller but closer region B.

Imagine a cavern at the center of a planet. The cavern would be gravity-free because of the cancellation of gravity in every direction. Amazingly, the size of the cavern doesn't change this fact—even if it constitutes most of the volume of the planet! A hollow planet, like a huge basketball, would have no gravitational field anywhere inside it. Complete cancellation of gravity occurs everywhere inside. To see why, consider the particle P in Figure 8.22, which is twice as far from the left side of the planet as it is from the right side. If gravity depended only on distance, P would be attracted only $\frac{1}{4}$ as much to the left side as to the right side (according to the inverse-square law). But gravity also depends on mass. Imagine a cone reaching out to the left from P to encompass region A in the figure, and a cone of equal angle reaching out to the right to encompass region B. Region A will have four times the area and therefore four times the mass of region B. Since $\frac{1}{4}$ of 4 is equal to 1, P is attracted to the farther but more massive region A with just as much force as it is to the closer but less massive region B. Cancellation occurs. More thought will show that cancellation will occur anywhere inside a planetary shell that has uniform thickness and composition. A gravitational field would exist at and beyond its outer surface and would behave as if all the mass of the planet were concentrated at its center—the center of gravity; but everywhere inside the hollow part, the gravitational field is zero. Anyone inside would feel weightless.

Questions

1. If you stepped into a hole bored clear through the center of the earth and made no attempt to grab the edges at either end, what kind of motion would you experience?
2. Halfway to the center of the earth, would you weigh more or weigh less than you weigh at the surface of the earth?

Although gravity can be canceled inside a body or between bodies, it cannot be shielded in the same way that electricity and magnetism can. We will see that electricity and magnetism have repelling as well as attracting forces that enable shielding, but gravitation only attracts and therefore cannot be shielded. Eclipses provide convincing evidence for this. The moon is in the gravitational field of both the sun and the earth. During a lunar eclipse the earth is directly between the moon and the sun, and any shielding of the sun's field by the earth would result in a deviation of the moon's orbit.

Answers

1. You would oscillate down and up. If the earth were an ideal sphere of uniform density (and if there were no air resistance!), your oscillation would be what is called *simple harmonic motion*. Each round trip would take nearly 90 min. Interestingly enough, we will see in the next chapter that an earth satellite in close orbit about the earth also takes the same 90 min to make a complete round trip. (This is no coincidence, for if you study physics further, you'll learn that "down and up" simple harmonic motion is simply the vertical component of uniform circular motion—interesting stuff.)
2. You would weigh less, because the part of the earth's mass under you is less and pulls you with less force. If the earth were a uniform sphere with uniform density, your weight halfway to the center would be exactly half your surface weight. But since the earth's core is so dense (about seven times the density of surface rock), your weight would be somewhat more than half surface weight. Exactly how much depends on how the earth's density varies with depth, information that is not known today.

Even a very slight shielding effect would accumulate over a period of years and show itself in the timing of subsequent eclipses. But there have been no such discrepancies; past and future eclipses are calculated to a high degree of accuracy using only the simple law of gravitation. No shielding effect in gravitation has ever been found.

Einstein's Theory of Gravitation

A model for gravity quite unlike Newton's was presented by Einstein in his general theory of relativity. Einstein perceived a gravitational field as a geometrical warping of four-dimensional space and time; he realized that bodies put dents in space and time somewhat like a massive ball placed in the middle of a large waterbed dents the two-dimensional surface (Figure 8.23). The more massive the ball, the greater the dent or warp. If we roll a marble across the top of the bed but away from the ball, the marble will roll in a straight-line path. But if we roll the marble near the ball, it will curve as it rolls across the indented surface of the waterbed. If the curve closes on itself, the marble will orbit the ball in an elliptical or circular path. From a Newtonian viewpoint, the marble curves because it is attracted to the ball. From an Einsteinian viewpoint, the marble curves not because of any force, but because the surface on which it moves curves.* In Chapter 35 we will treat Einstein's theory of gravitation in more detail.

FIGURE 8.23 Warped space-time. Space near a star is curved in four dimensions in a way similar to the two-dimensional surface of a waterbed when a heavy ball rests on it.

Black Holes

Suppose you were indestructible and could go in a spaceship to the surface of a star. Your weight would depend both on your mass and the star's mass and on the distance between the star's center and your belly button. If the star were to burn out and collapse to half radius with no change in its mass, your weight at its surface, determined by the inverse-square law, would be four times as much (Figure 8.24). If the star collapsed to a tenth its radius, your weight at its surface would be 100 times as much. If the star kept shrinking, the gravitational field at the surface would become stronger. It would be more and more difficult for a starship to leave. The velocity required to escape, *escape velocity,* would increase. If a star such as our sun collapsed to a radius of less than 3 kilometers, the escape velocity from its surface would exceed the speed of light, and

FIGURE 8.24 If a star collapses to half its radius and there is no change in its mass, gravitation at its surface is multiplied by 4.

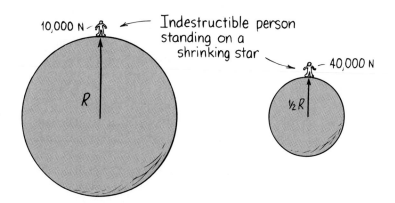

10,000 N — Indestructible person standing on a shrinking star — 40,000 N

R   ½R

---

* Don't be discouraged if you cannot visualize four-dimensional space-time. Einstein himself often told his friends, "Don't try. I can't do it either." Perhaps we are not too different from the great thinkers around Galileo who couldn't think of a moving earth!

nothing—not even light—could escape! The sun would be invisible. It would be a black hole.

The sun, in fact, probably has too little mass to experience such a collapse, but when some stars with greater mass—now estimated to be at least 1.5 solar masses or more—reach the end of their nuclear resources, they undergo collapse; and unless rotation is high enough, the collapse continues until the stars reach infinite densities. Gravitation near these shrunken stars is so enormous that light cannot escape from their vicinity. They have crushed themselves out of visible existence. The results are *black holes,* which are completely invisible.

A black hole is no more massive than the star from which it collapsed, so the gravitational field in regions at and greater than the original star's radius is no different after the star's collapse than before. But closer distances near the vicinity of a black hole are nothing less than the collapse of space itself, with a surrounding warp into which anything that passes too close—light, dust, or a spaceship—is drawn. Astronauts could enter the fringes of this warp and, with a powerful spaceship, still escape. After a certain distance, however, they could not, and they would disappear from the observable universe. Any object falling into a black hole would be torn to pieces. No feature of the object would survive except its mass, its angular momentum (if any), and its electric charge (if any).

FIGURE 8.25 Anything that falls into a black hole is crushed out of existence. Only mass, angular momentum, and electric charge are retained by the black hole.

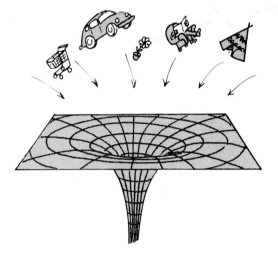

A theoretical entity with some similarity to a black hole is the "wormhole" (Figure 8.26). Like a black hole, a wormhole is an enormous distortion of space and time. But instead of collapsing toward an infinitely dense point, the wormhole opens out again in some other part of the universe—or even, conceivably, in some other universe! Whereas the existence of black holes has been confirmed through experiment, the wormhole remains speculative. But since the same laws of physics that explain the black hole also predict the possibility of wormholes, it won't be surprising if someday their existence is confirmed. Some physicists imagine that the wormhole opens the possibility of time travel.

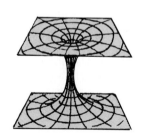

FIGURE 8.26 A wormhole may be the portal to another part of our universe or even to another universe.

How can a black hole be detected, if there is literally no way to "see" it? It makes itself felt by its gravitational influence on neighboring stars. There is now good evidence that some binary star systems consist of a luminous star and an invisible companion with black-hole-like properties orbiting around each other. Even stronger evidence for more massive black holes is observed in the central regions of some other galaxies.

There, stars are circling in a powerful gravitational field around an apparently empty center. These black holes may have masses more than one billion times the mass of our sun. Such a black hole possibly occupies the center of our own galaxy. Discoveries are coming faster than textbooks can report. Check with your astronomy instructor for the latest update.

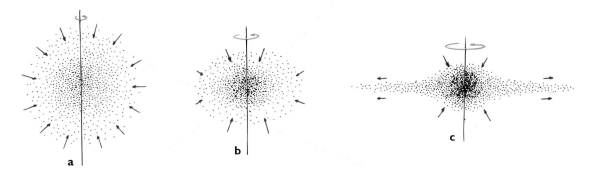

FIGURE 8.27 Solar system formation. A slightly rotating ball of interstellar gas (a) contracts due to mutual gravitation and (b) conserves angular momentum by speeding up. The increased momentum of individual particles and clusters of particles causes them to (c) sweep in wider paths about the rotational axis, producing an overall disk shape. The greater surface area of the disk promotes cooling and condensation of matter in swirling eddies—the birthplace of the planets.

Universal Gravitation

We all know that the earth is round. But why is the earth round? It is round because of gravitation. Everything attracts everything else, and so the earth has attracted itself together as far as it can! Any "corners" of the earth have been pulled in; as a result, every part of the surface is equidistant from the center of gravity. This makes it a sphere. Therefore, we see from the law of gravitation that the sun, the moon, and earth are spherical because they have to be (rotational effects make them slightly ellipsoidal).

If everything pulls on everything else, then the planets must pull on each other. The force that controls Jupiter, for example, is not just the force from the sun; there are also the pulls from the other planets. Their effect is small in comparison with the pull of the much more massive sun, but it still shows. When Saturn is near Jupiter, its pull disturbs the otherwise smooth ellipse traced by Jupiter. Both planets "wobble" about their expected orbits. The interplanetary forces causing this wobbling are called *perturbations*. By the 1840s, studies of the most recently discovered planet, Uranus, showed that the deviations of its orbit from a perfect ellipse could not be explained by perturbations from all other known planets. Either the law of gravitation was failing at this great distance from the sun or an unknown eighth planet was perturbing Uranus. An Englishman and a Frenchman, J. C. Adams and Urbain Leverrier, each assumed Newton's law to be valid and independently calculated where an eighth planet should be. At about the same time, Adams sent a letter to the Greenwich Observatory in England and Leverrier sent a letter to the Berlin Observatory in Germany, both suggesting that a certain area of the sky be searched for a new planet. The request by Adams was delayed by misunderstandings at Greenwich, but Leverrier's request was heeded immediately. The planet Neptune was discovered that very night!

Studies of the orbit of Neptune and further refinements in calculations of the orbit of Uranus led to the prediction and discovery of the ninth planet, Pluto, in 1930 at the Lowell Observatory in Arizona. Pluto takes 248 years to make a single revolution about the sun, so no one will see it in its discovered position again until the year 2178.

The shapes of distant galaxies provide further evidence that the law of gravitation applies to larger distances—even underlying the fate of the entire universe. Current scientific speculation is that the universe originated in the explosion of a primordial fireball some 8 to 15 billion years ago. This is the **Big Bang** theory of the origin of the universe. All the matter of the universe was hurled outward from this event and continues in an outward expansion. We find ourselves in an expanding universe. This expanding may go on indefinitely, or it may eventually be overcome by the combined gravitation of all the galaxies and come to a stop. Like a stone thrown upward, whose departure from the ground comes to an end when it reaches the top of its trajectory and which then begins its descent to the place of its origin, the universe may contract and fall back into a single unity. This would be the *Big Crunch*. After that, we can only speculate that the universe might re-explode to produce a new universe. The same course of action might repeat itself and the process may well be cyclic. If this speculation is true, we live in an oscillating universe.

If the universe does oscillate, who can say how many times this process has repeated? We know of no way a civilization could leave a trace of ever having existed, for all the matter in the universe would be reduced to bare subatomic particles or new entities during such an event. We can only speculate that formation of the elements, stars, galaxies, and life again takes place. All the laws of nature, such as the law of gravitation, might then be rediscovered by the higher-evolving life forms. Then students of these laws might read about them, as you are doing now. Think about that!

We do not know whether the expansion is indefinite because we are uncertain about whether enough mass exists to halt the expansion. If the expansion halts and is followed by contraction, the time from Big Bang to Big Crunch is estimated to be somewhat less than 100 billion years. Our universe is still young. But humankind is younger by far.

Few theories have affected science and civilization as much as Newton's theory of gravitation. The successes of Newton's ideas ushered in the so-called Age of Enlightenment, for Newton had demonstrated that by observation and reason and by employing mechanical models and deducing mathematical laws, people could uncover the very workings of the physical universe. How profound that all the moons and planets and stars and galaxies have such a beautifully simple rule to govern them; namely,

$$F = G \frac{m_1 m_2}{d^2}$$

The formulation of this simple rule is one of the major reasons for the success in science that followed, for it provided hope that other phenomena of the world might also be described by equally simple laws.

This hope nurtured the thinking of many scientists, artists, writers, and philosophers of the 1700s. One of these was the English philosopher John Locke, who argued that observation and reason, as demonstrated by Newton, should be our best judge and guide in all things and that all of nature and even society should be searched to discover any "natural laws" that might exist. Using Newtonian physics as a model of reason, Locke and his followers modeled a system of government that found adherents in the thirteen British colonies across the Atlantic. These ideas culminated in the Declaration of Independence and the Constitution of the United States of America.

Summary of Terms

Kepler's Laws of Planetary Motion

Law 1: Each planet moves in an elliptical orbit with the sun at one focus.

Law 2: The line from the sun to any planet sweeps out equal areas of space in equal time intervals.

Law 3: The squares of the times of revolution (days, months, or years) of the planets are proportional to the cubes of their average distances from the sun ($T^2 \sim R^3$ for all planets).

Law of universal gravitation Every body in the universe attracts every other body with a force that for two bodies is directly proportional to the product of their masses and inversely proportional to the square of the distance separating them:

$$F = G\frac{m_1 m_2}{d^2}$$

Inverse-square law A law relating the intensity of an effect to the inverse square of the distance from the cause:

$$\text{Intensity} \sim \frac{1}{\text{distance}^2}$$

Gravity follows the inverse-square law, as do the effects of electric, magnetic, light, sound, and radiation phenomena.

Weightlessness A condition encountered in free fall wherein a support force is lacking.

Spring tide A high or low tide that occurs when the sun, earth, and moon are all lined up so that the tides due to the sun and moon coincide, making the high tides higher than average and the low tides lower than average.

Neap tide A tide that occurs when the moon is midway between new and full, in either direction. Tides due to the sun and moon partly cancel, making the high tides lower than average and the low tides higher than average.

Gravitational field The influence that a massive body extends into the space around itself, producing a force on another massive body. It is measured in newtons per kilogram (N/kg).

Black hole A concentration of mass resulting from gravitational collapse, near which gravity is so intense that not even light can escape.

Big Bang The primordial explosion that is thought to have resulted in the expanding universe.

Suggested Reading

Einstein, A., and L. Infeld. *The Evolution of Physics.* New York: Simon & Schuster, 1938.

Gamow, G. *Gravity.* Science Study Series. Garden City, NY: Doubleday (Anchor), 1962.

Smoot, George, and Keay Davidson. *Wrinkles in Time.* New York: Morrow, 1993. A leading cosmologist's personal account of the discovery of tiny wrinkles in spacetime left over from the Big Bang.

Review Questions

Copernicus, Brahe, and Kepler

1. Who measured the positions of the planets so accurately that the data could be used to find details of the planets' motion? Who used the data to show that planets travel in elliptical orbits around the sun? Who explained these elliptical orbits?

Kepler's Laws

2. What relationship between the speed of planets and their distance from the sun did Kepler discover?

3. What relationship between the times of revolutions of the planets and their distance from the sun did Kepler discover?

4. In Kepler's thinking, what was the direction of force on a planet?

Newton's Law of Universal Gravitation

5. In Newton's thinking, what was the direction of force on a planet?

6. In Newton's insight, what did a falling apple have in common with the moon?

7. How can the moon "fall" without getting closer to the earth?

8. How does the force of gravity between two objects depend on their masses?

9. How does the force of gravity depend on the distance between two objects?

The Universal Gravitational Constant, G

10. How was *G* measured?

11. What is the magnitude of the gravitational force between the earth and a 1-kg body? (Hint: You don't have to use the law of universal gravitation to figure this out.)

12. What is the magnitude of the gravitational force between you and the earth?

Gravity and Distance: The Inverse-Square Law

13. If the earth were relocated to an orbit only half as far from the sun as its present orbit, how would the force of the sun on the earth change?

14. If the earth were relocated as in Question 13, how would the force of the earth on you change?

15. Can you escape from the earth's gravity by getting above the atmosphere? By going to the moon? Defend your answers.

Weight and Weightlessness

16. Would the springs inside a bathroom scale be more compressed or less compressed if you weighed yourself in an elevator that accelerated upward? Downward?

17. Would the springs inside a bathroom scale be more compressed or less compressed if you weighed yourself in an elevator that moved upward at constant velocity? Downward at constant velocity?

18. Can you have weight if you are in a spacecraft far removed from sources of gravitational force? Explain.

Ocean Tides

19. Does the moon circle the earth, does the earth circle the moon, or do both circle each other? Explain.

20. Why do both the sun and the moon exert a greater gravitational force on one side of the earth than on the other?

21. Do tides depend more on the strength of gravitational pull or on the *difference* in strengths? Explain.

22. (Fill in the blank.) Gravitational force (units N) depends on the inverse square of distance. Tidal force, which is the difference in gravitational forces per unit mass (units N/kg), depends on the inverse _____ of distance.

23. Distinguish between *spring tides* and *neap tides*.

Tides in the Earth and Atmosphere

24. How do the sizes of tides compare for the ocean and the "solid" earth?

25. What are *magnetic tides*?

26. Why are all tides greatest at the time of a full moon or new moon?

Tides on the Moon

27. Does the fact that one side of the moon always faces the earth mean the moon rotates about its axis (like a top), or doesn't rotate about its axis?

28. Why is there a torque about the moon's center of mass when the moon's long axis is not aligned with the earth's gravitational field?

29. Is there a torque about the moon's center of mass when the moon's long axis is aligned with the earth's gravitational field? Explain.

Gravitational Fields

30. What is a gravitational field, and how can its presence be detected?

The Gravitational Field Inside a Planet

31. What is the magnitude of the gravitational field at the center of the earth?

32. What would be the magnitude of the gravitational field halfway to the center of a planet with a uniform density, relative to the field at the planet's surface?

33. What would be the magnitude of the gravitational field anywhere inside a hollow planet?

34. We will see in later chapters that electricity and magnetism can be shielded. Why cannot gravity be shielded?

Einstein's Theory of Gravitation

35. Newton said that the path of a planet curves because a force acts on it. Why does a planet's path curve according to Einstein?

Black Holes

36. If the earth shrank with no change in its mass, what would happen to your weight at the surface?

37. What happens to the strength of the gravitational field at the surface of a star that shrinks?

38. How can a black hole be detected if it cannot emit light?

Universal Gravitation

39. Why is the earth "round"? Why isn't it *exactly* spherical?

40. Distinguish between the *Big Bang* and the *Big Crunch*.

Projects

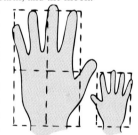

1. Hold up your thumb and first two fingers and make a V sign. Place a strong rubber band across your thumb and first finger. This represents the force of gravity between the sun and the earth. Place a medium-strength rubber band across your thumb and second finger to represent the force of gravity between the sun and moon. Then place a weak rubber band across your first two fingers to represent the force of gravity between the moon and earth. Note how all fingers pull on each other. Likewise for the gravitational pulls between the sun, earth, and the moon.

2. Hold your hands outstretched, one twice as far from your eyes as the other, and make a casual judgment as to which hand looks bigger. Most people see them to be about the same size, while many see the nearer hand as slightly bigger. Almost nobody upon casual inspection sees the nearer hand as four times as big. But by the inverse-square law, the nearer hand should appear twice as tall and twice as wide and therefore occupy four times as much of your visual field as the farther hand. Your belief that your hands are the same size is so strong that you likely overrule this information. Now if you overlap your hands slightly and view them with one eye closed, you'll see the nearer hand as clearly bigger. This raises an interesting question: What other illusions do you have that are not so easily checked?

Exercises

1. Comment on whether or not this label on a consumer product should be cause for concern. *CAUTION: The mass of this product pulls on every other mass in the universe, with an attracting force that is proportional to the product of the masses and inversely proportional to the square of the distance between them.*

2. Gravitational force acts on all bodies in proportion to their masses. Why, then, doesn't a heavy body fall faster than a light body?

3. Which weighs more, a sheet of aluminum foil or the same sheet crumpled into a tight wad?

4. The sun pulls on the moon with about twice the force of the earth's pull on it. Why, then, isn't the moon pulled away from the earth?

5. Somewhere between the earth and the moon, gravity from these two bodies on a space pod would cancel. Is this location nearer the earth or the moon?

6. Which planets, those closer to the sun than the earth or those farther from the sun than the earth, have a period greater than 1 earth year?

7. When a planet is at its farthest point from the sun, it moves more slowly than when it is at its closest point to the sun. (a) How does this fact relate to Kepler's second law (the law of areas)? (b) How does this fact relate to the law of angular momentum conservation?

8. What are the magnitude and direction of the gravitational force that acts on a man who weighs 700 N at the surface of the earth?

9. The weight of an apple near the surface of the earth is 1 N. What is the weight of the earth in the gravitational field of the apple?

10. The earth and the moon are attracted to each other by gravitational force. Does the more massive earth attract the less massive moon with a force that is greater, smaller, or the same as the force with which the moon attracts the earth? (With an elastic band stretched between your thumb and forefinger, which is pulled more strongly by the band, your thumb or your forefinger?)

11. If the mass of the earth somehow increased and nothing else changed, would your weight also increase? (*Hint*: Let the equation for gravitational force guide your thinking.)

12. The intensity of light from a central source varies inversely as the square of the distance. If you lived on a planet only half as far from the sun as our earth, how would the light intensity compare with that on earth? How about a planet ten times as far away as the earth?

13. A small light source located 1 m in front of a 1-m² opening illuminates a wall behind. If the wall is 1 m behind the opening (2 m from the light source), the illuminated area covers 4 m². How many square meters will be illuminated if the wall is 3 m from the light source? 5 m? 10 m?

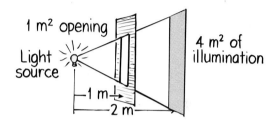

14. If you want to make a profit in buying precious material by weight at one altitude and selling it at another altitude for the same price per body weight, should you buy or sell at the higher altitude? (Assume weighing is done on a spring scale.)

15. The planet Jupiter is more than 300 times as massive as the earth, so it might seem that a body on the surface of Jupiter would weigh 300 times as much as it does on the earth. But it so happens that a body would scarcely weigh three times as much on the surface of Jupiter as it would on the surface of the earth. Can you think of an explanation for why this is so? (*Hint*: Let the terms in the equation for gravitational force guide your thinking.)

16. From the data in the preceding exercise, estimate the radius of Jupiter relative to the radius of the earth.

17. Why do the passengers of high-altitude jet planes feel the sensation of weight while passengers in an orbiting space vehicle such as the space shuttle do not?

18. Imagine a skydiver falling at terminal speed. Is she weightless? How about a scuba diver who is hovering motionless in the water? How about a "space-walking" astronaut hovering near his spacecraft? An astronaut within an accelerating spacecraft in deep space?

19. Why is a person in free fall weightless, while a person falling at terminal velocity is not?

20. If you were in a car that drove off the edge of a cliff, why would you be momentarily weightless? Would gravity still be acting on you?

21. If you were in a freely falling elevator and you dropped a pencil, you'd see the pencil hovering. Is the pencil falling? Explain.

22. Explain why the following reasoning is wrong. "The sun attracts all bodies on the earth. At midnight, when the sun is directly below, it pulls on an object in the same direction as the pull of the earth on that object; at noon, when the sun is directly overhead it pulls on an object in a direction opposite to the pull of the earth. Therefore, all objects should be somewhat heavier at midnight than they are at noon." (*Hint*: Relate this to the preceding two exercises.)

23. If the mass of the earth increased, your weight would correspondingly increase. But if the mass of the sun increased, your weight would not be affected at all. Why?

24. Most people today know that the ocean tides are caused principally by the gravitational influence of the moon. And most people, therefore, think that the gravitational pull of the moon on the earth is greater than the gravitational pull of the sun on the earth. What do you think?

25. If somebody tugged on your shirt sleeve, it would likely tear. But if all parts of your shirt were tugged equally, no tearing would occur. How does this relate to tidal forces?

26. Would ocean tides exist if the gravitational pull of the moon (and sun) were somehow equal on all parts of the world? Explain.

27. Why aren't high ocean tides exactly 12 h apart?

28. With respect to spring and neap ocean tides, when are the lowest tides? That is, when is it best for digging clams?

29. Whenever the ocean tide is unusually high, will the following low tide be unusually low? Defend your answer in terms of "conservation of water." (If you slosh water in a tub so it is extra deep at one end, will the other end be extra low?)

30. The Mediterranean Sea has very little sediment churned up and suspended in its waters, mainly because of the absence of any substantial ocean tides. Why do you suppose the Mediterranean Sea has practically no tides? Similarly, are there tides in the Black Sea? Great Salt Lake? Your county reservoir? A glass of water? Explain.

31. The human body is composed of mostly water. Why does the gravitational pull of the moon overhead cause appreciably less biological tides in the fluid compartment of the body than the gravitational pull of a 1-kg melon held over your head?

32. If the moon didn't exist, would the earth still have ocean tides? If so, how often?

33. What would be the effect on the earth's tides if the diameter of the earth were very much larger than it is? If the earth were as it presently is, but the moon very much larger with the same mass?

34. Does the strongest tidal force on our bodies come from the earth, the moon, or the sun?

35. Since a "tidal bulge" sweeps around the earth in approximately 24 hours, does this mean that the incoming tide at the seashore comes in faster than a jet plane (which can almost beat the time zones)?

36. Exactly why do tides occur in the earth's crust and in the earth's atmosphere?

37. The value of g at the earth's surface is about 9.8 m/s^2. What is the value of g at a distance of twice the earth's radius?

38. If the earth were of uniform density (same mass/volume throughout), what would the value of g be inside the earth at half its radius?

39. If the earth were of uniform density, would your weight increase or decrease at the bottom of a deep mine shaft? Defend your answer.

40. It so happens that an actual *increase* in weight is found even in the deepest mine shafts. What does this tell us about how the earth's density changes with depth?

41. Which requires more fuel—a rocket going from the earth to the moon or a rocket coming from the moon to the earth? Why?

42. If you could somehow tunnel inside a star, would your weight increase or decrease? If, instead, you somehow stood on the surface of a shrinking star, would your weight increase or decrease? Why are your answers different?

43. If our sun shrank in size to become a black hole, show from the gravitational force equation that the earth's orbit would not be affected.

44. If the earth were hollow but still had the same mass and same radius, would your weight in your present location be more, less, or the same as it is now? Explain.

45. Make up two multiple-choice questions: One that would check a classmate's understanding of the inverse-square law, and another that would check a distinction between weight and weightlessness.

46. Some people dismiss the validity of scientific theories by saying they are "only" theories. The law of universal gravitation is a theory. Does this mean that scientists still doubt its validity? Explain.

47. We say on page 156 that gravity cannot be shielded and on the same page that gravitational components cancel to zero inside a uniform shell. Why are these two statements not contradictory?

48. Strictly speaking, you weigh a tiny bit less when you are in the lobby of a massive skyscraper than you do at home. Why is this so?

49. A person falling into a black hole would probably be killed by tidal forces before ever encountering the hole itself. Explain why this is so.

50. Ultimately, the universe may expand without limit, or it may coast to a stop, or it may turn around and collapse to a "big crunch." What is the single most important quantity that will determine which of these fates is in store for the universe?

Problems

1. The value of g at the earth's surface is about 9.8 m/s². What is the value of g at a distance from the earth's center that is 4 times the earth's radius?

2. Show by algebraic reasoning that your gravitational acceleration toward an object of mass M a distance d away is $a = GM/d^2$.

3. The mass of a certain neutron star is 3.0×10^{30} kg (1.5 solar masses) and its radius is 8000 m (8 km). What is the acceleration of gravity at the surface of this condensed, burned-out star?

4. Many people mistakenly believe that the astronauts who orbit the earth are "above gravity." Calculate g for space shuttle territory, 200 kilometers above the earth's surface (dashed line in sketch). Earth's mass is 6×10^{24} kg, and its radius is 6.38×10^6 m (6380 km). Your answer is what percentage of 9.8 m/s²?

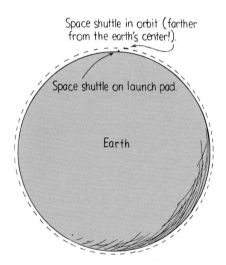

5. A 3-kg newborn baby at the earth's surface is gravitationally attracted to earth with a force of about 30 N. (a) Calculate the force of gravity with which the baby on earth is attracted to the planet Mars when Mars is closest to earth. (The mass of Mars is 6.4×10^{23} kg and its closest distance is 5.6×10^{10} m). (b) Calculate the force of gravity between the baby and the physician who delivers it. Assume the physician has a mass of 100 kg and is 0.5 m from the baby. (c) How do the forces compare?

6. Calculate the force of gravity between the earth (mass = 6×10^{24} kg) and the sun (mass = 2×10^{30} kg, distance = 1.5×10^{11} m).

7. To better comprehend the magnitude of the gravitational force between the earth and the sun, pretend gravity is turned off and the pull replaced by the tension in a steel cable joining them. How thick would such a cable need to be? You can estimate the diameter by knowing the tensile strength of steel cable is about 5.0×10^8 N/m² (each square meter cross section can support a 5.0×10^8-newton force).

8. The difference in force per mass (N/kg) across a body, tidal force T_F, is approximated by $T_F = 4GMR/d^3$, where G is the gravitational constant, M the mass of the body causing tides, R the radius of the body having tides, and d the center-to-center distance between the bodies. Make two calculations: (a) T_F that the moon exerts on you, and (b) T_F that a 1-kg melon 1 m above your head exerts on you. For simplicity, let R for you be 1 m (pretend you're 2 m tall). Let d to the moon be 3.8×10^8 m, and to the melon 2 m. Mass M of the moon is 7.4×10^{22} kg. (c) After you have made your calculations, compare them. Which is larger, and by how much? Then share this information with any friends who say that the tidal forces of planets on people influence their lives!

9

. . . .

Satellite Motion

: The astronaut falls around, rather than into, the earth.

Not too long ago, perhaps in the heyday of your grandparents, it was a fanciful and far-fetched idea that humans would soon be in comfortable spaceships high above the atmosphere orbiting the planet earth. As recently as 1969, when the first edition of this book was being written, humans first reached the moon. Today the news you see on TV from around the world is beamed to and from satellites. As you're reading this, astronauts and cosmonauts may be circling you. We are preparing for space exploration that will take us who knows where. Our spacefaring accomplishments, amazing by present-day standards yet perhaps quaint and antiquated by tomorrow's, had their beginnings back in 1665 on a farm in Woolsthorpe, England. It was there that Isaac Newton thought of universal gravitation and its role in the motions of the moon, planets, and satellites.

The Falling Apple

Perhaps the sight of a falling apple triggered Newton's idea that the pull of the earth on the apple extends also to the moon. The young Newton—still in his early 20s—wanted to explain the fact that the moon does not follow a straight-line path, but instead circles about the earth. He had the concept of inertia developed earlier by Galileo: without an outside force, a moving object continues its motion at constant speed in a straight line. He knew that if an object undergoes a change in speed or direction, then a force is responsible. Newton had the insight to see that the force that pulled the apple may extend to the moon and pull it into a circular path about the earth.

The Falling Moon

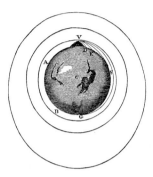

FIGURE 9.1   "The greater the velocity . . . with which (a stone) is projected, the farther it goes before it falls to the earth. We may therefore suppose the velocity to be so increased, that it would describe an arc of 1, 2, 5, 10, 100, 1000 miles before it arrived at the earth, till at last, exceeding the limits of the earth, it should pass into space without touching." —Isaac Newton, *System of the World.*

In the previous chapter we saw that the moon falls beneath the straight-line path it would follow if no force acted on it. Newton hypothesized that the moon is simply a projectile circling the earth under the attraction of gravity. This concept is illustrated in a drawing by Newton, shown in Figure 9.1. He compared the motion of the moon to that of a cannonball fired from the top of a high mountain. He imagined that the mountain top was above the earth's atmosphere, so air resistance would not slow the motion of the ball. If a cannonball were fired with a small horizontal speed, it would follow a parabolic path and soon hit the earth below. If it were fired faster, its path would be less curved and it would hit the earth farther away. If the cannonball were fired fast enough, Newton reasoned, the parabolic path would become a circle and the cannonball would circle indefinitely. It would be in orbit. So Newton visualized satellites nearly 300 years before they became reality.

Both cannonball and moon have "sideways" velocity, or *tangential velocity*—the velocity parallel to the earth's surface—sufficient to ensure motion *around* the earth rather than *into* it. Without air resistance to reduce its speed, the moon "falls" around and around the earth indefinitely.

For the idea to advance from hypothesis to theory, it would have to be tested. Newton's test was to see if the moon's fall beneath its otherwise straight-line path was in correct proportion to the fall of an apple or any object near the earth's surface. He reasoned that the mass of the moon should not affect how it falls, just as mass has no effect on the acceleration of freely falling objects on earth. How far the moon falls and how far an apple near the earth's surface falls should relate only to their respective distances from the earth's center. If the distance of fall for the moon and the apple are in correct proportion, then the hypothesis that earth gravity reaches to the moon must be taken seriously.

The distance from the center of the earth to the center of the moon was known to be 60 times the distance from the center of the earth to the center of the apple close to the earth's surface. The apple will fall nearly 5 meters (4.9 meters) in its first second of fall. Newton reasoned from Kepler's laws that gravitational attraction to the earth must

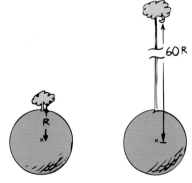

FIGURE 9.2   An apple falls 4.9 m during its first second of fall when it is near the earth's surface. Newton asked how far the moon would fall in the same time if it were 60 times as far from the center of the earth. His answer was (in today's units) 1.4 mm.

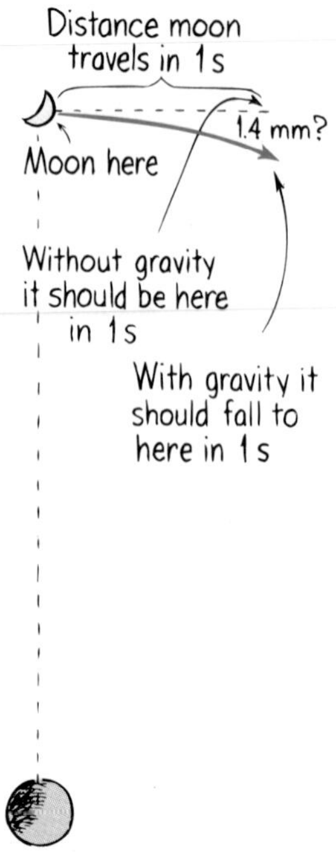

Distance moon
travels in 1s

1.4 mm?

Moon here

Without gravity
it should be here
in 1s

With gravity it
should fall to
here in 1s

FIGURE 9.3 If the force that pulls apples off trees also pulls the moon into orbit, the circle of the moon's orbit should fall 1.4 mm below a point along the straight line where the moon would otherwise be one second later.

be weakened by the inverse square of distance. So if the moon is 60 times as far away, its fall toward earth should be $1/(60)^2$ the fall of an apple near the earth's surface. In one second the moon should fall $1/(60)^2$ of 4.9 meters, which is 1.4 millimeters.*

Newton confirmed his hypothesis. Every second the moon falls 1.4 millimeters beneath the tangent line it would follow without gravity. Newton concluded that the earth and planets orbit the sun in the same way that the moon orbits the earth. The planets continually fall around the sun in closed paths. Why don't the planets crash into the sun? They don't because of their tangential velocities. What would happen if their tangential velocities were reduced to zero? The answer is simple enough: their motion would be straight toward the sun and they would indeed crash into it. Any objects in the solar system with insufficient tangential velocities have long ago crashed into the sun. What remains is the harmony we observe.

Satellite Motion

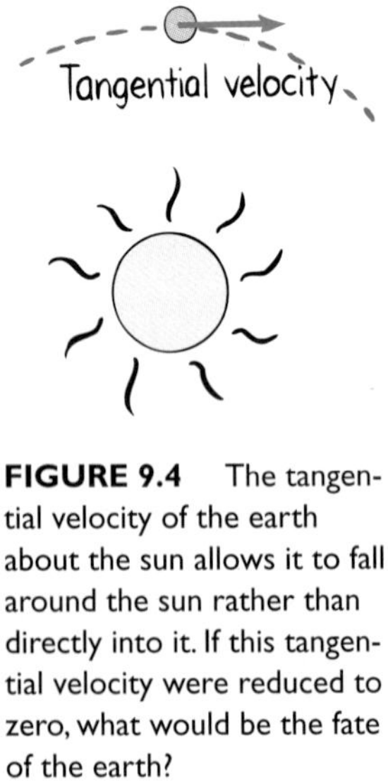

Tangential velocity

FIGURE 9.4 The tangential velocity of the earth about the sun allows it to fall around the sun rather than directly into it. If this tangential velocity were reduced to zero, what would be the fate of the earth?

Circular Orbits

Recall from Chapter 3 that the tangential speed a projectile needs to orbit the earth is 8 kilometers per second. In each second the projectile goes 8 kilometers horizontally (tangential), and falls 4.9 meters vertically (radial)—the same distance the earth curves for each 8-kilometer tangent. So an 8-kilometer-per-second cannonball fired horizontally from Newton's mountain would follow the earth's curvature and coast around the earth again and again (provided the cannoneer and the cannon got out of the way). Fired slower, the cannonball would strike the earth's surface; fired faster it would overshoot a circular orbit as we will discuss shortly. Newton calculated the speed for circular orbit, and since such a cannon-muzzle velocity was clearly impossible, he did not foresee people launching satellites (he did not foresee multistage rockets).

Note that in circular orbit the speed of a satellite is not changed by gravity; only the direction changes. Compare a satellite in circular orbit with a bowling ball rolling along a bowling alley. Why doesn't the gravitational force acting on the bowling ball change its speed? Because gravity is not pulling forward or backward; gravity pulls straight downward. There is no component of gravitational force in the direction of motion (Figure 9.5).

The same is true for a satellite in circular orbit; it is always moving perpendicularly to the earth's gravitational field. It does not move in the direction of the field, which would increase its speed, nor does it move in a direction against the field, which would

*Or, working backwards, $(.0014 \text{ m})(60^2) = 4.9 \text{ m}$.

FIGURE 9.5 (a) The force of gravity on the bowling ball is at 90° to its direction of motion, so it has no component of force to pull it forward or backward, and the ball rolls at constant speed. (b) The same is true even if the bowling alley is larger and remains "level" with the curvature of the earth. (c) If the ball moves at 8 km/s with no air resistance, would it need the alley?

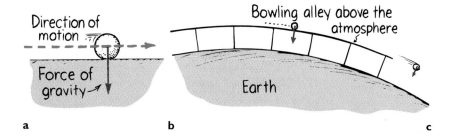

a   b   c

decrease its speed. Instead, the satellite moves at right angles to the earth's gravitational field, so no change in speed occurs—only change in direction. A satellite in circular orbit coasts at constant speed parallel to the earth's surface—a special form of free fall.

Questions Consider a ball rolling along a bowling alley that completely circles the earth, and is elevated high enough so air resistance can be neglected.

1. Why would the force of gravity not change the speed of the ball?
2. If a section of the alley were cut away to leave a large gap, how fast must the ball travel to clear the gap and continue its motion as usual? At this speed, what would be the maximum gap for unchanged motion?

For a satellite close to the earth, the period (the time for a complete orbit about the earth) is about 90 minutes. For higher altitudes, the orbital speed is less and the period is longer. For example, communication satellites located in orbit 5.5 earth radii above the surface of the earth have a period of 24 hours. This period matches the period of daily earth rotation. For an orbit around the equator, these satellites stay above the same point on the ground. The moon is even farther away and has a period of 27.3 days. The higher the orbit of a satellite, the less its speed and the longer its period.*

Answers

1. A change in speed requires a force or component of force along the direction of travel. In this case the alley and gravitational force on the ball are everywhere perpendicular to each other, so there is no force component in the direction of motion.
2. To clear the gap without bumping into the edge of the alley, the ball must have orbital speed. Then its curved path matches that of the alley's surface. In this case the alley can be completely removed to have in effect a 360° gap, because the ball would be in earth orbit anyway!

*The speed of a satellite in circular orbit is given by $v = \sqrt{d/GM}$ and the period of satellite motion is given by $T = 2\pi \sqrt{GM/d^3}$, where G is the universal gravitational constant (see Chapter 8), M is the mass of the earth (or whatever body the satellite orbits), and d is the altitude of the satellite measured from the center of the earth or parent body.

Questions

1. One of the beauties of physics is that there are usually different ways to view and explain a given phenomenon. Is the following explanation valid? Satellites remain in orbit instead of falling to the earth because they are beyond the main pull of earth's gravity.

2. Satellites in close circular orbit fall about 4.9 m during each second of orbit. Why doesn't this distance accumulate and send satellites crashing into the earth's surface?

Elliptical Orbits

If a projectile just above the drag of the atmosphere is given a horizontal speed somewhat greater than 8 kilometers per second, it will overshoot a circular path and trace an oval path called an **ellipse**.

An ellipse is a specific curve; the closed path taken by a point that moves in such a way that the sum of its distances from two fixed points (called *foci*) is constant. For a satellite orbiting a planet, one focus is at the center of the planet; the other focus could be inside or outside of the planet. An ellipse can be easily constructed by using a pair of tacks, one at each focus, a loop of string, and a pencil (Figure 9.6). The closer the foci are to each other, the closer the ellipse is to a circle. When both foci are together, the ellipse is a circle. So we see that a circle is a special case of an ellipse.

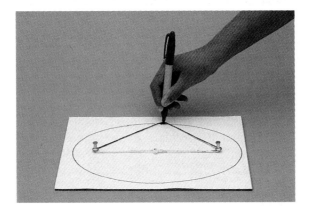

FIGURE 9.6 A simple method for constructing an ellipse.

Answers

1. No, no, a thousand times no! If any moving object were beyond the pull of gravity, it would move in a straight line and would not curve around the earth. Satellites remain in orbit because they are being pulled by gravity, not because they are beyond it. For the altitudes of most earth satellites, the earth's gravitational field is only a few percent weaker than at the earth's surface.

2. In each second, the satellite falls about 4.9 m below the straight-line tangent it would have taken if there were no gravity. The earth's surface also curves 4.9 m beneath a straight-line 8-km tangent. The process of falling with the curvature of the earth continues from tangent line to tangent line, so the curved path of the satellite and the curve of the earth's surface "match" all the way around the earth. Satellites in fact do crash to the earth's surface from time to time, but this is principally because they encounter air re-sistance in the upper atmosphere that decreases their orbital speed.

FIGURE 9.7 The shadows cast by the ball are all ellipses, one for each lamp in the room. The point at which the ball makes contact with the table is the common focus of all three ellipses.

Whereas the speed of a satellite is constant in a circular orbit, speed varies in an elliptical orbit. When the initial speed is greater than 8 kilometers per second, the satellite overshoots a circular path and moves away from the earth, against the force of gravity. It therefore loses speed. Like a rock thrown into the air, it slows to a point where it no longer recedes and then begins to fall back toward the earth. The speed it loses in receding is regained as it falls back toward the earth, and it finally rejoins its original path with the same speed it had initially (Figure 9.8). The procedure repeats over and over, and an ellipse is traced each cycle.

FIGURE 9.8 Elliptical orbit. An earth satellite that has a speed somewhat greater than 8 km/s overshoots a circular orbit (a) and travels away from the earth. Gravitation slows it to a point where it no longer moves farther from the earth (b). It falls toward the earth gaining the speed it lost in receding (c) and follows the same path as before in a repetitious cycle.

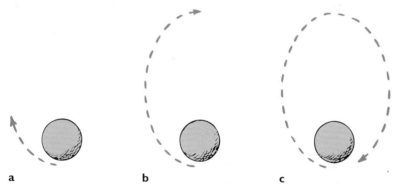

a   b   c

Interestingly enough, the parabolic path of a projectile such as a tossed baseball or a cannonball (Chapter 3) is actually a tiny segment of a thin ellipse that extends within and just beyond the center of the earth (Figure 9.9a). In Figure 9.9b, we see several paths of cannonballs fired from Newton's mountain. All these ellipses have the center of the earth as one focus. As muzzle velocity is increased, the ellipses are less eccentric (more nearly circular); and when muzzle velocity reaches 8 kilometers per second, the ellipse rounds into a circle and does not intercept the earth's surface. The cannonball

FIGURE 9.9 (a) The parabolic path of the cannonball is part of an ellipse that extends within the earth. The earth's center is the far focus. (b) All paths of the cannonball are ellipses. For less than orbital speeds, the center of the earth is the far focus; for circular orbit, both foci are the earth's center; for greater speeds, the near focus is the earth's center.

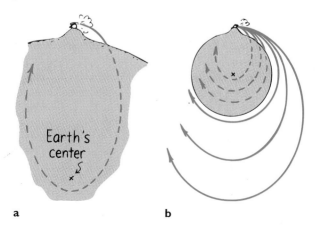

a   b

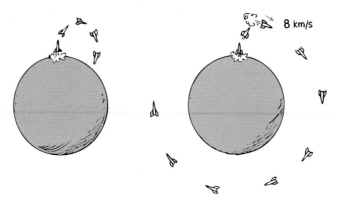

FIGURE 9.10   The initial thrust of the rocket pushes it up above the atmosphere. Another thrust to a horizontal speed of at least 8 km/s is required if the rocket is to fall around rather than into the earth.

coasts in circular orbit. At greater muzzle velocities, the orbiting cannonball traces the familiar external ellipse.

Putting a payload into earth orbit requires control over the speed and direction of the rocket that carries it above the atmosphere. A rocket initially fired vertically is intentionally tipped from the vertical course; then, once above the drag of the atmosphere, it is aimed *horizontally,* whereupon the payload is given a final thrust to 8 kilometers per second or more. This is shown in Figure 9.10, where for the sake of simplicity the payload is the entire single-stage rocket. We see that with the proper tangential velocity it falls around the earth, rather than into it, and becomes an earth satellite.

Question The orbital path of a satellite is shown in the sketch. In which of the marked positions A through D does the satellite have the greatest speed? Lowest speed?

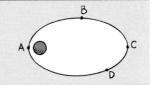

Energy Conservation and Satellite Motion

Recall from Chapter 6 that an object in motion possesses kinetic energy (KE) by virtue of its motion. An object above the earth's surface possesses potential energy (PE) by virtue of its position. Everywhere in its orbit, a satellite has both KE and PE. The sum of the KE and PE will be a constant all through the orbit. The simplest case occurs for a satellite in circular orbit.

In a circular orbit the distance between the body's center and the satellite does not change, which means the PE of the satellite is the same everywhere in orbit. Then, by

Answer The satellite has its greatest speed as it whips around A and has its lowest speed at position C. After passing C, it gains speed as it falls back to A to repeat its cycle.

the conservation of energy, the KE must also be constant. So a satellite in circular orbit coasts at an unchanging PE, KE, and speed (Figure 9.11).

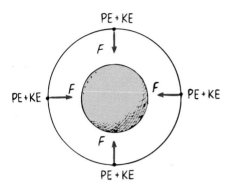

FIGURE 9.11 The force of gravity on the satellite is always toward the center of the body it orbits. For a satellite in circular orbit, no component of force acts along the direction of motion. The speed, and thus the KE, do not change.

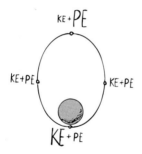

FIGURE 9.12 The sum of KE and PE for a satellite is a constant at all points along its orbit.

In an elliptical orbit the situation is different. Both speed and distance vary. PE is greatest when the satellite is farthest away (at the *apogee*) and least when the satellite is closest (at the *perigee*). Note that the KE will be least when the PE is most, and the KE will be most when the PE is least. At every point in the orbit, the sum of KE and PE is the same (Figure 9.12).

At all points along the orbit, except at the apogee and perigee, there is a component of gravitational force parallel to the direction of motion of the satellite. The component of force in the direction of motion changes the speed of the satellite. Or we can say that

$$\text{(This component of force)} \times \text{(distance moved)} = \Delta\text{KE}.$$

Either way, when the satellite gains altitude and moves against this component, its speed and KE decrease. The decrease continues to the apogee. Once past the apogee, the satellite moves in the same direction as the component, and the speed and KE increase. The increase continues until the satellite whips past the perigee and repeats the cycle.

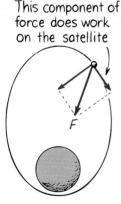

FIGURE 9.13 In elliptical orbit, a component of force exists along the direction of the satellite's motion. This component changes the speed and, thus, the KE. (The perpendicular component changes only the direction.)

World Monitoring by Satellite

Figure 9.14 shows the path traced in one period by a satellite in circular orbit launched in a northeastern direction from Cape Canaveral, Florida. The path is curved only because the map is flat. Note that the path crosses the equator twice in one period, for the path describes a circle whose plane passes through the earth's center. Note also that the path does not end where it begins. This is because the earth rotates beneath the satellite while it orbits. During the 90-minute period, the earth turns 22.6 degrees, so when the satellite makes a com-

plete orbit it begins its new sweep many kilometers to the west (about 2500 km at the equator). This turns out to be quite advantageous for earth-monitoring satellites. Figure 9.15 shows the area monitored over a week by successive sweeps for a typical satellite. A dramatic but typical example of such monitoring is the three-year worldwide watch of the distribution of ocean phytoplankton (Figure 9.16). Such extensive information would have been impossible to acquire before the advent of satellites.

FIGURE 9.14 The path of a typical satellite launched in a northeasterly direction from Cape Canaveral. Because the earth rotates while the satellite orbits, each sweep passes overhead some 2100 km farther west at the latitude of Cape Canaveral.

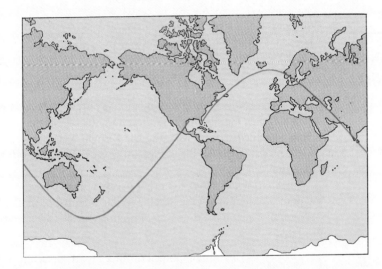

FIGURE 9.15 Typical sweep pattern for a satellite over a week.

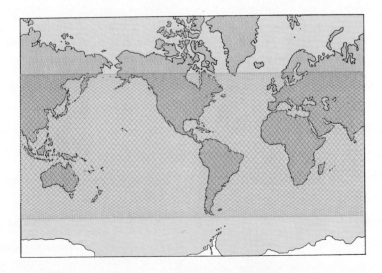

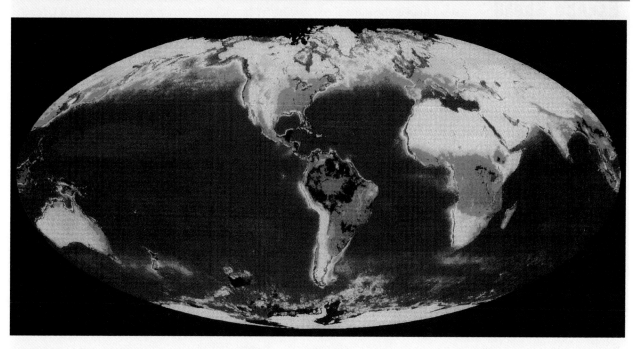

FIGURE 9.16 Colored land areas indicate different vegetation types, while the oceans are colored to indicate variations in surface chlorophyll content.

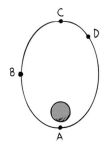

Questions

1. The orbital path of a satellite is shown in the sketch. In which marked positions A through D does the satellite have the greatest KE? Greatest PE? Greatest total energy?
2. Why does the force of gravity change the speed of a satellite when it is in an elliptical orbit, but not when it is in a circular orbit?

Answers

1. KE is maximum at the perigee A; PE is maximum at the apogee C; the total energy is the same everywhere in the orbit.
2. At any point on its path, the direction of motion of a satellite is always tangent to its path. If a component of force exists along this tangent, then the acceleration of the satellite will involve a change in speed as well as direction. In circular orbit the gravitational force is always perpendicular to the direction of motion of the satellite, just as every part of the circumference of a circle is perpendicular to the radius. So there is no component of gravitational force along the tangent, and only the direction of motion changes, not the speed. But when the satellite moves in directions that are not perpendicular to the force of gravity, as in an elliptical path, there is a component of force along the direction of mo-tion or opposite to that direction, which changes the speed of the satellite. A component of force parallel to the direction the satellite moves does work to change its KE.

Escape Speed

We know that a cannonball fired horizontally at 8 kilometers per second from Newton's mountain would find itself in orbit. But what would happen if the cannonball were instead fired at the same speed *vertically*? It would rise to some maximum height, reverse direction, and then fall back to earth. Then the old saying "What goes up must come down" would hold true, just as surely as a stone tossed skyward will be returned by gravity (unless, as we shall see, its speed is too great).

In today's spacefaring age, it is more accurate to say "What goes up may come down," for there is a critical speed at which a projectile is able to outrun gravity and escape the earth. This critical speed is called the **escape speed** or, if direction is involved, the *escape velocity*. From the surface of the earth, escape speed is 11.2 kilometers per second. Launch a projectile at any speed greater than that and it will leave the earth, traveling slower and slower, never stopping due to earth gravity.* We can understand the magnitude of this speed from an energy point of view.

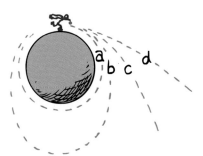

FIGURE 9.17 If Superman tosses a ball 8 km/s horizontally from the top of a mountain high enough to be just above air drag (a), then about 90 minutes later he can turn around and catch it (neglecting the earth's rotation). Tossed slightly faster (b), it will take an elliptical orbit and return in a slightly longer time. Tossed at more than 11.2 km/s (c), it will escape the earth. Tossed at more than 42.5 km/s (d), it will escape the solar system.

How much work would be required to lift a payload against the force of earth gravity to a distance very very far ("infinitely far") away? We might think PE would be infinite because the distance is infinite. But gravity diminishes with distance by the inverse-square law. The force of gravity on the payload would be strong only close to the earth. Most of the work done in launching a rocket occurs near the earth. It turns out that the PE of a 1-kilogram body infinitely far away is 62 million joules (62 MJ). So to put a payload infinitely far from the earth's surface requires at least 62 million joules of energy per kilogram of load. We won't go through the calculation here, but 62 million joules per kilogram corresponds to a speed of 11.2 kilometers per second, whatever the total mass involved. This is the escape speed from the surface of the earth.†

If we give a payload any more energy than 62 million joules per kilogram at the surface of the earth or, equivalently, any more speed than 11.2 kilometers per second, then, neglecting air resistance, the payload will escape from the earth never to return. As it continues outward, its PE increases and its KE decreases. Its speed becomes less and less, though it is never reduced to zero. The payload outruns the gravity of the earth. It escapes.

The escape speeds of various bodies in the solar system are shown in Table 9.1. Note that the escape speed from the sun is 620 kilometers per second at the surface of the sun. Even at a distance equaling that of the earth's orbit, the escape speed from the sun is 42.5 kilometers per second, considerably more than the escape speed from the

---

*Escape speed from any planet or any body is given by $v = \sqrt{d/2GM}$, where G is the universal gravitational constant, M is the mass of the attracting body, and d is the distance from its center. (At the surface of the body, d would simply be the radius of the body.)

†Interestingly enough, this might well be called the *maximum falling speed*. Any object, however far from earth, released from rest and allowed to fall to earth only under the influence of the earth's gravity would not exceed 11.2 km/s (with air friction, it would be less).

Table 9.1 Escape Speeds at the Surface of Bodies in the Solar System

Astronomical Body	Mass (earth masses)	Radius (earth radii)	Escape Speed (km/s)
Sun	333,000	109	620
Sun (at a distance of the earth's orbit)		23,500	42.2
Jupiter	318	11	60.2
Saturn	95.2	9.2	36.0
Neptune	17.3	3.47	24.9
Uranus	14.5	3.7	22.3
Earth	1.00	1.00	11.2
Venus	0.82	0.95	10.4
Mars	0.11	0.53	5.0
Mercury	0.055	0.38	4.3
Moon	0.0123	0.27	2.4

earth. An object projected from the earth at a speed greater than 11.2 kilometers per second but less than 42.5 kilometers per second will escape the earth but not the sun. Rather than recede forever, it will take up an orbit around the sun.

The first probe to escape the solar system, Pioneer 10, was launched from earth in 1972 with a speed of only 15 kilometers per second. The escape was accomplished by directing the probe into the path of oncoming Jupiter. It was whipped about by Jupiter's great gravitational field, picking up speed in the process—similar to the increase in the speed of a ball encountering an oncoming bat when it departs from the bat. Its speed of departure from Jupiter was increased enough to exceed the sun's escape speed at the distance of Jupiter. Pioneer 10 passed the orbit of Pluto in 1984. Unless it collides with another body, it will wander indefinitely through interstellar space. Like a note in a bottle cast into the sea, Pioneer 10 contains information about the earth that might be of interest to extraterrestrials, in hopes that it will one day wash up and be found on some distant "seashore."

It is important to point out that the escape speed from a body is the initial speed given by a brief thrust, after which there is no force to assist motion. One could escape

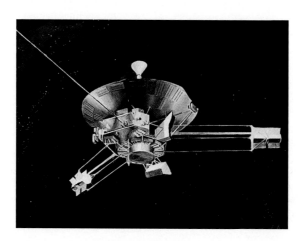

FIGURE 9.18 Pioneer 10, launched from earth in 1972, escaped from the solar system in 1984 and is wandering in interstellar space.

the earth at any sustained speed more than zero, given enough time. For example, suppose a rocket is launched to a destination such as the moon. If the rocket engines burn out when still close to the earth, the rocket needs a minimum speed of 11.2 kilometers per second. But if the rocket engines can be sustained for long periods of time, the rocket could go to the moon without ever attaining 11.2 kilometers per second.

It is interesting to note that the accuracy with which an unmanned rocket reaches its destination is not accomplished by staying on a preplanned path or by getting back on that path if it strays off course. No attempt is made to return the rocket to its original path. Instead, the control center in effect asks, "Where is it now with respect to where it ought to go? What is the best way to get there from here, given its present situation?" With the aid of high-speed computers, the answers to these questions are used in finding a new path. Corrective thrusters put the rocket on this new path. This process is repeated over and over again all the way to the goal.*

Summary of Terms

Satellite   A projectile or small celestial body that orbits a larger celestial body.

Ellipse   The oval path followed by a satellite. The sum of the distances from any point on the path to two points inside called foci is a constant. When the foci are together at one point, the ellipse is a circle. As the foci get farther apart, the ellipse gets more "eccentric."

Escape speed   The speed that a projectile, space probe, or similar object must reach to escape the gravitational influence of the earth or celestial body to which it is attracted.

Suggested Reading

Dyson, Freeman. *From Eros to Gaia.* New York: Pantheon Books, 1992.

Review Questions

The Falling Apple

1. According to Newton, what does a falling apple have in common with the moon?

The Falling Moon

2. What exactly is *tangential velocity?*
3. How far does the moon fall beneath a straight-line tangent to its motion in 1 s?
4. Why don't the planets fall into the sun?

Satellite Motion

Circular Orbits

5. Why doesn't the force of gravity change the speed of a satellite in circular orbit?
6. How much time is taken for a complete revolution of a satellite in close orbit about the earth?
7. For orbits of greater altitude, is the period greater or less?

Elliptical Orbits

8. Why does the force of gravity change the speed of a satellite in an elliptical orbit?
9. At what part of an elliptical orbit does a satellite have the greatest speed? The least speed?

Energy Conservation and Satellite Motion

10. Why is kinetic energy a constant for a satellite in circular orbit?
11. Why is kinetic energy a variable for a satellite in an elliptical orbit?
12. With respect to the apogee and perigee of an elliptical orbit, where is the gravitational potential energy greatest? Least?
13. Is the sum of kinetic and potential energies a constant for satellites in circular orbits, elliptical orbits, or both?

Escape Speed

14. What is the minimum speed for orbiting the earth in close orbit? The maximum speed? What happens above this speed?
15. How was Pioneer 10 able to escape the solar system with an initial speed less than escape speed?

*Is there a lesson to be learned here? Suppose you find that you are off course. You may, like the rocket, find it more fruitful to take a course that leads to your goal as best plotted from your present position and circumstances, rather than try to get back on the course you plotted from a previous position and under, perhaps, different circumstances.

Exercises

1. The astronaut on page 166 appears to be floating free of gravity. Is he? Explain.

2. Since the moon is gravitationally attracted to the earth, why doesn't it simply crash into the earth?

3. Does the speed of a falling object depend on its mass? Does the speed of a satellite in orbit depend on its mass? Defend your answers.

4. If you have ever watched the launching of an earth satellite, you may have noticed that the rocket starts vertically upward, then departs from a vertical course and continues its climb at an angle. Why does it start vertically? Why does it not continue vertically?

5. If a cannonball is fired from a tall mountain, gravity changes its speed all along its trajectory. But if it is fired fast enough to go into circular orbit, gravity does not change its speed at all. Explain.

6. From atop a mountain high enough to be above air drag, Superman throws a ball fast enough to circle the earth. In order to simply turn around and catch the ball some 90 minutes later, should he be located at the north pole, between the pole and the equator, or at the equator? Does the earth's spin make a difference?

7. A satellite can orbit at 5 km above the moon, but not at 5 km above the earth. Why?

8. Apollo Command Modules in low orbits about the moon had much lower orbital speeds than satellites in low earth orbit have. Why?

9. If a projectile were launched vertically from the surface of the earth at a speed of 10 km/s, would it orbit the earth?

10. Would the speed of a satellite in close circular orbit about Jupiter be greater than, equal to, or less than 8 km/s?

11. Why are satellites normally sent into orbit by firing them in an easterly direction, the direction in which the earth spins?

12. If the earth rotated once around its axis every 90 min instead of every 24 h, would you press against the earth's surface if you were at the equator? At the poles? In the middle of the United States? Explain.

13. Of all the United States, why is Hawaii the most efficient launching site for non-polar satellites? (*Hint:* Look at the spinning earth from above either pole and compare it to a spinning turntable.)

14. Relative to the sun, when do you move faster in earth orbit, at day or at night? Explain.

15. Two planets are never seen at midnight. Which two, and why?

16. Does a satellite coast in a plane that necessarily intersects the earth's center? Explain.

17. Why does a satellite burn up when it descends into the atmosphere, but doesn't burn up when it ascends through the atmosphere?

18. Neglecting air drag, could a satellite be put into orbit in a circular tunnel beneath the earth's surface? Discuss.

19. In the sketch on the left, a ball gains KE when rolling down a hill because work is done by the component of weight (*F*) that acts in the direction of motion. Sketch in the similar component of gravitational force that does work to change the KE of the satellite on the right.

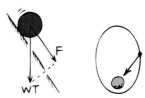

20. Why is work done by the force of gravitation on a satellite when it moves from one part of an elliptical orbit to another, but not when it moves from one part of a circular orbit to another?

21. What is the shape of the orbit when the velocity of the satellite is everywhere perpendicular to the force of gravity?

22. A communications satellite with a 24-h period hovers over a fixed point on the earth. Why is this fixed point always on the equator? (*Hint:* Think of the satellite's orbit as a ring around the earth.)

23. If a pair of satellites at different altitudes but moving in the same direction are seen from the earth to pass overhead, which of the two is overtaking the other?

24. A communications satellite is in a circular orbit far above the earth. A spy satellite of equal mass is in a circular orbit closer to the earth. Which satellite has the greater (a) speed? (b) period? (c) kinetic energy? (d) potential energy?

25. If a flight mechanic drops a wrench from a high-flying jumbo jet, it crashes to earth. If an astronaut on the orbiting space shuttle drops a wrench, does it crash to earth also? Defend your answer.

26. The orbiting space shuttle travels at 8 km/s with respect to the earth. Suppose it projects a capsule rearward at 8 km/s with respect to the shuttle. Describe the path of the capsule with respect to the earth.

27. A satellite in circular orbit about the moon fires a small projectile in a direction opposite to the velocity of the satellite. If the speed of the projectile is the same as the satellite's speed, describe the motion of the probe. If the speed is twice the speed of the satellite, why would this pose a danger to the satellite?

28. The orbital velocity of the earth about the sun is 30 km/s. If the earth were suddenly stopped in its tracks, it would simply fall radially into the sun. Devise a plan

whereby a rocket loaded with radioactive wastes could be fired into the sun for permanent disposal. How fast and in what direction with respect to the earth's orbit should the rocket be fired?

29. If you stopped an earth satellite dead in its tracks, it would simply crash into the earth. Why, then, don't the communications satellites that "hover motionless" above the same spot on earth crash into the earth?

30. What would be the consequences of the sun momentarily stopping in its apparent path across the sky?

31. The sun appears to move faster across the background of stars in December than in June. How could you interpret this fact?

32. A space explorer lands on a small rotating asteroid, only to find that its gravity is not strong enough to hold her on the surface. Why is this? Shouldn't any gravity, no matter how weak, attract the explorer to the surface and hold her there?

33. A rocket coasts in an elliptical orbit around the earth. To attain the greatest amount of KE for escape using a given amount of fuel, should it fire its engines at the apogee or the perigee? (*Hint:* Let the formula $Fd = \Delta KE$ be your guide to thinking. Suppose the thrust F is brief and of the same duration in either case. Then consider the distance d the rocket would travel during this brief burst at the apogee and at the perigee.)

34. Escape speed from the surface of the earth is 11.2 km/s, but a space vehicle could escape from the earth at half this speed and less. Explain.

35. What is the maximum possible speed of impact upon the surface of the earth for a faraway body initially at rest that falls to the earth by virtue of the earth's gravity only?

36. If Pluto were somehow stopped short in its orbit, it would fall into rather than around the sun. How fast would it be moving when it hit the sun?

37. If the earth shrank in size, all other factors remaining the same, would escape velocity from its surface be greater, less, or the same as it presently is? Explain.

38. At which of the indicated positions does the satellite in elliptical orbit experience the greatest gravitational force? Have the greatest speed? The greatest velocity? The greatest momentum? The greatest kinetic energy? The greatest gravitational potential energy? The greatest total energy? The greatest angular momentum? The greatest acceleration?

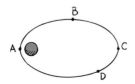

39. Many schools and hospitals could have been constructed with the funds used in the last decade to probe and explore the solar system. How many of these schools and hospitals do you speculate would in fact have been constructed if the various space missions were not funded?

40. Make up two multiple-choice questions that test for a distinction between the characteristics of a satellite in a circular and an elliptical orbit.

Problems

1. The force of gravity between the earth and an earth satellite is given by $F = GmM/r^2$, where m is the mass of the satellite, M the mass of the earth, and r the radial distance between the satellite and the center of the earth. If the satellite follows a circular orbit, the force is a centripetal force, given by $F = mv^2/r$. Equate the two expressions for force to show that the speed $v = \sqrt{GM/r}$. (Here we substitute r for d).

2. Calculate the speed in m/s at which the earth revolves about the sun. You may assume the orbit is nearly circular.

3. The moon is about 3.8×10^5 km from the earth. Find its average orbital speed about the earth.

4. A certain satellite has a kinetic energy of 8 billion joules at perigee (closest to earth) and 5 billion joules at apogee (farthest from earth). As the satellite travels from apogee to perigee, how much work does the gravitational force do on it? Does its potential energy increase or decrease during this time, and by how much?

5. The escape speed at a distance d from the center of a body of mass M is given by

$$v_{esc} = \sqrt{\frac{2GM}{d}}$$

Calculate the escape speed from the surface of Mars, which has a radius of 3.4×10^6 m and a mass of 6.6×10^{23} kg.

6. Referring to the equation in the preceding problem, we see that for planets of equal radius, the escape speed is proportional to the square root of the planet's mass. Suppose a planet made of kryptonite has the same radius as the earth but is 16 times more massive. What is the escape speed from its surface?

7. The planet Saturn has a mass 95 times that of the earth and a radius 9.2 times the earth's. Find the escape speed for objects from the surface of Saturn.

P a r t

PROPERTIES OF MATTER

Atoms are so tiny that I inhale billions of trillions with each breath, nearly a trillion times more atoms than the total population of people since time zero! In each breath, I inhale billions of atoms exhaled by every person whose breaths are now evenly mixed in the air. So there are atoms in my body from every person who ever lived, excepting newborn babies far away. But atoms I've recycled are now part of them. From a physics point of view, we're all one!

10

The Atomic Nature of Matter

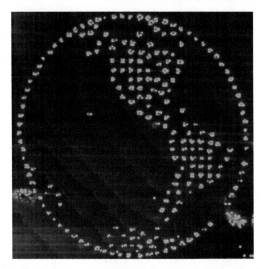

Each dot in the world's tiniest map consists of a few thousand atoms of gold. The atoms were moved into place by a scanning tunneling microscope.

Sometime when you have a few quiet moments, take a fantasy trip. The only ticket required is a fertile imagination. Imagine that you fall off your chair in slow motion, and while falling to the floor, you also slowly shrink in size. What would such a trip be like? What would you see? As you topple off the chair and approach the floor, you brace yourself for impact against the smooth, solid surface. And as you get nearer and nearer to it, becoming smaller and smaller all the while, you note that the floor is not as smooth as you supposed it to be, for great cracks appear. These are the microscopic irregularities found in all apparently smooth surfaces. In falling into one of these cracks, which look like canyons as you continue to shrink in size, you again brace yourself for impact against the canyon floor only to find that the bottom of the canyon is itself a myriad of cracks and crevices. Falling into one of these crevices and becoming still smaller, you note the solid walls have given way to nebulous surfaces that throb and pucker. If you could see greater detail, you would notice that the throbbing surfaces consist of hazy blobs, mostly spherical, some egg-shaped, some larger than others, and all oozing into each other, making up long chains of complicated structures. Falling still farther, you again brace yourself for impact as you approach one of these cloudy spheroids closer and closer, smaller and smaller, and—wow!—you find you have penetrated into a new universe. You fall into a sea of emptiness, occupied by occasional specks that whirl past at unbelievably high speeds. You are in an atom, as empty of matter as the solar system. You have found that the solid floor you have fallen into is, except for specks of matter here and there, empty space. If you continue falling, you might fall many meters through "solid" matter before making a direct hit with a subatomic speck.

All matter, however solid it appears, is made up of tiny building blocks, which themselves are mostly empty space. These are atoms—atoms that combine to form molecules that in turn combine to form the compounds and substances of matter. In this chapter we will study the atomic nature of matter. We will follow this up in succeeding chapters by investigating the properties of matter in the solid, liquid, gaseous, and plasma phases.

Atoms

All manner of things—shoes, and ships, and sealing wax; cabbages and kings—anything we can think of is composed of atoms. One might think that an incredible number of different kinds of atoms exist to account for the rich variety of substances we find around us. But the number is surprisingly small. The great variety of substances results not from any great variety of atoms, but from the many ways a few types of atoms can be combined—just as in a color print three colors can be combined to form almost every conceivable color. To date (1997) we know of 112 distinct **atoms**. These are the chemical *elements*. Only 90 elements are found naturally; the others are made in the laboratory with high-energy atomic accelerators and nuclear reactors. These heaviest elements are too unstable (radioactive) to occur naturally in appreciable amounts.

Hydrogen was apparently the original element and still makes up over 90 percent of the atoms in the known universe. Elements heavier than hydrogen are manufactured in the deep interiors of stars, where enormous temperatures and pressures fuse the cores of hydrogen atoms together to form heavier elements. With the exception of some of the hydrogen and trace amounts of other light elements, all the elements that occur naturally on earth are remnants of stars that exploded long before the solar system came into being.

These star remnants are the building blocks of all matter. And all matter, however complex, living or nonliving, is some combination of these elements. From a pantry having about 100 bins, each containing a different element, we have all the materials needed to make up any substance occurring in the universe. Slightly more than a dozen elements compose more than 99 percent of the earth; the majority of elements are relatively rare.* Living things, for example, are composed primarily of five elements: oxygen (O), carbon (C), hydrogen (H), nitrogen (N), and calcium (Ca). The letters in the parentheses represent the chemical symbols for these elements.

Atoms are small and numerous. There are about 10^{23} atoms in a gram of water (a thimbleful). The number 10^{23} is an enormous number, more than the number of drops of water in all the lakes and rivers of the world. So there are more atoms in a thimbleful of water than there are drops of water in the world's lakes and rivers.

Atoms are ageless. Many atoms in your body are nearly as old as the universe itself, cycling and recycling among innumerable hosts, both nonliving and living. When

---

* Most common substances are formed out of combinations of two or more of these elements: hydrogen (H), carbon (C), nitrogen (N), oxygen (O), sodium (Na), magnesium (Mg), aluminum (Al), silicon (Si), phosphorus (P), sulfur (S), chlorine (Cl), potassium (K), calcium (Ca), and iron (Fe).

FIGURE 10.1 Leslie is made of stardust—in the sense that the carbon, oxygen, nitrogen, and other atoms that make up her body originated in the deep interiors of ancient stars that have long since exploded.

you breathe, for example, only some of the atoms that you inhale are exhaled in your next breath. The remaining atoms are taken into your body to become part of you, and they later leave your body by various means. You don't "own" the atoms that make up your body; you borrow them. We all share from the same atom pool as atoms forever migrate around, within, and among us. So some of the atoms in the nose you scratch today could have been part of your neighbor's ear yesterday!

Atoms get around. Just as a drop of ink put into a glass of water soon colors the entire glassful, the atoms making up a cupful of DDT or any material thrown into an ocean spread around and are later found in every part of the world's oceans. The same is true of materials released into the atmosphere.

Interestingly, there are about as many atoms of air in your lungs at any moment as there are breathfuls of air in the atmosphere of the whole world. (There are about 10^{22} atoms in a liter of air at atmospheric pressure, and about 10^{22} liters of air in the atmosphere.) When you exhale a single breath of air, in about 6 years that breath becomes evenly mixed in the atmosphere. Then, every person in the world who inhales a breath of air gets, on average, one of the many atoms you exhaled in that single breath! And this occurs for *each* breath you exhale! Considering the many thousands of breaths that people exhale, at any moment you have hordes of atoms in your lungs that were once in the lungs of every person who ever lived. We are literally breathing one another.

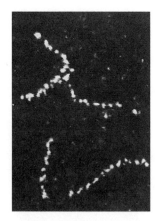

FIGURE 10.2 The first photo of individual atoms made in 1970 with a scanning electron microscope by researchers at the University of Chicago's Enrico Fermi Institute. Strings of dots are chains of thorium atoms.

Atoms are too small to be seen—at least with visible light. You could stack microscope atop microscope and never "see" an atom. This is because light is made up of waves, and atoms are smaller than the wavelengths of visible light. The size of a particle visible under the highest magnification must be larger than the wavelength of light. (More about this in Chapter 28.) The first "photo" of individual atoms is the historic image in Figure 10.2. It was taken in 1970 with a scanning electron microscope, a nonlight imaging device that bypasses light and optics altogether. The higher-resolution images of atoms in Figure 10.3 were obtained with an improved device called a scanning tunneling microscope. Even better detail can be seen with newer types of imaging devices that are presently revolutionizing microscopy.

FIGURE 10.3 An image of graphite obtained with a scanning tunneling microscope. The "bumps" indicate the location of individual carbon atoms.

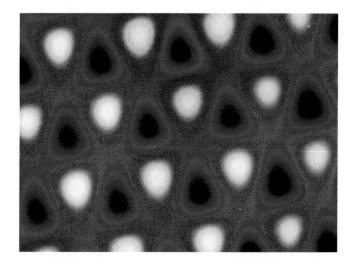

The first somewhat direct evidence for the existence of atoms was unknowingly discovered in 1827 by a Scottish botanist, Robert Brown, while studying the spores of pollen under a microscope. He noticed that the spores were in a constant state of agitation, always moving, always jumping about. At first he thought that the spores were some sort of moving life forms. But later he found that inanimate dust particles and grains of soot also showed this motion. This perpetual jiggling of particles—called **Brownian motion**—results from bombardment of the visible particles by invisible molecules.

Questions

1. World population grows each year. Does this mean the mass of the earth increases each year?

2. Are there really atoms that were once a part of Albert Einstein in the brain matter of all your classmates?

Answers

1. The mass of the earth does increase by the incidence of roughly 40,000 tons of interplanetary dust each year. But the increasing number of people increases the mass of the earth by zero. The atoms that make up our bodies are the same atoms that were here before we were born—we are but dust, and unto dust we shall return. Human cells are merely rearrangements of material previously present. The atoms that make up a baby forming in its mother's womb must be supplied by the food she eats. And those atoms were formed in stars that long since exploded (Figure 10.1).

2. Yes, and from Charlie Chaplin, too, although the configurations of these atoms with respect to others are now quite different! The next time you have one of those days when you feel like you'll never amount to anything, take comfort in the thought that many of the atoms that now compose you will live forever in the bodies of all the people on earth who are yet to be.

Molecules

Atoms combine to form **molecules** (Figure 10.4). Two atoms of hydrogen (H_2) combine with a single atom of oxygen (O) to produce a water molecule (H_2O). A molecule may be as simple as the two-atom combination of oxygen (O_2) or nitrogen (N_2) or as complex as the double helix of deoxyribonucleic acid (DNA), which consists of millions of atoms—the basic building block of life.

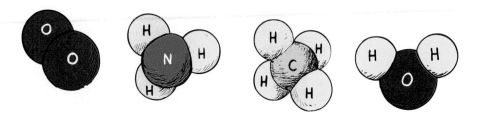

FIGURE 10.4 Models of simple molecules. The atoms in a molecule are not just mixed together but are joined in a well-defined way.

A molecule can be divided into atoms that have chemical properties of their own. You can do a simple experiment to see this for yourself. Place two wires that are connected to the terminals of an ordinary battery into a glass of salted water (Figure 10.5). Position the wires on opposite sides of the glass so they don't touch, and you will see bubbles of gas forming on the wires. If you collect these gases, you will find that they have entirely different chemical properties from water vapor. One of the gases is hydrogen, and the other is oxygen. If the gases are mixed and ignited with a match, a quick fire or explosion results. The explosion is the violent combination of hydrogen and oxygen to form water again. The amount of energy that is suddenly released is the same amount of energy that was gradually drawn from the battery to separate the gases.

FIGURE 10.5 When electric current passes through salted water, bubbles of oxygen form at the left wire and bubbles of hydrogen form at the right wire.

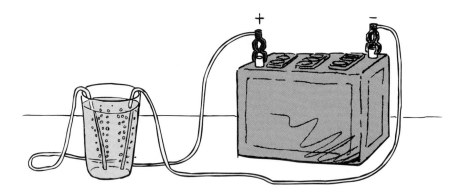

The energy released in common chemical burning comes mainly from carbon and hydrogen atoms combining with oxygen atoms to form carbon dioxide (CO_2) and water (H_2O). (In incomplete combustion, carbon monoxide (CO) may also be formed.) Oxygen atoms are attracted strongly to carbon and hydrogen atoms. When these attracting atoms snap together, they produce energy—much as a slamming screen door produces sound. This energy is ordinarily the kinetic energy of emerging molecules—often so great that light is generated in their violent collisions. This is how we get flames.

It is not only things containing carbon and hydrogen, such as natural gas and gasoline, that can burn. Iron can "burn" (oxidize) too. That's what rusting is—the slow combination of oxygen atoms with iron atoms, releasing energy. When the rusting of iron is speeded up, it makes a nice hand warmer for skiers and winter hikers. Any process in which atoms rearrange to form different molecules is called a **chemical reaction**.

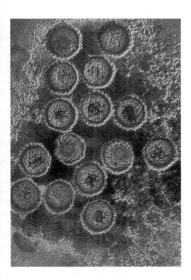

FIGURE 10.6 An electron microscope photo of virus molecules.

Molecules are too small to be seen with optical microscopes, but exceedingly small quantities of them are evident to our sense of smell. Our olfactory organs clearly discern noxious gases such as sulfur dioxide, ammonia, and ether. The smell of perfume is the result of molecules that rapidly evaporate (change from the liquid to the gaseous phase; see Chapter 16) and bumble around haphazardly in the air until some of them get close enough to our nose to be inhaled. They are just a few of the billions of jostling molecules that, in their aimless wanderings, happen to bumble up into the nose. You can get an idea of the speed of molecular diffusion in the air when you are in your bedroom and smell food almost immediately after the oven door in the kitchen has been opened.

More direct evidence of molecules is shown in the electron microscope photograph of virus molecules which are composed of thousands of atoms (Figure 10.6). These giant molecules are still too small to be seen with visible light. We will see in Chapter 30 that an atom or a molecule that is invisible to light may be visible to a shorter-wavelength electron beam.

We often say that seeing is believing. But we don't always have to see something to believe in its existence. We don't have to look inside a closed cardboard box that is too heavy to lift to know that something is in it. We could measure its mass without benefit of light. Atoms and molecules are too small to be seen with light, but this doesn't prevent us from measuring their masses with great precision.

Molecular and Atomic Masses

The earliest determination of molecular and atomic masses was made with measurements of gases in the early 1800s. As an example of how this is done, consider the hydrogen and oxygen gases that result when water is decomposed by electric current. If you do an experiment like the one shown in Figure 10.7, you find that the volume of hydrogen gas produced is twice the volume of oxygen gas produced. Chemists have learned that equal volumes of gases (at equal temperature and pressure) contain equal numbers of molecules. So twice as many hydrogen as oxygen molecules are produced when water is decomposed. That's not surprising, because every molecule of water (H_2O) that breaks apart releases two hydrogen atoms and one oxygen atom. (These atoms then combine in pairs to make hydrogen and oxygen molecules, H_2 and O_2.) Now

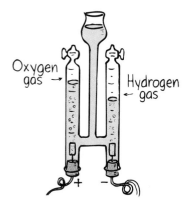

Oxygen gas →

Hydrogen gas ←

FIGURE 10.7 In the electrolysis of water, twice the volume of hydrogen gas as oxygen gas is collected.

if you measure the masses of the hydrogen and oxygen gases that are produced, you find that the oxygen is eight times as massive as the hydrogen. So you conclude that a single molecule of oxygen is 16 times as massive as a single molecule of hydrogen. Correspondingly, the mass of an oxygen atom is 16 times that of a hydrogen atom.

The rule that equal volumes of gases at the same temperature and pressure contain equal numbers of molecules, regardless of the size of the molecules, was first proposed in 1811 by the Italian physicist Amedeo Avogadro and is called **Avogadro's principle**. This rule makes sense only when you realize that in a gas the molecules are far apart. For steam (water vapor) the distance between molecules is about twelve times the molecular diameter, so the molecules occupy about 1700 times the volume of the same mass of liquid ($12^3 = 1728$) where the molecules are almost "touching."

The theory of relative masses of molecules follows from Avogadro's principle. We simply weigh equal volumes of two gases and compare the masses. The ratio of the masses of gas is identical to the ratio of the masses of the individual molecules making up the gases. For example, if gas A is twice as heavy as an equal volume of gas B, then molecule A has twice the mass of molecule B. If the chemical formula for the molecule is known, the relative masses of individual atoms can be determined.

To compare the masses of elements, scientists in the nineteenth century chose one atom of hydrogen (the simplest element) to be *exactly* one **atomic mass unit** (abbreviated **amu**).* The mass of each other kind of atom was expressed relative to it. Thus the mass of the most common kind of carbon atom is 12 amu; that of oxygen is 16 amu; and that of uranium, one of the heaviest, is 238 amu.

Elements, Compounds, and Mixtures

Certain solids such as gold, liquids such as mercury, and gases such as neon are composed of a single kind of atom. These substances are called *elements*. Certain other solids such as crystals of common table salt, liquids such as water, and gases such as methane are made up of elements that are chemically combined. These are called **compounds**. A compound may or may not be made of molecules. Water and carbon dioxide are compounds made of molecules. On the other hand, table salt (Nacl) is made of different kinds of atoms arranged in a regular pattern. Each chlorine atom is surrounded by six sodium atoms (Figure 10.8). In turn, every sodium atom is surrounded by six chlorine atoms. (More accurately, the atoms in table salt are *ions,* charged either positive or negative by deficiency or excess of electrons). As a whole, there is one sodium atom for each chlorine atom, but there are no separate sodium-chlorine groups that can be labeled as molecules.

Compounds often have properties quite different from those of the elements of which they are composed. Sodium is a metal that reacts violently with water. Chlorine is a poisonous greenish gas. Yet the compound of these two elements is the harmless white crystal (NaCl) you sprinkle on your potatoes. At ordinary temperatures water (H_2O) is a liquid, yet at these temperatures hydrogen and oxygen are both gases.

Compounds are formed only when elements react chemically and bond with each other. But not all elements, atoms, or molecules react with one another chemically when they are brought close together. Substances that are mixed together without chemically

---

* Later, for reasons that do not concern us now (including greater precision), the atom of a particular carbon isotope was chosen as *exactly* 12.000 amu. The masses of other atoms then turn out to be *almost* integers. The most common isotope of hydrogen, for example, has a mass of 1.008; the most abundant form of oxygen is 15.995; and what we call uranium 238 has an atomic mass of 238.051.

FIGURE 10.8 Table salt (NaCl) is a compound that is not made of molecules. The sodium and chlorine atoms are arranged in a repeating pattern where each atom is surrounded by six atoms of the other kind.

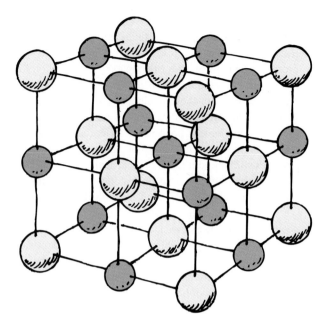

combining are called **mixtures**. Sand combined with salt is a mixture. Hydrogen and oxygen gases form a mixture until ignited, whereupon they form the compound water. A common mixture on which we all depend is nitrogen and oxygen together with a little argon and small amounts of carbon dioxide and other gases. It is the air we breathe.

Question A mixture of hydrogen and oxygen gases when ignited will form the compound water, H_2O. How many grams of oxygen should be mixed with 2 g of hydrogen to get pure water?

Atomic Structure

Nearly all the mass of an atom is concentrated in its *nucleus,* which occupies less than a trillionth of the volume of the atom. The nucleus therefore is extremely dense. If bare atomic nuclei could be packed against each other into a lump 1 centimeter in diameter (about the size of a large pea), the lump would weigh about a billion tons! Huge electrical forces of repulsion prevent such close packing of atomic nuclei. Each nucleus is electrically charged and repels other nuclei. Only under special circumstances can the nuclei of two or more atoms be "squashed into contact," and when this happens, a violent reaction may take place. This is what happens in a hydrogen bomb, which will be discussed in Chapter 33.

Answer 16 g. Each H_2O molecule has two atoms of H for each atom of O. The relative masses of H and O atoms are 1 and 16. So water has 2 g of H for every 16 grams of O. When two or more elements combine to form a compound, they always do so in the same proportion by mass. For water it's always 8 g O to 1 g H. (If you study chemistry, you'll call this the *law of definite proportions.*)

The principal building block of the nucleus is the *nucleon*. Nucleons, in turn, are composed of fundamental particles called *quarks*. When the nucleon is in its electrically neutral state, it is a *neutron;* when it is in its electrically charged state, it is a *proton*. All protons are identical; they are copies of one another. Likewise with neutrons; each neutron is like every other neutron. The known atomic nuclei are composed of between 1 and 260 or so of these nucleons. The lighter nuclei have roughly equal numbers of protons and neutrons; more massive nuclei have more neutrons than protons.

Protons have a positive electric charge. Particles with positive electric charge repel other particles with positive charge but attract particles with negative charge. So particles with the same kind of electrical charge repel one another, and particles with unlike charge attract one another. Protons in the nucleus attract a surrounding cloud of negatively charged particles called *electrons* to make an atom.

Moving electrons make up electric current in electrical circuits. Electrons are exceedingly light, less than $1/2000$ the mass of a nucleon, so they contribute very little mass to the atom. An electron in one atom is identical to any electron in or out of any other atom. Electrons have negative charges so they repel other electrons, and are attracted to the positively charged nucleus.

The number of protons in the nucleus is electrically balanced by an equal number of electrons whirling in the surrounding cloud. The atom itself is electrically neutral, so it normally doesn't attract or repel other atoms. But when atoms get close together, their electron clouds interact, causing the atoms to exert forces on one another. Too close, and the atoms repel. That's why solid matter doesn't squash down to greater density.

Like the solar system, the atom is mostly empty space. The nucleus is tiny and the surrounding cloud of electrons is easily penetrated by individual high-energy particles. Nevertheless, the electron cloud is surprisingly resilient. It pushes back against any force trying to compress or deform it. It's like a tough truck tire. A sharp needle can penetrate it, but it takes a great force to squeeze it. "Keep out of my space" is the message that electrons in one atom give to electrons in another atom. So, even though you and the floor are mostly empty space, you don't fall through the floor. The electron clouds in the atoms of the floor defend their space against the electron clouds in the atoms of your shoe, and you are supported. Ultimately, these forces exerted by the electron clouds are electrical in nature. So it is electricity that is holding you up, just as it is electricity that is holding the electrons in the atom.

Attractions between atoms can be summed up in a few very simple rules of thumb: If the centers of two atoms get closer together than one atomic diameter, the atoms repel each other strongly. If the atoms pull apart to a slightly greater separation, they attract each other. (This is why some atoms combine to form molecules and why solids don't crumble and fall apart.) If the atoms are separated by two atomic diameters or more, they have practically no influence on each other.

The main characteristic that distinguishes one atom from another is the number of protons in the nucleus (equivalently, the number of electrons about the nucleus). The hydrogen atom is the simplest. It consists of a single proton with a single orbiting electron. The helium atom is next. It consists of two protons and two neutrons surrounded by two orbiting electrons. The lithium atom has three protons and three electrons, and the list continues with the atom of each successive element having an additional proton in the

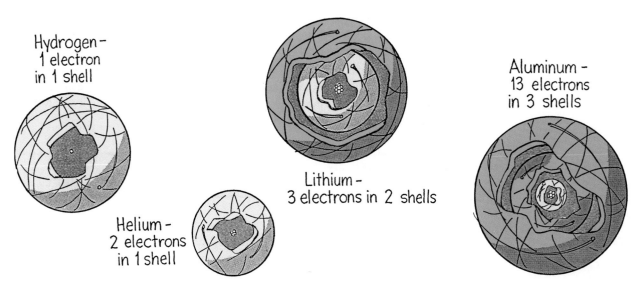

FIGURE 10.9 The classical model of the atom consists of a tiny nucleus surrounded by electrons that orbit within spherical shells. As the size and charge of nuclei increase, electrons are pulled closer, and the shells become smaller.

nucleus and an additional electron in the cloud of orbiting electrons (Figure 10.9). Equal numbers of protons and electrons make each atom electrically neutral.

Atoms are classified by their **atomic number**, which is the same as the number of protons in the nucleus. Hydrogen has atomic number 1, helium 2, and so on in sequence to naturally occurring uranium with atomic number 92. The numbers continue through the artificially produced transuranic elements, at this writing up to 110. The arrangement of elements by their atomic numbers makes up the **periodic table**.* See the fold-out page following page 202.

We find that electrons revolve about the nucleus in clouds, or waves, that spread out in concentric shells at various distances from the nucleus.† In the innermost shell there are at most two electrons; in the second shell, eight electrons; in the third shell, eighteen. In atoms having more than eighty-six electrons, there are seven shells altogether. These shells are indicated in the periodic table by rows. Note that the uppermost row consists of only two elements, hydrogen and helium. The electrons in helium complete the innermost shell. The second row contains eight elements, beginning with lithium and ending with neon, whose electrons complete the second shell, and so on. The order in which successive shells (and subshells) are filled becomes more complicated for the heavier elements.

* The enormous human effort and ingenuity that went into finding regularity and system in the chemical elements make a fascinating atomic detective story that can be found in the books suggested at the end of the chapter.
† An atomic "shell" cannot be visualized as a thin layer like an eggshell. Because of the electron's wave nature, discussed in Chapter 31, the atomic shell is more like the spread-out dough of a doughnut.

In the periodic table the elements are arranged vertically by similarity in electronic configuration, which is then reflected in similarity in the physical and chemical properties of the elements and their compounds. It is the configuration of electrons in the outer shell of the atom that quite literally gives life and color to the world. The configuration of the outer electrons dictates whether and how atoms bond to become molecules, the melting and freezing temperatures, and the electrical conductivity, as well as the taste, texture, appearance, and color of substances.

Question How many atomic nuclei are in an oxygen atom? An oxygen molecule?

Antimatter

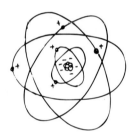

FIGURE 10.10 An atom of antimatter has a negatively charged nucleus surrounded by positrons.

Whereas matter is composed of atoms with positively charged nuclei and negatively charged electrons, **antimatter** is composed of atoms with negative nuclei and positive electrons, or *positrons* (Figure 10.10).

Positrons were first discovered in 1932, in cosmic rays bombarding the earth's atmosphere. Today antiparticles of all types are regularly produced in laboratory experiments involving nuclear physics. The first complete anti-atom was constructed in 1995. A positron has the same mass as an electron and the same magnitude of charge but the opposite sign. Antiprotons have the same mass as protons but are negatively charged. Every charged particle has an antiparticle of the same mass and opposite charge. Neutral particles (such as the neutron) also have antiparticles, alike in mass and in some other properties, but opposite in some properties. For every particle there is an antiparticle. There are even antiquarks.

Gravitational force does not distinguish between matter and antimatter—both attract the other. Also, there is no way to tell whether something is made of matter or antimatter by the light it emits. Only through much subtler, hard-to-measure nuclear effects could we tell whether a distant galaxy is made of matter or antimatter. But if an antistar met a star, it would be a different story. They would mutually annihilate each other, with most of the matter converting to radiant energy. This process, more so than any other known, results in the maximum energy output per gram of substance— $E = mc^2$ with a 100 percent mass conversion. (Nuclear fission and fusion, in contrast, convert less than 1 percent of the matter involved.)*

There cannot be both matter and antimatter in our immediate environment, at least not in appreciable amounts or for appreciable times, for something made of antimatter would be completely transformed to radiant energy as soon as it touched matter, consuming an equal amount of normal matter in the process. If the moon were made of antimatter, for example, a flash of energetic radiation would result as soon as one of our spaceships touched it. Both the spaceship and an equal amount of the antimatter moon

Answer There is one nucleus in an oxygen atom (O), and two in the combination of two oxygen atoms—an oxygen molecule (O_2).

* Some physicists speculate that the early universe, right after the big bang, was billions of times more massive than its present state, and that near total extinction of matter-antimatter left only the amount now present.

would disappear in a burst of radiant energy. We know the moon is not antimatter because this didn't happen during moon missions. (The astronauts weren't in this kind of danger, for other previous evidence showed the moon is made of matter.) But what about other galaxies? There is strong reason to believe that in our part of the universe, galaxies are made only of normal matter. But what of more distant parts of the universe? We don't know.

Phases of Matter

Matter exists in four phases (or states): *solid, liquid, gaseous,* and *plasma.* In all phases, the atoms are perpetually moving. In the solid phase the atoms and molecules vibrate about fixed positions. If the rate of molecular vibration is increased sufficiently, molecules will shake apart and wander throughout the material, jostling each other as they go. The shape of the material is no longer fixed but takes the shape of its container. This is the liquid phase. If more energy is put into the material and the molecules gain even more energy, they may break away from one another and assume the gaseous phase. H_2O provides a common example of this changing of phases. When solid, it is ice. If we heat the ice, the increased molecular motion jiggles the molecules out of their fixed positions, and we have water. If we heat the water, we can reach a stage where faster molecular motion results in a separation between water molecules, and we have steam. Continued heating causes the molecules to separate into atoms—a process called dissociation. If we heat the steam to temperatures exceeding 2000°C, the atoms themselves will be jostled so violently that some or all of their electrons shake loose, making a gas of free electrons and positive ions called **plasma.**

Although the plasma phase is less common to our everyday experience, it may be considered the normal phase of matter in the universe; the sun and other stars as well as much of the intergalactic matter are in the plasma phase. All substances can be transformed from any phase to another.

The following three chapters treat these phases of matter in turn. We will continue the story of the atom in Chapter 31.

Summary of Terms

Atom The smallest particle of an element that has all of the element's chemical properties.

Brownian motion The haphazard movement of tiny particles suspended in a gas or liquid resulting from bombardment by the fast-moving molecules of the gas or liquid.

Molecule The smallest particle of any substance that has all the substance's chemical properties. Atoms combine to form molecules.

Chemical reaction The process of rearrangement of atoms transforming one molecule into another.

Avogadro's principle Equal volumes of all gases at the same temperature and pressure contain the same number of molecules.

Atomic mass unit (amu) The standard unit of atomic mass. It is based on the mass of the common carbon atom, which is arbitrarily given the value of exactly 12. One amu is one-twelfth the mass of the common carbon atom.

Compound A chemical substance made of atoms of two or more different elements combined in a fixed proportion.

Mixtures Substances mixed together without combining chemically.

Atomic number A number designating an atom, which is the same as the number of protons in the nucleus or the same as the number of extranuclear electrons in a neutral atom.

Periodic table A chart that lists elements by their atomic number and by their electron arrangements, so that elements with similar chemical properties are in the same column. (See fold-out page.)

Antimatter Matter composed of atoms with negative nuclei and positive electrons.

Plasma The fourth phase of matter, in addition to the solid, liquid, and gaseous phases, which consists of positive ions (atoms missing some or all of their electrons) and free electrons.

Suggested Reading

Asimov, I. *A Short History of Chemistry*. Chaps. 5 and 8. Science Studies Series. Garden City, N.Y.: Doubleday (Anchor), 1965.

Feynman, R. P., R. B. Leighton, and M. Sands. *The Feynman Lectures on Physics*. Vol. I, Chap. 1. Reading, Mass.: Addison-Wesley, 1963.

Lederman, Leon. *The God Particle*. Boston: Houghton Mifflin Company, 1993.

Review Questions

Atoms

1. What is the most abundant element in the known universe?
2. How many elements are known today?
3. How does the age of most atoms compare with the age of the solar system?
4. From where do elements originate?
5. What are the five most common elements in living things?
6. How does the approximate number of atoms in the air in your lungs compare to the number of breaths of air in the atmosphere of the whole world?
7. Why cannot individual atoms be seen with visible light?
8. What causes dust particles and tiny grains of soot to move with Brownian motion?

Molecules

9. Distinguish between an atom and a molecule.
10. How many elements compose a water molecule? How many individual atoms compose a water molecule?
11. Compared to the energy it takes to separate oxygen and hydrogen from water, how much energy is given off when they recombine? (What general physics principle is illustrated here?)
12. Is energy given off or taken in when carbon and oxygen atoms combine?
13. When the oven is opened in the kitchen, how do you know that molecules from the hot apple pie are moving?

Molecular and Atomic Masses

14. When water is decomposed, why is the volume of hydrogen gas collected twice as much as the volume of oxygen gas?
15. When water is decomposed, for every 2 g of H formed, 16 g of O are formed. Since H has twice the volume, does this mean that if a certain volume of O has a mass of 16 g, the same volume of H would have a mass of 1 g?

16. If the same volumes of gas at the same temperature and pressure have equal numbers of molecules, does this mean that an oxygen molecule is 16 times as massive as a molecule of hydrogen?
17. Oxygen and hydrogen molecules, O_2 and H_2, are composed of double atoms. Does it follow that a single oxygen atom has sixteen times the mass of a single hydrogen atom?
18. What does Avogadro's principle have to do with the last two questions?
19. If a volume of gas is three times as heavy as another gas of equal volume, how do the molecular masses compare?
20. A meter is the standard of length, a kilogram the standard of mass, and a second the standard of time. Similarly, what is the standard atomic mass unit?

Elements, Compounds, and Mixtures

21. What is a compound? Give three examples.
22. What is a mixture? Give three examples.

Atomic Structure

23. Which part of an atom contributes most to an atom's mass?
24. Distinguish between a proton and a neutron. What are both called?
25. What is the rule for attractive and repelling forces between particles with positive and negative electric charges?
26. How does the mass of an electron compare to the mass of a nucleon?
27. How do the number of electrons in an electrically neutral atom compare to the number of protons in the nucleus?
28. Why do we say an atom is mostly "empty space"?
29. If atoms are mostly empty space, why don't materials made of atoms simply fall through each other?
30. What distinguishes one kind of atom from another?
31. What is meant by the *atomic number* of an atom? What is the atomic number of hydrogen? Iron? Gold? Uranium?
32. What is the *periodic table*?
33. What are *atomic shells*?

Antimatter

34. How do matter and antimatter differ?
35. What happens when a particle of matter and a particle of antimatter meet?
36. If a 1-gram body made of antimatter meets a 1-kilogram body made of matter, what would be the mass of surviving matter?

Phases of Matter

37. What is the role of temperature in the four states of matter?

38. What are the relative amounts of energy in a quantity of H_2O when it is ice, water, and vapor?

39. Which of the four states of matter is most common in the universe?

40. How does a plasma differ from a gas?

Project

A candle will burn only if oxygen is present. Will a candle burn twice as long in an inverted liter jar as it will in an inverted half-liter jar? Try it and see.

Exercises

1. A cat strolls across your backyard. An hour later a dog with his nose to the ground follows the trail of the cat. Explain this occurrence from a molecular point of view.

2. If no molecules in a body could escape, would the body have any odor?

3. The average speed of a perfume vapor molecule at room temperature may be about 300 m/s, but you'll find the speed at which the scent travels across the room is much less. Why?

4. Which are older, the atoms in the body of an elderly person or those in the body of a baby?

5. Where were the atoms that make up a newborn "manufactured"?

6. A class of meteorites called chondrites contain a relative abundance of elements identical to the relative abundance observed in the sun (except for the volatile gases hydrogen and helium). What does this scientific finding suggest about the origin of the solar system?

7. Why is Brownian motion apparent only for microscopic particles?

8. Why are atoms visible with electron microscopes yet invisible with even ideal optical microscopes?

9. If the elements were arranged in increasing order by atomic mass rather than by atomic number, would the sequence be the same? (To answer this, check the periodic table on the fold-out.)

10. Which of the following are pure elements: H_2, H_2O, He, Na, NaCl, H_2SO_4, U?

11. A particular atom has 29 electrons, 34 neutrons, and 29 protons. What is the atomic number of this element, and what is its name?

12. A radical is a group of atoms that acts like a single particle in combining with atoms or other radicals. The hydroxyl radical is OH. What common compound does it form?

13. How do the number of protons in an atomic nucleus dictate the chemical properties of the element?

14. Which would be the more valued result: taking one proton from each nucleus of a sample of gold or adding one proton to each gold nucleus? Explain.

15. If two protons and two neutrons are removed from the nucleus of an oxygen atom, what nucleus remains?

16. What element results if you add a pair of protons to the nucleus of mercury? (See periodic table.)

17. What element results if two protons and two neutrons are ejected from a radium nucleus?

18. You could swallow a capsule of germanium without ill effects. But if a proton were added to each of the germanium nuclei, you would not want to swallow the capsule. Why? (Consult the periodic table of the elements.)

19. What element results if one of the neutrons in a nitrogen nucleus is converted by radioactive decay into a proton?

20. Which of the following elements would you predict to have properties most like those of silicon (Si): aluminum (Al), phosphorus (P), or germanium (Ge)? (Consult the periodic table of the elements.)

21. Carbon, with a half-full outer shell of electrons—four in a shell that can hold eight—readily shares its electrons with other atoms and forms a vast number of molecules, many of which are the organic molecules that form the backbone of living matter. Looking at the periodic table, what other element might play a role like carbon in life forms on some other planet?

22. Which contributes more to an atom's mass; electrons or protons? Which contributes more to an atom's size?

23. The atoms that comprise your body are mostly empty space, and structures such as the chair you're sitting on are composed of atoms that are also mostly empty space. So why don't you fall through the chair?

24. When 50 cubic centimeters (cm^3) of alcohol are mixed with 50 cm^3 of water, the volume of the mixture is only 98 cm^3. Can you offer an explanation for this?

25. (a) From an atomic point of view, why do you have to heat a solid to melt it? (b) If you have a solid and a liquid at room temperature, what conclusion can you draw about the relative strengths of their interatomic forces?

26. At high temperature, the energy of thermal agitation breaks apart organic molecules and destroys living matter. Does this mean that low temperature is most favorable for life? What might work against life when it gets too cold?

27. In what sense can you truthfully say that you are a part of every person in history? In what sense can you say that you will tangibly contribute to every person on earth who will follow?

28. What are the chances that at least one of the atoms exhaled by your very first breath will be in your next breath?

29. Somebody told your friend that if an antimatter alien ever set foot upon the earth, the whole world would explode into pure radiant energy. Your friend looks to you for verification or refutation of this claim. What do you say?

30. Make up a multiple-choice question that will test your classmates on the distinction between any two terms in the Summary of Terms list.

Problems

1. How many grams of oxygen are there in 18 g of water?

2. How many grams of hydrogen are there in 16 g of methane gas? The chemical formula for methane is CH_4.

3. Gas A is composed of diatomic molecules (two atoms to a molecule) of a pure element. Gas B is composed of monatomic molecules (one atom to a "molecule") of another pure element. Gas A has three times the mass of an equal volume of gas B at the same temperature and pressure. How do the atomic masses of elements A and B compare?

4. A teaspoon of an organic oil dropped on the surface of a quiet pond spreads out to cover almost an acre. The oil film has a thickness equal to the size of a molecule. In the lab when you drop 0.001 milliliter (10^{-9} m^3) of the organic oil on the still surface of water, you find that it spreads to cover an area of 1.0 m^2. If the layer is one molecule thick, what is the size of a single molecule?

The problems that follow require some knowledge of exponents.

5. The diameter of an atom is about 10^{-10} m. (a) How many atoms make a line a millionth of a meter (10^{-6} m) long? (b) How many atoms cover a square a millionth of a meter on a side? (c) How many atoms fill a cube a millionth of a meter on a side? (d) If a dollar were attached to each atom, what could you buy with your line of atoms? With your square of atoms? With your cube of atoms?

6. There are approximately 10^{23} H_2O molecules in a thimbleful of water and 10^{46} H_2O molecules in the earth's oceans. Suppose that Columbus had thrown a thimbleful of water into the ocean and that the water molecules have by now mixed uniformly with all the water molecules in the oceans. Can you show that if you dip a sample thimbleful of water from anywhere in the ocean, you'll probably scoop up at least one of the molecules that was in Columbus's thimble? (*Hint:* The ratio of the number of molecules in a thimble to the number of molecules in the ocean will equal the ratio of the number of molecules in question to the number of molecules the thimble will hold.)

7. There are approximately 10^{22} molecules in a single medium-sized breath of air and approximately 10^{44} molecules in the atmosphere of the whole world. Now the number 10^{22} squared is equal to 10^{44}. So how many breaths of air are there in the world's atmosphere? How does this number compare with the number of molecules in a single breath? If all the molecules from Julius Caesar's last dying breath are now thoroughly mixed in the atmosphere, how many of these on the average do we inhale with each single breath?

8. Assume that the present world population of about 4×10^9 people is about one-thirtieth the number of people who ever lived on earth. How does the number of people who ever lived compare to the number of air molecules in a single breath?

11

....

Solids

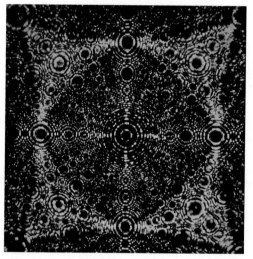

: **A**tomic arrangement at the tip of a needle.

Humans have been using solid materials for many thousands of years. The names Stone Age, Bronze Age, and Iron Age tell us the importance of solid materials in the development of civilization. Wood and clay were also important to early peoples, and gems were put to use for art and adornment. The number and uses of materials multiplied over the centuries, yet there was little progress in understanding the nature of solids. Not until the twentieth century did discoveries concerning atoms make it possible to understand the structure of materials. Now chemists, metallurgists, and materials scientists routinely design and produce new materials to meet specific needs. Solid-state physicists explore solids such as semiconductors and tailor them to meet the needs of this information age.

Müller's Micrograph

A striking example showing the structure of solid matter is shown above. The picture is a micrograph produced back in 1958 by Erwin Müller. Dr. Müller used an extremely fine platinum needle with a hemisphere-shaped tip that was 40-millionths of a centimeter in diameter. The needle was enclosed in a tube of rarefied helium and subjected to a very high positive voltage (25,000 volts). This voltage created such an intense electric force that any helium atoms "settling" on the atoms of the needle tip were stripped of electrons to become *ions* (electrically charged atoms). Positively charged helium ions streamed away from the needle tip in a direction almost perpendicular to its surface at every point. They then struck a fluorescent screen, producing this picture of the platinum needle tip, which magnifies the spacing of the atoms by approximately 750,000 times. Clearly, the platinum is crystalline, with the atoms arranged like oranges in a grocer's display. Although the photograph is not of the atoms themselves, it shows the positions of the atoms and reveals the microarchitecture of one of the solids that make up our world.

Crystal Structure

The existence of crystals (salt and quartz, for example) has been known for centuries, but it was not until the development of atomic theory in the twentieth century that crystals could be interpreted as regular arrays of atoms. The crystal as a latticework of atoms was confirmed in 1912 by using X rays. Atoms in a crystal are very close together; most are about 0.3 nanometers apart (1 nanometer = 10^{-9} meter), about the same as the wavelength of X rays. The German physicist Max von Laue discovered that a beam of X rays directed upon a crystal is diffracted into a characteristic pattern called a *diffraction pattern* (Figure 11.1). The diffraction patterns made by X rays on photographic film show the crystal to be a neat mosaic of atoms sited on a regular lattice, like a three-dimensional chessboard or a child's jungle gym. Metals such as iron, copper, and gold have relatively simple crystal structures. Tin and cobalt are only a little more complex. All metals contain a jumble of many crystals, each almost perfect, each with the same regular lattice but at some inclination to the crystal nearby. These metal crystals can be seen when a metal surface is cleaned (etched) with acid (Figure 11.2). You can see them on the surface of galvanized iron that has been exposed to the weather or on brass doorknobs that have been etched by the perspiration of hands.

Von Laue's photographs of the X-ray diffraction patterns fascinated English scientists William Henry Bragg and his son William Lawrence Bragg. They developed a mathematical formula that showed just how X rays should scatter from the various regularly spaced atomic layers in a crystal. With this formula and an analysis of the pattern of spots in a diffraction pattern, they could determine distances between the atoms in a crystal.

A solid can be amorphous or crystalline. In the *amorphous* state, the atoms and molecules are distributed randomly. In the more common crystalline state, the atoms or ions are in a three-dimensional orderly arrangement—a crystalline latticework of atoms

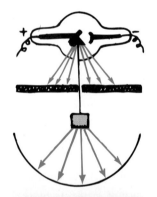

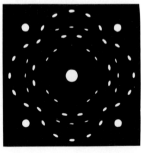

FIGURE 11.1 X-ray determination of crystal structure. The photograph of salt is a product of X-ray diffraction. Rays from the X-ray tube are blocked by a lead screen except for a narrow beam that hits the crystal of sodium chloride (common table salt). The radiation that penetrates the crystal and reaches the photographic film makes the pattern shown. The white spot in the center is the main unscattered beam of X-rays. The size and arrangement of the other spots result from the latticework structure of sodium and chlorine ions in the crystal. Every crystal line structure has its own unique X-ray diffraction picture. A crystal of sodium chloride always produces this same design. (Adapted from *Experimental Chemistry* 6/e by M. J. Sienko, et al. 1984 McGraw-Hill, Inc. Reprinted by permission of McGraw-Hill.)

FIGURE 11.2   Silver crystals.

(Figure 11.3). Each atom or ion vibrates about its own fixed position and is unable to move through the lattice. Atoms are tied together in this lattice by what are called bonding forces. These bonding forces are electrical in nature, and are electrical pushes and pulls between electrical charges in atoms.

We will not go into **atomic bonding** except to say there are four principal types in solids: ionic, covalent, metallic, and the weakest, van der Waals'. More can be found about them in almost any chemistry text.

FIGURE 11.3   Sodium chloride crystal: the large spheres represent chlorine atoms (actually ions).

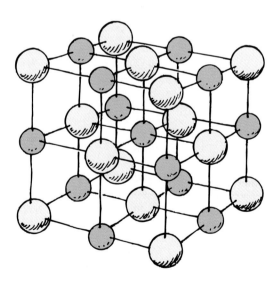

Density

The masses of atoms and the spacing between atoms determine the density of materials. We think of **density** as the "lightness" or "heaviness" of materials of the same size. It is a measure of the compactness of matter, of how much mass is squeezed into a given space; it is the amount of mass per unit volume:*

$$\text{Density} = \frac{\text{mass}}{\text{volume}}$$

The densities of a few materials are given in Table 11.1. Mass is measured in grams or kilograms and volume in cubic centimeters (cm^3) or cubic meters (m^3). A gram of material is defined as equal to the mass of 1 cubic centimeter of water at a temperature of 4°C.[†] Therefore, water has a density of 1 gram per cubic centimeter. Gold, density 19.3 grams per cubic centimeter, is therefore 19.3 times as massive as an equal volume of water. Osmium, a hard, bluish-white metallic element, is the densest substance on earth. Although the individual osmium atom is less massive than individual atoms of gold, mercury, lead, and uranium, the close spacing of osmium atoms in the crystalline

FIGURE 11.4 When the volume of the bread is reduced, its density increases.

Table 11.1 Densities of Various Substances

Material	Grams per Cubic Centimeter	Kilograms per Cubic Meter
Solids		
Osmium	22.6	22,600
Platinum	21.5	21,500
Gold	19.3	19,300
Uranium	19.0	19,000
Lead	11.3	11,300
Silver	10.5	10,500
Copper	8.9	8900
Brass	8.6	8600
Iron	7.9	7860
Tin	5.8–7.3	5800–7300
Aluminum	2.7	2700
Ice	0.92	917
Liquids		
Mercury	13.6	13,600
Glycerin	1.26	1260
Seawater	1.03	1025
Water at 4°C	1.00	1000
Benzene	0.81	810
Ethyl alcohol	0.79	791

* Mass density is related to weight density, which is defined as weight per unit volume:

$$\text{Weight density} = \frac{\text{weight}}{\text{volume}}$$

Weight density is measured in N/m^3. As a 1-kg body has a weight of 9.8 N, weight density is numerically 9.8 times greater than mass density. For example, weight density of water is 9800 N/kg^3. In the British system, 1 cubic foot (ft^3) of fresh water (almost 7.5 gallons) weighs 62.4 pounds. Thus, fresh water has a weight density of 62.4 lb/ft^3.

[†] A cubic meter is a sizable volume and contains a million cubic centimeters, so there are a million grams of water in a cubic meter (or, equivalently, a thousand kilograms of water in a cubic meter). Hence, 1 g/cm^3 = 1000 kg/m^3.

form gives it the greatest density. More osmium atoms fit into a cubic centimeter than other more massive and more widely spaced atoms would fit into the same volume.

Questions

1. What is the density of 100 kg of water?
2. Which has the greater density—10 kg of lead or 100 kg of aluminum?
3. Which has the greater density—the microscopic tip of a platinum needle or the whole world?

Elasticity

When an object is subjected to external forces, it undergoes changes in size or shape or both. The changes depend on the arrangement and bonding of the atoms in the material.

A weight hanging on a spring stretches the spring. Additional weight stretches it still more. If the weights are removed, the spring returns to its original length. We say the spring is *elastic*. When batters hit baseballs, the baseballs temporarily change their shape. An archer, about to shoot an arrow, first bends the bow, which springs back to its original form when the arrow is released. The spring, the baseball, and the bow are examples of elastic objects. **Elasticity** is the property of changing shape when a deforming force acts on an object and returning to the original shape when the deforming force is removed. Not all materials return to their original shape when a deforming force is applied and then removed. Materials that do not resume their original shape after being deformed are said to be *inelastic*. Clay, putty, and dough are inelastic materials. Lead is also inelastic, since it's easy to distort lead permanently.

FIGURE 11.5   A baseball is elastic.

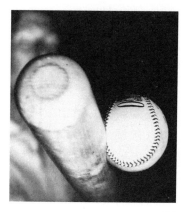

Answers

1. The density of any amount of water is 1000 kg/m³, which means that the mass of water that would exactly fill a cubic-meter tank would be 1000 kg. Can you see that the volume of 100 kg of water must be 1/10 m³? And that 100 kg/0.1 m³ = 1000 kg/1 m³?

2. Density is a *ratio* of mass and volume (or weight and volume), and this ratio is greater for any amount of lead than for any amount of aluminum—see Table 11.1.

3. The platinum needle tip has a greater density. Only if the world were composed entirely of platinum atoms and not more closely packed in the compressed core, would the densities of each be the same. (The average density of the earth is about 5,500 kg/m³, intermediate between the densities of aluminum and iron.)

FIGURE 11.6 The stretch of the spring is directly proportional to the applied force. If the weight is doubled, the spring stretches twice as much.

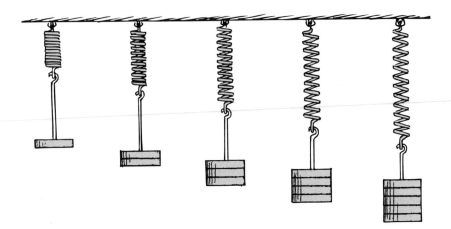

When a weight is hung on a spring, a force (gravity) acts on it. The stretch (or compression) is directly proportional to the applied force (Figure 11.6). This relationship, noted in the mid-seventeenth century by the British physicist Robert Hooke, a contemporary of Isaac Newton, is called **Hooke's law**. The amount of stretch or compression (change in length), Δx, is directly proportional to the applied force F. In shorthand notation,

$$F \sim \Delta x$$

If an elastic material is stretched or compressed beyond a certain amount, it will not return to its original state and will remain distorted. The distance beyond which permanent distortion occurs is called the *elastic limit*. Hooke's law holds only as long as the force does not stretch or compress the material beyond its elastic limit.

Questions

1. A 2-kg load is hung from the end of a spring. The spring then stretches a distance of 10 cm. If, instead, a 4-kg load is hung from the same spring, how much will the spring stretch? How about if a 6-kg load were hung from the same spring? (Assume that none of these loads stretches the spring beyond its elastic limit.)

2. If a force of 10 N stretches a certain spring 4 cm, how much stretch will occur for an applied force of 15 N?

Answers

1. A 4-kg load has twice the weight of a 2-kg load. In accord with Hooke's law, $F \sim \Delta x$, two times the applied force will result in two times the stretch, so the spring should stretch 20 cm. The weight of the 6-kg load will make the spring stretch three times as far, 30 cm. (When the elastic limit is exceeded, then the amount of stretch cannot be predicted with the information given.)

2. The spring will stretch 6 cm. By ratio and proportion, $\dfrac{10\ N}{4\ cm} = \dfrac{15\ N}{6\ cm}$, which is read: 10 newtons is to 4 centimeters as 15 newtons is to 6 centimeters. If taking a lab, you'll learn that the ratio of force to stretch is called the spring constant k (in this case $k = 2.5$ N/cm), and Hooke's law is expressed as the equation $F = k\ \Delta x$.

The Periodic Table
Guide to the Elements

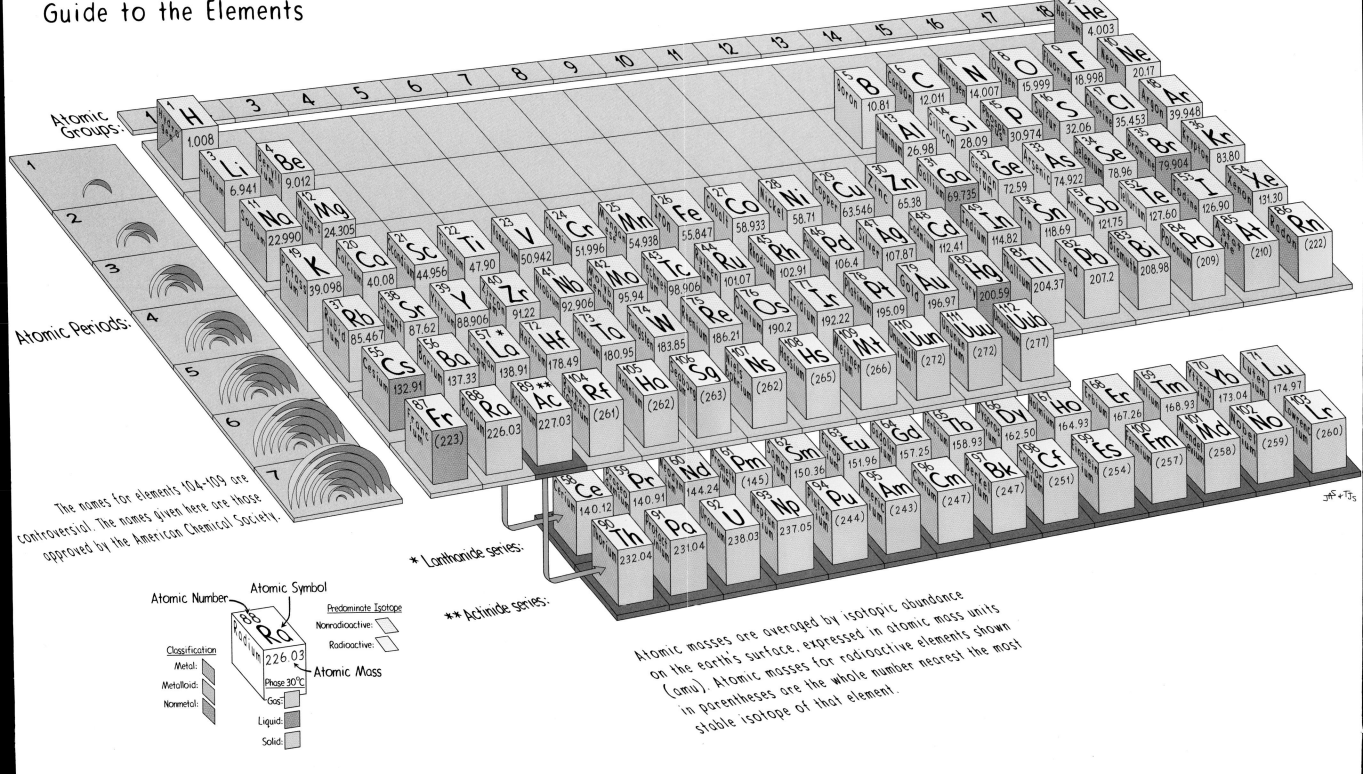

Atomic Groups:

Atomic Periods:

The names for elements 104-109 are controversial. The names given here are those approved by the American Chemical Society.

* Lanthanide series:

** Actinide series:

Atomic Number
Atomic Symbol
Predominate Isotope
Nonradioactive:
Radioactive:

88
Ra
Radium
226.03
Phase 30°C

Classification
Metal:
Metalloid:
Nonmetal:

Atomic Mass

Gas:
Liquid:
Solid:

Atomic masses are averaged by isotopic abundance on the earth's surface, expressed in atomic mass units (amu). Atomic masses for radioactive elements shown in parentheses are the whole number nearest the most stable isotope of that element.

JAF + TJs

The Periodic Table

The Periodic Table

Tension and Compression

When something is pulled on (stretched), it is said to be in *tension*. When it is pushed in (squashed up), it is in *compression*. Compression causes things to get shorter and fatter, whereas tension causes them to get longer and thinner. This is not obvious for rigid materials because the shortening or lengthening is very small.

Steel is an excellent elastic material, for it can withstand large forces and then return to its original size and shape. Because of its strength and elastic properties, it is used to make not only springs but construction girders as well. Vertical girders of steel used in the construction of tall buildings undergo only slight compression. A typical 25-meter-long vertical girder (column) used in high-rise construction is compressed about a millimeter when it carries a 10-ton load. More deformation occurs when girders are used horizontally, where the tendency is to sag under heavy loads.

FIGURE 11.7 The top part of the beam is stretched and the bottom part is compressed. What happens in the middle portion, between top and bottom?

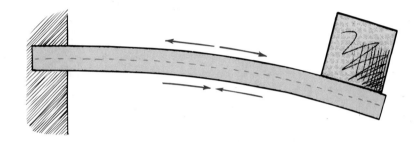

When a horizontal beam is supported at one or both ends, it is under both a tension and a compression from its weight and the load it supports. Consider the horizontal beam supported at one end (known as a cantilever beam) in Figure 11.7. It sags because of its own weight and because of the load it carries at its end. A little thought will show that the top part of the beam tends to be stretched. Atoms tend to be pulled apart. The top part is slightly longer because of its deformation. The top part is under tension. More thought shows that the bottom part of the beam is under compression. Atoms there are squeezed together. The bottom part is slightly shorter because of the way it is bent. So the top part is under tension and the bottom part of the beam is under compression. Can you see that somewhere in between the top and bottom there will be a region where nothing happens, where there is neither tension nor compression? This is the *neutral layer.*

The horizontal beam shown in Figure 11.8, known as a simple beam, is supported at both ends, and carries a load in the middle. This time there is compression in the top of the beam and tension in the bottom part. Again, there is a neutral layer along the middle portion of the thickness of the beam all along its length.

FIGURE 11.8 The top part of the beam is compressed and the bottom part is stretched. Where is the neutral layer (the part that is not under stress due to tension or compression)?

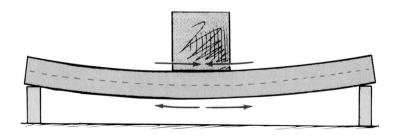

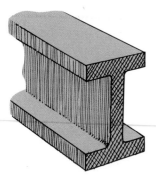

FIGURE 11.9 An I-beam is like a solid bar with some of the steel scooped from its middle where it is needed least. The beam is therefore lighter for nearly the same strength.

We can see why the cross-section of steel girders has the form of the letter I (Figure 11.9). Most of the material in these I-beams is concentrated in the top and bottom flanges; the piece joining the flanges, called the *web*, is much less wide. Thus when the beam is used horizontally in construction, the stress is predominantly in the top and bottom flanges—not in the central portion. One flange is squeezed while the other is stretched. Between the top and bottom flanges is a relatively stress-free region that acts principally to hold the top and bottom flanges well apart. For this purpose, comparatively little material is needed. The flanges carry virtually all stresses in the beam. An I-beam is nearly as strong as a solid rectangular bar of the same overall dimensions, and its weight is considerably less. A large rectangular steel beam on a certain span might collapse under its own weight, whereas an I-beam of the same depth could carry much added load.

Question If you had to make a hole horizontally through one of the tree branches shown in Figure 11.10 in a location that would weaken it the least, would you bore it through the upper portion, the middle, or the lower portion?

FIGURE 11.10 The upper half of each horizontal branch is under tension because of its weight, while the lower half is under compression. In what location is the wood neither stretched nor compressed?

Answer It would be best to drill the hole in the middle, through the neutral layer. Wood fibers in the top part of the branch are being stretched, and if you drilled the hole there, that part of the branch may pull apart. Fibers in the lower part are being compressed, and if you drilled the hole there, that part of the branch might crush under compression. In between, in the neutral layer, the hole will hardly affect the strength of the branch because fibers there are neither stretched nor compressed.

Arches

Stones break more easily under tension than compression. The roofs of stone structures built by the Egyptians during the time of the pyramids were constructed with many horizontal stone slabs. Because of the weakness of these slabs under the tension forces produced by gravity, many vertical columns had to be erected to hold the roofs up. Likewise for the temples in ancient Greece. Then came arches—and fewer vertical columns.

FIGURE 11.11 Common semi-circular stone arches, which have stood for centuries.

FIGURE 11.12 The horizontal slabs of stone of the roof cannot be too long for stone easily breaks under tension. That is why many vertical columns are needed to support the roof.

FIGURE 11.13 Both the curve of a sagging chain and the arch are catenaries.

Look at the tops of the windows in old brick buildings. Chances are the tops are arched. Likewise for the shape of the tops of doorways and corridors of stone structures and of old stone bridges. When a load is placed on a properly arched structure, the compression strengthens rather than weakens the structure. The stones are pushed together more firmly and are held together by compressive forces. With just the right shape of the arch, the stones do not even need cement to hold them together. When the load being supported is uniform horizontally, as with a bridge, the proper shape is a parabola, the same curve followed by a thrown ball. The cables of a suspension bridge provide an example of an "upside-down" parabolic arch. If, on the other hand, the arch is supporting only its own weight, the curve that gives it maximum strength is called a *catenary*. A catenary is the curve formed by a rope or chain hung between two points of support. Tension along every part of the rope or chain is parallel to the curve. So when a freestanding arch takes the shape of an inverted cantenary, compression due to its weight is everywhere parallel to the arch, just as tension between adjacent links of a hanging chain is everywhere parallel to the chain. The arch that graces the city of St. Louis is a catenary (Figure 11.13).

Twirl an arch through a complete circle and you have a dome. The weight of the dome, like that of an arch, produces compression. Modern domes, such as the Astrodome in Houston, cover vast areas without the interruption of columns. There are shallow domes (the Jefferson Monument) and tall ones (the United States Capitol). And before these there were the famous Eskimo igloos.

FIGURE 11.14 The weight of the dome produces compression, not tension, so no support columns are needed in the middle.

Question Why is it easier for a chicken inside an eggshell to poke its way out than it is for a chicken on the outside to poke its way in?

Scaling*

Did you ever notice how strong an ant is for its size? An ant can carry the weight of several ants on its back, whereas a strong elephant couldn't even carry one elephant on its back. How strong would an ant be if it were scaled up to the size of an elephant? Would this "super ant" be several times stronger than an elephant? Surprisingly, the answer is no. Such an ant would not be able to lift its own weight off the ground. Its legs would be too thin for its greater weight and they would likely break.

There is a reason for the thin legs of the ant and the thick legs of an elephant. As the size of a thing increases, it grows heavier much faster than it grows stronger. You can

Answer To poke its way into a shell, a chicken on the outside must contend with compression, which greatly resists shell breakage. Only the weaker shell tension must be overcome when poking from the inside. To see that the compression of the shell is strong, try squashing an egg along its long axis by squeezing it between your thumb and fingers. Surprised? Try it along its shorter diameter. Surprised? (Do this over a sink with some protection such as gloves for possible shell splinters.)

*Material in this section is based on two delightful and informative essays: "On Being the Right Size," from *Possible Worlds* by J. B. S. Haldane. Copyright 1928 by Harper & Row, Publishers, Inc. Renewed 1956 by J. B. S. Haldane. "On Magnitude," from *On Growth and Form* by Sir D'Arcy Wentworth Thompson (New York: Cambridge University Press).

support a toothpick horizontally at its ends and notice no sag. But support a tree of the same kind of wood horizontally at its ends and you'll see a noticeable sag. Relative to its weight, the toothpick is much stronger than the tree. **Scaling** is the study of how the volume and shape (size) of any object affect the relationship of its weight, strength, and surface area.

Strength comes from the area of the cross-section (the "square" of the linear size), whereas weight depends on volume (the "cube" of the linear size). To understand this square-cube relationship, consider the simplest case of a solid cube of matter, 1 centimeter on a side. A 1-cubic centimeter cube has a cross-section of 1 square centimeter. That is, if we sliced through the cube parallel to one of its faces, the sliced area would be 1 square centimeter. Compare this to a cube that has double the linear dimensions, a cube 2 centimeters on each side. Its cross-section area will be 2×2 or 4 square centimeters and its volume will be $2 \times 2 \times 2$ or 8 cubic centimeters. For the same density it would be 8 times heavier. Careful investigation of Figure 11.15 shows that for increases of linear dimensions the cross-sectional area (as well as the total area) grows as the square of the linear dimension, whereas volume and weight grow as the cube of the linear dimension.

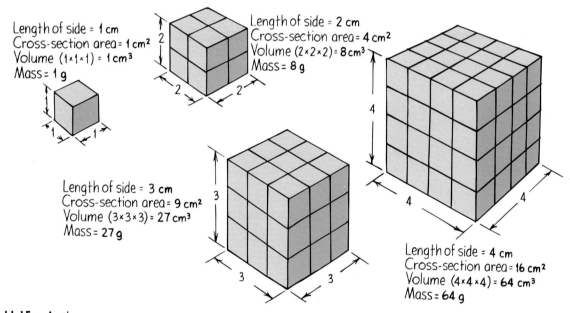

Length of side = 1 cm
Cross-section area = 1 cm²
Volume (1×1×1) = 1 cm³
Mass = 1 g

Length of side = 2 cm
Cross-section area = 4 cm²
Volume (2×2×2) = 8 cm³
Mass = 8 g

Length of side = 3 cm
Cross-section area = 9 cm²
Volume (3×3×3) = 27 cm³
Mass = 27 g

Length of side = 4 cm
Cross-section area = 16 cm²
Volume (4×4×4) = 64 cm³
Mass = 64 g

FIGURE 11.15 As the linear dimensions of an object change by some factor, the cross-section area changes as the square of this factor, and the volume (and hence the weight) changes as the cube of this factor. We see that when the linear dimensions are doubled (factor = 2), the area grows by $2^2 = 4$, and the volume grows by $2^3 = 8$.

Volume (and thus weight) increase much faster than the corresponding increase of cross-sectional area. Although the figure demonstrates the simple example of a cube, the principle applies to an object of any shape. Consider a football player who can do many pushups. Suppose he could somehow be scaled up to twice his size—that is, twice as tall and twice as broad, with his bones twice as thick and every linear dimension increased by a factor of two. Would he be twice as strong and be able to lift himself with twice the ease? The answer is no. Although his twice-as-thick arms would have four times the cross-sectional area and be four times as strong, he would be eight times as heavy. For comparable effort he would be able to lift only half his weight. Relative to his new weight, he would be weaker than before.

We find in nature that large animals have disproportionately thick legs compared with those of small animals. This is because of the relationship between volume and area; the fact that volume (and weight) grow as the cube of the factor by which the linear dimension increases, while strength (and area) grow as the square of the increase factor. So we see there is a reason for the thin legs of a daddy longlegs, deer, or antelope, and for the thick legs of a rhino, hippo, or elephant.

Questions

1. Consider a 1-cubic centimeter cube scaled up to a cube 10 centimeters long on each edge.

   (a) What would be the volume of the scaled-up cube?

   (b) What would be its cross-sectional surface area?

   (c) What would be its total surface area?

2. If you were somehow scaled up to twice your size while retaining your present proportions, would you be stronger or weaker? Explain your reasoning.

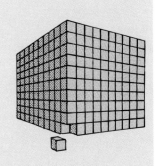

So the great strengths attributed to King Kong and other fictional giants cannot be taken seriously. The fact that the consequences of scaling are conveniently omitted is one of the differences between science and science fiction.

Important also is the comparison of total surface area to volume (Figure 11.16). Total surface area, just like cross-sectional area, grows in proportion to the square of an object's linear size, whereas volume grows in proportion to the cube of the linear size. So as an object grows, its surface area and volume grow at different rates, with the result that the surface area to volume ratio *decreases*. In other words, both the surface area and the volume of a growing object increase, but the growth of surface area *relative* to the growth of volume decreases. Not many people really understand this concept. The following examples may be helpful.

Answers

1. (a) The volume of the scaled-up cube would be (length of side)^3 = (10 cm)^3, or 1000 cm^3.

   (b) Its cross-sectional surface area would be (length of side)^2 = (10 cm)^2, or 100 cm^2.

   (c) Its total surface area = 6 sides x area of a side = 600 cm^2.

2. Your scaled-up self would be four times stronger, because the cross-sectional area of your twice-as-thick bones and muscles increase by four. You could lift a load four times as heavy. But your weight would be eight times as much as before, so you would not be stronger relative to your greater weight. Having four times the strength carrying eight times the weight gives you a strength-to-weight ratio of only half its former value. So if you can barely lift your own weight now, when scaled up you could only lift half your new weight. Your strength would increase, but your strength-to-weight ratio would decrease. Stay as you are!

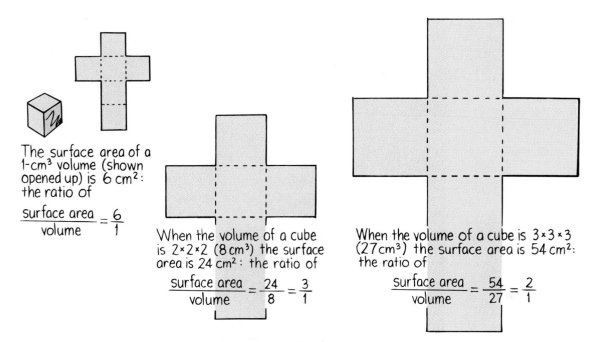

The surface area of a 1-cm³ volume (shown opened up) is 6 cm²: the ratio of

$$\frac{\text{surface area}}{\text{volume}} = \frac{6}{1}$$

When the volume of a cube is 2×2×2 (8 cm³) the surface area is 24 cm²: the ratio of

$$\frac{\text{surface area}}{\text{volume}} = \frac{24}{8} = \frac{3}{1}$$

When the volume of a cube is 3×3×3 (27 cm³) the surface area is 54 cm²: the ratio of

$$\frac{\text{surface area}}{\text{volume}} = \frac{54}{27} = \frac{2}{1}$$

FIGURE 11.16 As the size of an object increases, there is a greater factor of increase in volume than in surface area; as a result, the ratio of surface area to volume decreases.

The large ears of elephants are nature's way of making up for the small ratio of surface area to volume for these large creatures. The large ears are not for better hearing, but for cooling. The rate at which a creature dissipates heat is proportional to its surface area. If an elephant didn't have large ears, it would not have enough surface area to cool its huge mass. The large ears of the African elephant greatly increase overall surface area, which facilitates cooling in hot climates.

At the biological level, living cells must contend with the fact that the growth of volume is faster than the growth of surface area. Cells obtain nourishment by diffusion through their surfaces. As cells grow their surface area increases, but not fast enough to keep up with volume. For example, if the surface area increases by four, the corresponding volume increases by eight. Eight times as much mass must be sustained by only four times as much nourishment. This puts a limit on the growth of a living cell. So cells divide, and there is life as we know it. That's nice.

Not so nice is the fate of large creatures when they fall. The statement "The bigger they are, the harder they fall" holds true and is a consequence of the small ratio of surface area to weight. Resistance of the air to movement through it is proportional to the surface area of the moving object. If you fall out of a tree, even in the presence of air resistance, your rate of fall is very nearly 1 g. You don't have enough surface area relative to your weight—unless you wear a parachute. Small creatures, on the other hand, need no parachute. They have plenty of surface area relative to their small weight. An insect can fall from the top of a tree to the ground below without harm. The surface area per weight ratio is in the insect's favor, for the insect in effect *is* its own parachute.

The fact that small things have great surface areas relative to their volume, mass, or weight is evident in the kitchen. A cook knows that more skin results when peeling 5 kilograms of small potatoes than when peeling 5 kilograms of large potatoes. Smaller objects have more surface area per kilogram. Crushed ice will cool a drink much faster than a single ice cube of the same mass, because crushed ice presents more surface area to the beverage. Steel wool rusts away at the sink while steel sink fixtures don't. Rusting is a surface phenomenon. Iron rusts when exposed to air, but it rusts much faster and is soon eaten away if it is in the form of small strands or filings.

FIGURE 11.17 The African elephant has less surface area relative to its weight than other animals. It compensates with its large ears, which significantly increase its radiating surface area and promote cooling.

FIGURE 11.18 The long tail of the monkey not only helps maintain balance but also effectively radiates heat.

Chunks of coal burn, while coal dust explodes when ignited. Thin french fries cook faster in oil than fat fries. Flat hamburgers cook faster than meatballs of the same mass. Large raindrops fall faster than small raindrops, and large fish swim faster than small fish. These are all consequences of the fact that volume and area are not in direct proportion to each other.

It is interesting to note that the rate of heartbeat in a mammal is related to the size of the mammal. The heart of a tiny shrew beats about twenty times faster than the heart of an elephant. In general, small mammals live fast and die young; larger animals live at a leisurely pace and live longer. Don't feel bad about a pet hamster who doesn't live as long as a dog. All warm-blooded animals have about the same life span—not in terms of years, but in the average number of heartbeats (about 800 million). Humans are the exception: we live two to three times longer than other mammals of our size.

Summary of Terms

Atomic bonding   The linking together of atoms to form larger structures, including solids.

Density   The mass of a substance per unit volume:

$$\text{Density} = \frac{\text{mass}}{\text{volume}}$$

Or we may define *weight density* as weight/volume:

$$\text{Weight density} = \frac{\text{weight}}{\text{volume}}$$

Elasticity   The property of a material wherein it changes shape when a deforming force acts on it, and returns to its original shape when the force is removed.

Hooke's law   The amount of stretch or compression of an elastic material is directly proportional to the applied force.

$$F \sim \Delta x$$

Scaling   The study of how size affects the relationships among weight, strength, and surface.

Suggested Reading

For more about the relationships among the size, area, and volume of objects, read these essays:"On Being the Right Size" by J. B. S. Haldane, and "On Magnitude" By Sir D'Arcy Wentworth Thompson. Both are in J. R. Newman, ed. *The World of Mathematics,* Vol. II. New York: Simon & Schuster, 1956.

Morrison, P. *Powers of Ten.* New York: W. H. Freeman, 1982.

Review Questions

Müller's Micrograph

1. Distinguish between an atom and an ion.

Crystal Structure

2. How does the arrangement of atoms in a crystalline substance differ from that in a non-crystalline substance?

3. What evidence can you cite for the microscopic crystal nature of certain solids? For macroscopic crystal nature?

Density

4. What happens to the volume of a loaf of bread that is squeezed? The mass? The density? (Why are all three answers different?)

5. What happens to the density of a chocolate bar when you cut it in half?

6. The uranium atom is the heaviest atom found in nature. Why is uranium metal not the densest material?

7. Which has the greater density, a heavy bar of pure gold or a pure gold ring? Defend your answer.

8. Distinguish between mass density and weight density.

Elasticity

9. What is the evidence for the claim that a baseball is elastic?

10. Cite a substance that is inelastic. What is the evidence for your answer?

11. What is Hooke's law, and how does it apply to things that can be squeezed or stretched?

12. What is meant by the elastic limit for a particular object?

13. If the weight of a 1-kg body stretches a spring by 2 cm, how much will the spring be stretched when it supports a 3-kg load? (Assume the spring does not reach its elastic limit.)

Tension and Compression

14. Distinguish between *tension* and *compression.*

15. What is the neutral layer in a beam that supports a load?

16. Why are the cross-sections of metal beams in the shape of the letter I instead of solid rectangles?

Arches

17. Why were so many vertical columns needed to support the roofs of stone buildings in ancient Egypt and Greece?

18. Is it *tension* or *compression* that strengthens an arch that supports a load?

19. Why are no vertical columns needed to support the middle of the Houston Astrodome?

Scaling

20. What is the square-cube relationship in scaling?

21. Distinguish among *linear dimension, area,* and *volume.*

22. If the linear dimensions of an object are doubled, by how much does the surface area increase? By how much does the volume increase?

23. As the volume of an object is increased, its surface area also increases. Does the *ratio* of square meters to cubic meters decrease? Defend your answer.

24. Which will rust faster, a piece of iron or the same amount of iron in "steel wool"? Why?

25. Which has more skin, an elephant or a mouse? Which has more skin *per bodyweight,* an elephant or a mouse?

Projects

1. If you live in a region where snow falls, collect some snowflakes on black cloth and examine them with a magnifying glass. You'll see that all the many shapes are hexagonal crystalline structures, among the most beautiful sights in nature.

2. Simulate atomic close packing with a couple of dozen or so pennies. Arrange them in a square pattern so each penny inside the perimeter makes contact with four others. Then arrange them hexagonally, so contact with six others occurs. Compare the areas occupied by the same number of pennies close-packed both ways.

3. Are you slightly longer while lying down than you are tall when standing up? Make measurements and see.

4. If you nail four sticks together to form a rectangle, they can be deformed into a parallelogram without too much effort. But if you nail three sticks together to form a triangle, the shape cannot be changed without actually breaking the sticks or dislodging the nails. The triangle is the strongest of all the geometrical figures, which is why you see triangular shapes in bridges. Try it and see, and then look at the triangles used in strengthening structures of many kinds.

Exercises

1. Silicon is the chief ingredient of both glass and semiconductor devices, yet the physical properties of glass are different from those of semiconductor devices. Explain.

2. What evidence can you cite to support the claim that crystals are composed of atoms that are arranged in specific patterns?

3. Is iron necessarily heavier than cork? Explain.

4. When water freezes, it expands. What does this say about the density of ice relative to the density of water?

5. In a deep dive, a whale is appreciably compressed by the pressure of the surrounding water. What happens to the whale's density?

6. When you are floating in water with your lungs full of air and you exhale, you sink lower in the water. Which is more significant in causing you to sink, your change of mass or your change of density?

7. The uranium atom is the heaviest and most massive atom among the naturally occurring elements. Why then, isn't a solid bar of uranium the densest metal?

8. Which has more volume—a kilogram of gold or a kilogram of aluminum?

9. Which has more weight—a liter of ice or a liter of water?

10. Why does the suspended spring stretch more at the top than at the bottom?

11. If the spring in the sketch above were supporting a heavy weight, how would the sketch be changed?

12. A thick rope is stronger than a thin rope of the same material. Is a long rope stronger than a short rope?

13. Tension and compression occur in a partially supported horizontal beam when it sags due to gravity or supports a load. Make a simple sketch to show a means of supporting the beam so that tension occurs at the top part and compression at the bottom. Sketch another case where compression is at the top and tension occurs at the bottom.

14. Can a horizontal I-beam support a greater load when the web is horizontal or when the web is vertical? Explain.

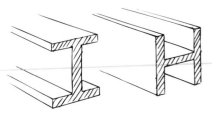

15. The sketches are of top views of a dam to hold back a lake. Which of the two designs is better? Why?

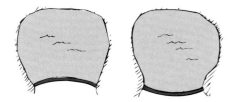

16. Consider a very large wooden barrel, such as the kind used in wineries. Should the opposite ends be concave (bending inward) or convex (bending outward)? Why?

17. Archie designs an arch to serve as an outdoor sculpture in a park. The arch is to be a certain width and a certain height. To get the size and shape for the strongest arch, he suspends a chain from two equally elevated supports as far apart as the arch is wide and lets the chain hang as low as the arch is high. He then designs the arch to have exactly the inverted shape of the hanging chain. Explain why.

18. The photo shows a circular arch of stone. Note that it must be held together with steel rods to prevent outward movement. If the shape of the arch were not a semicircle, but the shape used by Archie in the previous exercise, would the steel rods be necessary? Explain.

19. Why do you suppose that girders are so often arranged to form triangles in the construction of bridges and other structures? (Compare the stability of three sticks nailed together to form a triangle with four sticks nailed to form a rectangle, or any number of sticks to form multi-legged geometric figures. Try it and see!)

20. Consider two bridges that are exact replicas of each other except that every dimension in the larger is exactly twice that of the other—that is, twice as long, structural elements twice as thick, etc. Which bridge is more likely to collapse under its own weight?

21. A candymaker making taffy apples decides to use 100 kg of large apples rather than 100 kg of small apples. Will the candymaker need to make more or less taffy to cover the apples?

22. Will a person twice as heavy as another use more or less than twice as much suntan lotion at the beach?

23. Why is it easier to start a fire with kindling rather than large sticks and logs of the same kind of wood?

24. Why does a chunk of coal burn when ignited, whereas coal dust explodes?

25. Why does a two-story house that is roughly a cube suffer less heat loss than a rambling one-story house of the same volume?

26. Why is heating more efficient in large apartment buildings than in single-family dwellings?

27. If you are grilling hamburgers and get impatient, why is it a good idea to flatten the burgers to make them wider and thinner?

28. Why do thin French fries cook faster than fat fries?

29. Some environmentally conscious people build their homes in the shape of domes. Why is there less heat loss in a dome-shaped dwelling?

30. If you use a batch of cake batter for cupcakes instead of cake and bake them for the time suggested for baking a cake, what will be the result?

31. Why are mittens warmer than gloves on a cold day? And which parts of the body are most susceptible to frostbite? Why?

32. Why are gymnasts usually short in stature?

33. How does scaling relate to the fact that the heartbeat of large creatures is generally slower than the heartbeat of smaller creatures?

34. Nourishment is obtained from food through the inner surface area of the intestines. Why is it that a small organism, such as a worm, has a simple and relatively straight intestinal tract, while a large organism, such as a human being, has a complex and many-folded intestinal tract?

35. The human lungs have a volume of only about 4 liters (L), yet an internal surface area of nearly 100 m^2. Why is this important, and how is this accomplished?

36. What does the concept of scaling have to do with the fact that living cells in a whale are about the same size as those in a mouse?

37. Which fall faster, large or small raindrops?

38. Who has more need for drink in a dry desert climate—a child or an adult?

39. Why doesn't a hummingbird soar like an eagle and an eagle flap its wings like a hummingbird?

40. Can you relate the idea of scaling to the governance of small versus large groups of citizens? Explain.

Problems

1. Find the density of a 5-kg solid cylinder. The cylinder is 10 cm tall and has a radius of 3 cm.
2. What is the weight of a cubic meter of cork? Could you lift it? (For the density of cork, use 400 kg/m^3.)
3. A certain spring stretches 3 cm when a load of 15 N is suspended from it. How much will the spring stretch if 45 N is suspended from it (and it doesn't reach its elastic limit)?
4. A certain spring stretches 4 cm when a load of 10 N is suspended from it. How much will the spring stretch if an identical spring also supports the load as shown in a and b? (Neglect the weights of the springs.)

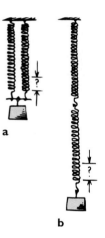

a

b

5. If a certain spring stretches 4 cm when a load of 10 N is suspended from it, how much will the spring stretch if it is cut in half and 10 N is suspended from it?
6. If the linear dimensions of a storage tank are reduced to half their former values, by how much does the overall surface area of the tank decrease? By how much does its volume decrease?
7. A cube 2 cm on a side is cut into cubes 1 cm on a side. (a) How many cubes result? (b) What was the surface area of the original cube and what is the total surface area of the eight smaller cubes? What is the ratio of surface areas? (c) What are the surface-to-volume ratios of the original cube and the combination of all the smaller cubes?
8. Consider eight little spheres of mercury, each with a diameter of 1 mm. When they coalesce to form a single sphere, how big will it be? How does its surface area compare with the total surface area of the previous eight little spheres?

12

Liquids

Molecules beautifully sliding over one another.

Unlike a solid, a liquid can flow. Molecules that make up a liquid are not confined to fixed positions as in solids, but can move freely from position to position by sliding over one another, and the liquid takes the shape of its container. Molecules of a liquid are close together and greatly resist compressive forces. Liquids, like solids, are hard to compress.

Pressure

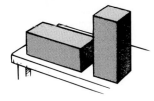

FIGURE 12.1 Although the weight of both blocks is the same, the upright block exerts greater pressure against the table.

A liquid contained in a vessel exerts forces against the walls of the vessel. To discuss the interaction between the liquid and the walls, it is convenient to introduce the concept of **pressure**. Pressure is defined as force per unit area and is obtained by dividing the force by the area on which the force acts:*

$$\text{Pressure} = \frac{\text{force}}{\text{area}}$$

As an illustration of the distinction between pressure and force, consider the two blocks in Figure 12.1. The blocks are identical, but, one stands on its end and the other

* Pressure may be measured in any unit of force divided by any unit of area. The standard international (SI) unit of pressure, the newton per square meter, is called the *pascal* (Pa) after the seventeenth-century theologian and scientist, Blaise Pascal. A pressure of 1 Pa is very small and approximately equals the pressure exerted by a dollar bill resting flat on a table. Science types more often use kilopascals (1 kPa = 1000 Pa).

215

on its side. Both blocks are of equal weight and therefore exert the same force on the surface, but the upright block exerts a greater pressure against the surface. This is because the force is distributed over a smaller area, which increases the pressure. If the block were tipped up so that it made contact with a single corner, the pressure would be greater still.

Pressure in a Liquid

When you swim under water, you can feel the water pressure acting against your eardrums. The deeper you swim, the greater the pressure. What causes this pressure? It is simply the weight of the water (and air) above you pushing against you. If you swim twice as deep, there is twice the weight of water above, and twice the water pressure. The pressure of the air above is always transmitted down through the water and adds to the water pressure. But, because air pressure near earth's surface is nearly constant, the only thing that changes with depth is water pressure. That's what we consider here.

Liquid pressure depends on density as well as depth. If you were submerged in a liquid more dense than water, the pressure would be correspondingly greater. The pressure in a liquid is equal to the product of weight density and depth:*

$$\text{Liquid pressure} = \text{weight density} \times \text{depth}$$

Simply put, the pressure that a liquid exerts against the sides and bottom of a container depends only on the density and depth of the liquid. At twice the depth, the liquid pressure against the bottom is twice as great; at three times the depth, the liquid pressure is threefold; and so on. Or, if the liquid is two or three times as dense, liquid pressure is correspondingly two or three times as great for any given depth. Liquids are practically

FIGURE 12.2 The dependence of liquid pressure on depth is not a problem for the giraffe because of its large heart and intricate system of valves and elastic, absorbent blood vessels in the brain. Without these structures, it would faint when suddenly raising its head and would be subject to brain hemorrhaging when lowering it.

* This is derived from the definitions of pressure and density. Consider an area at the bottom of a vessel of liquid. The weight of the column of liquid directly above this area produces pressure. From the definition Weight density = weight/volume, we can express this weight of liquid as Weight = weight density × volume, where the volume of the column is simply the area multiplied by the depth. Then we get

$$\text{Pressure} = \frac{\text{force}}{\text{area}} = \frac{\text{weight}}{\text{area}} = \frac{\text{weight density} \times \text{volume}}{\text{area}} = \frac{\text{weight density} \times (\text{area} \times \text{depth})}{\text{area}}$$
$$= \text{weight density} \times \text{depth}$$

(Strictly speaking, we should add to this equation the pressure due to the atmosphere on the liquid surface.)

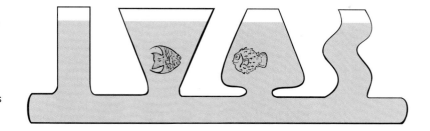

FIGURE 12.3 Liquid pressure is the same for any given depth below the surface, regardless of the shape of the containing vessel. Liquid pressure = weight density × depth (plus the air pressure at the top).

incompressible—that is, their volume can hardly be changed at all by pressure. So, except for small changes produced by temperature, the density of a liquid is practically the same at all depths.*

It is important to note that the pressure does not depend on the amount of liquid. For example, you feel the same pressure whether you dunk your head a meter under water in a small pool or a meter under water in the middle of a large lake. The same is true for a fish. Refer to the connecting vases in Figure 12.3. If we hold a goldfish by its tail and dunk its head a couple of centimeters under the surface, water pressure on the fish's head will be the same in any of the vases. If we release the fish and it swims a few centimeters deeper, the pressure on the fish will increase with depth but be the same no matter what vase the fish is in. If the fish swims to the bottom, pressure will be greater, but it makes no difference what vase it is in. All vases are filled to equal depths, so the water pressure is the same at the bottom of each vase, regardless of its shape or volume. If water pressure at the bottom of a vase were greater than water pressure at the bottom of a neighboring narrower vase, the greater pressure would force water sideways and then up the narrower vase to a higher level until the pressures at the bottom were equalized. But this doesn't happen. Pressure is depth dependent and not volume dependent, so we see there is a reason why water seeks its own level.

FIGURE 12.4 The average water pressure acting against the dam depends on the average depth of the water and not on the volume of water held back. The large shallow lake exerts only one-half the average pressure that the small deep pond exerts.

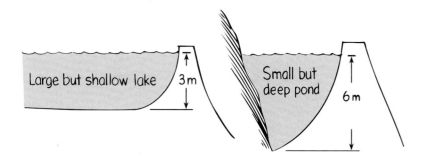

Large but shallow lake   3 m

Small but deep pond   6 m

The fact that water seeks its own level can be demonstrated by filling a garden hose with water and holding the two ends at the same height. The water levels will be equal. If one end is raised higher than the other, water will flow out of the lower end, even if it has to flow "uphill" part of the way to get there. This fact was not fully appreciated by some of the Romans, who built elaborate aqueducts with tall arches and roundabout routes to ensure that water would always flow slightly downward every

---

* The density of fresh water is 1000 kg/m³. Since the weight (*mg*) of 1000 kg is 1000 × 9.8 = 9800 N, the weight density of water is 9800 newtons per cubic meter (9800 N/m³). The water pressure beneath the surface of a lake is simply equal to this density multiplied by the depth in meters. For example, water pressure is 9800 N/m² at a 1-m depth and 98,000 N/m² at a depth of 10 m. In SI units pressure is measured in pascals, so this would be 9800 Pa and 98,000 Pa, respectively; or, in kilopascals, 9.8 kPa and 98 kPa, respectively. For the *total pressure* in these cases, add the pressure of the atmosphere, 101.3 kPa.

FIGURE 12.5 Roman aqueducts assured that water flowed slightly downhill from reservoir to city.

place along its route from the reservoir to the city. If pipes were laid in the ground and followed the natural contour of the land, in some places the water would have to flow uphill, and the Romans were skeptical of this. Careful experimentation was not yet the mode, so with plentiful slave labor the Romans built unnecessarily elaborate aqueducts.

Another interesting fact about liquid pressure is that it is exerted equally in all directions. For example, if we are submerged in water, no matter which way we tilt our heads we feel the same amount of water pressure on our ears. Because a liquid can flow, the pressure isn't only downward. We know pressure acts sideways when we see water spurting sideways from a leak in the side of an upright can. We know pressure also acts upward when we try to push a beach ball beneath the surface of the water. The bottom of a boat is certainly pushed upward by water pressure. Pressure in a liquid at any location is exerted in equal amounts in all directions.

FIGURE 12.6 The forces of a liquid pressing against a surface add up to a net force that is perpendicular to the surface.

When liquid presses against a surface, there is a net force directed perpendicular to the surface (Figure 12.6). If there is a hole in the surface, the liquid spurts at right angles to the surface before curving downward due to gravity (Figure 12.7). At greater depths the pressure is greater and the speed of the escaping liquid is greater.*

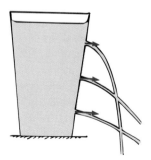

FIGURE 12.7 Water pressure acts perpendicular to the sides of a container, and increases with increasing depth.

* The speed of liquid out of the hole is $\sqrt{2gh}$, where h is the depth below the free surface. Interestingly enough, this is the same speed the water or anything else would have if freely falling the same vertical distance h.

Buoyancy

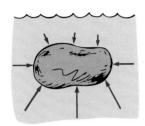

FIGURE 12.8 The greater pressure against the bottom of a submerged object produces an upward buoyant force.

Anyone who has ever lifted a submerged object out of water is familiar with buoyancy, the apparent loss of weight of objects submerged in a liquid. For example, lifting a large boulder off the bottom of a riverbed is a relatively easy task as long as the boulder is below the surface. When it is lifted above the surface, however, the force required to lift it is considerably more. This is because when the boulder is submerged, the water exerts an upward force on it that is exactly opposite in direction to gravity. This upward force is called the **buoyant force** and is a consequence of pressure increasing with depth. Figure 12.8 shows why the buoyant force acts upward. Force due to water pressure is exerted everywhere against the object in a direction perpendicular to its surface. The arrows represent the magnitude's and directions of the forces at different places. Forces that produce pressures against the sides due to equal depths cancel one another. Pressure is greatest against the bottom of the boulder because the bottom of the boulder is at a greater depth. Since the upward forces against the bottom are greater than the downward forces against the top, the forces do not cancel, and there is a net force upward. This net upward force is the buoyant force.

If the weight of the submerged object is greater than the buoyant force, the object will sink. If the weight is equal to the buoyant force on the submerged object, it will remain at any level, like a fish. If the buoyant force is greater than the weight of the completely submerged object, it will rise to the surface and float.

FIGURE 12.9 When a stone is submerged, it displaces water that has a volume equal to the volume of the stone.

FIGURE 12.10 The raised level equals that which would occur if, instead of putting the stone in the container, water equal to the stone's volume were poured in.

Understanding buoyancy requires understanding the meaning of the expression "volume of water displaced." If a stone is placed in a container that is brimful of water, some water will overflow (Figure 12.9). Water is *displaced* by the stone. A little thought will tell us that the *volume of the stone*—that is, the amount of space it takes up or the number of its cubic centimeters—is equal to the *volume of water displaced*. Place any object in a container partly filled with water, and the level of the surface rises (Figure 12.10). By how much? Exactly the same as if a volume of water were poured in that was equal to the volume of the submerged object. This is a good method for determining the volume of irregularly shaped objects: *A completely submerged object always displaces a volume of liquid equal to its own volume.*

Archimedes' Principle

The relationship between buoyancy and displaced liquid was first discovered in the third century BC by the Greek scientist Archimedes. It is stated as follows:

> **An immersed body is buoyed up by a force equal to the weight of the fluid it displaces.**

This relationship is called **Archimedes' principle**. It is true of liquids and gases, which are both fluids. If an immersed body displaces 1 kilogram of fluid, the buoyant force

FIGURE 12.11 A liter of water occupies a volume of 1000 cm³, has a mass of 1 kg, and weighs 9.8 N. Its density may therefore be expressed as 1 kg/L and its weight density as 9.8 N/L. (Seawater is slightly denser, about 10.0 N/L.)

acting on it is equal to the weight of 1 kilogram.* By *immersed,* we mean either *completely* or *partially submerged.* If we immerse a sealed 1-liter container halfway into the water, it will displace a half-liter of water and be buoyed up by a force equal to the weight of a half-liter of water. If we immerse it completely (submerge it), it will be buoyed up by a force equal to the weight of a full liter of water (having a mass of 1 kilogram). Unless the container is compressed, the buoyant force will equal the weight of 1 kilogram of water at *any* depth, as long as it is completely submerged. This is because at any depth the container can displace no greater volume of water than its own volume. And the weight of this displaced water (not the weight of the submerged object!) is equal to the buoyant force.

If a 25-kilogram object displaces 20 kilograms of fluid upon immersion, its apparent weight will be equal to the weight of 5 kilograms. Note that in Figure 12.12 the 3-kilogram block has an apparent weight equal to the weight of 1 kilogram when submerged. The apparent weight of a submerged object is its weight in air minus the buoyant force.

Questions

1. Does Archimedes' principle tell us that if an immersed object displaces liquid weighing 10 N, the buoyant force on the object is 10 N?
2. A 1-L container completely filled with lead has a mass of 11.3 kg and is submerged in water. What is the buoyant force acting on it?
3. A boulder is thrown into a deep lake. As it sinks deeper and deeper into the water, does the buoyant force on it increase? Decrease?
4. Since buoyant force is the net force that a fluid exerts on a body and we learned in Chapter 4 that net forces produce accelerations, why doesn't a submerged body accelerate?

Answers

1. Yes. Looking at it another way, the immersed object pushes 10 N of fluid aside. The displaced fluid reacts by pushing back on the immersed object with 10 N.
2. The buoyant force is equal to the weight of 1 kg (9.8 N) because the volume of water displaced is 1 L, which has a mass of 1 kg and a weight of 9.8 N. The 11.3 kg of the lead is irrelevant; 1 L of anything submerged in water will displace 1 L and be buoyed upward with a force of 9.8 N, the weight of 1 kg.
3. Buoyant force does not change as the boulder sinks because the boulder displaces the same volume of water at any depth. Since water is practically incompressible, its density is nearly the same at all depths; hence, the weight of water displaced, or the buoyant force, is practically the same at all depths.
4. It does accelerate if the buoyant force is not balanced by other forces that act on it—the force of gravity and fluid resistance. The net force on a submerged body is the result of the net force the fluid exerts (buoyant force), the weight of the body, and, if moving, the force of fluid friction.

---

* A kilogram is not a unit of force but a unit of mass. So, strictly speaking, the buoyant force is not 1 kg, but the *weight* of 1 kg, which is 9.8 N. We could as well say that the buoyant force is 1 *kilogram weight,* not simply 1 kg.

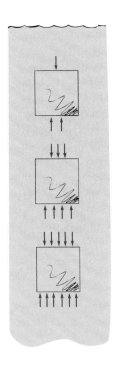

FIGURE 12.12 A 3-kg block weighs more in air than in water. When submerged, its loss in weight is the buoyant force, which is equal to the weight of water displaced.

Perhaps your instructor will summarize Archimedes' principle by way of a numerical example to show that the difference between the upward-acting and the downward-acting forces due to the similar pressure differences on a submerged cube is numerically identical to the weight of fluid displaced. It makes no difference how deep the cube is placed, for although the pressures are greater with increasing depths, the *difference* in pressure up against the bottom of the cube and down against the top of the cube is the same at any depth (Figure 12.13). Whatever the shape of the submerged body, the buoyant force is equal to the weight of fluid displaced.

FIGURE 12.13 The difference in the upward and downward force acting on the submerged block is the same at any depth.

Floating Mountains

The tip of a floating iceberg above the water surface comprises only 10% of the whole iceberg. That's because ice is 0.9 times the density of water, so 90% of it extends into the water. A mountain similarly floats on the earth's semi-liquid mantle with only its tip showing. That's because the earth's continental crust is about 0.85 times the density of the mantle it floats upon; thus about 85% of a mountain extends into the mantle. So like floating icebergs, mountains are appreciably deeper than they are high.

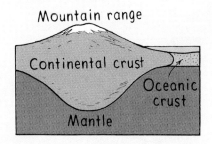

FIGURE 12.14 The continental crust is deeper beneath mountains.

There is an interesting gravitational side-light to this: Recall in Chapter 8 that the gravitational field at the earth's surface varies slightly with varying densities of underlying rock—valued information to geologists and oil prospectors. Recall also that gravitation is less at the top of a mountain because of the greater distance to the earth's center. Combining these ideas we see that because the bottom of a mountain pushes so deep into the earth's mantle, there is increased distance between a mountaintop and the more dense parts of the earth. This increased "gap" further reduces gravitation at the top of mountains.

Another interesting fact about mountains: If you could shave off the top of an iceberg, the iceberg would be lighter and be buoyed up to nearly its original height, before its top was shaved. Similarly, when mountains erode they are lighter, and are pushed up from below to float to nearly their original heights. So when a kilometer of mountain erodes away, some 85% of a kilometer of mountain comes back. That's why it takes so long for mountains to weather away.

What Makes an Object Sink or Float?

It is important to emphasize that the buoyant force that acts on a submerged object depends on the volume of the object. Small objects displace small amounts of water and are acted on by small buoyant forces. Large objects displace large amounts of water and are acted on by larger buoyant forces. It is the *volume* of the submerged object—not its *weight*—that determines buoyant force. Buoyant force is equal to the weight of fluid displaced. (A misunderstanding of this idea is at the root of a lot of confusion that you or your friends may have about buoyancy!)

The weight of an object does play a role, for whether an object will sink or float in a liquid depends on the relative magnitudes of buoyant force and weight. Can you see that if the buoyant force is exactly equal to the weight of the object, then the weight of the object must be the same as the weight of the water displaced? What would its density be in this case? Since the weights are the same and the volumes of object and displaced water are the same, their densities must be the same. The density of the object equals the density of water. This is true for a fish, which is "at one" with the water—it doesn't sink and it doesn't float. If the fish were somehow bloated up, it would be less dense than water and it would float to the top. If the fish swallowed a stone and became more dense than water, it would sink to the bottom. From this we see three simple rules.

1. If an object is denser than the fluid in which it is immersed, it will sink.
2. If an object is less dense than the fluid in which it is immersed, it will float.
3. If an object has a density equal to the density of the fluid in which it is immersed, it will neither sink nor float.

From these rules, what can you say about people who, try as they may, cannot float?* They're simply too dense! To float more easily, you must reduce your density. The formula: Weight density = weight/volume tells you that you must either reduce your weight or increase your volume. The purpose of a life jacket is to increase volume while correspondingly adding very little to your weight.

The weight, not the volume, of a submarine is varied to achieve the desired density. Water is taken into and out of its ballast tanks. A fish regulates its density by expanding and contracting an air sac that changes its volume. The fish can move upward by increasing its volume (which decreases its density) and downward by contracting its volume (which increases its density). The overall density of a crocodile increases when it swallows stones. From 4 to 5 kilograms of stones have been found lodged in the front part of the stomach in large crocodiles. Because of this increased density, the crocodile swims lower in the water, thus exposing less of itself to its prey (Figure 12.15).

FIGURE 12.15 (left) A crocodile coming toward you in the water. (right) A stoned crocodile coming toward you in the water.

* Interestingly, the people who can't float are, nine times out of ten, males. Most males are more muscular and slightly denser than females.

Flotation

Iron is much denser than water and therefore sinks, but an iron ship floats. Why is this so? Consider a solid 1-ton block of iron. Iron is nearly eight times as dense as water, so when it is submerged it will displace only $\frac{1}{8}$ ton of water, which is certainly not enough to keep it from sinking. Suppose we reshape the same iron block into a bowl, as shown in Figure 12.16. It still weighs 1 ton. When we place it in the water, it settles into the water, displacing a greater volume of water than before. The deeper it is immersed, the more water it displaces and the greater the buoyant force that acts on it. When the buoyant force equals 1 ton, it will sink no further.

FIGURE 12.16 An iron block sinks, while the same quantity of iron shaped like a bowl floats.

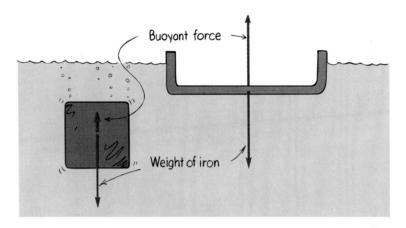

FIGURE 12.17 The weight of a floating object equals the weight of the water displaced by the submerged part.

When the iron boat displaces water that weighs as much as the boat, the boat floats. This is sometimes called the **principle of flotation**, which states:

A floating object displaces a weight of fluid equal to its own weight.

FIGURE 12.18 A floating object displaces a weight of fluid equal to its own weight.

Every ship, every submarine, and every dirigible must be designed to displace a weight of fluid equal to its own weight. Thus, a 10,000-ton ship must be built wide enough to displace 10,000 tons of water before it sinks too deep in the water. The same holds true for vessels in air. A dirigible that weighs 100 tons displaces at least 100 tons of air. If it displaces more, it rises; if it displaces less, it falls. If it displaces exactly its weight, it hovers at constant altitude.

Since the buoyant force upon a body equals the weight of the fluid it displaces, denser fluids will exert a greater buoyant force upon a body than less dense fluids of the same volume. A ship therefore floats higher in salt water than in fresh water because salt water is slightly denser than fresh water. In the same way, a solid chunk of iron will float in mercury even though it sinks in water.

FIGURE 12.19 The same ship empty and loaded. How does the weight of its load compare to the weight of extra water displaced?

Questions

Fill in the blanks for these statements:

1. The volume of a submerged body is equal to the _____ of the fluid displaced.
2. The weight of a floating body is equal to the _____ of the fluid displaced.

Answers

1. Volume.
2. Weight.

Pascal's Principle

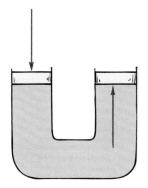

FIGURE 12.20 The force exerted on the left piston increases the pressure in the liquid and is transmitted to the right piston.

One of the most important facts about fluid pressure is that a change in pressure at one part of the fluid will be transmitted undiminished to other parts. For example, if the pressure of city water is increased at the pumping station by 10 units of pressure, the pressure everywhere in the pipes of the connected system will be increased by 10 units of pressure (providing the water is at rest). This rule is called **Pascal's principle**:

> **A change in pressure at any point in an enclosed fluid at rest is transmitted undiminished to all points in the fluid.**

Pascal's principle was discovered by Blaise Pascal (1623–1662), a French mathematician, physicist, and theologian. The SI unit of pressure, the pascal (1 Pa = 1 N/m^2), is named after him.

Fill a U-tube with water and place pistons at each end, as shown in Figure 12.20. Pressure exerted against the left piston will be transmitted throughout the liquid and against the bottom of the right piston. (The pistons are simply "plugs" that can slide freely but snugly inside the tube.) The pressure that the left piston exerts against the water will be exactly equal to the pressure the water exerts against the right piston at the same height. This is nothing to write home about. But suppose you make the tube on the right side wider and use a piston of larger area; then the result is impressive. In Figure 12.21 the piston on the right has 50 times the area of the piston on the left (say the left has 100 square centimeters and the right 5000 square centimeters). Suppose a 10-kg load is placed on the left piston. Then an additional pressure (nearly 1000 N/cm^2) due to the weight of the load is transmitted throughout the liquid and up against the larger piston. Here is where the difference between force and pressure comes in. The additional pressure is exerted against every square centimeter of the larger piston. Since there is 50 times the area, 50 times as much force is exerted on the larger piston. Thus, the larger piston will support a 500-kg load—fifty times the load on the smaller piston!

FIGURE 12.21 A 10-kg load on the left piston will support 500 kg on the right piston.

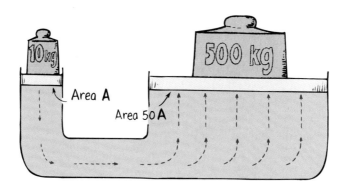

Area A

Area 50 A

This *is* something to write home about, for we can multiply forces with such a device. One newton input produces 50 newtons output. By further increasing the area of the larger piston (or reducing the area of the smaller piston), we can multiply force, in principle, by any amount. Pascal's principle underlies the operation of the hydraulic press.

The hydraulic press does not violate energy conservation, for the increase in force is compensated by a decrease in distance moved. When the small piston in Figure 12.21 is moved downward 10 centimeters, the large piston will be raised only one-fiftieth of this, or 0.2 centimeter. The input force multiplied by the distance moved by the smaller piston is equal to the output force multiplied by the distance moved by the larger piston; this is one more example of a simple machine, operating on the same principle as a mechanical lever.

FIGURE 12.22 Pascal's principle in a gas station.

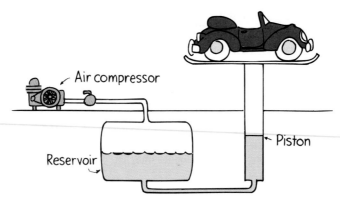

Pascal's principle applies to all fluids, gases as well as liquids. A typical application of Pascal's principle for gases and liquids is the automobile lift seen in many service stations (Figure 12.22). Increased air pressure produced by an air compressor is transmitted through the air to the surface of oil in an underground reservoir. The oil in turn transmits the pressure to a piston, which lifts the automobile. The relatively low pressure that exerts the lifting force against the piston is about the same as the air pressure in the tires of the automobile.

Question As the automobile in Figure 12.22 is being lifted, how does the change in oil level in the reservoir compare to the distance the automobile moves?

Surface Tension

Suppose we suspend a bent piece of clean wire from a sensitive spiral spring (Figure 12.23), lower the wire into water, and then raise it. As we attempt to free the wire from the water surface, we see from the stretched spring that the water surface exerts an

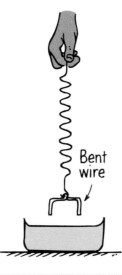

FIGURE 12.23 When the bent wire is lowered into the water and then raised, the spring will stretch because of surface tension.

Answer The car moves up a greater distance than the oil level drops, since the area of the piston is smaller than the surface area of the oil in the reservoir.

appreciable force on the wire. The water surface resists being stretched, for it has a tendency to contract. This can be seen when a fine-haired paintbrush is wet. When the brush is under water, the hairs are fluffed pretty much as they are when the brush is dry, but when the brush is lifted out, the surface film of water contracts and pulls the hairs together (Figure 12.24). We call this contractive tendency of the surface of liquids **surface tension**.

Surface tension accounts for the spherical shape of liquid drops. Raindrops, drops of oil, and falling drops of molten metal are all spherical because their surfaces tend to contract and force each drop into the shape having the least surface area. This is a sphere, the geometrical figure that has the least surface area for a given volume. For this reason the mist and dewdrops on spider webs or on the downy leaves of plants are nearly spherical blobs.

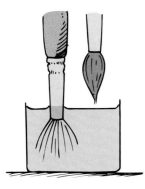

FIGURE 12.24 When the brush is taken out of the water, the hairs are held together by surface tension.

FIGURE 12.25 Small blobs of water are drawn by surface tension into spherelike shapes. Can you apply the ideas of Figure 11.15 in the previous chapter to an explanation of why larger drops are more flattened by gravity, while the tinier drops are more spherical?

Surface tension is caused by molecular attractions. Beneath the surface, each molecule is attracted in every direction by neighboring molecules, with the result that there is no tendency to be pulled in any preferred direction. A molecule on the surface of a liquid, however, is pulled only by neighbors to each side and downward from below; there is no pull upward (Figure 12.26). These molecular attractions thus tend to pull the molecule from the surface into the liquid. This tendency to pull surface molecules into the liquid causes the surface to become as small as possible. The surface behaves as if it were tightened into an elastic film. This is evident when dry steel needles or razor blades seem to float on water; actually, they are supported by the surface molecules opposing an increase of surface area. The same effect lets certain bugs run across the surface of a pond.

Surface tension of water is greater than that of other common liquids, and pure water has a stronger surface tension than soapy water. We can see this when a little soap film on the surface of water is effectively pulled out over the entire surface. This minimizes the surface area of the water. The same thing happens for oil or grease floating on water. Oil has less surface tension than cold water and is drawn out into a film covering the whole surface. But hot water has less surface tension than cold water because the faster-moving molecules are not held as cohesively. This allows the grease or oil in hot soups to float in little bubbles on the surface of the soup. When the soup cools and the surface tension of water increases, the grease or oil is dragged out over the surface of the soup. The soup becomes "greasy." Hot soup tastes different than cold soup primarily because the surface tension of water changes with temperature.

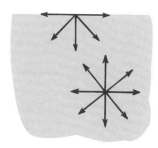

FIGURE 12.26 A molecule at the surface is pulled only sideways and downward by neighboring molecules. A molecule beneath the surface is pulled equally in all directions.

Capillarity

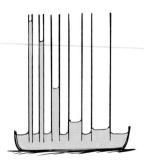

FIGURE 12.27 Capillary tubes.

If the end of a thoroughly clean glass tube with a small inside diameter is dipped into water, the water will wet the inside of the tube and rise in it. In a tube with a bore of about $\frac{1}{2}$ millimeter in diameter, for example, the water will rise slightly higher than 5 centimeters. With a still smaller bore, the water will rise much higher (Figure 12.27). This rise of a liquid in a fine, hollow tube or in a narrow space is **capillarity**.

Water molecules are attracted to glass more than to each other. The attraction between unlike substances is called *adhesion*. The attraction between like substances is called *cohesion*. When a glass tube is dipped into water, the adhesion between the glass and the water causes a thin film of water to be drawn up over the inner and outer surfaces of the tube (Figure 12.28a). Surface tension causes this film to contract (Figure 12.28b). The film on the outer surface contracts enough to make a rounded edge. The film on the inner surface contracts more by raising water with it until the adhesive force is balanced by the weight of the water lifted (Figure 12.28c). In a small tube, the weight of the water in the tube is small and the water is lifted higher than if the tube were large.

FIGURE 12.28 Hypothetical stages of capillary action, as seen in a cross-sectional view of a capillary tube.

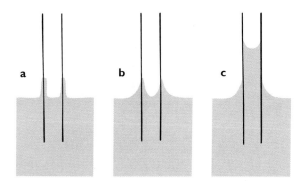

If a paintbrush is dipped partway into water, the water will rise up into the narrow spaces between the bristles by capillary action. If your hair is long, let it hang into the sink or bathtub, and water will seep up to your scalp in the same way. This is how oil soaks upward in a lamp wick and water into a bath towel when one end hangs in water. Dip one end of a lump of sugar in coffee, and the entire lump is quickly wet. Capillary action is essential for plant growth. It brings water to the roots of plants and carries sap and nourishment to high branches of trees. Just about everywhere we look, we can see capillary action at work. How nice.

Summary of Terms

Pressure The ratio of force to the area over which that force is distributed:

$$\text{Pressure} = \frac{\text{force}}{\text{area}}$$

$$\text{Liquid pressure} = \text{weight density} \times \text{depth}$$

Buoyant force The net upward force that a fluid exerts on an immersed object.

Archimedes' principle An immersed body is buoyed up by a force equal to the weight of the fluid it displaces.

Principle of flotation A floating object displaces a weight of fluid equal to its own weight.

Pascal's principle The pressure applied to a motionless fluid confined in a container is transmitted undiminished throughout the fluid.

Surface tension The tendency of the surface of a liquid to contract in area and thus behave like a stretched elastic membrane.

Capillarity The rise of a liquid in a fine, hollow tube or in a narrow space.

Suggested Reading

Rogers, E. *Physics for the Inquiring Mind*. Princeton, NJ: Princeton University Press, 1960. Chapter 6 of this textbook, an oldie but goodie, treats surface tension in interesting detail.

Review Questions

Pressure

1. Distinguish between *force* and *pressure*.

Pressure in a Liquid

2. What is the relationship between liquid pressure and the depth of a liquid? Between liquid pressure and density?

3. (a) If you swim twice as deep in water, how much more water pressure is exerted on your ears? (b) If you swim in salt water, will the pressure be greater than in fresh water at the same depth? Why or why not?

4. How does water pressure one meter below the surface of a small pond compare to water pressure one meter below the surface of a huge lake?

5. If you punch a hole in a container filled with water, in what direction does the water initially flow outward from the container?

Buoyancy

6. Why does buoyant force act upward on an object submerged in water?

7. How does the volume of a completely submerged object compare to the volume of water displaced?

Archimedes' Principle

8. How does the buoyant force that acts on a fish compare to the weight of the fish?

9. If a 1-L container is immersed halfway into water, what is the volume of water displaced? What is the buoyant force on the container?

10. How does the buoyant force on a submerged object compare with the weight of water displaced?

11. What is the mass of 1 L of water? What is its weight in newtons?

12. Does the buoyant force on a submerged object depend on the weight of the object itself or on the weight of the fluid displaced by the object? Does it depend on the volume of the object? Defend your answers.

What Makes an Object Sink or Float?

13. If the buoyant force on a fully submerged object is equal to the weight of the object, how do the densities of the object and water compare?

14. If the buoyant force on a fully submerged object is greater than the weight of the object, how do the densities of the object and water compare?

15. If the buoyant force on a fully submerged object is less than the weight of the object, how do the densities of the object and water compare?

16. How is the density of a submarine controlled? How is the density of a fish controlled?

Flotation

17. What weight of water is displaced by a 100-ton ship? What is the buoyant force that acts on a 100-ton ship?

18. Does the buoyant force on a floating object depend on the weight of the object itself or on the weight of the fluid displaced by the object? Or are these both the same for the special case of floating? Defend your answer.

Pascal's Principle

19. What will happen to the pressure in all parts of a confined fluid if you increase the pressure in one part?

20. If the pressure in a hydraulic press is increased by an additional 10 N/cm^2, how much extra load will the output piston support if its cross-sectional area is 50 cm^2?

Surface Tension

21. Why does the surface of a blob of water contract?

22. What geometric shape has the least surface area for a given volume?

23. Why is cold soup greasy?

Capillarity

24. Distinguish between *adhesive* and *cohesive* forces.

25. What determines how high water will climb in a capillary tube?

Projects

1. Try to float an egg in water. Then dissolve salt in the water until the egg floats. How does the density of an egg compare to that of tap water? To salt water?

2. Punch a couple of holes in the bottom of a water-filled container, and water will spurt out because of water pressure. Now drop the container, and as it freely falls note that the water no longer spurts out! If your friends don't understand this, could you figure it out and then explain it to them?

3. Float a water-soaked Ping-Pong ball in a can of water held more than a meter above a rigid floor. Then drop the can. What happens to the ball as both drop (and what does this say about surface tension)? More dramatically, what happens to the ball, and why, when the can makes impact with the floor? Try it and you'll be astounded! (Caution: Unless you're wearing safety goggles, keep your head away from above the can when it makes impact.)

4. Soap greatly weakens the cohesive forces between water molecules. You can see this by putting some oil in a bottle of water and shaking it so that the oil and water mix. Notice that the oil and water quickly separate as soon as you stop shaking the bottle. Now add some soap to the mixture. Shake the bottle again and you will see that the soap makes a fine film around each little oil bead and that a longer time is required for the oil to gather after you stop shaking the bottle.

   This is how soap works in cleaning. It breaks the surface tension around each particle of dirt so that the water can reach the particles and surround them. The dirt is carried away in rinsing. Soap is a good cleaner only in the presence of water.

Exercises

1. Stand on a bathroom scale and read your weight. When you lift one foot off so you're standing on one foot, does the reading change? Does a scale read force or pressure?

2. Why are persons confined to bed less likely to develop bedsores on their bodies if they use a waterbed rather than an ordinary mattress?

3. You know that a sharp knife cuts better than a dull knife. Do you know why this is so?

4. If water faucets upstairs and downstairs are turned fully on, will more water per second come out of the upstairs or downstairs faucet?

5. Which do you suppose exerts more pressure on the ground—an elephant or a lady standing on spike heels? (Which will be more likely to make dents in a linoleum floor?) Can you approximate a rough calculation for each?

6. The photo at the top of the next column shows physics teacher Marshall Ellenstein walking barefoot on broken glass bottles in his class. What physics concept is Marshall demonstrating, and why is he careful that the broken pieces are small and numerous? (The Band-Aids on his feet are for humor!)

7. Why does your body get more rest when you're lying than when you're sitting? And why is blood pressure measured on the upper arm, at the elevation of your heart? Is blood pressure in your legs greater?

8. Which teapot holds more liquid?

9. The sketch shows the reservoir that supplies water to a farm. It is made of wood and reinforced with metal hoops. (a) Why is it elevated? (b) Why are the hoops closer together near the bottom part of the tank?

10. A block of aluminum with a volume of 10 cm^3 is placed in a beaker of water filled to the brim. Water overflows. The same is done in another beaker with a 10-cm^3 block of lead. Does the lead displace more, less, or the same amount of water?

11. A block of aluminum with a mass of 1 kg is placed in a beaker of water filled to the brim. Water overflows. The same is done in another beaker with a 1-kg block of lead. Does the lead displace more, less, or the same amount of water?

12. A block of aluminum with a weight of 10 N is placed in a beaker of water filled to the brim. Water overflows. The same is done in another beaker with a 10-N block of lead. Does the lead displace more, less, or the same amount of water? (Why are your answers to this exercise and Exercise 11 different from your answer to Exercise 10?)

13. At a depth of 10 m below the surface of a deep lake, the "liquid pressure," which is the extra pressure due to the liquid supported above this point, is 1 atmosphere. What is the total pressure at this point, and why is it different?

14. There is a legend of a Dutch boy who bravely held back the whole Atlantic Ocean by plugging a hole in a dike with his finger. Is this possible and reasonable? (See also Problem 4.)

15. If you've wondered about the flushing of toilets on the upper floors of city skyscrapers, how do you suppose the plumbing is designed so that there is not an enormous impact of sewage arriving at the basement level? (Check your speculations with someone who is into architecture.)

16. Why does water "seek its own level"?

17. Suppose you wish to lay a level foundation for a home on hilly and bushy terrain. How can you use a garden hose filled with water to determine equal elevations for distant points?

18. When you are bathing on a stony beach, why do the stones hurt your feet less when you get in deep water?

19. If you cut your finger, why does holding it above your head reduce bleeding?

20. If liquid pressure were the same at all depths, would there be a buoyant force on an object submerged in the liquid? Explain.

21. A can of diet drink floats in water, whereas a can of regular soda sinks. What is your explanation?

22. How does the fraction of the submerged part of a floating object relate to its density relative to the density of the liquid in which it floats?

23. The Himalayan Mountains are slightly less dense than the mantle material upon which they "float." Do you suppose that, like floating icebergs, they are deeper than they are high?

24. Why is a high mountain composed mostly of lead impossible for the planet earth?

25. How much force is needed to push a nearly weightless but rigid 1-L carton beneath a surface of water?

26. Why is it inaccurate to say that heavy objects sink and that light objects float? Give exaggerated examples to support your answer.

27. A piece of iron placed on a block of wood makes it float lower in the water. If the iron were instead suspended beneath the wood, would it float as low, lower, or higher? Defend your answer.

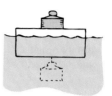

28. Compared to an empty ship, would a ship loaded with a cargo of Styrofoam sink deeper into water or rise in water? Defend your answer.

29. If a submarine starts to sink, will it continue to sink to the bottom of the sea if no changes are made? Explain.

30. A barge filled with scrap iron is in a canal lock. If the iron is thrown overboard, does the water level at the side of the lock rise, fall, or remain unchanged? Explain.

31. Would the water level in a canal lock go up or down if a battleship in the lock sank?

32. A balloon is weighted so that it is barely able to float in water. If it is pushed beneath the surface, will it come back to the surface, stay at the depth to which it is pushed, or sink? Explain. (*Hint:* Does the balloon's density change?)

33. The density of a rock doesn't change when it is submerged in water, but your density changes when you are submerged. Why is this so?

34. In answering the question of why bodies float higher in salt water than fresh water, your friend replies that the reason is that salt water is denser than fresh water. (Does your friend often answer questions by reciting only factual statements that relate to the answers but don't provide any concrete reasons?) How would you answer the same question?

35. A ship sailing from the ocean into a fresh-water harbor sinks slightly deeper into the water. Does the buoyant force on it change? If so, does it increase or decrease?

36. Suppose you wear two life preservers that are identical in size, first a light one filled with Styrofoam and then a very heavy one filled with lead pellets. If you submerge these life preservers in the water, upon which will the buoyant force be greater? Upon which will the buoyant force be ineffective? Why are your answers different?

37. The weight of the human brain is about 15 N. The buoyant force supplied by fluid around the brain is about 14.5 N. Does this mean that the weight of fluid surrounding the brain is at least 14.5 N? Defend your answer.

38. The relative densities of water, ice, and alcohol are 1.0, 0.9, and 0.8 respectively. Do ice cubes float higher or lower in a mixed alcoholic drink compared to plain water? What can you say about a cocktail in which the ice cubes lie submerged at the bottom of the glass?

39. When an ice cube in a glass of water melts, does the water level in the glass rise, fall, or remain unchanged? Does your answer change if the ice cube has many air bubbles? How about if the ice cube contains many grains of heavy sand?

40. A half-filled bucket of water is on a spring scale. Will the reading of the scale increase or remain the same if a fish is placed in the bucket? (Will your answer be different if the bucket is initially filled to the brim?)

41. The weight of the container of water is equal to the weight of the stand and suspended solid iron ball as shown below in the sketch (a). When the suspended ball is lowered into the water, the balance is upset (b). Will the additional weight that must be put on the right side to restore balance be greater than, equal to, or less than the weight of the ball?

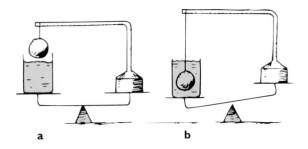

a   **b**

42. Would a fish float to the surface, sink, or stay at the same depth if the gravitational field of the earth increased?

43. What would we experience when swimming in water in an orbiting space habitat where simulated gravity is $\frac{1}{2} g$? Would we float in the water as we do on earth?

44. We say that the shape of a liquid is that of its container. But with no container and no gravity, what is the natural shape of a blob of water?

45. If you release a Ping-Pong ball beneath the surface of water, it will rise to the surface. Would it do the same if submerged in a big blob of water at the hub of an orbiting structure in space? Explain.

46. So you're on a run of bad luck and you fall quietly into a small quiet pool as hungry crocodiles lurking at the bottom are appreciating Pascal's principle. What does Pascal's principle have to do with their delight?

47. In the hydraulic arrangement shown, the larger piston has an area which is fifty times that of the smaller piston. The strongman hopes to exert enough force on the large piston to raise the 10 kg that rest on the small piston. Do you think he will be successful? Defend your answer.

48. In the hydraulic arrangement shown in Figure 12.21, the multiplication of force is equal to the ratio of areas of the large and small pistons. Some people are surprised to learn that the area of the liquid surface in the reservoir of the arrangement shown in Figure 12.22 is immaterial. What is your explanation to clear up this confusion?

49. Why will hot water leak more readily than cold water through small leaks in a car radiator?

50. On the surface of a pond it is common to see water striders, insects that can "walk" on the surface of water without sinking. What physics concept explains this act? Explain.

Problems

1. The piston cross-section area is 400 cm² for the hydraulic lift shown in Figure 12.22. The mass of the car and the lift itself is 2000 kg. How much pressure must be applied to the surface of the fluid in the reservoir to lift the car?

2. How much pressure do you experience when you balance a 5-kg ball on the tip of your finger, say of area 1 cm²? How does this compare with your answer to the previous question?

3. A rectangular barge, 5 m long and 2 m wide, floats in fresh water. (a) Find how much deeper it floats when its load is a 400-kg horse. (b) If the barge can only be pushed 15 cm deeper into the water before water overflows to sink it, how many 400-kg horses can it carry?

4. A dike in Holland springs a leak through a hole of area 1 cm² at a depth of 2 m below the water surface. With what force would a boy have to push on the hole with his thumb to stop the leak? Could he do it?

5. A merchant in Katmandu sells you a solid gold 1-kg statue for a very reasonable price. When you get home, you wonder whether or not you got a bargain, so you lower the statue into a measuring cup and measure its volume. What volume will verify that it's pure gold?

6. When a 2.0-kg object is suspended in water, it "masses" 1.5 kg. What is the density of the object?

7. An ice cube measures 10 cm on a side, and floats in water. One cm extends above water level. If you shaved off the 1-cm part, how many cm of the remaining ice would extend above water level?

8. A swimmer wears a heavy belt to make her average density exactly equal to the density of water. Her mass, including the belt, is 60 kg. (a) What is the swimmer's weight in newtons? (b) What is the swimmer's volume in m³? (c) At a depth of 2 m below the surface of a pond, what buoyant force acts on the swimmer? What net force acts on her?

9. A vacationer floats lazily in the ocean with 90% of his body below the surface. The density of the ocean water is 1025 kg/m³. What is the vacationer's average density?

10. A salvage ship is able to raise a chunk of steel from the ocean floor to the water surface, but cannot raise it above the water. The ship's captain knows that the density of steel is about four times the density of water, and he wonders whether one other ship with a crane of equal capacity will be enough to help him lift the steel above the water, or whether he will need the help of more than one other ship. Explain to the captain why one other ship will be enough.

13

·············

Gases and Plasmas

: **C**olorful parts of the air.

Gases as well as liquids flow; hence, both are called *fluids*. The primary difference between a gas and a liquid is the distance between molecules. In a gas, the molecules are far apart and free from the cohesive forces that dominate their motions when in the liquid and solid phases. Their motions are less restricted. A gas expands indefinitely and fills all space available to it. Only when the quantity of gas is very large, such as in the earth's atmosphere or in a star, do gravitational forces limit the size or determine the shape of the mass of a gas.

The Atmosphere

The thickness of our atmosphere is determined by two competing factors: the kinetic energy of its molecules, which tends to spread the molecules apart; and gravity, which tends to hold them near the earth. If somehow the earth's gravity were shut off, atmospheric molecules would dissipate and disappear. Or if gravity acted but the molecules moved too slowly to form a gas (as might occur on a remote, cold planet), our "atmosphere" would be a liquid or solid layer, just so much more matter lying on the ground. There would be nothing to breathe. Again, no atmosphere.

But our atmosphere is a happy compromise between energetic molecules that tend to fly away and gravity that holds them back. Without the heat of the sun, air molecules would lie on the earth's surface the way settled popcorn lies at the bottom of a popcorn machine. But add heat to the popcorn and to atmospheric gases, and both will bumble their way up to higher altitudes. Pieces of popcorn in a popper attain speeds of a few kilometers per hour and reach altitudes up to a meter or two; molecules in the air move at speeds of about 1600 kilometers per hour and bumble up to many kilometers in altitude. Fortunately there is an energizing sun, and there is gravity, and we have an atmosphere.

The exact height of the atmosphere has no real meaning, for the air gets thinner and thinner the higher one goes. Eventually it thins out to emptiness in interplanetary space. Even in the vacuous regions of interplanetary space, however, there is a gas density of about 1 molecule per cubic centimeter. This is primarily hydrogen, the most plentiful element in the universe. About 50 percent of the atmosphere is below an altitude of 5.6 kilometers (18,000 ft), 75 percent is below 11 kilometers (36,000 ft), 90 percent is below 18 kilometers, and 99 percent is below about 30 kilometers (Figure 13.1). A detailed description of the atmosphere can be found in any encyclopedia.

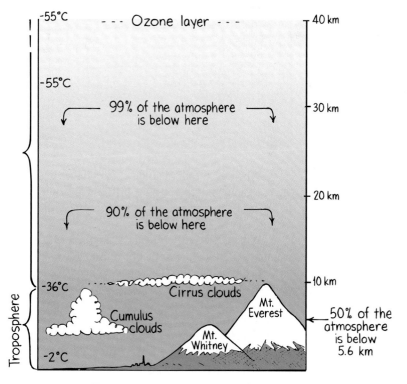

FIGURE 13.1 The atmosphere. Air is more compressed at sea level than at higher altitudes. Like feathers in a huge pile, what's at the bottom is more squashed than what's nearer the top.

Atmospheric Pressure

We live at the bottom of an ocean of air. The atmosphere, much like the water in a lake, exerts pressure. One of the most celebrated experiments demonstrating the pressure of the atmosphere was conducted in 1654 by Otto von Guericke, burgermeister of Magdeburg and inventor of the vacuum pump. Von Guericke placed together two copper hemispheres about $\frac{1}{2}$ meter in diameter to form a sphere, as shown in Figure 13.2. He fashioned an airtight joint with an oil-soaked leather gasket. When he evacuated the sphere with his vacuum pump, two teams of eight horses each were unable to pull the hemispheres apart.

FIGURE 13.2 The famous "Magdeburg hemispheres" experiment of 1654, demonstrating atmospheric pressure. Two teams of horses couldn't pull the evacuated hemispheres apart. Were the hemispheres sucked together or pushed together? By what?

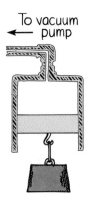

To vacuum pump

FIGURE 13.3 Is the piston pulled up or pushed up?

When the air pressure inside a cylinder like that shown in Figure 13.3 is reduced, there is an upward force on the piston. This force is large enough to lift a heavy weight. If the inside diameter of the cylinder is 10 centimeters or greater, a person can be lifted by this force.

What do the experiments of Figures 13.2 and 13.3 demonstrate? Do they show that air exerts pressure or that there is a "force of suction"? If we say there is a force of suction, then we assume that a vacuum can exert a force. But what is a vacuum? It is an absence of matter; it is a condition of nothingness. How can nothing exert a force? The hemispheres are not sucked together, nor is the piston holding the weight sucked upward. The hemispheres and the piston are *pushed* by the weight of the atmosphere.

Just as water pressure is caused by the weight of water, **atmospheric pressure** is caused by the weight of air. We have adapted so completely to the invisible air that we sometimes forget it has weight. Perhaps a fish "forgets" about the weight of water in the same way. The reason we don't feel this weight crushing against our bodies is that the pressure inside our bodies equals that of the surrounding air. There is no net force for us to sense.

At sea level, 1 cubic meter of air has a mass of about $1\frac{1}{4}$ kilograms. So the air in your kid sister's small bedroom weighs about as much as she does! The density of air

Table 13.1 Densities of Various Gases

Gas	Density(kg/m³)*
Dry air	
0°C	1.29
10°C	1.25
20°C	1.21
30°C	1.16
Hydrogen	0.090
Helium	0.178
Nitrogen	1.25
Oxygen	1.43

*At sea-level atmospheric pressure and at 0°C (unless otherwise specified)

decreases with altitude. At 10 kilometers, for example, 1 cubic meter of air has a mass of about 0.4 kilograms. To compensate for this, airplanes are pressurized; the additional air needed to fully pressurize a 747 jumbo jet, for example, is more than 1000 kilograms. Air is heavy if you have enough of it. If your kid sister doesn't believe air has weight, you can show her why she falsely perceives weight-free air. Hand her a plastic bag of water and she'll tell you it has weight. But hand her the same bag of water while she's submerged in a swimming pool, and she won't feel its weight. That's because she and the bag are surrounded by water. Likewise with the air that is around us.

FIGURE 13.4 You don't notice the weight of a bag of water while you're submerged in water. Similarly, you don't notice the weight of air.

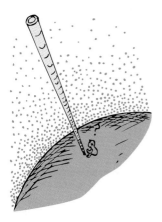

FIGURE 13.5 The mass of air that would occupy a bamboo pole that extends to the "top" of the atmosphere is about 1 kg. This air has a weight of 10 N.

Consider the mass of air in an upright 30-kilometer-tall bamboo pole that has an inside cross-sectional area of 1 square centimeter. If the density of air inside the pole matches the density of air outside, the mass of enclosed air would be about one kilogram. The weight of this much air is about 10 newtons. So air pressure at the bottom of the bamboo pole would be about 10 newtons per square centimeter (10 N/cm²). Of course, the same is true without the bamboo pole. There are 10,000 square centimeters in 1 square meter, so a column of air 1-square meter in cross section that extends up through the atmosphere has a mass of about 10,000 kilograms. The weight of this air is about 100,000 newtons (10^5 N). This weight produces a pressure of 100,000 newtons per square meter—or, equivalently, 100,000 pascals, or 100 kilopascals. To be more exact, the average atmospheric pressure at sea level is 101.3 kilopascals (101.3 kPa).*

The pressure of the atmosphere is not uniform. Besides altitude variations, there are variations in atmospheric pressure at any one locality due to moving fronts and storms. Measurement of changing air pressure is important to meteorologists when predicting weather.

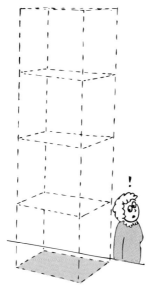

FIGURE 13.6 The weight of air that bears down on a one-square-meter surface at sea level is about 100,000 newtons. So atmospheric pressure is about 10^5 N/m², or about 100 kPa.

* The pascal is the SI unit of measurement. The average pressure at sea level (101.3 kPa) is often called 1 atmosphere. In British units, the average atmospheric pressure at sea level is 14.7 lb/in².

Questions

1. About how many kilograms of air occupy a classroom that has a 200-m² floor area and a 4-m-high ceiling? (Assume a chilly 10°C temperature.)

2. Why doesn't the pressure of the atmosphere break windows?

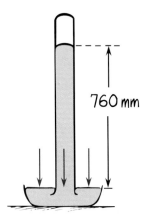

FIGURE 13.7 A simple mercury barometer.

Barometers

Instruments used for measuring the pressure of the atmosphere are called **barometers**. A simple mercury barometer is illustrated in Figure 13.7. A glass tube, longer than 76 centimeters (760 millimeters) and closed at one end, is filled with mercury and tipped upside down in a dish of mercury. The mercury in the tube runs out of the submerged open bottom until the level in the tube is 76 centimeters above the level in the dish. The empty space trapped above, except for some mercury vapor, is a vacuum. The vertical height of the mercury column remains constant even when the tube is tilted, unless the top of the tube is less than 76 centimeters above the level in the dish—in which case the mercury completely fills the tube.

Why does mercury behave this way? The explanation is similar to the reason a simple seesaw will balance when the weights of people at its two ends are equal. The barometer "balances" when the weight of liquid in the tube exerts the same pressure as the atmosphere outside. Whatever the width of the tube, a 76-centimeter column of mercury weighs the same as the air that would fill a super-tall 30-kilometer tube of the same width. If the atmospheric pressure increases, then the atmosphere pushes down harder on the mercury and the column is pushed higher than 76 centimeters. The mercury is literally pushed up into the tube of a barometer by the weight of the atmosphere.

Could water be used to make a barometer? The answer is yes, but the glass tube would have to be much longer—13.6 times as long, to be exact. You may recognize this number as the density of mercury relative to that of water. A volume of water 13.6 times that of mercury is needed to provide the same weight as the mercury in the tube. So the tube would have to be at least 13.6 times taller than the mercury column. A water barometer would have to be 13.6 × 0.76 meter, or 10.3 meters high—too tall to be practical.

What happens in a barometer is similar to what happens during the process of drinking through a straw. By sucking, you reduce the air pressure in the straw that is placed in a drink. The weight of the atmosphere on the drink pushes liquid up into the reduced-pressure region. Strictly speaking, the liquid is not sucked up; it is pushed up by the atmosphere. If the atmosphere is prevented from pushing on the surface of the drink, as in the party trick bottle with the straw through the air-tight cork stopper, one can suck and suck and get no drink.

FIGURE 13.8 Strictly speaking, they do not suck the soda up the straw. They instead reduce pressure in the straw and allow the weight of the atmosphere to press the liquid up into the straws. Could they drink a soda this way on the moon?

Answers

1. The mass of air is 1000 kg. The volume of air is 200 m² × 4 m = 800 m³; each cubic meter of air has a mass of about 1.25 kg, so 800 m³ × 1.25 kg/m³ = 1000 kg.

2. Atmospheric pressure is exerted on both sides of a window, so no net force is exerted on the window. If for some reason the pressure is reduced or increased on one side only, like when a tornado passes by, then watch out!

FIGURE 13.9 The atmosphere pushes water from below up into a pipe that is evacuated of air by the pumping action.

If you understand these ideas, you can understand why there is a 10.3-meter limit on the height that water can be lifted with vacuum pumps. The old fashioned farm-type pump, Figure 13.9, operates by producing a partial vacuum in a pipe that extends down into the water below. The weight of the atmosphere on the surface of the water simply pushes the water up into the region of reduced pressure inside the pipe. Can you see that even with a perfect vacuum, the maximum height to which water can be lifted is 10.3 meters?

FIGURE 13.10 The aneroid barometer.

Question What is the maximum height to which water could be drunk through a straw?

A small portable instrument that measures atmospheric pressure is the aneroid barometer (Figure 13.10). It uses a metal box that is partially exhausted of air and has a slightly flexible lid that bends in or out with changes in atmospheric pressure. Motion of the lid is indicated on a scale by a mechanical spring-and-lever system. Since the atmospheric pressure decreases with increasing altitude, a barometer can be used to determine elevation. An aneroid barometer calibrated for altitude is called an *altimeter* (altitude meter). Some altimeters are sensitive enough to indicate a change in elevation less than a meter.

The pressure (or vacuum) inside a television picture tube is about 1 ten-thousandth pascal (10^{-4} Pa). At an altitude of about 500 kilometers, artificial satellite territory, gas pressure is about one ten-thousandth of this (10^{-8} Pa). This is a pretty good vacuum by earthbound standards. Still greater vacuums exist in the wakes of satellites orbiting at this altitude, and they can reach 10^{-13} pascals. This is a "hard vacuum."

Answer However strong your lungs may be, or whatever device you use to make a vacuum in the straw, at sea level the water could not be pushed up by the atmosphere higher than 10.3 m.

Vacuums on the earth are produced by pumps, which work by virtue of a gas tending to fill its container. If a space with less pressure is provided, gas will flow from the region of higher pressure to the one of lower pressure. A vacuum pump simply provides a region of lower pressure into which the normally fast-moving gas molecules randomly move. The air pressure is repeatedly lowered by piston and valve action (Figure 13.11). The best vacuums attainable with mechanical pumps are about one pascal. Better vacuums, down to 10^{-8} pascals, are attainable with vapor diffusion pumps. Vacuums greater than this are very difficult to attain. Technologists requiring hard vacuums are looking more to the prospects of orbiting laboratories in space.

FIGURE 13.11 A mechanical vacuum pump. When the piston is lifted, the intake valve opens and air moves in to fill the empty space. When the piston is moved downward, the outlet valve opens and the air is pushed out. What changes would you make to convert this pump into an air compressor?

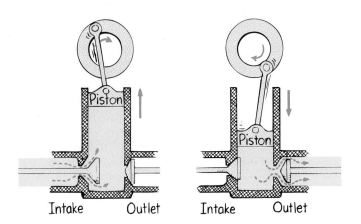

Boyle's Law

The air pressure inside the inflated tires of an automobile is considerably more than the atmospheric pressure outside. The density of air inside is also more than the density of air outside. To understand the relation between pressure and density, think of the molecules of air (primarily nitrogen and oxygen) inside the tire, which behave like tiny Ping-Pong balls—perpetually moving helter-skelter and banging against the inner walls. Their impacts produce a jittery force that appears to our coarse senses as a steady push. This pushing force averaged over a unit of area provides the pressure of the enclosed air.

Suppose there are twice as many molecules in the same volume (Figure 13.12). Then the air density is doubled. If the molecules move at the same average speed—or, equivalently, if they have the same temperature—then the number of collisions will be doubled. This means the pressure is doubled. So pressure is proportional to density.

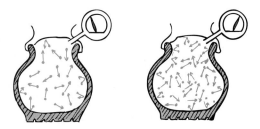

FIGURE 13.12 When the density of gas in the tire is increased, pressure is increased.

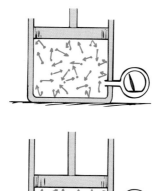

FIGURE 13.13 When the volume of gas is decreased, density and therefore pressure are increased.

We can also double the air density by compressing a *fixed* amount of air to half its volume. Consider the cylinder with the movable piston in Figure 13.13. If the piston is pushed downward so that the volume is half the original volume, the density of molecules will be doubled, and the pressure will correspondingly be doubled. Decrease the volume to a third its original value, and the pressure will be multiplied by three, and so forth (providing the temperature remains the same).

Notice in these examples with the piston that the product of pressure and volume is the same. For example, a doubled pressure multiplied by a halved volume gives the same value as a tripled pressure multiplied by a one-third volume. In general, we can say that the product of pressure and volume for a given mass of gas is a constant as long as the temperature does not change. "Pressure × volume" for a quantity of gas at one time is equal to "different pressure × different volume" at any other time. In shorthand notation,

$$P_1V_1 = P_2V_2$$

where P_1 and V_1 represent the original pressure and volume, respectively, and P_2 and V_2 the second pressure and volume. This relationship is called **Boyle's Law**, after Robert Law, the seventeenth-century physicist who is credited with its discovery.* Note the similarity of this relationship to the conservation of work in a machine, where the product of force and distance is constant: $F_1d_1 = F_2d_2$.

Boyle's law applies to ideal gases. An ideal gas is one in which the disturbing effects of the forces between molecules and the finite size of the individual molecules can be neglected. Air and other gases under normal pressures approach ideal gas conditions.[†]

Questions

1. A piston in an airtight pump is withdrawn so that the volume of the air chamber is increased three times. What is the change in pressure?
2. A scuba diver 10.3 m deep breathes compressed air. If she were to hold her breath while returning to the surface, by how much would the volume of her lungs tend to increase?

Answers

1. The pressure in the piston chamber is reduced to one-third. This is the principle that underlies a mechanical vacuum pump.
2. Atmospheric pressure can support a column of water 10.3 m high, so the pressure in water due to the weight of the water alone equals atmospheric pressure at a depth of 10.3 m. Taking the pressure of the atmosphere at the water's surface into account, the total pressure at this depth is twice atmospheric pressure. Unfortunately for the scuba diver, her lungs tend to inflate to twice their normal size if she holds her breath while rising to the surface. A first lesson in scuba diving is not to hold your breath when ascending. To do so can be fatal.

---

* Humor aside, Boyle's law is named after Robert Boyle.

[†] A general law that takes temperature changes into account is $P_1V_1/T_1 = P_2V_2/T_2$, where T_1 and T_2 represent the first and second *absolute* temperatures, measured in the SI unit called kelvin (Chapter 17).

Buoyancy of Air

A crab lives at the bottom of its ocean of water and looks upward at jellyfish drifting above it. Similarly, we live at the bottom of our ocean of air and look upward at balloons drifting above us. A balloon is suspended in air and a jellyfish is suspended in water for the same reason: each is buoyed upward by a displaced weight of fluid equal to its own weight. In one case the displaced fluid is air, and in the other case it is water. In water, immersed objects are buoyed upward because the pressure acting up against the bottom of the object exceeds the pressure acting down against the top. Likewise, air pressure acting up against an object immersed in air is greater than the pressure above pushing down. The buoyancy in both cases is numerically equal to the weight of fluid displaced. **Archimedes' principle** holds for air just as it does for water:

> **An object surrounded by air is buoyed up by a force equal to the weight of the air displaced.**

We know that a cubic meter of air at ordinary atmospheric pressure and room temperature has a mass of about 1.2 kilograms, so its weight is about 12 newtons. Therefore any 1-cubic-meter object in air is buoyed up with a force of 12 newtons. If the mass of the 1-cubic-meter object is greater than 1.2 kilograms (so that its weight is greater than 12 newtons), it falls to the ground when released. If this size object has a mass less than 1.2 kilograms, it rises in the air. Any object that has a mass less than the mass of an equal volume of air will rise in air. Another way to say this is that any object less dense than air will rise in air. Gas-filled balloons that rise in air are less dense than air.

Gas is used in balloons simply because its presence prevents the atmosphere from collapsing the balloon. In sport balloons, the gas is simply heated air. In balloons intended to reach very high altitudes or to stay up a long time, helium is usually used. Its density is small enough so that the combined weight of helium, balloon, and whatever the cargo happens to be is less than the weight of air it displaces.* Low-density gas is used in a balloon for the same reason cork is used in a swimmer's life preserver. The cork possesses no strange tendency to be drawn toward the surface of water, and the gas possesses no strange tendency to rise. Both are buoyed upward like anything else. They are simply light enough for the buoyancy to be significant.

Unlike water, there is no sharp surface at the "top" of the atmosphere. Furthermore, unlike water, the atmosphere becomes less dense with altitude. Whereas cork will float to the surface of water, a released helium-filled balloon does not rise to any atmospheric surface. How high will a balloon rise? We can state the answer in several ways. A balloon will rise only so long as it displaces a weight of air greater than its own weight. Since air becomes less dense with altitude, a lesser weight of air is displaced per given volume as the balloon rises. When the weight of displaced air equals the total weight of the balloon, upward acceleration of the balloon will cease. We can also say that when the buoyant force on the balloon equals its weight, the balloon will cease rising. Equivalently, when the average density of the balloon (including its load) equals the density of the surrounding air, the balloon will cease rising. Helium-filled toy balloons usually break when released in the air because as the balloon rises to regions of less pressure, the helium in the balloon expands, increasing the volume and stretching the rubber until it ruptures.

FIGURE 13.14 All bodies are buoyed up by a force equal to the weight of the air they displace. Why, then, don't all objects float like this balloon?

*Hydrogen is the least dense gas but is highly flammable, so it is seldom used.

Questions

1. Is there a buoyant force acting on you? If there is, why are you not buoyed up by this force?
2. Two balloons that have the same weight and volume contain equal amounts of helium. One is rigid and the other is free to expand as the pressure outside decreases. When released, which balloon will rise higher?

Large dirigible airships are designed so that when loaded they will slowly rise in air; that is, their total weight is a little less than the weight of air displaced. When in motion, the ship may be raised or lowered by means of horizontal "elevators."

Thus far we have treated pressure only as it applies to stationary fluids. Motion produces an additional influence.

Bernoulli's Principle

Mark this statement true or false: Atmospheric pressure increases in a gale, tornado, or hurricane. If you answered true, sorry; the statement is false. High-speed winds may blow the roof off your house, but the pressure within the winds is actually less than for still air of the same density inside the house. As strange as it may first seem, when the speed of a fluid increases, the internal pressure decreases. This is true whether the fluid is a gas or liquid.

Daniel Bernoulli, a Swiss scientist of the eighteenth century, studied the relationship of fluid speed and pressure. When a fluid flows through a narrow constriction, its speed increases. This is easily noticed by the increased speed of the water coming out of the nozzle of a garden hose when you narrow the opening of the nozzle. The fluid must speed up in the constricted region if the flow is to be continuous.

Bernoulli wondered how the fluid got the energy for this extra speed. He reasoned that it is acquired at the expense of a lowered internal pressure. His discovery, now called **Bernoulli's principle**, states:

When the speed of a fluid increases, pressure in the fluid decreases.

Bernoulli's principle is a consequence of the conservation of energy. The full energy picture for a fluid in motion is quite complicated. It includes energy associated with changes in temperature, density, and with heat energy generated by friction. But if temperature and density remain nearly constant and friction is small, only three energy terms need to be considered: kinetic energy due to motion, gravitational potential energy due to elevation, and work done by pressure forces. In the steady flow of an incompressible fluid without friction, the sum of three terms—kinetic energy, potential

Answers

1. There is a buoyant force acting on you, and you are buoyed upward by it. You don't notice it only because your weight is so much greater.
2. The balloon that is free to expand will displace more air as it rises than the balloon that is restrained. Hence, the balloon that is free to expand will experience more buoyant force and rise higher.

FIGURE 13.15 The pressure in the spout reduces when the plug is removed.

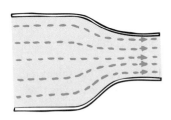

FIGURE 13.16 Water speeds up when it flows into the narrower pipe. The constricted streamlines indicate increased speed and decreased internal pressure.

FIGURE 13.17 The paper rises when air is blown across its top surface.

energy, and work remains constant.* If the elevation of the flowing fluid does not change, then potential energy is constant and only the terms representing kinetic energy and work done by pressure forces remain. When one of these terms increases, the other term decreases. So more speed and kinetic energy mean less pressure, and more pressure means less speed and kinetic energy.

The decrease of fluid pressure with increasing speed may at first seem surprising, particularly if we fail to distinguish between the pressure *in* the fluid and the pressure *by* the fluid on something that interferes with its flow. The pressure within fast-moving water in a fire hose is relatively low, whereas the pressure that the water can exert on anything in its path to slow it down may be huge.

In steady flow, one small bit of fluid follows along the same path as a bit of fluid in front of it. The motion of a fluid in steady flow follows *streamlines,* which are represented by dashed lines in Figure 13.15 and later figures. Streamlines are the smooth paths, or trajectories, of the bits of fluid. The lines are closer together in the narrower regions, where the flow speed is greater and the pressure within the fluid is less.

Bernoulli's principle holds only for steady flow. If the flow speed is too great, the flow may become turbulent and follow a changing, curling path known as an *eddy.* This exerts friction on the fluid and causes some of its energy to be transformed to heat. Then Bernoulli's principle does not hold.

Applications of Bernoulli's Principle

Hold a sheet of paper in front of your mouth, as shown in Figure 13.17. When you blow across the top surface, the paper rises. This is because the pressure of moving air against the top of the paper is less than the pressure of air at rest against the lower surface.

We began our discussion of Bernoulli's principle by stating that atmospheric pressure decreases in a strong wind, tornado, or hurricane. As it turns out, an unvented building with airtight closed windows is in more danger of losing its roof than a well-vented building. This is because the air pressure inside may be appreciably greater than the reduced atmospheric pressure outside, and the roof is more likely to be pushed off by the higher pressure air in the building than blown off by the wind. When the wind is blowing over a peaked roof as shown in Figure 13.18, the effect is even more pronounced.

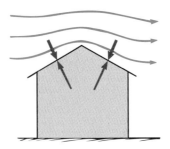

FIGURE 13.18 Air pressure above the roof is less than air pressure beneath the roof.

* In mathematical form: $\frac{1}{2}mv^2 + mgy + pV =$ constant; where m is the mass of some small volume V, v its speed, g the acceleration due to gravity, y its elevation, and p its pressure. If mass m is expressed in terms of density ρ, where $\rho = m/V$, and each term is divided by V, Bernoulli's equation takes the form:

$$\frac{1}{2}\rho v^2 + \rho gy + p = \text{constant.}$$

Then all three terms have units of pressure. If y does not change, then an increase in v means a decrease in p, and vice versa.

The crowding of the streamlines shows this. The difference in outside and inside pressure need not really be very much. A small pressure difference over a large area can be formidable. So if you're ever caught in an unvented building in a tornado or hurricane, consider opening the windows a bit so that the pressures inside and outside are more nearly equal.*

If we think of the blown-off roof as an airplane wing, we can better understand the lifting force that supports a heavy airliner. In both cases a greater pressure below pushes the roof and wing into a region of lesser pressure above. When a wing moves through the air, the pressure difference is established by causing air to move faster over the top of the wing than across the bottom of the wing. This difference in speed is achieved by having the wing tilt slightly upward in front (the tilt is called "angle of attack"). Some wings are flatter on the bottom and more curved on the top to accentuate the difference in speeds for the air going over and under the wing.

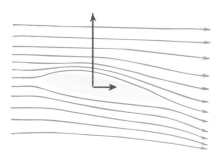

FIGURE 13.19 The vertical arrow represents the net upward force (lift) that results from less air pressure above the wing than below the wing. The horizontal arrow represents air drag.

The net upward pressure on the wing multiplied by the surface area of the wing gives the net lifting force. The lift is greater when there is a large wing area and when the plane is traveling fast. Gliders have a very large wing area relative to their weight so that they do not have to be going very fast for sufficient lift. At the other extreme, fighter planes designed for high speed have small wing areas relative to their weight. Consequently, they must take off and land at high speeds.

FIGURE 13.20 Where is air pressure greater—on the top or bottom surface of the hang glider?

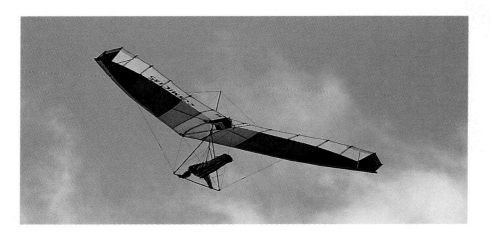

---

* A word of caution: In a building that has venting adequate to tolerate the sudden pressure drop without the need to open windows, opening a window may actually increase damage.

We all know that a baseball pitcher can throw a ball in such a way that it will curve off to one side of its trajectory. This is accomplished by imparting a large spin to the ball. Similarly, a tennis player can hit a ball that will curve. A thin layer of air is dragged around the spinning ball by friction, which is enhanced by the baseball's threads or the tennis ball's fuzz. The moving layer produces a crowding of streamlines on one side. Note in Figure 13.21b that the streamlines are more crowded at B than at A for the direction of spin shown. Air pressure is greater at A, and the ball curves as shown.

FIGURE 13.21   (a) The streamlines are the same on either side of a nonspinning baseball. (b) A spinning ball produces a crowding of streamlines. The resulting "lift" (red arrow) causes the ball to curve as shown by the blue arrow.

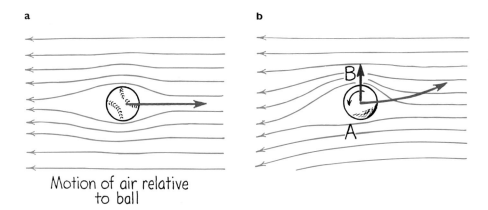

a   **b**

Motion of air relative to ball

The net force produced by unequal pressures is not restricted to airplane wings and spinning balls. Bernoulli's principle has been applied with varying degrees of success to sailboats without sails since the 1920s. In the place of masts, these ships have large vertical cylinders that are made to rotate about their vertical axes by motors. Like the case of the spinning baseball, a net force is produced that moves the ship, in most cases, more efficiently than canvas sails. A more recent design is Jacques Cousteau's ship, the *Alcyone* (Figure 13.22). Instead of rotating cylinders, it has fixed cylinders with special venting and an internal fan to produce the unequal pressures. A pair of conventional diesel engines power the ship when wind speed is insufficient. With a good wind, *Alcyone* can save over 50 percent in fuel, and with winds over 25 knots (12.5 meters per second), the diesel engines can be shut off altogether for a cruising speed of nine knots.

Bernoulli's principle accounts for the fact that passing ships run the risk of a sideways collision. Water flowing between the ships travels faster than water flowing past the outer sides. The streamlines are more compressed between the ships than outside. Water pressure acting against the hulls is therefore reduced between the ships. Unless the ships are steered to compensate for this, the greater pressure against the outer sides of the ships then forces them together. Figure 13.23 shows how this can be demonstrated in your kitchen sink or bathtub.

FIGURE 13.22   Jacques Cousteau's *Alcyone* is a wind-powered ship that employs a Turbosail™ system instead of sails. Like a spinning baseball, the vertical cylinders deflect the airstream. In this design the cylinders do not spin, but instead suck in air through perforated lateral vents on one or the other side by means of an internal fan. This action simulates spinning cylinders and deflects the airstream. The airstream reacts by exerting a net force on the cylinders. The component of this net force in the forward direction propels the ship.

FIGURE 13.23 Try this in your sink. Loosely moor a pair of toy boats side by side. Then direct a stream of water between them. The boats will draw together and collide. Why?

Question If you take a barometer outside on a windy day, will it give a lower reading than if you keep it indoors?

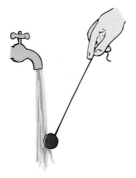

FIGURE 13.24 Pressure is greater in the stationary fluid (air) than in the moving fluid (water stream). The atmosphere pushes the ball into the region of reduced pressure.

Here 's another way to demonstrate Bernoulli's principle in your kitchen sink (Figure 13.24). Tape a Ping-Pong ball to a string and allow the ball to swing into a stream of running water. You'll see that it will remain in the stream even when tugged slightly to the side, as shown. This is because the pressure of the stationary air next to the ball is greater than the pressure of the moving water. The ball is pushed into the region of reduced pressure by the atmosphere.

A similar thing happens to a bathroom shower curtain when the shower water is turned on full blast. The pressure in the shower stall is reduced, and the relatively greater pressure outside the curtain pushes it inward. Although this effect is small compared with the convection produced by temperature differences, the next time you're taking a shower and the curtain swings in against your legs, think of Daniel Bernoulli!

Rats to you too, Daniel Bernoulli!

FIGURE 13.25 The curved shape of an umbrella can be disadvantageous on a windy day.

Answer Atmospheric pressure is reduced where the wind is blowing, so the barometer should give a lower reading. (The effect, however, is very small so you might not detect it unless the wind is very strong or your barometer is very sensitive.)

Plasma

In addition to solids, liquids, and gases, there is a fourth phase of matter, the least common phase in our everyday environment—**plasma** (not to be confused with the clear liquid part of blood, also called plasma).

A plasma is an electrified gas. The atoms and molecules that make it up are positively *ionized,* stripped of one or more electrons, with a corresponding number of free electrons. Recall that a neutral atom has as many positive protons inside the nucleus as it has negative electrons outside the nucleus. When one or more of these electrons is stripped from the atom, the atom has more positive charge than negative charge and is called a positive *ion* (under some conditions, it may have extra electrons, in which case it is called a *negative ion*). Although the electrons and ions are themselves electrically charged, the plasma as a whole is electrically neutral because there are still equal numbers of positive and negative charges, just as there are in an ordinary gas. Nevertheless, a plasma and a gas have very different properties. The plasma readily conducts electric current, it absorbs certain kinds of radiation that pass unhindered through a gas, and it can be shaped, molded, and moved by electric and magnetic fields.

In laboratories on earth, a plasma is often created by heating a gas to very high temperatures, making it so hot that electrons are "boiled" off the atoms. The center of our sun consists of such a hot plasma. Plasmas may also be created at lower temperatures by bombarding atoms with high-energy particles or radiation.

Plasma in the Everyday World

If you happen to be reading this by the light emitted by a fluorescent lamp, you don't have to look far to see plasma in action. Within the glowing tube of the lamp is a plasma that contains argon and mercury ions (as well as many neutral atoms of these elements). When you turn the lamp on, a high voltage between electrodes at the ends of the tube causes electrons to flow. These electrons ionize some of the atoms, forming a plasma, which then provides a conducting path to keep the current flowing. The current activates some mercury atoms, causing them to emit radiation, mostly in the invisible ultraviolet. This radiation, in turn, causes the phosphor coating on the inner surface of the tube to glow with visible light.

The neon gas in an advertising sign similarly becomes a plasma when its atoms are ionized by electron bombardment as a current flows through it. Some neon atoms, after being activated by the current, emit predominantly red light. "Neon lights" do not always contain just neon. The different colors seen in these signs correspond to different kinds of atoms.

Vapor lamps used in street lighting also emit light stimulated by glowing plasmas (Figure 13.26). The bluish-green, almost white light from some of these lamps comes from mercury atoms; in other lamps, excited sodium atoms emit yellow light.

The aurora borealis (or the northern lights) shown on the cover of this book is glowing plasmas in the upper atmosphere. Layers of low-temperature plasma encircle the whole earth. Occasionally, showers of electrons from outer space and from radiation belts enter the "magnetic windows" near the earth's poles, crash into the layers of plasma, and produce light.

These layers, which extend upward from about 80 kilometers, make up the ionosphere and act as mirrors to low-frequency radio waves. Higher-frequency radio and TV waves pass through the ionosphere. This is why you can pick up radio stations from long distances on your lower-frequency AM radios, but you have to be in the "line of sight"

FIGURE 13.26 Streets are illuminated at night by vapor lamps containing plasma.

of broadcasting or relay antennas to pick up higher-frequency FM and TV signals. Ever notice that at nighttime you can pick up very distant stations on your AM set? This is because the plasma layers settle closer together in the absence of the energizing sunlight and act as better radio reflectors.

Plasma Power

A higher-temperature plasma is the exhaust of a jet engine. It is a weakly ionized plasma, but when small amounts of potassium salts or cesium metal are added, it becomes a very good conductor, and when it is directed into a magnet, electricity is generated! This is MHD power, the **m**agneto**h**ydro**d**ynamic interaction between a plasma and a magnetic field. (We will treat the mechanics of how electricity is generated in this way in Chapter 24.) Low-pollution MHD power is now in the developmental stage and is in operation at a few places in the world already. We can expect to see more plasma power with MHD.

An even more promising achievement will be plasma power of a different kind—the controlled fusion of atomic nuclei. We will treat the physics of fusion in Chapter 33. Equally important as the physics of thermonuclear fusion is the social impact of its controlled use. We have already seen the early effects of uncontrolled thermonuclear fusion, the hydrogen bomb. Like all technology, fusion can be applied for humanity's benefit as well as for its destruction. The benefits of controlled fusion may well be more far-reaching than the benefits that have come from harnessing electrical power in the nineteenth century. Fusion plants may ultimately not only make abundant electrical energy but provide the energy and means to recycle and even synthesize elements as well. The control of fusion may provide the setting for a new age.

We have come a long way with our mastery of the first three phases of matter. Our mastery of the fourth phase may bring us ever so much farther.

Summary of Terms

Atmospheric Pressure The pressure exerted against bodies immersed in the atmosphere. It results from the weight of air pressing down from above. At sea level, atmospheric pressure is about 101 kPa.

Barometer Any device that measures atmospheric pressure.

Boyle's law The product of pressure and volume is a constant for a given mass of confined gas, as long as temperature remains unchanged:

$$P_1 V_1 = P_2 V_2$$

Archimedes' principle for air An object surrounded by air is buoyed up with a force equal to the weight of displaced air.

Bernoulli's principle The pressure in a fluid moving steadily without friction or outside energy input decreases when the fluid velocity increases.

Plasma An electrified gas containing charged atoms (ions) and free electrons. Most of the matter in the universe is in the plasma phase.

Review Questions

1. Distinguish among a *gas*, a *liquid*, and a *fluid*.

The Atmosphere

2. What is the energy source for the motion of gas in the atmosphere? What prevents atmospheric gases from flying off into space?

3. How high would you have to go in the atmosphere for half of the mass of air to be below you?

Atmospheric Pressure

4. What is the cause of atmospheric pressure?

5. What is the mass of a cubic meter of air at room temperature (20°C)?

6. What is the approximate mass of a column of air 1 cm^2 in area that extends from sea level to the upper atmosphere? What is the weight of this amount of air?

7. What is the SI unit of atmospheric pressure?

8. What is the pressure at the bottom of the column of air discussed in question 6?

Barometers

9. How does the downward pressure of the 76-cm column of mercury in a barometer compare to air pressure at the bottom of the atmosphere?

10. How does the weight of mercury in a barometer compare to the weight of an equal cross-section of air from sea level to the top of the atmosphere?

11. Why would a water barometer have to be 13.6 times taller than a mercury barometer?

12. When you drink liquid through a straw, is it more accurate to say the liquid is pushed up the straw rather than sucked up the straw? What exactly does the pushing? Defend your answer.

13. Why will a vacuum pump not operate for a well that is more than 10.3 m deep?

14. Why is it that an aneroid barometer is able to measure altitude as well as atmospheric pressure?

15. How does a mechanical pump produce a vacuum?

Boyle's Law

16. By how much does the density of air increase when it is compressed to half its volume?

17. What happens to the air pressure inside a balloon when it is squeezed to half its volume at constant temperature?

18. What is an ideal gas?

Buoyancy of Air

19. A balloon that weighs 1 N is suspended in air, drifting neither up nor down. How much buoyant force acts on it? What happens if the buoyant force decreases? If it increases?

20. Does the air exert buoyant force on all objects in air or only on objects such as balloons that are very light for their size? Why does the air support only things with a very low density?

21. What usually happens to a toy helium-filled balloon that rises high into the atmosphere?

Bernoulli's Principle

22. What happens to the internal pressure in a fluid flowing in a horizontal pipe when its speed increases?

23. What are streamlines? Is pressure greater or less in regions where streamlines are crowded?

Applications of Bernoulli's Principle

24. What does Bernoulli's principle have to do with the flight of airplanes?

25. What does the tilt of the wing have to do with the flight of airplanes?

26. Why does a spinning ball curve in its flight?

27. Does atmospheric pressure increase or decrease on a windy day?

Plasma

28. How does a plasma differ from a gas?

29. (a) In what sense is a plasma electrically charged? (b) In what sense is a plasma electrically neutral?

Plasma in the Everyday World

30. Cite at least three examples of plasma in your daily environment.

31. What does plasma have to do with the aurora borealis (northern lights)?

32. Why is AM radio reception better at night?

Plasma Power

33. What is MHD power?

34. To what use might a future fusion power plant be put besides generating electricity?

Projects

1. You can find the pressure exerted by the tires of your car on the road and compare it with the air pressure in the tires. For this project, you need to get the weight of your car from the manual or a dealer, and divide by four to get the approximate weight held up by one tire. You can closely approximate the area of contact of a tire with the road by tracing the edges of tire contact on a sheet of paper marked with 1-in^2 squares beneath the tire. After you get the pressure of the tire on the road, compare it with the air pressure in the tire. Are they nearly equal? If not, which one is greater?

2. Try this in the bathtub or when you're washing dishes. Lower a drinking glass, mouth downward, over a small floating object. What do you observe? How deep will the glass have to be pushed in order to compress the enclosed air to half its volume? (You won't be able to do this in your bathtub unless it's 10.3 m deep!)

3. You ordinarily pour water from a full glass to an empty glass by simply placing the full glass above the empty glass and tipping. Have you ever poured air from one glass to the other? The procedure is similar. Lower two glasses in water, mouths downward. Let one fill with water by tilting its mouth upward. Then hold the water-filled glass mouth downward above the air-filled glass. Slowly tilt the lower glass and let the air escape, filling the upper glass. You will be pouring air from one glass to another!

4. Raise a filled glass of water above the waterline, but with its mouth beneath the surface. Why does the water not run out? How tall would a glass have to be before water began to run out? (You won't be able to do this indoors unless you have a 10.3-m ceiling.)

5. Place a card over the open top of a glass filled to the brim with water, and invert it. Why does the card stay intact? Try it sideways.

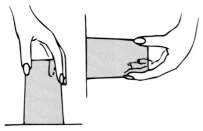

6. Invert a water-filled pop bottle or small-necked jar. Notice that the water doesn't simply fall out, but gurgles out of the container. Air pressure won't let it get out until some air has pushed its way up inside the bottle to occupy the space above the liquid. How would an inverted, water-filled bottle empty if this were done on the moon?

7. Pour about a half cup of water in a 5-or-so-L metal can with a screw top. Place the can *open* on a stove and heat until the water boils and steam comes out of the opening. Quickly remove the can and screw the cap on tightly. Allow the can to stand and observe the results. The effect can be hastened by cooling the can with a dousing of cold water. Explain your observations. (Don't do this with a can you expect to use again. The can better be expendable!)

8. Heat a small amount of water to boiling in an aluminum soda-pop can and invert it quickly into a dish of cold water. Surprisingly dramatic!

9. Make a small hole near the bottom of an open tin can. Fill it with water, which proceeds to spurt from the hole. Cover the top of the can firmly with the palm of your hand and the flow stops. Explain.

10. Lower a narrow glass tube or drinking straw into water and place your finger over the top of the tube. Lift the tube from the water and then lift your finger from the top of the tube. What happens? (You'll do this often if you enroll in a chemistry lab.)

11. Fold the ends of a filing card down so that you make a little bridge. Stand it on the table and blow through the arch as shown. No matter how hard you blow, you will not succeed in blowing the card off the table (unless you blow against the side of it). Try this with your non-physics friends. Then explain it to them!

12. Push a pin through a small card and place it in the hole of a thread spool. Try to blow the card from the spool by blowing through the hole. Try it in all directions.

13. Hold a spoon in a stream of water as shown and feel the effect of the differences in pressure.

Exercises

1. It is said that a gas fills all the space available to it. Why then doesn't the atmosphere go off into space?

2. Why is there no atmosphere on the moon?

3. Count the tires on a large tractor trailer that is unloading food at your local supermarket, and you may be surprised to count 18 tires. Why so many tires? (*Hint:* See Project 1.)

4. What is the purpose of the ridges that prevent the funnel from fitting tightly in the mouth of a bottle?

5. How does the density of air in a deep mine compare with the air density at the earth's surface?

6. At the bottom of an imaginary hole drilled to the center of the earth, why would air pressure be greater than at the earth's surface even though gravity has diminished to zero?

7. When an air bubble rises in water, what happens to its mass, volume, and density?

8. The valve stem on a tire must exert a certain force on the air within to prevent any of that air from leaking out. If the diameter of the valve stem were doubled, by how much would the force exerted by the valve stem increase?

9. Two teams of eight horses each were unable to pull the Magdeburg hemispheres apart (Figure 13.2). Why? Suppose two teams of nine horses each could pull them apart. Then would one team of nine horses succeed if the other team were replaced with a strong tree? Defend your answer.

10. Before boarding an airplane, you buy a roll of camera film or any item packaged in an airtight foil package, and while in flight you notice that it is puffed up. Explain why this happens.

11. Why do you suppose airplane windows are smaller than bus windows?

12. A half cup or so of water is poured into a 5-L can, which is put over a source of heat until most of the water has boiled away. Then the top of the can is screwed on tightly and the can is removed from the source of heat and allowed to cool. What happens to the can and why?

13. Will breaking a TV picture tube cause the tube to implode or to explode? Explain.

14. We can understand how pressure in water depends on depth by considering a stack of bricks. The pressure below the bottom brick is determined by the weight of the entire stack. Halfway up the stack, the pressure is half because the weight of the bricks above is half. To explain atmospheric pressure, we should consider compressible bricks, like foam rubber. Why is this so?

15. The "pump" in a vacuum cleaner is merely a high-speed fan. Would a vacuum cleaner pick up dust from a rug on the moon? Explain.

16. Suppose the pump shown in Figure 13.9 worked with a perfect vacuum. From how deep a well could water be pumped?

17. If a liquid only half as dense as mercury were used in a barometer, how high would its level be on a day of normal atmospheric pressure?

18. Why does the size of the cross-sectional area of a mercury barometer not affect the height of the enclosed mercury column?

19. From how deep a container could mercury be drawn with a siphon?

20. If you could somehow replace the mercury in a mercury barometer with a denser liquid, would the height of the liquid column be greater or less than with mercury? Why?

21. Would it be slightly more difficult to draw soda through a straw at sea level or on top of a very high mountain? Explain.

22. The pressure exerted against the ground by an elephant's weight distributed evenly over its four feet is less than 1 atmosphere. Why, then, would you be crushed beneath the foot of an elephant, while you're unharmed by the pressure of the atmosphere?

23. Your friend says that the buoyant force of the atmosphere on an elephant is significantly greater than the buoyant force of the atmosphere on a small helium-filled balloon. What do you say?

24. On a sensitive balance, weigh an empty flat thin plastic bag. Then weigh the bag when air is in it. Will the readings differ? Explain.

25. Why is it so difficult to breathe when snorkeling at a depth of 1 m, and practically impossible at a 2-m depth? Why can't a diver simply breathe through a hose that extends to the surface?

26. Why does the weight of an object in air differ from its weight in a vacuum (remembering that weight is the force exerted against a supporting surface)? Cite an example where this would be an important consideration.

27. A little girl sits in a car at a traffic light holding a helium-filled balloon. The windows are up and the car is relatively airtight. When the light turns green and the car accelerates forward, her head pitches backward but the balloon pitches forward. Explain why.

28. Would a bottle of helium gas weigh more or less than an identical bottle filled with air at the same pressure? Than an identical bottle with the air pumped out?

29. When you replace helium in a balloon of given size by less-dense hydrogen, does the buoyant force on the balloon change? Explain.

30. A steel tank filled with helium gas doesn't rise in air, but a balloon containing the same helium easily does. Why?

31. A balloon filled with air falls to the ground, but a balloon filled with helium rises. Why is this so?

32. The gas pressure inside an inflated rubber balloon is always greater than the air pressure outside. Why is this so?

33. Two identical balloons of the same volume are pumped up with air to more than atmospheric pressure and suspended on the ends of a stick that is horizontally balanced. One of the balloons is then punctured. Is the balance of the stick upset? If so, which way does it tip?

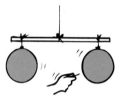

34. Two balloons that have the same weight and volume are filled with equal amounts of helium. One is rigid and the other is free to expand as the pressure outside decreases. When released, which will rise higher? Explain.

35. Imagine a huge space colony that consists of a rotating air-filled cylinder. How would the density of air at "ground level" compare to the air densities "above"?

36. Would a helium-filled balloon "rise" in the atmosphere of a rotating space habitat? Defend your answer.

37. The force of the atmosphere at sea level against the outside of a 10-m^2 store window is about a million N. Why does this not shatter the window? Why might the window shatter in a strong wind?

38. Why does the fire in a fireplace burn more briskly on a windy day?

39. In a department store, an airstream from a hose connected to the exhaust of a vacuum cleaner blows upward at an angle and supports a beach ball in midair. Does the air blow under or over the ball to provide support?

40. What provides the lift to keep a Frisbee in flight?

41. When a steadily flowing gas flows from a larger-diameter pipe to a smaller-diameter pipe, what happens to (a) its speed, (b) its pressure, and (c) the spacing between its streamlines?

42. Why is it easier to throw a curve with a tennis ball than a baseball?

43. How is an airplane able to fly upside down?

44. When a jet plane is cruising at high altitude, the flight attendants have more of a "hill" to climb as they walk forward along the aisle than when the plane is cruising at a lower altitude. Why does the pilot have to fly with a greater "angle of attack" at high altitude than at low?

45. What physics concept underlies these three observations? When passing an oncoming truck on the highway, your car tends to sway toward the truck. The canvas roof of a convertible automobile bulges upward when the car is traveling at high speeds. The windows of older trains sometimes break when a high-speed train passes by on the next track.

46. A steady wind blows over the waves of an ocean. Why does the wind increase the humps and troughs of the waves?

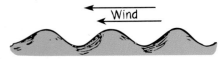

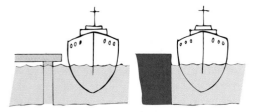

47. In answering the question of why a flag flaps in the wind, your friend replies that it flaps because of Bernoulli's principle. (Not really a convincing explanation, is it?) How would you answer the same question?

48. Wharves are made with pilings that permit the free passage of water. Why would a solid-walled wharf be disadvantageous to ships attempting to pull alongside?

49. Is lower pressure the result of fast-moving air, or is fast-moving air the result of lower pressure? Give one example supporting each point of view. (In physics, when two things are related—such as force and acceleration or speed and pressure—it is usually arbitrary which one we call cause and which one we call effect.)

50. Why can you pick up far-away radio stations better at nighttime on your AM radio?

Problems

1. Estimate the buoyant force that air exerts on you. (To do this, you can estimate your volume by knowing your weight and by assuming that your weight density is a bit less than that of water.)

2. Nitrogen and oxygen in their liquid states have densities only 0.8 and 0.9 that of water. Atmospheric pressure is due primarily to the weight of nitrogen and oxygen gas in the air. If the atmosphere liquefied, would its depth be greater or less than 10.3 m?

3. Air in a cylinder is compressed to one-tenth its original volume with no change in temperature. (a) What happens to its pressure? (b) If a valve is then opened to let out enough air to bring the pressure back down to its original value, what percentage of the molecules escape?

4. On a perfect fall day, you are hovering at low altitude in a hot-air balloon, accelerated neither upward nor downward. The total weight of the balloon, including its load and the hot air in it, is 20,000 N. (a) What is the weight of the displaced air? (b) What is the volume of the displaced air?

5. Just after it is launched, a helium-filled research balloon resembles a thin vertical cigar. When it reaches an altitude of 50,000 feet, it looks like a fat peach. The material of the balloon exerts no appreciable force on the gas within, but simply serves as a boundary separating air from helium. Explain carefully why the buoyant force on the balloon remains constant as it rises.

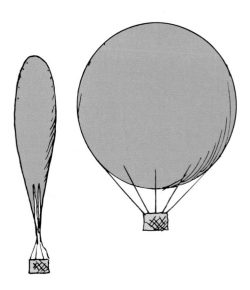

6. A mercury barometer reads 760 mm at sea level. When it is carried to an altitude of 5.6 km, the height of the mercury column is half, 380 mm. What is the air pressure at this altitude relative to sea-level pressure? If the barometer is taken up another 5.6 km to an altitude of 11.2 km, will the height of its mercury column fall another 380 mm and be zero? Why or why not?

P a r t

3

HEAT

· ·

Although the temperature of these sparks exceeds 2000°C, the heat they impart in striking my skin is very small -- which illustrates that *temperature* and *heat* are different concepts. Learning to distinguish between closely related concepts is the challenge and essence of *Conceptual Physics*.

14

. . . .

Temperature, Heat, and Expansion

What is the source of thermal energy in the earth's
interior? (See page 577.)

All matter is composed of continually jiggling atoms or molecules. Whether the atoms and molecules combine to form solids, liquids, gases, or plasmas depends on how fast the molecules are moving. By virtue of their motion, the molecules or atoms in matter possess kinetic energy. The average kinetic energy of the individual particles is directly related to a property you can sense: how hot something is. Whenever something becomes warmer, we know that the kinetic energy of its particles increases. Strike a solid penny with a hammer and it becomes warm because the hammer's blow causes the atoms in the metal to jostle faster. Put a flame to a liquid and it too becomes warmer. Rapidly compress air in a tire pump and the air becomes warmer. When a solid, liquid, or gas gets warmer, its atoms or molecules move faster. The atoms or molecules have more kinetic energy.

Temperature

The quantity that tells how warm or cold an object is with respect to some standard is called **temperature**. We express the temperature of matter by a number that corresponds to the degree of hotness on some chosen scale.

256

FIGURE 14.1 Can we trust our sense of hot and cold? Will both fingers feel the same temperature when they are put in the warm water?

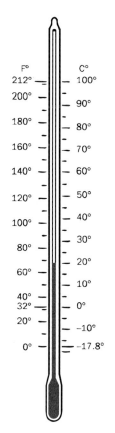

FIGURE 14.2 Fahrenheit and Celsius scales on a thermometer.

Nearly all materials expand when their temperature is raised and contract when it is lowered. A thermometer is a common instrument that measures temperature by means of the expansion and contraction of a liquid, usually mercury or colored alcohol.

On the scale commonly used in laboratories, the number 0 is assigned to the temperature at which water freezes and the number 100 to the temperature at which water boils (at standard atmospheric pressure). The space between is divided into 100 equal parts called *degrees;* hence, a thermometer so calibrated has been called a *centigrade thermometer* (from *centi,* "hundredth," and *gradus,* "degree"). It is now called a *Celsius thermometer* in honor of the man who first suggested the scale, the Swedish astronomer Anders Celsius (1701–1744).

In the United States, the number 32 is assigned to the temperature at which water freezes, and the number 212 is assigned to the temperature at which water boils. Such a scale makes up a Fahrenheit thermometer, named after its illustrious originator, the German physicist G. D. Thermometer (1686–1736). The Fahrenheit scale will become obsolete if and when the United States goes metric.*

Still another temperature scale, favored by scientists, is the Kelvin scale, named after the British physicist Lord Kelvin (1824–1907). This scale is calibrated not in terms of the freezing and boiling points of water, but in terms of energy itself. The number 0 is assigned to the lowest possible temperature—**absolute zero**, at which a substance has absolutely no kinetic energy to give up.† Absolute zero corresponds to $-273°C$ on the Celsius scale. Units on the Kelvin scale are the same size as degrees on the Celsius scale, so the temperature of melting ice is $+273$ kelvins. There are no negative numbers on the Kelvin scale. We won't treat this scale further until we return to it when we study thermodynamics in Chapter 17.

Arithmetic formulas are used for converting from Fahrenheit to Celsius and Celsius to Fahrenheit and are popular in classroom exams. Such arithmetic exercises are not really physics, and the probability of your having the occasion to do this task elsewhere is small, so we will not be concerned with it here. Besides, this conversion can be very closely approximated by simply reading the corresponding temperature from the side-by-side scales in Figure 14.2.

Temperature is related to the random motion of the molecules in a substance. More specifically, it is proportional to the average kinetic energy of molecular translational motion (that is, molecular motion along a straight or curved path). It is important to understand that temperature is not a measure of the *total* kinetic energy of molecules in a substance. For example, there is twice as much molecular kinetic energy in 2 liters

---

* The conversion to Celsius will put the United States in step with the rest of the world, where the Celsius scale is the standard. Americans are slow to convert. Changing any long-established custom is difficult, and the Fahrenheit scale does have some advantages in everyday use. For example, its degrees are smaller ($1°F = \frac{5}{9}°C$), which gives greater accuracy when reporting the weather in whole-number temperature readings. Then, too, people somehow attribute a special significance to numbers increasing by an extra digit, so that when the temperature of a hot day is reported to reach 100°F, the idea of heat is conveyed more dramatically than by saying it is 38°C. Like so much of the British system of measure, the Fahrenheit scale is geared to human beings. (And, for the record, the Fahrenheit scale is named after Gabriel Daniel Fahrenheit.)

† Even at absolute zero, a substance has what is called "zero-point energy," which is unavailable energy that cannot be transferred to a different substance.

FIGURE 14.3 There is more molecular kinetic energy in the container filled with warm water than in the small cupful of higher-temperature water.

of boiling water as in 1 liter of boiling water—but the temperatures of both amounts of water are the same because the *average* kinetic energy per molecule in each is the same. This concept is further illustrated in Figure 14.3.

Interestingly enough, a thermometer registers its own temperature. When a thermometer is in thermal contact with something whose temperature we wish to know, energy will flow between the two until their temperatures are equal and thermal equilibrium is established. If we know the temperature of the thermometer, we then know the temperature of the something. A thermometer should be small enough that it doesn't appreciably alter the temperature of the something being measured. If you are measuring the room air temperature, then your thermometer is small enough. But if you are measuring the temperature of a drop of water, contact between the drop and the thermometer may change the drop's temperature—a classic case of changing what is being measured by the measuring process.

Heat

If you touch a hot stove, energy enters your hand because the stove is warmer than your hand. When you touch a piece of ice, on the other hand, energy passes out of your hand and into the colder ice. The direction of spontaneous energy transfer is always from a warmer thing to a neighboring cooler thing. The energy transferred from one thing to another because of a temperature difference between the things is called **heat**.

It is important to point out that matter does not *contain* heat. Matter contains molecular kinetic energy and possibly potential energy, *not heat*. Heat is energy in transit from a body of higher temperature to one of lower temperature. Once transferred, the energy ceases to be heat. (As an analogy, work is also energy in transit. A body does not *contain work*. It *does* work or has work done on it.) In previous chapters we called the energy resulting from heat flow *thermal energy,* to make clear its link to heat and temperature. In this chapter, we will use the term that scientists prefer, *internal energy.*

Internal energy is the grand total of all energies inside a substance. In addition to the translational kinetic energy of jostling atoms in a substance, there is energy in other forms. There is rotational kinetic energy of molecules and kinetic energy due to internal movements of atoms within molecules. There is also potential energy due to the forces between molecules. So a substance does not contain heat—it contains internal energy.

When a substance absorbs or gives off heat, internal energy in the substance changes. Thus, as a substance absorbs heat, this energy may or may not make the molecules or atoms jostle faster. In some cases, as when ice is melting, a substance absorbs heat without an increase in molecular kinetic energy. The substance undergoes a change of phase, which we will cover in detail in Chapter 16.

For things in thermal contact, heat flow is from the substance at a higher temperature to a substance at a lower temperature, but not necessarily a flow from a substance with more internal energy to a substance with less internal energy. There is more internal energy in a bowl of warm water than there is in a red-hot thumbtack; if the tack is immersed in the water, heat flow is not from the warm water to the tack. Instead, heat flow is from the hot tack to the cooler water. Heat never flows of itself from a low-temperature substance into a higher-temperature substance.

FIGURE 14.4 Just as water in the pipes seeks a common level (where the pressures at any given depth are the same), the thermometer and its immediate surroundings reach a common temperature (where the average molecular KE for both is the same).

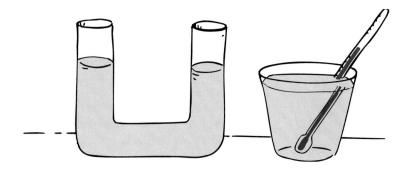

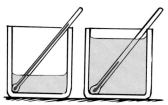

Hot stove

FIGURE 14.5 Although the same quantity of heat is added to both containers, the temperature increases more in the container with the smaller amount of water.

How much heat flows depends not only on the temperature difference between substances but on the amount of material as well. For example, a barrelful of hot water will transfer more heat to a cooler substance than a cupful of water at the same temperature. There is more internal energy in the larger amount of water.

Questions

1. Suppose you apply a flame to 1 L of water for a certain amount of time and its temperature rises by 2°C. If you apply the same flame for the same time to 2 L of water, by how much will its temperature rise?
2. If a fast marble hits a random scatter of slow marbles, does the fast marble usually speed up or slow down? Which lose(s) kinetic energy and which gain(s) kinetic energy, the initially fast-moving marble or the initially slow ones? How do these questions relate to the direction of heat flow?

Measurement of Heat

So heat is molecular energy being transferred from one thing to another because of a temperature difference. Since heat is a form of energy, it can be measured in joules. A more common unit of heat, however, is the calorie. The calorie is defined as the amount of heat required to change the temperature of 1 gram of water by 1 Celsius degree.*

Answers

1. Its temperature will rise by only 1°C, because there are twice as many molecules in 2 L of water and each molecule receives only half as much energy on the average.
2. A fast-moving marble slows when it hits slower-moving marbles. It gives up some of its kinetic energy to the slower ones. Likewise with the flow of heat. Molecules with more kinetic energy that are in contact with molecules having less kinetic energy give up some of their excess energy to the less energetic ones. The direction of energy transfer is from hot to cold. For both the marbles and the molecules, however, the total energy before and after contact is the same.

---

* Another common unit of heat is the British thermal unit (BTU). The BTU is defined as the amount of heat required to change the temperature of 1 lb of water by 1 Fahrenheit degree.

The energy ratings of foods and fuels are determined by burning them and measuring the energy released. (Digestion is really "burning"—at a slow rate). The heat unit used to label foods is actually the kilocalorie, which is 1000 calories (the heat required to change 1 kilogram of water by 1°C). To distinguish this unit from the smaller calorie, the food unit is sometimes called a *Calorie* (written with a capital *C*). It is important to remember that the calorie and Calorie are units of energy. These names are historical carry-overs from the early idea that heat was an invisible fluid called *caloric*. This view persisted into the nineteenth century. We now know that heat is a form of energy, so it doesn't need its own separate unit. Someday the calorie may give way to the SI unit, the joule, as the common unit for measuring heat. (The relationship between calories and joules is that 1 calorie = 4.184 joules.)

FIGURE 14.6 To the weight watcher, the peanut contains 10 calories; to the physicist, it releases 10,000 calories (or 41,840 joules) of energy when burned or digested.

Specific Heat Capacity

Have you ever noticed that some foods remain hotter much longer than others? Boiled onions and squash on a hot dish, for example, are often too hot to eat when mashed potatoes—if not too moist—may be eaten comfortably. The filling of hot apple pie can burn your tongue while the crust will not, even when the pie has just been taken out of the oven. The aluminum covering on a hot frozen dinner can be peeled off with your bare fingers as soon as it is removed from the oven. A piece of toast may be comfortably eaten a few seconds after coming from the hot toaster, whereas we must wait several minutes before eating soup from a stove as hot as the toaster.

Different substances have different capacities for storing internal energy. If we heat a pot of water on a stove, we might find that it requires 15 minutes to raise it from room temperature to its boiling temperature. But if we put an equal mass of iron on the same flame, we would find that it would rise through the same temperature range in only about 2 minutes. For silver, the time would be less than a minute. We find that different materials require different quantities of heat to raise the temperature of a given mass of the material by a specified number of degrees. Different materials absorb energy in different ways. The energy may increase the jiggling motion of molecules, which raises the temperature. Or it may increase the amount of internal vibration or rotation within the molecules and go into potential energy, which does not raise the temperature. Generally there is a combination of both.

A gram of water requires 1 calorie of energy to raise the temperature 1 Celsius degree. It takes only about one-eighth as much energy to raise the temperature of a gram of iron by the same amount. Water absorbs more heat than iron for the same change in temperature. We say water has a higher **specific heat capacity** (sometimes simply called *specific heat*).

FIGURE 14.7 The filling of hot apple pie may be too hot to eat, whereas the crust is not.

> **The specific heat capacity of any substance is defined as the quantity of heat required to change the temperature of a unit mass of the substance by 1 degree Celsius.**

We can think of specific heat capacity as thermal inertia. Recall that inertia is a term used in mechanics to signify the resistance of an object to a change in its state of motion. Specific heat capacity is like a thermal inertia since it signifies the resistance of a substance to a change in temperature.

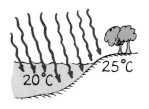

FIGURE 14.8 Water has a high specific heat capacity and is transparent, so it takes more energy to heat up than land. Solar energy incident upon land is concentrated at the surface, but upon water it extends and dilutes beneath the surface.

Question Which has a higher specific heat capacity, water or sand?

Water has a much higher capacity for storing energy than all but a few uncommon materials. A relatively small amount of water absorbs a great deal of heat for a correspondingly small temperature rise. Because of this, water is a very useful cooling agent and is used in the cooling systems of automobiles and other engines. If a liquid of lower specific heat capacity were used in cooling systems, its temperature would rise higher for a comparable absorption of heat.

Water also takes a long time to cool, a fact that explains why hot-water bottles used to be employed on cold winter nights. (Electric blankets have, for the most part, taken their place.) This tendency on the part of water to resist changes in temperature improves the climate in many places. The next time you are looking at a world globe, notice the high latitude of Europe. If water did not have a high specific heat capacity, the countries of Europe would be as cold as the northeastern regions of Canada, for both Europe and Canada get about the same amount of sunlight per square kilometer. The Atlantic current known as the Gulf Stream carries warm water northeast from the Caribbean. It holds much of its internal energy long enough to reach the North Atlantic off the coast of Europe, where it then cools. The energy released, 1 calorie per degree for each gram of water that cools, is carried by the westerly winds over the European continent. A similar effect occurs in the United States. The winds in the latitudes of North America are westerly. On the West Coast, air moves from the Pacific Ocean to the land. Because of water's high specific heat capacity, an ocean does not vary much in temperature from summer to winter. The water is warmer than the air in the winter and cooler than the air in the summer. In winter the water warms the air that moves over it and warms the coastal regions of North America. In summer, the water cools the air and the coastal regions are cooled. On the east coast, air moves from the land to the Atlantic Ocean. Land, with a lower specific heat capacity, gets hot in the summer but cools rapidly in the winter. As a result of water's high specific heat capacity and the wind directions, the West Coast city of San Francisco is warmer in the winter and cooler in the summer than the East Coast city of Washington, D.C., which is at about the same latitude.

Islands and peninsulas that are more or less surrounded by water do not have the same extremes of temperatures that are observed in the interior of a continent. The high summer and low winter temperatures common in Manitoba and the Dakotas, for example, are largely due to the absence of large bodies of water. Europeans, islanders, and people living near ocean air currents should be glad that water has such a high specific heat capacity. San Franciscans are!

Answer Water has the higher specific heat capacity. The temperature of water increases less than the temperature of sand in the same sunlight. Sand's low specific heat capacity, as evidenced by how quickly the surface warms in the morning sun and how quickly it cools at night, affects local climates.

Expansion

When the temperature of a substance is increased, its molecules or atoms jiggle faster and tend to move farther apart, on the average. The result is an expansion of the substance. With few exceptions, all forms of matter—solids, liquids, gases, and plasmas—generally expand when they are heated and contract when they are cooled.

In many cases the changes in the size of substances are not very noticeable, but careful observation will usually detect them. Telephone wires are longer and sag more on a hot summer day than they do on a cold winter day. Metal lids on glass fruit jars can often be loosened by heating them under hot water. If one part of a piece of glass is heated or cooled more rapidly than adjacent parts, the expansion or contraction that results may break the glass. This is especially true with thick glass. Pyrex glass is specially formulated to expand very little with increasing temperature.

The expansion of substances must be allowed for in structures and devices of all kinds. A dentist uses filling material that has the same rate of expansion as teeth. The aluminum pistons of some automobile engines are just smaller enough in diameter than the steel cylinders to allow for the much greater expansion rate of aluminum. A civil engineer uses reinforcing steel of the same expansion rate as concrete. Long steel bridges commonly have one end fixed while the other rests on rockers (Figure 14.9). The roadway itself is segmented with tongue-and-groove type gaps called *expansion joints* (Figure 14.10). Similarly, concrete roadways and sidewalks are intersected by gaps, sometimes filled with tar, so that the concrete can expand freely in summer and contract in winter.

FIGURE 14.9 One end of the bridge is fixed, while the end shown rides on rockers to allow for thermal expansion.

FIGURE 14.10 This gap is called an expansion joint; it allows the bridge to expand and contract.

Different substances expand at different rates. When two strips of different metals, say one of brass and the other of iron, are welded or riveted together, the greater expansion of one metal results in the bending shown in Figure 14.11. Such a compound thin bar is called a *bimetallic strip*. When the strip is heated, one side of the double strip becomes longer than the other, causing the strip to bend into a curve. On the other hand,

FIGURE 14.11 A bimetallic strip. Brass expands (or contracts) more when heated (or cooled) than iron does, so the strip bends as shown.

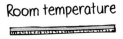

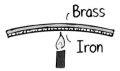

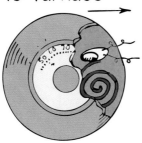

To furnace

FIGURE 14.12 A thermostat. When the coil expands, the drop of liquid mercury rolls away from the electrical contacts and breaks the circuit. When the coil contracts, the mercury rolls against the contacts and completes the electrical circuit.

when the strip is cooled, it tends to bend in the opposite direction, because the metal that expands more also shrinks more. The movement of the strip may be used to turn a pointer, regulate a valve, or close a switch.

A practical application of this is the thermostat (Figure 14.12). The back-and-forth bending of the bimetallic coil opens and closes an electric circuit. When the room becomes too cold, the coil bends toward the brass side, and in so doing activates an electrical switch that turns on the heat. When the room becomes too warm, the coil bends toward the iron side, which activates an electrical contact that turns off the heating unit. Refrigerators are equipped with thermostats to prevent them from becoming either too warm or too cold. Bimetallic strips are used in oven thermometers, electric toasters, automatic chokes on carburetors, and various other devices.

Liquids expand appreciably with increases in temperature. In most cases the expansion of liquids is greater than the expansion of solids. The gasoline overflowing a car's tank on a hot day is evidence for this. If the tank and contents expanded at the same rate, they would expand together and no overflow would occur. Similarly, if the expansion of the glass of a thermometer were as great as the expansion of the mercury, the mercury would not rise with increasing temperature. The reason the mercury in a thermometer rises with increasing temperature is that the expansion of liquid mercury is greater than the expansion of glass.

FIGURE 14.13 Place a dented Ping-Pong ball in boiling water, and you'll remove the dent. Why?

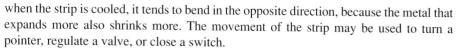

Questions

1. The Concorde supersonic airplane is 20 cm longer when in flight. Offer an explanation.
2. How would a thermometer be different if glass expanded more with increasing temperature than mercury?

Expansion of Water

Increase the temperature of any common liquid and it will expand. But not water at temperatures near the freezing point: ice-cold water does just the opposite! Water at the temperature of melting ice, 0°C (or 32°F), *contracts* when the temperature is increased. This is most unusual. As the water is heated and its temperature rises, it continues to *contract* until it reaches a temperature of 4°C. With further increase in temperature, the

Answers

1. At cruising speed (faster than the speed of sound), air friction against the Concorde raises its temperature dramatically, resulting in this significant thermal expansion.
2. The scale would have to be upside down. Do you see why?

FIGURE 14.14 The contraction and then expansion of water with increasing temperature.

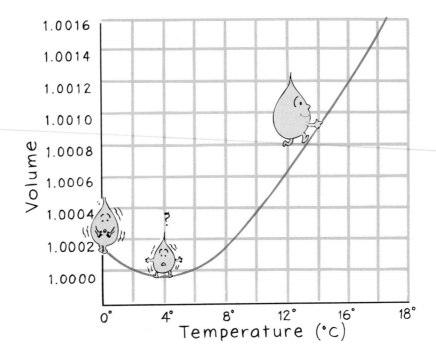

water then begins to expand; the expansion continues all the way to the boiling point, 100°C. This odd behavior is shown graphically in Figure 14.14.

A given amount of water has its smallest volume—and thus its greatest density—at 4°C. Just below 0°C, when water has become solid ice, its volume is considerably larger—and its density smaller. (Remember that ice floats in water, evidence that it is less dense than water.) The volume of ice at 0°C is not shown in Figure 14.14. (If it were plotted to the same exaggerated scale, the graph would extend far beyond the top of the page.) After water has turned to ice, further cooling causes it to contract.

Question What was the precise temperature at the bottom of Lake Michigan on New Year's Eve in 1901?

Ice has a crystalline structure. The crystals of most solids are arranged in such a way that the solid state occupies a smaller volume than the liquid state. Ice, however, has open-structured crystals (Figure 14.15). This structure results from the angular shape of the water molecules and the fact that the forces binding water molecules

Answer The temperature at the bottom of any body of water that has 4°C water in it is 4°C at the bottom, for the same reason that rocks are at the bottom. Both 4°C water and rocks are more dense than water at any other temperature. Water is a poor heat conductor, so if the body of water is deep and in a region of long winters and short summers, the water at the bottom is likely a constant 4°C year round.

FIGURE 14.15 Water molecules in their crystal form have an open-structured hexagonal arrangement that accounts for the expansion of water upon freezing. Ice therefore is less dense than water.

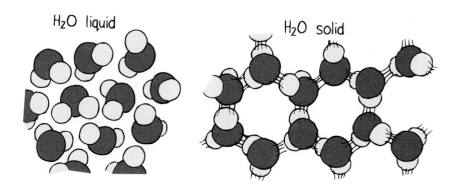

together are strongest at certain angles. Water molecules in this open structure occupy a greater volume than they do in the liquid state. Consequently, ice is less dense than water.

The reason for the dip in the curve of Figure 14.14 is that two types of volume changes take place in ice-cold water. The open-structured crystals that make up solid ice are present, to a small extent, in ice-cold water—a "microscopic slush." These crystals are buffeted by neighboring molecules and have momentary life spans—some are broken apart while others form. At any moment there are enough of them to alter the density of water. At about 10°C all the ice crystals have collapsed. The left-hand graph in Figure 14.16 indicates how the volume of cold water changes as more and more microscopic ice crystals collapse. At the same time that crystals are collapsing due to rising temperature, increased molecular motion results in expansion. This effect is shown in the center graph in Figure 14.16. Whether ice crystals are in the water or not, increased kinetic energy of the molecules increases the volume of the water. When we combine the effects of contraction and expansion, the curve looks like the right-hand graph in Figure 14.16 (or Figure 14.14).

FIGURE 14.16 The collapsing of ice crystals plus increased molecular motion with increasing temperature produce the overall effect of water being most dense at 4°C.

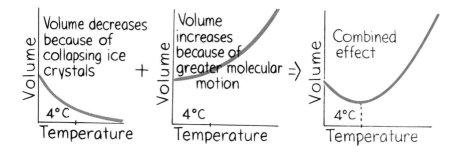

This behavior of water is of great importance in nature. Suppose that the greatest density of water were at its freezing point and that it shrank upon freezing, as is true of most liquids. Then the coldest water would settle to the bottom, and ponds would freeze from the bottom up. Pond organisms would then be destroyed in winter months. Fortunately, this does not happen. The densest water, which settles at the bottom of a pond, is 4° above the freezing temperature. Water at the freezing point, 0°C, is less dense and "floats," so ice forms at the surface while the pond remains liquid below the ice.

Let's examine this in more detail. Most of the cooling in a pond takes place at its surface when the surface air is colder than the water. As the surface water is cooled, it becomes denser and sinks to the bottom. Water will "float" at the surface for further cooling only if it is equally as dense as or less dense than the water below.

Consider a pond that is initially at, say, 10°C. It cannot possibly be cooled to 0°C without first being cooled to 4°C. And water at 4°C cannot remain at the surface for further cooling unless all the water below has at least an equal density—that is, unless all the water below is at 4°C. If the water below the surface is any temperature other than 4°C, any surface water at 4°C will be denser and will sink before it can be further cooled. So before any ice can form, all the water in a pond must be cooled to 4°C. Only when this condition is met can the surface water be cooled to 3°, 2°, 1°, and 0°C without sinking. Then ice can form.

So we see that the water at the surface is first to freeze. Continued cooling of the pond results in the freezing of the water just below the ice, so a pond freezes from the surface downward. In a cold winter the ice will be thicker than in a milder winter. Very deep bodies of water are not ice-covered even in the coldest of winters. This is because all the water in a lake must be cooled to 4°C before lower temperatures can be reached and because the winter is not long enough for all the water to be cooled to 4°C. If only some of the water is at 4°C, it lies on the bottom. Because of water's high specific heat and poor ability to conduct heat, the bottom of deep lakes in cold regions is a constant 4°C the year round. Fish should be glad that this is so.

FIGURE 14.17 As water is cooled, it sinks until the entire pond is 4°C. Then, as water at the surface is cooled further, it floats on top and can freeze. Once ice is formed, temperatures lower than 4°C can extend down into the lake.

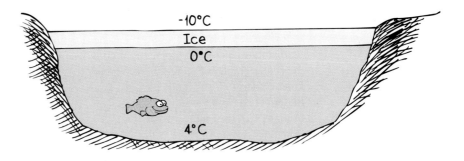

Summary of Terms

Temperature   A measure of the average translational kinetic energy per molecule in a substance, measured in degrees Celsius or Fahrenheit or in kelvins.

Absolute zero   The lowest possible temperature that a substance may have—the temperature at which molecules of the substance have their minimum kinetic energy.

Heat   The energy that flows from a substance of higher temperature to a substance of lower temperature, commonly measured in calories or joules.

Internal energy   The total of all molecular energies, kinetic plus potential, that are internal to a substance.

Specific heat capacity   The quantity of heat per unit mass required to raise the temperature of a substance by 1 Celsius degree.

Review Questions

1. Why does a penny become warmer when it is struck by a hammer?
2. What happens to the temperature of air when it is rapidly compressed?

Temperature

3. What are the temperatures for freezing water on the Celsius and Fahrenheit scales? For boiling water?
4. What are the temperatures for freezing water and boiling water on the Kelvin temperature scale?
5. Why are there no negative numbers on the Kelvin scale?
6. What is meant by the statement that a thermometer measures its own temperature?

Heat

7. When you touch a cold surface, does cold travel from the surface to your hand or does energy travel from your hand to the cold surface? Explain.

8. Distinguish between temperature and heat.

9. Distinguish between heat and internal energy.

10. What determines the direction of heat flow?

Measurement of Heat

11. How is the energy value of foods determined?

12. Distinguish between a calorie and a Calorie.

13. Distinguish between a calorie and a joule.

Specific Heat Capacity

14. Which warms up faster when heat is applied—iron or silver?

15. Does a substance that heats up quickly have a high or a low specific heat capacity?

16. Does a substance that cools off quickly have a high or a low specific heat capacity?

17. How does the specific heat of water compare to the specific heats of other common materials?

18. Northeastern Canada and much of Europe receive about the same amount of sunlight per unit area. Why then is Europe generally warmer in the winter?

19. By energy conservation: If ocean water cools, then should something else warm? What is it that warms?

20. Why is the temperature fairly constant for land masses surrounded by large bodies of water?

Expansion

21. Why will hot water poured into a drinking glass be more likely to break the glass if the glass is thick?

22. How can a bimetallic strip be used to regulate temperature?

23. Which generally expands more for an equal increase in temperature—solids or liquids?

Expansion of Water

24. When the temperature of ice-cold water is increased slightly, does it undergo a net expansion or net contraction?

25. What is the reason for ice being less dense than water?

26. Does "microscopic slush" in water tend to make it more dense or less dense?

27. What happens to the amount of "microscopic slush" in cold water when its temperature is increased?

28. What happens to the kinetic energy of molecular motion in water when its temperature is increased?

29. At what temperature do the combined effects of contraction and expansion produce the smallest volume for water?

30. Why does ice form at the surface of a body of water instead of at the bottom?

Exercises

1. In your room there are things such as tables, chairs, other people, and so forth. Which things in the room you are in have a temperature (1) lower than, (2) greater than, and (3) equal to the temperature of the air?

2. Which is greater, an increase in temperature of 1 C° or one of 1 F°?

3. Why wouldn't you expect all the molecules in a gas to have the same speed?

4. Why can't you establish whether you are running a high temperature by touching your own forehead?

5. Which has the greater amount of internal energy, an iceberg or a cup of hot coffee? Explain.

6. Would a common mercury thermometer be feasible if glass and mercury expanded at the same rates for changes in temperature? Explain.

7. A thermometer outdoors on a sunny day shows a temperature higher than the temperature of the air. Why? Does this mean that the thermometer is "wrong"?

8. What is *temperature* a measurement of?

9. Why is there a minimum temperature (absolute zero) but no maximum temperature?

10. If you drop a hot rock into a pail of water, the temperature of the rock and the water will change until both are equal. The rock will cool and the water will warm. Does this hold true if the hot rock is dropped into the Atlantic Ocean? Explain.

11. Would you expect the temperature of water at the bottom of Niagara Falls to be slightly higher than the temperature at the top of the falls? Why?

12. Why does the pressure of gas enclosed in a rigid container increase as the temperature increases?

13. Adding the same amount of heat to two different objects does not necessarily produce the same increase in temperature. Why not?

14. In addition to the random motions of a molecule from place to place that are associated with temperature, some molecules can absorb large amounts of energy that go into vibrations and rotations of the molecule itself. Would you expect materials composed of such molecules to have a high or a low specific heat? Explain.

15. Why will a watermelon stay cool for a longer time than sandwiches when both are removed from a cooler on a hot day?

16. Bermuda is about as far north of the equator as North Carolina, but unlike North Carolina it has a tropical climate year round. Why is this so?

17. Iceland, so named to discourage conquest by expanding empires, is not ice-covered like Greenland and parts of Siberia, even though it is nearly on the Arctic Circle. The average winter temperature of Iceland is considerably higher than regions at the same latitude in eastern Greenland and central Siberia. Why is this so?

18. Why does the presence of large bodies of water tend to moderate the climate of nearby land—make it warmer in cold weather, and cooler in hot weather?

19. If the winds at the latitude of San Francisco and Washington, D.C., were from the east rather than from the west, why might San Francisco be able to grow only cherry trees and Washington, D.C., only palm trees?

20. In the old days, on a cold winter night it was common to bring a hot object to bed with you. Which would be better to keep you warm through the cold night—a 10-kg iron brick or a 10-kg jug of hot water at the same temperature? Explain.

21. The desert sand is very hot in the day and very cool at night. What does this tell you about its specific heat?

22. Cite an exception to the claim that all substances expand when heated.

23. Would a bimetallic strip function if the two different metals happened to have the same rates of expansion? Is it important that they expand at different rates? Explain.

24. Steel plates are commonly attached to each other with rivets, which are slipped into holes in the plates and rounded over with hammers. The hotness of the rivets makes them easier to round over, but their hotness has another important advantage in providing a tight fit. What is it?

25. A method for breaking boulders used to be putting them in a hot fire, then dousing them with cold water. Why would this fracture the boulders?

26. After a car has been driven for some distance, why does the air pressure in the tires increase?

27. Will a grandfather pendulum clock run faster or slower on a hot day? Explain.

28. Creaking noises are often heard in the attics of old houses on cold nights. Give an explanation in terms of thermal expansion.

29. An old remedy for a pair of drinking glasses that stick together is to run water at different temperatures into the inner glass and over the surface of the outer glass. Which water should be hot, and which should be cold?

30. Any architect will tell you that chimneys are never used as a weight-bearing part of a wall. Why?

31. Looking at the expansion joint in the photo of Figure 14.10, would you say it was taken on a warm day or a cold day? Why?

32. Would you or the gas company gain by having gas warmed before it passed through your gas meter?

33. A metal ball is just able to pass through a metal ring. When the ball is heated, however, it will not pass through the ring. What would happen if the ring, rather than the ball, were heated? Does the size of the hole increase, stay the same, or decrease?

34. After a machinist slips a hot, snugly fitting iron ring over a very cold brass cylinder very quickly, there is no way that the two can be separated intact. Can you explain why this is so?

35. Suppose you cut a small gap in a metal ring. If you heat the ring, will the gap become wider or narrower?

36. When a mercury thermometer is warmed, the mercury level momentarily goes down before it rises. Can you give an explanation for this?

37. When something is heated, it expands. Is the air in your house an exception to this rule? You can heat the air without increasing the volume of the house.

38. Why are incandescent bulbs typically made of very thin glass?

39. One of the reasons the first light bulbs were expensive is that the electrical lead wires into the bulb were made of platinum, which expands at about the same rate as glass when heated. Why is it important that the metal leads and the glass have the same coefficient of expansion?

40. After you measure the dimensions of a plot of land with a steel tape on a hot day, you come back and re-measure on a cold day. On which day do you determine the larger area for the land?

41. What was the precise temperature at the bottom of Lake Superior at 12:01 AM on October 31, 1894?

42. Suppose that water is used in a thermometer instead of mercury. If the temperature is at 4°C and then changes, why can't the thermometer indicate whether the temperature is rising or falling?

43. A piece of solid iron sinks in a container of molten iron. A piece of solid aluminum sinks in a container of molten aluminum. Why does a piece of solid water (ice) not sink in a container of "molten" (liquid) water? Explain in molecular terms.

44. How does the combined volume of the billions and billions of hexagonal open spaces in the structures of ice crystals in a piece of ice compare to the portion of ice that floats above the water line?

45. How would the shape of the curve in Figure 14.14 differ if density instead of volume were plotted against temperature? Make a rough sketch.

46. State whether water at the following temperatures will expand or contract when warmed a little: 0°C; 4°C; 6°C.

47. Why is it important to protect water pipes so they don't freeze?

48. If cooling occurred at the bottom of a pond instead of at the surface, would a lake freeze from the bottom up? Explain.

49. If water had a lower specific heat, would ponds be more likely to freeze or less likely to freeze?

50. Cite some quantity other than temperature that equalizes when two systems are brought together.

Problems

1. What would be the final temperature of a mixture of 50 g of 20°C water and 50 g of 40°C water?

2. If you wish to warm 100 kg of water by 20°C for your bath, how much heat is required? (Give your answer in calories and joules.)

3. Suppose a bar 1 m long expands 0.5 cm when heated. By how much will a bar 100 m long of the same material expand when similarly heated?

4. Steel expands 1 part in 100,000 (fractional expansion of 10^{-5}) for each Celsius degree increase in temperature. Suppose the 1.3-km main span of the Golden Gate Bridge had no expansion joints. How much longer would it be for an increase in temperature of 10°C?

5. Consider a 40,000-km steel pipe that forms a ring to fit snugly all around the circumference of the world. Suppose people along its length breathe on it so as to raise its temperature 1°C. The pipe gets longer. It also is no longer snug. How high does it stand above ground level? (To simplify, consider only the expansion of its radial distance from the center of the earth, and apply the geometry formula that relates circumference C and radius r, $C = 2\pi r$. The result is surprising!)

6. The specific heat capacity of copper is 0.092 calories per gram per degree Celsius. How much heat is required to raise the temperature of a 10-g piece of copper from 0°C to 100°C? How does this compare with the heat needed to raise the temperature of the same mass of water through the same temperature difference?

7. A cook pours a liter of ice water at 0°C into a pan of hot water at 80°C and finds that the mixture reaches a temperature of 60°C. How much hot water was in the pan?

8. In a laboratory, a student pours 1 kg of boiling water at 100°C into a 1-kg pan which has an initial temperature of 23°C. The specific heat capacity of the pan is 0.10 calories per gram per degree Celsius, one-tenth the specific heat capacity of water. If all the heat lost by the water goes into heating the pan, what will be the final temperature of the water and pan?

15

. . . .

Heat Transfer

: **W**hy are the firewalker's feet not burned?

Heat transfers from warmer to cooler things. If several objects with different temperatures are in contact, those that are warm become cooler and those that are cool become warmer. They tend to reach a common temperature. This equalizing of temperature occurs in three ways: *conduction, convection,* and *radiation.*

Conduction

Hold one end of an iron nail in a flame. It will quickly become too hot to hold. The heat enters the metal nail at the end that is kept in the flame and the heat is transmitted along the nail's whole length. The transmission of heat in this manner is called **conduction**. The fire causes the molecules at the heated end of the nail to move more rapidly. Because of their increased motion, these molecules and free electrons collide with their neighbors, and so on. This bumping process continues until the increased motion has been transmitted to all the molecules and the entire body has become hot. Heat conduction occurs by electron and molecular collisions.

How well a solid object conducts heat depends on the bonding of its molecular structure. Solids whose molecules have one or more "loose" outer electrons conduct heat (and electricity) well. Metals have the "loosest" outer electrons and are the best conductors of heat and electricity for this reason. Silver is the best, copper is next, and,

FIGURE 15.1 The tile floor feels colder than the wooden floor, even though both floor materials are the same temperature. This is because tile is a better conductor than wood, and heat is more readily conducted from the foot that makes contact with the tile.

among the common metals, aluminum and then iron are next in order. Wool, wood, straw, paper, cork, and Styrofoam are poor conductors of heat. The outer electrons in the molecules of these materials are firmly attached. Poor conductors are called *insulators*.

Because wood is a good insulator it is used for handles of cookware. Even when it is hot, you can quickly grasp the wooden handle of a pot with your bare hand and remove the pot from a hot oven without harm. An iron handle of the same temperature would surely burn your hand. Wood is a good insulator even when it's red hot, which is why firewalking professor John Suchocki can walk barefoot on red-hot wood coals without burning his feet (chapter opener photo, facing page). (CAUTION: Don't try this on your own; even experienced firewalkers sometimes receive bad burns when conditions aren't just right.) The principal factor here is the low conductivity of wood—even red-hot wood. Although its temperature is high, relatively little heat is conducted to the feet, just as little heat is conducted by air when you put your hand briefly in a hot pizza oven. If you touch metal in the hot oven, OUCH! Similarly, a firewalker who steps on a hot piece of metal or another good conductor will be burned. Evaporation can play a role in firewalking too, as we'll see in the next chapter.

Liquids and gases, in general, are poor conductors. Air is a very poor conductor, which, as previously mentioned, is why your hand isn't harmed if put briefly in a hot pizza oven. The good insulating properties of such things as wool, fur, and feathers are largely due to the air spaces they contain. Porous substances are likewise good insulators because of their many small air spaces. Be glad that air is a poor conductor; if it weren't, you'd feel quite chilly on a 20°C (68°F) day!

Snow is a poor conductor (good insulator)—about the same as dry wood. Hence a blanket of snow can literally keep the ground warm in winter. Snowflakes are formed of crystals, which collect into feathery masses, imprisoning air and thereby interfering with the escape of heat from the earth's surface. Traditional Eskimo winter dwellings are shielded from the cold by their snow covering. Animals in the forest find shelter from the cold in snowbanks and in holes in the snow. The snow doesn't provide them with heat; it simply slows down the loss of heat they generate.

Heat is transmitted from a higher to a lower temperature. We often hear people say they wish to keep the cold out of their homes. A better way to put this is to say that they want to prevent the heat from escaping. There is no "cold" that flows into a warm home. If the home becomes colder, it is because heat flows out. Homes are insulated with rock

FIGURE 15.2 Snow lasts longer on the non-conducting lawn, and on parts of the roofs that are well-insulated. Snow patterns reveal the conduction, or lack of conduction, of heat through the roofs.

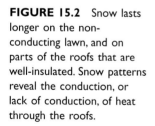

wool or spun glass to prevent heat from escaping rather than to prevent cold from entering. Interestingly enough, insulation of whatever kind does not actually prevent heat from getting through it; it simply slows the rate at which heat penetrates. Even a well-insulated warm home in winter will gradually cool. Insulation delays the transfer of heat.

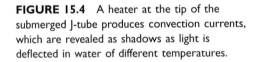

Question In desert regions that are hot in the daytime and cold at nighttime, the walls of houses are often made of mud. Why is it important that the mud walls be thick?

Convection

FIGURE 15.3 (a) Convection currents in air. (b) Convection currents in liquid.

Liquids and gases transmit heat mainly by **convection**, which is heat transfer by the actual motion of the fluid—by currents. Convection can occur in all fluids, whether liquids or gases. Whether we heat water in a pan or heat air in a room, the process is the same (Figure 15.3). If the fluid is heated from below, the molecules at the bottom increase in speed; the heated fluid becomes less dense and is pushed up by the denser cooler fluid that takes its place at the bottom. In this way, convection currents keep the fluid stirred up as it heats. Convection currents occur in the atmosphere, affecting weather. To understand the atmosphere's convection currents, we must first understand why warm air rises, why certain gases such as helium rise in air, and why rising air cools.

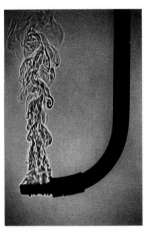

FIGURE 15.4 A heater at the tip of the submerged J-tube produces convection currents, which are revealed as shadows as light is deflected in water of different temperatures.

Answer A wall of appropriate thickness keeps the house warm at night by slowing the flow of heat from inside to outside, and keeps the house cool in the daytime by slowing the flow of heat from outside to inside. Such a wall has "thermal inertia."

Question You can hold your fingers beside the candle flame without harm, but not above the flame. Why?

Why Fast-Moving Molecules Rise in Air

We all know from common experience that warm air rises. From our study of buoyancy we understand why this is so. Warm air expands, becomes less dense than the surrounding air, and is buoyed upward like a balloon. As the rising warm air cools and reaches the density of the surrounding air, it no longer rises. This is evident when we see smoke from a fire rise, and then settle off as it cools to the temperature of the surrounding air.

Any molecule that has higher average speed than the surrounding air tends to rise. To see why, consider a fairly large region of identical gas molecules. Because of gravity, more molecules are near the bottom than near the top; the gas is denser toward the ground. If the region has a uniform temperature, then each molecule has the same average kinetic energy and the same average speed. Each molecule, therefore, has the same tendency to migrate up, down, or sideways throughout the region. Now consider a single faster-moving molecule. Can you see that it will migrate farther and more rapidly than its neighbors? If placed in the middle of our region, it will bump into and rebound from molecules in all directions—but when moving upward, it rebounds from a smaller number of molecules than when moving downward. This is because the density of molecules is less above; there is less opposition to an upward migration. So when our molecule happens to move upward, it travels farther before making a collision than when moving downward. It has a "wider window" above. When this window more than compensates for the downward pull of gravity, a faster-moving molecule bumbles upward in its random jostling.

So we can visualize how a fast-moving molecule rises in air. Our individual fast-moving-molecule description is particularly well-suited to helium or other lightweight molecules. All air molecules in the same locality have the same temperature, and therefore the same average kinetic energy. Average *speed,* however, depends on mass; smaller-mass molecules with the same kinetic energy have higher speeds. Helium's average speed is always considerably greater than that of neighboring heavier nitrogen and oxygen molecules. Helium therefore migrates more than its heavier neighbors and bumbles its way to the top of the atmosphere. Most of it, interestingly, escapes into outer space. That's why helium, the seventh most common gas in the earth's atmosphere, doesn't normally exist in the lower atmosphere. But our description isn't a good picture for the rising of warm air because a single fast-moving molecule too soon shares its excess energy with its less energetic neighbors. Only a large cluster of energetic molecules would appreciably rise before dissipating excess energy—and for a large cluster, buoyancy explains the upward motion.

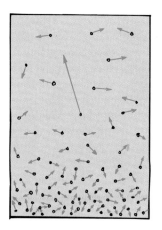

FIGURE 15.5 A fast-moving molecule tends to migrate toward the region of least obstruction—upward.

Answer Heat travels upward by air convection. Since air is a poor conductor, very little heat travels sideways.

FIGURE 15.6 The hot steam expands from the pressure cooker and is cool to Millie's touch.

Why Rising Air Cools

Rising warm air, like a rising balloon, expands. Why? Because less atmospheric pressure squeezes on it as it rises to higher altitudes. As the air expands, it cools. Do the following experiment right now. With your mouth open, blow on your hand. Your breath is warm. Now repeat, but pucker your lips to make a small hole so your breath expands as it leaves your mouth. Note that your breath is appreciably cooler! Expanding air cools. This is just the opposite of what occurs when air is compressed. If you've ever compressed air with a tire pump, you probably noticed that the air and pump became quite hot.

A dramatic example of cooling by expansion occurs with steam expanding through the nozzle of a pressure cooker (Figure 15.6). The cooling effect of both expansion and rapid mixing with cooler air allows you to hold your hand comfortably in the jet of condensed vapor. (CAUTION: If you try this, be sure to place your hand high above the nozzle at first and then lower it to a comfortable distance. If you put your hand at the nozzle where no steam appears, watch out! Steam is invisible, and it's near the nozzle before it has sufficiently expanded and cooled. The cloud of "steam" you see is actually condensed water vapor—much cooler.)

We can understand the cooling of expanding air by thinking of molecules of air as tiny Ping Pong balls bouncing against one another. A ball picks up speed when it is hit by another that approaches with a greater speed. But when a ball collides with one that is receding, its rebound speed is reduced. Likewise for a Ping-Pong ball moving toward a paddle; it picks up speed when it hits an approaching paddle, but loses speed when it hits a receding paddle. The same idea applies to a region of air that is expanding; molecules collide, on the average, with more molecules that are receding than are approaching (Figure 15.7). Thus, in expanding air, the average speed of the molecules decreases and the air cools.*

FIGURE 15.7 Molecules in a region of expanding air collide more often with receding molecules than with approaching ones. Their rebound speeds therefore tend to decrease and, as a result, the expanding air cools.

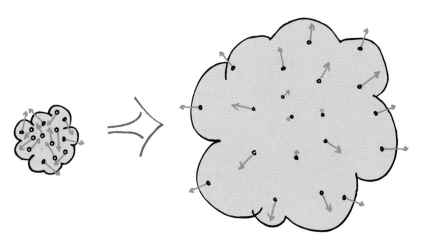

Convection currents stirring the atmosphere result in winds. Some parts of the earth's surface absorb heat from the sun more readily than others, and as a result the air near the surface is heated unevenly and convection currents form. This is evident at the seashore. In the daytime the shore warms more easily than the water; air over the shore is pushed up (we say it rises) by cooler air from above the water taking its place. The result is a sea breeze. At night the process reverses because the shore cools off more

---

* Where does the energy go in this case? We will see in Chapter 17 that it goes into work done on the surrounding air as the expanding air pushes outward.

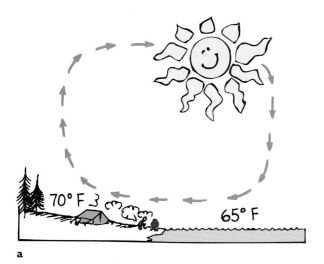

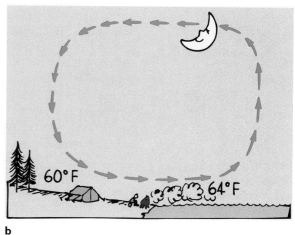

a

b

FIGURE 15.8 Convection currents produced by unequal heating. Warmed air rises and cools as it expands. It then sinks and flows toward the warmed area to replace the air that has risen. The land is (a) warmer than the water in the day (lower specific heat) and (b) cooler than the water at night, so the direction of air flow reverses.

quickly than the water, and then the warmer air is over the sea (Figure 15.8). Build a fire on the beach and you'll notice that the smoke sweeps inward during the day and seaward at night.

Radiation

Heat from the sun passes through space and then through the atmosphere before it warms the earth's surface. This heat does not pass through the atmosphere by conduction, for air is a poor conductor. Nor does it pass through by convection, for convection begins only after the earth is warmed. We also know that neither convection nor conduction is possible in the empty space between our atmosphere and the sun. We can see that heat must be transmitted some other way—by **radiation**.* The energy so radiated is called *radiant energy.*

Radiant energy is in the form of *electromagnetic waves.* It includes radio waves, microwaves, infrared radiation, visible light, ultraviolet radiation, X rays, and gamma rays. These types of radiant energy are listed in order of wavelength, from longest to shortest. Infrared (below-the-red) radiation has longer wavelengths than those of visible light. The longest visible wavelengths are for red light, and the shortest are for violet light. Ultraviolet (beyond-the-violet) radiation has shorter wavelengths. (Wavelength is treated in more detail in Chapter 18, and electromagnetic waves are covered in more detail in Chapters 24 and 25).

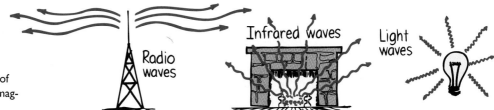

FIGURE 15.9 Types of radiant energy (electromagnetic waves).

* Do not confuse radiation with radioactivity. The radiation we are talking about here is electromagnetic radiation, including visible light. We will return to it in Part 6. Radioactivity is a process of the atomic nucleus, which we shall study in Part 7.

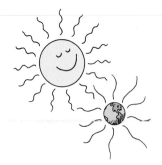

FIGURE 15.10 Both the sun and earth emit the same kind of radiant energy. The sun's glow is visible to the eye; the earth's glow is of longer waves not visible to the eye.

Most people know that the sun glows and emits radiant energy, and most educated people know that the source of the sun's radiant energy involves nuclear reactions in its deep interior. But relatively few people know that the earth also "glows" and emits radiant energy of the same nature, whose source also involves nuclear reactions in its interior (we'll treat nuclear reactions in detail in Chapters 32 and 33). Visit the depths of any mine and you'll find it's warm down there—year round. Radioactivity in the earth's interior warms the earth. The earth's glow, however, is of non-visible infrared waves. Electromagnetic waves emitted by both the sun and the earth differ only in range of wavelengths, and the amount. The sun emits enormously more radiant energy than earth, and at visible wavelengths.

All objects continually emit radiant energy in a mixture of wavelengths. Objects at low temperatures emit long waves, just as long lazy waves are produced when you shake a rope with little energy (Figure 15.11). Higher-temperature objects emit waves of shorter wavelengths. Objects of everyday temperatures emit waves mostly in the long-wavelength end of the infrared region. Shorter-wavelength infrared waves absorbed by our skin produce the sensation of heat. So it is common to refer to infrared radiation as *heat radiation.*

FIGURE 15.11 Shorter wavelengths are produced when the rope is shaken more rapidly.

When an object is hot enough, some of the radiant energy it emits is in the range of visible light. At a temperature of about 500°C an object begins to emit the longest waves we can see, red light. Higher temperatures produce a yellowish light. At about 1200°C all the different waves to which the eye is sensitive are emitted and we see an object as "white hot."

Common sources that give the sensation of heat are the burning embers in a fireplace, a lamp filament, and the sun. All of these emit both infrared radiation and visible light. When this radiant energy falls on other objects, it is partly reflected and partly absorbed. The part that is absorbed increases the internal energy of the objects.

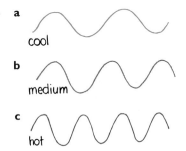

FIGURE 15.12 (a) A low-temperature source emits primarily long waves. (b) A medium-temperature source emits primarily medium waves. (c) A high-temperature source emits primarily short waves.

Absorption and Reflection of Radiant Energy

Absorption and reflection are opposite processes. Therefore, a good absorber of radiant energy reflects very little radiant energy, including the range we call light. So a good absorber appears dark. A perfect absorber reflects no radiant energy and appears perfectly black. The pupil of the eye, for example, allows radiant energy to enter with no reflection, which is why it appears black. (The pink "pupils" that appear in some flash portraits are from direct light reflected off the red retina at the back of the eyeball.)

Look at the open ends of pipes in a stack. The holes appear black. Look at open doorways or windows of distant houses in the daytime, and they too look black. Openings appear black because the radiant energy that enters is reflected from the inside walls many times and is partly absorbed at each reflection until very little or none remains to come back out (Figure 15.13).

Good reflectors, on the other hand, are poor absorbers. Clean snow is a good reflector and therefore does not melt rapidly in sunlight. If the snow is dirty, it absorbs radiant energy from the sun and melts faster. Dropping black soot by aircraft on snowed-

FIGURE 15.13 The hole looks perfectly black and indicates a black interior, when in fact the interior has been painted a bright white.

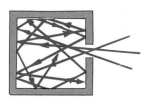

FIGURE 15.14 Radiation that enters the cavity has little chance of leaving before it is completely absorbed.

FIGURE 15.15 When the containers are filled with hot (or cold) water, the blackened one cools (or warms) faster.

in mountainsides is a technique sometimes used in flood control. Controlled melting at favorable times rather than a sudden runoff of melted snow is thereby accomplished.

Light-colored buildings stay cooler in summer because they reflect much of the oncoming radiant energy. Light-colored buildings are also poor emitters, so they retain more of their internal energy and stay warm. Paint your house a light color.

Emission of Radiant Energy

Emission is the reradiation of absorbed electromagnetic waves. Good absorbers are also good emitters; poor absorbers are poor emitters. For example, a radio antenna that is constructed to be a good emitter of radio waves will also, by its very design, be a good receiver (absorber) of radio waves. A poorly designed transmitting antenna will also be a poor receiver.

Interestingly enough, if a good absorber were not also a good emitter, black objects would remain warmer than lighter-colored objects and never reach a common temperature with them. But objects in thermal contact, given sufficient time, do reach the same temperature. Then each object emits as much energy as it absorbs. When time is not sufficient, like when a blacktop pavement or dark automobile body remains hotter than the surroundings on a hot day, thermal equilibrium is not reached. But at nightfall the dark objects cool faster! Given sufficient time, all objects come to thermal equilibrium. So a dark object that absorbs a lot must emit a lot as well.

You can check this out with a pair of metal containers of the same size and shape, one with a white or mirrorlike surface and the other with a blackened surface (Figure 15.15). Fill the containers with hot water, and place thermometers in the water. You will find that the black container cools faster. The blackened surface is a better emitter. Coffee or tea will stay hot longer in a shiny mirrorlike pot than in a blackened one.

The same experiment can be done in reverse. This time fill each container with ice water and place the containers in front of a fireplace or outside on a sunny day—wherever there is a good source of radiant energy. You'll find the black container will warm up faster. A good emitter of radiant energy is also a good absorber.

Whether a surface plays the role of net emitter or net absorber depends on whether its temperature is above or below the surroundings. If the surface is hotter than the surroundings, for example, it will be a net emitter and will cool. If the surface is colder than the surroundings, it will be a net absorber and will become warmer. Every surface, hot or cold, both absorbs and emits radiant energy. If the surface absorbs more than it emits it is a net absorber; if it emits more than it absorbs it is a net emitter.

Questions

1. If a good absorber of radiant energy were a poor emitter, how would its temperature compare with the temperature of its surroundings?
2. Is it more efficient to paint a heating radiator black or silver?

Cooling at Night by Radiation

Bodies that radiate more energy than they receive become cooler. This happens at night when solar radiation is absent. Objects out in the open radiate energy into the night and, because of the absence of warmer bodies, may receive very little energy in return. They give out more energy than they receive and become cooler. If the object is a good conductor of heat—like metal, stone, or concrete—heat from the ground will be conducted

FIGURE 15.16 Patches of frost crystals betray the hidden entrances to mouse burrows. Each cluster of crystals is frozen mouse breath!

Answers

1. If a good absorber were not also a good emitter, there would be a net absorption of radiant energy and the temperature of good absorbers would remain higher than the temperature of the surroundings. Things around us approach a common temperature only because good absorbers are, by their very nature, also good emitters.
2. Most of the heat provided by a heating radiator is accomplished by convection, so the color is not really that important. For optimum efficiency, however, the radiators should be painted a dull black so that the contribution by radiation is increased.

to it, somewhat stabilizing its temperature. But materials such as wood, straw, and grass are poor conductors, and little heat is conducted into them from the ground. These insulating materials are net radiators and get *colder than the air.* It is common for frost to form on these kinds of materials even when the temperature of the air does not go down to freezing. Have you ever seen a frost-covered lawn or field on a chilly but above-freezing morning before the sun is up? The next time you see this, notice that the frost forms only on the grass, straw, or on other poor conductors, while none forms on the cement, stone, or other good conductors.

Snow is a good example. During the day the snow gains very little heat energy from the sun because of its reflectivity, and at night it loses energy rapidly by radiating infrared radiation to space. Because of snow's poor conductivity, the ground conducts very little heat to it, and the surface of the snow becomes cooler than the surrounding air. Snow therefore has a significant cooling effect on the earth and atmosphere. The bottom of the snow remains at about ground temperature. That's why Midwestern wheat farmers like deep snows in winter—to protect their wheat fields from the harsh, cold weather.

Questions

1. Which is likely to be colder; a night when the stars are out or a night with no stars?
2. In winter, why do road surfaces on bridges tend to be more icy than the road surfaces on either side?

Newton's Law of Cooling

FIGURE 15.17 The long stem of a wine glass helps to prevent heat from the hand from warming the wine.

An object at a different temperature from its surroundings will ultimately come to a common temperature with its surroundings. A relatively hot object cools as it warms its surroundings; a cool object warms as it cools its surroundings.

The rate of cooling of an object depends on how much hotter the object is than the surroundings. The temperature change per minute of a hot apple pie will be more if the hot pie is put in a cold freezer than if put on the kitchen table. When the pie cools in the freezer, the temperature difference between it and its surroundings is greater. A warm home will leak heat to the cold outside at a greater rate when there is a large difference in the inside and outside temperatures. Keeping the inside of your home at a high temperature on a cold day is more costly than keeping it at a lower temperature. If you keep the temperature difference small, the rate of cooling will be correspondingly low.

The rate of cooling of an object—whether by conduction, convection, or radiation—is approximately proportional to the temperature difference ΔT between the object and its surroundings.

$$\text{Rate of cooling} \sim \Delta T$$

Answers

1. It is colder on the starry night, when the earth radiates directly to frigid deep space, where the temperature is only 3 K (3 degrees above absolute zero). On a cloudy night, net radiation is less, for the clouds radiate energy back to the earth's surface.
2. Energy radiated by road surfaces is partly replaced by heat conducted from the warmer earth below. But bridges make very little thermal contact with the earth, with very little energy conducted to the bridges as they radiate energy away. So the bridges get colder, with more chance of ice formation.

This is known as **Newton's law of cooling**. (Guess who is credited with discovering this?)

The law holds also for heating. If an object is cooler than its surroundings, its rate of warming up is also proportional to ΔT. Frozen food will warm up faster in a warm room than in a cold room.

The rate of cooling we experience on a cold day can be increased by the added convection of wind. We speak of this in terms of *wind chill*. For example, a wind chill of $-20°C$ means we are losing heat at the same rate as if the temperature were $-20°C$ without wind.

Question Since a hot cup of tea loses heat more rapidly than a lukewarm cup of tea, would it be correct to say that a hot cup of tea will cool to room temperature before a lukewarm cup of tea will?

The Greenhouse Effect and Global Warming

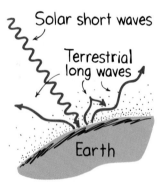

FIGURE 15.18 The green house effect due to the earth's atmosphere. Carbon dioxide and other "greenhouse gases" in the atmosphere act to absorb and retain heat that would otherwise be radiated from earth into space.

The earth and its atmosphere gain energy when they absorb radiant energy from the sun. This warms the earth. The earth, in turn, emits what is called **terrestrial radiation**, much of which escapes to outer space. Absorption and emission go on at equal rates to produce an average equilibrium temperature. Over the last 500,000 years the average temperature of the earth has fluctuated between 19°C and 27°C and is presently at the high point, 27°C. The earth's temperature increases when either the radiant energy coming in increases, or there is a decrease in the escape of terrestrial radiation.

The **greenhouse effect** is the effect of atmospheric gases on the balance of terrestrial radiation and radiant energy from the sun. Hot sources emit short waves, while cooler sources emit longer waves. Thus, radiant energy from the hot sun is composed of short waves—ultraviolet, visible light, and short-wavelength infrared waves. The atmosphere is transparent to much of this radiation, especially the visible light, so solar energy reaches the earth's surface and is absorbed. The surface in turn reradiates part of this energy. But since the earth's surface is relatively cool, it reradiates the energy as long waves—mainly long-wavelength infrared. Atmospheric gases (mainly carbon dioxide and water vapor) absorb and re-emit much of this long-wave radiation back to earth. So the long-wave radiation that cannot escape the earth's atmosphere helps to keep the earth warm. This process is very nice, for the earth would be a frigid $-18°C$ otherwise. Our present environmental concern is that excess carbon dioxide and other atmospheric gases will trap too much energy and make the earth too warm.

The atmospheric greenhouse effect gets its name from the glass structures used by farmers and florists to "trap" solar energy. Glass is transparent to waves of visible light and opaque to ultraviolet and infrared. Glass acts as a sort of one-way valve. It allows visible light to enter but prevents longer waves from leaving. So short waves of sunlight

---

Answer No! Although the rate of cooling is greater for the hotter cup, it has farther to cool to reach thermal equilibrium. The extra time is equal to the time it takes to cool to the initial temperature of the lukewarm cup of tea. Cooling *rate* and cooling *time* are not the same thing.

FIGURE 15.19 Glass is transparent to short-wavelength radiation but opaque to long-wavelength radiation. Reradiated energy from the plant is long wavelength because the plant has a relatively low temperature.

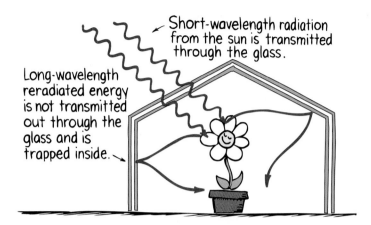

Short-wavelength radiation from the sun is transmitted through the glass.

Long-wavelength reradiated energy is not transmitted out through the glass and is trapped inside.

enter through the glass roof and are absorbed by soil and plants inside. The soil and plants, in turn, emit long infrared waves. This energy cannot get through the glass and the greenhouse warms up.

Interestingly enough, in the farmer's or florist's greenhouse, heating is mainly due to the ability of glass to prevent convection currents from mixing the cooler outside air with the warmer inside air. The greenhouse effect plays a bigger role in the warming of the earth than it does in the warming of greenhouses.

Question What does it mean to say that the greenhouse effect is like a one-way valve?

Solar Power

Step from the shade into the sunshine and you're noticeably warmed. The warmth you feel isn't so much because the sun is hot, for its surface temperature of 6000°C is no hotter than the flames of some welding torches; we are warmed principally because the sun is so big. As a result, it emits enormous amounts of radiant energy, less than one part in a billion of which reaches the earth. The amount of radiant energy received each second over each square meter that is at right angles to the sun's rays at the top of the atmosphere is 1400 joules (1.4 kJ) (Figure 15.20). This amount of energy is called the **solar**

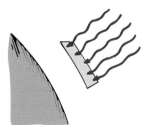

FIGURE 15.20 Over each square meter that is perpendicular to the sun's rays at the top of the atmosphere, the sun pours 1400 J of radiant energy each second. Hence the solar constant is 1.4 kJ/s/m², or 1.4 kW/m².

Answer The transparent material, atmosphere for the earth and glass for the greenhouse, passes only incoming short waves and blocks outgoing long waves. As a result, radiant energy is trapped within the "greenhouse."

constant. Expressed in terms of power, this is 1.4 kilowatts per square meter (1.4 kW/m²). The amount of **solar power** that reaches the ground is attenuated by the atmosphere and reduced by nonperpendicular elevation angles of the sun—and, of course, it is cut off at night. The solar power received in the United States, averaged over day and night, summer and winter, is about 13 percent of the solar constant (0.18 kW/m²). This amount of power, falling on the roof area of a typical American house, is twice the power needed to comfortably heat and cool the house year round. It isn't surprising that we see more and more homes using solar energy for space heating and water heating. (Harnessing solar energy for cooling is not yet practical—except in very dry climates where evaporating water cools houses.)

FIGURE 15.21 Solar water heaters are covered with glass to provide a greenhouse effect, which further heats the water. Why are the collectors painted black?

Solar heating needs a distribution system to move solar energy from the collector to the storage or living space. When the distribution system requires external energy to operate fans or pumps, we have an active system. When the distribution is by natural means (conduction, convection, or radiation), we have a passive system. At the present time passive systems are essentially problem-free and serve as an economical supplement to conventional heating—even in the northern states.

On a larger scale, the problems of utilizing solar power are greater. First, there is the fact that no energy arrives at night. This calls for supplemental sources of energy or efficient solar-energy storage devices. Variations in weather, particularly cloud cover, produce a variable energy supply from day to day and from season to season. Even in clear daylight hours, the sun is high in the sky only part of the day. Solar-energy collecting and concentration systems, whether arrays of mirrors or photovoltaic cells, at this writing are not yet competitive in cost with electrical power generated by conventional power sources. Projections indicate the story may be different within a decade or two.

FIGURE 15.22 Solar power yesterday and today.

FIGURE 15.23 Solar power tomorrow: A Boeing conception of a photovoltaic power satellite being constructed in low earth orbit. The weightlessness in orbit allows the use of large weblike structures of a kind that would be crushed if used on earth. The satellite can be deployed in geosynchronous orbit after completion. Electric energy, produced continuously, could be converted to microwaves and beamed to earth or wherever needed.

The Thermos Bottle

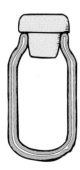

We can briefly summarize the ways in which heat is transferred by considering a device that inhibits these three methods—the common vacuum, or thermos, bottle. The thermos bottle is double-walled glass, with a vacuum between the walls. The two inner glass surfaces facing each other are silvered. When a hot liquid is poured into such a bottle, it remains at very nearly the same temperature for many hours. This is because the transfer of heat by conduction, convection, and radiation is severely inhibited.

1. Heat transfer by *conduction* through the vacuum is impossible. Some heat escapes by conduction through the glass and stopper, but this is a slow process as glass and plastic or cork are poor conductors.
2. The vacuum also prevents heat loss through the walls by *convection*.
3. Heat loss by *radiation* is prevented by the silvered surfaces of the walls, which reflect heat waves back into the bottle.

FIGURE 15.24 A thermos bottle.

Summary of Terms

Conduction   The transfer of heat energy by molecular and electron collisions within a substance (especially a solid).

Convection   The transfer of heat energy in a gas or liquid by means of currents in the heated fluid. The fluid moves, carrying energy with it.

Radiation   The transfer of energy at the speed of light by means of electromagnetic waves.

Newton's law of cooling   The rate of loss of heat from an object is proportional to the temperature difference between the object and its surroundings.

Greenhouse effect   The heating effect of a medium such as glass or the earth's atmosphere that is transparent to the short-wavelength radiation of sunlight but opaque to long-wavelength terrestrial radiation. Energy of sunlight that enters the glass of a florist's greenhouse or the atmosphere of the earth is absorbed and reradiated at a longer wavelength that is consequently trapped, producing heating.

Solar constant   1400 J/m^2 received from the sun each second at the top of the earth's atmosphere on an area perpendicular to the sun's rays; expressed in terms of power, 1.4 kW/m^2.

Solar power   Energy per unit time derived from the sun.

Review Questions

1. What are the three common ways in which heat is transferred?

Conduction

2. What is the role of "loose" electrons in heat conductors?
3. Distinguish between a conductor and an insulator.
4. Why does room-temperature tile feel cooler to the bare feet than a wooden floor?
5. Why are materials such as wood, fur, feathers, and even snow good insulators?
6. How does a blanket keep you warm on a cold night, even though it is not really a source of energy?
7. Why do we say that cold is not a tangible thing?

Convection

8. How is heat transferred from one place to another by convection?

Why Fast-Moving Molecules Rise in Air

9. How does buoyancy relate to convection?
10. Why are fast-moving air molecules more likely to migrate upward than downward?

11. Why do helium atoms in the atmosphere keep their higher average speeds, while molecules of hot air in the atmosphere lose speed?

Why Rising Air Cools

12. As a given mass of air rises, what happens to its pressure? What happens to its volume? What happens to its temperature?
13. How are the speeds of molecules of air affected when the air is compressed by the action of a tire pump?
14. How are the speeds of molecules of air affected when the air expands rapidly?
15. Why is Millie's hand not burned when she holds it above the escape valve of the pressure cooker (Figure 15.6)?
16. Why does the direction of coastal winds change from day to night?

Radiation

17. What exactly is radiant energy?
18. How do the wavelengths of radiant energy vary with the temperature of the radiating source?

Absorption and Reflection of Radiant Energy

19. Since all objects are absorbing energy from their surroundings, why don't the temperatures of all objects continuously increase?
20. Since all objects are emitting energy to their surroundings, why don't the temperatures of all objects continuously decrease?
21. What determines whether an object is a net absorber or net emitter at a given time?
22. Why does the pupil of the eye appear black?

Emission of Radiant Energy

23. Which will normally cool faster, a black pot of hot water or a silvered pot of hot water? Explain.
24. Which will normally warm faster, a black pot of cold water or a silvered pot of cold water? Explain.
25. Is a good absorber of radiation a good emitter or a poor emitter?

Cooling at Night by Radiation

26. What happens to the temperature of something that radiates energy without absorbing the same amount in return?
27. Will a good conductor that is in contact with the relatively warm earth become significantly colder than the earth when it radiates energy? Why or why not?
28. Will a good insulator that is in contact with the relatively warm earth become significantly colder than the earth when it radiates energy? Why or why not?

Newton's Law of Cooling

29. If you want a room-temperature can of beverage to cool quickly, should you put it in the freezer compartment or in the main part of your refrigerator? Or does it not matter?

30. Which will undergo the greater rate of cooling, a red-hot poker in a warm oven or a red-hot poker in a cold room (or do both cool at the same rate)?

31. Does Newton's law of cooling apply to warming as well as cooling?

The Greenhouse Effect and Global Warming

32. What is meant by *terrestrial radiation*?

33. Why is radiant energy from the sun composed of short waves and terrestrial radiation composed of relatively longer waves?

Solar Power

34. Distinguish between *active* and *passive* solar heating systems.

The Thermos Bottle

35. What is the function of the silver surfaces of the thermos bottle?

36. Heat cannot readily escape a thermos bottle, so hot things inside stay hot. Will cold things inside a thermos bottle likewise stay cold? Explain.

Projects

1. Hold the bottom end of a test tube full of cold water in your hand. Heat the top part in a flame until it boils. The fact that you can still hold the bottom shows that water is a poor conductor of heat and that convection does not move the hot water downward. This is even more dramatic when you wedge chunks of ice at the bottom; then the water above can be brought to a boil without melting the ice. Try it and see.

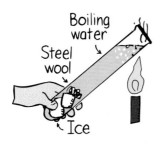

2. If you live where there is snow, do as Benjamin Franklin did nearly two centuries ago: Lay samples of light and dark cloth on the snow on a sunny day. Note the difference in the rate of melting beneath the cloths.

3. Wrap a piece of paper around a thick metal bar and place it in a flame. Note that the paper will not catch fire. Can you figure out why? (Paper generally will not ignite until its temperature reaches about 230°C.)

Exercises

1. Why is it difficult to estimate the temperature of things by touching them?

2. Wrap a fur coat around a thermometer. Will its temperature rise?

3. If 70°F air feels warm and comfortable to us, why does 70°F water feel cool when we swim in it?

4. At what common temperature will both a block of wood and a block of metal feel neither hot nor cold to the touch?

5. If you hold one end of a metal nail against a piece of ice, the end in your hand soon becomes cold. Does cold flow from the ice to your hand? Explain.

6. Why do restaurants serve baked potatoes wrapped in aluminum foil?

7. Many tongues have been injured by licking a piece of metal on a very cold day. Why would no harm result if a piece of wood were licked on the same day?

8. Wood is a better insulator than glass. Yet fiberglass is commonly used as an insulator in wooden buildings. Explain.

9. Visit a snow-covered cemetery and note the snow does not slope upward against the gravestones, but instead forms depressions as shown. Can you think of a reason for this?

10. If you were caught in freezing weather with only your own body heat for a heat source, would you be warmer in an igloo or in a wooden shack?

11. Heat transfer is important on a hot mid-summer day when air takes heat from the earth and prevents the earth's surface from becoming unbearably hot. Describe how conduction and convection play a role.

12. You can bring water in a paper cup to a boil by placing it over a hot flame. Why doesn't the paper cup burn?

13. Why can you comfortably hold your fingers close beside a candle flame, but not very close above the flame?

14. Why is it that you can safely hold your bare hand in a hot pizza oven for a few seconds, but if you momentarily touch the metal insides you'll burn yourself?

15. Wood conducts heat very poorly — it has a very low conductivity. Does wood still have a low conductivity if it is hot? Could you quickly and safely grab the wooden handle of a pan from a hot oven with your bare hand? Although the pan handle is hot, does much heat conduct from it to your hand if you do it quickly? Could you do the same with an iron handle? Explain.

16. Wood has a very low conductivity. Does it still have a low conductivity if it is very hot — that is, in the stage of smoldering red-hot coals? Could you safely walk across a bed of red-hot wood coals with bare feet? Although the coals are hot, does much heat conduct from them to your feet if you step quickly? Could you do the same on red-hot iron coals? Explain. (*Caution:* Coals can stick to your feet, so OUCH — don't try it!)

17. On page 255, the opener to Part 3, Terrence Jones with a fireworks sparkler cites the distinction between temperature and heat. Suppose instead that the photo of his brother Marques, shown here, were used as the part opener. Note that Marques holds a frying pan by both its wooden handle and by its iron rim. Do two things: (1) Fill in the balloon with a statement that illustrates good physics; (2) Make a statement that illustrates poor physics, but which may seem to be good physics to somebody ignorant about conductivity.

18. A friend says that in a mixture of gas in thermal equilibrium that the molecules have the same average kinetic energy. Do you agree or disagree? Explain.

19. Why would you not expect all molecules in a gas to have the same average speed?

20. What happens to the gas pressure within a sealed gallon can when it is heated? Cooled? Why?

21. In a still room, smoke from a cigar will sometimes rise only so far, not reaching the ceiling. Explain why.

22. Why would you expect a single helium atom to continuously rise in an atmosphere of nitrogen and oxygen? Why doesn't it stop rising like the smoke in the preceding question?

23. Release a single molecule in an evacuated region. Will it fall differently than a baseball in the same region? Explain.

24. Why do most molecules not travel in perfectly straight lines between collisions in a gas? Some do; which?

25. Does gravity tend to decrease upward movement of air, and increase downward movement? And is there a greater "window" above any point in air than below, which favors upward migration? Guess how these two opposite effects compare when air is in thermal equilibrium and convection zero.

26. In a mixture of hydrogen and oxygen gases at the same temperature, which molecules move faster? Why?

27. One container is filled with argon gas and the other with krypton gas. If both gases have the same temperature, in which container are the atoms moving faster? Why?

28. Which atoms have the greatest average speed in a mixture, U-238 or U-235? How would this affect diffusion through a porous membrane of otherwise identical gases made from these isotopes?

29. Should you expect the ratio of nitrogen molecules to oxygen molecules in the atmosphere to increase or decrease with increasing altitude? Explain.

30. If we warm a volume of air, it expands. Does it then follow that if we expand a volume of air, it warms? Explain.

31. How could you change the drawing in Figure 15.7 to make it illustrate the heating of air when air is compressed? Make a sketch of this case.

32. A snow-making machine used for ski areas blows a mixture of compressed air and water through a nozzle. The temperature of the mixture may initially be well above the freezing temperature of water, yet crystals of snow are formed as the mixture is ejected from the nozzle. Explain how this happens.

33. What does the high specific heat of water have to do with convection currents in the air at the seashore?

34. What would be the most efficient color for steam radiators?

35. Turn an incandescent lamp on and off quickly while you are standing near it. You feel its heat but find that when you touch the bulb, it is not hot. Explain why you felt heat from it.

36. Why does a good *emitter* of heat radiation appear black at room temperature?

37. A number of bodies at different temperatures placed in a closed room share radiant energy and ultimately come to the same temperature. Would this thermal equilibrium be possible if good absorbers were poor emitters and poor absorbers were good emitters? Explain.

38. From the rules that a good absorber of radiation is a good radiator and a good reflector is a poor absorber, state a rule relating the reflecting and radiating properties of a surface.

39. Suppose you are served coffee at a restaurant before you are ready to drink it. In order that it be hottest when you are ready for it, would you be wiser to add cream to it right away or when you are ready to drink it?

40. Even though metal is a good conductor, frost can be seen on parked cars in the early morning even when the air temperature is above freezing. Can you explain this?

41. When there is morning frost on the ground in an open park, why is it likely that none is on the ground beneath park benches?

42. It's commonly known that a white-painted house is a good idea in hot summer weather, because more sunlight is reflected from the exterior, making for a cooler interior. But a white-painted house is also a good idea in the cold winter. Why so?

43. Outer space is not "nothingness." It is full of radiation with a temperature of about 3 K. Since this radiation is shining on the earth at night, why does the earth get cold at night?

44. Why is whitewash sometimes applied to the glass of florists' greenhouses in the summer?

45. On a very cold sunny day you wear a black coat and a transparent plastic coat. Which should be worn on the outside for maximum warmth?

46. If the composition of the upper atmosphere were changed so that it permitted a greater amount of terrestrial radiation to escape, what effect would this have on the earth's climate? How about if the atmosphere reduced the escape of terrestrial radiation?

47. Is it important to convert temperatures to the Kelvin scale when we use Newton's law of cooling? Why or why not?

48. If you wish to save fuel and you're going to leave your warm house for a half hour or so on a very cold day, should you turn your thermostat down a few degrees, turn it off altogether, or let it remain at the room temperature you desire?

49. If you wish to save fuel and you're going to leave your cool house for a half hour or so on a very hot day, should you turn your air conditioning thermostat up a bit, turn it off altogether, or let it remain at the room temperature you desire?

50. As more energy from fossil and other non-renewable fuels is consumed on earth, the overall temperature of the earth tends to rise. Regardless of the increase in energy, however, the temperature does not rise indefinitely. By what process is an indefinite rise prevented? Explain your answer.

Problems

1. A container of hot water at 80°C cools to 79°C in 15 seconds when it is placed in a room that is at 20°C. Use Newton's law of cooling to estimate the time it will take for the container to cool from 50°C to 49°C? And still later, from 40°C to 39°C?

2. In a 25°C room, hot coffee in a vacuum flask cools from 75°C to 50°C in eight hours. What will its temperature be after another eight hours?

3. At a certain location, the solar power per unit area reaching the earth's surface is 200 W/m^2, averaged over a 24-hour day. If you live in a house with an average power requirement of 3 kW and you can convert solar power to electric power with 10% efficiency, how large a collector area will you need to meet all your household energy requirements from solar energy? Will the collector fit in your yard?

16

· · · · · · · · · · · ·

Change of Phase

H$_2$O as ice, water, and vapor.

The matter in our environment exists in four common *phases* (or *states*). Ice, for example, is the *solid* phase of H$_2$O. Add energy, and you add motion to the rigid molecular structure, which breaks down to form H$_2$O in the *liquid* phase, water. Add more energy, and the liquid changes to the *gaseous* phase. Add still more energy, and the molecules break into ions and electrons, giving the *plasma* phase. The phase of matter depends on its temperature and the pressure that is exerted on it. Changes of phase are often accompanied by changes in temperature or pressure—and almost always require a transfer of energy.

Evaporation

Water in an open container will eventually evaporate, or dry up. The liquid that disappears becomes water vapor in the air. **Evaporation** is a change of phase from liquid to gas that takes place at the surface of a liquid.

The temperature of anything is related to the average kinetic energy of its particles. Molecules in liquid water move about in all directions and bump into one another while moving at different speeds. Some gain kinetic energy while others lose kinetic energy. Molecules at the surface that gain kinetic energy by being bumped from below may have enough energy to break free of the liquid. They can leave the surface and fly into the space above the liquid. In this way they become molecules of vapor.

The increased kinetic energy of molecules bumped free of the liquid comes from molecules remaining in the liquid. This is "billiard-ball physics": When balls bump into one another and some gain kinetic energy, the others lose the same amount. Molecules going from liquid to gas are the gainers, while the losers of energy remain in the liquid. Thus the average kinetic energy of the molecules remaining in the liquid is lowered—evaporation is a cooling process.

The canteen shown in Figure 16.1 keeps cool because of evaporation when the cloth covering the sides is kept wet. As the faster-moving water molecules leave the cloth, the temperature of the cloth decreases. The cool cloth in turn cools the metal canteen by conduction, which in turn cools the water inside. So heat is transferred from the water in the canteen to the air outside. In this way the water is cooled appreciably below outside air temperature.

The cooling effect of evaporation is strikingly evident when rubbing alcohol is poured on your back. The alcohol evaporates very rapidly, cooling the surface of the body quickly. The more rapid the evaporation, the faster the cooling.

When our bodies tend to overheat, our sweat glands produce perspiration. This is part of nature's thermostat, for the evaporation of perspiration cools us and helps us maintain a stable body temperature. Many animals do not have sweat glands and must cool themselves by other means (Figures 16.2 and 16.3).

FIGURE 16.1 The cloth covering on the sides of the canteen promotes cooling when it is wet. As the fastest-moving molecules evaporate from the wet cloth, the temperature of the cloth decreases and cools the metal, which in turn cools the water within. In this way the water in the canteen is cooled appreciably below air temperature.

FIGURE 16.2 Dogs have no sweat glands (except between the toes). They cool themselves by panting. In this way evaporation occurs in the mouth and within the bronchial tract.

FIGURE 16.3 Pigs have no sweat glands and therefore cannot cool by the evaporation of perspiration. Instead, they wallow in the mud to cool themselves.

Question   Would evaporation be a cooling process if each molecule at the surface of a liquid had the same kinetic energy before and after bumping into other molecules?

Condensation

FIGURE 16.4   Heat is given up by steam when it condenses inside the radiator.

The process that is the opposite of evaporation is **condensation**—the changing of a gas into a liquid. When gas molecules near the surface of a liquid are attracted to the liquid, they strike the surface with increased kinetic energy and become part of the liquid. In collisions with low-energy molecules in the liquid, they share excess kinetic energy and increase the liquid temperature. We say that condensation is a warming process.

A dramatic example of the warming that results from condensation is the energy given up by steam when it condenses—a painful experience if it condenses on you. That's why a steam burn is much more damaging than a burn from boiling water of the same temperature; the steam gives up considerable energy when it condenses to a liquid and wets the skin. This energy release by condensation is utilized in steam-heating systems.

Water vapor need not be as hot as steam to give up energy when it condenses. You are warmed by condensation after you have taken a shower—even a cold shower—if you remain in the moist shower area. You quickly sense the difference if you step outside. Away from the moisture, net evaporation takes place quickly and you feel chilly. When you remain in the shower stall, even with the water off, the warming effect of condensation counteracts the cooling effect of evaporation. If as much moisture condenses as evaporates, you feel no change in body temperature. If condensation exceeds evaporation, you are warmed. If evaporation exceeds condensation, you are cooled. So now you know why you can dry yourself with a towel much more comfortably if you remain in the shower area. To dry yourself thoroughly, you can finish the job in a less moist area.

Spend a July afternoon in dry Tucson or Phoenix where evaporation is appreciably greater than condensation. The result of this pronounced evaporation is a much cooler feeling than you would experience in a same-temperature July afternoon in New York City or New Orleans. In these humid locations, condensation noticeably counteracts evaporation, and you feel the warming effect as vapor in the air condenses

FIGURE 16.5   If you're chilly outside the shower stall, step back inside and be warmed by the condensation of the excess water vapor there.

Answer   No. If there were no transfer of kinetic energy during molecular bumping, there would be no change in temperature. The liquid cools only when there is a lowering of average kinetic energy of molecules in the liquid. This occurs when some molecules (like billiard balls) gain speed at the expense of others that lose speed. Those that leave (evaporate) are gainers while losers remain behind in the liquid and lower the average kinetic energy.

on your skin. You are quite literally being "tattooed" by the impact of H_2O molecules in the air that slam into you. Put more mildly, you are warmed by the condensation of vapor in the air upon your skin.

Question If the water level in a dish of water remains unchanged from one day to the next, can you conclude that no evaporation or condensation is taking place?

a   b   c   d

FIGURE 16.6 The toy drinking bird operates by the evaporation of ether inside its body and by the evaporation of water from the outer surface of its head. The lower body contains liquid ether, which evaporates rapidly at room temperature. As it (a) vaporizes, it (b) creates pressure (inside arrows), which pushes ether up the tube. Ether in the upper part does not vaporize because the head is cooled by the evaporation of water from the outer felt-covered beak and head. When the weight of ether in the head is sufficient, the bird (c) pivots forward, permitting the ether to run back to the body. Each pivot wets the felt surface of the beak and head, and (d) the cycle is repeated.

Condensation in the Atmosphere

There is always some water vapor in the air. The warmer the day, the more water vapor the air can hold. We can understand this by thinking of a fly making a grazing contact with flypaper. At high speed it has enough momentum and energy to rebound from the flypaper into the air without sticking, but at low speed it more likely gets stuck. Similarly, when water vapor molecules collide at high speed, they are more likely to bounce off one another and remain in the vapor phase. At low speeds they are more likely to

---

Answer Not at all, for there is much activity taking place at the molecular level. Both evaporation and condensation occur continuously. The fact that the water level remains constant simply indicates equal rates of evaporation and condensation. As many molecules leave the surface by evaporation as return by condensation, so no *net* evaporation or condensation takes place. The two processes cancel each other.

FIGURE 16.7
Condensation of water
vapor.

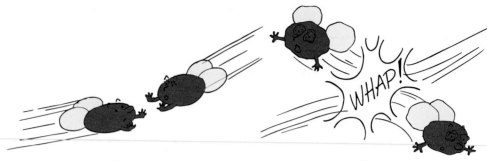

Fast-moving H₂O molecules rebound upon collision

Slow-moving H₂O molecules coalesce upon collision

stick together and become part of a liquid (Figure 16.7). The faster that water molecules in the vapor phase move, the less chance there is that they will condense to form droplets. So we see why more water molecules can exist in the vapor phase in warmer air than in cooler air.

There is a limit to the humidity of air at a given temperature. Beyond this limit, the water vapor condenses to a liquid. Maximum humidity occurs when the maximum amount of vapor in air is reached—that is, when the air is *saturated*. When meteorologists report the *relative humidity,* they are comparing the amount of vapor in the air to the amount at saturation for the given air temperature. A relative humidity of 100 percent means that the air is saturated. A relative humidity of 50 percent, on the other hand, means there is half the amount of vapor in the air as when it is saturated at the same temperature. For the average person, weather is ideal when the temperature is about 20°C and the relative humidity is about 50 to 60 percent. When the relative humidity is high, condensation counteracts evaporation of perspiration, and we say the weather is muggy.

Fog and Clouds

FIGURE 16.8 Why is it common for clouds to form where there are updrafts of warm moist air?

Warm air rises. As it rises, it expands. As it expands, it chills. As it chills, water vapor molecules begin sticking together after colliding rather than bouncing off one another. If there are larger and slower-moving particles or ions present, water vapor condenses upon these particles, and with sufficient buildup we have a cloud. If these particles are not present, we can stimulate cloud formation by "seeding" the air with appropriate particles or ions.

Warm breezes blow over the ocean. When the moist air moves from warmer to cooler waters or from warm water to cool land, it chills. As it chills, water vapor molecules begin sticking rather than bouncing off one another. Condensation takes place near ground level, and we have fog. The difference between fog and a cloud is basically altitude. Fog is a cloud that forms near the ground. Flying through a cloud is much like driving through fog.

Boiling

Under the right conditions, evaporation can take place beneath the surface of a liquid, forming bubbles of vapor that are buoyed to the surface where they escape. This change of phase throughout a liquid rather than only at the surface is called **boiling**. Bubbles in the liquid can form only when the pressure of the vapor within the bubbles is great enough to resist the pressure of the surrounding liquid. Unless the vapor pressure is great enough, the surrounding pressure will collapse any bubbles that may form. At temperatures below the boiling point, the vapor pressure in bubbles is not great enough, so bubbles do not form until the boiling point is reached. At this temperature, 100°C for water at atmospheric pressure, molecules are energetic enough to exert a vapor pressure as great as the pressure of the surrounding water (due mainly to atmospheric pressure).

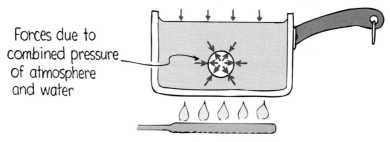

Forces due to combined pressure of atmosphere and water

FIGURE 16.9 The motion of molecules in the bubble of steam (much enlarged) creates a gas pressure that counteracts the water pressure against the bubble. Most of the water pressure is due to the air pressing down from above.

If pressure is increased, the molecules in the vapor must move faster to exert enough pressure to keep the bubble from collapsing. Extra pressure can be provided either by going deeper below the surface of the liquid (as in geysers, discussed below) or by increasing the air pressure above the liquid surface—which is the way a pressure cooker works. A pressure cooker has a tight-fitting lid that does not allow vapor to escape until it reaches a certain pressure greater than normal air pressure. As the evaporating vapor builds up inside the sealed pressure cooker, pressure on the surface of the liquid increases, which at first prevents boiling. Bubbles that would normally form are crushed. Continued heating increases the temperature beyond 100°C. Boiling does not occur until the vapor pressure within the bubbles overcomes the increased pressure on the water. The boiling point is raised. Conversely, lowered pressure (as at high altitudes) decreases the boiling point of the liquid. So we see that boiling depends not only on temperature but on pressure as well.

FIGURE 16.10 A pressure cooker.

At high altitudes, water boils at a lower temperature. For example, in Denver, Colorado, the "mile-high city," water boils at 95°C instead of the 100°C boiling temperature characteristic of sea level. If you try to cook food in boiling water of a lower temperature, you must wait a longer time for proper cooking. A 3-minute boiled egg in Denver is yucky. If the temperature of the boiling water is too low, food will not cook at all. Note that the high temperature of water cooks food, not the boiling process itself.

Geysers

A geyser is a periodically erupting pressure cooker. It consists of a long, narrow, vertical hole into which underground streams seep (Figure 16.11). The column of water is heated by volcanic heat below to temperatures exceeding 100°C. This can happen

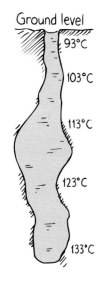

Ground level

93°C
103°C
113°C
123°C
133°C

FIGURE 16.11 An Old Faithful type of geyser.

because the vertical column of water exerts pressure on the deeper water, thereby increasing the boiling point. The narrowness of the shaft shuts off convection currents, which allows the deeper portions to become considerably hotter than the water surface. Water at the surface is less than 100°C, but the water temperature below, where it is being heated, is more than 100°C, high enough to permit boiling before water at the top reaches the boiling point. Boiling begins near the bottom, where the rising bubbles push out the column of water above, and the eruption starts. As the water gushes out, the pressure on the remaining water is reduced. It rapidly boils and erupts with great force.

Boiling Is a Cooling Process

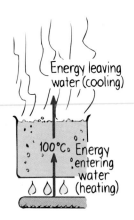

FIGURE 16.12 Heating warms the water, and boiling cools it.

Evaporation is a cooling process. So is boiling. At first thought, this may seem surprising—perhaps because we usually associate boiling with heating. But heating water is one thing; boiling is another. When 100°C water at atmospheric pressure is boiling, its temperature remains constant. It is being cooled by boiling as fast as it is being heated by energy from the high-temperature source (Figure 16.12). If cooling did not take place, continued input of energy to a pot of boiling water would result in a continued increase in temperature. The reason a pressure cooker reaches higher temperatures is because it prevents normal boiling, which in effect prevents cooling.

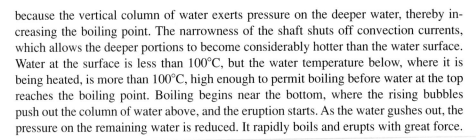

Question Since boiling is a cooling process, would it be a good idea to cool your hot and sticky hands by dipping them into boiling water?

Boiling and Freezing at the Same Time

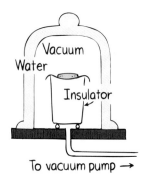

FIGURE 16.13 Apparatus to demonstrate that water will freeze and boil at the same time in a vacuum. A gram or two of water is placed in a dish that is insulated from the base by a polystyrene cup.

We usually boil water by the application of heat. But we can boil water by the reduction of pressure. We can dramatically show the cooling effect of evaporation and boiling when room-temperature water is placed in a vacuum jar (Figure 16.13). If the pressure in the jar is slowly reduced by a vacuum pump, the water will start to boil. The boiling process takes heat away from the water left in the dish, which cools to a lower temperature. As the pressure is further reduced, more and more of the slower-moving molecules boil away. Continued boiling results in a lowering of temperature until the freezing point of approximately 0°C is reached. Continued cooling by boiling causes ice to form over the surface of the bubbling water. Boiling and freezing are taking place at the same time! This must be witnessed to be appreciated. Frozen bubbles of boiling water are a remarkable sight.

Spray some drops of coffee into a vacuum chamber, and they, too, will boil until they freeze. Even after they are frozen, the water molecules will continue to evaporate into the vacuum until little crystals of coffee solids are left. This is how freeze-dried coffee is made. The low temperature of this process tends to keep the chemical structure of coffee solids from changing. When hot water is added, much of the original flavor of the coffee is restored. Boiling really is a cooling process!

Answer No, no, no! When we say boiling is a cooling process, we mean that the water (not your hands!) is being cooled relative to the higher temperature it would attain otherwise. Because of cooling, it remains at 100°C instead of getting hotter. A dip in 100°C water would be most uncomfortable for your hands!

Melting and Freezing

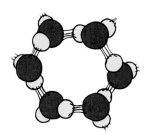

FIGURE 16.14 The open structure of pure ice crystals that normally fuse at 0°C. When other kinds of molecules or ions are introduced, crystal formation is interrupted, and the freezing temperature is lowered.

Suppose you held hands with someone, and each of you started jumping around randomly. The more violently you jumped, the more difficult keeping hold would be. If you jumped violently enough, keeping hold would be impossible. Something like this happens to the molecules of a solid when it is heated. As heat is absorbed, the molecules vibrate more and more violently. If enough heat is absorbed, the attractive forces between the molecules will no longer be able to hold them together. The solid melts.

Freezing is the converse of this process. As energy is withdrawn from a liquid, molecular motion diminishes until finally the molecules, on the average, are moving slowly enough so that the attractive forces between them are able to cause cohesion. The molecules then vibrate about fixed positions and form a solid.

At atmospheric pressure, water freezes at 0°C—unless substances such as sugar or salt are dissolved in it. Then the freezing point is lower. In the case of salt, chlorine ions grab electrons from the hydrogen atoms in H_2O and impede crystal formation. The result of this interference by "foreign" ions is that slower motion is required for the formation of the six-sided ice-crystal structures. As ice crystals do form, the interference is intensified because the proportion of "foreign" molecules or ions among nonfused water molecules increases. Connections become more and more difficult. Only when the water molecules move slowly enough for attractive forces to play an unusually large part in the process can freezing be completed. The ice first formed is almost always pure H_2O. In general, adding anything to water lowers the freezing temperature. Antifreeze is a practical application of this process.

Regelation

Because H_2O molecules form open structures in the solid phase (Figure 16.14), the application of pressure lowers the melting point of ice. The crystals are simply crushed to the liquid phase. (The temperature of the melting point is only slightly lowered, 0.007°C for each additional atmosphere of added pressure.) When the pressure is removed, molecules crystallize and refreezing occurs. This phenomenon of melting under pressure and freezing again when the pressure is reduced is called **regelation**. It is one of the properties of water that make it different from other materials.

Regelation is nicely illustrated in Figure 16.15. A fine copper wire with weights attached to its ends is hung over a block of ice.* The wire will slowly cut its way through the ice, but its track will be left full of ice. So the wire and weights will fall to the floor, leaving the ice a solid block.

The making of snowballs is a good example of regelation. When we compress the snow with our hands, we cause a slight melting of the ice crystals; when pressure is removed, refreezing occurs and binds the snow together. Making snowballs is difficult in very cold weather because the pressure we can apply is not enough to melt the snow.

A skater skates on a thin film of water between the blade and the ice, which is produced by blade pressure and friction. As soon as the pressure is released, the water refreezes.

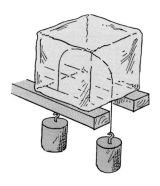

FIGURE 16.15 Regelation. The wire gradually passes through the ice without cutting it in half.

* Changes of phase are occurring as ice melts and the water refreezes, and, as will be stressed in the next section and the rest of the chapter, energy is needed for these changes. When the water immediately above the wire refreezes, it gives up energy. How much? Enough to melt an equal amount of ice immediately under the wire. This energy must be conducted through the thickness of the wire. Hence this demonstration requires that the wire be an excellent conductor of heat. String, for example, won't do.

Energy and Changes of Phase

If we continuously heat a solid or liquid, the solid or liquid will eventually change phase. A solid will liquefy, and a liquid will vaporize. Energy is required for both liquefaction of a solid and vaporization of a liquid. Conversely, energy must be extracted from a substance to change its phase in the direction from gas to liquid to solid (Figure 16.16).

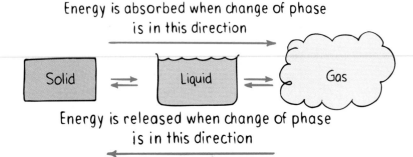

Energy is absorbed when change of phase is in this direction

Solid ⇄ Liquid ⇄ Gas

Energy is released when change of phase is in this direction

FIGURE 16.16 Energy changes with change of phase.

The cooling cycle of a refrigerator nicely employs the concepts shown in Figure 16.16. A liquid of low boiling point (one of the newer refrigerants replacing environmentally harmful Freon) is pumped into the cooling unit, where it turns into a gas. In order to vaporize, it draws heat from the food items stored there. The gas with its added energy is directed outside the cooling unit to coils, appropriately called condensation coils, located in the back of the refrigerator. At the coils, energy is given off to the air as the gas condenses to a liquid again. A motor pumps the fluid through the system, where it is made to undergo the cyclic process of vaporization and condensation. The next time you're near a refrigerator, place your hand near the condensation coils in the back and you'll feel the warm air that has been heated by the energy extracted from inside.

An air conditioner employs the same principles and simply pumps heat energy from one part of the unit to another location—outside. When the direction of energy flow is reversed, the air conditioner becomes a heat pump—a heater.

So we see that a solid must absorb energy to melt, and a liquid must absorb energy to vaporize. Conversely, a gas must release energy to liquefy, and a liquid must release energy to solidify.

Question When H_2O in the vapor phase condenses, is the surrounding air warmed or cooled?

Answer The surrounding air is warmed; energy is released by the vapor when it turns to liquid (Figure 16.16). Another way to see why the air is warmed is to think of molecules in the air (a mixture of gases, including H_2O) as tiny billiard balls bouncing off one another. The total kinetic energy before and after collisions always remains the same. If one molecule gains kinetic energy in a collision, the other loses the same amount. Some molecules gain speed; some lose speed. What happens when the losers are H_2O that are near other slow-moving H_2O that have also just given much of their kinetic energy to other molecules? The answer is, they stick to one another (Figure 16.7). They condense from the air. And what is the last act of the H_2O molecules before they condense? Their last act in the vapor phase is to transfer much of their kinetic energy to other molecules. So the air is warmed. How much? By about 540 calories for each gram of H_2O that condenses.

The general behavior of many substances can be illustrated by a description of the changes of phase that occur in H_2O. To make the numbers simple, consider a 1-gram piece of ice at a temperature of $-50°C$ in a closed container that is put on a stove to heat. A thermometer in the container reveals a slow increase in temperature up to $0°C$. At $0°C$, the temperature stops rising, even though energy is continually added. This energy melts the ice. In order for the whole gram of ice to melt, 80 calories are absorbed by the ice. The temperature will not begin rising until all the ice melts. Only when all the ice melts will each additional calorie absorbed by the water increase its temperature by $1°C$ until the boiling temperature, $100°C$, is reached. Again, as energy is added, the temperature remains constant while more and more of the gram of water is boiled away and becomes steam. The water must absorb 540 calories of heat energy to vaporize the whole gram. Finally, when all the water has become steam at $100°C$, the temperature begins to rise once more. It will continue to rise as long as energy is added. This process is graphed in Figure 16.17.

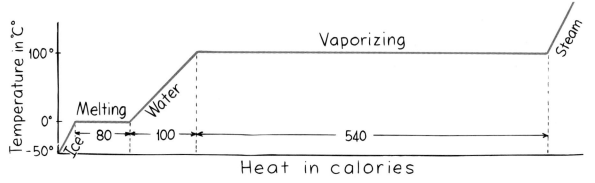

FIGURE 16.17 A graph showing the energy involved in the heating and the changes of phase of 1 g of H_2O.

The 540 calories required to vaporize a gram of water is a relatively large amount of energy—much more than is required to transform a gram of ice at absolute zero to boiling water at $100°C$. So 540 calories is a relatively large amount of energy. Although the molecules in steam and boiling water at $100°C$ have the same average kinetic energy, and therefore the same temperature, steam has more potential energy because the molecules are relatively free of each other and are not bound together as in the liquid phase. Steam contains a vast amount of energy that can be released during condensation.

The change of phase sequence is reversible. If the molecules in a gram of steam condense to form boiling water, they transfer 540 calories of energy to the environment. So, for example, when water is cooled from $100°C$ to $0°C$, 100 additional calories are transferred to the environment. When ice water fuses to become solid ice, 80 more calories of energy are transferred by the water to the environment. The condensation of water releases 540 calories per gram, and the fusion (freezing) of water releases 80 calories per gram.*

* These values are known as the *heat of vaporization* and the *heat of fusion,* respectively. In SI units, the heat of vaporization of water is 2.26 megajoules per kilogram (MJ/kg), and the heat of fusion of water is 0.335 MJ/kg.

Questions

1. How much energy is transferred when a gram of steam at 100°C condenses to water at 100°C?

2. How much energy is transferred when a gram of boiling water at 100°C cools to ice water at 0°C?

3. How much energy is transferred when a gram of ice water at 0°C freezes to ice at 0°C?

4. How much energy is transferred when a gram of steam at 100°C turns to ice at 0°C?

The large value of 540 calories per gram explains why under some conditions hot water will freeze faster than warm water.* This phenomenon is evident when a thin layer of water is distributed over a large area—like when you wash a car with hot water on a cold winter day or flood a skating rink with hot water, which melts, smooths out the rough spots, and refreezes quickly. The rate of cooling by rapid evaporation is very high because each evaporating gram draws at least 540 calories from the water left behind. This is an enormous amount of energy compared to the 1 calorie per Celsius degree that is drawn from each gram of water that cools by thermal conduction. Evaporation is truly a cooling process.

FIGURE 16.18 On a cold day, hot water freezes faster than warm water because of the energy that leaves the hot water during rapid evaporation.

Answers

1. One gram of steam at 100°C transfers 540 calories of energy when it condenses to become water at the same temperature.

2. One gram of boiling water transfers 100 calories when it cools 100°C to become ice water.

3. One gram of ice water at 0°C transfers 80 calories to become ice at 0°C.

4. One gram of steam at 100°C transfers to the surroundings a grand total of the above values, 720 calories, to become ice at 0°C.

* Boiling hot water will not freeze before cold water does, but will freeze before moderately hot water does. Water that is boiling hot, for example, will freeze before water warmer than about 60°C, but not before water cooler than 60°C. Try it and see.

FIGURE 16.19 Paul Ryan tests the hotness of molten lead by dragging his wetted finger through it.

Question Suppose 4 g of boiling water is poured onto a cold surface and 1 g rapidly evaporates. If evaporation takes 540 calories from the remaining 3 g of water and no other heat transfer takes place, what will be the temperature of the remaining 3 g?

You dare not touch your dry finger to a hot skillet on a hot stove, but you can certainly do so without harm if you first wet your finger and touch the skillet briefly. You can even touch it a few times in succession as long as your finger remains wet. This is because energy that ordinarily would go into burning your finger goes instead into changing the phase of the moisture on your finger. The energy converts the moisture to a vapor, which then provides an insulating layer between your finger and the skillet. Similarly, you are able to judge the hotness of a hot clothes iron.

Supervisor Paul Ryan of the Department of Public Works in Malden, Massachusetts, has for years used molten lead to seal pipes in certain plumbing operations. He startles onlookers by dragging his finger through molten lead to judge its hotness (Figure 16.19). He is sure that the lead is very hot and his finger is thoroughly wet before he does this. (Do not try this on your own, for if the lead is not hot enough it will stick to your finger and you could be seriously burned!) Similarly, firewalkers who walk barefoot on red-hot coals often prefer wet feet (others prefer dry feet, claiming hot coals more readily stick to wet feet—ouch!). The low conductivity of wood coals, however, (as discussed in the previous chapter) is the principal reason the feet of bare-foot firewalkers are not burned.

FIGURE 16.20 Professor Dave Willey walks with wetted bare feet across red-hot wood coals without harm.

Answer The remaining 3 g will turn to 0°C ice under conditions where all 540 calories are taken from the remaining water (for example when the surroundings are below freezing and don't contribute energy). 540 calories from 3 g means each gram gives up 180 calories. 100 calories from a gram of boiling water reduces its temperature to 0°C, and 80 more calories taken away turns it to ice. This is why hot water so quickly turns to ice in a freezing-cold environment.

Summary of Terms

Evaporation   The change of phase at the surface of a liquid as it passes to the gaseous phase. This is caused by the random motion of molecules that occasionally escape from the liquid surface. Cooling of the liquid results.

Condensation   The change of phase from gas to liquid; the opposite of evaporation. Warming of the liquid results.

Boiling   A rapid evaporation that takes place within the liquid as well as at its surface. As with evaporation, cooling of the liquid results.

Regelation   The process of melting under pressure and the subsequent refreezing when the pressure is removed.

Review Questions

1. What are the four common phases of matter?

Evaporation

2. Do all the molecules or atoms in a liquid have about the same speed or much different speeds?

3. What is evaporation, and why is it a cooling process? Exactly what is it that cools?

4. Why does a hot dog pant?

Condensation

5. What is condensation, and why is it a warming process? Exactly what is it that warms?

6. Why is a steam burn more damaging than a burn from boiling water of the same temperature?

7. Why do we feel uncomfortably warm on a hot and humid day?

Condensation in the Atmosphere

8. Why is there usually more water vapor in warm air than in cool air?

9. Why does water vapor in the air condense when the air is chilled?

Fog and Clouds

10. Why does warm moist air form clouds when it rises?

11. What is the basic difference between a cloud and fog?

Boiling

12. Distinguish between evaporation and boiling.

13. Why does water not boil at 100°C when it is under a pressure of more than one atmosphere?

14. Why is the boiling point of water lower at high altitude?

15. What causes food to cook faster in a pressure cooker than in an open pot?

Geysers

16. How is a geyser similar to a pressure cooker?

17. Why will water at the bottom of a geyser not boil when it is at 100°C?

18. Why does water at the bottom of a geyser get hotter and boil before water at the top?

19. What happens to the water pressure at the bottom of a geyser when some of the water above gushes out?

20. Why does the remaining water in a geyser erupt violently when some of the water above gushes out?

Boiling Is a Cooling Process

21. What is it that cools when boiling occurs—the water that leaves in the form of vapor or the water left behind?

22. We observe that the temperature of boiling water doesn't increase with continued energy input. Explain how this observation is evidence that boiling is a cooling process.

Boiling and Freezing at the Same Time

23. When will water boil at a temperature less than 100°C?

24. We can add energy to water and boil it to form steam. Isn't it a contradiction to say we can boil water to form ice? Explain.

Melting and Freezing

25. Why does increasing the temperature of a solid make it melt?

26. Why does decreasing the temperature of a liquid make it freeze?

27. Why does freezing of water not occur at 0°C when molecules other than H_2O are present?

28. Which freezes at the lower temperature, pure water or salt water?

Regelation

29. What happens to the hexagonal open structure of ice when sufficient pressure is applied to it?

30. What happens to the water that results from crushing crystals when the pressure is removed?

31. Why does a wire not simply cut a block of ice in half when it passes through the ice?

Energy and Changes of Phase

32. Does a liquid give off or absorb energy when it turns into a gas? When it turns into a solid?

33. Does a gas give off or absorb energy when it turns into a liquid?

34. Does a solid give off or absorb energy when it turns into a liquid?

35. Is Figure 16.16 consistent with or contrary to the statement that both evaporation and boiling are cooling processes and that condensation is a warming process? Defend your answer.

36. Is the food compartment in a refrigerator cooled by vaporization or by condensation of the refrigerating fluid?

37. How many calories are needed to change the temperature of 1 g of water by 1°C? To melt 1 g of ice at 0°C? To vaporize 1 g of boiling water at 100°C?

38. Why does the temperature of melting ice not rise when energy is added?

39. Why does the temperature of boiling water not rise when energy is added?

40. Why is the rate of cooling of hot water by evaporation greater than the rate of cooling of lukewarm water by thermal conduction?

41. Why is it important that your finger be wet if you touch it briefly to a hot clothes iron?

42. When we discussed in Chapter 4 the acceleration and high speeds you would encounter when falling from great heights, no words of caution told you of the dangers of trying this on your own. Why are there words of caution in this chapter with respect to walking barefoot on hot coals and dipping a finger into molten lead?

Projects

1. Place a Pyrex funnel mouth down in a saucepan full of water so the straight tube of the funnel sticks above the water. Rest a part of the funnel on a nail or coin so water can get under it. Place the pan on a stove and watch the water as it begins to boil. Where do the bubbles form first? Why? As the bubbles rise, they expand rapidly and push water ahead of them. The funnel confines the water, which is forced up the tube and driven out at the top. Now do you know how a geyser and a coffee percolator work?

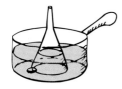

2. Watch the spout of a teakettle of boiling water. Notice that you cannot see the steam that issues from the spout. The cloud that you see farther away from the spout is not steam, but condensed water droplets. Now hold the flame of a candle in the cloud of condensed steam. Can you explain your observations?

3. You can make rain in your kitchen. Put a cup of water in a Pyrex saucepan or Silex coffeemaker and heat it slowly over a low flame. When the water is warm, place a saucer filled with ice cubes on top of the container. As the water below is heated, droplets form at the bottom of the cold saucer and combine until they are large enough to fall, producing a steady "rainfall" as the water below is gently heated. How does this resemble and how does it differ from the way natural rain is formed?

4. Measure the temperature of boiling water and the temperature of a boiling solution of salt and water. How do they compare?

5. Suspend a heavy weight by copper wire over an ice cube. In a matter of minutes the wire will be pulled through the ice. The ice will melt beneath the wire and refreeze above it, leaving a visible path if the ice is clear.

6. Place a tray of boiling hot water and a tray of water from a hot-water tap in a freezer. The trays should be filled to about the same depth. See which tray of water will freeze first.

7. If you suspend a container of water in a pan of boiling water, the water in the inner container will reach 100°C but will not boil. Can you figure out why this is so?

Exercises

1. You can determine wind direction by wetting your finger and holding it up in the air. Explain.

2. When you step out of a swimming pool on a hot, dry day in the Southwest, you feel quite chilly. Why?

3. Why does blowing over hot soup cool the soup?

4. Can you give two reasons why pouring a cup of hot coffee into the saucer results in faster cooling?

5. A covered glass of water sits for days with no drop in water level. Strictly speaking, can you say that nothing has happened, that no evaporation or condensation has taken place? Explain.

6. We say that evaporation is a cooling process. (a) What cools and what warms? We say that condensation is a warming process. (b) What warms and what cools?

7. If all the molecules in a liquid had the same speed, and some were able to evaporate, would the remaining liquid be cooled? Explain.

8. Where does the energy come from that keeps the dunking bird in Figure 16.6 operating?

9. An inventor claims to have developed a new perfume that lasts a long time because it doesn't evaporate. Comment on this claim.

10. Porous canvas bags filled with water are used by travelers in hot weather. When the bags are slung on the outside of a fast-moving car, the water inside is cooled considerably. Explain.

11. Why will wrapping a bottle in a wet cloth at a picnic often produce a cooler bottle than placing the bottle in a bucket of cold water?

12. The human body can maintain its customary temperature of 37°C on a day when the temperature is above 40°C. How is this done?

13. Double-pane windows have nitrogen gas or very dry air between the panes. Why is ordinary air a poor idea?

14. Why are icebergs often surrounded by fog?

15. How does Figure 16.7 help explain the moisture that forms on the inside of car windows when you're parking on a cool night?

16. You know that the windows in your warm house get wet on a cold day. But can moisture form on the windows if the inside of your house is cold on a hot day? How is this different?

17. Why do clouds often form above mountain peaks? (*Hint*: Consider the updrafts.)

18. Why will clouds tend to form above either a flat or a mountainous island in the middle of the ocean? (*Hint*: Compare the specific heat of the land with that of the water and the subsequent convection currents in the air.)

19. A great amount of water vapor changes phase to become liquid water droplets in the clouds that form a thunderstorm. Does this phase change release energy or absorb it?

20. Why does the temperature of boiling water remain the same as long as the heating and boiling continue?

21. Why do vapor bubbles in a pot of boiling water get larger as they rise in the water?

22. Why does the boiling temperature of water increase when the water is under increased pressure?

23. Why does the boiling temperature of water decrease when the water is under reduced pressure, such as at higher altitude?

24. Place a jar of water on a small stand within a saucepan of water so the bottom of the jar is held above the bottom of the pan. When the pan is put on a stove, the water in the pan will boil, but not the water in the jar. Why?

25. Why should you not pick up a hot skillet with a wet dish cloth?

26. Water will boil spontaneously in a vacuum—on the moon, for example. Could you cook an egg in this boiling water? Explain.

27. Our inventor friend proposes a design of cookware that will allow boiling to take place at a temperature less than 100°C so that food can be cooked with less energy consumption. Comment on this idea.

28. If water that boils due to reduced pressure is not hot, then is ice formed by the reduced pressure not cold? Explain.

29. Your instructor hands you a closed flask of room-temperature water. When you hold it, the heat transfer between your bare hands and the flask causes the water to boil. Quite impressive! How is this accomplished?

30. When you boil potatoes, will your cooking time be reduced with vigorously boiling water instead of gently boiling water? (Directions for cooking spaghetti call for vigorously boiling water—not to lessen cooking time, but to prevent something else. If you don't know what it is, ask a cook.)

31. Why does putting a lid over a pot of water on a stove shorten the time it takes for the water to come to a boil, whereas after the water is boiling, use of the lid only slightly shortens the cooking time?

32. In a nuclear submarine power plant, the temperature of the water in the reactor is above 100°C. How is this possible?

33. Explain why the eruptions of many geysers repeat with notable regularity.

34. Why does the water in a car radiator often boil explosively when the radiator cap is removed?

35. Why is very cold ice "sticky"?

36. Would regelation occur if ice crystals were not open-structured? Explain.

37. You'll slip on a smoothly polished floor much more readily than on an unpolished floor, but you'll slip more readily on bumpy ice than on smooth ice. Why is this so? (*Hint*: Why is ice slippery?)

38. People who live where snowfall is common will tell you that air temperatures are higher on snowy days than on clear days. Some misinterpret this by stating that snowfall can't occur on very cold days. Explain this misinterpretation.

39. A piece of metal and an equal mass of wood are both removed from a hot oven at equal temperatures and dropped onto blocks of ice. The metal has a lower specific heat capacity than the wood. Which will melt more ice before cooling to 0°C?

40. How does melting ice change the temperature of surrounding air?

41. Why does dew form on a cold soft drink can?

42. Air-conditioning units contain no water whatsoever, yet it is common to see water dripping from them when they're running on a hot day. Explain.

43. The "dew point" is the temperature at which moisture would condense out of the air if the temperature were lowered to that value. Would you expect the dew point to be higher on a moist summer day or a dry summer day?

44. Some old-timers found that when they wrapped newspaper around the ice in their iceboxes, melting was inhibited. Discuss the advisability of this.

45. Why is it that during cold winters a tub of water placed in a farmer's canning cellar helps prevent canned food from freezing?

46. Why will spraying fruit trees with water before a frost help to protect the fruit from freezing?

47. There are several theories to explain how an ice age might begin. Consider the following one: If the temperature of the world increases, evaporation of the oceans increases, precipitation increases, and snowfall increases. This results in more snow left in cooler places at the end of each summer, with successively increasing depths each winter that become hard-packed from the bottom up to become ice. All the while, the ice reflects more solar radiation that would otherwise be absorbed. This, in turn, cools the earth further and allows for more freezing. Can you continue this sequence and explain how the process reverses itself and how the ice age withdraws?

48. It so happens that drops of water placed in a hot skillet will quickly vaporize away, but drops in a *very* hot skillet will dance around for a longer time before vaporizing away. Can you think of a reason for this?

49. You have a friend who tells you that in exchange for a given sum of money, some people he met taught him techniques of mind over matter that allow him to walk harmlessly with bare feet on red-hot coals. This is no jive, for after walking on wetted grass and then over smoldering wood embers his feet are completely un-harmed. Furthermore, your friend urges that you raise money and learn these wonderful techniques yourself. What is your response?

50. Why does a hot dog pant?

Problems

1. The quantity of heat Q that changes the temperature ΔT of a mass m of a substance is given by $Q = mc\Delta T$, where c is the specific heat capacity of the substance. For example, for H_2O, $c = 1$ cal/g°C. And for a change of phase the quantity of heat Q that changes the phase of a mass m is $Q = mL$, where L is the heat of fusion or heat of vaporization of the substance. For example, for H_2O the heat of fusion is 80 cal/g or 80 kcal/kg, and the heat of vaporization is 540 cal/g or 540 kcal/kg. Use these relationships to determine the number of calories to change (a) 1 kg of 0°C ice to 1 kg 0°C ice water; (b) 1 kg of 0°C ice water to 1 kg 100°C boiling water; (c) 1 kg of 100°C boiling water to 1 kg 100°C steam; and (d) 1 kg of 0°C ice to 1 kg 100°C steam.

2. The specific heat capacity of ice is about 0.5 cal/g°C. Calculate the number of calories it would take to change a 1-gram ice cube at absolute zero (-273°C) to 1 gram of boiling water. How does this number of calories compare to the number of calories required to change the same gram of 100°C boiling water to 100°C steam?

3. Find the mass of 0°C ice that 10 g of 100°C steam will completely melt.

4. Calculate the height that a block of ice at 0°C must be dropped to completely melt upon impact. Assume no air resistance and that all the energy goes into melting the ice. (*Hint:* Equate the joules of gravitational potential energy to the product of the mass of ice and its heat of fusion (in SI units, 335,000 J/kg). Do you see why the answer doesn't depend on mass?)

5. A 10-kg iron ball is dropped onto the pavement from a height of 100 m. If half of the heat generated goes into warming the ball, find the temperature increase of the ball. (In SI units the specific heat capacity of iron is 450 J/kg°C.) Why is the answer the same for any mass ball?

6. The heat of vaporization of ethyl alcohol is about 200 cal/g. If 2 kg of it were allowed to vaporize in a refrigerator, how many grams of ice would be formed from 0°C water?

17

. . . .

Thermodynamics

The colors in the thermogram indicate where heat is escaping from the buildings. Red and orange indicate warm spots; blue and purple indicate cooler areas.

The study of heat and its transformation to mechanical energy is called **thermodynamics** (which stems from Greek words meaning "movement of heat"). The science of thermodynamics was developed in the nineteenth century, before the atomic and molecular nature of matter was understood. Whereas in previous chapters we related heat to the change in energy of jiggling atoms and molecules, in this chapter we see that thermodynamics invokes only macroscopic notions—such as mechanical work, pressure, and temperature—and their roles in energy transformations. Thermodynamics is a powerful theoretical science that bypasses the molecular details of the system altogether. Its foundation is the conservation of energy and the fact that heat flows spontaneously from hot to cold and not the other way around. It provides the basic theory of heat engines, from steam turbines to fusion reactors, and the basic theory of refrigerators and heat pumps. We begin our study of thermodynamics with a look at one of its early concepts—a lowest limit of temperature.

Absolute Zero

In principle, there is no upper limit of temperature. As thermal motion increases, a solid object first melts; with further increase in motion it then vaporizes; as the temperature is further increased, molecules break up into atoms. Still further increase in temperature

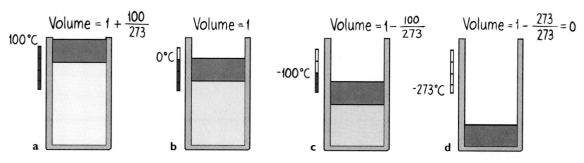

FIGURE 17.1 The volume of a gas changes by $\frac{1}{273}$ its volume at 0°C with each 1°C change in temperature when the pressure is held constant. (a) At 100°C the volume is $\frac{100}{273}$ greater than it is at (b) 0°C. (c) When the temperature is reduced to −100°C, the volume is reduced by $\frac{100}{273}$. (d) At −273°C the volume of the gas would be reduced by $\frac{273}{273}$ and so would be zero.

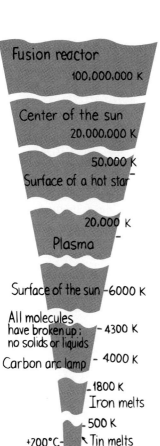

finds atoms losing some or all of their electrons, thereby forming a cloud of electrically charged particles—a plasma. This situation exists in stars, where the temperature is many millions of degrees Celsius.

In contrast, there is a definite limit at the other end of the temperature scale. Gases expand when heated and contract when cooled. Nineteenth-century experiments found that all gases, regardless of their initial pressures or volumes, change by $\frac{1}{273}$ of their volume at 0°C for each degree Celsius change in temperature, provided the pressure is held constant. So if a gas at 0°C were cooled down by 273°C, it would contract, according to this rule, by $\frac{273}{273}$ of its volume and be reduced to zero volume. Clearly, we cannot have a substance with zero volume. Scientists also found that the pressure of any gas in any container of *fixed* volume changes by $\frac{1}{273}$ of its pressure at 0°C for each degree Celsius change in temperature. So a gas in a container of fixed volume cooled 273°C below zero would have no pressure whatsoever. In practice, every gas liquefies before it gets this cold. Nevertheless, these decreases by $\frac{1}{273}$ increments suggested the idea of a lowest temperature: −273°C. So there is a limit to coldness. When atoms and molecules lose all available kinetic energy, they reach the **absolute zero** of temperature.* At absolute zero, no more energy can be extracted from a substance and no further lowering of its temperature is possible. This limiting temperature is actually 273.15° below zero on the Celsius scale (and 459.7° below zero on the Fahrenheit scale).

On the absolute temperature scale, called the Kelvin scale,† absolute zero is 0 K (short for "0 kelvin," rather than "0 degrees kelvin"). There are no minus numbers on the Kelvin scale. Degrees on the Kelvin scale are calibrated with the same-sized divisions as on the Celsius scale. The melting point of ice thus is 273.15 K, and the boiling point of water is 373.15 K.

FIGURE 17.2 Some absolute temperatures.

* Even at absolute zero, molecules still have a small kinetic energy, called the *zero-point energy*. Helium, for example, has enough motion at absolute zero to keep it from freezing. The explanation for this involves quantum theory.

† Named after the famous British physicist Lord Scale, who coined the word *thermodynamics* and was the first to suggest this thermodynamic temperature scale. (Lord *Scale*, really?)

Internal Energy

We all know that there is a vast amount of energy locked in all materials—this book, for example. The pages of this book are composed of molecules that are in constant motion. They have kinetic energy. Due to interactions with neighboring molecules, they also have potential energies. The pages can be easily burned, so we know they store chemical energy, which is really electric potential energy at the molecular level. We know there are vast amounts of energy associated with atomic nuclei. Then there is the "energy of being," described by the celebrated equation $E = mc^2$ (mass energy). Energy within a substance is found in these and other forms, which, when taken together, are called **internal energy**.* The internal energy in even the simplest substance can be quite complex. But in our study of heat changes and heat flow, we will be concerned only with the *changes* in the internal energy of a substance. Changes in temperature indicate these changes in internal energy.

First Law of Thermodynamics

Some two hundred years ago heat was thought to be an invisible fluid called *caloric,* which flowed like water from hot objects to cold objects. Caloric appeared to be conserved—that is, it seemed to flow from one place to another without being created or destroyed. This idea was the forerunner of the law of conservation of energy. By the middle of the nineteenth century, it became apparent that the flow of heat was nothing more than the flow of energy itself. The caloric theory of heat was gradually abandoned.† Today we view heat as energy being transferred from one place to another, usually by molecular collisions. Heat is energy in transit.

When the law of energy conservation is enlarged to include heat, we call it the **first law of thermodynamics**. We state it generally in the following form:

When heat flows to or from a system, the system gains or loses an amount of energy equal to the amount of heat transferred.

Answers

1. Neither. They are equal.
2. A container of helium twice as hot has twice the absolute temperature, or two times 273 K. This would be 546 K, or 273°C. (Simply subtract 273 from the Kelvin temperature to convert to Celsius. Can you see why?)

---

* If this book were poised at the edge of a table and ready to fall, it would possess gravitational potential energy; if it were tossed into the air, it would possess kinetic energy. But these are not examples of internal energy, because they involve more than just the elements of which the book is composed. They include gravitational interactions with the earth and motion with respect to the earth. If we wish to include these forms, we must do so in terms of a larger "system"—a system enlarged to include both the book and the earth. They are not part of the internal energy of the book itself.

† Popular ideas, when proved wrong, are seldom suddenly discarded. People tend to identify with the ideas that characterize their time; hence, it is often the young who are more prone to discover and accept new ideas and push the human adventure forward.

By *system,* we mean a well-defined group of atoms, molecules, particles, or objects. The system may be the steam in a steam engine or it may be the whole earth's atmosphere. It can even be the body of a living creature. The important point is that we must be able to define what is contained within the system and what is outside it. If we add heat to the steam in a steam engine, to the earth's atmosphere, or to the body of a living creature, we will increase the internal energies of these things. And because of this added energy, these things may do work on external things. This added energy does one or both of two things: (1) it increases the internal energy of the system if it remains in the system or (2) it does work on things external to the system if it leaves the system. So, more specifically, the first law states:

Heat added to a system = increase in internal energy + external work done by the system.

Note that the first law is an overall principle that is not concerned with the inner workings of the system itself. Whatever the details of molecular behavior in the system, the added heat energy serves only two functions: It increases the internal energy of the system or it enables the system to do external work (or both). Our ability to describe and predict the behavior of systems that may be too complicated to analyze in terms of atomic and molecular processes is one of the beauties of thermodynamics. Thermodynamics bridges the microscopic and macroscopic worlds.

Put an airtight can of air on a hot stove and heat it up. Since the can is of fixed size, enclosed air does no work on it, and all the heat that goes into the can increases the internal energy of the enclosed air. Its temperature rises. If the can is fitted with a movable piston, then the heated air can do work as it expands and pushes the piston outward. Can you see that this work done means less temperature rise? So when energy is added to a system that does no external work, the amount of heat added is equal to the increase in the internal energy of the system. If the system does external work, the increase in internal energy is correspondingly less. The first law of thermodynamics is simply the thermal version of the law of conservation of energy.

Consider a given amount of energy supplied to a steam engine. The amount supplied will be evident in the increase in the internal energy of the steam and in the mechanical work done. The sum of the increase in internal energy and the work done will equal the energy input. Energy output cannot exceed energy input.

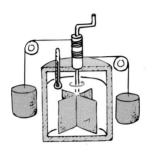

FIGURE 17.3 Paddle-wheel apparatus first used to compare heat with mechanical energy. As the weights fall, they give up potential energy, which is transformed to heat that warms the water. This was first demonstrated by James Joule, for whom the unit of energy is named.

Questions

1. If 100 J of heat is added to a system that does no external work, by how much is the internal energy of that system increased?
2. If 100 J of heat is added to a system that does 40 J of external work, by how much is the internal energy of that system increased?

Answers

1. 100 J.
2. 60 J. We see from the first law that 100 J = 60 J + 40 J.

FIGURE 17.4 Do work on the pump by pressing down on the piston and you compress the air inside. What happens to the temperature of the enclosed air? What happens to its temperature if the air expands and pushes the piston back upward?

There are other ways to add energy to a system than by heat. We can, for instance, do work on a system and produce a result similar to adding heat. A bicycle pump provides a good example. When we pump on the handle, the pump becomes hot. Why? Because we are putting mechanical work into the system and raising its internal energy. If the process happens quickly enough so that very little heat is conducted away, then most of the work input will go into increasing the internal energy, and the system becomes hotter.

Adiabatic Processes

The process of compression or expansion of a gas so that no heat enters or leaves a system is said to be *adiabatic* (from the Greek for "impassable"). **Adiabatic processes** can be achieved either by thermally insulating a system from its surroundings (with Styrofoam, for example) or by performing the process rapidly so that energy has little time to enter or leave. Pumping a bicycle pump is a familiar example. If we set the "heat added" part of the first law to zero, we see that changes in internal energy are equal to the work done on or by the system.* If work is done on a system—compressing it, for example—the internal energy increases. We raise its temperature. If work is done by the system—expanding against its surroundings, for example—the internal energy decreases. It cools.

A common example of an adiabatic process is the compression or expansion of the gases in the cylinders of an automobile engine (Figure 17.5). Compression and expansion occur in only hundredths of a second, too short a time for any appreciable

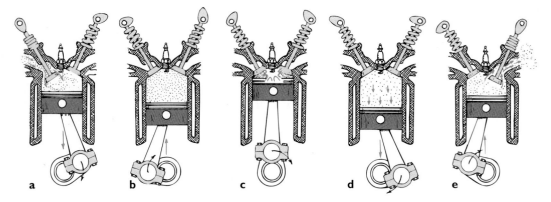

FIGURE 17.5 A four-cycle internal-combustion engine. (a) A fuel-air mixture from the carburetor fills the cylinder as the piston moves down. (b) The piston moves upward and compresses the mixture—adiabatically, since no appreciable heat transfer occurs. (c) The spark plug fires, igniting the mixture, which reaches a high temperature. (d) Adiabatic expansion pushes the piston downward, the power stroke. (e) The burned gases are pushed out the exhaust pipe. Then the intake valve opens and the cycle repeats.

---

* ΔHeat = Δ internal energy + work

            0 = Δ internal energy + work

Then we can say

      $-$Work = Δ internal energy

amount of heat to flow into or out of the combustion chamber. For very high compressions, like those that occur in diesel engines, the temperatures achieved are sufficient to ignite a fuel mixture without the use of spark plugs. Diesel engines have no spark plugs.

So when work is done on a gas by adiabatically compressing it, the gas gains internal energy and becomes warmer. We note this by the warmth of a bicycle pump when air is compressed. When a gas adiabatically expands, it does work on its surroundings, gives up internal energy, and becomes cooler. This can be noted by an experiment you can do right now: With your mouth wide open, blow on your hand. Your breath is warm. Now repeat, but pucker your lips to make a small hole so your breath must expand as it leaves your mouth. Your breath is noticeably cooler! Expanding air cools.

Meteorology and the First Law

Thermodynamics is useful to meteorologists when analyzing weather. Meteorologists express the first law of thermodynamics in the following form:

Air temperature increases as heat is added or as pressure is increased.

The temperature of the air may be changed by adding or subtracting heat, by changing the pressure of the air, or by both. Heat is added by solar radiation, by long-wave earth radiation, by moisture condensation, or by contact with the warm ground. An increase in air temperature results. The atmosphere may lose heat by radiation to space, by evaporation of rain falling through dry air, or by contact with cold surfaces. The result is a drop in air temperature.

There are some atmospheric processes in which the amount of heat added or subtracted is very small—small enough so that the process is nearly adiabatic. Then we have the adiabatic form of the first law:

Air temperature increases (or decreases) as pressure increases (or decreases).

Adiabatic processes in the atmosphere are characteristic of parts of the air, which we'll call *blobs,* which have dimensions on the order of tens of meters to kilometers. These blobs are large enough that outside air doesn't appreciably mix with the air inside them during the minutes to hours that they exist. Blobs behave as if they are enclosed in giant, tissue-light garment bags. As a blob of air flows up the side of a mountain, its pressure lessens, allowing it to expand and cool. The reduced pressure results in reduced temperature.* Measurements show that the temperature of a blob of dry air will decrease by 10°C for a decrease in pressure that corresponds to an increase in altitude of 1 kilometer. So dry air cools 10°C for each kilometer it rises (Figure 17.7). Air flowing over tall mountains or rising in thunderstorms or cyclones may change elevation by several kilometers. Thus, if a blob of dry air at ground level with a comfortable temperature of 25°C were lifted to 6 kilometers, the temperature would be a frigid −35°C. On the other hand, if air at a typical temperature of −20°C at an altitude of 6 kilometers descends to the ground, its temperature would be a whopping 40°C. A dramatic example of this adiabatic warming is the *chinook*—a wind that blows down from the Rocky Mountains across the Great Plains. Cold air moving down the slopes of the mountains

FIGURE 17.6 Blow warm air onto your hand from your wide-open mouth. Now reduce the opening between your lips so the air expands as you blow. Adiabatic expansion—the air is cooled.

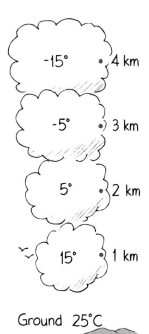

FIGURE 17.7 The temperature of a blob of dry air that expands adiabatically decreases by about 10°C for each kilometer of elevation.

* Recall that in Chapter 15 we treated the cooling of rising air at the microscopic level by considering the behavior of colliding molecules. With thermodynamics, we consider only the macroscopic measurements of temperature and pressure to get the same results. It's nice to analyze things from more than one point of view.

FIGURE 17.8 Chinooks, warm dry winds, occur when high-altitude air descends and is adiabatically warmed.

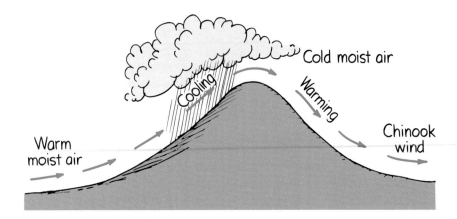

FIGURE 17.9 A thunder head is the result of the rapid adiabatic cooling of a rising mass of moist air. It gets energy from condensation of water vapor.

FIGURE 17.10 The layer of campfire smoke over the lake indicates a temperature inversion. The air near the ground is cooler than the air above it.

is compressed into a smaller volume and is appreciably warmed (Figure 17.8). The effect of expansion or compression on gases is quite impressive.*

A rising blob cools as it expands. But the surrounding air is cooler at increased elevations also. The blob will continue to rise as long as it is warmer (less dense) than the surrounding air. If it gets cooler (denser) than its surroundings, it will sink. Under some conditions, large blobs of cold air sink and remain at a low level, with the result that the air above is warmer. When the upper regions of the atmosphere are warmer than the lower regions, we have a **temperature inversion**. If any rising warm air is less dense than this upper layer of warm air, it will rise no farther. It is common to see evidence of this over a cold lake where visible gases and particles, such as smoke, spread out in a flat layer above the lake rather than rise and dissipate higher in the atmosphere (Figure 17.10). Temperature inversions trap smog and other thermal pollutants. The smog of Los Angeles often is trapped by such an inversion, caused by low-level cold air from the ocean over which is a layer of hot air that moves over the mountains from the hotter Mojave Desert. The mountains aid in holding the trapped air (Figure 17.11). The mountains on the edge of Denver play a similar role in trapping smog beneath a temperature inversion.†

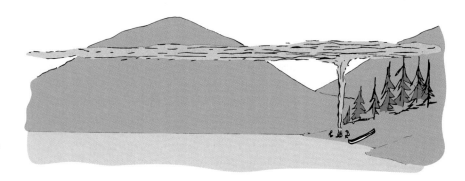

* Interestingly enough, when you're flying at high altitudes where outside air temperature is typically −35°C, you're quite comfortable in your warm cabin—but not because of heaters. The process of compressing outside air to a cabin pressure of nearly sea level would normally heat the air to a roasting 55°C (131°F). So air conditioners must be used to extract heat from the pressurized air.

† Strictly speaking, meteorologists call any temperature profile that thwarts upward convection a *temperature inversion*—including instances where the upper regions of air are cooler rather than warmer.

FIGURE 17.11 Smog in Los Angeles is trapped by the mountains and a temperature inversion caused by warm air from the Mojave Desert overlying cool air from the Pacific Ocean.

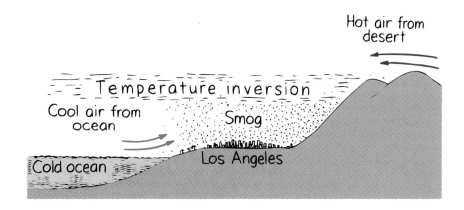

Adiabatic blobs are not restricted to the atmosphere, and changes in these blobs do not necessarily happen quickly. Convection of some deep ocean currents takes thousands of years for circulation. The water masses are so huge and conductivities are so low that no appreciable quantities of heat are transferred to or from these blobs over these long periods of time. They are warmed or cooled adiabatically by changes in internal pressure. Changes in adiabatic ocean convection, as evidenced by the recurring El Niño a few years ago, have a great effect on the earth's climate.* Ocean convection is influenced by the temperature of the ocean floor, which in turn is influenced by convection currents in the molten material that lies beneath the earth's crust (Figure 17.12). Knowledge of the behavior of molten material in the earth's mantle is harder to come by. Once a blob of hot liquid material deep within the mantle begins to rise, will it continue to rise to the earth's crust? Or will its rate of adiabatic cooling find it cooler and denser than its surroundings, at which point it will sink? Is convection self-perpetuating? Geophysical types are currently pondering these questions.

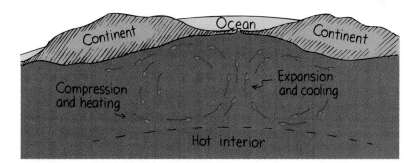

FIGURE 17.12 Do convection currents in the earth's mantle drive the continents as they drift across the global surface? Do rising blobs of molten material cool faster or slower than the surrounding material? Do sinking blobs heat to temperatures above or below those of the surroundings? The answers to these questions are not known at this writing.

---

* El Niño is a recurrent warming (3–7 year cycle) of the ocean surface off the west coast of South America caused by the *lack* of upwelling of cold, nutrient-rich water from the ocean floor. The El Niño results in the dying of plankton and fish and affects weather over the Pacific, North America, and sometimes even worldwide.

Second Law of Thermodynamics

Suppose we place a hot brick next to a cold brick in a thermally insulated region. We know that the hot brick will cool as it gives heat to the cold brick, which will warm. They will arrive at a common temperature: thermal equilibrium. No energy will be lost, in accordance with the first law of thermodynamics. But pretend the hot brick extracts heat from the cold brick and becomes hotter. Would this violate the first law of thermodynamics? Not if the cold brick becomes correspondingly colder so that the combined energy of both bricks remains the same. If this were to happen, it would not violate the first law. But it would violate the **second law of thermodynamics**. The second law distinguishes the direction of energy transformation in natural processes. The second law of thermodynamics can be stated in many ways, but most simply it is this:

Heat will never of itself flow from a cold object to a hot object.

In winter, heat flows from inside a warm heated home to the cold air outside. In summer, heat flows from the hot air outside into the cooler interior. The direction of spontaneous heat flow is from hot to cold. Heat can be made to flow the other way, but only by putting in work or other energy from another source—as occurs with heat pumps and air conditioners, both of which cause heat to flow from a cooler to a warmer place. The huge amount of internal energy in the ocean cannot be used to light a single flashlight lamp without external effort. Energy will not of itself flow from the lower temperature ocean to the higher temperature lamp filament. Without external effort, the direction of heat flow is from hot to cold.

Heat Engines

It is easy to change work completely into heat—simply rub your hands together briskly. The heat that is created then adds to the internal energy of your hands, making them warmer. Or push a crate at constant speed along a floor. All the work you do in over-

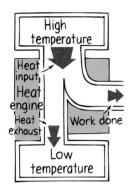

FIGURE 17.13 When heat flows in any heat engine from a high-temperature place to a low-temperature place, part of this heat can be turned into work. (If work is put into a heat engine, the flow of energy may go from a low-temperature to a high-temperature place, as in a refrigerator or air conditioner. Then the arrows in the diagram are reversed.)

coming friction is completely converted to heat, which warms the crate and the floor. But the reverse process, changing heat completely into work, can never occur. The best that can be done is the conversion of some heat to mechanical work. The first heat engine to do this was the steam engine, invented in about 1700.

A **heat engine** is any device that changes internal energy into mechanical work. The basic idea behind a heat engine, whether a steam engine, internal combustion engine, or jet engine, is that mechanical work can be obtained only when heat flows from a high temperature to a low temperature. In every heat engine only some of the heat can be transformed into work.

In considering heat engines, we talk about *reservoirs*. Heat flows out of a high-temperature reservoir and into a low-temperature reservoir. Every heat engine will (1) gain heat from a reservoir of higher temperature, increasing the engine's internal energy; (2) convert some of this energy into mechanical work; and (3) expel the remaining energy as heat to some lower-temperature reservoir, usually called a *sink* (Figure 17.13). In a gasoline engine, for example, (1) the products of burning fuel in the combustion chamber provide the high-temperature reservoir, (2) the hot gases do mechanical work on the piston, and (3) heat is expelled to the environment via the cooling system and the exhaust. The second law tells us that no heat engine can convert all the heat supplied into mechanical energy. Only some of the heat can be transformed into work, with the remainder expelled in the process. Applied to heat engines, the second law may be stated:

> **When work is done by a heat engine running between two temperatures, T_{hot} and T_{cold}, only some of the input heat at T_{hot} can be converted to work, and the rest is expelled at T_{cold}.**

Every heat engine discards some heat, which may be desirable or undesirable. Hot air expelled in a laundry on a cold winter day may be quite desirable, while the same hot air on a hot summer day is something else. When expelled heat is undesirable, we call it *thermal pollution*.

Before scientists understood the second law, many people thought that a very low friction heat engine could convert nearly all the input heat energy to useful work. But not so. In 1824 the French engineer Sadi Carnot analyzed the functioning of a heat engine and made a fundamental discovery. He showed that the greatest fraction of energy input that can be converted to useful work, even under ideal conditions, depends on the temperature difference between the hot reservoir and the cold sink. His equation is

$$\text{Ideal efficiency} = \frac{T_{hot} - T_{cold}}{T_{hot}}$$

where T_{hot} is the temperature of the hot reservoir and T_{cold} the temperature of the cold.* Ideal efficiency depends only on the temperature difference between input and exhaust. Whenever ratios of temperatures are involved, the absolute temperature scale must be used. So T_{hot} and T_{cold} are expressed in kelvins. For example, when the hot reservoir in a steam turbine is 400 K (127°C), and the sink is 300 K (27°C), the ideal efficiency is

$$\frac{400 - 300}{400} = \frac{1}{4}$$

---

* Efficiency = work output/heat input. From energy conservation, heat input = work output + heat that flows out at low temperature (See Figure 17.13). So work output = heat input − heat output. So efficiency = (heat input − heat output)/(heat input). In the ideal case, it can be shown that the ratio (heat out)/(heat in) = T_{cold}/T_{hot}. Then we can say ideal efficiency = $(T_{hot} - T_{cold})/T_{hot}$.

This means that even under ideal conditions, only 25% of the heat provided by the steam can be converted into work, while the remaining 75% is expelled as waste. This is why steam is superheated to high temperatures in steam engines and power plants. The higher the steam temperature driving a motor or turbogenerator, the higher the efficiency of power production. (Increasing operating temperature in the example to 600 K yields an efficiency $(600 - 300)/600 = \frac{1}{2}$; twice the efficiency at 400 K.)

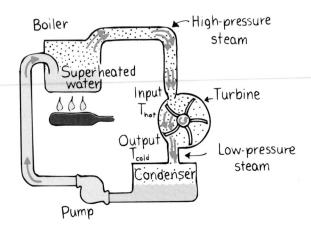

FIGURE 17.14 A simplified steam turbine. The turbine turns because high-temperature steam from the boiler exerts more pressure on the front side of the turbine blades than the low-temperature steam exerts on the back side of the blades. Without a pressure difference, the turbine would not turn and deliver energy to an external load (an electric generator, for example). The presence of steam pressure on the back side of the blades, even in the absence of friction, prevents a perfectly efficient engine.

We can see the role of temperature difference between heat reservoir and sink in the operation of the steam-turbine engine in Figure 17.14. The hot reservoir is steam from the boiler and the cold sink is the exhaust region in the condenser. The hot steam exerts pressure and does work on the blades when it pushes on their front sides. This is nice. But steam pressure is not confined to the front sides of the blades; steam pressure is also exerted on the backs of the blades—counter-effective and not so nice. Pressure on the back sides is reduced most importantly because steam cools after giving much of its energy to the blades (and in practice pressure is further reduced by lowering exhaust temperature via condensation outside the turbine). Even if friction were absent, the turbine's net work output would be the difference in work done on the blades by hot steam and work done by the blades on the cooler steam in exhausting it. We know that with confined steam, temperature and pressure go hand in hand; increase temperature and you increase pressure—decrease temperature and you decrease pressure. So the pressure difference necessary for the operation of a heat engine is directly related to the temperature difference between source and sink. The greater the temperature difference, the greater the efficiency.

Carnot's equation states the upper limit of efficiency for all heat engines. The higher the operating temperature (relative to exhaust temperature) of any heat engine, whether in an ordinary automobile, nuclear-powered ship, or jet aircraft, the higher the efficiency of that engine. In practice, friction is always present in all engines, and efficiency is always less than ideal.* So whereas friction is solely responsible for the inefficiencies of many devices, in the case of heat engines the overriding concept is the second law of thermodynamics; only some of the heat input can be converted to work—even without friction.

---

* The ideal efficiency of an ordinary automobile engine is more than 50%, but in practice the actual efficiency is about 25%. Engines of higher operating temperatures (compared to sink temperatures) would be more efficient, but the melting point of engine materials limits the upper temperatures at which they can operate. Higher efficiencies await engines made with new materials with higher melting points. Watch for ceramic engines!

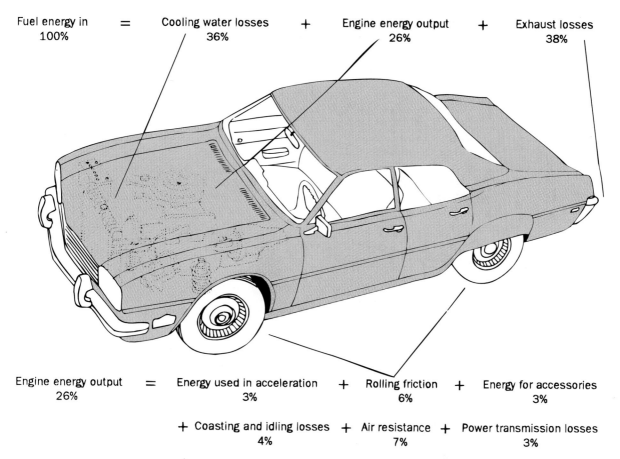

Fuel energy in 100% = Cooling water losses 36% + Engine energy output 26% + Exhaust losses 38%

Engine energy output 26% = Energy used in acceleration 3% + Rolling friction 6% + Energy for accessories 3%

+ Coasting and idling losses 4% + Air resistance 7% + Power transmission losses 3%

FIGURE 17.15 Only 26% of the heat energy produced in burning the gasoline in an automobile is transformed to mechanical energy, and most of that is lost in friction and in overcoming air resistance. The losses shown here are for a typical American automobile and are averages over different driving situations. (Courtesy of the Exploratorium, San Francisco.)

Questions

1. What would be the ideal efficiency of an engine if its hot reservoir and exhaust were the same temperature—say 400 K?
2. What would be the ideal efficiency of a machine having a hot reservoir at 400 K and a cold reservoir somehow maintained at absolute zero, 0 K?

Answers

1. Zero efficiency; $\dfrac{400 - 400}{400} = 0$. So no work output is possible for any heat engine unless a temperature difference exists between the reservoir and the sink.
2. $\dfrac{400 - 0}{400} = 1$; only in this idealized case is an ideal efficiency of 100% possible.

OTEC Power

A non-polluting power plant, presently in the research and development stage off the Kona Coast in Hawaii, is OTEC, the acronym for Ocean Thermal Energy Conversion. An OTEC power plant produces electric power by exploiting the temperature difference between warm surface waters and cold deep waters. More specifically, it is a heat engine that runs on the temperature difference between solar-lit 26°C tropical surface waters and deep dark 4°C waters. Because of the small temperature difference, the efficiency is low.

The low efficiency requires that large amounts of water be circulated. But that can have some advantages. Cold deep waters, saturated with nitrogen and other nutrients locked away from the surface food chain in darkness for centuries, are circulated to the surface. This upwelling of nutrient-rich water exposed to sunlight should produce an explosion of phytoplankton growth comparable to that obtained when fertilizers are added to land crops. So the promise of OTEC is not only pollution-free electric energy, but increased food from the sea.

The operation of OTEC is similar to that of a conventional steam-powered plant. Warm seawater is admitted into a partially evacuated chamber where it boils (recall that the boiling point of water depends on pressure). The expanding "steam," actually a vapor, turns a turbogenerator, after which it is condensed by cold water pumped up from the depths. In an alternative closed-cycle system, a working fluid such as ammonia is vaporized and condensed through heat exchange with the warm and cold seawater, providing a greater pressure difference across the turbine, which allows more straightforward scale-up to commercial-sized plants. Like wind power and hydropower, OTEC power comes ultimately from sunlight. It needs no other fuel. Watch for OTEC in the new century!

FIGURE 17.16 The mini-OTEC power plant pictured here is a heat engine that operates on the temperature difference between warm surface water and cold deep water.

Order Tends to Disorder

The first law of thermodynamics states that energy can be neither created nor destroyed. The second law qualifies this by adding that in transformations, energy "deteriorates" from more useful forms to less useful forms. Energy becomes more diffuse and ultimately degenerates into waste. Another way to say this is that organized energy (concentrated and therefore usable energy) degenerates into disorganized energy (nonusable energy). The energy of gasoline is organized and usable energy. When gasoline burns in a car engine, part of its energy does useful work and part is "thrown away" as wasted heat in the cooling system and exhaust. Even the part that goes into useful work ultimately makes its way into disorganized heat energy as the car encounters friction with the road and the air and the brakes.

Organized energy in the form of electricity that goes to the electric lights in homes and office buildings degenerates into heat energy. This is a principal source of heating in many office buildings where the climate is moderate, such as the Transamerica Pyramid in San Francisco. All the electrical energy in the lamps, even the part that briefly exists in the form of light, is turned into heat energy, which is used to warm the building. This explains why the lights are on most of the time. The energy so generated has no further use. All the heat in the buildings cannot be reused to light a single lamp. Heat, diffused into the environment as thermal energy, is the graveyard of useful energy.

We see that the "quality" of energy is lowered with each transformation. Organized forms of energy tend to turn into disorganized forms. With this broader perspective, the second law can be stated another way:

Natural systems tend to proceed toward a state of greater disorder.

FIGURE 17.17 The Transamerica Pyramid and some other buildings are heated by electric lighting, which is why the lights are on most of the time.

FIGURE 17.18 Push a heavy crate across a rough floor and all your work will go into heating the floor and crate. You have transformed ordered energy into disordered energy.

You finally straighten up your room. Everything is dust-free and in its proper place. A week later it is cluttered again. You stack a pile of pennies on your table, all heads up. Somebody walks by and accidentally bumps into the table and the pennies topple to the floor below, certainly not all heads up. You're on the job at a large company and a co-worker puts up a joke sign that reads, "If you thing thinks are confused now, just wait." These are examples of the tendency of the universe and all that is in it to become disordered. In the broadest sense, that is the message of the second law of thermodynamics. By disordered, we mean more random. For example, molecules of gas all moving in harmony make up an orderly state—and also an unlikely state. On the other hand, molecules of gas moving with haphazard directions and speeds make up a disorderly state—a more probable state. If you remove the lid of a bottle containing some colored gas, the gas molecules escape into the room and disappear from view as they mix with the air molecules (Figure 17.19). They have changed from a more orderly to a less orderly state. You would not expect the reverse to happen; that is, you would not expect the molecules that were released to spontaneously fly back into the bottle and thereby return to the more ordered containment. Processes in which disorder changes spontaneously to order are simply not observed to happen.

FIGURE 17.19 Molecules of gas go readily from the bottle to the air but they won't return spontaneously.

Disordered energy can be changed to ordered energy only at the expense of some organizational effort or work input. For example water in a refrigerator freezes and becomes more ordered because work is put into the refrigeration cycle; gas can be ordered into a small region if a compressor supplied with outside energy does work. But without some outside energy input, processes in which the net effect is an increase in order are not observed in nature.

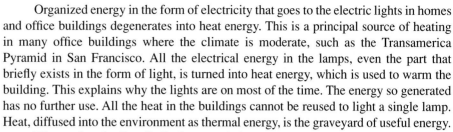

Entropy

The idea of lowering the "quality" of energy is embodied in the idea of **entropy**.* Entropy is the measure of the *amount of disorder.* The second law states that, in the long run, entropy always increases. Gas molecules escaping from a bottle move from a relatively orderly state to a disorderly state. Disorder increases; entropy increases. Whenever a physical system is allowed to distribute its energy freely, it always does so in a manner such that entropy increases while the energy of the system that remains available for doing work decreases.

Consider the old riddle, "How do you unscramble an egg?" The answer is simple: "Feed it to a chicken." But even then you won't get all your original egg back—egg making has its inefficiencies too. All living organisms, from bacteria to trees to human beings, extract energy from their surroundings and use it to increase their own organization. In living organisms, entropy decreases. But the order in life forms is maintained by increasing entropy elsewhere; life forms plus their waste products have a net increase in entropy.[†] Energy must be transformed into the living system to support life. When it isn't, the organism soon dies and tends toward disorder.

FIGURE 17.20 Entropy.

The first law of thermodynamics is a universal law of nature for which no exceptions have been observed. The second law, however, is a probabilistic statement. Given enough time, even the most improbable states may occur; entropy may sometimes decrease. For example, the haphazard motions of air molecules could momentarily be-

* Entropy can be expressed mathematically. The increase in entropy ΔS of a thermodynamic system is equal to the amount of heat added to the system ΔQ divided by the temperature T at which the heat is added: $\Delta S = \Delta Q/T$.

[†] Interestingly enough, the American writer Ralph Waldo Emerson, who lived during the time the second law of thermodynamics was the new science topic of the day, philosophically speculated that not everything becomes more disordered with time and cited the example of human thought. Ideas about the nature of things grow increasingly refined and better organized as they pass through the minds of succeeding generations. Human thought is evolving toward more order.

FIGURE 17.21 Why is the motto of this contractor — "Increasing entropy is our business"—so appropriate?

come harmonious in a corner of the room, just as a barrelful of pennies dumped on the floor could all come up heads. These situations are possible—but not probable. The second law tells us the most probable course of events, not the only possible one.

The laws of thermodynamics are often stated this way: You can't win (because you can't get any more energy out of a system than you put into it), you can't break even (because you can't get as much useful energy out as you put in), and you can't get out of the game (entropy in the universe is always increasing).

Summary of Terms

Thermodynamics The study of heat and its transformation to different forms of energy.

Absolute zero The lowest possible temperature that a substance can have; the temperature at which particles of a substance have their minimum kinetic energy.

Internal energy The total energy (kinetic plus potential) of the submicroscopic particles that make up a substance. *Changes* in internal energy are of principal concern in thermodynamics.

First law of thermodynamics A restatement of the law of energy conservation, applied to systems in which energy is transferred by heat and/or work. The heat added to a system equals its increase in internal energy plus the external work it does on its environment.

Ideal efficiency $= \dfrac{T_{\text{hot}} - T_{\text{cold}}}{T_{\text{hot}}}$

where T_{hot} is the temperature of the hot reservoir and T_{cold} the temperature of the cold sink.

Adiabatic process A process, often of fast expansion or compression, wherein no heat enters or leaves a system.

Temperature inversion A condition wherein upward convection of air is stopped, sometimes because an upper region of the atmosphere is warmer than the region below it.

Second law of thermodynamics Thermal energy never spontaneously flows from a cold object to a hot object. Also, no machine can be completely efficient in converting heat to work; some of the heat supplied to the machine at high temperature is dissipated as waste heat at lower temperature. And finally, all systems tend to become more and more disordered as time goes by.

Heat engine A device that uses heat as input and supplies mechanical work as output, or that uses work as input and moves heat "uphill" from a cooler to a warmer place.

Entropy A measure of the disorder of a system. Whenever energy freely transforms from one form to another, the direction of transformation is toward a state of greater disorder and therefore toward one of greater entropy.

Review Questions

1. From where does the word *thermodynamics* stem?
2. Is the study of thermodynamics concerned primarily with microscopic or macroscopic processes?

Absolute Zero

3. By how much does the volume of gas at 0°C contract for each 1 Celsius degree decrease in temperature when the pressure is held constant?

4. By how much does the pressure of gas at 0°C decrease for each 1 Celsius degree decrease in temperature when the volume is held constant?

5. If we assume the gas does not condense to a liquid, what volume is approached for a gas at 0° cooled by 273 Celsius degrees?

6. What is the lowest possible temperature on the Celsius scale? On the Kelvin scale?

Internal Energy

7. What else besides molecular kinetic energy contributes to the internal energy of a substance?

8. In the study of thermodynamics, is the principal concern the amount of internal energy in a substance or the changes in internal energy in a substance?

First Law of Thermodynamics

9. How does the law of the conservation of energy relate to the first law of thermodynamics?

10. What happens to the internal energy of a system when mechanical work is done on it? What happens to its temperature?

11. What is the relationship between heat added to a system, change in its internal energy, and external work done by the system?

Adiabatic Processes

12. What condition is necessary for a process to be adiabatic?

13. If work is done *on* a system, does the internal energy of the system increase or decrease? If work is done *by* a system, does the internal energy of the system increase or decrease?

Meteorology and the First Law

14. Name at least six ways in which the temperature of the air can change.

15. What generally happens to the temperature of rising air?

16. What generally happens to the temperature of sinking air?

17. What is a chinook?

18. What is a temperature inversion?

19. Do adiabatic processes apply to water masses in the ocean? Defend your answer.

Second Law of Thermodynamics

20. How does the second law of thermodynamics relate to the direction of heat flow?

Heat Engines

21. What three processes occur in every heat engine?

22. What exactly is thermal pollution?

23. If all friction could be removed from a heat engine, would it be 100% efficient? Explain.

Order Tends to Disorder

24. What does it mean to say that when energy is transformed, it becomes less useful?

25. Give at least two examples to distinguish between organized energy and disorganized energy.

26. How much of the electrical energy transformed by a common light bulb becomes heat energy?

27. With respect to orderly and disorderly states, what do natural systems tend to do? Can a disorderly state ever transform to an orderly state? Explain.

Entropy

28. What is the physicist's term for *measure of messiness?*

29. What is the relationship between the second law of thermodynamics and entropy?

30. Distinguish between the first and second laws of thermodynamics in terms of whether or not exceptions occur.

Exercises

1. A friend said the temperature inside a certain oven is 500 and the temperature inside a certain star is 50,000. You're unsure about whether your friend meant Celsius degrees or kelvins. How much difference does it make in each case?

2. The temperature of the sun's interior is about 10^7 degrees. Does it matter whether this is degrees Celsius or kelvins? Explain.

3. Helium has the special property that its internal energy is directly proportional to its absolute temperature. Consider a flask of helium with a temperature of 10°C. If it is heated until it has twice the internal energy, what will its temperature be?

4. If you vigorously shake a can of liquid back and forth for more than a minute, will the temperature of the liquid increase? (Try it and see.)

5. When air is quickly compressed, why does its temperature increase?

6. When you pump a tire with a bicycle pump, the cylinder of the pump becomes hot. Give two reasons why this is so.

7. Is it possible to wholly convert a given amount of mechanical energy into thermal energy? Is it possible to wholly convert a given amount of thermal energy into mechanical energy? Cite examples to illustrate your answers.

8. Why do diesel engines need no spark plugs?

9. Everybody knows that warm air rises. So it might seem that the air temperature should be higher at the top of mountains than down below. But the opposite is most often the case. Why?

10. What is the original source of the energy in coal, oil, and wood?

11. What is the original source of the energy in a hydroelectric power plant?

12. What is the original source of the energy in an OTEC power plant?

13. We say that energy "deteriorates" in energy transformations. Does this contradict the first law of thermodynamics? Defend your answer.

14. The combined molecular kinetic energies of molecules in a very large container of cold water are greater than the combined molecular kinetic energies in a cup of hot tea. Pretend you partially immerse the teacup in the cold water and that the tea absorbs 10 units of energy from the water and becomes hotter, while the water that gives up 10 units of energy becomes cooler. Would this energy transfer violate the first law of thermodynamics? The second law of thermodynamics? Explain.

15. Why is *thermal pollution* a relative term?

16. Is it possible to construct a heat engine that produces no thermal pollution? Defend your answer.

17. Why is it advantageous to use steam as hot as possible in a steam-driven turbine?

18. How does the ideal efficiency of an automobile relate to the temperature of the engine and the temperature of the environment in which it runs? Be specific.

19. Will the efficiency of a car engine increase, decrease, or remain the same if the muffler is removed? If driven on a very cold day? Defend your answers.

20. What happens to the efficiency of a heat engine when the temperature of the reservoir into which thermal energy is transferred is lowered?

21. To increase the efficiency of a heat engine, would it be better to produce the same temperature increment by increasing the temperature of the reservoir while holding the temperature of the sink constant, or to decrease the temperature of the sink while holding the temperature of the reservoir constant? Explain.

22. Under what conditions would a heat engine be 100% efficient?

23. Could you cool a kitchen by leaving the refrigerator door open and closing the kitchen door and windows? Explain.

24. Could you warm a kitchen by leaving the hot oven door open? Explain.

25. An electric fan not only doesn't decrease the temperature of air, but actually increases air temperature. How then, are you cooled by a fan on a hot day?

26. Why will a refrigerator with a fixed amount of food consume more energy in a warm room than in a cold room?

27. Although the efficiency of an OTEC power plant is very small compared with the efficiency of fossil fuel and nuclear power plants, this isn't a serious shortcoming. Why not?

28. In buildings that are being electrically heated, is it at all wasteful to turn all the lights on? Is turning all the lights on wasteful if the building is being cooled by air conditioning?

29. Using both the first and the second laws of thermodynamics, defend the statement that 100% of the electrical energy that goes into lighting a lamp is converted to thermal energy.

30. Molecules in the combustion chamber of a rocket engine are in a high state of random motion. When the molecules leave through a nozzle in a more ordered state, will their temperature be more, less, or the same as their temperature in the chamber before being exhausted?

31. A wet bathing suit spontaneously chills itself (and its occupant). How can this happen without violating the second law of thermodynamics?

32. Is the total energy of the universe becoming more unavailable with time? Explain.

33. Comment on this statement: The second law of thermodynamics is one of the most fundamental laws of nature, yet it is not an exact law at all.

34. Water evaporates from a salt solution and leaves behind salt crystals that have a higher degree of molecular order than the more randomly moving molecules in the salt water. Has the entropy principle been violated? Why or why not?

35. Water put into a freezer compartment in your refrigerator goes to a state of less molecular disorder when it freezes. Is this an exception to the entropy principle? Explain.

36. As a chicken grows from an egg, it becomes more ordered with time. Does this violate the principle of entropy? Explain.

37. The Patent Office rejects claims for perpetual motion machines (in which the energy out is as great or greater than the energy in) without even studying them. Why is this?

38. (a) If you spent ten minutes repeatedly shaking and throwing down a pair of coins, would you expect to see two heads come up at least once? (b) If you spent an hour shaking a handful of ten coins and throwing them down, would you expect to see all ten come up heads at least once? (c) If you stirred a box of 10,000 coins and dumped them repeatedly on the floor all day long, would you expect to see all 10,000 come up heads at least once?

39. In your bedroom are probably some 10^{27} air molecules. If they all happened to congregate on the opposite side of the room, you could suffocate. But this is unlikely. Is such a circumstance less likely, more likely, or the same if there are many times fewer molecules in the room?

40. Make up two multiple-choice questions that would check a classmate's understanding of the distinction between heat and internal energy.

Problems

1. Calculate the Carnot efficiency of an OTEC power plant that operates on the temperature difference between deep 4°C water and 25°C surface water.

2. On a chilly 10°C day, your friend who likes cold weather says she wishes it were twice as cold. Taking this literally, what temperature would this be?

3. Imagine a giant dry-cleaner's bag full of air at a temperature of −35°C floating like a balloon with a string hanging from it 10 km above the ground. Estimate its temperature if you were able to yank it suddenly to the earth's surface.

4. A power station with an efficiency of 0.4 generates 10^8 W of electric power and dissipates 1.5×10^8 W of thermal energy to the cooling water that flows through it. Knowing the specific heat of water in SI units is 4184 J/kg°C, calculate how many kg of water flow through the plant each second if the water is heated through 3 Celsius degrees.

5. A "heat pump" moves heat from a cooler to a warmer place, and is the heart of a refrigerator or air conditioner; it is sometimes used to heat houses, too. The minimum work input needed to move energy "uphill" from T_{cold} to T_{hot} is

$$\text{Min work} = (\text{Energy moved}) \times \frac{T_{hot} - T_{cold}}{T_{cold}}.$$

Calculate the minimum work needed to move 1 J of energy (a) from inside a room at $T_{cold} = 295$ K to the outdoors at $T_{hot} = 308$ K, (b) from inside a lab freezer at $T_{cold} = 173$ K to a room at $T_{hot} = 293$ K, and (c) from a helium refrigerator with an inside temperature at $T_{cold} = 4$ K to a room at $T_{hot} = 300$ K. Comment on the differences.

6. Construct a table of all the possible combinations of numbers that can come up when you throw two dice. Your friend says, "Yes, I know that seven is the most likely total number when two dice are thrown. But *why* seven?" Based on your table, answer your friend, and explain that in thermodynamics the situations likely to be observed are those that can be formed in the most different ways.

P a r t

4

SOUND

This CD is the pits -- billions of them, carefully inscribed in an array that is scanned by a laser beam millions of pits per second. It's the sequence of pits detected as light and dark spots that forms a binary code that is converted into a continuous audio wave form. Digitized music! Who ever thought that something as complex as Beethoven's Fifth Symphony could be reduced to a series of ones and zeros?

18

Vibrations and Waves

Overlapping waves.

Many things in nature wiggle and jiggle. We call a wiggle in time a *vibration*. A vibration cannot exist in an instant, for the body that vibrates needs time to move to and fro. When we strike a bell the vibrations continue for some time before they die down. We call a wiggle in space and time a *wave*. A wave cannot exist in one place but must extend from one place to another. Light and sound are both vibrations that propagate through space as waves, but of two very different kinds. Sound is the propagation of vibrations through a material medium—a solid, liquid, or gas. If there is no medium to vibrate, then no sound is possible. On the other hand, light is a vibration not of matter, but of nonmaterial electric and magnetic fields. Light can pass through many materials, but it needs none; it can propagate through a vacuum, for example between the sun and the earth. But the source of all waves—sound, light, or whatever—is something that is vibrating. For sound it can be the vibrating prongs of a tuning fork and for light the vibrations of electrons in an atom. Waves may be periodic vibrations, as in music, or nonperiodic wave pulses, as in the cracking sound of a firecracker or the snapping of one's fingers. We shall begin our study of vibrations and waves by considering the motion of a simple pendulum.

Vibration of a Pendulum

If we suspend a stone at the end of a piece of string, we have a simple pendulum. Pendulums swing to and fro with such regularity that for a long time they were used to control the motion of most clocks. They can still be found in grandfather clocks and cuckoo clocks. Galileo discovered that the time a pendulum takes to swing to and fro through small distances depends only on the *length of the pendulum.** The time of a to-and-fro swing, called the **period**, does not depend on the mass of the pendulum or on the size of the arc through which it swings.

In addition to length, the period of a pendulum depends on the acceleration of gravity. Oil and mineral prospectors use very sensitive pendulums to detect slight differences in this acceleration, which is affected by the densities of underlying formations.

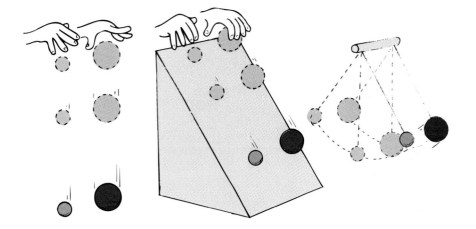

FIGURE 18.1 Drop two balls of different mass and they accelerate at *g*. Let them slide without friction down the same incline and they slide together at the same fraction of *g*. Tie them to strings of the same length so they are pendulums, and they swing to and fro in unison. In all cases, the motions are independent of mass.

A long pendulum has a longer period than a shorter pendulum; that is, it swings to and fro less frequently than a short pendulum. When walking, we allow our legs to swing with the help of gravity, like a pendulum. In the same way that a long pendulum has a greater period, a person with long legs tends to walk with a slower stride than a person with short legs. This is most noticeable in long-legged animals such as giraffes, horses, and ostriches, which run with a slower gait than do short-legged animals such as dachshunds, hamsters, and mice.

Wave Description

The to-and-fro vibratory motion (often called *oscillatory* motion) of a swinging pendulum in a small arc is called **simple harmonic motion.**† The pendulum bob filled with sand in Figure 18.2 exhibits simple harmonic motion above a conveyor belt. When the

---

* The exact equation for the period of a simple pendulum is $T = 2\pi\sqrt{l/g}$, where T is the period, l is the length of the pendulum, and g is the acceleration of gravity.

† The condition for simple harmonic motion is that the restoring force is proportional to the displacement from equilibrium. This condition is nearly always met, at least approximately, for most vibrations. The component of weight that restores a displaced pendulum to its equilibrium position is directly proportional to the pendulum's displacement (for small angles)—likewise for a bob attached to a spring. Recall from page 202 in Chapter 11, Hooke's law for a spring: $F = k\Delta x$, where the force to stretch (or compress) a spring is directly proportional to the distance stretched (or compressed).

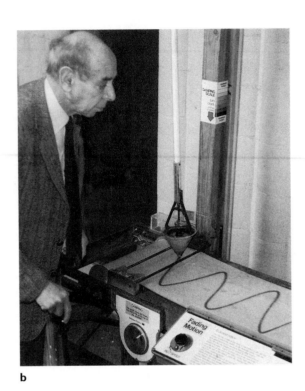

a   b

FIGURE 18.2 Frank
Oppenheimer at the San
Francisco Exploratorium
demonstrates (a) a straight
line traced by a swinging
pendulum bob that leaks
sand on the stationary
conveyor belt. (b) When the
conveyor belt is uniformly
moving, a sine curve is
traced.

conveyor belt is not moving (left), the sand traces out a straight line. More interestingly,
when the conveyor belt is moving at constant speed (right), the sand traces out a special
curve known as a **sine curve**.

A sine curve can also be traced by a bob attached to a spring undergoing vertical
simple harmonic motion (Figure 18.3). A marking pen attached to the bob traces a sine
curve on a sheet of paper that is moved horizontally at constant speed. A sine curve is a
pictorial representation of a wave. Like a water wave, the high points of a sine wave are
called *crests,* and the low points are called *troughs.* The straight dashed line represents
the "home" position, or midpoint of the vibration. The term **amplitude** refers to the distance from the midpoint to the crest (or trough) of the wave. So the amplitude equals the
maximum displacement from equilibrium.

FIGURE 18.3 A sine
curve.

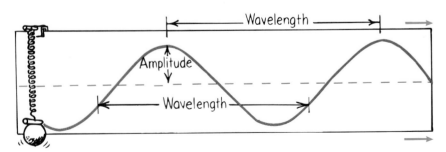

The **wavelength** of a wave is the distance from the top of one crest to the top of
the next one. Or equivalently, the wavelength is the distance between any successive
identical parts of the wave. The wavelengths of waves at the beach are measured in meters, the wavelengths of ripples in a pond in centimeters, and the wavelengths of light in
billionths of a meter (nanometers).

How frequently a vibration occurs is described by its **frequency**. The frequency of a vibrating pendulum, or object on a spring, specifies the number of to-and-fro vibrations it makes in a given time (usually one second). A complete to-and-fro oscillation is one vibration. If it occurs in one second, the frequency is one vibration per second. If two vibrations occur in one second, the frequency is two vibrations per second, and so forth.

The unit of frequency is called the **hertz** (Hz), after Heinrich Hertz, who demonstrated radio waves in 1886. One vibration per second is 1 hertz; two vibrations per second is 2 hertz, and so on. Higher frequencies are measured in kilohertz (kHz, thousands of hertz), and still higher frequencies in megahertz (MHz, millions of hertz) or gigahertz (GHz—billions of hertz). AM radio waves are measured in kilohertz, while FM radio waves are measured in megahertz; radar and microwave ovens operate at gigahertz frequencies. A station at 960 kHz on the AM radio dial, for example, broadcasts radio waves that have a frequency of 960,000 vibrations per second. A station at 101.7 MHz on the FM dial broadcasts radio waves with a frequency of 101,700,000 hertz. These radio-wave frequencies are the frequencies at which electrons are forced to vibrate in the antenna of a radio station's transmitting tower. The source of all waves is something that vibrates. The frequency of the vibrating source and the frequency of the wave it produces are the same.

FIGURE 18.4 Electrons in the transmitting antenna vibrate 940,000 times each second and produce 940-kHz radio waves.

If an object's frequency is known, its period can be calculated, and vice versa. Suppose, for example, that a pendulum makes two vibrations in one second. Its frequency is 2 Hz. The time needed to complete one vibration—that is, the period of vibration—is $\frac{1}{2}$ second. Or if the vibration frequency is 3 Hz, then the period is $\frac{1}{3}$ second. The frequency and period are the inverse of each other:

$$\text{Frequency} = \frac{1}{\text{period}}$$

or vice versa:

$$\text{Period} = \frac{1}{\text{frequency}}$$

Questions

1. What is the frequency in vibrations per second of a 60-Hz wave? What is its period?
2. Gusts of wind make the Sears Building in Chicago sway back and forth at a vibration frequency of about 0.1 Hz. What is its period of vibration?

Answers

1. A 60-Hz wave vibrates 60 times per second and has a period of $\frac{1}{60}$ second.
2. The period is 1/frequency = 1/(0.1 Hz) = 1/(0.1 vibration/s) = 10 s. Each vibration therefore takes 10 seconds.

Wave Motion

Most information about our surroundings comes to us in some form of waves. It is through wave motion that sounds come to our ears, light to our eyes, and electromagnetic signals to our radios and television sets. Through *wave motion,* energy can be transferred from a source to a receiver without the transfer of matter between the two points.

Wave motion can be most easily understood by first considering its simplest case in a horizontally stretched rope. If one end of such a rope is shaken up and down, a rhythmic disturbance travels along the rope. Each particle of the rope moves up and down, while at the same time the disturbance moves along the length of the rope. The medium, rope or whatever, returns to its initial condition after the disturbance has passed.

Perhaps a more familiar example of wave motion is provided by a water wave. If a stone is dropped into a quiet pond, waves will travel outward in expanding circles, the centers of which are at the source of the disturbance. In this case we might think that water is being transported with the waves, since water is splashed onto previously dry ground when the waves meet the shore. We should realize, however, that barring obstacles the water will run back into the pond, and things will be much as they were in the beginning: The surface of the water will have been disturbed, but the water itself will have gone nowhere. A leaf on the surface will bob up and down as the waves pass, but will end up where it started. Again, the medium returns to its initial condition after the disturbance has passed.

Let us consider another example of a wave to illustrate that what is transported from one place to another is a disturbance in a medium, not the medium itself. If you view a field of tall grass from an elevated position on a gusty day, you will see waves travel across the grass. The individual stems of grass do not leave their places; instead, they swing to and fro. Furthermore, if you stand in a narrow footpath, the grass that blows over the edge of the path, brushing against your legs, is very much like the water that doused the shore in our earlier example. While wave motion continues, the tall grass swings back and forth, vibrating between definite limits but going nowhere. When the wave motion stops, the grass returns to its initial position.

Wave Speed

The speed of periodic wave motion is related to the frequency and wavelength of the waves. We can understand this by considering the simple case of water waves (Figures 18.5 and 18.6). Imagine that we fix our eyes at a stationary point on the surface of water and observe the waves passing by this point. We can measure how much time passes between the arrival of one crest and the arrival of the next one (the period), and also observe the distance between crests (the wavelength). We know that speed is defined as distance divided by time. In this case, the distance is one wavelength and the time is one period, so wave speed = wavelength/period.

FIGURE 18.5 Water waves.

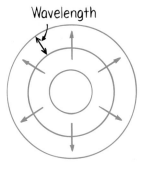

FIGURE 18.6 A top view of water waves.

FIGURE 18.7 If the wavelength is 1 m, and one wavelength per second passes the pole, then the speed of the wave is 1 m/s.

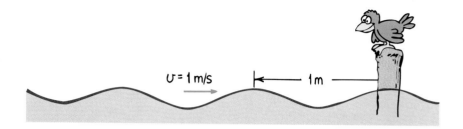

For example, if the wavelength is 10 meters and the time between crests at a point on the surface is 0.5 second, the wave moves 10 meters in 0.5 seconds and its speed is 10 meters divided by 0.5 seconds, or 20 meters per second.

Since period is equal to the inverse of frequency, the formula wave speed = wavelength/period can also be written

Wave speed = wavelength × frequency

This relationship holds true for all kinds of waves, whether they are water waves, sound waves, or light waves.

Questions

1. If a train of freight cars, each 10 m long, rolls by you at the rate of three cars each second, what is the speed of the train?

2. If a water wave oscillates up and down three times each second and the distance between wave crests is 2 m, what is its frequency? Its wavelength? Its wave speed?

Transverse Waves

Fasten one end of a rope to a wall and hold the free end in your hand. If you suddenly twitch the free end up and then down, a pulse will travel along the rope and back (Figure 18.8). In this case the motion of the rope (up and down arrows) is at right angles to the direction of wave speed. The right-angled, or sideways, motion is called *transverse motion*. Now shake the rope with a regular, continuing up-and-down motion, and the series

FIGURE 18.8 A transverse wave.

Answers

1. 30 m/s. We can see this in two ways. (a) According to the speed definition from Chapter 2, $v = d/t = 3 \times 10$ m/1 s = 30 m/s, since 30 m of train passes you in 1 s. (b) If we compare our train to wave motion, where wavelength corresponds to 10 m and frequency is 3 Hz, then Speed = wavelength × frequency = 10 m × 3 Hz = 10 m × 3/s = 30 m/s.

2. The frequency of the wave is 3 Hz, its wavelength is 2 m, and its wave speed = wavelength × frequency = 2 m × 3/s = 6 m/s. It is customary to express this as the equation $v = \lambda f$ where v is wave speed, λ (the Greek letter lambda) is wavelength, and f is wave frequency.

of pulses will produce a wave. Since the motion of the medium (the rope in this case) is transverse to the direction the wave travels, this type of wave is called a **transverse wave**.

Waves in the stretched strings of musical instruments and upon the surfaces of liquids are transverse. We will see later that electromagnetic waves, which make up radio waves and light, are also transverse.

Longitudinal Waves

Not all waves are transverse. Sometimes parts that make up a medium move to and fro in the same direction in which the wave travels. Motion is *along* the direction of the wave rather than at right angles to it. This produces a **longitudinal wave**.

Both a transverse and a longitudinal wave can be demonstrated with a spring or a Slinky stretched out on the floor, as shown in Figure 18.9. A transverse wave is demonstrated by shaking the end of a Slinky up and down. A longitudinal wave is demonstrated by rapidly pulling and pushing the end of the Slinky toward and away from you. In this case we see that the medium vibrates parallel to the direction of energy transfer. Part of the Slinky is compressed, and a wave of *compression* travels along the spring. Between successive compressions is a stretched region, called a *rarefaction*. Both compressions and rarefactions travel in the same direction along the Slinky. Sound waves are longitudinal waves.

FIGURE 18.9 Both waves transfer energy from left to right. (a) When the end of the Slinky is pushed and pulled forward and back, a longitudinal wave is produced. (b) When it's shaken up and down, a transverse wave is produced.

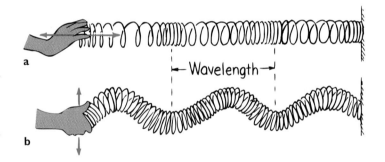

Waves that travel in the ground generated by earthquakes are of two main types: longitudinal P waves, and transverse S waves. (Geology students often remember P waves as "push-pull" waves, and S waves as "side-to-side" waves.) S waves cannot travel through liquid matter, while P waves can travel through both molten and solid parts of the earth's interior. Study of these waves reveals much about the earth's interior.

FIGURE 18.10 Waves generated by an earthquake. P waves are longitudinal and travel through both molten and solid materials. S waves are transverse and travel only through solid materials. Reflections and refractions of the waves provide information about the earth's interior.

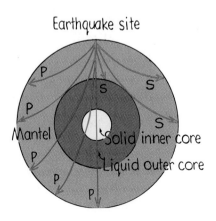

The wavelength of a longitudinal wave is the distance between successive compressions or equivalently, the distance between successive rarefactions. The most common example of longitudinal waves is sound in air. Elements of air vibrate to and fro about some equilibrium position as the waves move by. We will treat sound waves in detail in the next chapter.

Interference

Whereas a material object like a rock will not share its space with another rock, more than one vibration or wave can exist at the same time in the same space. If we drop two rocks in water, the waves produced by each can overlap and form an **interference pattern**. Within the pattern, wave effects may be increased, decreased, or neutralized.

When more than one wave occupies the same space at the same time, the displacements add at every point. This is the *superposition principle*. So when the crest of one wave overlaps the crest of another, their individual effects add together to produce a wave of increased amplitude. This is called *constructive interference* (Figure 18.11). When the crest of one wave overlaps the trough of another, their individual effects are reduced. The high part of one wave simply fills in the low part of another. This is called *destructive interference*.

FIGURE 18.11
Constructive and destructive interference in a transverse wave.

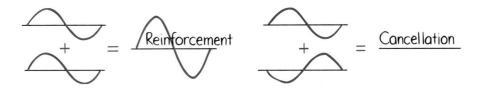

Wave interference is easiest to see in water. In Figure 18.12 we see the interference pattern made when two vibrating objects touch the surface of water. We can see the regions where the crest of one wave overlaps the trough of another to produce regions of zero amplitude. At points along these regions, the waves arrive out of step. We say they are *out of phase* with each other.

Interference is characteristic of all wave motion, whether the waves are water waves, sound waves, or light waves. We will treat the interference of sound in the next chapter and the interference of light in Chapter 28.

FIGURE 18.12 Two sets of overlapping water waves produce an interference pattern. The left diagram is an idealized drawing of the expanding waves from the two sources. The right diagram is a photograph of an actual interference pattern.

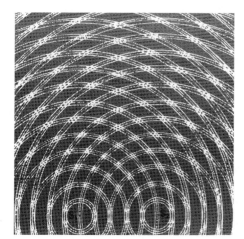
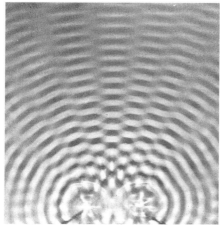

Standing Waves

If we tie a rope to a wall and shake the free end up and down, we produce a train of waves in the rope. The wall is too rigid to shake, so the waves are reflected back along the rope. By shaking the rope just right, we can cause the incident and reflected waves to form a **standing wave**, where parts of the rope, called the *nodes,* are stationary. Nodes are the regions of minimal or zero energy, whereas *antinodes* (not labeled in Figure 18.13) are the regions of maximum displacement and maximum energy. You can hold your fingers on either side of the nodes and the rope doesn't touch them. Other parts of the rope, especially the antinodes, would make contact with your fingers. Antinodes occur halfway between nodes.

FIGURE 18.13 The incident and reflected waves interfere to produce a standing wave.

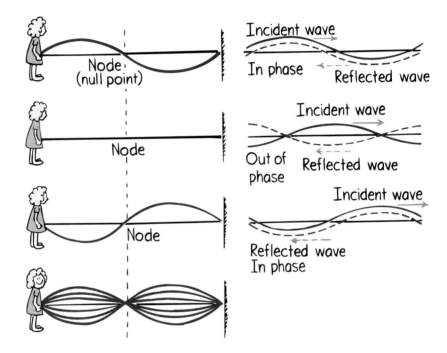

Standing waves are the result of interference (and as we will see in the next chapter, *resonance*). When two sets of waves of equal amplitude and wavelength pass through each other in opposite directions, the waves are steadily in and out of phase with each other. They produce stable regions of constructive and destructive interference.

It is easy to make standing waves yourself. Tie a rope, or better, a rubber tube between two firm supports. Shake the tube up and down with your hand near one of the supports. If you shake the tube with the right frequency, you will set up a standing wave as shown in Figure 18.14a. Shake the tube with twice the frequency, and a standing wave of half the previous wavelength, having two loops, will result. (The distance between successive nodes is a half wavelength; two loops make up a full wavelength.) Triple the frequency, and a standing wave with one-third the original wavelength, having three loops, results, and so forth.

FIGURE 18.14 (a) Shake the rope until you set up a standing wave of one segment ($\frac{1}{2}$ wavelength). (b) Shake with twice the frequency and produce a wave with two segments (1 wavelength). (c) Shake with three times the frequency and produce three segments ($1\frac{1}{2}$ wavelengths).

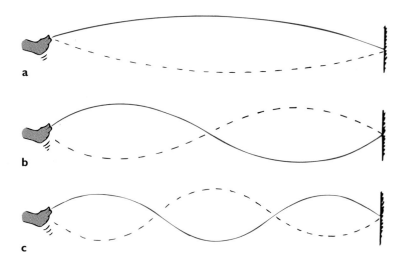

Standing waves are set up in the strings of musical instruments when plucked, bowed, or struck. They are set up in the air in an organ pipe, a trumpet, or a clarinet, and the air of a soda-pop bottle when air is blown over the top. Standing waves can be set up in a tub of water or a cup of coffee by sloshing it back and forth with the right frequency. Standing waves can be produced with either transverse or longitudinal vibrations.

Questions

1. Is it possible for one wave to cancel another wave so that no amplitude remains?
2. Suppose you set up a standing wave of three segments, as shown in Figure 18.14c. If you shake with twice as much frequency, how many wave segments will occur in your new standing wave? How many wavelengths?

Doppler Effect

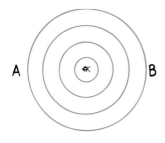

FIGURE 18.15 Top view of water waves made by a stationary bug jiggling in still water.

A pattern of water waves produced by a bug jiggling its legs and bobbing up and down in the middle of a quiet puddle is shown in Figure 18.15. The bug is not going anywhere but is merely treading water in a fixed position. The waves it makes spread over the surface of the water. The waves form concentric circles, because wave speed is the same in all directions. If the bug bobs in the water at a constant frequency, the distance between wave crests (the wavelength) is the same in all directions. Waves encounter point A as frequently as they encounter point B. This means that the frequency of wave motion is

Answers

1. Yes. This is called destructive interference. In a standing wave in a rope, for example, parts of the rope have no amplitude—the nodes.
2. If you impart twice the frequency to the rope, you'll produce a standing wave with twice as many segments. You'll have six segments. Since a full wavelength has two segments, you'll have three complete wavelengths in your standing wave.

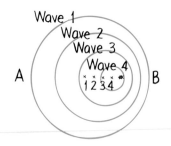

FIGURE 18.16 Water waves made by a bug swimming in still water.

the same at points A and B, or anywhere in the vicinity of the bug. This wave frequency is the same as the bobbing frequency of the bug.

Suppose the jiggling bug moves across the water at a speed less than the wave speed. In effect, the bug chases part of the waves it has produced. The wave pattern is distorted and is no longer made of concentric circles (Figure 18.16). The center of the outer wave was made when the bug was at the center of that circle. The center of the next smaller wave was made when the bug was at the center of that circle, and so forth. The centers of the circular waves move in the direction of the swimming bug. Although the bug maintains the same bobbing frequency as before, an observer at B would see the waves coming more often. The observer would measure a higher frequency. This is because each successive wave has a shorter distance to travel and therefore arrives at B more frequently than if the bug weren't moving toward B. An observer at A, on the other hand, measures a *lower* frequency because of the longer time between wave-crest arrivals. This is because, to reach A, each crest has to travel farther than the one ahead of it due to the bug's motion. This change in frequency due to the motion of the source (or receiver) is called the **Doppler effect** (after the Austrian scientist Christian Doppler, 1803–1853).

Water waves spread over the flat surface of the water. Sound and light waves, on the other hand, travel in three-dimensional space in all directions like an expanding balloon. Just as circular waves are closer together in front of the swimming bug, spherical sound or light waves ahead of a moving source are closer together and reach a receiver more frequently.

The Doppler effect is evident when you hear the changing pitch of a car horn as the car passes you. When the car approaches, the pitch is higher than normal (higher like a higher note on a musical scale). This is because the crests of the sound waves are hitting your ear more frequently. And when the car passes and moves away, you hear a drop in pitch because the crests of the waves are hitting your ear less frequently.

FIGURE 18.17 The pitch of sound gets higher as the source moves toward you; the pitch gets lower as the source moves away.

The Doppler effect also occurs for light. When a light source approaches, there is an increase in its measured frequency; and when it recedes, there is a decrease in its frequency. An increase in frequency is called a *blue shift*, because the increase is toward the high frequency, or blue, end of the color spectrum. A decrease in frequency is called a *red shift*, referring to a shift toward the lower-frequency, or red, end of the color spectrum. Distant galaxies, for example, show a red shift in the light they emit. A measurement of this shift permits a calculation of their speeds of recession. A rapidly spinning star shows a red shift on the side turning away from us and a blue shift on the side turning toward us. This enables a calculation of the star's spin rate.

Question When a sound source moves toward you, do you measure an increase or decrease in wave speed?

Bow Waves

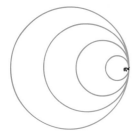

FIGURE 18.18 Wave pattern made by a bug swimming at wave speed.

When the speed of a source is as great as the speed of the waves it produces, something interesting happens. The waves pile up in front of the source. Consider the bug in our previous example when it swims as fast as the wave speed. Can you see that the bug will keep up with the waves it produces? Instead of the waves moving ahead of the bug, they superimpose and hump up on one another directly in front of the bug (Figure 18.18). The bug moves right along with the leading edge of the waves it is producing.

A similar thing happens when an aircraft travels at the speed of sound. In the early days of jet aircraft, it was believed that this pile-up of sound waves in front of the air-plane imposed a "sound barrier" and that to go faster than the speed of sound, the plane would have to "break the sound barrier." What actually happens is that the overlapping wave crests disrupt the flow of air over the wings, making it more difficult to control the craft. But the barrier is not real. Just as a boat can easily travel faster than the waves it produces, with sufficient power an aircraft easily travels faster than the speed of sound. Then we say that it is *supersonic*. A supersonic airplane flies into smooth, undisturbed air because no sound wave can propagate out in front of it. Similarly, a bug swimming faster than the speed of water waves finds itself always entering into water with a smooth, unrippled surface.

When the bug swims faster than wave speed, ideally it produces a wave pattern as shown in Figure 18.19. It outruns the waves it produces. The waves overlap at the edges, and the pattern made by these overlapping waves is a V shape, called a **bow wave**, which appears to be dragging behind the bug. The familiar bow wave generated by a speedboat knifing through the water is a non-periodic wave produced by the overlapping of many periodic circular waves.

FIGURE 18.19 Wave pattern made by a bug swimming faster than wave speed.

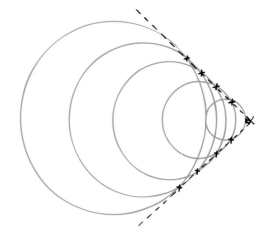

Answer Neither! It is the *frequency* of a wave that undergoes a change where there is motion of the source, not the *wave speed*. Be clear about the distinction between frequency and speed. How frequently a wave vibrates is altogether different from how fast the disturbance moves from one place to another.

FIGURE 18.20 Patterns made by a bug swimming at successively greater speeds. Overlapping at the edges occurs only when the bug swims faster than wave speed.

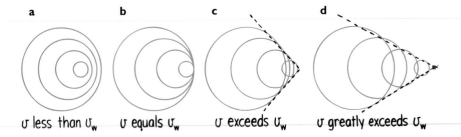

a   b   c   d

υ less than υ_w υ equals υ_w υ exceeds υ_w υ greatly exceeds υ_w

Some wave patterns made by sources moving at various speeds are shown in Figure 18.20. Note that after the speed of the source exceeds wave speed, increased speed of the source produces a narrower V shape.*

Shock Waves

A speedboat knifing through the water generates a two-dimensional bow wave. A supersonic aircraft similarly generates a three-dimensional **shock wave**. Just as a bow wave is produced by overlapping circles that form a V, a shock wave is produced by overlapping spheres that form a cone. And just as the bow wave of a speedboat spreads until it reaches the shore of a lake, the conical wake generated by a supersonic craft spreads until it reaches the ground.

The bow wave of a speedboat that passes by can splash and douse you if you are at the water's edge. In a sense, you can say that you are hit by a "water boom." In the same way, when the conical shell of compressed air that sweeps behind a supersonic aircraft reaches listeners on the ground below, the sharp crack they hear is described as a **sonic boom**.

We don't hear a sonic boom from slower-than-sound, or subsonic, aircraft because the sound waves reach our ears one at a time and are perceived as one continuous tone. Only when the craft moves faster than sound do the waves overlap to reach the listener in a single burst. The sudden increase in pressure is much the same in effect as the sudden expansion of air produced by an explosion. Both processes direct a burst of high-pressure air to the listener. The ear is hard pressed to distinguish between the high pressure from an explosion and the high pressure from many overlapping waves.

FIGURE 18.21 Shock wave of a bullet piercing a sheet of Plexiglas. Light is deflected as it passes through the compressed air that makes up the shock wave, which makes it visible. Note the slight bending of the wave (not shown in following figures) due to air temperature and density changes.

* Bow waves generated by boats in water are more complex than is indicated here. Our idealized treatment serves as an analogy for the production of the less complex shock waves in air.

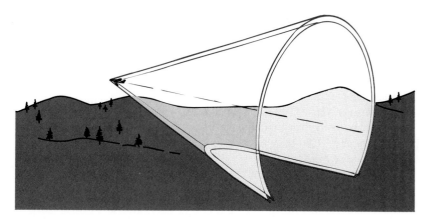

FIGURE 18.22 A shock wave.

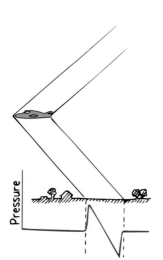

FIGURE 18.23 The shock wave is actually made up of two cones—a high-pressure cone with the apex at the bow and a low-pressure cone with the apex at the tail. A graph of the air pressure at ground level between the cones takes the shape of the letter N.

A water skier is familiar with the fact that next to the high hump of the V-shaped bow wave is a V-shaped depression. The same is true of a shock wave, which usually consists of two cones: a high-pressure cone generated at the bow of the supersonic aircraft and a low-pressure cone that follows at the tail of the craft.* The edges of these cones are visible in the photograph of the supersonic bullet in Figure 18.21. Between these two cones the air pressure rises sharply to above atmospheric pressure, then falls below atmospheric pressure before sharply returning to normal beyond the inner tail cone (Figure 18.23). This overpressure suddenly followed by underpressure intensifies the sonic boom.

A common misconception is that sonic booms are produced when an aircraft flies through the "sound barrier"—that is, just as the aircraft surpasses the speed of sound. This is the same as saying that a boat produces a bow wave when it first overtakes its own waves. This is not so. The fact is that a shock wave and its resulting sonic boom are swept continuously behind an aircraft traveling faster than sound, just as a bow wave is swept continuously behind a speedboat. In Figure 18.24, listener B is in the process of hearing a sonic boom. Listener C has already heard it, and listener A will hear it shortly. The aircraft that generated this shock wave may have broken through the sound barrier hours ago!

It is not necessary that the moving source be "noisy" to produce a shock wave. Once an object is moving faster than the speed of sound, it will *make* sound. A supersonic bullet passing overhead produces a crack, which is a small sonic boom. If the bullet were larger and disturbed more air in its path, the crack would be more boomlike. When the a lion tamer cracks a circus whip, the cracking sound is actually a sonic boom produced by the tip of the whip when it travels faster than the speed of sound. Both the bullet and the whip are not in themselves sound sources, but when traveling at supersonic speeds they produce their own sound as they generate shock waves.

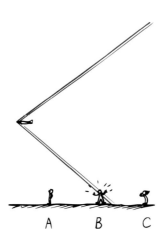

FIGURE 18.24 The shock wave has not yet reached listener A, but is now reaching listener B and has already reached listener C.

---

* Shock waves are often more complex and involve multiple cones.

Summary of Terms

Sine curve A wave form traced by simple harmonic motion, which can be made visible on a moving conveyor belt by a pendulum swinging at right angles above the moving belt.

Amplitude For a wave or vibration, the maximum displacement on either side of the equilibrium (midpoint) position.

Wavelength The distance between successive crests, troughs, or identical parts of a wave.

Frequency For a vibrating body or medium, the number of vibrations per unit time. For a wave, the number of crests that pass a particular point per unit time.

Hertz The SI unit of frequency. One hertz (symbol Hz) equals one vibration per second.

Period The time in which a vibration is completed. The period of a wave equals the period of the source, and is equal to 1/frequency.

Wave speed The speed with which waves pass a particular point:

$$\text{Wave speed} = \text{wavelength} \times \text{frequency}$$

Transverse wave A wave in which the vibration is in a direction perpendicular (transverse) to the direction in which the wave travels. Light consists of transverse waves.

Longitudinal wave A wave in which the medium vibrates in a direction parallel (longitudinal) to the direction in which the wave travels. Sound consists of longitudinal waves.

Interference pattern The pattern formed by superposition of different sets of waves that produces mutual reinforcement in some places and cancellation in others.

Standing wave A stationary wave pattern formed in a medium when two sets of identical waves pass through the medium in opposite directions.

Doppler effect The shift in received frequency due to motion of a vibrating source toward or away from a receiver.

Bow wave The V-shaped wave made by an object moving across a liquid surface at a speed greater than the wave speed.

Shock wave The cone-shaped wave made by an object moving at supersonic speed through a fluid.

Sonic boom The loud sound resulting from the incidence of a shock wave.

Review Questions

1. What is a *wiggle in time* called? A *wiggle in space and time?*
2. What is the source of all waves?
3. Distinguish between sound waves and light waves.

Vibration of a Pendulum

4. What feature about a pendulum makes it useful in a grandfather clock?
5. What is the period of a pendulum?
6. If a pendulum takes one second to make a complete to and fro swing, how great is its period?
7. The period of a certain pendulum is 2 s, and the period of another is 1 s. Which pendulum is longer?

Wave Description

8. How is a sine curve related to a wave?
9. Distinguish between these different parts of a wave: amplitude, wavelength, frequency, and period.
10. How many vibrations per second are represented in a radio wave of 101.7 MHz?
11. How do *frequency* and *period* relate to each other?

Wave Motion

12. Exactly what is it that moves from source to receiver in wave motion?
13. Does the medium in which a wave moves travel along with the wave itself? Give examples to support your answer.

Wave Speed

14. What is the relationship among frequency, wavelength, and wave speed?
15. As the frequency of a wave of constant speed is increased, does the wavelength increase or decrease?

Transverse Waves

16. In a transverse wave, in which direction do the vibrations move when compared with the direction of wave travel?

Longitudinal Waves

17. In a longitudinal wave, in which direction do the vibrations move when compared with the direction of wave travel?
18. How do compressions and rarefactions of longitudinal waves compare to crests and troughs of transverse waves?

Interference

19. Distinguish between *constructive interference* and *destructive* interference.
20. What does it mean to say one wave is out of phase with another?
21. What kinds of waves can show interference?

Standing Waves

22. What causes a standing wave?

23. What is a *node?* What is an *antinode?*

Doppler Effect

24. In the Doppler effect, does frequency change when the source moves? Does wavelength change? Does wave speed change?

25. Can the Doppler effect be observed with longitudinal waves, transverse waves, or both?

26. What is meant by a blue shift and a red shift for light?

Bow Waves

27. How fast must a bug swim to keep up with the waves it is producing? How fast must a boat move to produce a bow wave?

28. How do the speed of a source of waves and the speed of the waves themselves compare for the production of a bow wave?

29. How does the V shape of a bow wave depend on the speed of the source?

Shock Waves

30. How is a bow wave similar to a shock wave? How are they different?

31. How does the V shape of a shock wave depend on the speed of the source?

32. True or false: A sonic boom occurs only when an aircraft is breaking through the sound barrier.

33. True or false: In order for an object to produce a sonic boom, it must be "noisy."

Projects

1. Tie a rubber tube, a spring, or a rope to a fixed support and produce standing waves. See how many nodes you can produce.

2. Wet your finger and rub it slowly around the rim of a thin-rimmed, stemmed glass while you hold the base of the glass firmly to a tabletop with your other hand. The friction of your finger will excite standing waves in the glass, much like the wave made on the strings of a violin by the friction from a violin bow. Try it with a metal bowl.

Exercises

1. A grandfather pendulum clock keeps perfect time. Then it is brought to a summer home high in the mountains. Does it run faster, slower, or the same? Explain.

2. If a pendulum is shortened, does its frequency increase or decrease? What about its period?

3. You let an empty suitcase swing to and fro at its natural frequency. If the case were filled with books, would the natural frequency be lower, greater, or the same as before?

4. Is the time required to swing to and fro (the period) on a playground swing longer or shorter when you stand rather than sit on the swing? Explain.

5. Why doesn't the frequency of a simple pendulum depend on its mass?

6. You clamp one end of a hacksaw blade in a vise and twang the free end. It vibrates. Now repeat, but first put a wad of clay on the free end. How, if at all, will the frequency of vibration differ? Would it make a difference if the wad of clay were stuck to the middle? Explain. (Why could this question have been asked back in Chapter 7?)

7. The needle of a sewing machine moves up and down in simple harmonic motion. Its driving force comes from a rotating wheel that is powered by an electric motor. How do you suppose the period of the up-and-down needle compares to the period of the rotating wheel?

8. What kind of motion should you impart to the nozzle of a garden hose so that the resulting stream of water approximates a sine curve?

9. What kind of motion should you impart to a stretched coiled spring (or Slinky) to provide a transverse wave? A longitudinal wave?

10. What kind of wave is each of the following: (a) An ocean wave rolling toward Waikiki beach? (b) The sound of one whale calling another whale under water? (c) A pulse sent down a stretched rope by snapping one end of it?

11. If a gas tap is turned on for a few seconds, someone a couple of meters away will hear the gas escaping long before she smells it. What does this indicate about the speed of sound and the motion of molecules in the sound-carrying medium?

12. If we double the frequency of a vibrating object, what happens to its period?

13. Red light has a longer wavelength than violet light. Which has the greater frequency?

14. What is the frequency of the second hand of a clock? The minute hand? The hour hand?

15. What is the source of wave motion?

16. You dip your finger repeatedly into a puddle of water and make waves. What happens to the wavelength if you dip your finger more frequently?

17. How does the frequency of vibration of a small object floating in water compare to the number of waves passing it each second?

18. How far, in terms of wavelength, does a wave travel in one period?

19. A rock is dropped in water, and waves spread over the flat surface of the water. What becomes of the energy in these waves when they die out?

20. The wave patterns seen in Figure 18.5 are composed of circles. What does this tell you about the speed of waves moving in different directions?

21. Why is lightning seen before thunder is heard?

22. A pair of loudspeakers on two sides of a stage are emitting identical pure tones (tones of a fixed frequency and fixed wavelength in air). When you stand in the center aisle, equally distant from the two speakers, you hear the sound loud and clear. Why does the intensity of the sound diminish considerably when you step to one side? Suggestion: Use a diagram to make your point.

23. A banjo player plucks the middle of an open string. Where are the nodes of the standing wave in the string? What is the wavelength of the vibrating string?

24. Violinists sometimes bow a string to produce maximum vibration (antinodes) at one-quarter and three-quarters of the string length rather than at the middle of the string. Then the string vibrates with a wavelength equal to the string length rather than twice the string length. (See Figures 18.14 a and b.) What is the effect on frequency when this occurs?

25. Why is there a Doppler effect when the source of sound is stationary and the listener is in motion? In which direction should the listener move to hear a higher frequency? A lower frequency?

26. A railroad locomotive is at rest with its whistle shrieking, then starts moving toward you. (a) Does the frequency that you hear increase, decrease, or stay the same? (b) How about the wavelength reaching your ear? (c) How about the speed of sound in the air between you and the locomotive?

27. When you blow your horn while driving toward a stationary listener, an increased frequency of the horn is heard by the listener. Would the listener hear an increase in horn frequency if he were also in a car traveling at the same speed in the same direction as you are? Explain.

28. Is there an appreciable Doppler effect when the motion of the source is at right angles to a listener? Explain.

29. How does the Doppler effect aid police in detecting speeding motorists?

30. Astronomers find that light coming from one edge of the sun has a slightly higher frequency than light from the opposite edge. What do these measurements tell us about the sun's motion?

31. Would it be correct to say that the Doppler effect is the apparent change in the speed of a wave due to motion of the source? (Why is this question a test of reading comprehension as well as a test of physics knowledge?)

32. How does the phenomenon of interference play a role in the production of bow or shock waves?

33. What can you say about the speed of a boat that makes a bow wave?

34. Does the conical angle of a shock wave open wider, narrow down, or remain constant as a supersonic aircraft increases its speed?

35. If the sound of an airplane does not come from the part of the sky where the plane is seen, does this imply that the airplane is traveling faster than the speed of sound? Explain.

36. Does a sonic boom occur only at the moment when an aircraft exceeds the speed of sound? Explain.

37. Why is it that a subsonic aircraft, no matter how loud it may be, cannot produce a sonic boom?

38. Imagine a super-fast fish that is able to swim faster than the speed of sound in water. Would such a fish produce a "sonic boom"?

39. Make up a multiple-choice question that would check a classmate's understanding of the distinction between a transverse and longitudinal wave.

40. Make up two multiple-choice questions that would check a classmate's understanding of the terms that describe a wave.

Problems

1. A skipper on a boat notices wave crests passing his anchor chain every 5 s. He estimates the distance between wave crests to be 15 m. He also correctly estimates the speed of the waves. What is this speed?

2. A weight suspended from a spring is seen to bob up and down over a distance of 20 centimeters twice each second. What is its frequency? Its period? Its amplitude?

3. Radio waves travel at the speed of light—300,000 km/s. What is the wavelength of radio waves received at 100 MHz on your radio dial?

4. A mosquito flaps its wings 600 vibrations per second, which produces the annoying 600-Hz buzz. The speed of sound is 340 m/s. How far does the sound travel between wing beats? In other words, find the wavelength of the mosquito's sound.

5. On a keyboard, you strike middle C, which has a frequency of 256 Hz. (a) What is the period of one vibration of this tone? (b) As the sound leaves the instrument at a speed of 340 m/s, what is its wavelength in air?

6. (a) If you were so foolish as to play your keyboard instrument under water, where the speed of sound is 1500 m/s, what would be the wavelength of the middle-C tone in water? (b) Explain *why* middle C (or any other tone) has a longer wavelength in water than in air.

7. The wavelength of the signal from TV Channel 6 is 3.42 m. Does Channel 6 broadcast on a frequency above or below the FM radio band, which is 88 to 108 MHz?

8. As shown in the drawing at the right, the half-angle of the shock wave cone generated by a supersonic transport is 45°. What is the speed of the plane relative to the speed of sound?

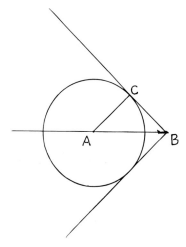

19

Sound

Resonance!

If a tree fell in the middle of a deep forest hundreds of kilometers away from any living being, would there be a sound? Different people will answer this question in different ways. "No," some will say, "sound is subjective and requires a listener. If there is no listener, there will be no sound." "Yes," others will say, "a sound is not something in a listener's head. A sound is an objective thing." Discussions like this one often are beyond agreement because the participants fail to realize that they are arguing not about the nature of sound but about the definition of the word. Either side is right, depending on which definition is taken, but investigation can proceed only when a definition has been agreed on. The physicist usually takes the objective position and defines sound as a form of energy that exists whether or not it is heard and goes on from there to investigate its nature.

Origin of Sound

Most sounds are waves produced by the vibrations of material objects. In a piano, violin, and guitar, the sound is produced by the vibrating strings; in a saxophone, by a vibrating reed; in a flute, by a fluttering column of air at the mouthpiece. Your voice results from the vibration of your vocal chords.

In each of these cases, the original vibration stimulates the vibration of something larger or more massive, such as the sounding board of a stringed instrument, the air column within a reed or wind instrument, or the air in the throat and mouth of a singer. This vibrating material then sends a disturbance through the surrounding medium, usually air, in the form of longitudinal waves. Under ordinary conditions, the frequency of the vibrating source and the frequency of the sound waves produced are the same.

We describe our subjective impression about the frequency of sound by the word *pitch*. A high-pitched sound like that from a piccolo has a high frequency of vibration, while a low-pitched sound like that from a foghorn has a low frequency of vibration. The human ear of a young person can normally hear pitches corresponding to the range of frequencies between about 20 and 20,000 hertz. As we grow older, the limits of this human hearing range shrink, especially at the high-frequency end. Sound waves with frequencies below 20 hertz are **infrasonic**, and those with frequencies above 20,000 hertz are called **ultrasonic**. We cannot hear infrasonic and ultrasonic sound waves.

Nature of Sound in Air

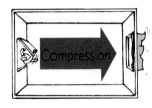

a

b

FIGURE 19.1 (a) When the door is opened, a compression travels across the room. (b) When the door is closed, a rarefaction travels across the room. (From *The New College Physics: A Spiral Approach* by A. V. Baez. Copyright © 1967 by W. H. Freeman and Company. Reprinted with permission.)

FIGURE 19.2 Compressions and rarefactions traveling from the tuning fork through the tube.

When we clap our hands, the sound produced is nonperiodic. It consists of a wave *pulse* that travels out in all directions. The pulse disturbs the air in the same way that a similar pulse would disturb a coiled spring or a Slinky. Each particle moves to and fro along the direction of the expanding wave.

For a clearer picture of this process, consider a long room as shown in Figure 19.1a. At one end is an open window with a curtain over it. At the other end is a door. When we open the door, we can imagine the door pushing the molecules next to it away from their initial positions and into their neighbors. The neighboring molecules, in turn, push into their neighbors, and so on, like a compression traveling along a spring, until the curtain flaps out the window. A pulse of compressed air has moved from the door to the curtain. This pulse of compressed air is called a **compression**.

When we close the door (Figure 19.1b), the door pushes some air molecules out of the room. This produces an area of low pressure behind the door. Neighboring molecules then move into it, leaving a zone of lower pressure behind them. We say this zone of lower-pressure air is *rarefied*. Other molecules farther away from the door, in turn, move into these rarefied regions, and a disturbance again travels across the room. This is evidenced by the curtain, which flaps inward. This time the disturbance is a **rarefaction**.

As with all wave motion, it is not the medium itself that travels across the room, but the energy-carrying pulse. In both cases the pulse travels from the door to the curtain. We know this because in both cases the curtain moves after the door is opened or closed. If you continually swing the door open and closed in periodic fashion, you can set up a wave of periodic compressions and rarefactions that will make the curtain swing in and out of the window. On a much smaller but more rapid scale, this is what happens when a tuning fork is struck. The periodic vibrations of the tuning fork and the waves it produces are considerably higher in frequency and lower in amplitude than those caused by the swinging door. You don't notice the effect of sound waves on the curtain, but you are well aware of them when they meet your sensitive eardrums.

Consider sound waves in the tube shown in Figure 19.2. For simplicity, only the waves that travel in the tube are depicted. When the prong of the tuning fork next to the

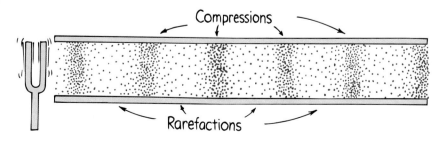

FIGURE 19.3 Waves of compressed and rarefied air, produced by the vibrating cone of the loudspeaker, make up the pleasing sound of music.

tube moves toward the tube, a compression enters the tube. When the prong swings away in the opposite direction, a rarefaction follows the compression. It's like a Ping-Pong paddle moving to and fro in a room packed with Ping-Pong balls. As the source vibrates, a periodic series of compressions and rarefactions are produced. The frequency of the vibrating source and the frequency of the wave it produces are the same.

Pause to reflect on the physics of sound while you are listening to your radio sometime. The radio loudspeaker is a paper cone that vibrates in rhythm with an electrical signal. Air molecules next to the vibrating cone of the speaker are themselves set into vibration. This air in turn vibrates against neighboring particles, which in turn do the same, and so on. As a result, rhythmic patterns of compressed and rarefied air emanate from the loudspeaker, showering the whole room with undulating motions. The resulting vibrating air sets your eardrum into vibration, which in turn sends cascades of rhythmic electrical impulses along the cochlear nerve and into the brain. And you listen to the sound of music.

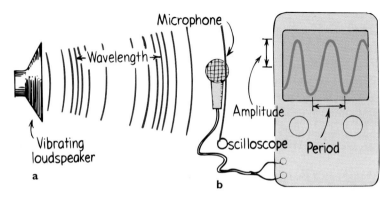

FIGURE 19.4 (a) The radio loudspeaker is a paper cone that vibrates in rhythm with an electric signal. The sound that is produced sets up similar vibrations in the microphone, which are displayed on an oscilloscope. (b) The shape of the pattern on the screen of the oscilloscope reveals information about the sound.

Media That Transmit Sound

Most sounds that we hear are transmitted through the air. However, any elastic substance—whether solid, liquid, gas, or plasma—can transmit sound. Elasticity is the ability of a material that has changed shape in response to an applied force to resume its initial shape once the distorting force is removed. Steel is an elastic substance. In contrast, putty is inelastic.* In elastic liquids and solids, the atoms are relatively close together and respond quickly to each other's motions and transmit energy with little loss. Sound travels about four times faster in water than in air and about fifteen times faster in steel than in air.

Relative to solids and liquids, air is a poor conductor of sound. You can hear the sound of a distant train more clearly if your ear is placed against the rail. Similarly, a watch placed on a table beyond hearing distance can be heard if you place your ear to the table. Or click some rocks together under water while your ear is submerged. You'll hear the clicking sound very clearly. If you've ever been swimming in the presence of motorized boats, you probably noticed that you can hear the boats' motors much more clearly under water than above water. Liquids and crystalline solids are generally excellent conductors of sound—much better than air. The speed of sound is generally greater

* Elasticity is not "stretchiness," like a rubber band. Some very stiff materials are elastic—like steel.

in liquids than in gases, and still greater in solids. Sound won't travel in a vacuum, for the transmission of sound requires a medium. If there is nothing to compress and expand, there can be no sound.

Speed of Sound in Air

If we watch a person at a distance chopping wood, or see a far-away baseball player hit a ball, we can easily see that it takes an appreciable time for the sound of the blow to reach our ears. Thunder is heard after a flash of lightning is seen. These common experiences show that sound requires a recognizable time to travel from one place to another. The speed of sound depends on wind conditions, temperature, and humidity. It does not depend on the loudness or the frequency of the sound; all sounds travel at the same speed. The speed of sound in dry air at 0°C is about 330 meters per second, nearly 1200 kilometers per hour (a little more than one-millionth the speed of light). Water vapor in the air increases this speed slightly. Sound travels faster through warm air than cold air. This is to be expected because the faster-moving molecules in warm air bump into each other more often and therefore can transmit a pulse in less time.* For each degree rise in temperature above 0°C, the speed of sound in air increases by 0.6 meter per second. So in air at a normal room temperature of about 20°C, sound travels at about 340 meters per second.

Questions

1. Do compressions and rarefactions in a sound wave travel in the same direction or in opposite directions from one another?
2. What is the approximate distance of a thunderstorm when you note a 3-s delay between the flash of lightning and the sound of thunder?

Reflection of Sound

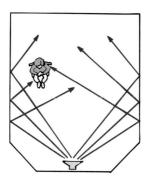

FIGURE 19.5 The angle of incident sound is equal to the angle of reflected sound.

We call the reflection of sound an *echo*. The fraction of energy carried by the reflected sound wave is large if the surface is rigid and smooth and less if the surface is soft and irregular. Sound energy not carried by the reflected sound wave is carried by the "transmitted" (absorbed) wave.

Sound reflects from a smooth surface the same way that light does—the angle of incidence is equal to the angle of reflection (Figure 19.5). Sometimes when sound reflects from the walls, ceiling, and floor of a room, the reflecting surfaces are too reflective and the sound becomes garbled. This is due to multiple reflections called **reverberations**. On the other hand, if the reflective surfaces are too absorbent, the sound level would be low and the hall would sound dull and lifeless. Reflection of sound in a room makes it sound lively and full, as you have probably found out while singing in the

Answers

1. They travel in the same direction.
2. Assuming the speed of sound in air is about 340 m/s, in 3 s it will travel (340 × 3) = 1020 m. There is no appreciable delay for the light, so the storm is slightly more than 1 km away.

---

* The speed of sound in a gas is about $\frac{3}{4}$ the average speed of the gas molecules.

shower. In the design of an auditorium or concert hall, a balance must be found between reverberation and absorption. The study of sound properties is called *acoustics*.

It is often advantageous to place highly reflective surfaces behind the stage to direct sound out to an audience. Reflecting surfaces are suspended above the stage in some concert halls. The ones in the opera hall in San Francisco are large shiny plastic surfaces that also reflect light (Figure 19.6). A listener can look up at these reflectors and see the reflected images of the members of the orchestra. (The plastic reflectors are somewhat curved, which increases the field of view.) Both sound and light obey the same law of reflection, so if a reflector is oriented so that you can see a particular musical instrument, rest assured that you will hear it also. Sound from the instrument will follow the line of sight to the reflector and then to you.

FIGURE 19.6 The plastic plates above the orchestra reflect both light and sound. Adjusting them is quite simple: what you see is what you hear.

Refraction of Sound

Sound waves bend when parts of the wave fronts travel at different speeds. This happens in uneven winds or when sound is traveling through air of uneven temperatures. This bending of sound is called **refraction**. On a warm day, the air near the ground may be appreciably warmer than the rest of the air, so the speed of sound near the ground increases. Sound waves therefore tend to bend away from the ground, resulting in sound that does not seem to carry well.

We hear thunder when the lightning is reasonably close, but we often fail to hear the thunder for distant lightning because of refraction. The sound travels more slowly at higher altitudes and bends away from the ground. The opposite often occurs on a cold day or at night when the layer of air near the ground is colder than the air above. Then the speed of sound near the ground is reduced. The higher speed of the wave fronts above causes a bending of the sound toward the earth, resulting in sound that can be heard over considerably longer distances (Figure 19.7).

The refraction of sound occurs under water, where the speed of sound varies with temperature. This poses a problem for surface vessels that bounce ultrasonic waves off

FIGURE 19.7 The wave fronts of sound are bent in air of uneven temperatures.

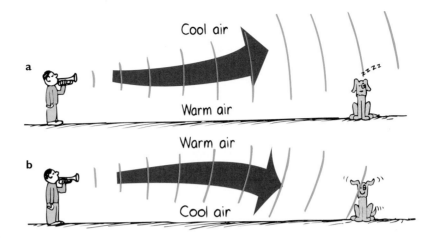

a

Cool air

Warm air

b

Warm air

Cool air

FIGURE 19.8 A 5-month fetus displayed on a viewing screen by ultrasound.

FIGURE 19.9 A dolphin emits ultrahigh-frequency sound to locate and identify objects in its environment. Distance is sensed by the time delay between sending sound and receiving its echo, and direction is sensed by differences in time for the echo to reach its two ears. A dolphin's main diet is fish and, since hearing in fish is limited to fairly low frequencies, they are not alerted to the fact they are being hunted.

the bottom of the ocean to chart the sea bottom's features. Refraction is a blessing to submarines that wish to escape detection. Because of thermal gradients and layers of water at different temperatures, the refraction of sound leaves gaps or "blind spots" in the water. This is where submarines hide. If it weren't for refraction, submarines would be easier to detect.

The multiple reflections and refractions of ultrasonic waves are used by physicians in a technique for harmlessly "seeing" inside the body without the use of X rays. When high-frequency sound (ultrasound) enters the body, it is reflected more strongly from the outside of organs than from their interior, and a picture of the outline of the organs is obtained. When ultrasound is incident upon a moving object, the reflected sound has a slightly different frequency. Using this Doppler effect, a physician can "see" the beating heart of a fetus as early as 11 weeks (Figure 19.8).

The ultrasound echo technique may be relatively new to humans, but not to bats or dolphins. It is well known that bats emit ultrasonic squeaks and locate objects by their echoes. Dolphins do this and more.* The ultrasonic waves emitted by a dolphin enable

* The primary sense of the dolphin is acoustic, for vision is not a very useful sense in the often murky and dark depths of the ocean. Whereas sound is a passive sense for us, it is an active sense for the dolphin that sends out sounds and then perceives its surroundings on the basis of the echoes that come back. What's more interesting, the dolphin can reproduce the sonic signals that paint the mental image of its surroundings; thus the dolphin probably communicates its experience to other dolphins by communicating the full acoustic image of what is "seen," placing it directly in the minds of other dolphins. It needs no word or symbol for "fish," for example, but communicates an image of the real thing—perhaps with emphasis highlighted by selective filtering, as we similarly communicate a musical concert to others via various means of sound reproduction. Small wonder that the language of the dolphin is very unlike our own!

it to "see" through the bodies of other animals and people. Skin, muscle, and fat are almost transparent to dolphins, so they "see" a thin outline of the body—but the bones, teeth, and gas-filled cavities are clearly apparent. Physical evidence of cancers, tumors, heart attacks, and even emotional state can all be "seen" by the dolphin—as humans have only recently been able to do with ultrasound.

Question An oceanic depth-sounding vessel surveys the ocean bottom with ultrasonic sound that travels 1530 m/s in seawater. How deep is the water if the time delay of the echo from the ocean floor is 2 s?

Energy in Sound Waves

Wave motion of all kinds possesses energy of varying degrees. Electromagnetic waves from the sun, for instance, bring us the enormous quantities of energy that are necessary for life on earth. By comparison, the energy in sound is extremely small. That's because producing sound requires only a small amount of energy. For example, 10,000,000 people talking at the same time would produce sound energy equal only to the energy needed to light a common flashlight. Hearing is possible only because of our remarkably sensitive ears. Only the most sensitive microphone can detect sound that is softer than what we can hear.

Sound energy dissipates to thermal energy while sound travels in air. For waves of higher frequency, the sound energy is transformed into internal energy more rapidly than for waves of lower frequencies. As a result, sound of low frequencies will travel farther through air than sound of higher frequencies. That's why the foghorns of ships are of a low frequency.

Forced Vibrations

If we strike an unmounted tuning fork, the sound from it may be rather faint. If we hold the same fork against a table after striking it, the sound is louder. This is because the table is forced to vibrate, and its larger surface sets more air in motion. The table will be forced into vibration by a fork of any frequency. This is a case of **forced vibration**.

The mechanism in a music box is mounted on a sounding board. Without the sounding board, the sound the music-box mechanism makes is barely audible. Sounding boards are important in all stringed musical instruments.

Natural Frequency

When someone drops a wrench on a concrete floor, we are not likely to mistake its sound for that of a baseball bat hitting the floor. This is because the two objects vibrate differently when they are struck. Tap a wrench and the vibrations it makes are different from the vibrations of a baseball bat, or of anything else. Any object composed of an elastic material when disturbed will vibrate at its own special set of frequencies, which together form its special sound. We speak of an object's **natural frequency**, which depends on factors such as the elasticity and shape of the object. Bells and tuning forks, of course, vibrate at their own characteristic frequencies. And interestingly enough, most things from planets to atoms and almost everything else in between have a springiness to them and vibrate at one or more natural frequencies.

FIGURE 19.10 The natural frequency of the smaller bell is higher than that of the larger bell, and it rings at a higher pitch.

Answer 1530 m

Resonance

FIGURE 19.11 Pumping a swing in rhythm with its natural frequency produces a large amplitude.

When the frequency of forced vibrations on an object matches the object's natural frequency, a dramatic increase in amplitude occurs. This phenomenon is called **resonance**. Literally, *resonance* means "resounding," or "sounding again." Putty doesn't resonate because it isn't elastic, and a dropped handkerchief is too limp. In order for something to resonate, it needs a force to pull it back to its starting position and enough energy to keep it vibrating.

A common experience illustrating resonance occurs on a swing. When pumping a swing, we pump in rhythm with the natural frequency of the swing. More important than the force with which we pump is the timing. Even small pumps or small pushes from someone else, if delivered in rhythm with the frequency of the swinging motion, produce large amplitudes. A common classroom demonstration of resonance is illustrated with a pair of tuning forks adjusted to the same frequency and spaced a meter or so apart. When one of the forks is struck, it sets the other fork into vibration. This is a small-scale version of pushing a friend on a swing—it's the timing that's important. When a series of sound waves impinge on the fork, each compression gives the prong of the fork a tiny push. Since the frequency of these pushes corresponds to the natural frequency of the fork, the pushes will successively increase the amplitude of vibration. This is because the pushes occur at the right time and repeatedly occur in the same direction as the instantaneous motion of the fork. The motion of the second fork is often called a *sympathetic vibration.*

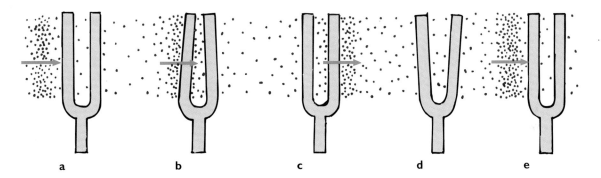

| a | b | c | d | e |

FIGURE 19.12 Stages of resonance. (a) The first compression meets the fork and gives it a tiny and momentary push; (b) the fork bends and then (c) returns to its initial position just at the time a rarefaction arrives and (d) overshoots in the opposite direction. Just when it returns to its initial position (e) the next compression arrives to repeat the cycle. Now it bends farther because it is moving.

If the forks are not adjusted for matched frequencies, the timing of pushes is off, and resonance will not occur. When you tune your radio set, you are similarly adjusting the natural frequency of the electronics in the set to match one of the many surrounding signals. The set then resonates to one station at a time instead of playing all stations at once.

Resonance is not restricted to wave motion. It occurs whenever successive impulses are applied to a vibrating object in rhythm with its natural frequency. Cavalry troops marching across a footbridge near Manchester, England, in 1831 inadvertently caused the bridge to collapse when they marched in rhythm with the bridge's natural frequency. Since then, it is customary to order troops to "break step" when crossing bridges. A more recent bridge disaster was caused by wind-generated resonance (Figure 19.13).

FIGURE 19.13 In 1940, four months after being completed, the Tacoma Narrows Bridge in the state of Washington was destroyed by wind-generated resonance. The mild gale produced a fluctuating force in resonance with the natural frequency of the bridge, steadily increasing the amplitude until the bridge collapsed.

The effects of resonance are all about us. Resonance underscores not only the sound of music, but the color of autumn leaves, the height of ocean tides, the operation of lasers, and a vast multitude of phenomena that add to the beauty of the world about us.

Interference

Sound waves, like any waves, can be made to exhibit interference. Recall that wave interference was discussed in the last chapter. A comparison of interference for transverse waves and longitudinal waves is shown in Figure 19.14. In either case, when the crests of one wave overlap the crests of another wave, increased amplitude results. Or when the crest of one wave overlaps the trough of another wave, decreased amplitude results. In the case of sound, the crest of a wave corresponds to a compression, and the trough of a wave corresponds to a rarefaction. Interference occurs for both transverse and longitudinal waves.

An interesting case of sound interference is illustrated in Figure 19.15. If you are an equal distance from two sound speakers that emit identical tones of fixed frequency, the sound is louder because the effects of the two speakers add. The compressions and rarefactions of the tones arrive in step, or in *phase*. However, if you move to the side so that the paths from the speakers to you differ by a half wavelength, then the rarefactions from one speaker will be filled in by the compressions from the other speaker. This is destructive interference. It is just as if the crest of one water wave exactly filled in the trough of another water wave. If the region is devoid of any reflecting surfaces, little or no sound will be heard!

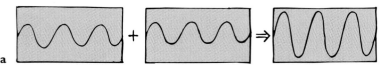

a

The superposition of two identical transverse waves in phase produces a wave of increased amplitude.

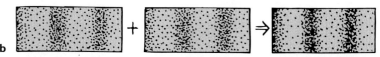

b

The superposition of two identical longitudinal waves in phase produces a wave of increased intensity.

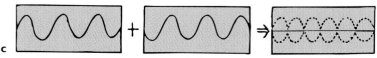

c

Two identical transverse waves that are out of phase destroy each other when they are superimposed.

d

Two identical longitudinal waves that are out of phase destroy each other when they are superimposed.

FIGURE 19.14 Wave interference for transverse and longitudinal waves.

If the speakers emit a whole range of sounds with different frequencies, only some waves destructively interfere for a given difference in path lengths. So interference of this type is usually not a problem, because there is usually enough reflection of sound to fill in canceled spots. Nevertheless, "dead spots" are sometimes evident in poorly designed theaters or music halls, where sound waves reflect off walls and interfere with nonreflected waves to produce zones of low amplitude. When you move your head a few centimeters in either direction, you may hear a noticeable difference.

FIGURE 19.15

Interference of sound waves. (a) Waves arrive in phase and interfere constructively when the path lengths from the speakers are the same. (b) Waves arrive out of phase and interfere destructively when the path lengths differ by half a wavelength (or $\frac{3}{2}$, $\frac{5}{2}$, etc.).

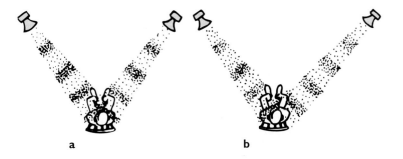

a   **b**

Sound interference is dramatically illustrated when monaural sound is played by stereo speakers that are out of phase. Speakers are out of phase when the input wires to one speaker are interchanged (positive and negative wire inputs reversed). For a monaural signal this means that when one speaker is sending a compression of sound, the other is sending a rarefaction. The sound produced is not as full and not as loud as from speakers properly connected in phase, because the longer waves are being canceled by

FIGURE 19.16 The positive and negative wire inputs to one of the stereo speakers have been interchanged, resulting in speakers that are out of phase. When the speakers are far apart, monaural sound is not as loud as from properly phased speakers. When they are brought face to face, very little sound is heard. Interference is nearly complete as the compressions of one speaker fill in the rarefactions of the other!

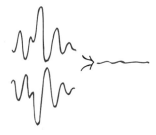

FIGURE 19.17 When a mirror image of a sound signal combines with the sound itself, the sound is canceled.

interference. All waves that are appreciably longer than the distance between the speakers arrive out of phase to a listener's ear. The preponderance of surviving shorter waves makes this sound "tinny." Shorter waves are canceled as the speakers are brought closer together, and when the pair of speakers are brought face to face, against each other, very little sound is heard! Only the sound waves having the highest frequencies survive cancellation. You must try this to appreciate it. And then share this impressive physics demonstration with your friends!

Destructive sound interference is a useful property in *anti-noise technology*. Noisy devices such as jackhammers are being equipped with microphones that send the sound of the device to electronic microchips, which create mirror-image wave patterns of the sound signals. For the jackhammer, this mirror-image sound signal is fed to earphones worn by the operator. Sound compressions (or rarefactions) from the hammer are canceled by mirror image rarefactions (or compressions) in the earphones. The combination of signals cancels the jackhammer noise. Noise-canceling earphones are already common for pilots. Watch for this principle applied to electronic mufflers in cars, where anti-noise is blasted through loudspeakers, canceling about 95% of the original noise.

FIGURE 19.18 Ken Ford tows gliders in quiet comfort when he wears his noise-canceling earphones.

Beats

When two tones of slightly different frequency are sounded together, a fluctuation in the loudness of the combined sounds is heard; the sound is loud, then faint, then loud, then faint, and so on. This periodic variation in the loudness of sound is called **beats** and is due to interference. Strike two slightly mismatched tuning forks, and because one fork vibrates at a different frequency than the other, the vibrations of the forks will be momentarily in step, then out of step, then in again, and so on. When the combined waves reach our ears in step—say when a compression from one fork overlaps a compression from the other—the sound is a maximum. A moment later, when the forks are out of step, a compression from one fork is met with a rarefaction from the other, resulting in a minimum. The sound that reaches our ears throbs between maximum and minimum loudness and produces a tremolo effect.

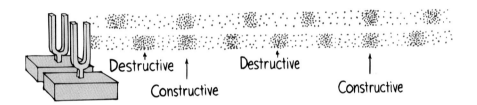

FIGURE 19.19 The interference of two sound sources of slightly different frequencies produces beats.

We can understand beats by considering the analogous case of two people walking side by side with different strides. At some moment they will be in step, a little later out of step, then in step again, and so on. Imagine that one person, perhaps with longer legs, takes exactly seventy steps in 1 minute, and the shorter person takes seventy-two steps in the same time. The shorter person gains two steps per minute on the taller person. A little thought will show that they will both be momentarily in step twice each minute. In general, if two people with different strides walk together, the number of times they are in step in each unit of time is equal to the difference in the frequencies of the steps. This applies also to the pair of tuning forks. If one fork undergoes 264 vibrations each second and the other fork vibrates 262 times per second, they will be in step twice each second. A beat frequency of 2 hertz will be heard. The overall tone will correspond to the average frequency, 263 hertz.

FIGURE 19.20 The unequal spacings of the combs produce a moiré pattern that is similar to beats.

> **Question** What is the beat frequency when a 262-Hz and a 266-Hz tuning fork are sounded together? A 262-Hz and a 272-Hz fork?

If we overlap two combs of different teeth spacings, we'll see a moiré pattern that is related to beats (Figure 19.20). The number of beats per length will equal the difference in the number of teeth per length for the two combs.

> **Answer** For the 262-Hz and 266-Hz forks, the ear will hear 264 Hz, which will beat at 4 Hz (266 − 262). For the 272-Hz and 262-Hz forks, 267 Hz will be heard, and some people will hear it throb ten times each second. Beat frequencies greater than 10 Hz are too rapid to be heard normally.

Radio Broadcasts

A radio receiver emits sound but, interestingly enough, it doesn't receive sound waves. A radio receiver, like a television set, receives *electromagnetic waves*—actually low-frequency light waves. These waves, which we will treat in detail in Part 6, are fundamentally different than sound waves—not only in their completely different nature, but in their extremely high frequencies, which are way beyond the range of human hearing.

Every radio station has an assigned frequency at which it broadcasts. The electromagnetic wave transmitted at this frequency is the **carrier wave**. The relatively low-frequency sound signal to be communicated is superimposed on the much higher frequency carrier wave in two principal ways: by slight variations in amplitude that match the audio frequency or by slight variations in frequency. This impression of the sound wave on the higher-frequency radio wave is **modulation**. When the amplitude of the carrier wave is modulated we call it AM, or **amplitude mod-**

ulation. AM stations broadcast in the range of 535 to 1605 kilohertz. When the frequency of the carrier wave is modulated, we call it FM, or **frequency modulation**. FM stations broadcast in the higher-frequency range of 88 to 108 megahertz. Amplitude modulation is like rapidly changing the brightness of a constant-color light bulb. Frequency modulation is like rapidly changing the color of a constant-intensity light bulb.

Turning the knob of a radio receiver to select a particular station is like adjusting movable masses on the prongs of a tuning fork to make it resonate to the sound produced by another fork. In choosing a radio station you adjust the frequency of an electrical circuit inside the radio receiver to match and resonate with the frequency of the station you want. You sort out one carrier wave from many. Then the impressed sound signal is separated from the carrier wave, amplified, and fed to the loudspeaker. It's nice to hear only one station at a time!

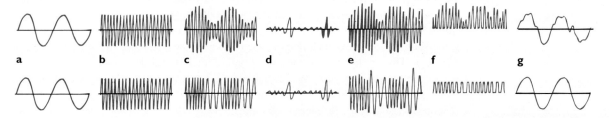

FIGURE 19.21 AM and FM radio signals. (a) Sound waves that enter a microphone. (b) Radio-frequency carrier wave produced by transmitter without sound signal. (c) Carrier wave modulated by signal. (d) Static interference. (e) Carrier wave and signal affected by static. (f) Radio receiver cuts out negative half of carrier wave. (g) Signal remaining is rough for AM because of static but is smooth for FM because the tips of the wave form are clipped without loss to the signal.

Beats can occur with any kind of wave and provide a practical way to compare frequencies. To tune a piano, for example, a piano tuner listens for beats produced between a standard tuning fork and those of a particular string on the piano. When the frequencies are identical, the beats disappear. The members of an orchestra tune up by listening for beats between their instruments and a standard tone produced by a piano or some other instrument.

Beats are utilized by dolphins in surveying the motions of things around them. When a dolphin sends out sound signals, beats may be produced when the echoes it receives interfere with the sound it sends. When there is no relative motion between the dolphin and the object returning the sound, the sending and receiving frequencies are the same and no beats occur. But when there is relative motion, the echo has a different frequency due to the Doppler effect, and beats are produced when the echo and emitted sound combine. The same principle is applied by the radar guns used by police officers. The beats between the signal that is sent and the one that is reflected are used to determine how fast the car that reflected the signal is moving.

Question Is it correct to say that in every case, without exception, any radio wave travels faster than any sound wave?

Answer Yes, because any radio wave travels at the speed of light. A radio wave is an electromagnetic wave—in a very real sense, a low-frequency light wave (or we can say a light wave is a high-frequency radio wave!). A sound wave, on the other hand, is a mechanical disturbance propagated through a material medium by material particles that vibrate against one another. In air, the speed of sound is about 340 m/s, about one-millionth the speed of a radio wave. Sound travels faster in other media, but in no case at the speed of light. No sound wave can travel as fast as light.

Summary of Terms

Infrasonic   Describes a sound that has a frequency too low to be heard by the normal human ear; below 20 hertz.

Ultrasonic   Describes a sound that has a frequency too high to be heard by the normal human ear; above 20 kilohertz.

Compression   Condensed region of the medium through which a longitudinal wave travels.

Rarefaction   Rarefied region, or region of lessened pressure, of the medium through which a longitudinal wave travels.

Forced vibration   The setting up of vibrations in an object by a vibrating force.

Natural frequency   A frequency at which an elastic object naturally tends to vibrate if it is disturbed and the disturbing force is removed.

Resonance   The response of a body when a forcing frequency matches its natural frequency.

Beats   A series of alternate reinforcements and cancellations produced by the interference of two waves of slightly different frequencies, heard as a throbbing effect in sound waves.

Carrier wave   A wave, usually of radio frequency, whose characteristics are modified in the process of modulation.

Modulation   The process of impressing one wave system upon another of higher frequency.

Amplitude modulation (AM)   A type of modulation in which the amplitude of the carrier wave is varied above and below its normal value by an amount proportional to the amplitude of the impressed signal.

Frequency modulation (FM)   A type of modulation in which the frequency of the carrier wave is varied above and below its normal frequency by an amount that is proportional to the amplitude of the impressed signal. In this case, the amplitude of the modulated carrier wave remains constant.

Review Questions

1. How does a physicist usually define sound?

Origin of Sound

2. What is the source of all sounds?

3. What is the relationship between *frequency* and *pitch*?

4. What is the average range of a young person's hearing?

5. Distinguish between *infrasonic* and *ultrasonic* sound waves.

Nature of Sound in Air

6. Distinguish between a *compression* and a *rarefaction.*

7. Do compressions and rarefactions travel in the same or opposite directions in a wave? Cite evidence to support your answer.

8. How does the paper cone of a radio loudspeaker emit sound?

Media That Transmit Sound

9. Relative to solids and liquids, how does air rank as a conductor of sound?

10. Why will sound not travel in a vacuum?

Speed of Sound in Air

11. What factors does the speed of sound depend upon? What are some factors that it does *not* depend upon?

12. What is the speed of sound in dry air at 0°C?

13. Does sound travel faster in warm air than in cold air? Defend your answer.

14. How does the speed of sound in water compare with the speed of sound in air? How does the speed in steel compare with the speed in air?

Reflection of Sound

15. What is an *echo*?

16. What is the law of reflection for sound?

17. What exactly is a *reverberation*?

Refraction of Sound

18. What is the cause of refraction?

19. Does sound tend to bend upward or downward when its speed is less near the ground?

20. Why does sound sometimes refract under water?

21. There is a difference between the way we passively see our surroundings in daylight and the way we actively probe our surroundings with a searchlight in the darkness. Which of these ways of perceiving our surroundings is more like the way a dolphin perceives its environment?

Energy in Sound Waves

22. Which is normally greater, the energy in ordinary sound or the energy in ordinary light?

23. What ultimately becomes of the energy of the sound in the air?

24. Why will sound of low frequencies travel farther in air than sound of high frequencies?

25. Why are foghorns low in frequency?

Forced Vibrations

26. Why will a struck tuning fork sound louder when it is held against a table?

27. Give at least three examples of forced vibration.

Natural Frequency

28. Give at least two factors that determine the natural frequency of an object.

29. Does a blob of putty have a natural frequency? Explain.

Resonance

30. Distinguish between *forced vibrations* and *resonance*.

31. What is required to make an object resonate?

32. When you listen to a radio, why do you hear only one station at a time instead of hearing all stations at once?

33. Why do troops "break step" when crossing a bridge?

Interference

34. Is it possible for one wave to cancel another? Defend your answer.

35. What kind of waves exhibit interference?

36. Distinguish between *constructive interference* and *destructive interference*.

37. What is responsible for "dead spots" in poorly designed theaters or concert halls?

Beats

38. What physical phenomenon underlies the production of beats?

39. What beat frequency will be heard when a 370-Hz and a 374-Hz tuning fork are sounded together?

40. How is the phenomenon of beats useful for tuning musical instruments?

Radio Broadcasts

41. How does a radio wave differ from a sound wave?

42. Distinguish between a carrier wave and a sound wave.

43. Distinguish between AM and FM.

44. How is tuning a radio like adjusting a pair of tuning forks for resonance?

Projects

1. Suspend the wire grille from a refrigerator or an oven from a string, the ends of which you hold to your ears. Let a friend gently stroke the grille with pieces of broom straw and other objects. The effect is best appreciated in a relaxed condition with your eyes closed. Be sure to try this!

2. In the bathtub, submerge your head and listen to the sound you make when clicking your fingernails together or tapping the tub beneath the water surface. Compare the sound with what you hear when both the source and your ears are above the water. At the risk of getting the floor wet, slide back and forth in the tub at different frequencies and see how the amplitude of the sloshing waves quickly builds up when you slide in rhythm with the waves. (The latter of these projects is most effective when you are alone in the tub.)

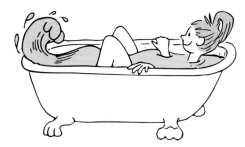

3. Stretch a piece of balloon rubber not too tightly over a radio loudspeaker. Glue a small, very lightweight piece of mirror, aluminum foil, or polished metal near one edge. Project a narrow beam of light on the mirror while your favorite music is playing and observe the beautiful patterns that are reflected on a screen or wall.

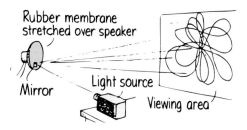

Exercises

1. Toss a stone in still water and concentric circles are formed. What form will waves have if a stone is tossed into smoothly flowing water?

2. Why do flying bees buzz?

3. A cat can hear sound frequencies up to 70,000 Hz. Bats send and receive ultrahigh-frequency squeaks up to 120,000 Hz. Which hears sound of shorter wavelengths, cats or bats?

4. Scientifically speaking, what does it mean to say that a radio station is "at 101.1 on your FM dial"?

5. Sound from Source A has twice the frequency of sound from Source B. Compare the wavelengths of sound from the two sources.

6. Suppose a sound wave and an electromagnetic wave have the same frequency. Which has the longer wavelength?

7. At the stands of a race track you notice smoke from the starter's gun before you hear it fire. Explain.

8. In Olympic competition, a microphone picks up the sound of the starter's gun and sends it electrically to speakers at every runner's starting block. Why?

9. When a sound wave moves past a point in air, are there changes in the density of air at this point? Explain.

10. At the instant that a high-pressure region is created just outside the prongs of a vibrating tuning fork, what is being created inside between the prongs?

11. Why is it so quiet after a snowfall?

12. If a bell is ringing inside a bell jar, we can no longer hear it when the air is pumped out, but we can still see it. What differences in the properties of sound and light does this indicate?

13. Why is the moon described as a "silent planet"?

14. As you pour water into a glass, you repeatedly tap the glass with a spoon. As the tapped glass is being filled, does the pitch of the sound increase or decrease? (What should you do to answer this question?)

15. If the speed of sound depended on its frequency, would you enjoy a concert when you were sitting in the second balcony?

16. If the frequency of sound is doubled, what change will occur in its speed? In its wavelength?

17. Why does sound travel faster in warm air?

18. Why does sound travel faster in moist air? (*Hint:* At the same temperature, water vapor molecules have the same average kinetic energy as the heavier nitrogen and oxygen molecules in the air. How, then, do the average speeds of H_2O molecules compare with those of N_2 and O_2 molecules?)

19. Would the refraction of sound be possible if the speed of sound were unaffected by wind, temperature, and other conditions? Defend your answer.

20. Why can the tremor of the ground from a distant explosion be felt before the sound of the explosion can be heard?

21. What kinds of wind conditions would make sound more easily heard at long distances? Less easily heard at long distances?

22. Ultrasonic waves have many applications in technology and medicine. One advantage is that large intensities can be used without danger to the ear. Cite another advantage of their short wavelength. (*Hint*: Why do microscopists use blue light rather than white light to see detail?)

23. Why is an echo weaker than the original sound?

24. What two physics mistakes occur in a science fiction movie that shows a distant explosion in outer space, where you see and hear the explosion at the same time?

25. A rule of thumb for estimating the distance in kilometers between an observer and a lightning stroke is to divide the number of seconds in the interval between the flash and the sound by 3. Is this rule correct?

26. If a single disturbance some unknown distance away sends out both transverse and longitudinal waves that travel with distinctly different speeds in the medium, such as in the ground during earthquakes, how could the distance to the disturbance be determined?

27. Why will marchers at the end of a long parade following a band be out of step with marchers near the front?

28. Why do soldiers break step in marching over a bridge?

29. Why is the sound of a harp soft in comparison with the sound of a piano?

30. Apartment dwellers will testify that bass notes are more distinctly heard from music played in nearby apartments. Why do you suppose lower-frequency sounds get through walls, floors, and ceilings more easily?

31. If the handle of a tuning fork is held solidly against a table, the sound from the tuning fork becomes louder. Why? How will this affect the length of time the fork keeps vibrating? Explain.

32. The sitar, an Indian musical instrument, has a set of strings that vibrate and produce music, even though they are never plucked by the player. These "sympathetic strings" are identical to the plucked strings and are mounted below them. What is your explanation?

33. Why are you not troubled by sound interference as shown in Figure 19.15 when you listen to stereo or quadraphonic music?

34. A special device uses earphones to transmit sound out of phase from a noisy jackhammer to its operator. Over the noise of the jackhammer, the operator can easily hear your voice while you are unable to hear his. Explain.

35. When Meidor brings the pair of out-of-phase speakers together as shown in Figure 19.16, which waves are most canceled, long waves or short waves? Why?

36. An object resonates when the frequency of a vibrating force either matches its natural frequency or is a sub-multiple of its natural frequency. Why will it not resonate to multiples of its natural frequency? (Think of pushing a child in a swing.)

37. Two sound waves of the same frequency can interfere, but to make beats, two sound waves have to have different frequencies. Why?

38. Walking beside you, your friend takes 50 strides per minute while you take 48 strides per minute. If you start in step, when will you be in step again?

39. Suppose a piano tuner hears 3 beats per second when listening to the combined sound from his tuning fork and the piano note being tuned. After slightly tightening the string, he hears 5 beats per second. Should the string be loosened or tightened?

40. A human cannot hear sound at a frequency of 100 kHz, or sound at 102 kHz. But if you walk into a room in which two sources are emitting sound waves at 100 kHz and 102 kHz, you'll hear sound. Explain.

Problems

1. What is the wavelength of a 340-Hz tone in air? What is the wavelength of a 34,000-Hz ultrasonic wave in air?

2. An oceanic depth-sounding vessel surveys the ocean bottom with ultrasonic waves that travel 1530 m/s in seawater. How deep is the water directly below the vessel if the time delay of the echo to the ocean floor and back is 6 s?

3. A bat flying in a cave emits a sound and receives its echo 1 s later. How far away is the cave wall?

4. You watch a distant woman driving nails into her front porch at a regular rate of 1 stroke per second. You hear the sound of the blows exactly synchronized with the blows you see. And then you hear one more blow after you see her stop hammering. How far away is she?

5. Two speakers are wired to emit identical sounds in unison. The wavelength in air of the sounds is 6 m. Do the sounds interfere constructively or destructively (a) at a distance of 12 m from both speakers? (b) At a distance of 9 m from both speakers? (c) At a distance of 9 m from one speaker and 12 m from the other?

6. What is the frequency of the sound emitted by the speakers in the previous problem? Is this a low pitch or a high pitch relative to the range of human hearing?

7. A grunting porpoise emits a sound of the same frequency as the sounds in the previous two problems. What is the wavelength of this sound in water, where the speed of sound is 1500 m/s?

8. What beat frequencies are possible with tuning forks of frequencies 256, 259, and 261 Hz, respectively?

20

. . . .

Musical Sounds

: **P**leasant vibrations.

Most of the sounds we hear are noises. The impact of a falling object, the slamming of a door, the roaring of a motorcycle, and most of the sounds from traffic in city streets are noises. Noise corresponds to an irregular vibration of the eardrum produced by some irregular vibration. The sound of music has a different character, having periodic tones—or musical "notes." The line that separates music and noise, however, is thin and subjective. To some contemporary composers, it is nonexistent.

Noise Versus Music

Some people consider contemporary music and music from other cultures to be noise. Differentiating these types of music from noise becomes a problem of aesthetics. However, differentiating traditional music—that is, Western classical music and most types of popular music—from noise presents no problem. A person with total hearing loss could distinguish between these by using an oscilloscope. Recall from Figure 19.4 in the previous chapter that when an electrical signal from a microphone is fed into an oscilloscope, patterns of air pressure variations with time are nicely displayed—which easily distinguishes between noise and traditional music (Figure 20.1).

Musicians usually speak of musical tones in terms of three principal characteristics: pitch, loudness, and quality.

FIGURE 20.1 Graphical representations of noise and music.

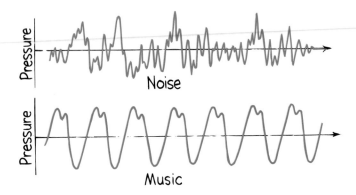

Pitch

The **pitch** of a sound relates to its frequency. Most sounds are composites of frequencies, in which case the pitch corresponds to the lowest frequency component. Rapid vibrations of the sound source produce a high note, whereas slow vibrations produce a low note. We speak of the pitch of a sound in terms of its position on the musical scale. When concert A is struck on a piano, a hammer strikes two or three strings, each of which vibrates 440 times in 1 second. The pitch of concert A corresponds to 440 hertz.*

Different musical notes are obtained by changing the frequency of the vibrating sound source. This is usually done by altering the size, the tightness, or the mass of the vibrating object. A guitarist or violinist, for example, adjusts the tightness, or tension, of the strings of the instrument when tuning them. Then different notes can by played by altering the length of each string by "stopping" it with the fingers.

In wind instruments, the length of the vibrating air column can be altered (trombone and trumpet) or holes in the side of the tube can be opened and closed in various combinations (saxophone, clarinet, flute) to change the pitch of the note produced.

High-pitched sounds used in music are most often less than 4000 hertz, but the average human ear can hear sounds with frequencies up to 18,000 hertz. Some people can hear tones of higher pitch than this, and so can most dogs. In general, the upper limit of hearing in people gets lower as they grow older. A high-pitched sound is often inaudible to an older person and yet may be clearly heard by a younger one. So by the time you can really afford that high-fidelity music system, you may not be able to appreciate the difference.

Sound Intensity and Loudness

The **intensity** of sound depends on the amplitude of pressure variations within the sound wave. (And, as with all waves, intensity is directly proportional to the square of the wave amplitude.) Intensity is measured in units of watts/meter^2. The human ear responds to intensities covering the enormous range from 10^{-12} W/m^2 (the threshold of hearing) to more than 1 W/m^2, (the threshold of pain). Because the range is so great, intensities are scaled by factors of ten, with the barely audible 10^{-12} W/m^2 as a reference intensity—called 0 *bel,* a unit named after Alexander Graham Bell. A sound ten times

*Interestingly enough, concert A varies from as low as 436 Hz to as high as 448 Hz in different symphony orchestras.

Table 20.1 Common sources and sound intensities

Source of sound	Intensity (W/m²)	Sound level (dB)
Jet airplane 30 m away	10^2	140
Air-raid siren, nearby	1	120
Disco music, amplified	10^{-1}	110
Riveter	10^{-3}	90
Busy street traffic	10^{-5}	70
Conversation in home	10^{-6}	60
Quiet radio in home	10^{-8}	40
Whisper	10^{-10}	20
Rustle of leaves	10^{-11}	10
Threshold of hearing	10^{-12}	0

more intense has an intensity of 1 bel (10^{-11} W/m²) or 10 *decibels*. Table 20.1 lists typical sounds and their intensities.

A sound of 10 decibels is 10 times as intense as 0 decibels, the threshold of hearing. 20 decibels is 100 or 10^2 times the intensity of the threshold of hearing. Accordingly, 30 decibels is 10^3 times the threshold of hearing and 40 decibels is 10^4 times. So 60 decibels represents sound intensity a million times (10^6) greater than 0 decibels; 80 decibels represents sound 10^2 times as intense as 60 decibels.*

Physiological hearing damage begins at exposure to 85 decibels, the degree depending on the length of exposure and on frequency characteristics. Damage from loud sounds can be temporary or permanent, depending on whether the organ of Corti, the receptor organ in the inner ear, is impaired or destroyed. A single burst of sound can produce vibrations in the organs intense enough to tear them apart. Less intense, but severe, noise can interfere with cellular processes in the organs and cause their eventual breakdown. Unfortunately, the cells of these organs do not regenerate.

Sound intensity is a purely objective and physical attribute of a sound wave and can be measured by various acoustical instruments (and the oscilloscope in Figure 20.2). **Loudness**, on the other hand, is a physiological sensation. The ear senses

FIGURE 20.2 James displays a sound signal on an oscilloscope.

*The decibel scale is called a logarithmic scale. The decibel rating is proportional to the logarithm of the intensity.

some frequencies much better than others. A 3500-Hz sound at 80 decibels, for example, sounds about twice as loud to most people as a 125-Hz sound at 80 decibels; humans are more sensitive to the 3500-Hz range of frequencies. The loudest sounds we can tolerate have intensities a million million times greater than the faintest sounds. The difference in perceived loudness, however, is much less than this amount.

Question Can hearing be permanently impaired when attending concerts, clubs, or functions that feature very loud music?

Quality

We have no trouble distinguishing between the tone from a piano and a like-pitched tone from a clarinet. Each of these tones has a characteristic sound that differs in **quality**, or timbre. Most musical sounds are composed of a superposition of many tones differing in frequency. The various tones are called **partial tones**, or simply *partials*. The lowest frequency, called the **fundamental frequency**, determines the pitch of the note. Partial tones whose frequencies are whole number multiples of the fundamental frequency are called **harmonics**. A tone that has twice the frequency of the fundamental is the second harmonic, a tone with three times the fundamental frequency is the third harmonic, and so on (Figure 20.3).* It is the variety of partial tones that give a musical note its characteristic quality.

FIGURE 20.3 Modes of vibration of a guitar string.

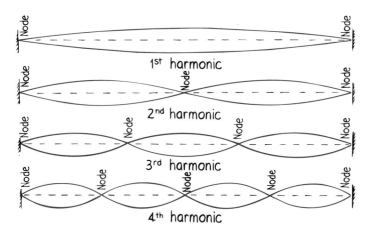

1st harmonic

2nd harmonic

3rd harmonic

4th harmonic

Answer Yes, depending on how loud, how long, how near, and how often. Some music groups have emphasized loudness over quality. Tragically, as hearing becomes more and more impaired, members of such groups (and their fans) require louder and louder sounds for stimulation. Hearing loss caused by sounds is particularly common in the frequency range of 2000–5000 Hz. Human hearing is normally most sensitive around 3000 Hz.

---

*In the terminology often used in music, the second harmonic is called the first *overtone,* the third harmonic the second overtone, and so on.

Not all partial tones present in a complex tone are integer multiples of the fundamental. Unlike the harmonics of woodwinds and brasses, stringed instruments such as a piano produce "stretched" partial tones that are nearly, but not quite, harmonics. This is an important factor in tuning pianos and happens because the stiffness of the strings adds a little bit of restoring force to the tension.

Thus, if we strike middle C on the piano, we produce a fundamental tone with a pitch of about 262 hertz and also a blending of partial tones of two, three, four, five, and so on times the frequency of middle C. The number and relative loudness of the partial tones determine the quality of sound associated with the piano. Sound from practically every musical instrument consists of a fundamental and partials. Pure tones, those having only one frequency, can be produced electronically. Electronic synthesizers produce pure tones and mixtures of these to give a vast variety of musical sounds.

FIGURE 20.4 A composite vibration of the fundamental mode and the third harmonic.

The quality of a tone is determined by the presence and relative intensity of the various partials. The sound produced by a certain tone from the piano and the one produced by one of the same pitch from a clarinet have different qualities that the ear recognizes because their partials are different. A pair of tones of the same pitch with different qualities have either different partials or a difference in the relative intensity of the partials.

FIGURE 20.5 Sounds from the piano and clarinet differ in quality.

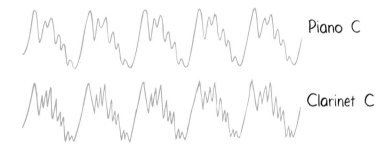

Piano C

Clarinet C

Musical Instruments

Conventional musical instruments can be grouped into one of three classes: those in which the sound is produced by vibrating strings, those in which the sound is produced by vibrating air columns, and those in which the sound is produced by *percussion*—the vibrating of a two-dimensional surface.

In a stringed instrument, the vibration of the strings is transferred to a sounding board and then to the air, but with low efficiency. To compensate for this, we find relatively large string sections in orchestras. A smaller number of the high-efficiency wind instruments sufficiently balances a much larger number of violins.

In a wind instrument, the sound is a vibration of an air column in the instrument. There are various ways to set air columns into vibration. In brass instruments such as trumpets, French horns, and trombones, vibrations of the player's lips interact with standing waves that are set up by acoustic energy reflected within the instrument by the flared bell. The lengths of the vibrating air columns are manipulated by pushing valves that add or subtract extra segments or by extending the length of the tube.* In woodwinds such as clarinets, oboes, and saxophones, a stream of air produced by the musician sets a reed vibrating, whereas in fifes, flutes, and piccolos, the musician blows air

---

*A bugle has neither valves nor variable length. A bugler must be adept in creating different overtones to get different notes.

against the edge of a hole to produce a fluttering stream that sets the air column into vibration.

Some instruments, such as a cello or a trombone, can produce a continuous range of possible fundamental frequencies. Other instruments, such as a piano or a clarinet, are limited to only certain fundamental frequencies. Most music is based on a particular set of frequencies called a *scale*. In the "well-tempered scale," each note has a frequency that is the twelfth root of 2 ($\sqrt[12]{2}$), or 1.05946, times the frequency of the note just below it. You can see that if you start at a certain note, and advance through 12 successive notes in the scale, you will reach a note that has twice the frequency of the original note. A factor-of-two increase in frequency is called an *octave*. Twelve more steps and you will advance another octave. The seven white keys and five black keys on a piano span one octave.

In percussion instruments such as drums and cymbals, a two-dimensional membrane or elastic surface is struck to produce sound. The fundamental tone produced depends on the geometry, the elasticity, and, in some cases, the tension of the surface. Changes in pitch result from changing the tension in the vibrating surface; depressing the edge of a drum membrane with the hand is one way of accomplishing this. Different modes of vibration can be set up by striking the surface in different places. In the kettledrum, the shape of the kettle changes the frequency of the drum. As in all musical sounds, the quality depends on the number and relative loudness of the partial tones.

Electronic musical instruments differ markedly from conventional musical instruments. Instead of strings that must be bowed, plucked, or struck, or reeds over which air must be blown, or diaphragms that must be tapped to produce sounds, some electronic instruments use electrons to generate the signals that make up musical sounds. Others start with sound from an acoustical instrument and then modify it. Electronic music demands of the composer and player an expertise beyond the knowledge of musicology. It brings a powerful new tool to the hands of the musician.

Fourier Analysis

Did you ever look closely at the grooves in an old phonograph record? Variations in the width of the grooves, seen in Figure 20.6, cause the phonograph needle (stylus) that rides in the groove to vibrate. These mechanical vibrations, in turn, are transformed into electrical vibrations to produce the sound you hear from a record. Isn't it remarkable that all the distinct vibrations made by the various pieces of an orchestra are captured by

FIGURE 20.6 A microscopic view of the grooves in a phonograph record.

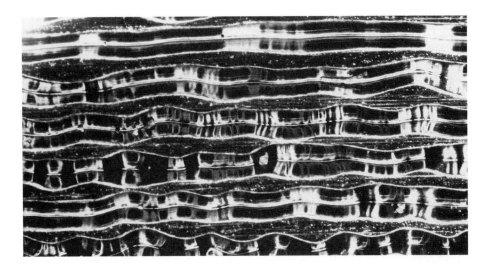

FIGURE 20.7 Wave forms of (a) an oboe, (b) a clarinet, and (c) the oboe and clarinet sounded together.

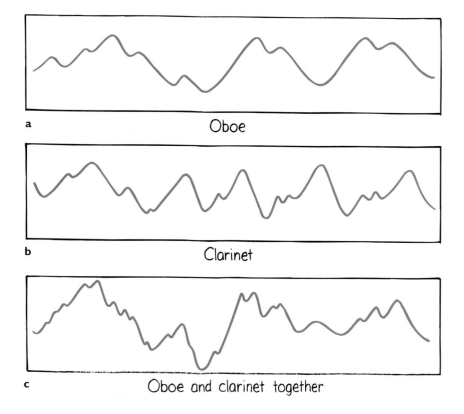

a   Oboe

b   Clarinet

c   Oboe and clarinet together

the single-wave groove of the record? The sound of an oboe when captured by the groove of phonograph records and displayed on an oscilloscope screen looks like Figure 20.7a. This wave corresponds to the electronic signal produced by the vibrating needle. It also corresponds to the amplified signal that activates the loudspeaker of the sound system and to the amplitude of air vibrating against the eardrum. Figure 20.7b shows the wave appearance of a clarinet. When oboe and clarinet are sounded together, the principle of superposition is evident as their individual waves combine to produce the wave form shown in Figure 20.7c.

The shape of the wave in Figure 20.7c is the net result of shapes a and b superposing (interfering). If we know a and b, it is a simple thing to create c. But it is a far different problem to discern in c the shapes of a and b that make it up. Looking only at shape c, we cannot unscramble the oboe from the clarinet.

But play the record on the phonograph, and our ears will at once know what instruments are being played, what notes they are playing, and what their relative loudness is. Our ears break the overall signal into its component parts automatically.

In 1822 the French mathematician Joseph Fourier discovered a mathematical regularity to the component parts of periodic wave motion. He found that even the most complex periodic wave motion is created by simple sine waves that add together. A sine wave is the simplest of waves, having a single frequency (Figure 20.8). Fourier found that all periodic waves may be broken down into constituent sine waves of different amplitudes and frequencies. The mathematical operation for doing this is called **Fourier analysis**. We will not explain the mathematics here but simply point out that by such analysis one can find the pure sine waves that add to compose the tone of, say, a violin. When these pure tones with the appropriate amplitudes are sounded together, as by striking a number of tuning

FIGURE 20.8 A sine wave.

forks or by selecting the proper keys on an electric organ, they combine to give the tone of the violin. The lowest-frequency sine wave is the fundamental and determines the pitch of the note. The higher-frequency sine waves are the partials that give the characteristic quality. Thus, the wave form of any musical sound is no more than a sum of simple sine waves.

Since the wave form of music is a multitude of various sine waves, to duplicate sound accurately by radio, record player, or tape recorder we should be able to process as large a range of frequencies as possible. The notes of a piano keyboard range from 27 hertz to 4200 hertz, but to duplicate the music of a piano composition accurately, the sound system must have a range of frequencies up to 20,000 hertz. The greater the range of the frequencies of an electrical sound system, the closer the musical output approximates the original sound, hence the wide range of frequencies in a high-fidelity sound system.

FIGURE 20.9 The fundamental and its harmonics combine to produce a composite wave.

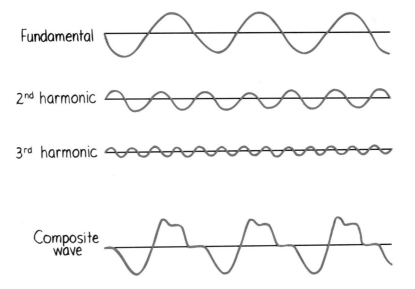

Our ear performs a sort of Fourier analysis automatically. It sorts out the complex jumble of air pulsations that reach it and transforms them into pure tones made up of sine waves. And we recombine various groupings of these pure tones when we listen.

FIGURE 20.10 Does each hear the same music?

What combinations of tones we have learned to focus our attention on determines what we hear when we listen to a concert. We can direct our attention to the sounds of the various instruments and discern the faintest tones from the loudest; we can delight in the intricate interplay of instruments and still detect the extraneous noises of others around us. This is a most incredible feat.

Compact Discs

We now experience the rich, full sounds of a string quartet or a symphony orchestra in our own rooms on a CD (compact disc) by means of the remarkable sound recording and reproduction technique called *digital audio*. Mom and Dad may have had their music on LP records that utilized a conventional stylus that vibrated in the record groove. Today the digital player utilizes a laser beam, which is directed onto a plastic reflective disc. The input to the amplifier comes from a light sensor rather than a stylus.

On yesterday's phonograph record, the stylus was made to vibrate when it rode in the squiggly phonograph groove. The output was a signal like those shown in Figure 20.7. This type of continuous wave form is called an *analog* signal. It can be changed to a *digital* signal by measuring the numerical value of its amplitude during each split second (Figure 20.11). This numerical value can be expressed in a number system that is convenient for computers, called *binary*. In the binary code, any number can be expressed as a succession of ones and zeros; for example, the number 1 is 1, 2 is 10, 3 is 11, 4 is 100, 5 is 101, 17 is 10001, etc. So the shape of the analog wave form can be expressed as a series of "on" and "off" pulses that corresponds to a series of ones and zeros in binary code. That's where the laser disc comes in.

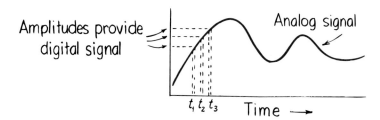

FIGURE 20.11 The amplitude of the analog wave form is measured at successive split seconds to provide digital information that is recorded in binary form on the reflective surface of the laser disc.

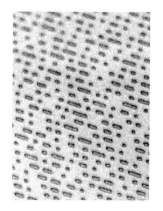

FIGURE 20.12 A microscopic view of the pits on a laser disc.

Instead of a squiggly phonograph groove, a CD has a series of microscopic pits about thirty times thinner than a human hair (Figure 20.12). When the laser beam falls on a flat portion of the reflective surface, it is reflected directly into the player's optical system; this gives an "on" pulse. When the beam is incident upon a passing pit, very little of the laser beam returns to the optical sensor; this gives an "off" pulse. A flickering of "on" and "off" pulses generates the "one" and "zero" digits of the binary code.

The rate at which these tiny pits on the disc are sampled is 44,100 times per second. A single audio disc is $\frac{1}{6}$ the size of a conventional LP and contains billions of bits of information. All this information is encoded on the reflective surface, which is covered with a protective layer of clear plastic. Since the laser beam is focused onto the signal surface below, the CD is immune to dust, scratches, and fingerprints. No more of the snap, crackle, and pop so characteristic of yesterday's LP. And since the laser beam does not touch the disc, the disc never wears out—no matter how many times you play it.

FIGURE 20.13   A tightly focused laser beam reads digital information represented by a series of pits on the laser disc.

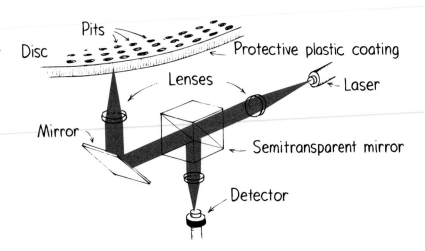

The most extraordinary feature of the laser disc, however, is the quality of the sound. You can hear the difference.

Summary of Terms

Pitch   The "highness" or "lowness" of a tone, as on a musical scale, which is principally governed by frequency. A high-frequency vibrating source produces a sound of high pitch; a low-frequency vibrating source produces a sound of low pitch.

Intensity   The power per square meter carried by a sound wave, often measured in decibels.

Loudness   The physiological sensation directly related to sound intensity or volume.

Quality   The characteristic timbre of a musical sound, governed by the number and relative intensities of partial tones.

Partial tone   Single-frequency component sound wave of a complex tone. When the frequency of a partial tone is an integer multiple of the lowest frequency, it is a harmonic.

Fundamental frequency   The lowest frequency of vibration, or first harmonic, in a musical tone.

Harmonic   A partial tone whose frequency is an integer multiple of the fundamental frequency. The second harmonic has twice the frequency of the fundamental, the third harmonic three times the frequency, and so on in sequence.

Fourier analysis   A mathematical method that resolves any periodic wave form into a combination of simple sine waves.

Suggested Reading

Rossing, Thomas D. "The Compact Disc Digital Audio System." *The Physics Teacher* 25, no. 9 (Dec. 1987): 556-562. A close look at CD technology.

Review Questions

Noise Versus Music

1. Distinguish between noise and music.
2. What are the three principal characteristics of musical tones?

Pitch

3. How can the pitch of a guitar string be increased?
4. What is the range of normal human hearing? What happens to this range with age?

Sound Intensity and Loudness

5. How do the *intensities* of the loudest sounds we can tolerate compare to the lowest intensities we can hear?
6. Is the sound of 30 dB 30 times greater than the threshold of hearing, or 10^3 (a thousand) times greater?

Quality

7. What exactly determines the pitch of a note?
8. If the fundamental frequency of a note is 200 Hz, what is the frequency of the second harmonic? The third harmonic?
9. What exactly determines the musical quality of a note?
10. Why do the same notes plucked on a banjo and a guitar have distinctly different sounds?

Musical Instruments

11. What are the three principal classes of musical instruments?

12. Why do orchestras generally have a greater number of stringed instruments than wind instruments?

Fourier Analysis

13. What did Fourier discover about complex periodic wave patterns?
14. A high-fidelity sound system may have a frequency range that extends up to or beyond 20,000 hertz. Of what use is this extended range?

Compact Discs

15. How is the sound signal captured on a conventional phonograph record? How is the sound signal captured on a laser disc?
16. Why does a laser disc not wear out like a conventional phonograph record?

Projects

1. Test to see which ear has the better hearing by covering one ear and finding how far away your open ear can hear the ticking of a clock; repeat for the other ear. Notice also how the sensitivity of your hearing improves when you cup your ears with your hands.
2. With a strong magnifying glass, examine the grooves in phonograph records. If you have an old 78-RPM disk, compare the grooves with those of a $33\frac{1}{3}$-RPM disk.
3. Make the lowest-pitched sound you are capable of; then keep doubling the pitch to see how many octaves your voice can span. If you are a singer, what is your range?
4. On a sheet of graph paper, construct one full cycle (one period of the fundamental) of the composite wave of Figure 20.9 by superposing various vertical displacements of the fundamental and first two partial tones. Your instructor can show you how this is done. Then find the composite waves of partial tones of your own choosing.

Exercises

1. The yellow-green light emitted by street lights matches the yellow-green color to which the human eye is most sensitive. Consequently, a street light that emits light of this color is better seen at night. Similarly, the monitored sound intensity of television commercials is louder than the sound from regular programming, yet doesn't exceed the regulated intensities. At what frequencies do advertisers concentrate the commercials' sound?

2. Explain how you can lower the pitch of a note on a guitar by altering (a) the length of the string, (b) the tension of the string, or (c) the thickness or the mass of the string.
3. Why is the thickness greater for the bass strings of a guitar than for the treble strings?
4. Would a plucked guitar string vibrate for a longer or a shorter time if the instrument had no sounding board? Why?
5. If you touch a guitar string very lightly at its midpoint, you can hear a tone that is one octave above the fundamental for that string. (A tone that is an octave above another tone has twice the frequency.) Explain.
6. If a guitar string vibrates in two segments, where can a tiny piece of folded paper be supported without flying off? How many pieces of folded paper could similarly be supported if the wave form were of three segments?
7. If a vibrating string is made shorter (as by holding a finger on it), what effect does this have on the frequency of vibration and on the pitch?
8. A violin string playing the note A oscillates at 440 Hz. What is the period of the string's oscillation?
9. The amplitude of a transverse wave in a stretched string is the maximum displacement of the string from its equilibrium position. What does the amplitude of a longitudinal sound wave in air correspond to?
10. Which of the two musical notes displayed one at a time on an oscilloscope screen has the higher pitch? Which is the louder (if they are being detected by equivalent microphones)?

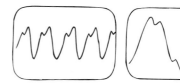

11. In a hi-fi speaker system, why is the woofer (low-frequency speaker) larger than the tweeter (high-frequency speaker)?
12. One person has a threshold of hearing of 5 dB, and another of 10 dB. Which person has the more acute hearing?
13. How much more intense is sound at 110 dB than sound at 50 dB?
14. How is an electronic organ able to imitate the sounds made by various musical instruments?
15. A person talking after inhaling helium gas has a high-pitched voice. One of the reasons for this is the higher speed of sound in helium than in air. Why does sound travel faster in helium?

16. Why does your voice sound fuller in the shower?

17. The frequency range for a telephone is between 500 and 4000 Hz. Why does a telephone not do a very good job of transmitting music?

18. How many octaves does normal human hearing span? How many octaves are on a common piano keyboard? (If you're not sure, look and see.)

19. If the fundamental frequency of a violin string is 440 Hz, what is the frequency of the second harmonic? The third?

20. At an outdoor concert, will the pitch of musical tones be affected on a windy day? Explain.

21. A trumpet has keys and valves that let the trumpeter change the length of the vibrating air column and the position of the nodes. A bugle has no such keys and valves, yet it can sound different notes. How do you think the bugler achieves different notes?

22. The human ear is sometimes called a Fourier analyzer. What does this mean and why is it an apt description?

23. Most compact discs are "read-only." They can't easily be erased and re-recorded the way the floppy disk or hard disk in your computer can. Why is the normal CD "read-only"?

24. Do all the people in a group hear the same music when they listen to it attentively? (Do all see the same sight when looking at a painting? Do all taste the same flavor when sipping the same wine? Do all perceive the same aroma when smelling the same perfume? Do all feel the same texture when touching the same fabric? Do all come to the same conclusion when listening to a logical presentation of ideas?)

Problems

1. How much more intense than the threshold of hearing is a sound of 10 dB? 30 dB? 60 dB?

2. How much more intense is a sound of 40 dB than a sound of 30 dB?

3. A certain note has a frequency of 1000 Hz. What is the frequency of a note one octave above it? Two octaves above it? One octave below it? Two octaves below it?

4. Starting with a fundamental tone, how many harmonics are there between the first and second octaves? Between the second and third octaves? (Look at Figure 20.3 to get started.)

5. A cello string 0.75 m long has a 220-Hz fundamental frequency. Find the wave speed along the vibrating string.

Part

5

ELECTRICITY
AND MAGNETISM

···························

This simple electric circuit illustrates some intriguing physics. The battery provides *voltage*, an electric pressure that pushes electrons through the wire and lamp. Electrons flow easily through the relatively thick wire, but with difficulty through the lamp filament, which has *resistance* to electron flow. *Current* squeezed through it shakes the atoms so vigorously that they glow. That's why the filament emits *light* while the connecting wire doesn't. Even the light is electrical in nature -- *magnetic* too, as Part 5 will show

Onward!

21

. . . .

Electrostatics

⋮ **O**ne cloud provides more voltage than five million car batteries.

Electricity is the name given to a wide range of phenomena that in one form or another underlie just about everything around us—from lightning in the sky, sparks beneath our feet when we scuff across a rug, to what holds atoms together to form molecules. The control of electricity is evident in technological devices of many kinds, from lamps to computers. In this technological age it is important to have an understanding of the basics of electricity and of how these basics can be manipulated to produce a prosperity unknown before recent times.

In this chapter we will investigate electricity at rest, or **electrostatics**, as it is called. This involves electric charges, the forces between them, the aura that surrounds them, and their behavior in materials. In the next chapter we will investigate the motion of electric charges, or *electric currents,* and the voltages that produce them and how they can be controlled. In Chapter 23 we will study the relationship of electric currents to magnetism and in Chapter 24 how magnetism and electricity can be controlled to operate motors and other electrical devices.

An understanding of electricity requires a step-by-step approach, for one concept is the building block for the next, and so on. So please put extra care into the study of this material. It can be difficult, confusing, and frustrating if you're hasty. But with careful effort, it can be comprehensible and rewarding. Onward!

Electrical Forces

Consider a universal force which, like gravity, varies inversely as the square of distance but which is billions upon billions of times stronger than gravity. If there were such a force and if it were everywhere attractive like gravity, the universe would be pulled together into a tight ball with all the matter pulled as close together as it could get. But suppose this force were a repelling force, with every bit of matter repelling every other bit of matter. What then? The universe would be an ever-expanding gaseous cloud. Suppose, however, that the universe consisted of two kinds of particles, say, positives and negatives. Suppose that positives repelled positives but attracted negatives, and that negatives repelled negatives but attracted positives. Like kinds repel and unlike kinds attract (Figure 21.1). And suppose that there were equal numbers of each—and some neutrals unaffected by this force. What would the universe be like? The answer is simple: It would be like the one we are living in. For there are such particles and there is such a force—the *electrical force*.

FIGURE 21.1 Like charges repel. Unlike charges attract.

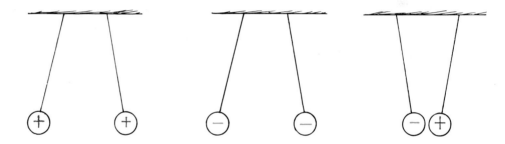

Clusters of positive and negative particles have been pulled together by the enormous attraction of the electrical force. By forming compact and evenly mixed clusters of positives and negatives, the huge forces have balanced themselves out almost perfectly. These clusters are the atoms of matter. When two or more atoms join to form a molecule, the molecule also contains balancing positives and negatives. And when trillions of molecules combine to form a speck of matter, the electrical forces balance again. Between two pieces of ordinary matter, there is scarcely any electrical attraction or repulsion at all, because each piece contains equal numbers of positives and negatives. Between the earth and the moon, for example, the electrical forces cancel out. The much weaker gravitational force, which only attracts, is left as the predominant force between these bodies.

Electric Charges

The terms *positive* and *negative* refer to electric *charge,* the fundamental quantity that underlies all electrical phenomena. Protons are positively charged and electrons are negatively charged. The attractive force between protons and electrons holds atoms together. Between neighboring atoms the negative electrons of one atom may at times be closer to the positive protons of another atom, so the attractive force between these charges is greater than the repulsive force, and the atoms combine to form a molecule. In fact, all the chemical bonding forces that hold atoms together to form molecules are electrical forces acting in small regions where the balance of attractive and repelling forces is not perfect. Anyone planning to study chemistry should first know something about electrical attraction and repulsion, and before studying electrical phenomena

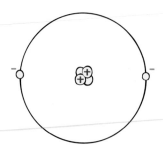

FIGURE 21.2 Model of a helium atom. The atomic nucleus is made up of two protons and two neutrons. The positively charged protons attract two negative electrons.

should know something about atoms. Here are some important facts about atoms:

1. Every atom has a positively charged nucleus surrounded by negatively charged electrons.

2. All electrons are identical; that is, they all have the same mass and the same quantity of negative charge.

3. The nucleus is composed of protons and neutrons. (The common form of hydrogen, which has no neutron, is the only exception.) All protons are identical; similarly, all neutrons are identical. A proton has nearly 2000 times the mass of an electron but its positive charge is equal in magnitude to the negative charge of the electron. A neutron has slightly greater mass than a proton and has no charge.

4. Atoms usually have as many electrons as protons, so the atom has zero *net* charge.

Why don't protons pull the oppositely charged electrons into the nucleus? Prior to quantum mechanics, no one had a satisfactory answer. A popular answer, now known to be incorrect, was that electrons aren't pulled into the nucleus by electrical force for the same reason planets aren't pulled into the sun by gravitational force—electrons are held in orbit by the pull of the protons. This oversimplified explanation served as a starter for understanding the electrical nature of the atom. The terms *orbit* and *orbital* are still used, although a better word is *shell,* which suggests that the electrons are spread out over a spherical region. The answer today has to do with the wave nature of electrons. An electron behaves like a wave and has to have a certain amount of space related to its wavelength. We'll see in Chapter 31, when we treat quantum mechanics, that atomic size is set by the minimum amount of "elbow room" that an electron requires.

Why don't the protons in the nucleus mutually repel and fly apart? What holds the nucleus together? The answer is that in addition to electrical forces in the nucleus, there are even stronger non-electrical nuclear forces that are able to hold the protons together in spite of the electrical repulsion. We will treat this further in Chapter 32.

Questions

1. Beneath the complexities of electrical phenomena, there lies a fundamental rule from which nearly all other effects stem. What is this fundamental rule?

2. How does the charge of an electron differ from the charge of a proton?

Conservation of Charge

In a neutral atom there are as many electrons as protons, so there is no net charge. The positive balances the negative exactly. If an electron is removed from an atom, then it is no longer neutral. The atom then has one more positive charge (proton) than negative charge (electron) and is said to be positively charged.* A charged atom is called an *ion.*

Answers

1. Like charges repel; opposite charges attract.

2. The charge of an electron is equal in magnitude, but opposite in sign, to the charge of a proton.

---

* Each proton has a charge $+e$, equal to $+ 1.6 \times 10^{-19}$ coulomb. Each electron has a charge $-e$, equal to -1.6×10^{-19} coulomb. Why such different particles have the same magnitude of charge is an unanswered question in physics. The equality of the magnitudes has been tested to high accuracy.

FIGURE 21.3 Electrons are transferred from the fur to the rod. The rod is then negatively charged. Is the fur charged? How much compared to the rod? Positively or negatively?

A *positive ion* has a net positive charge. A *negative ion,* an atom with one or more extra electrons, is negatively charged.

Material objects are made of atoms, which means they are composed of electrons and protons (and neutrons as well). Objects ordinarily have equal numbers of electrons and protons and are therefore electrically neutral. But if there is a slight imbalance in the numbers, the object is electrically charged. An imbalance comes about when electrons are added or removed. Although the innermost electrons in an atom are bound very tightly to the oppositely charged atomic nucleus, the outermost electrons of many atoms are bound very loosely and can be easily dislodged. How much work is required to tear an electron away from an atom varies for different substances. The electrons are held more firmly in plastic than in your hair, for example. Hence, when a comb is rubbed through your hair, electrons transfer from the hair to the comb. The plastic then has an excess of electrons and is said to be *negatively charged*. The hair, in turn, has a deficiency of electrons and is said to be *positively charged.* If you rub a glass or plastic rod with silk, you'll find that the rod becomes positively charged. The silk has a greater affinity for electrons than the glass or plastic rod. Electrons are rubbed off the rod and onto the silk.

So we see that an object having unequal numbers of electrons and protons is electrically charged. If it has more electrons than protons, it is negatively charged. If it has fewer electrons than protons, it is positively charged.

It is important to note that when we charge something, no electrons are created or destroyed. Electrons are simply transferred from one material to another. Charge is conserved. In every event, whether large-scale or at the atomic and nuclear level, the principle of **conservation of charge** applies. No case of the creation or destruction of electric charge has ever been found. The conservation of charge is a cornerstone in physics, ranking with the conservation of energy and momentum.

Any object that is electrically charged has an excess or deficiency of some whole number of electrons—electrons cannot be divided into fractions of electrons. This means that the charge of the object is a whole-number multiple of the charge of an electron. It cannot have a charge equal to the charge of $1\frac{1}{2}$ or $1000\frac{1}{2}$ electrons, for example. Charge is "grainy," or made up of elementary units called quanta. We say that charge is *quantized,* with the smallest quantum of charge being that of the electron (or proton). No smaller units of charge have ever been found.* All charged objects to date have a charge that is a whole-number multiple of the charge of a single electron.

Question If you scuff electrons onto your feet while walking across a rug, are you negatively or positively charged?

Answer You have more electrons after you scuff your feet, so you are negatively charged (and the rug is positively charged).

* Within the atomic nucleus, however, elementary particles called *quarks* carry charges $\frac{1}{3}$ and $\frac{2}{3}$ of the magnitude of the electron's charge. Each proton and each neutron is made up of three quarks. Since quarks always exist in such combinations and have never been found separated, the whole-number-multiple rule of electron charge holds for nuclear processes as well.

Coulomb's Law

The electrical force, like gravitational force, decreases inversely as the square of the distance between charged bodies. This relationship was discovered by Charles Coulomb in the eighteenth century and is called **Coulomb's law**. It states that for two charged objects that are much smaller than the distance between them, the force between the two objects varies directly as the product of their charges and inversely as the square of the separation distance. (Review the inverse-square law in Figure 8.4 back on page 147.) The force acts along a straight line from one charged object to the other. Coulomb's law can be expressed as

$$F = k \frac{q_1 q_2}{d^2}$$

where d is the distance between the charged particles, q_1 represents the quantity of charge of one particle, q_2 represents the quantity of charge of the other particle, and k is the proportionality constant.

The unit of charge is called the **coulomb**, abbreviated C. It turns out that a charge of 1C is the charge associated with 6.25 billion billion electrons. This might seem like a great number of electrons, but it represents only the amount of charge that passes through a common 100-watt light bulb in a little over a second.

The proportionality constant k in Coulomb's law is similar to G in Newton's law of gravitation. Instead of being a very small number like G (6.67×10^{-11}), the electrical proportionality constant k is a very large number. It is approximately

$$k = 9,000,000,000 \text{ N·m}^2/\text{C}^2$$

or, in scientific notation, $k = 9 \times 10^9$ N·m^2/C^2. The unit N·m^2/C^2 is not central to our interest here; it simply converts the right-hand side of the equation to the unit of force, the newton (N). What is important is the large magnitude of k. If, for example, a pair of like charged particles of 1 coulomb each were 1 meter apart, the force of repulsion between them would be 9 billion newtons.* That would be more than ten times the weight of a battleship! Obviously, such amounts of net charge do not usually exist in our everyday environment.

So Newton's law of gravitation for massive bodies is similar to Coulomb's law for electrically charged bodies.† Whereas the gravitational force of attraction between particles such as an electron and a proton is extremely small, the electrical force between these particles is relatively enormous. Other than the big difference in strength, the most important difference between gravitational and electrical forces is that electrical forces may be either attractive or repulsive, whereas gravitational forces are only attractive.

---

* In comparing the magnitudes of G and k, we must note that they depend on the units chosen for mass and electric charge, which could have been chosen differently. So our comparison only reminds us that electric forces are usually enormous compared with gravitational forces. Contrast the 9 billion newtons between the two unit charges 1 meter apart with the gravitational force of attraction between two unit masses (kilograms) 1 m apart: 6.67×10^{-11} N—an extremely small force. For the force to be 1 N, the masses at 1 m apart would have to be nearly 122,000 kg each! Gravitational forces between ordinary objects are much too small to be detected except in delicate experiments. But electrical forces between ordinary objects can be relatively huge. Even for highly charged objects, however, the imbalance of electrons to protons is normally less than one part in a trillion.

† According to quantum theory, a force varies inversely as the square of the distance if it involves the exchange of particles with no mass. Exchange of massless photons is responsible for the electrical force and exchange of massless gravitons accounts for the gravitational force. Some scientists have looked for an even deeper relationship between gravity and electricity. Albert Einstein spent the latter part of his life searching with little success for a "unified field theory." More recently, the electrical force has been unified with one of the two nuclear forces, the *weak force*, which plays a role in radioactive decay.

Questions

1. The proton that is the nucleus of the hydrogen atom attracts the electron that orbits it. Relative to this force, does the electron attract the proton with less force, more force, or the same amount of force?

2. If a proton at a particular distance from a charged particle is repelled with a given force, by how much will the force decrease when the proton is three times as distant from the particle? Five times as distant?

3. What is the sign of charge of the particle in this case?

Conductors and Insulators

It is easy to establish an electric current in metals because one or more of the electrons in the outer shell of the atoms in a metal are not anchored to the nuclei of particular atoms, but are free to wander in the material. Such materials are called good **conductors**. Metals are good conductors of electric current for the same reason they are good heat conductors. Electrons in their outer atomic shell are "loose."

The electrons in other materials, rubber and glass for example, are tightly bound and belong to particular atoms. They are not free to wander about among other atoms in the material. Consequently, it isn't easy to make them flow. These materials are poor conductors of electric current for the same reason they are generally poor heat conductors. Such materials are called good **insulators**.

FIGURE 21.4 It is easier to establish an electric current through hundreds of kilometers of metal wire than through a few centimeters of insulating material.

Answers

1. The same amount of force, in accord with Newton's third law—basic mechanics! Recall that a force is an interaction between two things, in this case between the proton and the electron. They pull on each other—equally.

2. It decreases to $\frac{1}{9}$ its original value; to $\frac{1}{25}$.

3. Positive.

All substances can be arranged in order of their ability to conduct electric charges. Those at the top of the list are the conductors and those at the bottom are the insulators. The ends of the list are very far apart. The conductivity of a metal, for example, can be more than a million trillion times greater than the conductivity of an insulator such as glass. In a common appliance cord, electrons flow through several meters of wire rather than flowing directly across from one wire to the other through a small fraction of a centimeter of rubber insulation.

Semiconductors

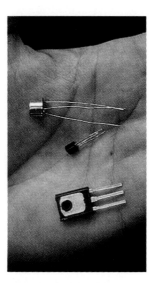

FIGURE 21.5 Transistors.

Whether a substance is classified as a conductor or an insulator depends on how tightly the atoms of the substance hold their electrons. Some materials, such as germanium and silicon, are neither good conductors nor good insulators, and fall in the middle of the range of electrical resistivity. These materials are fair insulators in their pure crystalline form but increase tremendously in conductivity when even one atom in 10 million is replaced with an impurity that adds or removes an electron from the crystal structure. These materials can be made to behave sometimes as insulators and sometimes as conductors, and are called **semiconductors**. Thin layers of semiconducting materials sandwiched together make up *transistors,* which are used to control the flow of currents in circuits, to detect and amplify radio signals, and to produce oscillations in transmitters; they also act as digital switches. These tiny solids were the first electrical components in which materials with different electrical characteristics were not interconnected by wires but were physically joined in one structure. They require very little power and in normal use last indefinitely.

A semiconductor will also conduct when light of the proper color shines on it. A pure selenium plate is normally a good insulator, and any electric charge built up on its surface will remain there for extended periods in the dark. If the plate is exposed to light, however, the charge leaks away almost immediately. If a charged selenium plate is exposed to a pattern of light, such as the pattern of light and dark that makes up this page, the charge will leak away only from the areas exposed to light. If a black plastic powder were brushed across its surface, the powder would stick only to the charged areas where the plate had not been exposed to light. Now if a piece of paper with an electric charge on the back were put over the plate, the black plastic powder would be drawn to the paper to form the same pattern as, say, the one on this page. If the paper were then heated to melt the plastic and fuse it to the paper, you might pay a nickel for it and call it a Xerox copy.

Superconductors

At temperatures near absolute zero, certain metals acquire infinite conductivity (zero resistance to the flow of charge). These are **superconductors**. In 1987, superconductivity at "high" temperature (above 100 K) was discovered in a non-metallic compound. Once electric current is established in a superconductor, the electrons flow indefinitely. With no electrical resistance, current passes through a superconductor without losing energy; no heat loss occurs when charges flow. At this writing superconductivity at both "high" and low temperature is being intensely investigated. Potential applications include long-distance transmission of power without loss, and high-speed magnetically levitated vehicles to replace trains.

Charging

We charge things by transferring electrons from one place to another. We can do this by physical *contact*, as occurs when substances are rubbed together or simply touched. Or we can redistribute the charge on an object simply by putting a charged object near it—this is called *induction*.

Charging by Friction and Contact

We are all familiar with the electrical effects produced by friction. We can stroke a cat's fur and hear the crackle of sparks that are produced, or comb our hair in front of a mirror in a dark room and see as well as hear the sparks. We can scuff our shoes across a rug and feel a tingle as we reach for the doorknob, or feel a surprising shock after sliding across a plastic seat cover while parked in an automobile (Figure 21.6). In all these cases electrons are being transferred by friction when one material rubs against another.

Electrons can be transferred from one material to another by simply touching. For example, when a negatively charged rod is placed in contact with a neutral object, some electrons will move to the neutral object. This method of charging is simply called *charging by contact*. If the object is a good conductor, electrons will spread to all parts of its surface because the transferred electrons repel one another. If it is a poor conductor, it may be necessary to touch the rod at several places on the object in order to get a more or less uniform distribution of charge.

Charging by Induction

If you bring a charged object *near* a conducting surface, you will induce electrons to move in the material even though there is no physical contact. Consider the two insulated metal spheres, A and B, in Figure 21.7. (a) They touch each other, so in effect they form a single noncharged conductor. (b) When a negatively charged rod is brought near A, electrons in the metal, being free to move, are repelled as far as possible until their mutual repulsion is big enough to balance the influence of the rod. Charge is redistributed. (c) If A and B are separated while the rod is still present, (d) they will each be equally and oppositely charged. This is **charging by induction**. The charged rod has never touched them, and it retains the same charge it had initially.

We can similarly charge a single sphere by induction if we touch it when different parts of it are differently charged. Consider the metal sphere that hangs from a noncon-

FIGURE 21.6 Charging by friction and then by contact.

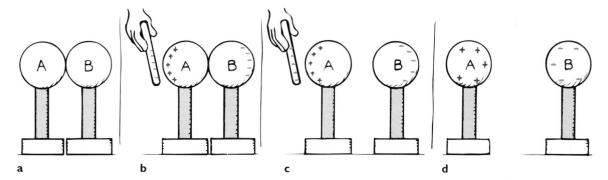

a   b   c   d

FIGURE 21.7 Charging by induction.

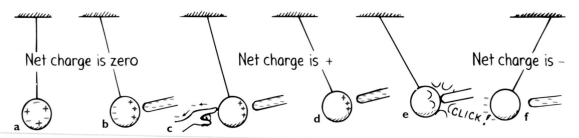

FIGURE 21.8 Stages of charge induction by grounding. (a) Net charge on the metal ball is zero. (b) Charge redistribution is induced on the ball by the presence of the charged rod. The net charge on the ball is still zero. (c) Touching the negative side of the ball removes electrons by contact. (d) This leaves the ball positively charged. (e) The ball is more strongly attracted to the negative rod, and when it touches, charging by contact occurs as electrons move from the rod to the sphere. (f) The negative ball is repelled by the still somewhat negatively charged rod.

ducting string, as shown in Figure 21.8. When we touch the metal surface with a finger, we are providing a path for charge to flow to or from a very large reservoir for electric charge—the ground. We say that we are *grounding* the sphere, a process that may leave it with a net charge. We will return to this idea of grounding in the next chapter when we discuss electric currents.

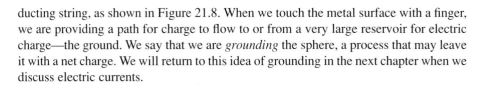

Questions

1. Would the charges induced on the spheres A and B of Figure 21.7 necessarily be exactly equal and opposite?
2. Why does the negative rod in Figure 21.7 have the same charge before and after the spheres are charged, but not when charging takes place as in Figure 21.8?

Charging by induction occurs during thunderstorms. The negatively charged bottoms of clouds induce a positive charge on the surface of the earth below. Benjamin Franklin was the first to demonstrate this when his famous kite-flying experiment proved that lightning is an electrical phenomenon.* Lightning is an electrical discharge

FIGURE 21.9 The negative charge at the bottom of the cloud induces a positive charge at the surface of the ground below.

Answers

1. The charges must be equal and opposite on both spheres, because each single positive charge on sphere A is the result of a single electron being taken from A and moved to B. This is like taking bricks from the surface of a brick road and putting them all on the sidewalk. The number of bricks on the sidewalk will be exactly matched by the number of holes in the road. Likewise, the number of extra electrons on B will exactly match the number of "holes" (positive charges) left in A. Remember that a positive charge is the equivalent of an absent electron.
2. In the charging process of Figure 21.7, no contact was made between the negative rod and either of the spheres. In Figure 21.8, however, the rod touched the positively charged sphere. A transfer of charge by contact reduced the negative charge on the rod.

* Benjamin Franklin was careful to isolate himself from his apparatus and keep out of the rain when he conducted this experiment, so he wasn't electrocuted as were others who attempted to duplicate his experiment. In addition to being a great statesman, Franklin was a first-rate scientist. He introduced the terms *positive* and *negative* as they relate to electricity, but nevertheless supported the one-fluid theory of electric charge, and contributed to our understanding of grounding and insulation. He also published a newspaper, formed the first fire insurance company, and invented a safer, more efficient stove—a very busy man! Only a task as important as helping to form the United States' system of government kept him from spending even more of his time on his favorite activity—the scientific investigation of nature.

between a cloud and the oppositely charged ground or between oppositely charged parts of clouds.

Franklin also found that charge flows readily to or from sharp metal points, and fashioned the first lightning rod. If the rod is placed above a building and connected to the ground, the point of the rod collects electrons from the air, preventing a large buildup of positive charge on the building by induction. This continual "leaking" of charge prevents a charge buildup that might otherwise lead to a sudden discharge between the cloud and the building. The primary purpose of the lightning rod, then, is to prevent a lightning discharge from occurring. If for any reason sufficient charge does not leak from the air to the rod, and lightning strikes anyway, it may be attracted to the rod and short-circuited to the ground, thereby sparing the building. The overall purpose of the lightning rod is to prevent a fire caused by lightning.

FIGURE 21.10 (a) Lightning rods atop a building serve double purpose; first to lower the likelihood of a lightning strike by discharging the structure, and second, to provide a path to ground in the event of a strike. (b) Note the heavy-duty grounding cable, which can conduct large currents in the event of a strike.

Charge Polarization

Charging by induction is not restricted to conductors. When a charged rod is brought near an insulator, there are no free electrons that can migrate throughout the insulating material. Instead, there is a rearrangement of charges within the atoms and molecules themselves (Figure 21.11). One side of the atom or molecule is induced into becoming more negative (or positive) than the opposite side. The atom or molecule is said to be

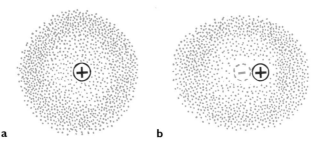

a   b

FIGURE 21.11 An electron buzzing around the atomic nucleus makes up an electron cloud. (a) The center of the negative electron cloud coincides with the center of the positive nucleus in an atom. (b) When an external negative charge is brought nearby to the right, the electron cloud is distorted so the centers of negative and positive charge no longer coincide. The atom is electrically polarized.

FIGURE 21.12 All the atoms or molecules near the surface become electrically polarized. Surface charges of equal magnitude and opposite sign are induced on opposite surfaces of the material.

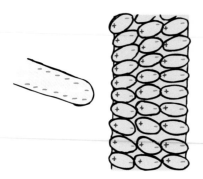

electrically polarized. If the charged rod is negative, say, then the positive part of the atom or molecule is tugged in a direction toward the rod, and the negative side of the atom or molecule is pushed in a direction away from the rod. The positive and negative parts of the atoms and molecules become aligned. They are electrically polarized.

We can see why electrically neutral bits of paper are attracted to a charged object. Molecules are polarized in the paper, with the sides of molecules opposite in sign to the charged object closest to it. Closeness wins, and the bits of paper experience a net attraction. Sometimes they will cling to the charged object and then suddenly fly off. This repulsion occurs because the paper bits acquire the same sign of charge as the charged object when they are in contact.

FIGURE 21.13 A charged comb attracts an uncharged piece of paper because the force of attraction for the closer charge is greater than the force of repulsion for the farther charge.

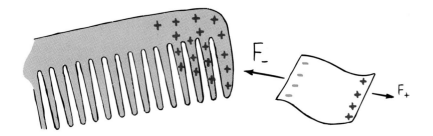

Rub an inflated balloon on your hair, and it becomes charged. Place the balloon against the wall, and it sticks. This is because the charge on the balloon induces an opposite surface charge on the wall. Closeness wins, for the charge on the balloon is slightly closer to the opposite induced charge than to the charge of same sign (Figure 21.14). Many molecules, H_2O for example, are electrically polarized in their normal

FIGURE 21.14 The negatively charged balloon polarizes molecules in the wooden wall and creates a positively charged surface, so the balloon sticks to the wall.

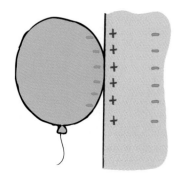

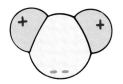

FIGURE 21.15 An H$_2$O molecule is an electric dipole.

states. The distribution of electric charge is not perfectly even. There is a little more negative charge on one side of the molecule than the other (Figure 21.15). Such molecules are said to be *electric dipoles*.

Question A negatively charged rod is brought close to some small pieces of neutral paper. The positive sides of molecules in the paper are attracted to the rod and the negative sides of the molecules are repelled. Since negative and positive sides are equal in number, why don't the attractive and repulsive forces cancel out?

Electric Field

Electrical forces, like gravitational forces, act between things that are not in contact with each other. For both electricity and gravitation, a *force field* exists that influences charged and massive bodies respectively. The properties of space surrounding any massive body can be considered to be so altered that another massive body introduced to this region will experience a force. The "alteration in space" caused by a massive body is called its *gravitational field*. We can think of any other massive body as interacting with the field and not directly with the massive body producing the field. For example, when an apple falls from a tree, we say it is interacting with the earth, but we can also think of the apple as interacting with the gravitational field of the earth. The field plays an intermediate role in the force between bodies. It is common to think of distant rockets and the like as interacting with gravitational fields rather than with the masses of the earth and other bodies responsible for the fields. Just as the space around a planet and every other massive body is filled with a gravitational field, the space around every electrically charged body is filled with an **electric field**—a kind of aura that extends through space.

FIGURE 21.16 A gravitational force holds the satellite in orbit about the planet (a), and an electrical force holds the electron in orbit about the proton (b). In both cases there is no contact between the bodies. We say that the orbiting bodies interact with the *force fields* of the planet and proton and are everywhere in contact with these fields. Thus, the force that one electric charge exerts on another can be described as the interaction between one charge and the field set up by the other.

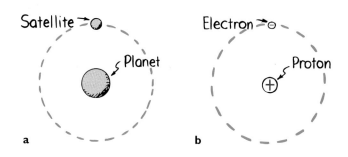

Answer The positive sides are simply closer to the rod. In accord with Coulomb's law, they therefore experience a greater electrical force than the farther-away negative sides. Hence we say that closeness wins. This greater force between positive and negative is attractive, so the neutral paper is attracted to the charged rod. Can you see that if the rod were positive, attraction would still occur?

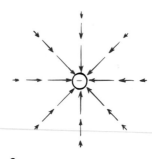

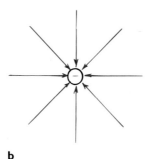

a

b

FIGURE 21.17 Electric field representations about a negative charge. (a) A vector representation. (b) A lines-of-force representation.

An electric field has both magnitude (strength) and direction. The magnitude of the field at any point is simply the force per unit of charge. If a body with charge q experiences a force F at some point in space, then the electric field E at that point is

$$E = \frac{F}{q}$$

The electric field is depicted with vector arrows in Figure 21.17a. The direction of the field is shown by the vectors and is defined to be the direction in which a small positive test charge at rest would be moved.* The direction of the force and that of the field at any point are the same. In the figure we see that all the vectors therefore point to the center of the negatively charged ball. If the ball were positively charged, the vectors would point away from its center because a positive test charge in the vicinity would be repelled.

A more useful way to describe an electric field is with electrical lines of force (Figure 21.17b). The lines of force shown in the figure represent a small number of the infinitely numerous possible lines that indicate the direction of the field. Where the lines are farther apart, the field is weaker. For an isolated charge, the lines extend to infinity; for two or more opposite charges, we represent the lines as emanating from a positive charge and terminating on a negative charge. Some electric field configurations are shown in Figure 21.18, and photographs of field patterns are shown in Figure 21.19. The photographs show bits of thread that are suspended in an oil bath surrounding charged conductors. The ends of the thread bits are charged by induction and tend to line up end to end with the field lines, like iron filings in a magnetic field.

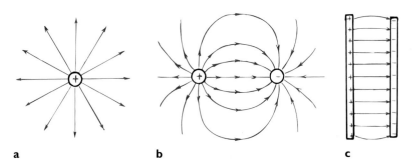

a   **b**   **c**

FIGURE 21.18 Some electric field configurations. (a) Lines of force about a single positive charge. (b) Lines of force for a pair of equal but opposite charges. Note that the lines emanate from the positive charge and terminate on the negative charge. (c) Uniform lines of force between two oppositely charged parallel plates.

The electric field concept helps us understand not only the forces between isolated stationary charged bodies, but also what happens when charges move. When charges move, their motion is communicated to neighboring charged bodies in the form of a field disturbance. The disturbance emanates from the accelerating charged body at the

---

* The test charge is small so that it does not appreciably influence the sources of the field being measured. Recall in our study of heat the similar need for a thermometer of small mass when measuring the temperature of bodies of larger masses.

FIGURE 21.19 Bits of thread suspended in an oil bath surrounding charged conductors line up end-to-end along the direction of the field. (a) Equal and opposite charges. (b) Equal like charges. (c) Oppositely charged plates. (d) Oppositely charged cylinder and plate.

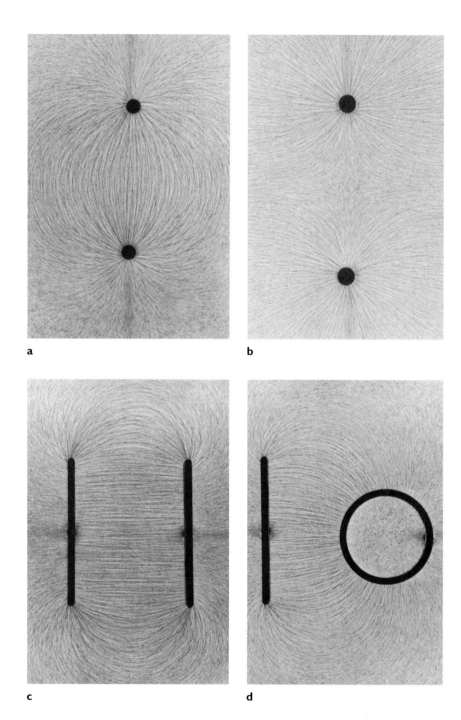

speed of light. We will learn that the electric field is a storehouse of energy and that energy can be transported over long distances in an electric field. Energy that is traveling in an electric field may be directed through and guided by metal wires. Or, it may be teamed up with a magnetic field to move through empty space. We will return to this idea in the next chapter and later when we learn about electromagnetic radiation.

Microwave Ovens

Imagine an enclosure with Ping-Pong balls among a few batons, all at rest. Now imagine the batons suddenly flipping back and forth like semi-rotating propellers, striking neighboring Ping-Pong balls. The balls are energized, moving in all directions. A microwave oven operates similarly. The batons are water molecules made to flip back and forth in rhythm with microwaves in the enclosure. The Ping-Pong balls are non-water molecules that make up the bulk of the food being cooked.

Each H_2O molecule is an electric dipole that aligns with an electric field, like a compass needle aligns with a magnetic field. When the electric field is made to oscillate, the H_2O molecules oscillate also. The H_2O molecules move quite energetically when the oscillation frequency matches their natural frequency—at resonance. Food is cooked by a sort of "kinetic friction" as flip-flopping H_2O molecules impart thermal motion to surrounding molecules. The metal enclosure reflects microwaves to and fro throughout the oven for rapid cooking.

Dry paper, foam plates, or other materials recommended for use in microwave ovens contain no water or other polar molecules, so microwaves pass through them with no effect. Likewise for ice, because H_2O molecules are locked in position and can't rotate to and fro.

Electric Shielding

An important difference between electric and gravitational fields is that electric fields can be shielded by various materials, while gravitational fields cannot. The amount of shielding depends on the material used for shielding. For example, air makes the electric field between two charged objects slightly less than it would be in a vacuum, while oil placed between the objects can diminish the field nearly a hundredfold. Metal can completely shield an electric field. When no current is flowing, the electric field inside metal is zero, regardless of the field strength outside.

Consider, for example, electrons on a spherical metal ball. Because of mutual repulsion the electrons will spread out uniformly over the outer surface of the ball. It is not difficult to see that the electrical force exerted on a sample test charge placed at the exact center of the ball is zero because opposing forces balance in every direction. Interestingly enough, complete cancellation occurs anywhere inside a conducting sphere. Understanding why this is true requires more thought and involves the inverse-square law and a bit of geometry. Consider the test charge at point P in Figure 21.20. The test charge is twice as far from the left side of the charged sphere as it is from the right side. If the electrical force between the test charge and the charges depended only on distance, then the test charge would be attracted only $\frac{1}{4}$ as much to the left side as to the right side. (Remember the inverse-square law: twice as far away means only $\frac{1}{4}$ the effect, three times as far away means only $\frac{1}{9}$ the effect, and so on.) But the force depends also on the amount of charge. In the figure, the cones extending from point P to areas A and B have the same apex angle, but one has twice the altitude of the other. This means that area A at the base of the longer cone is four times area B at the base of the shorter cone, true for any apex angle. Since $\frac{1}{4}$ of 4 is equal to 1, a test charge at P is attracted equally to each side. Cancellation occurs. A similar argument applies if the cones emanating from point P are oriented in any direction. Complete cancellation occurs at all points within the sphere. (Recall this argument back in Chapter 8 for the cancellation of gravity inside a hollow planet. The metal ball behaves the same way whether hollow or not because all of its charge gathers on its outer surface.)

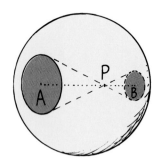

FIGURE 21.20 The test charge at P is attracted just as much to the greater number of farther charges at A as it is to the smaller number of closer charges at B. The net force on the test charge anywhere inside the conductor is zero. The electric field inside is therefore zero.

FIGURE 21.21 The charges are distributed on the surface of all conductors in a way such that the electric field inside the conductor is zero.

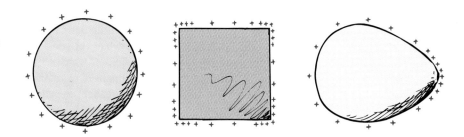

If the conductor is not spherical, then the charge distribution will not be uniform. The charge distribution over conductors of various shapes is shown in Figure 21.21. Most of the charge on a conducting cube, for example, is mutually repelled toward the corners. The remarkable thing is this: The exact charge distribution over the surface of a conductor is such that the electric field everywhere inside the conductor is zero. Look at it this way. If there were an electric field inside a conductor, then free electrons inside the conductor would be set in motion. How far would they move? Until equilibrium is established—which is to say, when the positions of all the electrons produce a zero field inside the conductor.

FIGURE 21.22 Electrons from the lightning bolt mutually repel to the outer metal surface. Although the electric field they set up may be great *outside* the car, the net electric field *inside* the car is practically zero.

We cannot shield ourselves from gravity, because gravity only attracts. There are no repelling parts of gravity to offset attracting parts. Shielding electric fields, however, is quite simple. Surround yourself or whatever you wish to shield with a conducting surface. Put this surface in an electric field of whatever field strength. The free charges in the conducting surface will arrange themselves on the surface of the conductor in such a way that all field contributions inside cancel one another. That's why certain electronic components are encased in metal boxes and why certain cables have a metal covering—to shield them from outside electrical activity.

Question Small bits of aligned thread vividly show the electric fields in the four photos of Figure 21.19. But the threads are not aligned inside the cylinder in Figure 21.19d. Why?

Electric Potential

When we studied energy in Chapter 6, we learned that an object has gravitational potential energy because of its location in a gravitational field. Similarly, a charged object has potential energy by virtue of its location in an electric field. Just as work is required to lift a massive object against the gravitational field of the earth, work is required to push a charged particle against the electric field of a charged body. This work changes the electric potential energy of the charged particle.* Consider the particle with the small positive charge located at some distance from a positively charged sphere in Figure 21.24. If you push the particle closer to the sphere, you will expend energy to overcome electrical repulsion; that is, you will do work in pushing the charged particle against the electric field of the sphere. This work done in moving the particle to its new location increases its energy. We call the energy the particle pos-

FIGURE 21.23 The sketch to the left shows the PE (potential energy) of a massive body held in a gravitational field. The sketch to the right shows the PE of a charged particle held in an electric field. When released, how does the KE (kinetic energy) acquired by each compare to the decrease in PE?

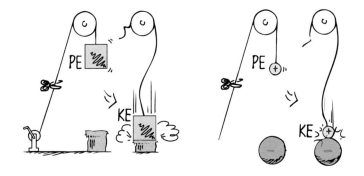

Answer The electric field is shielded inside the cylinder, shown as a circle in the two-dimensional photograph. Hence the threads are without alignment. The electric field inside any conductor is zero—so long as no electric current is flowing.

* This work is positive if it increases the electric potential energy of the charged particle and negative if it decreases it.

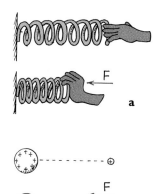

FIGURE 21.24 (a) The spring has more elastic PE when compressed. (b) The small charge similarly has more PE when pushed closer to the charged sphere. In both cases the increased PE is the result of work required to reach the closer location.

sesses by virtue of its location **electric potential energy**. If the particle is released, it accelerates in a direction away from the sphere, and its electric potential energy changes to kinetic energy.

If we push a particle with twice the charge instead, we do twice as much work pushing it, so the doubly charged particle in the same location has twice as much electric potential energy as before. A particle with three times the charge has three times as much potential energy; ten times the charge, ten times the potential energy; and so on. Rather than dealing with the total potential energy of a charged body, it is convenient when working with charged particles in an electric field to consider the electric potential energy *per unit of charge*. We simply divide the amount of energy by the amount of charge. For example, a particle with ten times as much charge as another in the same location will have ten times as much potential energy—but having ten times as much charge means that the energy per charge is the same. The concept of potential energy per unit charge is called **electric potential**; that is,

$$\text{electric potential} = \frac{\text{electric potential energy}}{\text{charge}}$$

The unit of measurement for electric potential is the **volt**, so electric potential is often called *voltage*. A potential of 1 volt (V) equals 1 joule (J) of energy per 1 coulomb (C) of charge.

$$1 \text{ volt} = 1 \frac{\text{joule}}{\text{coulomb}}$$

Thus a 1.5-volt battery gives 1.5 joules of energy to every 1 coulomb of charge passing through the battery. Both the names *electric potential* and *voltage* are common, so either may be used. In this book the names will be used interchangeably.

FIGURE 21.25 The larger test charge has more PE in the field of the charged dome, but the *electric potential* of any charged particle at the same location is the same, no matter how much its charge. So each point in space has a certain potential whether or not a charge is there.

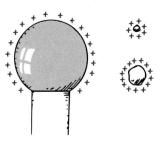

The significance of electric potential, or voltage, is that a definite value for it can be assigned to a location, whether or not a charged particle exists at that location (once the location of zero voltage has been specified). We can speak about electric potentials at different locations in an electric field whether or not charged particles occupy those locations. Likewise with voltages at various locations in an electric circuit. In the next chapter you will see that the location of the positive terminal of a 12-volt battery is maintained at a voltage 12 volts higher than the location of the negative terminal. When a conducting medium connects these voltage-differing terminals, charges in the medium will move between these voltage-differing locations.

FIGURE 21.26 Although the voltage of the charged balloon is high, the electric potential energy is low because of the small amount of charge.

Electric Energy Storage

FIGURE 21.27 Capacitor consisting of two closely spaced metal parallel plates. When connected to a battery, the plates acquire equal and opposite charges. The voltage between the plates then matches the electric potential difference between the battery terminals.

Questions

1. If there were twice as many coulombs in the test charge near the charged sphere in Figure 21.24, would the electric potential energy of the test charge with respect to the charged sphere be the same or would it be twice as great? Would the electric potential of the test charge be the same or would it be twice as great?
2. What does it mean to say that your car has a 12-volt battery?

Rub a balloon on your hair, and the balloon becomes negatively charged—perhaps to several thousand volts! That would be several thousand joules of energy if the charge were 1 coulomb. However, 1 coulomb is a very large amount of charge. The charge on a balloon rubbed on hair is more typically much less than a millionth of a coulomb. Therefore, the amount of energy associated with the charged balloon is very, very small. A high voltage means a lot of energy only if a lot of charge is involved. There is an important difference between electric potential energy and electric potential.

Electric energy can be stored in a common device called a **capacitor**, found in nearly all electronic circuits. Capacitors are used as low-energy on-off switches in computer motherboards, as a storehouse of larger amounts of energy in photoflash units that release the energy rapidly during the short duration of the flash. Similarly, but on a grander scale, enormous amounts of energy are stored in banks of capacitors that power giant lasers in national laboratories.

The simplest capacitor is a pair of conducting plates separated by a small distance, but not touching each other. When the plates are connected to a charging device such as the battery shown in Figure 21.27, electrons are transferred from one plate to the other. This occurs as the positive battery terminal pulls electrons from the plate connected to it. These electrons in effect are pumped through the battery and through the negative terminal to the opposite plate. The capacitor plates then have equal and opposite charges—the positive plate connected to the positive battery terminal, and the negative plate connected to the negative terminal. The charging process is complete when the potential

Answers

1. Twice as many coulombs would find the test charge with twice as much electric potential energy. (This is because it would have taken twice as much work to put the charge at that location.) But the electric potential would be the same. This is because the electric potential is total electric potential energy divided by total charge. For example, ten times the energy divided by ten times the charge will give the same value as two times the energy divided by two times the charge. Electric potential is not the same thing as electric potential energy. Be sure you understand this before you study further.
2. It means that one of the battery terminals is 12 V higher in potential than the other one. In the next chapter you will see that it also means that when a circuit is connected between these terminals, each coulomb of charge in the resulting current will be given 12 J of energy as it passes through the battery.

FIGURE 21.28 Practical capacitors.

difference between the plates equals the potential difference between the battery terminals—the battery voltage. The greater the battery voltage, and the larger and closer the plates, the greater the charge that can be stored. In practice, the plates may be thin metallic foils separated by a thin sheet of paper. This "paper sandwich" is then rolled up to save space and inserted into a cylinder. Such a practical capacitor is shown with others in Figure 21.28. (We will consider the role of capacitors in circuits in the next chapter.)

A charged capacitor is discharged when a conducting path is provided between the plates. Discharging a capacitor can be a shocking experience if you happen to be the conducting path. The energy transfer that occurs can be fatal where high voltages are present, such as the power supply in a TV set—even after the set has been turned off. This is the main reason for the warning signs on such devices.

FIGURE 21.29 Each key of the keyboard is part of a capacitor. When depressed, capacitor plates are pushed closer together to increase capacitance and signal the computer.

Question What is the net charge of a charged capacitor?

The energy stored in a capacitor comes from the work required to charge it. The energy is stored in the electric field between its plates. Between parallel plates the electric field is uniform, as indicated in Figures 21.18c and 21.19c on previous pages. So the energy stored in a capacitor is energy of its electric field. In Chapter 24 we will see how energy from the sun is radiated in the form of electric and magnetic fields. The fact that energy is contained in electric fields is truly far reaching.

Answer The net charge of a charged capacitor is zero because the charges on its two plates are equal in number and opposite in sign. If the capacitor is discharged, say by allowing the charge to flow from one plate to the other, the net charge of the capacitor remains zero, because then the net charge on each of its plates is zero.

Van de Graaff Generator

A common laboratory device for building up high voltages is the *Van de Graaff generator.* This is one of the lightning machines that mad scientists used in old science fiction movies. A simple model of the Van de Graaff generator is shown in Figure 21.30. A large, hollow metal sphere is supported by a cylindrical insulating stand. A motor-driven rubber belt inside the support stand passes a comb-like set of metal needles that are maintained at a large negative potential relative to ground. Discharge by the points deposits a continuous supply of electrons on the belt, which are carried up into the hollow conductor. Since the electric field inside the conductor is zero, the charge leaks onto metal points (tiny lightning rods) and is deposited on the inside of the sphere. The electrons repel each other to the outer surface of the conducting sphere. Static charge always lies on the outside surface of any conductor. This leaves the inside uncharged and able to receive more electrons as they are brought up by the belt. The process is continuous and the charge builds up until the negative potential on the sphere is much greater than at the voltage source at the bottom—on the order of millions of volts.

A sphere with a radius of 1 meter can be raised to a potential of 3 million volts before electrical discharge occurs through the air. The voltage can be further increased by increasing the radius of the sphere or by placing the entire system in a container filled with high-pressure gas. Van de Graaff generators can produce voltages as high as 20 million volts. These voltages are used to accelerate charged particles that can be used as projectiles for penetrating the nuclei of atoms. Touching one can be a hair-raising experience.

FIGURE 21.30   A simple model of a Van de Graaff generator

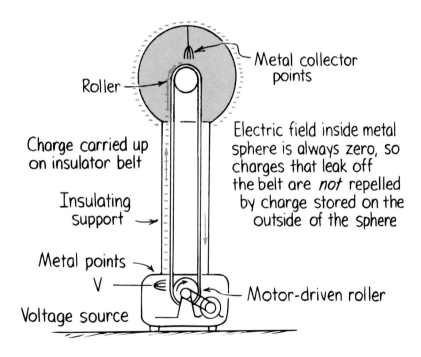

FIGURE 21.31 Both Meidor and the spherical dome of the Van de Graaff generator are charged to a high voltage. Why does her hair stand out?

Summary of Terms

Electricity   General term for electrical phenomena, much like gravity has to do with gravitational phenomena, or sociology with social phenomena.

Electrostatics   The study of electric charge at rest (not *in motion,* as in electric currents).

Conservation of charge   Electric charge is neither created nor destroyed. The total charge before an interaction equals the total charge after.

Coulomb's law   The relationship among electrical force, charge, and distance:

$$F = k\frac{q_1 q_2}{d^2}$$

If the charges are alike in sign, the force is repulsive; if the charges are unlike, the force is attractive.

Coulomb   The SI unit of electrical charge. One coulomb (symbol C) is equal to the total charge of 6.25×10^{18} electrons.

Conductor   Any material having free charged particles that easily flow through it when an electric force acts on them.

Insulator   A material without free charged particles and through which charge does not easily flow.

Electrically polarized   Term applied to an atom or molecule in which the charges are aligned so that one side has a slight excess of positive charge, the other side a slight excess of negative charge.

Electric field   Defined as force per unit charge, it can be considered to be an "aura" surrounding charged objects, and is a storehouse of electric energy. About a charged point, the field decreases with distance according to the inverse-square law, like a gravitational field. Between oppositely charged parallel plates, the electric field is uniform.

Electric potential energy   The energy a charged object possesses by virtue of its location in an electric field.

Electric potential   The electric potential energy per unit of charge, measured in volts, and often called *voltage:*

$$\text{Voltage} = \frac{\text{electric energy}}{\text{amount of charge}}$$

Volt   The unit of electric potential; 1 volt = 1 joule per coulomb.

Capacitor   An electrical device, in its simplest form a pair of parallel conducting plates separated by a small distance, that stores electric charge and energy.

Review Questions

Electrical Forces

1. Which is stronger, the electrical force between an electron and a proton or the gravitational force between these particles? Is the difference between these forces large or small?

2. Why does gravitational force predominate over electrical force for astronomical bodies?

Electric Charges

3. What part of an atom is *positively* charged and what part is *negatively* charged?

4. How does the charge of one electron compare to that of another electron?

5. How do the masses of electrons compare to the masses of protons? Neutrons?

6. How do the numbers of protons in the atomic nucleus normally compare to the number of electrons that orbit the nucleus?

Conservation of Charge

7. What kind of charge does an object acquire when electrons are stripped from it?

8. What is a positive ion? A negative ion?

9. What is meant by saying charge is *conserved*?

10. What is meant by saying charge is *quantized*?

11. What particle has exactly one quantum unit of charge?

Coulomb's Law

12. How does a *coulomb* of charge compare to the charge of a *single* electron?

13. How is Coulomb's law similar to Newton's law of gravitation? How is it different?

14. The proportionality constant k in Coulomb's law is huge in ordinary units, whereas the proportionality constant G in Newton's law of gravitation is tiny. What does this mean for the relative strengths of these two forces?

15. How does the magnitude of electrical force between a pair of charged objects change when the objects are moved twice as far apart? Three times as far apart?

Conductors and Insulators

16. Why are metals good conductors of both heat and electricity?

17. Why are materials such as glass and rubber good insulators?

Semiconductors

18. How does a *semiconductor* differ from a *conductor* or an *insulator*?

19. What is a transistor, and what are some of its functions?

Superconductors

20. How much electrical resistance does a superconductor have to a flow of electric charge?

Charging

21. What happens to electrons in any charging process?

Charging by Friction and Contact

22. Give an example of something charged by friction.

23. Give an example of something charged by simple contact.

Charging by Induction

24. Give an example of something charged by induction.

25. What occurs when we "ground" an object?

26. What is the purpose of a lightning rod?

Charge Polarization

27. How does an electrically *polarized* object differ from an electrically *charged* object?

28. What is normally the net charge of a polarized object?

29. A piece of paper becomes polarized in the presence of, say, a negative charge. The positive side of the paper is attracted to the negative charge, and the negative side of the paper is repelled by the negative charge. So why don't these forces cancel out?

30. Why will a balloon that is charged either positively or negatively stick to a neutral wall?

Electric Field

31. Give two examples of common force fields.

32. How is the direction of an electric field defined?

33. How is the magnitude of an electric field defined?

Electric Shielding

34. Why is there no electric field in the middle of a charged spherical conductor?

35. Does an electric field exist within a charged spherical conductor at points other than its center?

36. When charges mutually repel and distribute themselves on the surface of conductors, what is the effect inside the conductor?

37. What is the electric field inside a car struck by lightning?

Electric Potential

38. Distinguish between *electric potential energy* and *electric potential.*

39. A balloon may easily be charged to several thousand volts. Does that mean it has several thousand joules of energy? Explain.

Electric Energy Storage

40. Why does a charged capacitor have a net charge of zero?

41. Where is the stored energy in a capacitor?

42. In order to store more energy in a capacitor, how might you change the capacitor?

Van de Graaff Generator

43. What is the magnitude of the electric field inside the dome of a charged Van de Graaff generator?

44. Why does Meidor's hair stand out in Figure 21.31?

Projects

1. Demonstrate charging by friction and discharging from points with a friend who stands at the far end of a carpeted room. Scuff your way across the rug until your noses are close together. This can be a delightfully tingling experience, depending on how dry the air is and how pointed your noses are. Leather soles on your shoes also help.

2. Briskly rub a comb through your hair or on a woolen garment and bring it near a small but smooth stream of running water. Is the stream of water deflected?

Exercises

1. We do not feel the gravitational forces between ourselves and the objects around us because these forces are extremely small. Electrical forces, in comparison, are extremely huge. Since we and the objects around us are composed of charged particles, why don't we usually feel electrical forces?

2. With respect to forces, how are electric charge and mass alike? How are they different?

3. Why do clothes often cling together after tumbling in a clothes dryer?

4. When you remove your wool suit from the dry cleaner's garment bag, the bag becomes positively charged. Explain how this occurs.

5. When combing your hair, you scuff electrons from your hair onto the comb. Is your hair then positively or negatively charged? How about the comb?

6. At some automobile toll-collecting stations, a thin metal wire sticks up from the road and makes contact with cars before they reach the toll collector. What is the purpose of this wire?

7. Why are the tires for trucks carrying gasoline and other flammable fluids manufactured to conduct electricity?

8. An electroscope is a simple device consisting of a metal ball that is attached by a conductor to two thin leaves of metal foil protected from air disturbances in a jar, as shown. When the ball is touched by a charged body, the leaves that normally hang straight down spread apart. Why? (Electroscopes are useful not only as charge detectors, but also for measuring the quantity of charge: the more charge transferred to the ball, the more the leaves diverge.)

9. The leaves of a charged electroscope collapse in time. At higher altitudes they collapse more rapidly. Why is this true? (*Hint:* The existence of cosmic rays was first indicated by this observation.)

10. Is it necessary for a charged body to actually touch the ball of the electroscope for the leaves to diverge? Defend your answer.

11. Strictly speaking, when an object acquires a positive charge by the transfer of electrons, what happens to its mass? When it acquires a negative charge? Think small!

12. Strictly speaking, will a penny be slightly more massive if it has a negative charge or a positive charge? Explain.

13. In a crystal of salt are electrons and positive ions. How does the net charge of the electrons compare with the net charge of the ions? Explain.

14. How can you charge an object negatively with only the help of a positively charged object?

15. It is relatively easy to strip the outer electrons from a heavy atom like that of uranium (which then becomes a uranium ion) but very difficult to remove the inner electrons. Why do you suppose this is so?

16. When one material is rubbed against another, electrons jump readily from one to the other but protons do not. Why is this? (Think in atomic terms.)

17. If electrons were positive and protons negative, would Coulomb's law be written the same or differently?

18. The five thousand billion, billion freely moving electrons in a penny repel one another. Why don't they fly out of the penny?

19. How does the magnitude of electric force compare between a pair of charged particles when they are brought to half their original distance of separation? To one-quarter their original distance? To four times their original distance? (What law guides your answers?)

20. Two charged pellets are pulled apart to twice their original separation. (a) Which is likely to be larger, the gravitational or the electrical force between them? (b) Which will change by the greater factor when they are pulled apart, the gravitational or the electrical force between them?

21. Two equal charges exert equal forces on each other. What if one charge has twice the magnitude of the other. How do the forces they exert on each other compare?

22. Suppose that the strength of the electric field about an isolated point charge has a certain value at a distance of 1 m. How will the electric field strength compare at a distance of 2 m from the point charge? What law guides your answer?

23. Measurements show that there is an electric field surrounding the earth. Its magnitude is about 100 N/C at the earth's surface and points inward toward the earth's center. From this information, can you state whether the earth is negative or positive?

24. Why are metal-spiked shoes not a good idea for golfers on a stormy day?

25. If you are caught outdoors in a thunderstorm, why should you not stand under a tree? Can you think of a reason why you should not stand with your legs far apart? Or why lying down can be dangerous? (*Hint:* Consider electric potential difference.)

26. If a large enough electric field is applied, even an insulator will conduct an electric current, as is evident in lightning discharges through the air. Explain how this happens, taking into account the opposite charges in an atom and how ionization occurs.

27. Why is a good conductor of electricity also a good conductor of heat?

28. If you rub an inflated balloon against your hair and place it against a door, by what mechanism does it stick? Explain.

29. How can a charged atom (an ion) attract a neutral atom?

30. If you place a free electron and a free proton in the same electric field, how will the forces acting on them compare? Their accelerations? Their directions of travel?

31. Imagine a proton at rest a certain distance from a negatively charged plate. It is released and collides with the plate. Then imagine the similar case of an electron at rest, the same distance away from a plate of equal and opposite charge. In which case will the moving particle have the greater speed when the collision occurs? Why?

32. A gravitational field vector points toward the earth; an electric field vector points toward an electron. Why do electric field vectors point away from protons?

33. By what specific means do the bits of fine threads line up in the electric fields shown in Figure 21.19?

34. Suppose that a metal file cabinet is charged. How will the charge concentration at the corners of the cabinet compare with the charge concentration on the flat parts of the cabinet?

35. If you put in 10 joules of work to push 1 coulomb of charge against an electric field, what will be its voltage with respect to its starting position? When released, what will be its kinetic energy if it flies past its starting position?

36. You are not harmed by contact with a charged metal ball, even though its voltage may be very high. Is the reason similar to why you are not harmed by the greater-than-1000°C sparks from a 4th-of-July type sparkler? Defend your answer in terms of the energies that are involved.

37. What is the voltage at the location of a 0.0001 C charge that has an electric potential energy of 0.5 J (both measured relative to the same reference point)?

38. What safety is offered by staying inside an automobile during a thunderstorm? Defend your answer.

39. Would you feel any electrical effects if you were inside the charged sphere of a Van de Graaff generator? Why or why not?

40. A friend says that the reason one's hair stands out while touching a charged Van de Graaff generator is simply that the hair strands become charged and are light enough so that the repulsion between strands is visible. Do you agree or disagree?

Problems

1. Two point charges are separated by 6 cm. The attractive force between them is 20 N. Find the force between them when they are separated by 12 cm. (Why can you solve this problem without knowing the magnitudes of the charges?)

2. If the charges attracting each other in the previous problem have equal magnitude, what is the magnitude of each charge?

3. Two pellets, each with a charge of 1 microcoulomb (10^{-6} C), are located 3 cm (0.03 m) apart. What is the electric force between them? What mass object would experience this same force in the earth's gravitational field?

4. Electronic types neglect the force of gravity on electrons. To see why, compute the force of the earth's gravity on an electron and compare it with the force exerted on the electron by an electric field of magnitude 10,000 V/m (a relatively small field). The mass and charge of an electron are given on the inside back cover.

5. Atomic physicists ignore the effect of gravity within an atom. To see why, calculate and compare the gravitational and electrical forces between an electron and a proton separated by 10^{-10} m. The needed charges and masses are given on the inside back cover.

6. A droplet of ink in an industrial ink-jet printer carries a charge of 3.0×10^{-13} C and is deflected onto paper by a force of 2.4×10^{-7} N. Find the strength of the electric field to produce this force.

7. The potential difference between a storm cloud and the ground is 100 million volts. If a charge of 2 C flashes in a bolt from cloud to earth, what is the change of potential energy of the charge?

8. An energy of 0.1 J is stored in the metal ball on top of a Van de Graaff machine. A spark carrying 1 microcoulomb (10^{-6} C) discharges the ball. What was the ball's potential relative to ground?

9. In 1909 Robert Millikan was the first to find the charge of an electron in his now-famous oil drop experiment. In the experiment, tiny oil drops are sprayed into a uniform electric field between a horizontal pair of oppositely charged plates. The drops are observed with a magnifying eyepiece, and the electric field is adjusted so that the upward force on some negatively charged oil drops is just sufficient to balance the downward force of gravity. That is, when suspended, upward force qE just equals mg. Millikan accurately measured the charges on many oil drops and found the values to be whole-number multiples of 1.6×10^{-19} C—the charge of the electron. For this he won the Nobel prize. Questions: If a drop of mass 1.1×10^{-14} kg remains stationary in an electric field of 1.68×10^{5} N/C, (a) what is the charge of this drop? (b) How many extra electrons are on this particular oil drop (given the presently known charge of the electron)?

10. Find the voltage increase when (a) 12 J of work is done to push a 0.0001-C charge into an electric field, and (b) 24 J of work is done to push a 0.0002-C charge into the electric field.

22

Electric Current

: **P**ower is transferred quietly from one place to many others by tiny
: vibrations of electrons in wires.

In the previous chapter you were introduced to the concept of electric potential, measured in volts. Now we'll see that this voltage acts like an "electrical pressure" that can produce a flow of charge, or *current*, measured in amperes (or, simply, amps, and abbreviated A), and that the *resistance* that restrains this flow is measured in ohms (Ω). When the flow is in only one direction we'll call it *direct current* (dc), and for back-and-forth flow we'll call it *alternating current* (ac). Electric current can deliver electric *power*, measured, just like mechanical power, in watts (W) or in thousands of watts, kilowatts (kW). We see here many terms to be sorted out. This is easier to do when we have some understanding of the concepts these terms represent, which in turn are better understood if we know how they relate to one another. We begin with the flow of electric charge.

Flow of Charge

Recall from our study of heat and temperature that when the ends of a conducting material are at different temperatures, heat energy flows from the higher temperature to the lower temperature. The flow ceases when both ends reach the same temperature. Similarly, when the ends of an electrical conductor are at different electric potentials—when there is a **potential difference**—charge flows from one end to the other. The flow of charge persists for as long as there is a potential difference.* Without a potential difference, no charge flows. Connect one end of a wire to a charged Van de Graaff generator, for example, and the other end to the ground, and a surge of charge will flow through the wire. The flow will be brief, however, for the sphere will quickly reach a common potential with the ground.

To attain a sustained flow of charge in a conductor, some arrangement must be provided to maintain a difference in potential while charge flows from one end to the other. The situation is analogous to the flow of water from a higher reservoir to a lower one (Figure 22.1a). Water will flow in a pipe that connects the reservoirs only as long as a difference in water level exists. The flow of water in the pipe, like the flow of charge in the wire that connects the Van de Graaff generator to the ground, will cease when the pressures at each end are equal (we imply this when we say that water seeks its own level). A continuous flow is possible if the difference in water levels—hence water pressures—is maintained with the use of a suitable pump (Figure 22.1b).

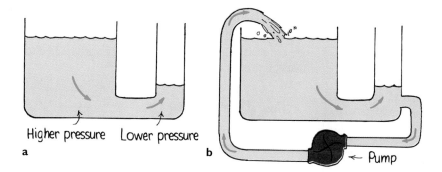

FIGURE 22.1 Water flows from the reservoir of higher pressure to the reservoir of lower pressure. (a) The flow will cease when the difference in pressure ceases. (b) Water continues to flow because a difference in pressure is maintained with the pump.

Higher pressure   Lower pressure

a   b   ← Pump

Electric Current

FIGURE 22.2 Each coulomb of charge that is made to flow in a circuit that connects the ends of this 1.5-V flash light cell is energized with 1.5 J.

Just as water current is the flow of H_2O molecules, **electric current** is simply the flow of electric charge. In circuits of metal wires, electrons make up the flow of charge. This is because one or more electrons from each metal atom are free to move throughout the atomic lattice. These charge carriers are called *conduction electrons*. Protons, on the other hand, do not move because they are bound inside the nuclei of atoms that are more or less locked in fixed positions. In conducting fluids, however—such as in a car battery—positive ions typically compose the flow of electric charge.

The electrical flow is measured in *amperes*. One ampere is a flow equal to 1 coulomb of charge per second. (Recall that 1 coulomb, the standard unit of charge, is the electric charge of 6.25 billion billion electrons.) In a wire that carries 5 amperes, for example, 5 coulombs of charge pass any cross section in the wire each second. So that's a lot of electrons! In a wire that carries 10 amperes, twice as many electrons pass any cross section each second.

It is interesting to note that a current-carrying wire is not electrically charged. Under ordinary conditions, negative conduction electrons swarm through the atomic lattice made up of positively charged atomic nuclei. So there are as many electrons as protons in the wire. Whether a wire carries a current or not, the net charge of the wire is normally zero at every moment.

* When we say that *charge flows,* we mean that charged *particles* flow. Charge is a property of particular particles, most significantly electrons, protons, and ions. When the flow is of negative charge, electrons or negative ions compose the flow. When the flow of charge is positive, protons or positive ions are flowing.

Voltage Sources

FIGURE 22.3 An unusual source of voltage. The electric potential between the head and tail of the electric eel (*Electrophorus electricus*) can be up to 600 V.

Charges flow only when they are "pushed" or "driven." A sustained current requires a suitable pumping device to provide a difference in electric potential—a voltage. An "electrical pump" is some sort of voltage source. If we charge one metal sphere positively and another negatively, we can develop a large voltage between the spheres. This voltage source is not a good electrical pump because when the spheres are connected by a conductor, the potentials equalize in a single brief surge of moving charges. It is not practical. Generators or chemical batteries, on the other hand, are sources of energy in electric circuits and are capable of maintaining a steady flow.

Batteries and electric generators do work to pull negative charges away from positive ones. In chemical batteries, this work is done by the chemical disintegration of zinc or lead in acid, and the energy stored in the chemical bonds is converted to electric potential energy.* Generators such as alternators in automobiles separate charge by electromagnetic induction, a process we will describe in Chapter 24. The work done by whatever means in separating the opposite charges is available at the terminals of the battery or generator. These different values of energy per charge create a difference in potential (voltage). This voltage provides the "electrical pressure" to move electrons through a circuit joined to these terminals.

The unit of electric potential difference (voltage) is the *volt.*[†] A common automobile battery will provide an electrical pressure of 12 volts to a circuit connected across its terminals. Then 12 joules of energy are supplied to each coulomb of charge that is made to flow in the circuit.

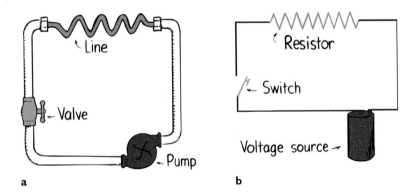

a   **b**

FIGURE 22.4 Analogy between a simple hydraulic circuit (a) and an electrical circuit (b).

There is often some confusion about charge flowing *through* a circuit and voltage placed, or impressed, *across* a circuit. We can distinguish between these ideas by considering a long pipe filled with water. Water will flow *through* the pipe if there is a difference in pressure *across* or between its ends. Water flows from the high-pressure end to the low-pressure end. Only the water flows, not the pressure. Similarly, electric charge flows because of the differences in electrical pressure (voltage). You say that charges flow *through* a circuit because of an applied voltage across the circuit.[‡] You don't say that voltage flows through a circuit. Voltage doesn't go anywhere, for it is the charges that move. Voltage produces current (if there is a complete circuit).

---

* You can find how this is done in almost any chemistry text.

[†] The terminology of this area of physics can be confusing, so here's a quick summary of terms: *electric potential* and *potential* mean the same thing—electrical potential energy per unit charge. Its units are volts. On the other hand, *potential difference* is the same as *voltage*—the difference in electrical potential between two points in a conducting path. Units of voltage are also volts.

[‡] We often say that current flows through a circuit, but don't say this around somebody who is picky about grammar, because the expression "current flows" is redundant. More properly, charge flows—which *is* current.

Electrical Resistance

FIGURE 22.5 More water flows through a thick hose than through a thin one connected to a city's water system (same water pressure). Likewise for electric current in thick and thin wires connected across the same potential difference.

A battery or generator of some kind is the prime mover and source of voltage in an electric circuit. How much current there is depends not only on the voltage but also on the **electrical resistance** the conductor offers to the flow of charge. This is similar to the rate of water flow in a pipe, which depends not only on the pressure difference between the ends of the pipe but also on the resistance offered by the pipe itself. The resistance of a wire depends on the conductivity of the material and also on its thickness and length. Thick wires have less resistance than thin wires. Longer wires have more resistance than short wires. In addition, electrical resistance depends on temperature. The greater the jostling about of atoms within the conductor, the greater resistance the conductor offers to the flow of charge. For most conductors, increased temperature means increased resistance.* The resistance of some materials reaches zero at very low temperatures. These are the superconductors discussed briefly in the previous chapter.

Electrical resistance is measured in units called *ohms*. The Greek letter *omega*, Ω, is commonly used as the symbol for the ohm. This unit is named after Georg Simon Ohm, a German physicist who in 1826 discovered a simple and very important relationship among voltage, current, and resistance.

FIGURE 22.6 Resistors in a circuit regulate the flow of charge. The symbol of resistance in an electric circuit is ─◢◣◢◣─ .

Ohm's Law

The relationship among voltage, current, and resistance is summarized by a statement called **Ohm's law**. Ohm discovered that the current in a circuit is directly proportional to the voltage established across the circuit, and is inversely proportional to the resistance of the circuit. In short,

$$\text{Current} = \frac{\text{voltage}}{\text{resistance}}$$

Or, in units form,

$$\text{Amperes} = \frac{\text{volts}}{\text{ohms}}$$

---

* Carbon is an interesting exception. As temperature increases, more carbon atoms shake loose an electron. This increases the electric current. So the resistance of carbon lowers with increasing temperature. This and (primarily) its high melting point are why carbon is used in arc lamps.

So for a given circuit of constant resistance, current and voltage are proportional to each other.* This means we'll get twice the current for twice the voltage. The greater the voltage, the greater the current. But if the resistance is doubled for a circuit, the current will be half what it would be otherwise. The greater the resistance, the smaller the current. Ohm's law makes good sense.

Ohm's law tells us that a potential difference of 1 volt established across a circuit that has a resistance of 1 ohm will produce a current of 1 ampere. If 12 volts are impressed across the same circuit, the current will be 12 amperes. The resistance of a typical lamp cord is much less than 1 ohm, while a typical light bulb has a resistance of more than 100 ohms. An iron or electric toaster has a resistance of 15–20 ohms. Remember that for a given potential difference, less resistance means more current. Inside electrical devices such as radio and television receivers, current is regulated by circuit elements called *resistors,* whose resistance may be a few ohms or millions of ohms.

Questions

1. How much current will flow through a lamp that has a resistance of 60 Ω when 12 V are impressed across it?
2. What is the resistance of an electric frying pan that draws 12 A when connected to a 120-V circuit?

Ohm's Law and Electric Shock

What causes electric shock in the human body—current or voltage? The damaging effects of shock are the result of current passing through the body. From Ohm's law, we can see that this current depends on the voltage that is applied, and also on the electrical resistance of the human body. The resistance of one's body depends on its condition and ranges from about 100 ohms if soaked with salt water to about 500,000 ohms if the skin is very dry. If we touch the two electrodes of a battery with dry fingers, completing the circuit from one hand to the other, we can expect to offer a resistance of about 100,000 ohms. We usually cannot feel the current produced by 12 volts, and 24 volts just barely tingles. If our skin is moist, 24 volts can be quite uncomfortable. Table 22.1 describes the effects of different amounts of current on the human body.

Answers

1. $\frac{1}{5}$ A. This is calculated from Ohm's law: 0.2 A = 12 V/60 Ω.
2. 10 Ω. Rearrange Ohm's law to read

$$\text{Resistance} = \frac{\text{voltage}}{\text{current}} = \frac{120 \text{V}}{12 \text{A}} = 10 \text{ Ω}.$$

* Many texts use V for voltage, I for current, and R for resistance, and express Ohm's law as $V = IR$. It then follows that $I = V/R$, or $R = V/I$, so if any two variables are known, the third can be found. Units are abbreviated V for volts, A for amperes, and Ω for ohms.

Table 22.1 Effect of Electric Currents on the Body

Current (A)	Effect
0.001	Can be felt
0.005	Is painful
0.010	Causes involuntary muscle contractions (spasms)
0.015	Causes loss of muscle control
0.070	If through the heart, serious disruption; probably fatal if current lasts for more than 1 s

Questions

1. If the resistance of your body were 100,000 ohms, what would be the current in your body when you touched the terminals of a 12-volt battery?
2. If your skin were very moist so that your resistance was only 1000 ohms, and you touched the terminals of a 24-volt battery, how much current would you draw?

Many people are killed each year by current from common 120-volt electric circuits. If you touch a faulty 120-volt light fixture with your hand while you are standing on the ground, there is a 120-volt "electrical pressure" between your hand and the ground, so the current would probably not be enough to do serious harm. But if you are standing barefoot in a wet bathtub connected through its plumbing to the ground, the resistance between you and the ground is very small. Your overall resistance is so low that the 120-volt potential difference may produce a harmful current in your body. Handling electrical devices while taking a bath is a definite no-no.

Drops of water that collect around the on-off switches of devices such as a hair dryer can conduct current to the user. Although distilled water is a good insulator, the ions in ordinary water greatly reduce the electrical resistance. These ions are contributed by dissolved materials, especially salts. There is usually a layer of salt left from perspiration on your skin, which when wet lowers your skin resistance to a few hundred ohms or less.

To receive a shock, there must be a *difference* in electric potential between one part of your body and another part. Most of the current will pass along the path of least electrical resistance connecting these two points. Suppose you fell from a bridge and managed to grab onto a high-voltage power line, halting your fall. So long as you touch nothing else of different potential, you will receive no shock at all. Even if the wire is thousands of volts above ground potential and even if you hang by it with two hands, no appreciable charge will flow from one hand to the other. This is because there is no appreciable difference in electric potential between your hands. If, however, you reach over with one hand and grab onto a wire of different potential . . . zap! We have all seen birds perched on high-voltage wires. Every part of their bodies is at the same high potential as the wire, so they feel no ill effects.

FIGURE 22.7 The bird can stand harmlessly on one wire of high potential, but it had better not reach over and grab a neighboring wire! Why not?

Answers

1. The current through your body would be $\frac{12}{100,000}$ A (0.00012 A).
2. You would draw $\frac{24}{1000}$ A (0.024 A). Ouch—loss of muscle control!

FIGURE 22.8 The third prong connects the body of the appliance directly to ground. Any charge that builds up on an appliance is therefore conducted to the ground.

Most electric plugs and sockets today are wired with three, instead of two, connections. The principal two flat prongs on a plug are for the current-carrying double wire, one part of which is "live" and the other neutral, while the third round prong is connected directly to the earth (Figure 22.8). Appliances that use a lot of power such as irons, stoves, washing machines, and dryers are connected with these three wires. If the live wire accidentally comes in contact with the metal surface of the appliance, the current will be directed to ground and won't shock anyone who handles it.

Electric shock can overheat tissues in the body and disrupt normal nerve functions. It can upset the nerve center that controls breathing. In rescuing shock victims, the first thing to do is find and turn off the power source. Then do CPR until help arrives.

Question What causes electric shock, current or voltage?

Direct Current and Alternating Current

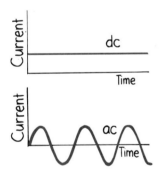

FIGURE 22.9 Time graphs of dc and ac.

Electric current may be dc or ac. By *dc,* we mean **direct current**, which refers to the flowing of charges in *one direction.* A battery produces direct current in a circuit because the terminals of the battery always have the same sign. Electrons move from the repelling negative terminal toward the attracting positive terminal, always moving through the circuit in the same direction. Even if the current occurs in unsteady pulses, so long as electrons move in one direction only, it is dc.

Alternating current (ac) acts as the name implies. Electrons in the circuit are moved first in one direction and then in the opposite direction, alternating to and fro about relatively fixed positions. This is accomplished by alternating the polarity of voltage at the generator or other voltage source. Nearly all commercial ac circuits in North America involve voltages and currents that alternate back and forth at a frequency of 60 cycles per second. This is 60-hertz current. In some places, 25-hertz, 30-hertz, or 50-hertz current is used. Throughout the world, most residential and commercial circuits are ac because electric energy in the form of ac can easily be stepped up to high voltage for long-distance transmission with small heat losses, then stepped down to convenient voltages where the energy is used. Why this is so will be discussed in Chapter 24.

Voltage of ac in North America is normally 120 volts.* In the early days of electricity, higher voltages burned out the filaments of electric light bulbs. Tradition has it that 110 volts was adopted as a first standard because it made bulbs of the day glow as brightly as a gas lamp. So the hundreds of power plants built in the U.S. prior to 1900 produced electricity at 110 volts (or 115 or 120 volts). By the time electricity became popular in Europe, engineers had figured out how to make light bulbs that would not burn out so fast at higher voltages. Power transmission is more efficient at higher

Answer Electric shock *occurs* when current is produced in the body, which is *caused* by an impressed voltage. So the initial *cause* is the voltage, but the current does the damage.

---

* 120 volts is what is called the "root-mean-square" average of the voltage. The actual voltage in a 120-volt ac circuit varies between +170 volts and −170 volts. It delivers the same power to an iron or a toaster as a 120-volt dc circuit.

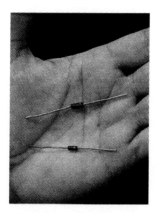

FIGURE 22.10 Diodes. As the symbol ──▶︎── suggests, current flows in the direction of the arrow but not in the reverse direction.

voltages, so Europe adopted 220 volts as their standard. The U.S. stayed with 110 volts (today officially 120 volts) because so much 110-volt equipment was already installed.

The primary use of electric current, whether dc or ac, is to transfer energy quietly, flexibly, and conveniently from one place to another.

Converting ac to dc

Household current is ac. The current in a battery-operated device such as a pocket calculator is dc. You can operate these devices on ac instead of batteries with an ac-dc converter. In addition to a transformer to lower the voltage (Chapter 24), the converter uses a *diode,* a tiny electronic device that acts as a one-way valve to allow electron flow in only one direction (Figure 22.10). Since alternating current changes its direction each half cycle, current passes through a diode only half of each period. The output is a rough dc, off half the time. To maintain continuous current while smoothing the bumps, a capacitor is used (Figure 22.11).

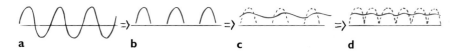

a   b   c   d

FIGURE 22.11 (a) When input to a diode is ac, (b) output is pulsating dc. (c) Slow charging and discharging of a capacitor provides continuous and smoother current. (d) In practice a pair of diodes are used so that there are no gaps in current output. The effect is to reverse the polarity of alternate half cycles instead of eliminating them.

Recall from the previous chapter that a capacitor acts as a storage reservoir for charge. Just as it takes time to raise the water level in a reservoir when water is added, it takes time to add or remove electrons from the plates of a capacitor. A capacitor therefore produces a retarding effect on changes in current flow. It smoothes the pulsed output.

FIGURE 22.12 Water input to the reservoir may be in repeated spurts or pulses, but the output is a fairly smooth stream. Likewise with a capacitor.

Speed and Source of Electrons in a Circuit

When we flip on the light switch on a wall and the circuit is completed, either ac or dc, the light bulb appears to glow immediately. When we make a telephone call, the electrical signal carrying our voice travels through the connecting wires at seemingly infinite speed. This signal is transmitted through the conductors at nearly the speed of light.

It is *not* the electrons that move at this speed.* Although electrons inside metal at room temperature have an average speed of a few million kilometers per hour, they make up no current because they are moving in random directions. There is no net flow

---

* Much effort and expense are expended in building particle accelerators that accelerate electrons and protons to speeds near that of light. If electrons in a common circuit traveled that fast, one would only have to bend a wire at a sharp angle and electrons traveling through the wire would possess so much momentum that they would fail to make the turn and fly off, providing a beam comparable to that produced by the accelerators!

FIGURE 22.13 The electric field lines between the terminals of a battery are directed through a conductor, which joins the terminals. A thick metal wire is shown here, but the path from one terminal to the other is usually an electric circuit. (If you touch this connecting wire, you won't be shocked, but the wire will heat quickly and may burn your hand!)

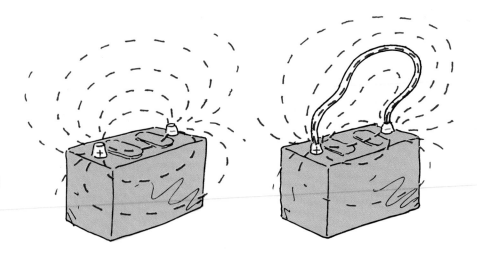

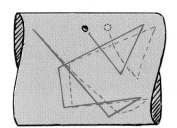

FIGURE 22.14 The solid lines depict a possible random path of an electron bounding about in an atomic lattice at a speed of about $\frac{1}{200}$ the speed of light. The dashed lines show an exaggerated view of how this path may be altered when an electric field is applied. The electron drifts toward the right with a speed much slower than a snail's pace. (This is quite idealized, for the electron wouldn't really encounter such similarly placed atoms and follow so similar a path when drifting.)

in any preferred direction. But when a battery or generator is connected, an electric field is established inside the conductor. The electrons continue their random motions while simultaneously being nudged by this field. It is the electric field that can travel through a circuit at nearly the speed of light. The conducting wire acts as a guide or "pipe" for electric field lines (Figure 22.13). In the space outside the wire, the electric field has a pattern determined by the location of electric charges, including charges in the wire. Inside the wire, the electric field is directed along the wire.

If the voltage source is dc, like the battery shown in Figure 22.13, the electric field lines are maintained in one direction in the conductor. Conduction electrons are accelerated by the field in a direction parallel to the field lines. Before they gain appreciable speed, they "bump into" the anchored metallic ions in their paths and transfer some of their kinetic energy to them. This is why current-carrying wires become hot. These collisions interrupt the motion of the electrons, so the actual *drift velocity* of electrons is extremely low. In a typical dc circuit—the electrical system of an automobile, for example—electrons have a drift velocity that averages about a hundredth of a centimeter per second. At this rate it would take about 3 hours for an electron to travel through 1 meter of wire! Large currents are possible because of large numbers of flowing electrons. So although an electric signal travels at nearly the speed of light in a wire, the electrons that move in response to this signal travel slower than a snail's pace.

In an ac circuit, the conduction electrons don't progress along the wire at all. They oscillate rhythmically to and fro about relatively fixed positions. When you talk to your friend on the telephone, it is the *pattern* of oscillating motion that is carried across town at nearly the speed of light. The electrons already in the wires vibrate to the rhythm of the traveling pattern.

A common misconception regarding electrical currents is that the current is propagated through the conducting wires by electrons bumping into one another—that an electrical pulse is transmitted in a manner similar to the way the pulse of a tipped domino is transferred along a row of closely spaced standing dominoes. This simply isn't true. The domino idea is a good model for the transmission of sound, but not for the transmission of electric energy. Electrons that are free to move in a conductor are accelerated by the electric field impressed upon them, but not because they bump into one another. True, they do bump into one another and other atoms, but this slows them down and offers resistance to their motion. Electrons throughout the entire closed path of a circuit all react simultaneously to the electric field.

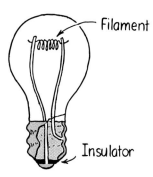

FIGURE 22.15 The conduction electrons that surge to and fro in the filament of the lamp do not come from the voltage source. They are in the filament to begin with. The voltage source simply provides them with surges of energy.

Another common misconception about electricity is the source of electrons. In a hardware store you can buy a water hose that is empty of water. But you can't buy a piece of wire, an "electron pipe," that is empty of electrons. The source of electrons in a circuit is the conducting circuit material itself. Some people think that the electrical outlets in the walls of their homes are a source of electrons. They think that electrons flow from the power utility through the power lines and into the wall outlets of their homes. This is not true. The outlets in homes are ac. Electrons make no net migration through a wire in an ac circuit.

When you plug a lamp into an outlet, *energy* flows from the outlet into the lamp, not electrons. Energy is carried by the pulsating electric field and causes vibratory motion of the electrons that already exist in the lamp filament. If 120 volts are impressed on a lamp, then an average of 120 joules of energy are dissipated by each coulomb of charge that is made to vibrate. Most of this electrical energy appears as heat, while some of it takes the form of light. Power utilities do not sell electrons. They sell *energy*. You supply the electrons.

So when you are jolted by an electric shock, the electrons making up the current in your body originate in your body. Electrons do not come out of the wire and through your body and into the ground; energy does. The energy simply causes free electrons in your body to vibrate in unison. Small vibrations tingle; large vibrations can be fatal.

Electric Power

FIGURE 22.16 The power and voltage on the light bulb read "100 W 120 V." How many amperes will flow through the bulb?

Unless it is in a superconductor, a charge moving in a circuit expends energy. This may result in heating the circuit or in turning a motor. The rate at which electric energy is converted into another form such as mechanical energy, heat, or light is called **electric power**. Electric power is equal to the product of current and voltage.*

$$\text{Power} = \text{current} \times \text{voltage}$$

If the voltage is expressed in volts and the current in amperes, then the power is expressed in watts. So, in units form,

$$\text{Watts} = \text{amperes} \times \text{volts}$$

If a lamp rated at 120 watts operates on a 120-volt line, you can see that it will draw a current of 1 ampere (120 watts = 1 ampere × 120 volts). A 60-watt lamp draws $\frac{1}{2}$ ampere on a 120-volt line. This relationship becomes a practical matter when you wish to know the cost of electrical energy, which is usually a few cents per kilowatt-hour depending on locality. A kilowatt is 1000 watts, and a kilowatt-hour represents the amount of energy consumed in 1 hour at the rate of 1 kilowatt.† Therefore, in a locality where electric energy costs 5 cents per kilowatt-hour, a 100-watt electric light bulb can be run for 10 hours at a cost of 5 cents, or a half cent for each hour. A toaster or iron, which draws much more current and therefore much more energy, costs about ten times as much to operate.

---

* Recall from Chapter 6 that Power = work/time; 1 Watt = 1J/s. Note that the units for mechanical power and electrical power check (work and energy are both measured in joules):

$$\text{Power} = \frac{\cancel{\text{charge}}}{\text{time}} \times \frac{\text{energy}}{\cancel{\text{charge}}} = \frac{\text{energy}}{\text{time}}$$

† Since Power = energy/time, simple rearrangement gives Energy = power × time; thus, energy can be expressed in the unit *kilowatt-hours* (kWh).

Questions

1. If a 120-V line to a socket is limited to 15 A by a safety fuse, will it operate a 1200-W hair dryer?
2. At 10¢/kWh, what does it cost to operate the 1200-W hair dryer for 1 h?

Electric Circuits

Any path along which electrons can flow is a *circuit*. For a continuous flow of electrons, there must be a complete circuit with no gaps. A gap is usually provided by an electric switch that can be opened or closed to either cut off or allow energy flow. Most circuits have more than one device that receives electric energy. These devices are commonly connected in a circuit in one of two ways, *series* or *parallel*. When connected in series, they form a single pathway for electron flow between the terminals of the battery, generator, or wall socket (which is simply an extension of these terminals). When connected in parallel, they form branches, each of which is a separate path for the flow of electrons. Both series and parallel connections have their own distinctive characteristics. We shall briefly treat circuits using these two types of connections.

Series Circuits

A simple **series circuit** is shown in Figure 22.17. Three lamps are connected in series with a battery. The same current exists almost immediately in all three lamps when the switch is closed. The charge does not "pile up" in any lamp but flows *through* each lamp. Electrons in all parts of the circuit begin to move at once. Some electrons move away from the negative terminal of the battery, some move toward the positive terminal,

FIGURE 22.17 A simple series circuit. The 6-V battery provides 2 V across each lamp.

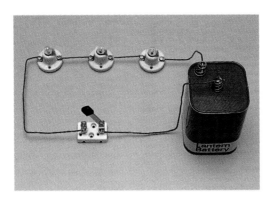
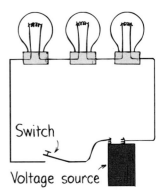

Answers

1. From the expression watts = amperes × volts, we see that current = 1200 W/120 V = 10 A, so the hair dryer will operate when connected to the circuit. But two hair dryers on the same circuit will blow the fuse.
2. 12¢ (1200 W = 1.2 kW; 1.2 kW × 1 h × 10¢/1 kWh = 12¢).

some move through the filament of each lamp. Eventually the electrons move all the way around the circuit (the same amount of current passes through the battery). This is the only path of the electrons through the circuit. A break anywhere in the path results in an open circuit, and the flow of electrons ceases. Burning out one of the lamp filaments or simply opening the switch could cause such a break.

The circuit shown in Figure 22.17 illustrates the following important characteristics of series connections:

1. Electric current has but a single pathway through the circuit. This means that the current passing through each electrical device is the same.

2. This current is resisted by the resistance of the first device, the resistance of the second, and that of the third also, so the total resistance to current in the circuit is the sum of the individual resistances along the circuit path.

3. The current in the circuit is numerically equal to the voltage supplied by the source divided by the total resistance of the circuit. This is in accord with Ohm's law.

4. Ohm's law also applies separately to each device. The *voltage drop*, or potential difference, across each device is proportional to its resistance. This follows from the fact that more energy is used to move a unit of charge through a large resistance than through a small resistance.

5. The total voltage impressed across a series circuit divides among the individual electrical devices in the circuit so that the sum of the voltage drops across each individual device is equal to the total voltage supplied by the source. This follows from the fact that the amount of energy used to move each unit of charge through the entire circuit equals the sum of the energies used to move that unit of charge through each electrical device in turn.

Questions

1. What happens to current in other lamps if one lamp in a series circuit burns out?
2. What happens to the light intensity of each lamp in a series circuit when more lamps are added to the circuit?

It is easy to see the main disadvantage of a series circuit: If one device fails, current in the whole circuit ceases. Some cheap Christmas tree lights are connected in series. When one bulb burns out, it's fun and games (or frustration) trying to find which one to replace.

Answers

1. If one of the lamp filaments burns out, the path connecting the terminals of the voltage source will break and current will cease. All lamps will go out.
2. The addition of more lamps in a series circuit results in a greater circuit resistance. This decreases the current in the circuit and therefore in each lamp, which causes dimming of the lamps. Energy is divided among more lamps, so the voltage drop across each lamp will be less.

Most circuits are wired so that it is possible to operate several electrical devices, each independently of the other. In your home, for example, a lamp can be turned on or off without affecting the operation of other lamps or electrical devices. This is because these devices are connected not in series, but in parallel with one another.

Parallel Circuits

A simple **parallel circuit** is shown in Figure 22.18. Three lamps are connected to the same two points A and B. Electrical devices connected to the same two points of an electrical circuit are said to be *connected in parallel*. The pathway for current from one terminal of the battery to the other is completed if only *one* lamp is lit. In this illustration, the circuit branches into three separate pathways from A to B. A break in any one path does not interrupt the flow of charge in the other paths. Each device operates independently of the other devices.

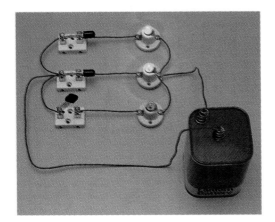
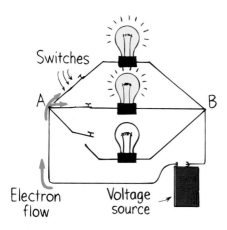

FIGURE 22.18 A simple parallel circuit. A 6-V battery provides 6 V across each lamp.

The circuit shown in Figure 22.18 illustrates the following major characteristics of parallel connections:

1. Each device connects the same two points A and B of the circuit. The voltage is therefore the same across each device.
2. The total current in the circuit divides among the parallel branches. Since the voltage across each branch is the same, the amount of current in each branch is inversely proportional to the resistance of the branch—Ohm's law applies separately to each branch.
3. The total current in the circuit equals the sum of the currents in its parallel branches.
4. As the number of parallel branches is increased, the overall resistance of the circuit is decreased. Overall resistance is lowered with each added path between any two points of the circuit. This means the overall resistance of the circuit is less than the resistance of any one of the branches.

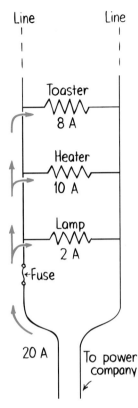

FIGURE 22.19 Circuit diagram for appliances connected to a household supply line.

Questions

1. What happens to the current in other lamps if one of the lamps in a parallel circuit burns out?
2. What happens to the light intensity of each lamp in a parallel circuit when more lamps are added in parallel to the circuit?

Parallel Circuits and Overloading

Electricity is usually fed into a home by way of two wires called *lines*. These lines are very low in resistance and are connected to wall outlets in each room—sometimes through two or more separate circuits. About 110–120 volts are impressed on these lines by a transformer in the neighborhood that steps down the higher voltage supply by generators at the power utility. This voltage is applied to appliances and other devices that are connected in parallel by plugs to the house circuits. As more devices are connected to the circuits, more pathways for current result in lowering of the combined resistance of each circuit. Therefore, a greater amount of current occurs in the circuits. Circuits that carry more than a safe amount of current are said to be *overloaded*.

We can see how overloading occurs by considering the circuit in Figure 22.19. The supply line is connected to an electric toaster that draws 8 amperes, to an electric heater that draws 10 amperes, and to an electric lamp that draws 2 amperes. When only the toaster is operating and drawing 8 amperes, the total line current is 8 amperes. When the heater is also operating, the total line current increases to 18 amperes (8 amperes to the toaster and 10 amperes to the heater). If you turn on the lamp, the line current increases to 20 amperes. Connecting any more devices increases the current still more. Connecting too many devices into the same circuit results in overheating that may cause a fire.

Answers

1. If one lamp burns out, the other lamps will be unaffected. The current in each branch, according to Ohm's law, is equal to voltage/resistance, and since neither voltage nor resistance is affected in the other branches, the current in those branches is unaffected. The total current in the overall circuit (the current through the battery), however, is decreased by an amount equal to the current drawn by the lamp in question before it burned out. But the current in any other single branch is unchanged.
2. The light intensity for each lamp is unchanged as other lamps are introduced (or removed). Only the total resistance and total current in the total circuit changes, which is to say the current in the battery changes. (There is resistance in a battery also, which we assume is negligible here.) As lamps are introduced, more paths are available between the battery terminals, which effectively decreases total circuit resistance. This decreased resistance is accompanied by an increased current, the same increase that feeds energy to the lamps as they are introduced. Although changes of resistance and current occur for the circuit as a whole, no changes occur in any individual branch in the circuit.

Safety Fuses

To prevent overloading in circuits, fuses are connected in series along the supply line. In this way the entire line current must pass through the fuse. The fuse shown in Figure 22.20 is constructed with a wire ribbon that will heat up and melt at a given current. If the fuse is rated at 20 amperes, it will pass 20 amperes, but no more. A current above 20 amperes will melt the fuse, which "blows out" and breaks the circuit. Before a blown fuse is replaced, the cause of overloading should be determined and remedied. Often, insulation that separates the wires in a circuit wears away and allows the wires to touch. This greatly reduces the resistance in the circuit, effectively shortening the circuit path, and is called a *short circuit*.

FIGURE 22.20 A safety fuse.

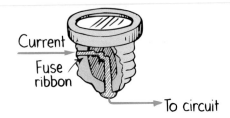

In modern buildings, fuses have been largely replaced by circuit breakers, which use magnets or bimetallic strips to open a switch when the current is too great. Utility companies use circuit breakers to protect their lines all the way back to the generators.

Summary of Terms

Potential difference The difference in voltage between two points, measured in volts. It can be compared to the difference in water pressure between two containers: If two containers having different water pressure are connected by a pipe, water will flow from the one with the higher pressure to the one with the lower pressure until the two pressures are equalized. Similarly, if two points with a difference in potential are connected by a conductor, charge will flow from the one with the greater potential to the one with the smaller potential so long as a potential difference is maintained.

Electric current The flow of electric charge that transports energy from one place to another. Measured in amperes, where 1 A is the flow of 6.25×10^{18} electrons per second, or 1 coulomb per second.

Electrical resistance The property of a material that resists electric current. Measured in ohms.

Superconductor A material in which the resistance to electric current drops to zero under special circumstances that usually include low temperatures.

Ohm's law The statement that the current in a circuit varies in direct proportion to the potential difference or voltage and inversely with resistance.

$$\text{Current} = \frac{\text{voltage}}{\text{resistance}}$$

A potential difference of 1 V across a resistance of 1 Ω produces a current of 1 A.

Direct current (dc) An electric current consisting of charges flowing in one direction only.

Alternating current (ac) Electric current that repeatedly reverses its direction; the electric charges vibrate about relatively fixed points. In the United States the vibrational rate is 60 Hz.

Electric power The rate of energy transfer, or the rate of doing work; the amount of energy per unit time, which electrically can be measured by the product of current and voltage.

$$\text{Power} = \text{current} \times \text{voltage}$$

Measured in watts (or kilowatts), where $1A \times 1V = 1W$.

Series circuit An electric circuit with devices connected in such a way that the electric current is the same through each of them.

Parallel circuit   An electric circuit with two or more devices connected in such a way that the same voltage acts across each one and any single one completes the circuit independently of all the others.

Review Questions

Flow of Charge

1. What condition is necessary for the flow of heat? What analogous condition is necessary for the flow of charge?
2. The flow of water is driven by a pressure difference. What drives the flow of charge?
3. What condition is necessary for the sustained flow of water in a pipe? What analogous condition is necessary for the sustained flow of charge in a wire?

Electric Current

4. Why are *electrons*, rather than *protons*, the principal charge carriers in metal wires?
5. What exactly is an *ampere*?
6. Why is a current-carrying wire normally not electrically charged?

Voltage Sources

7. Name two kinds of "electric pumps."
8. How much energy is given to each coulomb of charge passing through a 6-V battery?
9. Does charge flow *through* a circuit or *into* a circuit? Does voltage flow *through* a circuit, or is voltage established *across* a circuit?
10. Why do we usually think of voltage as a cause and current as an effect?

Electrical Resistance

11. Will water flow more easily through a wide pipe or a narrow pipe? Will current flow more easily through a thick wire or a thin wire?
12. Does heating a copper wire increase or decrease its electrical resistance?
13. How does a *superconductor* differ from an *ordinary conductor*?

Ohm's Law

14. If the voltage impressed across a circuit is held constant while the resistance doubles, what change occurs in the current?
15. If the resistance of a circuit remains constant while the voltage across the circuit decreases to half its former value, what change occurs in the current?
16. What role do resistors normally play in an electric circuit?

Ohm's Law and Electric Shock

17. How does wetness affect the resistance of your body?
18. What happens to the resistance of your skin when you perspire?
19. Why is it a very poor idea to handle electrical devices while in the bathtub?
20. High voltage by itself does not produce electric shock. What does?
21. What is the function of the third prong in a modern household electric plug?

Direct Current and Alternating Current

22. Distinguish between *dc* and *ac*.
23. Does a battery produce *dc* or *ac*? Does the generator at a power station produce *dc* or *ac*?
24. What does it mean to say that a certain current is 60 Hz?

Converting ac to dc

25. What property of a diode enables it to convert ac to pulsed dc?
26. A diode converts ac to pulsed dc. What electrical device smoothes the pulsed dc to a smoother dc?

Speed and Source of Electrons in a Circuit

27. True or false: Electrons in a common battery-driven circuit travel at about the speed of light. Defend your answer.
28. Exactly what does flow through a circuit at about the speed of light?
29. Why does a wire that carries electric current become hot?
30. What is meant by *drift velocity*?
31. A tipped domino sends a pulse along a row of standing dominoes. Is this a good analogy for the way electric current, sound, or both travel?
32. When you pay your household electric bill at the end of the month, which of the following are you paying for: voltage, current, power, energy? Explain.
33. Where do the electrons originate that produce an electric shock when you touch a charged conductor?

Electric Power

34. What is the relationship among electric power, current, and voltage?
35. Which of these is a unit of power and which is a unit of energy: a watt; a kilowatt; a kilowatt-hour?
36. Distinguish between a *kilowatt* and a *kilowatt-hour*.

Electric Circuits

37. What is an *electric circuit*?

Series Circuits

38. In a circuit of two lamps in series, if the current through one lamp is 1 A, what is the current through the other lamp?

39. If 6 V are impressed across the above circuit and the voltage across the first lamp is 2 V, what is the voltage across the second lamp?

40. What is a main shortcoming of a series circuit?

Parallel Circuits

41. In a circuit of two lamps in parallel, if there are 6 V across one lamp, what is the voltage across the other lamp?

42. If the current through each of the two branches of a parallel circuit is the same, what does this tell you about the resistance of the two branches?

43. How does the total current though the branches of a simple parallel circuit compare to the current that flows through the voltage source?

44. As more lines are added at a fast food restaurant, the resistance to getting people served is reduced. How is this similar to what happens when more branches are added to a parallel circuit?

45. Are household circuits normally wired in series or in parallel?

Parallel Circuits and Overloading

46. How does the amount of current in a home circuit differ from the amount of current in a reading lamp?

47. Why will too many electrical devices operating at one time often blow a fuse or trip a circuit breaker?

Safety Fuses

48. What is the function of fuses or circuit breakers in a circuit?

Projects

1. An electric cell is made by placing two plates of different materials that have different affinities for electrons in a conducting solution. You can make a simple 1.5-V cell by placing a strip of copper and a strip of zinc in a tumbler of salt water. The voltage of a cell depends on the materials used and the solution they are placed in, not the size of the plates. A battery is actually a series of cells.

Another easy cell to construct is the citrus cell. Stick a paper clip and a piece of copper wire into a lemon. Hold the ends of both close together, but not touching, and place the ends on your tongue. The slight tingle

that you feel and the metallic taste you experience result from a slight current of electricity pushed by the citrus cell through the wires when your moist tongue closes the circuit.

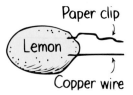

2. Examine the electric meter in your home. It is probably in the basement or on the outside of your house. You will see that in addition to the clocklike dials in the meter, there is a circular aluminum disk that spins between the poles of magnets when electric current goes into the house. The more electric current, the faster the disk turns. The speed of the disk is directly proportional to the number of watts used; for example, it will spin five times as fast for 500 W as for 100 W.

You can use the meter to determine how many watts an electrical device uses. First, see that all electrical devices in your home are disconnected (you may leave electric clocks connected—the 2 watts they use will hardly be noticeable). The disk will be practically stationary. Then connect a 100-W bulb and notice how many seconds it takes for the disk to make five complete rotations. The black spot painted on the edge of the disk makes this easy. Disconnect the 100-W bulb and plug in a device of unknown wattage. Again, count the seconds for five revolutions. If it takes the same time, it's a 100-W device; if it takes twice the time, it's a 50-W device; half the time, a 200-W device, and so forth. In this way you can estimate the power consumption of devices fairly accurately.

Exercises

1. If a wire connects two separated objects charged to different electric potential energies, can you say for certain which way charge will flow in the wire? How about if the objects are charged to different electric potentials?

2. If an electric current flows from one object to another, what can we say about the relative magnitudes of the electric potentials of the two objects?

3. One example of a water system is a garden hose that waters a garden. Another is the cooling system of an automobile. Which of these exhibits behavior more analogous to that of an electric circuit? Why?

4. What happens to the brightness of light emitted by a lightbulb when the current that flows in it increases?

5. Is a current-carrying wire electrically charged?

6. Your tutor tells you that an *ampere* and a *volt* really measure the same thing, and the different terms only serve to make a simple concept seem confusing. Why should you consider getting a different tutor?

7. In which of the circuits below does a current exist to light the bulb?

8. Does more current flow out of a battery than into it? Does more current flow into a lightbulb than out of it? Explain.

9. Energy is given to electric charge by pumping it from a low potential to a high potential. How do we get energy out of electricity?

10. Sometimes you hear someone say that a particular appliance "uses up" electricity. What is it that the appliance actually uses up and what becomes of it?

11. The engine in your automobile turns an alternator that supplies electricity to the car battery. Does this mean that the battery is an alternating-current device? Explain.

12. Suppose you leave your car lights on while at a movie. When you return, your battery is too "weak" to start your car. A friend comes and gives you a jump start with his battery and battery cables. What is happening when your friend gives you a jump start?

13. After your car is running, your friend disconnects the battery cables and you're on your way. Why is everything okay now? What about your weak battery?

14. An electron moving in a wire collides again and again with atoms and travels an average distance between collisions that is called the *mean free path*. If the mean free path is less in some metals, what can you say about the resistance of these metals? For a given conductor, what can you do to lengthen the mean free path?

15. Why will the resistance of a wire be slightly different immediately after you have held it in your hand?

16. Why is the current in an incandescent bulb greater immediately after it is turned on, than it is a few moments later?

17. A simple lie detector consists of an electric circuit, one part of which is part of your body—like from one finger to another. A sensitive meter shows the current that flows when a small voltage is applied. How does this technique indicate that a person is lying? (And when does this technique not tell when someone is lying?)

18. Only a small percentage of the electric energy fed into a common lightbulb is transformed into light. What happens to the rest?

19. Why are thick wires rather than thin wires usually used to carry large currents?

20. Will a lamp with a thick filament draw more current or less current than a lamp with a thin filament?

21. A 1-mile-long copper wire has a resistance of 10 ohms. What will be its new resistance when it is shortened by (a) cutting it in half; (b) doubling it over and using it as "one" wire?

22. What is the effect on current if both the voltage and the resistance are doubled? If both are halved?

23. Will the current in a lightbulb connected to a 220-V source be greater or less than when the same bulb is connected to a 110-V source?

24. Which will do less damage—plugging a 110-V appliance into a 220-V circuit or plugging a 220-V appliance into a 110-V circuit? Explain.

25. If a current of one- or two-tenths of an ampere flows into one of your hands and out the other, you will probably be electrocuted. But if the same current flows into your hand and out the elbow above the same hand, you can survive even though the current may be large enough to burn your flesh. Explain.

26. Would you expect to find dc or ac in the filament of a lightbulb in your home? How about in the headlight of an automobile?

27. Are automobile headlights wired in parallel or in series? What is your evidence?

28. A car's headlights consume 40 W on low beam, and 50 W on high beam. Is there more or less resistance in the high beam filament?

29. To connect a pair of resistors so their equivalent resistance will be more than the resistance of either one, should you connect them in series or in parallel?

30. To connect a pair of resistors so their equivalent resistance will be less than the resistance of either one, should you connect them in series or in parallel?

31. The damaging effects of electric shock result from the amount of current that flows in the body. Why, then, do we see signs that read "Danger—High Voltage" rather than "Danger—High Current"?

32. Comment on the warning sign shown in the sketch.

33. Is this label on a household product cause for concern? "Caution: This product contains tiny electrically charged particles moving at speeds in excess of 100,000,000 kilometers per hour."

34. Why is the wingspan of birds a consideration in determining the spacing between parallel wires in a power line?

35. Estimate the number of electrons that a power company delivers annually to the homes of a typical city of 50,000 people.

36. If electrons flow very slowly through a circuit, why does it not take a noticeably long time for a lamp to glow when you turn on a distant switch?

37. Why is the speed of an electric signal so much greater than the speed of sound?

38. If a glowing lightbulb is jarred and oxygen leaks inside, the bulb will momentarily brighten considerably before burning out. Putting excess current through a lightbulb will also burn it out. What physical change occurs when a lightbulb burns out?

39. In the circuit shown, how do the brightnesses of the identical lightbulbs compare? Which lightbulb draws the most current? What will happen if bulb A is unscrewed? If C is unscrewed?

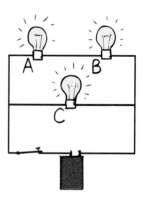

40. Consider a pair of flashlight bulbs connected to a battery. Will they glow brighter connected in series or in parallel? Will the battery run down faster if they are connected in series or in parallel?

41. If several bulbs are connected in series to a battery, they may feel warm to the touch but not visibly glow. What is your explanation?

42. As more and more bulbs are connected in series to a flashlight battery, what happens to the brightness of each bulb? Assuming heating inside the battery is negligible, what happens to the brightness of each bulb when more and more bulbs are connected in parallel?

43. What changes occur in the line current when more devices are introduced in a series circuit? In a parallel circuit? Why are your answers different?

44. When a pair of identical resistors are connected in series, which of the following is the same for both resistors: Voltage across each, power dissipated in each, current through each? Do any of your answers change if the resistors are different from each other?

45. When a pair of identical resistors are connected in parallel, which of the following is the same for both resistors: Voltage across each, power dissipated in each, current through each? Do any of your answers change if the resistors are different from each other?

46. A battery has internal resistance, so if the current it supplies goes up, the voltage it supplies goes down. If too many bulbs are connected in parallel across a battery, will their brightness diminish? Explain.

47. Why are devices in household circuits almost never connected in series?

48. Figure 22.19 shows a fuse placed in household circuitry. (a) Where else might a fuse be placed in this circuitry to serve a useful purpose, melting only if a problem arises? (b) Where would it be foolish to place a fuse in this circuitry because it would melt immediately?

49. Is the resistance of a 100-W bulb greater or less than the resistance of a 60-W bulb? Assuming the filaments in each bulb are of the same length and made of the same material, which bulb has the thicker filament?

50. If a 60-W bulb and a 100-W bulb are connected in series in a circuit, across which bulb will there be the greater voltage drop? How about if they are connected in parallel?

Problems

1. The wattage marked on a lightbulb is not an inherent property of the bulb but depends on the voltage to which it is connected, usually 110 or 120 V. How many amperes flow through a 60-W bulb connected in a 120-V circuit?

2. Rearrange the equation Current = voltage/resistance to express *resistance* in terms of current and voltage. Then solve the following: A certain device in a 120-V circuit has a current rating of 20 A. What is the resistance of the device (how many ohms)?

3. Using the formula Power = current x voltage, find the current drawn by a 1200-W hair dryer connected to 120 V. Then using the method you used in the previous problem, find the resistance of the hair dryer.

4. The total charge that an automobile battery can supply without being recharged is given in terms of ampere-hours. A typical 12-V battery has a rating of 60 ampere-hours (60 A for 1 h, 30 A for 2 h, and so on). Suppose you forget to turn off the headlights in your parked automobile. If each of the two headlight draws 3 A, how long will it be before your battery is "dead"?

5. How much does it cost to operate a 100-W lamp continuously for 1 week if the power utility rate is 10¢/kWh?

6. A 4-W night light is plugged into a 120-V circuit and operates continuously for 1 year. Find the following: (a) the current it draws, (b) the resistance of its filament, (c) the energy consumed in a year, and (d) the cost of its operation for a year at the utility rate of 10¢/kWh.

7. An electric iron connected to a 110-V source draws 9 A of current. How much heat (in joules) does it generate in a minute?

8. How many coulombs of charge flow through the iron in the previous problem in one minute?

9. A certain light bulb with a resistance of 95 ohms is labeled "150 W." Was this bulb designed for use in a 120-V circuit or a 220-V circuit?

10. In periods of peak demand, power companies lower their voltage. This saves them power (and saves you money!). To see the effect, consider a 1200-W toaster that draws 10 A when connected to 120 V. Suppose the voltage is lowered by 10% to 108 V. By how much does the current decrease? By how much does the power decrease? (*Caution*: The 1200-W label is valid only when 120 V is applied. When the voltage is lowered, it is the resistance of the toaster, not its power, that remains constant.)

23

············

Magnetism

: **"T**hey who control magnetism control the world." —Dick Tracy

Youngsters are fascinated with magnets, largely because they act at a distance. One can move a nail with a nearby magnet even when a piece of wood is in between. Likewise a neurosurgeon can guide a pellet through brain tissue to inoperable tumors, pull a catheter into position, or implant electrodes while doing little harm to brain tissue. The use of magnets grows daily.

The term *magnetism* comes from Magnesia, a province of Greece, where the Greeks found certain stones more than 2000 years ago. These stones, called *lodestones,* had the unusual property of attracting pieces of iron. Magnets were first fashioned into compasses and used for navigation by the Chinese in the twelfth century.

In the sixteenth century, William Gilbert, Queen Elizabeth's physician, made artificial magnets by rubbing pieces of iron against lodestone. He also suggested that a compass always points north and south because the earth itself has magnetic properties. Later, in 1750, John Michell in England found that magnetic poles obey the inverse-square law, and his results were confirmed by Charles Coulomb. Knowledge of magnetism and electricity developed almost independently of each other until 1820, when a Danish science professor named Hans Christian Oersted discovered in a classroom demonstration that an electric current affects a magnetic compass.* He

---

* We can only speculate about how often such relationships become evident when they "aren't supposed to" and are dismissed as "something wrong with the apparatus." Oersted, however, had the insight—characteristic of a good scientist—to see that nature was revealing another of its secrets.

saw confirming evidence that magnetism was related to electricity. Shortly thereafter, the French physicist André-Marie Ampère proposed that electric currents are the source of all magnetic phenomena.

Magnetic Forces

In Chapter 21 we discussed the forces that electrically charged particles exert on one another: The force between any two charged particles depends on the magnitude of the charge on each and their distance of separation, as specified in Coulomb's law. But Coulomb's law is not precisely true when the charged particles are moving with respect to each other. The force between electrically charged particles depends also, in a complicated way, on their motion. We find that in addition to the force we call *electrical*, there is a force due to the motion of the charged particles that we call the **magnetic force**. Both electrical and magnetic forces are actually different aspects of the same phenomenon of electromagnetism.

Magnetic Poles

The forces that magnets exert on one another are similar to electrical forces, for they can both attract and repel without touching, depending on which ends of the magnets are held near one another. Like electrical forces also, the strength of their interaction depends on the distance the two magnets are separated. Whereas electric charge is central to electrical forces, regions called *magnetic poles* give rise to magnetic forces.

A bar magnet that is suspended from its center by a piece of string will act as a compass. One end, called the *north-seeking pole,* points northward, and the opposite end, called the *south-seeking pole,* points southward. More simply, these are called the *north* and *south poles*. All magnets have both a north and a south pole. In a simple bar magnet these are located at the two ends. A common horseshoe magnet is simply a bar magnet that has been bent into a U shape. Its poles are also at its two ends (Figure 23.1).

When the north pole of one magnet is brought near the north pole of another magnet, they repel.* The same is true of a south pole near a south pole. If opposite poles are brought together, however, attraction occurs. We find that

Like poles repel; opposite poles attract.

This rule is similar to the rule for the forces between electric charges, where like charges repel one another and unlike charges attract. But there is a very important difference between magnetic poles and electric charges. Whereas electric charges can be isolated, magnetic poles cannot. Negatively charged electrons and positively charged protons are entities by themselves. A cluster of electrons need not be accompanied by a cluster of protons and vice versa. But a north magnetic pole never exists without the presence of a south pole and vice versa. If you break a bar magnet in half, each half still behaves as a complete magnet. Break the pieces in half again, and you have four complete magnets. You can continue breaking the pieces in half and never isolate a single pole.† Even when your piece is one atom thick, there are two poles. This suggests that atoms themselves are magnets.

FIGURE 23.1 A horseshoe magnet.

---

* The force of interaction between magnetic poles is given by $F \sim p_1 p_2 / d^2$, where p_1 and p_2 represent magnetic pole strengths and d represents the separation distance between the poles. Note the similarity of this relationship to Coulomb's law.

† Theoretical physicists have speculated for more than 65 years about the possible existence of discrete magnetic "charges," called *magnetic monopoles*. These tiny particles would carry either a single north or a single south magnetic pole and would be the counterparts to the positive and negative charges in electricity. Various attempts have been made to find monopoles, but none has proved successful. All known magnets always have at least one north and one south pole.

Question Does every magnet necessarily have a north and south pole?

Magnetic Fields

Recall that the space that surrounds a massive body contains a gravitational field, and the space that surrounds an electrically charged body contains an electric field. If the electrically charged body is moving, the region of space surrounding it is further altered. This alteration due to the motion of a charge is the **magnetic field**. We say a charged body in motion is surrounded by both an electric field and a magnetic field. Like the electric field, the magnetic field is a storehouse of energy. The greater the speed of the charged body, the greater the magnitude of the surrounding magnetic field. A magnetic field is produced by electric charge in motion.*

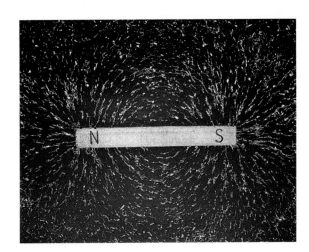

FIGURE 23.2 Top view of iron filings sprinkled on a sheet of paper on top of a magnet. The filings trace out a pattern of *magnetic field lines* in the surrounding space. The magnetic field lines continue inside the magnet (not revealed by the filings) and form closed loops. The source of the field is the motion of electrons in the iron atoms that compose the magnet.

So if the motion of electric charge produces magnetism, where is this motion in a common bar magnet? The answer is: in the electrons in the atoms. Although the magnet as a whole may be stationary, the magnet is composed of atoms whose electrons are in constant motion. Two kinds of electron motion contribute to magnetism: their spinning motion and their orbital motion. Electrons spin about their own axes like tops. A spinning electron is charge in motion. Electrons also revolve about the atomic nucleus—again, charge in motion. The field due to spinning predominates in most materials.

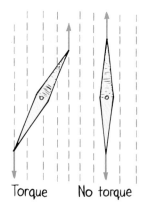

Torque   No torque

FIGURE 23.3 When the compass needle is not aligned with the magnetic field, the oppositely directed forces produce a pair of torques (called a *couple*) that twist the needle into alignment.

Answer Yes, just as every coin has two sides, a "head" and a "tail." Some "trick" magnets have more than two poles, but nevertheless their net "pole strength" is zero.

*Interestingly enough, since motion is relative, the magnetic field is relative. For example, when a charge moves by you, there is a definite magnetic field associated with the moving charge. But if you move along with the charge, so there is no motion relative to you, you will find no magnetic field associated with the charge. Magnetism is relativistic. In fact, it was Albert Einstein who first explained this when he published his first paper on special relativity, "On the Electrodynamics of Moving Bodies." (More on relativity in Chapters 34 and 35.)

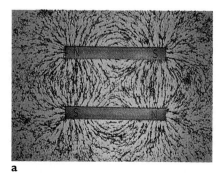
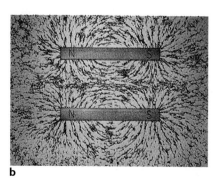

a   b

FIGURE 23.4   The magnetic field patterns for a pair of magnets. (a) Opposite poles are nearest each other, and (b) like poles are nearest each other.

FIGURE 23.5   The iron nails become induced magnets.

So every electron is a tiny electromagnet. A pair of electrons spinning in the same direction makes up a stronger electromagnet. A pair of electrons spinning in opposite directions, however, work against each other. Their magnetic fields cancel. This is why most substances are not magnets. In most atoms, the various fields cancel each other because the electrons spin in opposite directions. In materials such as iron, nickel, and cobalt, however, the fields do not cancel each other entirely. Each iron atom has four electrons whose spin magnetism is uncanceled. Each iron atom, then, is a tiny magnet. The same is true to a lesser degree for the atoms of nickel and cobalt.*

Most iron objects around you are magnetized to some degree. A filing cabinet, a refrigerator, or even cans of food on your pantry shelf, have north and south poles induced by the earth's magnetic field. Pass a compass from their bottoms to their tops and their poles are easily identified. Turn cans upside down and see how many days it takes for the poles to reverse themselves!

Magnetic Domains

Magnetic fields pass undiminished through most materials, including many metals. But not iron. The magnetic field of individual iron atoms is so strong that interaction among adjacent atoms causes large clusters of them to line up with each other. These clusters of aligned atoms are called **magnetic domains**. Each domain is perfectly magnetized and is made up of billions of aligned atoms. The domains are microscopic (Figure 23.6), and there are many of them in a crystal of iron.

FIGURE 23.6   A microscopic view of magnetic domains in a crystal of iron. Each domain consists of billions of aligned iron atoms.

* Most common magnets are made from alloys containing iron, nickel, cobalt, and aluminum in various proportions. In these the electron spin contributes virtually all the magnetic properties. In the rare earth metals like gadolinium, the orbital motion is more significant.

Not every piece of iron, however, is a magnet. This is because the domains in ordinary iron are not aligned. Consider a common iron nail: The domains in the nail are randomly oriented. They can be induced into alignment, however, when a magnet is brought nearby. (It is interesting to listen with an amplified stethoscope to the clickety-clack of domains undergoing alignment in a piece of iron when a strong magnet approaches.) The domains align themselves much as electrical charges in a piece of paper align themselves in the presence of a charged rod. When you remove the nail from the magnet, ordinary thermal motion causes most or all of the domains in the nail to return to a random arrangement. If the field of the permanent magnet is very strong, however, the nail may retain some permanent magnetism of its own after the two are separated.

FIGURE 23.7 Pieces of iron in successive stages of magnetism. The arrows represent domains; the head is a north pole and the tail a south pole. Poles of neighboring domains neutralize each other's effects, except at the ends.

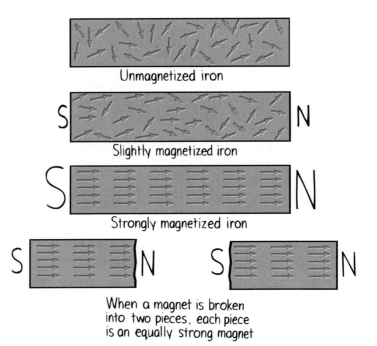

Unmagnetized iron

Slightly magnetized iron

Strongly magnetized iron

When a magnet is broken into two pieces, each piece is an equally strong magnet

Permanent magnets are made by simply placing pieces of iron or certain iron alloys in strong magnetic fields. Alloys of iron differ; soft iron is easier to magnetize than steel. It helps to tap the iron to nudge any stubborn domains into alignment. Another way of making a permanent magnet is to stroke a piece of iron with a magnet. The stroking motion aligns the domains in the iron. If a permanent magnet is dropped or heated, some of the domains are jostled out of alignment and the magnet becomes weaker.

Question How can a magnet attract a piece of iron that is not magnetized?

Answer Domains in the unmagnetized piece of iron are induced into alignment by the magnetic field of the nearby magnet. See the similarity of this with Figure 21.13. Like the pieces of paper that jump to the comb, pieces of iron will jump to a strong magnet when it is brought nearby. But unlike the paper, they are not then repelled. Can you think of the reason why?

A moving charge produces a magnetic field. A current of charges, then, also produces a magnetic field. The magnetic field that surrounds a current-carrying conductor can be demonstrated by arranging an assortment of compasses around a wire (Figure 23.8) and passing a current through it. The compasses line up with the magnetic field produced by the current and show it to be a pattern of concentric circles about the wire. When the current reverses direction, the compass needles turn around, showing that the direction of the magnetic field changes also. This is the effect that Oersted first demonstrated in his classroom.

Electric Currents and Magnetic Fields

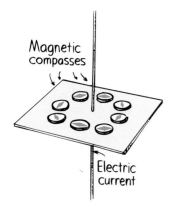

Magnetic compasses

Electric current

FIGURE 23.8 The compasses show the circular shape of the magnetic field surrounding the current-carrying wire.

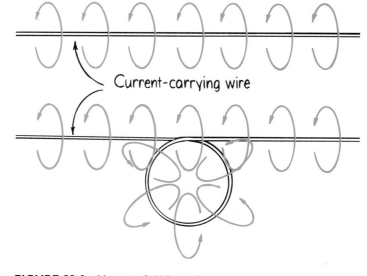

Current-carrying wire

FIGURE 23.9 Magnetic field lines about a current-carrying wire crowd up when the wire is bent into a loop.

FIGURE 23.10 Iron filings sprinkled on paper reveal the magnetic field configurations about (a) a current-carrying wire, (b) a current-carrying loop, and (c) a coil of loops.

If the wire is bent into a loop, the magnetic field lines become bunched up inside the loop (Figure 23.9). If the wire is bent into another loop, overlapping the first, the concentration of magnetic field lines inside the double loop is twice as much as in the single loop. It follows that the magnetic field intensity in this region is increased as the number of loops is increased. The magnetic field intensity is appreciable for a current-carrying coil of wire with many loops.

a
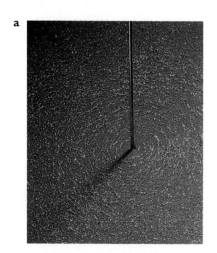

b
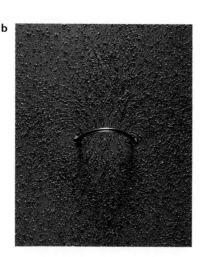

c
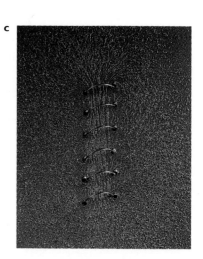

Superconducting Electromagnets

FIGURE 23.11 Scale model of a prototype magplane, a magnetically levitated vehicle. Whereas conventional trains vibrate as they ride on rails at high speeds, magplanes can travel vibration-free at high speeds because they make no physical contact with the guideway they float above.

Recall from Chapter 21 that in a superconductor there is no electrical resistance to limit the flow of electric charge and, therefore, no heating even for enormous currents. Electromagnets that utilize superconducting coils produce extremely strong magnetic fields—and do so very economically because there are no heat losses (although energy is used to keep the superconductors cold). At Fermilab near Chicago, superconducting electromagnets guide high-energy particles around the four-mile-circumference accelerator. Previously, when the accelerator used conventional elec-

tromagnets, the lab had to pay much greater electric bills every month to produce particles of less energy. Superconducting magnets can also be found in magnetic resonance imaging (MRI) devices in hospitals.

Another application to watch for is magnetically levitated, or "maglev" transportation. Figure 23.11 shows the scale model of a maglev system developed in the United States. The vehicle, called a magplane, carries superconducting coils on its underside. Moving along an aluminum trough, these coils generate currents in the aluminum that act as mirror-image magnets and repel the magplane. It floats six inches above the guideway, and its speed is limited only by air friction and passenger comfort. Someday you may ride swiftly and smoothly from one city to another in a magplane.

At this writing the strength of superconducting electromagnets is limited by the breakdown of superconductivity when the magnetic fields become too strong. Superconductors and superconducting magnets are presently generating a tremendous amount of interest and activity among physics types—for the stakes, both scientific and economic, are enormous.

Electromagnets

If a piece of iron is placed in a current-carrying coil of wire, the magnetic domains in the iron are induced into alignment. This further increases the magnetic field intensity, and we have an **electromagnet**! Strong electromagnets are used to control charged particle beams in high-energy accelerators. They also levitate and propel prototypes of high-speed trains.

Electromagnets powerful enough to lift automobiles are a common sight in junkyards. The strength of these electromagnets is limited by overheating of the current-carrying coils and saturation of magnetic domain alignment in the iron core. More powerful electromagnets omit the iron core altogether and use superconducting coils.

Magnetic Force on Moving Charged Particles

A charged particle at rest will not interact with a static magnetic field. But if the charged particle is moving in a magnetic field, the magnetic character of a charge in motion becomes evident. It experiences a deflecting force.* The force is greatest when the particle moves in a direction perpendicular to the magnetic field lines. At other angles, the

---

* When particles of electric charge q and velocity v move perpendicularly into a magnetic field of strength B, the force F on each particle is simply the product of the three variables: $F = qvB$. For nonperpendicular angles, v in this relationship must be the component of velocity perpendicular to B.

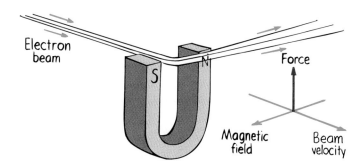

FIGURE 23.12 A beam of electrons is deflected by a magnetic field.

force is less and becomes zero when the particle moves parallel to the field lines. In any case, the direction of the force is always perpendicular both to the magnetic field lines and to the velocity of the charged particle (Figure 23.12). So a moving charge is deflected when it crosses through a magnetic field, but when it travels parallel to the field no deflection occurs.

This sideways deflection is very different from the forces that occur in other interactions, such as the gravitational forces between masses, the electric forces between charges, and the magnetic forces between magnetic poles. The force that acts on a moving charged particle does not act along the line that joins the sources of interaction, but instead acts perpendicularly to both the magnetic field and the particle velocity.

We are fortunate that charged particles are deflected by magnetic fields. This fact is employed to guide electrons onto the inner surface of a TV tube and provide a picture. The effect also works on a larger scale. Charged particles from outer space are deflected by the earth's magnetic field. The intensity of harmful cosmic rays striking the earth's surface would be more intense otherwise.

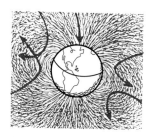

FIGURE 23.13 The magnetic field of the earth deflects many charged particles that make up cosmic radiation.

Magnetic Force on Current-Carrying Wires

Simple logic tells you that if a charged particle moving through a magnetic field experiences a deflecting force, then a current of charged particles moving through a magnetic field experiences a deflecting force also. If the particles are trapped inside a wire when they respond to the deflecting force, the wire will also be pushed (Figure 23.14).

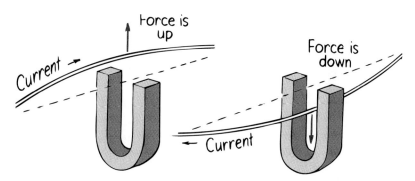

FIGURE 23.14 A current-carrying wire experiences a force in a magnetic field. (Can you see this is a simple extension of Figure 23.12?)

If we reverse the direction of current, the deflecting force acts in the opposite direction. The force is strongest when the current is perpendicular to the magnetic field lines. The direction of force is not along the magnetic field lines or along the direction of current. The force is perpendicular to both field lines and current. It is a sideways force.

We see that just as a current-carrying wire will deflect a magnet such as a compass needle, as discovered by Oersted in a classroom in 1820, a magnet will deflect a current-carrying wire. Discovering these complementary links between electricity and magnetism created much excitement, for almost immediately people began harnessing the electromagnetic force for useful purposes—with great sensitivity in electric meters and with great force in electric motors.

Question What law of physics tells you that if a current-carrying wire produces a force on a magnet, a magnet must produce a force on a current-carrying wire?

Electric Meters

The simplest meter used to detect electric current is simply a magnet that is free to turn—a compass. The next most simple meter is a compass in a coil of wires (Figure 23.15). When an electric current passes through the coil, each loop produces its own effect on the needle, so a very small current can be detected. A sensitive current-indicating instrument is called a *galvanometer*.*

FIGURE 23.15 A very simple galvanometer.

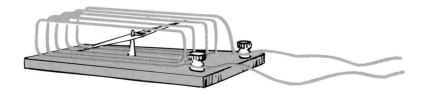

A more common design is shown in Figure 23.16. It employs more loops of wire and is therefore more sensitive. The coil is mounted for movement and the magnet is held stationary. The coil turns against a spring, so the greater the current in its windings, the greater its deflection. A galvanometer may be calibrated to measure current (amperes), in which case it is called an *ammeter*. Or it may be calibrated to measure electric potential (volts), in which case it is called a *voltmeter*.

FIGURE 23.17 Both the ammeter and the voltmeter are basically galvanometers. (The electrical resistance of the instrument is made to be very low for the ammeter, and very high for the voltmeter.)

FIGURE 23.16 A common galvanometer design.

Answer Newton's third law. It applies to *all* forces in nature.

* The galvanometer is named after Luigi Galvani (1737–1798), who discovered while dissecting a frog's leg that dissimilar metals touching the leg caused it to twitch. This chance discovery led to the invention of the chemical cell and the battery. The next time you pick up a galvanized pail, think of Luigi Galvani in his anatomy laboratory.

Electric Motors

If we modify the design of the galvanometer slightly, we have an electric motor. The principal difference is that the current is made to change direction every time the coil makes a half rotation. After being forced to turn one half rotation, it overshoots just in time for the current to reverse, whereupon it is forced to continue another half rotation, and so on in cyclic fashion to produce continuous rotation.

In Figure 23.18 we see the principle of the electric motor in bare outline. A permanent magnet produces a magnetic field in a region where a rectangular loop of wire is mounted to turn about the axis shown. When a current passes through the loop, it flows in opposite directions in the upper and lower sides of the loop (it has to do this because if charge flows into one end of the loop, it must flow out the other end). If the upper portion of the loop is forced to the left, then the lower portion is forced to the right, as if it were a galvanometer. But unlike a galvanometer, the current is reversed during each half revolution by means of stationary contacts on the shaft. The parts of the wire that brush against these contacts are called *brushes*. In this way, the current in the loop alternates so that the forces in the upper and lower regions do not change directions as the loop rotates. The rotation is continuous as long as current is supplied.

FIGURE 23.18 A simplified motor.

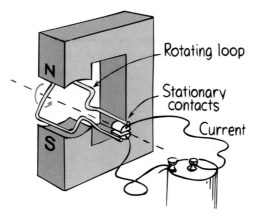

Rotating loop

Stationary contacts

Current

We have described here only a very simple dc motor. Larger motors, dc or ac, are usually made by replacing the permanent magnet by an electromagnet that is energized by the power source. Of course, more than a single loop is used. Many loops of wire are wound about an iron cylinder, called an *armature*, which then rotates when the wire carries current.

Question What is the major similarity between a galvanometer and a simple electric motor? What is the major difference?

Answer A galvanometer and a motor are similar in that they both employ coils positioned in a magnetic field. When a current passes through the coils, forces on the wires rotate the coils. The fundamental difference is that the maximum rotation of the coil in a galvanometer is one half turn, whereas in a motor the coil (wrapped on an armature) rotates through many complete turns. This is accomplished by alternating this current with each half turn of the armature.

The advent of electric motors brought to an end much human and animal toil in many parts of the world. Electric motors have greatly changed the way people live.

Earth's Magnetic Field

A suspended magnet or compass points northward because the earth itself is a huge magnet. The compass aligns with the magnetic field of the earth. The magnetic poles of the earth, however, do not coincide with the geographic poles—in fact, the magnetic and geographical poles are widely separated. The magnetic pole in the northern hemisphere, for example, is now located nearly 1800 kilometers from the geographic pole, somewhere in the Hudson Bay region of northern Canada. The other pole is located south of Australia (Figure 23.19). This means that compasses do not generally point to the true north. The discrepancy between the orientation of a compass and true north is known as the *magnetic declination*.

We do not know exactly why the earth itself is a magnet. The configuration of the earth's magnetic field is like that of a strong bar magnet placed near the center of the earth. But the earth is not a magnetized chunk of iron like a bar magnet. It is simply too hot for individual atoms to hold to a proper orientation. So the explanation must lie with electric currents deep in the earth. About 2000 kilometers below the outer rocky mantle (which itself is almost 3000 kilometers thick) lies the molten region that surrounds the solid center. Most earth scientists think that moving charges looping around within the molten part of the earth create the magnetic field. Some earth scientists speculate that the electric currents are the result of convection currents—from heat rising from the central core (Figure 23.20), and that such convection currents combined with the rotational effects of the earth produce the earth's magnetic field. Because of the earth's great size, the speed of moving charges need only be about a thousandth of a meter per second to account for the field. A firmer explanation awaits more study.

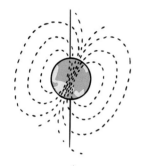

FIGURE 23.19 The earth is a magnet.

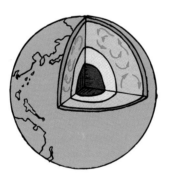

FIGURE 23.20 Convection currents in the molten parts of the earth's interior may drive electric currents to produce the magnetic field of the earth.

Whatever the cause, the magnetic field of the earth is not stable; it has wandered throughout geologic time. Evidence of this comes from analysis of the magnetic properties of rock strata. Iron atoms in a molten state are disoriented because of thermal motion, but a slight predominance of the iron atoms align with the magnetic field of the earth. When cooling and solidification occur, this predominance records the direction of the earth's magnetic field in the resulting igneous rock. It's similar for sedimentary rocks, where magnetic domains in grains of iron that settle in sediments tend to align themselves with the earth's magnetic field and become locked into the rock that forms. The slight magnetism that results can be measured with sensitive instruments. As samples of rock are tested from different strata formed throughout geologic time, the magnetic field of the earth for different periods can be charted. This evidence shows that there have been times when the magnetic field of the earth has diminished to zero, followed by reversal of the poles. More than twenty reversals have taken place in the past 5 million years. The most recent occurred 700,000 years ago. Prior reversals happened 870,000 and 950,000 years ago. Studies of deep sea sediments indicate the field was virtually switched off for 10,000 to 20,000 years just over 1 million years ago. We cannot predict when the next reversal will occur because the reversal sequence is not regular. But there is a clue in recent measurements that show a decrease of over 5 percent of the earth's magnetic field strength in the last 100 years. If this change is maintained, we may well have another reversal within 2000 years.

The reversal of magnetic poles is not unique to earth. The sun's magnetic field reverses regularly, with a period of 22 years. This 22-year magnetic cycle has been linked, through evidence in tree rings, to periods of drought on earth. Interestingly enough, the long-known 11-year sunspot cycle is just half the time during which the sun gradually reverses its magnetic polarity.

Varying ion winds in the earth's atmosphere cause more rapid but much smaller fluctuations in the earth's magnetic field. Ions in this region are produced by the energetic interactions of solar ultraviolet rays and X rays with atmospheric atoms. The motion of these ions produces a small but important part of the earth's magnetic field. Like the lower layers of air, the ionosphere is churned by winds. The variations in these winds are responsible for nearly all fast fluctuations in the earth's magnetic field.

The universe is a shooting gallery of charged particles. They are called *cosmic rays* and consist of protons and other atomic nuclei. The protons may be left over from the big bang; the heavier nuclei probably boiled off from exploding stars. In any event, they travel through space at fantastic speeds and make up the cosmic radiation that is hazardous to astronauts. Fortunately for those of us on the earth's surface, most of these charged particles are deflected away by the magnetic field of the earth. Some of them are trapped in the outer reaches of the earth's magnetic field and make up the Van Allen radiation belts (Figure 23.21).

FIGURE 23.21 The Van Allen radiation belts, shown here undistorted by the solar wind (cross section).

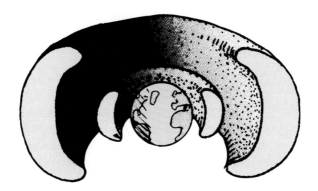

The Van Allen radiation belts consist of two doughnut-shaped rings, named after James A. Belts, who suggested their existence from data gathered by the U.S. satellite *Explorer I* in 1958.* The inner ring is centered about 3200 kilometers above the earth's surface, and the outer ring, which is a larger and wider doughnut, is centered about 16,000 kilometers overhead. Astronauts orbit at safe distances well below these belts of radiation. Most of the charged particles—protons and electrons—trapped in the outer belt probably come from the sun. Storms on the sun hurl charged particles out in great

---

* Humor aside, the name is actually James A. Van Allen (with his permission).

FIGURE 23.22 The aurora borealis (fluorescent lamp type) lighting of the sky caused by charged particles in the Van Allen belts striking atmospheric molecules.

fountains, many of which pass near the earth and are trapped by its magnetic field. The trapped particles follow corkscrew paths around the magnetic field lines of the earth and bounce between the earth's magnetic poles high above the atmosphere. Disturbances in the earth's field often allow the ions to dip into the atmosphere, causing it to glow like a fluorescent lamp. This is the beautiful *aurora borealis* (or northern lights); in the southern hemisphere, it is known as *aurora australis*. The aurora borealis in Finland skies is shown on the cover of this book.

The particles trapped in the inner belt probably originated from the earth's atmosphere. This belt gained newer electrons from high-altitude hydrogen bomb explosions in 1962.

In spite of the earth's protective magnetic field, many cosmic rays reach the earth's surface.* Cosmic ray bombardment is greatest at the magnetic poles, because charged particles that hit the earth there do not travel *across* the magnetic field lines, but rather *along* the field lines and are not deflected. Cosmic ray bombardment decreases away from the poles and is smallest in equatorial regions. At sea level, about three

---

* Some biological scientists speculate that the magnetic changes of the earth played a significant role in the evolution of life forms. One hypothesis is that in the early phases of primitive life, the earth's magnetic field was strong enough to shield the delicate life forms from high-energy charged particles. But during periods of zero strength, cosmic radiation and the spilling of the Van Allen belts increased the rate of mutation of more robust life forms—not unlike the mutations produced by X rays in the famous heredity studies of fruit flies. Coincidences between the dates of increased life changes and the dates of the magnetic pole reversals in the last few million years lend support to this hypothesis.

MRI: Magnetic Resonance Imaging

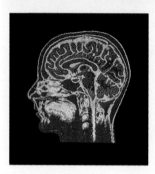

Magnetic resonance image scanners provide high-resolution pictures of the tissues inside a body. Superconducting coils produce a strong magnetic field up to 60,000 times stronger than the intensity of the earth's magnetic field, which is used to align the protons of hydrogen atoms in the body of the patient.

Like electrons, protons have a "spin" property, and will align with a magnetic field. Unlike a compass needle that aligns with the earth's magnetic field, the proton's axis wobbles about the applied magnetic field. Wobbling protons are slammed with a burst of radio waves tuned to push the proton's spin axis sideways, perpendicular to the applied magnetic field. When the radio waves pass and

the protons quickly return to their wobbling pattern, they emit faint electromagnetic signals whose frequencies depend slightly on the chemical environment in which the proton resides. The signals are picked up by sensors, and when analyzed by a computer reveal varying densities of hydrogen atoms in the body and their interactions with surrounding tissue. The images clearly distinguish fluid and bone.

It is interesting to note that MRI was formerly called NMR (nuclear magnetic resonance), because hydrogen nuclei resonate with the applied fields. Because of public phobia of anything "nuclear," the devices are now called MRI. Tell phobic friends that every atom in their bodies contains a nucleus!

---

particles strike each square centimeter each minute; this number increases rapidly with altitude. Most of these are secondary particles created when primary cosmic-ray particles strike nitrogen and oxygen nuclei high in the atmosphere. So cosmic rays are penetrating your body as you are reading this—and even when you aren't reading this!

Biomagnetism

Certain bacteria biologically produce single-domain grains of magnetite (a compound equivalent to iron ore) that they string together to form internal compasses. They then use these compasses to detect the dip of the earth's magnetic field. Equipped with a sense of direction, the organisms are able to locate food supplies. Amazingly, these bacteria south of the equator build the same single-domain magnets as their counterparts north of the equator, but then align them in opposite directions to coincide with the oppositely directed magnetic field in the southern hemisphere! Bacteria are not the only living organisms with built-in magnetic compasses: Pigeons have recently been found to have multiple-domain magnetite magnets within their skulls that are connected with a large number of nerves to the pigeon's brain. Pigeons have a magnetic sense, and not only can they discern longitudinal directions along the earth's magnetic field, but they can also detect latitude by the dip of the earth's field. Magnetic material has also been found in the abdomens of bees, whose behavior is affected by small magnetic fields. In 1992 researchers discovered tiny magnetite crystals in human brains, which resembled the crystals found in magnetic bacteria. No one knows if these are linked to our sensations. Like bacteria, pigeons, and bees, we may share a common magnetic sense.

FIGURE 23.23 The pigeon may well sense direction because of a built-in magnetic "compass" within its skull.

Summary of Terms

Magnetic force (1) Between magnets, it is the attraction of unlike magnetic poles for each other and the repulsion between like magnetic poles. (2) Between a magnetic field and a moving charged particle, it is a deflecting force due to the motion of the particle: The deflecting force is perpendicular to the velocity of the particle and perpendicular to the magnetic field lines. This force is greatest when the charged particle moves perpendicular to the field lines and is smallest (zero) when it moves parallel to the field lines.

Magnetic field The region of magnetic influence around a magnetic pole or a moving charged particle.

Magnetic domains Clustered regions of aligned magnetic atoms. When these regions themselves are aligned with one another, the substance containing them is a magnet.

Electromagnet A magnet whose field is produced by an electric current. It is usually in the form of a wire coil with a piece of iron inside the coil.

Review Questions

1. By whom, and in what setting, was the relationship between electricity and magnetism discovered?

Magnetic Forces

2. The force between electrically charged particles depends on the magnitude of charge, the distance of separation, and what else?
3. What is the origin of magnetic forces?

Magnetic Poles

4. Where are the magnetic poles located in a common bar magnet?
5. Where are the magnetic poles located in a common horseshoe magnet?
6. In what way is the rule for the interaction between magnetic poles similar to the rule for the interaction between electrically charged particles?
7. In what way are *magnetic poles* very different than *electric charges*?

Magnetic Fields

8. An electric field surrounds an electric charge. What additional field surrounds a moving electrically charged particle?
9. Why is *motion* a key word for magnetism?
10. What two kinds of rotational motion are exhibited by electrons in an atom?

Magnetic Domains

11. What is a magnetic domain?
12. Why is iron magnetic and wood not?
13. Why will dropping an iron magnet on a hard floor make it a weaker magnet?

Electric Currents and Magnetic Fields

14. How do the directions of magnetic field lines about a current-carrying wire differ from the directions of electric field lines about a charge?
15. What happens to the direction of the magnetic field about an electric current when the direction of the current is reversed?
16. Why is the magnetic field strength greater inside a current-carrying loop of wire than about a straight section of wire?

Electromagnets

17. Why does a piece of iron in a current-carrying loop increase the magnetic field strength?
18. Why are superconducting coils usually used to make very strong electromagnets?

Magnetic Force on Moving Charged Particles

19. In what direction relative to a magnetic field does a charged particle move in order to experience maximum deflecting force? Minimum deflecting force?
20. Both gravitational and electrical forces act along the direction of the force fields. How does the direction of the magnetic force on a moving charged particle differ?
21. What effect does the earth's magnetic field have on the intensity of cosmic rays striking the earth's surface?

Magnetic Force on Current-Carrying Wires

22. Since a magnetic force acts on a moving charged particle, does it make sense that a magnetic force also acts on a current-carrying wire? Defend your answer.
23. What relative direction between a magnetic field and a current-carrying wire results in the greatest force? Smallest force?
24. What happens to the direction of the force when the current in a wire is reversed?

Electric Meters

25. What is the function of a galvanometer?
26. What is a galvanometer called when calibrated to read current? Voltage?

Electric Motors

27. In what way is a galvanometer similar to an electric motor?

28. Why is it important that the current in the armature of a motor that uses a permanent magnet periodically change direction?

Earth's Magnetic Field

29. Why does a compass point northward? Why will the needle point in the same direction when in the southern hemisphere?

30. What is meant by magnetic declination?

31. Why are there probably no permanently aligned magnetic domains in the core of the earth?

32. How is the earth's magnetic field thought to be created?

33. What are *magnetic pole reversals*?

34. What are the *Van Allen radiation belts*?

35. What is the cause of the aurora borealis (northern lights)?

Biomagnetism

36. Name at least four creatures that are known to harbor tiny magnets within their bodies.

Projects

1. Find the direction and dip of the earth's magnetic field lines in your locality. Magnetize a large steel needle or straight piece of steel wire by stroking it a couple of dozen times in the same direction with a strong magnet. Run the needle or wire through a cork in such a way that when the cork floats your thin magnet remains horizontal (parallel to the water surface). Float the cork in a plastic or wooden container of water. The needle will point toward the magnetic pole. Then press a pair of unmagnetized common pins into the sides of the cork. Rest the pins on the rims of a pair of drinking glasses so that the needle or wire points toward the magnetic pole. It should dip in line with the earth's magnetic field.

2. An iron bar can be easily magnetized by aligning it with the magnetic field lines of the earth and striking it lightly a few times with a hammer. This works best if the bar is tilted down to match the dip of the earth's field. The hammering jostles the domains so they can better fall into alignment with the earth's field. The bar can be demagnetized by striking it when it is in an east-west direction.

3. Bring a magnetic compass near the tops of iron or steel objects in your home (radiators, refrigerators, stoves, lamps, etc.). You will find that the north pole of the compass needle points to the tops of these objects, and the south pole of the compass needle points to the bottoms. This shows that the objects are magnets, having a south pole on top and a north pole on the bottom. What is the explanation for this? You will find that even cans of food that have been in a vertical position in the pantry are magnetized. Turn one over and test to see how many days it takes to lose and then change its polarity.

Exercises

1. In what sense are all magnets electromagnets?

2. Since every iron atom is a tiny magnet, why aren't all iron materials themselves magnets?

3. If you place a chunk of iron near the north pole of a magnet, attraction will occur. Why will attraction also occur if you place the iron near the south pole of the magnet?

4. Do the poles of a horseshoe magnet attract each other? If you bend the magnet so that the poles get closer together, what happens to the force between the poles?

5. Why is it inadvisable to make a horseshoe magnet from a flexible material?

6. Small magnets such as those used to hold notes on refrigerator doors are often in the shape of a flat disk or square. Where do you think the poles of these magnets are located?

7. What surrounds a stationary electric charge? A moving electric charge?

8. "An electron always experiences a force in an electric field, but not always in a magnetic field." Defend this statement.

9. Why will a magnet attract an ordinary nail or paper clip, but not a wooden pencil?

10. A friend tells you that a refrigerator door, beneath its layer of white painted plastic, is made of aluminum. How could you check to see if this is true (without any scraping)?

11. Will either pole of a magnet attract a paper clip? Explain what is happening inside the attracted paper clip. (*Hint:* Consider Figure 21.12.)

12. Why does dropping a magnet on a hard surface weaken its magnetism?

13. One way to make a compass is to stick a magnetized needle into a piece of cork and float it in a glass bowl full of water. The needle will align itself with the magnetic field of the earth. Since the north pole of this compass is attracted northward, will the needle float toward the north side of the bowl? Defend your answer.

14. What is the net magnetic force on a compass needle? By what mechanism does a compass needle line up with a magnetic field?

15. A "dip needle" is a small magnet mounted on a horizontal axis so that it can swivel up or down (like a compass turned on its side). Where on the earth will a dip needle point most nearly vertically? Where will it point most nearly horizontally?

16. Since the iron filings that line up with the magnetic field of the bar magnet shown in Figure 23.2 are not little magnets themselves, by what mechanism do they align themselves with the field of the magnet?

17. The north pole of a compass is attracted to the north pole of the earth, yet like poles repel. Can you resolve this apparent dilemma?

18. Cans of food in your kitchen pantry are likely magnetized. Why?

19. One friend says that when a compass is taken across the equator, it turns around and points in the opposite direction. Another friend says this is not true, that southern-hemisphere types use the south pole of the compass to point toward the nearest pole. You're on; what do you say?

20. Why will a magnet placed in front of a television picture tube distort the picture? (*Note:* Do NOT try this with a color set. If you succeed in magnetizing the metal mask in back of the glass screen, you will have picture distortion even when the magnet is removed!)

21. The electrons in your TV set or your computer monitor are steered with magnetic fields. What is the direction of the magnetic field that pushes the electrons up or down? What is the direction of the field that pushes the electrons left and right?

22. Magnet A has twice the magnetic field strength of magnet B and at a certain distance pulls on magnet B with a force of 50 N. With how much force, then, does magnet B pull on magnet A?

23. Does a current-carrying wire exert a force on a magnet? Why or why not?

24. A strong magnet attracts a paper clip to itself with a certain force. Does the paper clip exert a force on the strong magnet? If not, why not? If so, does it exert as much force on the magnet as the magnet exerts on it? Defend your answers.

25. To make a compass, point an ordinary iron nail along the direction of the earth's magnetic field (which, in the northern hemisphere, is angled downward as well as northward) and repeatedly beat on it for a few seconds with a hammer or a rock. Then suspend it at its center of gravity by a string. Why does the beating magnetize the nail?

26. When iron naval ships are built, the location of the shipyard and the orientation in the ship while in the shipyard are recorded on a brass plaque permanently fixed to the ship. Why?

27. Can an electron at rest in a magnetic field be set into motion by the magnetic field? What if it were at rest in an electric field?

28. A cyclotron is a device for accelerating charged particles to high speed as they follow an expanding spiral path. The charged particles are subjected to both an electric field and a magnetic field. One of these fields increases the speed of the charged particles, and the other field causes them to follow a curved path. Which field performs which function?

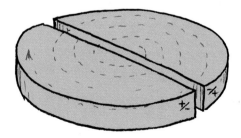

29. A magnetic field can deflect a beam of electrons, but it cannot do work on the electrons to change their speed. Why?

30. Two charged particles are projected into a magnetic field that is perpendicular to their velocities. If the charges are deflected in opposite directions, what does this tell you about them?

31. A beam of high-energy protons emerges from a cyclotron. Do you suppose there is a magnetic field associated with these particles? Why or why not?

32. Inside a laboratory room there is said to be either an electric field or a magnetic field, but not both. What experiments might be performed to establish what kind of field is in the room?

33. Why do astronauts keep to altitudes beneath the Van Allen radiation belts when doing space walks?

34. Residents of northern Canada are bombarded by more intense cosmic radiation than are residents of Mexico. Why is this so?

35. What changes in cosmic ray intensity at the earth's surface would you expect during periods in which the earth's magnetic field passed through a zero phase while undergoing pole reversals?

36. In a mass spectrometer, ions are directed into a magnetic field, where they are deflected and strike a detector. If a variety of singly ionized atoms travel at the same speed through the magnetic field, would you expect them all to be deflected by the same amount? Or would different ions be bent different amounts? What would you expect?

37. One way to shield a habitat in outer space from cosmic rays is with an absorbing blanket of some kind, which would function like the atmosphere that protects the earth. Speculate on a second way for shielding that is also similar to earth shielding.

38. If you had two bars of iron—one magnetized and the other not—and no other materials at hand, how could you tell which bar was the magnet?

39. Historically, replacing dirt roads by paved roads reduced friction on vehicles. Replacing paved roads by steel rails reduced friction further. What will be the next step to reduce friction of vehicles with the surface? What friction will remain after surface friction is eliminated?

40. Will a pair of parallel current-carrying wires exert forces on each other?

24

·············

Electromagnetic Induction

: **V**oltages are easily boosted or lowered via electromagnetic induction.

In the early 1800s, the only current-producing devices were voltaic cells, which produced small currents by dissolving metals in acids. These were the forerunners of present-day batteries. In 1820 Oersted found that magnetism was produced by current-carrying wires. The question arose as to whether electricity could be produced from magnetism. The answer was provided in 1831 by two physicists, Michael Faraday in England and Joseph Henry in the United States—each working independently and without knowledge of the other. Their discovery changed the world by making electricity commonplace—powering industries by day and lighting up cities at night.

Electro-magnetic Induction

Faraday and Henry both discovered that electric current can be produced in a wire by coiling the wire and simply moving a magnet in or out of it (Figure 24.1). No battery or other voltage source is needed—only the motion of a magnet in a wire loop. They discovered that voltage is caused or *induced* by the relative motion between a wire and a magnetic field. Voltage is induced whether the magnetic field of a magnet moves near a stationary conductor or the conductor moves in a stationary magnetic field (Figure 24.2). The results are the same for the same *relative* motion.

FIGURE 24.1 When the magnet is plunged into the coil, voltage is induced in the coil and charges in the coil are set in motion.

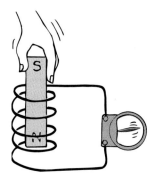

FIGURE 24.2 Voltage is induced in the wire loop whether the magnetic field moves past the wire or the wire moves through the magnetic field.

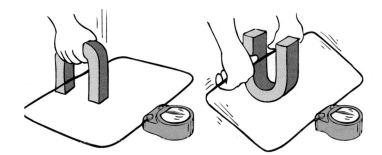

FIGURE 24.3 When a magnet is plunged into a coil of twice as many loops as another coil, twice as much voltage is induced. If the magnet is plunged into a coil with three times as many loops, then three times as much voltage is induced.

The greater the number of loops of wire that move in a magnetic field, the greater the induced voltage (Figure 24.3). Pushing a magnet into twice as many loops will induce twice as much voltage; pushing a magnet into ten times as many loops will induce ten times as much voltage; and so on. It may seem that we get something (energy) for nothing by simply increasing the number of loops in a coil of wire. But assuming the coil is connected to a resistor or other energy-dissipating device, we don't get energy for nothing: We find it is more difficult to push the magnet into a coil with more loops. This

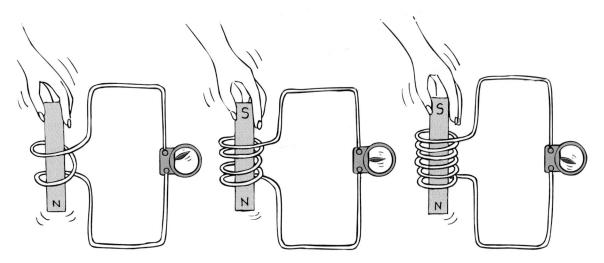

is because the induced voltage makes a current, which makes an electromagnet, which repels the magnet in our hand. So we do more work to induce more voltage (Figure 24.4). The amount of voltage induced depends on how fast the magnetic field lines are entering or leaving the coil. Very slow motion produces hardly any voltage at all. Quick motion induces a greater voltage. This phenomenon of inducing voltage by changing the magnetic field in a coil of wire is called **electromagnetic induction**.

FIGURE 24.4   It is more difficult to push the magnet into a coil with more loops if the coil is connected to a resistor. More current flows and the coil generates a stronger magnetic field that resists the motion of the magnet.

Faraday's Law

Electromagnetic induction is summarized by **Faraday's law**, which states,

> **The induced voltage in a coil is proportional to the product of the number of loops and the rate at which the magnetic field changes within those loops.**

The amount of *current* produced by electromagnetic induction depends not only on the induced voltage, but also on the resistance of the coil and the circuit to which it is connected.* For example, we can plunge a magnet in and out of a closed rubber loop and in and out of a closed loop of copper. The voltage induced in each is the same, providing the loops are the same size and the magnet moves with the same speed. But the current in each is quite different. The electrons in the rubber sense the same electric field as those in the copper, but their bonding to the fixed atoms prevents the movement of charge that so freely occurs in the copper.

Questions

1. What happens when a magnetically stored bit of information on a computer disk spins under a reading head that contains a small coil?
2. If you push a magnet into a coil connected to a resistor, as shown in Figure 24.4, you'll feel a resistance to your push. Why is this resistance greater in a coil with more loops?

Answers

1. The changing magnetic field in the coil induces voltage. In this way information stored magnetically on the disk is converted to electric signals.
2. Simply put, more work is required to provide more energy to be dissipated by more current in the resistor. You can also look at it this way: When you push a magnet into a coil, you cause the coil to become a magnet (an electromagnet). The more loops on the coil, the stronger is the electromagnet that you produce and the stronger it pushes back against the magnet you are moving. (If the coil's electromagnet attracted your magnet instead of repelling it, energy would be created from nothing and the law of energy conservation would be violated. So the coil has to repel your magnet.)

* Current also depends on the "inductance" of the coil. Inductance measures the tendency of a coil to resist a change in current because the magnetism produced by one part of the coil acts to oppose the change of current in other parts of the coil. We'll not cover this topic here.

We have mentioned two ways in which voltage can be induced in a loop of wire: by moving the loop near a magnet or by moving a magnet near the loop. There is a third way, by changing a current in a nearby loop. All three cases possess the same essential ingredient—a changing magnetic field in the loop.

Generators and Alternating Current

When one end of a magnet is repeatedly plunged into and back out of a coil of wire, the direction of the induced voltage alternates. As the magnetic field strength inside the coil is increased (magnet entering), the induced voltage in the coil is directed one way. When the magnetic field strength diminishes (magnet leaving), the voltage is induced in the opposite direction. The frequency of the alternating voltage induced is equal to the frequency of the changing magnetic field within the loop.

It is more practical to induce voltage by moving a coil than by moving a magnet. This can be done by rotating the coil in a stationary magnetic field (Figure 24.5). This arrangement is called a **generator**. The construction of a generator is in principle identical to that of a motor. Only the roles of input and output are reversed. In a motor, electric energy is the input and mechanical energy the output; in a generator, mechanical energy is the input and electric energy the output. Both devices simply transform energy from one form to another.

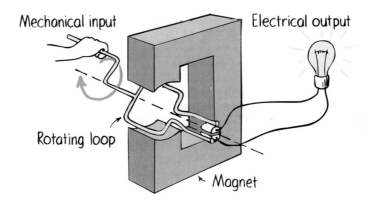

FIGURE 24.5 A simple generator. Voltage is induced in the loop when it is rotated in the magnetic field.

It is interesting to compare the physics of a motor and a generator and to see that both operate under the same underlying principle: that moving electrons experience a force that is perpendicular to both their velocity and to the magnetic field they traverse (Figure 24.6). We will call the deflection of the wire the *motor effect* and what happens as a result of the law of induction the *generator effect*. These effects are summarized in (a) and (b) of the figure. Study them. Can you see that the two effects are related?

FIGURE 24.6 (a) Motor effect: When a current moves along the wire, there is a perpendicular upward force on the electrons. Since there is no conducting path upward, the wire is tugged upward along with the electrons. (b) Generator effect: When a wire with no initial current is moved downward, the electrons in the wire experience a deflecting force perpendicular to their motion. There is a conducting path in the direction that the electrons follow, thereby constituting a current.

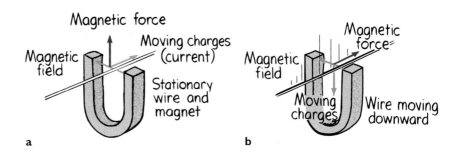

FIGURE 24.7 As the loop rotates, there is a change in the number of magnetic field lines it encloses. It varies from a maximum in (a) to a minimum at (c) and back to a maximum again at (e). The maximum induced voltage occurs at (c), where the greatest rate of change of enclosed field lines occurs.

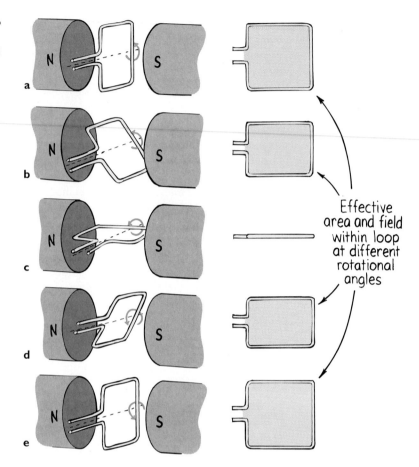

Effective area and field within loop at different rotational angles

We can see the details of electromagnetic induction in Figure 24.7. Note that when the loop of wire is rotated in the magnetic field, there is a change in the number of magnetic field lines within the loop. In (a) the loop has the largest number of lines inside it. As the loop rotates, (b), it encircles fewer of the field lines until at (c) the loop lies along the field lines and encloses none at all. As rotation continues, it encloses more field lines (d) and reaches a maximum of lines when it has made a half turn (e). As rotation continues, the magnetic field inside the loop changes in cyclic fashion, with the greatest rate of change of field lines occurring when the number of enclosed field lines goes through zero. Hence the induced voltage is greatest at these points (Figure 24.8). Because the voltage induced by the generator alternates, the current produced is ac, an alternating

FIGURE 24.8 As the loop rotates, the magnitude and direction of the induced voltage (and current) changes. One complete rotation of the loop produces one complete cycle in voltage (and current).

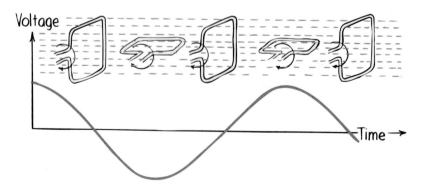

current.* It changes magnitude and direction periodically. The alternating current in our homes is produced by generators standardized so that the current goes through 60 full cycles of change in magnitude and direction each second—60 hertz.

Fifty years after Michael Faraday and Joseph Henry discovered electromagnetic induction, Nikola Tesla and George Westinghouse put those findings to practical use and showed the world that electricity could be generated reliably and in sufficient quantities to light entire cities.

Power Production

Nikola Tesla

Turbogenerator Power

Tesla built generators much like those still in use today—but quite a bit more complicated than the simple model we have discussed. Tesla's generators had armatures consisting of bundles of copper wires that were made to spin within strong magnetic fields by means of a turbine, which in turn was spun by the energy of falling water or steam. The rotating loops of wire in the armature cut through the magnetic field of the surrounding electromagnets, thereby inducing alternating voltage and current.

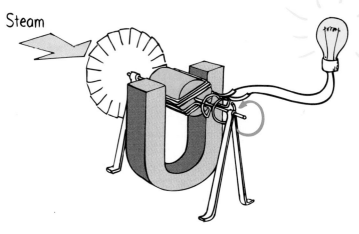

Steam

FIGURE 24.9 Steam drives the paddle wheel (turbine), which is connected to the armature of the generator.

We can look at this process from an atomic point of view. When the wires in the spinning armature cut through the magnetic field, oppositely directed electromagnetic forces act on the negative and positive charges. Electrons respond to this force by momentarily swarming relatively freely in one direction throughout the crystalline copper lattice; the copper atoms, which are actually positive ions, are forced in the opposite direction. But the ions are anchored in the lattice, so they hardly move at all. Only the electrons move, sloshing back and forth in alternating fashion with each rotation of the armature. The energy of this electronic sloshing is tapped at the electrode terminals of the generator.

MHD Power

An interesting device similar to the turbogenerator is the MHD (**m**agneto**h**ydro**d**ynamic) generator, which does away with a turbine and spinning armature altogether. Instead of making charges move in a magnetic field via a rotating armature, a plasma of

---

* With appropriate brushes and by other means, the ac in the loop(s) can be converted to dc to make a dc generator.

electrons and positive ions expands through a nozzle and moves at supersonic speed through a magnetic field. Like the armature in a turbogenerator, the motion of charges through a magnetic field gives rise to a voltage and flow of current in accordance with Faraday's law of induction. Whereas in a conventional generator "brushes" carry the current to the external load circuit, in the MHD generator the same function is performed by conducting plates, or *electrodes* (Figure 24.10). Unlike the turbogenerator, the MHD generator can operate at any temperature to which the plasma can be heated, either by combustion or nuclear processes. The high temperature results in a high thermodynamic efficiency, which means more power for the same amount of fuel and less waste heat. Efficiency is further boosted when the "waste" heat is used to turn water into steam and run a conventional steam-turbine generator.

FIGURE 24.10 A simplified MHD generator. Oppositely directed forces act on the positive and negative particles in the high-speed plasma moving through the magnetic field. The result is a voltage difference between the two electrodes. Current then flows from one electrode to the other through an external circuit. There are no moving parts; only the plasma moves. In practice, superconducting electromagnets are used.

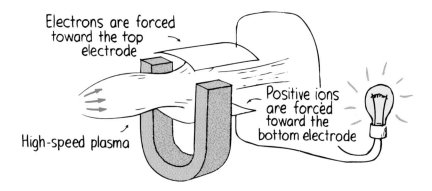

This substitution of a flowing plasma for rotating copper coils in a generator has become operational only recently because the technology to produce plasma of high enough temperatures is new. Current plants use a high-temperature plasma formed by combustion of fossil fuels in air or oxygen.*

Generators of whatever kind, of course, don't produce energy—they simply convert energy from some other form to electric energy. Some fraction of the energy from the source, whether fossil or nuclear fuel or wind or water, is converted to mechanical energy either to drive the turbine or to produce the plasma, and the generator converts most of this to electrical energy. The electricity that is produced simply carries this energy to distant places. Some people think that electricity is a primary source of energy. It is not. It is a form of energy that must have a source.

Transformers

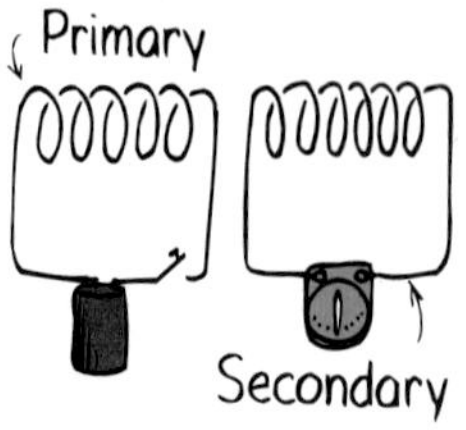

It is interesting to see that electric energy can be carried across empty space from one device to another with the simple arrangement shown in Figure 24.11. Note that one coil is connected to a battery and the other is connected to a galvanometer. It is customary to refer to the coil connected to the power source as the *primary* (input) and to the other as the *secondary* (output). As soon as the switch is closed in the primary and current passes through its coil, a current occurs in the secondary also—even though there is no material connection between the two coils. Only a brief surge of current occurs in the secondary, however. Then, when the primary switch is opened, a surge of current again registers in the secondary but in the opposite direction.

This is the explanation: A magnetic field builds up around the primary when the current begins to flow through the coil. This means that the magnetic field is growing

FIGURE 24.11 Whenever the primary switch is opened or closed, voltage is induced in the secondary circuit.

* The only large-scale MHD power plant that produces commercial electric power as of this writing is located in Russia—a joint collaboration of U.S. and Russian scientists and engineers.

(that is, *changing*) about the primary. But since the coils are near each other, this changing field extends into the secondary coil, thereby inducing a voltage in the secondary. This induced voltage is only temporary, for when the current and the magnetic field of the primary reach a steady state—that is, when the magnetic field is no longer changing—no further voltage is induced in the secondary. But when the switch is turned off, the current in the primary drops to zero. The magnetic field about the coil collapses, thereby inducing a voltage in the secondary coil, which senses the change. We see that voltage is induced whenever a magnetic field is *changing* through the coil, regardless of the reason.

Question When the switch of the primary in Figure 24.11 is opened or closed, the galvanometer in the secondary registers a current. But when the switch remains closed, no current is registered on the galvanometer of the secondary. Why?

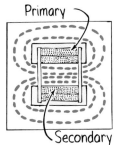

Primary

Secondary

FIGURE 24.13 A practical and more efficient transformer. Both primary and secondary coils are wrapped on the inner part of the iron core (yellow), which guides alternating magnetic lines (green) produced by ac in the primary. The alternating field induces ac voltage in the secondary. Thus power from the primary is transferred to the secondary.

If you place an iron core inside the primary and secondary coils of the arrangement of Figure 24.11, the magnetic field within the primary is intensified by the alignment of magnetic domains. The field is also concentrated in the core and extends into the secondary, which intercepts more of the field change. The galvanometer will show greater surges of current when the switch of the primary is opened or closed. Instead of opening and closing a switch to produce the change of magnetic field, suppose that alternating current is used to power the primary. Then the frequency of periodic changes in the magnetic field is equal to the frequency of the alternating current. Now we have a **transformer** (Figure 24.12). A more efficient arrangement is shown in Figure 24.13.

FIGURE 24.12 A simple transformer.

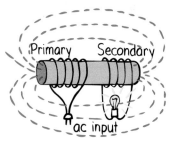

If the primary and secondary have equal numbers of wire loops (usually called *turns*), then the input and output alternating voltages will be equal. But if the secondary coil has more turns than the primary, the alternating voltage produced in the secondary coil will be greater than that produced in the primary. In this case, the voltage is said to be *stepped up*. If the secondary has twice as many turns as the primary, the voltage in the secondary will be double that of the primary.

Answer When the switch remains in the closed position, there is a steady current in the primary and a steady magnetic field about the coil. This field extends to the secondary, but unless there is a *change* in the field, electromagnetic induction does not occur.

FIGURE 24.14 (a) The voltage of 1 V induced in the secondary equals the voltage of the primary. (b) A voltage of 1 V is induced in the added secondary also because it intercepts the same magnetic field change from the primary. (c) The voltages of 1 V each induced in the two one-turn secondaries are equivalent to a voltage of 2 V induced in a single two-turn secondary.

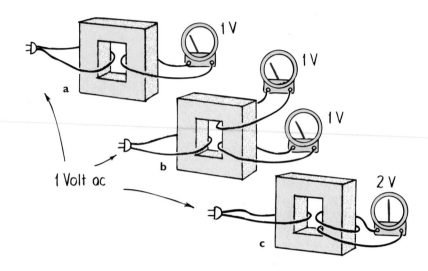

We can see this with the arrangements in Figure 24.14. First consider the simple case of a single primary loop connected to a 1-volt alternating source and a single secondary loop connected to the ac voltmeter (a). The secondary intercepts the changing magnetic field of the primary, and a voltage of 1 volt is induced in the secondary. If another loop is wrapped around the core so the transformer has two secondaries (b), it intercepts the same magnetic field change. We see that 1 volt is induced in it also. There is no need to keep both secondaries separate, for we could join them (c) and still have a total induced voltage of 1 volt +1 volt, or 2 volts. This is equivalent to saying that a voltage of 2 volts will be induced in a single secondary that has twice the number of loops as the primary. If the secondary is wound with three times as many loops, then three times as much voltage will be induced. Stepped-up voltage may light a neon sign or operate the picture tube in a television receiver or send power over long distance.

If the secondary has fewer turns than the primary, the alternating voltage produced in the secondary will be *lower* than that produced in the primary. The voltage is said to be *stepped down*. This stepped-down voltage may safely operate a toy electric train. If the secondary has half as many turns as the primary, then only half as much voltage is induced in the secondary. So electric energy can be fed into the primary at a given alternating voltage and taken from the secondary at a greater or lower alternating voltage, depending on the relative number of turns in the primary and secondary coil windings.

The relationship between primary and secondary voltages with respect to the relative number of turns is given by

$$\frac{\text{Primary voltage}}{\text{Number of primary turns}} = \frac{\text{secondary voltage}}{\text{number of secondary turns}}$$

It might seem that we get something for nothing with a transformer that steps up the voltage. Not so, for energy conservation always regulates what can happen. When voltage is stepped up, current in the secondary is less than in the primary. The transformer actually transfers energy from one coil to the other. The rate at which energy is transferred is called *power*. The power used in the secondary is supplied by the primary. The primary gives no more than the secondary uses, in accord with the law of conservation of energy. If the slight power losses due to heating of the core are neglected, then

$$\text{Power into primary} = \text{power out of secondary}$$

Electric power is equal to the product of voltage and current, so we can say

$$(\text{Voltage} \times \text{current})_{\text{primary}} = (\text{voltage} \times \text{current})_{\text{secondary}}$$

We see that if the secondary has more voltage than the primary, it will have less current than is in the primary. The ease with which voltages can be stepped up or down with a transformer is the principal reason that most electric power is ac rather than dc.

Questions

1. If 100 V of ac are put across a 100-turn transformer primary, what will be the voltage output if the secondary has 200 turns?
2. Assuming the answer to the last question is 200 V, and the secondary is connected to a floodlamp with a resistance of 50 Ω, what will be the ac current in the secondary circuit?
3. What is the power in the secondary coil?
4. What is the power in the primary coil?
5. What is the ac current drawn by the primary coil?
6. The voltage has been stepped up, and the current has been stepped down. Ohm's law says that increased voltage will produce increased current. Is there a contradiction here, or does Ohm's law not apply to circuits that have transformers?

Self-Induction

Current-carrying loops in a coil interact not only with loops of other coils but also with loops of the same coil. Each loop in a coil interacts with the magnetic field around the current in other loops of the same coil. This is *self-induction*. A self-induced voltage is produced. This voltage is always in a direction opposing the changing voltage that produces it and is commonly called the "back electromotive force," or simply "back emf."* We won't treat self-induction and back emfs here, except to acknowledge a common and dangerous effect. Suppose that a coil with a large number of turns is used as an electromagnet and is powered with a dc source, perhaps a small battery. Current in the coil is then accompanied by a strong magnetic field. When we disconnect the battery by

Answers

1. From 100 V/100 primary turns =(?) V/200 secondary turns, you can see that the secondary puts out 200 V.
2. From Ohm's law, 200 V/50 Ω = 4 A.
3. Power = 200 V $\times$ 4 A = 800 W.
4. By the law of conservation of energy, the power in the primary is the same, 800 W.
5. 800 W = 100 V $\times$ (?) A, so you see the primary draws 8 A. (Note that the voltage is stepped up from primary to secondary and that the current is correspondingly stepped down.)
6. Ohm's law still holds in the secondary circuit. The voltage induced across the secondary circuit, divided by the load (resistance) of the secondary circuit, equals the current in the secondary circuit. In the primary circuit, on the other hand, there is no conventional resistance. What "resists" the current in the primary is the energy transfer to the secondary.

---

* The opposition of an induced effect to an inducing cause is called *Lenz's law;* it is a consequence of the conservation of energy.

opening a switch, we had better be prepared for a surprise. When the switch is opened, the current in the circuit falls rapidly to zero and the magnetic field in the coil undergoes a sudden decrease (Figure 24.15). What happens when a magnetic field suddenly changes in a coil—even if it is the same coil that produced it? The answer is that a voltage is induced. The rapidly collapsing magnetic field with its store of energy may induce an enormous voltage, large enough to develop a strong spark across the switch—or to you, if you are opening the switch! For this reason, electromagnets are connected to a circuit that absorbs excess charge and prevents the current from dropping too suddenly. This reduces the self-induced voltage. This is also, by the way, why you should disconnect appliances by turning off the switch, not by pulling out the plug. The circuitry in the switch may prevent a sudden change in current.

FIGURE 24.15 When the switch is opened, the magnetic field of the coil collapses. This sudden change in the field can induce a huge voltage.

Power Transmission

Almost all electric energy sold today is in the form of ac, traditionally because of the ease with which it can be transformed from one voltage to another.* Large currents in wires produce heat and energy losses, so power is transmitted great distances at high voltages and correspondingly low currents (power = voltage × current). Power is generated at 25,000 V or less and is stepped up near the power station to as much as 750,000 V for long-distance transmission, then stepped down in stages at substations and distribution points to voltages needed in industrial applications (often 440 V or more) and for the home (240 and 120 V).

Energy, then, is transferred from one system of conducting wires to another by electromagnetic induction. It is but a short step further to find that the same principles account for eliminating wires and sending energy from a radio-transmitter antenna to a radio receiver many kilometers away. Extend these principles just a tiny step further to the transformation of energy of vibrating electrons in the sun to life energy on earth. The effects of electromagnetic induction are very far reaching.

FIGURE 24.16 Power transmission.

---

* Nowadays, power utilities can transform dc voltages using semiconductor technology. Keep an eye on the present advances in superconductor technology, and watch for resulting changes in the way that power is transmitted.

Field Induction

Electromagnetic induction has thus far been discussed in terms of the production of voltages and currents. Actually, the more fundamental way to look at it is in terms of the induction of electric *fields*. The electric fields, in turn, give rise to voltages and currents. Induction takes place whether or not a conducting wire or any material medium is present. In this more general sense, Faraday's law states

> **An electric field is created in any region of space in which a magnetic field is changing with time. The magnitude of the induced electric field is proportional to the rate at which the magnetic field changes. The direction of the induced electric field is at right angles to the changing magnetic field.**

If electric charge is present where the electric field is created, this charge will experience a force—which can cause charge to flow as current, or if the charge is in a wire, could cause the wire to be pushed to one side.

A second effect, which is the counterpart to Faraday's law, is very similar to Faraday's law, with only the roles of electric and magnetic fields interchanged. The symmetry between electric and magnetic fields revealed by this pair of laws is one of the many beautiful symmetries in nature. The companion to Faraday's law, advanced by the British physicist James Clerk Maxwell in the 1860s, states:

> **A magnetic field is created in any region of space in which an electric field is changing with time. The magnitude of the induced magnetic field is proportional to the rate at which the electric field changes. The direction of the induced magnetic field is at right angles to the changing electric field.**

These statements are two of the most important statements in physics. They underlie an understanding of the nature of light and of electromagnetic waves in general.

In Perspective*

The ancient Greeks discovered that when a piece of amber (a natural plastic-like mineral) was rubbed, it picked up little pieces of papyrus. They found strange rocks in the province of Magnesia that attracted iron. Probably because the air in Greece was relatively humid, they never noticed or studied the static electric charge effects common in dry climates. Further development of our knowledge of electrical and magnetic phenomena did not take place until 400 years ago. The human world shrank as more was learned about electricity and magnetism. It became possible first to signal by telegraph over long distances, then to talk to another person many kilometers away through wires, then not only to talk but also to send pictures over many kilometers with no physical connections in between.

Energy, so vital to civilization, could be transmitted over hundreds of kilometers. The energy of elevated rivers was diverted into pipes that fed giant "waterwheels" connected to assemblages of twisted and interwoven copper wires that rotated about specially designed chunks of iron-revolving monsters called generators. Out of these, energy was pumped through copper rods as thick as your wrist and sent to huge coils wrapped around transformer cores, boosting it to high voltages for efficient long-distance transmission to cities. Then the transmission lines split into branches—then to more transformers—then more branching and spreading, until finally the energy of the river was spread throughout whole cities—turning motors, making heat, making light, working gadgetry. There was the miracle of hot lights from cold water hundreds of kilometers away—a miracle made possible by specially designed bits of copper and iron that turned because people had discovered the laws of electromagnetism.

* Adapted from R. P. Feynman, R. B. Leighton, and M. Sands, *The Feynman Lectures on Physics*, Vol. II, pp. 1–10 and 1–11, 1964 by California Institute of Technology. Reading, Mass: Addison-Wesley. Richard P. Feynman, who was a Nobel laureate in physics and professor of physics at California Institute of Technology, is considered by many physicists to be among the most brilliant and inspirational physicists of his time—as well as the most colorful. He died in 1988.

These laws were discovered at about the time the American Civil War was being fought. From a long view of human history, there can be little doubt that events such as the American Civil War will pale into provincial insignificance in comparison with the more significant event of the nineteenth century: the discovery of the electromagnetic laws.

Summary of Terms

Electromagnetic induction   The induction of voltage when a magnetic field changes with time. If the magnetic field within a closed loop changes in any way, a voltage is induced in the loop:

$$\text{Voltage induced} \sim \text{no. of loops} \times \frac{\text{mag. field change}}{\text{time}}$$

This is a statement of Faraday's law. The induction of voltage is actually the result of a more fundamental phenomenon: the induction of an electric *field*, as defined for the more general case below.

Faraday's law   An electric field is created in any region of space in which a magnetic field is changing with time. The magnitude of the induced electric field is proportional to the rate at which the magnetic field changes. The direction of the induced electric field is at right angles to the changing magnetic field.

Generator   A device that produces electric current by rotating a coil of wire within a stationary magnetic field.

Transformer   A device for transferring electric power from one coil of wire to another by means of electromagnetic induction.

Maxwell's counterpart to Faraday's law   A magnetic field is created in any region of space in which an electric field is changing with time. The magnitude of the induced magnetic field is proportional to the rate at which the electric field changes. The direction of the induced magnetic field is at right angles to the changing electric field.

Review Questions

Electromagnetic Induction

1. Exactly what was it that Michael Faraday and Joseph Henry discovered?

2. When a magnet is thrust into a coil of wire that is connected to a lamp or some other device, voltage is induced in the wire, which in turn produces a current. Is each current-carrying loop of wire in the coil then an electromagnet?

3. What kind of interaction occurs between the end of the *magnet* thrust into a coil and the *electromagnet* that the coil becomes? (Attraction or repulsion?)

4. Exactly what is it that must change for the occurrence of electromagnetic induction?

Faraday's Law

5. Upon what two things does the current produced by electromagnetic induction depend?

6. What are the three ways that voltage can be induced in a wire?

Generators and Alternating Current

7. How does the frequency of induced voltage compare to how frequently a magnet is plunged in and out of a coil of wire?

8. What is the basic *difference* between a generator and an electric motor?

9. What is the basic *similarity* between a generator and an electric motor?

10. Where in the rotation cycle of a simple generator is the greatest rate of change of enclosed field lines? Where, then, is induced voltage at a maximum?

11. Why does the voltage induced in a generator alternate?

Power Production

12. Who discovered electromagnetic induction, and who put it to practical use?

Turbogenerator Power

13. What commonly supplies the energy input to a turbine?

MHD Power

14. What are the principal differences between an MHD *generator* and a *conventional generator*?

Transformers

15. What exactly does a transformer "transform" or, more exactly, transfer?

16. Why does a transformer require alternating current?

17. If a transformer is efficient enough, can it step up energy? Explain.

18. If 10 V ac is impressed across the primary coil of a transformer, how much voltage is induced in the secondary if it has five times the number of turns?

19. What name is given to the rate at which energy is transferred?

20. What is the principal advantage of *ac* over *dc*?

Self-Induction

21. When the magnetic field changes in a coil of wire, voltage in each loop of the coil is induced. Will voltage be induced in a loop if the source of the magnetic field is the coil itself?

Power Transmission

22. Why is power transmitted at high voltages over long distances?

23. Does the transmission of electric energy require electrical conductors between the source and receiver? Defend your answer.

Field Induction

24. What is induced by the rapid alternation of a *magnetic field*?

25. What is induced by the rapid alternation of an *electric field*?

In Perspective

26. How can part of the energy of a cold river become the energy of a hot lamp hundreds of kilometers away?

Exercises

1. A common pickup for an electric guitar consists of a coil of wire around a permanent magnet. The permanent magnet induces magnetism in the nearby guitar string. When the string is plucked, it oscillates above the coil, thereby changing the magnetic field that passes through the coil. The rhythmic oscillations of the string produce the same rhythmic changes in the magnetic field in the coil, which in turn induce the same rhythmic voltages in the coil, which when amplified and sent to a speaker produce music! Why will this type of pickup not work with nylon strings?

2. Why does an iron core increase the magnetic induction of a coil of wire?

3. Why are the armature and field windings of an electric motor usually wound on an iron core?

4. Why is a generator armature harder to rotate when it is connected to a circuit and supplying electric current?

5. Will a cyclist coast farther if the lamp connected to his generator is turned off? Explain.

6. If your metal car moves over a wide, closed loop of wire embedded in a road surface, will the magnetic field of the earth within the loop be altered? Will this produce a current pulse? Can you think of a practical application for this at a traffic intersection?

7. At the security area of an airport, you walk through a weak ac magnetic field inside a coil of wire. What is the result of a small piece of metal on your person that slightly alters the magnetic field in the coil?

8. A piece of plastic tape coated with iron oxide is magnetized more in some parts than in others. When the tape is moved past a small coil of wire, what happens in the coil? What is a practical application of this?

9. When a strip of magnetic material, variably magnetized, is embedded in a plastic card that is moved past a small coil of wire, what happens in the coil. What is a practical application of this?

10. A certain earthquake detector consists of a little box that contains a massive magnet suspended by sensitive springs. The magnet is surrounded by stationary coils of wire that are fastened to the box, which is firmly anchored to the earth. Explain how this device works, using two important principles of physics—one studied in Chapter 4, and the other in this chapter.

11. What is the primary difference between an electric *motor* and an electric *generator*?

12. Your friend says that if you crank the shaft of a dc motor manually, the motor becomes a dc generator. Do you agree or disagree?

13. Does the voltage output increase when a generator is made to spin faster? Explain.

14. An electric saw running at normal speed draws a relatively small current. But if a piece of wood being sawed jams, and the motor shaft is prevented from turning, the current dramatically increases and the motor overheats. Why?

15. If you place a metal ring in a region where a magnetic field is rapidly alternating, the ring may become hot to your touch. Why?

16. A magician places an aluminum ring on a table, underneath which is hidden an electromagnet. When the magician says "abracadabra" (and pushes a switch that starts current flowing through the coil under the table), the ring jumps into the air. Explain his "trick."

17. How could a lightbulb near, yet not touching, an electromagnet be lit? Is ac or dc required? Defend your answer.

18. A length of wire is bent into a closed loop and a magnet is plunged into it, inducing a voltage and, consequently, a current in the wire. A second length of wire, twice as long, is bent into two loops of wire and a magnet is similarly plunged into it. Twice the voltage is induced, but the current is the same as that produced in the single loop. Why?

19. Two separate but similar coils of wire are mounted close to each other, as shown below. The first coil is connected to a battery and has a direct current flowing through it. The second coil is connected to a galvanometer. What do the galvanometer readings show when the current in the first coil is increasing? Decreasing? Remaining steady?

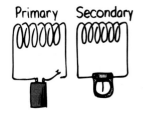

20. Why will more voltage be induced with the apparatus shown in the previous problem if an iron core is inserted in the coils?

21. Why does a transformer require alternating voltage?

22. How does the current in the secondary of a transformer compare with the current in the primary when the secondary voltage is twice the primary voltage?

23. In what sense can a transformer be thought of as an electrical lever? What does it multiply? What does it *not* multiply?

24. Why can a hum usually be heard when a transformer is operating?

25. Why is it important that the core of a transformer pass through both coils?

26. In the circuit shown, how many volts are impressed across and how many amps flow through the lightbulb?

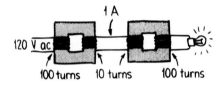

27. In the circuit shown, how many volts are impressed across and how many amps flow through the meter?

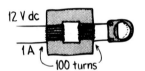

28. How would you answer the previous question if the input were 12-V ac?

29. Can an efficient transformer step up energy? Defend your answer.

30. Your friend says that, according to Ohm's law, high voltage produces high current. Then your friend asks, so how can power be transmitted at high voltage and low current in a power line? What is your illuminating response?

31. If a bar magnet is thrown into a coil of high-resistance wire, it will slow down. Why?

32. When a bar magnet is dropped through a vertical length of copper pipe, it falls noticeably more slowly than when dropped through a vertical length of plastic pipe. If the pipe is long enough, the magnet will reach a terminal falling speed. Why?

33. The metal wing of an airplane acts like a "wire" flying through the earth's magnetic field. A voltage is induced between the wing tips and a current flows along the wing, but only for a short time. Why does the current stop even though the airplane keeps flying through the earth's field?

34. What is wrong with this scheme? To generate electricity without fuel, arrange a motor to run a generator that will produce electricity that is stepped up with transformers so that the generator can run the motor and simultaneously furnish electricity for other uses.

35. With no magnets around, why will current flow in a coil of wire that is waved around in the air?

36. We know the source of a sound wave is a vibrating object. What is the source of an electromagnetic wave?

37. What does an incident radio wave do to the electrons in a receiving antenna?

38. How do you suppose the frequency of an electromagnetic wave compares with the frequency of the electrons it sets into oscillation in a receiving antenna?

39. A friend says that changing electric and magnetic fields generate one another, and this gives rise to visible light when the frequency of change matches the frequencies of light. Do you agree? Explain.

40. Would electromagnetic waves exist if changing magnetic fields could produce electric fields, but changing electric fields could not in turn produce magnetic fields? Explain.

Problems

1. The primary coil of a step-up transformer draws 100 W. Find the power provided by the secondary coil.

2. An ideal transformer has 50 turns in its primary and 250 turns in its secondary. 12 V ac is connected to the primary. Find: (a) volts ac available at the secondary; (b) current in a 10-ohm device connected to the secondary; (c) power supplied to the primary.

3. A model electric train requires 6 V to operate. If the primary coil of its transformer has 240 windings, how many windings should the secondary have if the primary is connected to a 120-V household circuit?

4. Neon signs require about 12,000 V for their operation. What should be the ratio of the number of loops in the secondary to the number of loops in the primary for a neon-sign transformer that operates off 120-V lines?

5. 100 kW (10^5 W) of power is delivered to the other side of a city by a pair of power lines between which the voltage is 12,000 V. (a) What current flows in the lines? (b) Each of the two lines has a resistance of 10 ohms. What is the voltage change *along* each line? (Think carefully. This voltage change is along each line, not between the lines.) (c) What power is expended as heat in both lines together (distinct from power delivered to customers)? Do you see why it is so important to step voltages up with transformers for long-distance transmission?

P a r t

LIGHT

Zowie, energetic photons in the sunlight are stimulating the vibrations of zillions of electrons in the molecular structure of this leaf. Some of these vibrations are vigorous and produce heat, and some are subtle and in turn send out new photons, which reveals a pattern of not only the overall shape of the leaf, but its translucence and delicate structure in intricate detail. And the radiating electrons don't vibe at any old frequency, by golly! They dance to an average rhythm of 6×10^{14} vibes per second, which is why the leaf is green!

25

$\cdots\cdots\cdots\cdots$

Properties of Light

Everything we see is a composite of electromagnetic waves.

Light is the only thing we can really see. But what *is* light? We know that during the day the primary source of light is the sun, and the secondary source is the brightness of the sky. Other common sources are flames, white-hot filaments in light bulbs and glowing gas in glass tubes. All light originates from the accelerated motion of electric charges. Light is an electromagnetic phenomenon, a tiny part of a larger whole—the *electromagnetic spectrum*. We begin our study of light by investigating its electromagnetic properties. In the next chapter we'll discuss its appearance—color. In Chapter 27 we'll learn how light behaves—how it reflects and refracts. Then we'll learn about its wave nature in Chapter 28 and its quantum nature in Chapters 29 and 30.

Electromagnetic Waves

Shake the end of a stick back and forth in still water, and you'll produce waves on the water surface. Similarly, shake an electrically charged rod to and fro in empty space, and you'll produce electromagnetic waves in space. This is because the moving charge is actually an electric current. What surrounds an electric current? The answer is a magnetic field. What surrounds a changing electric current? The answer is a changing magnetic field. Recall from the previous chapter that a changing magnetic field generates an electric field, in accordance with Faraday's law. If the magnetic field is oscillating, the electric field that it generates will be oscillating, too. And what does an oscillating electric field do? In accordance with Maxwell's counterpart to Faraday's law, it generates an oscillating magnetic field. The vibrating electric and magnetic fields regenerate each other to make up an **electromagnetic wave**, which emanates (moves outward) from the vibrating charge. There is only one speed, it turns out, for which the electric and magnetic fields remain in perfect balance, reinforcing each other as they carry energy through space. Let's see why this is so.

FIGURE 25.1 Shake an electrically charged object to and fro, and you produce an electromagnetic wave.

Electromagnetic Wave Velocity

A spacecraft cruising through space may gain or lose speed, even if its engines are shut off, because gravity can accelerate it forward or backward. But an electromagnetic wave traveling through space never changes its speed. Not because gravity doesn't act on light, for it does. Gravity can change the frequency of light or deflect light—but it can't change the speed of light. What keeps light moving always at the same, unvarying speed in empty space? The answer has to do with electromagnetic induction and energy conservation.

If light were to slow down, its changing electric field would generate a weaker magnetic field, which, in turn, would generate a weaker electric field, and so on, until the wave died out. But what would happen to the energy in the fields? If the fields faded away with no means of transferring energy to some other form, energy would be lost. That would be incompatible with the law of energy conservation. So the light can't slow down.

FIGURE 25.2 The electric and magnetic fields of an electromagnetic wave are perpendicular to each other and to the direction of motion of the wave.

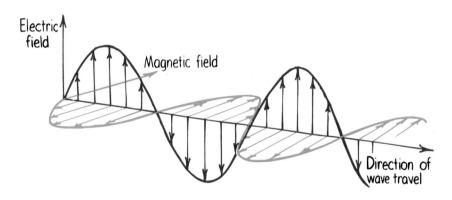

If light were to speed up, a similar argument prevails. The changing electric field would generate a stronger magnetic field, which, in turn, would generate a stronger electric field, and so on, a crescendo of ever-increasing field strength and ever-increasing energy, clearly a no-no with respect to energy conservation. At only one speed does mutual induction continue indefinitely, with neither a loss nor a gain in energy. From his equations of electromagnetic induction, Maxwell calculated the value of this critical speed and found it to be 300,000 kilometers per second. In his calculation he used only the constants in his equations determined by simple laboratory experiments with electric and magnetic fields. He didn't *use* the speed of light. He *found* the speed of light!

Maxwell quickly realized that he had discovered the solution to one of the greatest mysteries of the universe—the nature of light. He discovered that light is simply electromagnetic radiation within a particular frequency range, 4.3×10^{14} to 7×10^{14} vibrations per second! Such waves activate the "electrical antennae" in the retina of the

James Maxwell

eye. The lower frequency waves appear red, and the higher frequency waves appear violet.* Maxwell realized at the same time that electromagnetic radiation of any frequency propagates at the same speed as light.

On the evening of Maxwell's discovery, he had a date with a young lady he was later to marry. While they were walking in a garden, she remarked about the beauty and wonder of the stars. Maxwell asked how she would feel to know that she was walking with the only person in the world who knew what the starlight really was. For it was true. At that time, James Clerk Maxwell was the only person in the world to know that light of any kind is energy-carrying waves of electric and magnetic fields that continually regenerate each other and travel at one fixed speed.

Question The unvarying speed of electromagnetic waves in space is a remarkable consequence of what central principle in physics?

The Electromagnetic Spectrum

In a vacuum, all electromagnetic waves move at the same speed and differ from one another in their frequency. The classification of electromagnetic waves according to frequency is the **electromagnetic spectrum** (Figure 25.3). Electromagnetic waves have been detected with a frequency as low as 0.01 hertz (Hz). Electromagnetic waves with frequencies of several thousand hertz (kHz) are classified as very low frequency radio waves. One million hertz (MHz) lies in the middle of the AM radio band. The very high frequency (VHF) television band of waves starts at about 50 MHz and FM radio is found from 88 to 108 MHz. Then come ultrahigh frequencies (UHF), followed by microwaves, beyond which are infrared waves, often called "heat waves." Further still is visible light, which makes up less than a millionth of 1 percent of the measured electromagnetic spectrum. The lowest frequency light we can see with our eyes appears red.

FIGURE 25.3 The electromagnetic spectrum is a continuous range of waves extending from radio waves to gamma rays. The descriptive names of the sections are merely a historical classification, for all waves are the same in nature, differing principally in frequency and wavelength; all have the same speed.

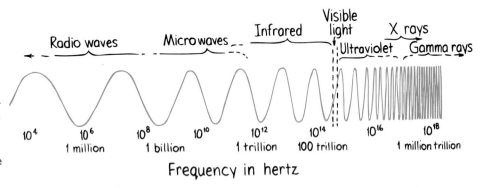

Answer The underlying principle that makes light and all other electromagnetic waves travel at one fixed speed is the conservation of energy.

---

* It is common to describe sound and radio by *frequency* and light by *wavelength*. In this book, however, we favor the single concept of frequency in describing light.

The highest frequencies of visible light are nearly twice the frequency of red and appear violet. Still higher frequencies are ultraviolet. These higher frequency waves cause sunburns. Higher frequencies beyond ultraviolet extend into the X-ray and gamma-ray regions. There are no sharp boundaries between the regions, which actually overlap each other. The spectrum is broken up into these arbitrary regions for classification.

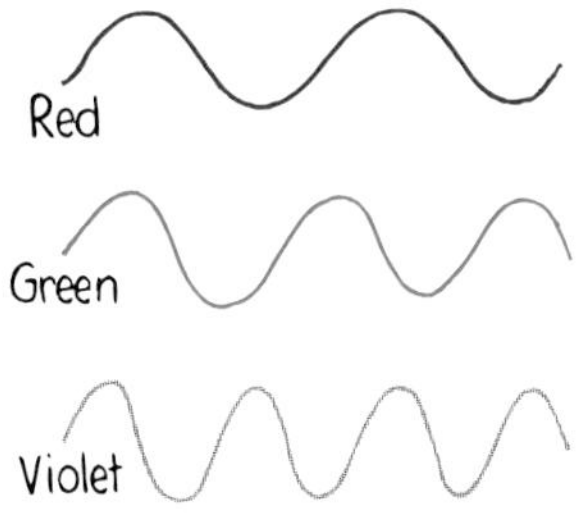

The concepts and relationships we treated earlier in our study of wave motion (Chapter 18) apply here. Recall that the frequency of a wave is the same as the frequency of the vibrating source. The same is true here: The frequency of an electromagnetic wave as it vibrates through space is identical to the frequency of the oscillating electric charge generating it.* Different frequencies correspond to different wavelengths—waves of low frequency have long wavelengths and waves of high frequencies have short wavelengths. For example, since the speed of the wave is 300,000 kilometers per second, an electric charge oscillating once per second (1 hertz) will produce a wave with a wavelength of 300,000 kilometers. This is because only one wavelength is generated in 1 second. If the frequency of oscillation were 10 hertz, then 10 wavelengths would be formed in 1 second, and the corresponding wavelength would be 30,000 kilometers. A frequency of 10,000 hertz would produce a wavelength of 30 kilometers. So the higher the frequency of the vibrating charge, the shorter the wavelength of radiation.[†]

We tend to think of space as empty—but only because we cannot see the montages of electromagnetic waves that permeate every part of our surroundings. We see some of these waves, of course, as light. These waves constitute only a microportion of the electromagnetic spectrum. We are unconscious of radio waves, which engulf us every moment. Free electrons in every piece of metal on the earth's surface continually dance to the rhythms of these waves. They jiggle in unison with the electrons being driven up and down along radio- and television-transmitting antennae. A radio or television receiver is simply a device that sorts and amplifies these tiny currents. There is radiation everywhere. Our first impression of the universe is one of matter and void, but actually the universe is a dense sea of radiation in which occasional concentrates are suspended.

FIGURE 25.4 Relative wavelengths of red, green, and violet light. Violet light has nearly twice the frequency of red light and half the wavelength.

Question Is it correct to say that a radio wave is a low-frequency light wave? Is a radio wave also a sound wave?

Answer Both a radio wave and a light wave are electromagnetic waves, which originate in the vibrations of electrons. Radio waves have lower frequencies than light waves, so a radio wave may be considered to be a low-frequency light wave (and a light wave a high-frequency radio wave). But a sound wave is a mechanical vibration of matter and is not electromagnetic. A sound wave is fundamentally different from an electromagnetic wave. So a radio wave is definitely not a sound wave.

---

* This is a rule of classical physics, valid when charges are oscillating over dimensions that are large compared with the size of a single atom (for instance, in a radio antenna). Quantum physics permits exceptions. Radiation emitted by a single atom or molecule can differ in frequency from the frequency of the oscillating charge within the atom or molecule.

[†] The relationship is $c = f\lambda$, where c is the wave speed (constant), f is the frequency, and λ is the wavelength.

Transparent Materials

Light is an energy-carrying electromagnetic wave that emanates from vibrating electrons in atoms. When light is incident upon matter, some of the electrons in the matter are forced into vibration. In this way, vibrations in the emitter are transmitted to vibrations in the receiver. This is similar to the way that sound is transmitted (Figure 25.5).

FIGURE 25.5 Just as a sound wave can force a sound receiver into vibration, a light wave can force electrons in materials into vibration.

Thus the way a receiving material responds when light is incident upon it depends on the frequency of the light and the natural frequency of the electrons in the material. Visible light vibrates at a very high rate, more than 100 trillion times per second (10^{14} hertz). If a charged object is to respond to these ultrafast vibrations, it must have very, very little inertia. Electrons are light enough to vibrate at this rate.

Materials such as glass and water allow light to pass through in straight lines. We say they are **transparent** to light. To understand how light gets through a transparent material, visualize the electrons in an atom as if they were connected by springs (Figure 25.6).* When a light wave is incident upon them, they are set into vibration.

Materials that are springy (elastic) respond more to vibrations at some frequencies than others (Chapter 18). Bells ring at a particular frequency, tuning forks vibrate at a particular frequency, and so do the electrons of atoms and molecules. The natural vibration frequencies of an electron depend on how strongly it is attached to a nearby nucleus. Different materials have different "spring strengths." Electrons in glass have a natural vibration frequency in the ultraviolet range. When ultraviolet rays shine on glass, resonance occurs as the wave builds and maintains a large amplitude of vibration of the electrons, just as pushing someone at the resonant frequency on a swing builds a large amplitude. The energy the atom receives may be passed on to neighboring atoms by collisions, or it may be re-emitted. Resonating atoms in the glass can hold onto the energy of the ultraviolet light for quite a long time (about 100 millionths of a second). During this time the atom makes about 1 million vibrations, and it collides with neighboring atoms and gives up its energy as heat. Thus, glass is not transparent to ultraviolet.

At lower wave frequencies, like those of visible light, electrons in the glass are forced into vibration, but at less amplitude. The atom holds the energy for less time, with less chance of collision with neighboring atoms, and less energy transformed to

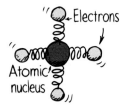

FIGURE 25.6 The electrons of atoms in glass have certain natural frequencies of vibration and can be modeled as particles connected to the atomic nucleus by springs.

---

* Electrons, of course, are not really connected by springs. Their "vibration" is actually orbital as they move around the nucleus, but the "spring model" helps us understand the interaction of light with matter. Physicists devise such conceptual models to understand nature, particularly at the submicroscopic level. The worth of a model lies not in whether it is "true," but whether it is useful. A good model not only is consistent with and explains observations, but also predicts what may happen. If predictions of the model are contrary to what happens, the model is usually either refined or abandoned. The simplified model that we present here—of an atom whose electrons vibrate as if on springs, with a time interval between absorbing energy and re-emitting energy—is quite useful for understanding how light passes through transparent solids.

heat. The energy of vibrating electrons is re-emitted as light. Clear glass is transparent to all visible light waves. The frequency of the re-emitted light that is passed from atom to atom is identical to the frequency of the light that produced the vibration in the first place. However, there is a slight time delay between absorption and re-emission.

It is this time delay that results in a lower average speed of light through a transparent material (Figure 25.7). Light travels at different average speeds through different materials. We say *average speeds,* for the speed of light in a vacuum, whether in interstellar space or in the space between atoms or molecules in a piece of glass, is a constant 300,000 kilometers per second. We call this speed of light c.* The speed of light in the atmosphere is slightly less than in a vacuum, but is usually rounded off as c. In water, light travels at 75% of its speed in a vacuum, or $0.75\ c$. In glass, light travels about $0.67\ c$, depending on the type of glass. In a diamond, light travels at less than half its speed in a vacuum, only $0.41\ c$. When light emerges from these materials into the air, it travels at its original speed, c.

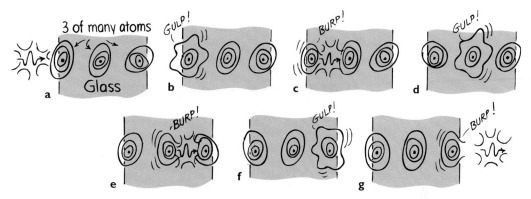

FIGURE 25.7 A light wave incident upon a pane of glass sets up vibrations in the atoms that produce a chain of absorptions and re-emissions, which pass the light energy through the material and out the other side. Because of the time delay between absorptions and re-emissions, the light travels through the glass more slowly than through empty space.

Infrared waves, with frequencies lower than those of visible light, vibrate not only the electrons, but entire atoms or molecules in the structure of the glass. This vibration increases the internal energy and temperature of the structure, which is why infrared waves are often called *heat waves.* Glass is transparent to visible light, but not to ultraviolet and infrared light.

FIGURE 25.8 Glass blocks both infrared and ultraviolet, but is transparent to visible light.

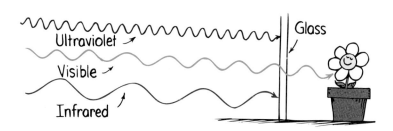

* The presently accepted value is 299,792 km/s, rounded to 300,000 km/s. (This corresponds to 186,000 mi/s.)

Questions

1. Why is glass transparent to visible light but opaque to ultraviolet and infrared?
2. Pretend that while you walk across a room, you make several momentary stops along the way to greet people who are "on your wavelength." How is this analogous to light traveling through glass?
3. In what way is it not analogous?

Opaque Materials

Most things around us are **opaque**—they absorb light without re-emission. Books, desks, chairs and people are opaque. Vibrations given by light to their atoms and molecules are turned into random kinetic energy—into internal energy. They become slightly warmer.

FIGURE 25.9 Metals are shiny because light that shines on them forces their free electrons into vibration. The electrons then emit their "own" light waves as reflection.

Answers

1. The natural frequency of vibration for electrons in glass is the same as the frequency of ultraviolet light, so resonance in the glass occurs when ultraviolet waves shine on it. The absorbed energy is passed on to other atoms as heat, not re-emitted as light, so the glass is opaque at ultraviolet frequencies. In the range of visible light, the forced vibrations of electrons in the glass are at smaller amplitudes—vibrations are more subtle, re-emission of light rather than the generation of heat occurs, and the glass is transparent. Lower-frequency infrared causes whole atoms and molecules, rather than electrons, to resonate; again heat is generated and the glass is opaque.

2. Your average speed across the room would be less because of the time delays associated with your momentary stops. Likewise, the speed of light in glass is less because of the time delays in interactions with atoms along its path.

3. In the case of walking across the room, it is you who begin the walk and you who complete the walk. This is not analogous to the similar case of light, for according to our model for light passing through a transparent material, the light that is absorbed by an electron made to vibrate is not the same light that is re-emitted—even though the two, like identical twins, are indistinguishable.

Metals are opaque. Interestingly enough, the outer electrons of atoms in metals are not bound to any particular atom. They are free to wander with very little restraint throughout the material (which is why metal conducts electricity and heat so well). When light shines on metal and sets these free electrons into vibration, their energy does not "spring" from atom to atom in the material, but instead goes into reflection. That's why metals are shiny.

The earth's atmosphere is transparent to visible light and some infrared, but, fortunately, quite opaque to high-frequency ultraviolet waves. The small amount of ultraviolet that does get through is responsible for sunburns. If it all got through we'd be in serious trouble without special protection. Clouds are semitransparent to ultraviolet, which is why you can get a sunburn on a cloudy day. Ultraviolet rays are not only harmful to the skin, but are also damaging to tar roofs. Now you know why tarred roofs are covered with gravel.

Have you noticed that things look darker when they are wet than when dry? Light incident on a dry surface bounces directly to your eye, while light incident on a wet surface bounces around inside the transparent wet region before it reaches your eye. What happens with each bounce? Absorption! So more absorption of light occurs in a wet surface, and the surface looks darker.

Shadows

A thin beam of light is often called a *ray*. When we stand in the sunlight, some of the light is stopped while other rays pass on in a straight-line path. We cast a **shadow**—a region where light rays cannot reach. If we are close to our shadow, it is sharp-edged because the sun is so far away. Either a large faraway light source or a small nearby light source will produce a sharp shadow. A large nearby light source produces a somewhat blurry shadow (Figure 25.10). There is usually a dark part on the inside and a lighter part around the edges of a shadow. A total shadow is called an **umbra** and a partial shadow a **penumbra**. A penumbra appears where some of the light is blocked but where other

FIGURE 25.10 A small light source produces a sharper shadow than a larger source.

FIGURE 25.11 An object held close to a wall casts a sharp shadow because light coming from slightly different directions does not spread much behind the object. As the object is moved farther away from the wall, penumbras are formed and the umbra becomes smaller. When the object is farther away, the shadow is less distinct. When the object is very far away (not shown), no shadow is evident because all the penumbras mix together into a big blur.

light fills it in. This can happen where light from one source is blocked and light from another source fills in (Figure 25.11). A penumbra also occurs where light from a broad source is only partially blocked.

Both the earth and the moon cast shadows when sunlight is incident upon them. When the path of either of these bodies crosses into the shadow cast by the other, an eclipse occurs (Figure 25.12). A dramatic example of the umbra and penumbra occurs when the shadow of the moon falls on the earth—during a **solar eclipse**. Because of the large size of the sun, the rays taper to provide an umbra and a surrounding penumbra (Figure 25.13). If you stand in the umbra part of the shadow you experience darkness during the day—a total eclipse. If you stand in the penumbra, you experience a partial eclipse, for you see a crescent of the sun.* In a **lunar eclipse**, the moon passes into the shadow of the earth.

Questions

1. Which type of eclipse—a solar eclipse, a lunar eclipse, or both—is dangerous to view with unprotected eyes?
2. Why are lunar eclipses more commonly seen than solar eclipses?

Answers

1. Only a solar eclipse is harmful when viewed directly because one views the sun directly. During a lunar eclipse, one views a very dark moon. It is not completely dark because the earth's atmosphere acts as a lens and bends some light into the shadow region. Interestingly enough, this is the light of red sunsets and sunrises all around the world, which is why the moon appears a faint deep red during a lunar eclipse.
2. The shadow of the relatively small moon on the large earth covers a very small part of the earth's surface. So only a relatively few people are in the shadow of the moon in a solar eclipse. But the shadow of the earth completely covers the moon during a total lunar eclipse, so everybody who views the nighttime sky can see the shadow of the earth on the moon.

---

* People are cautioned not to look at the sun at the time of a solar eclipse because the brightness and the ultraviolet light of direct sunlight are damaging to the eyes. This good advice is often misunderstood by those who then think that sunlight is more damaging at this special time. But staring at the sun when it is high in the sky is harmful whether or not an eclipse occurs. In fact, staring at the bare sun is more harmful than when part of the moon blocks it! The reason for special caution at the time of an eclipse is simply that more people are interested in looking at the sun during this time.

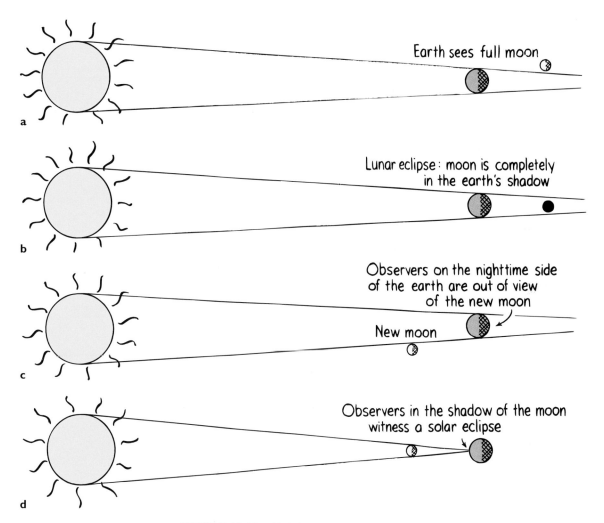

FIGURE 25.12 (a) A full moon is seen when the earth is between the sun and the moon. (b) When this alignment is perfect, the moon is in the earth's shadow, and a lunar eclipse is produced. (c) A new moon occurs when the moon is between the sun and the earth. (d) When this alignment is perfect, the moon's shadow falls on part of the earth to produce a solar eclipse.

FIGURE 25.13 Detail of a solar eclipse. A total eclipse is seen by observers in the umbra, and a partial eclipse is seen by observers in the penumbra. Most earth observers see no eclipse at all.

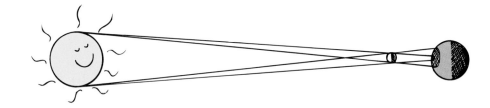

Seeing Light— The Eye

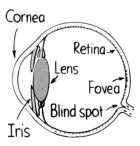

Cornea

Retina

Lens

Fovea

Blind spot

Iris

FIGURE 25.14 The human eye.

Light is the only thing we see with the most remarkable optical instrument known—the eye. A diagram of the human eye is shown in Figure 25.14.

Light enters the eye through the transparent cover called the *cornea,* which does about 70% of the necessary bending of the light before the light passes through the pupil (which is an aperture in the iris). The light then passes through the lens, which is used only to provide the extra bending power needed to focus images of nearby objects on the layer at the back of the eye. This layer—the *retina*—is extremely sensitive, and until very recently it was more sensitive to light than any artificial detector made. Different parts of the retina receive light from different parts of the visual field outside. The retina is not uniform. There is a spot in the center of our field of view called the *fovea,* or region of most distinct vision. Much greater detail can be seen here than at the side parts of the eye. There is also a spot in the retina where the nerves carrying all the information exit; this is the *blind spot.* You can demonstrate that you have a blind spot in each eye if you hold this book at arm's length, close your left eye, and look at Figure 25.15 with your right eye only. You can see both the round dot and the X at this distance. If you now move the book slowly toward your face, with your right eye fixed upon the dot, you'll reach a position about 20–25 centimeters from your eye where the X disappears. Now repeat with only the left eye open, looking this time at the X, and the dot will disappear. When you look with both eyes open, you are not aware of the blind spot, mainly because one eye "fills in" the part to which the other eye is blind. Amazingly, the brain fills in the "expected" view even with one eye closed. Repeat the exercise of Figure 25.15 with small objects on various backgrounds. Note that instead of seeing *nothing,* your brain gratuitously fills in the appropriate background. So you not only see what's there—you see what's not there!

FIGURE 25.15 The blind-spot experiment. Close your left eye and look with your right eye at the round dot. Adjust your distance and find the blind spot that erases the X. Switch eyes and look at the X and the dot disappears. Does your brain fill in crossed lines where the dot was?

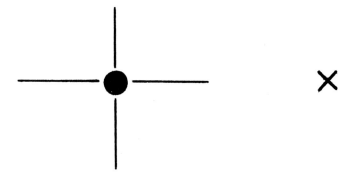

The retina is composed of tiny antennae that resonate to the incoming light. There are two basic kinds of antennae, the rods and the cones (Figure 25.16). As the names imply, some of the antennae are rod-shaped and some cone-shaped. There are three types of cones: those that are stimulated by low-frequency light, those stimulated by light of intermediate frequencies, and those stimulated by light of higher frequencies. The rods predominate toward the periphery of the retina, while the three types of cones are denser toward the fovea. The cones are very dense in the fovea itself, and since they are packed so tightly, they are much finer or narrower there than elsewhere in the retina. Color vision is possible because of the cones. Hence we see color most acutely by focusing an image on the fovea, where there are no rods. Primates and a species of ground squirrel are the only mammals that have the three types of cones and experience full color vision. The retinas of other mammals consist primarily of rods, which are sensitive only to lightness or darkness, like a black-and-white photograph or movie.

FIGURE 25.16 Magnified view of the rods and cones in the human eye.

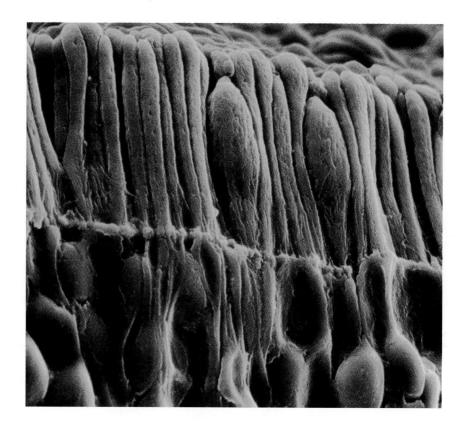

FIGURE 25.17 On the periphery of your vision you can see an object only if it is moving; you cannot see its color at all.

In the human eye, the number of cones decreases as we move away from the fovea. It's interesting that the color of an object disappears if viewed on the periphery of vision. This can be tested by having a friend enter your periphery of vision with some brightly colored objects. You will find that you can see the objects before you can see what color they are.

Another interesting fact is that the periphery of the retina is very sensitive to motion. Although our vision is poor from the corner of our eye, we are sensitive to anything moving there. We are "wired" to look for something jiggling to the side of our visual field, a feature that must have been important in our evolutionary development. So have your friend shake those brightly colored objects when she brings them into the periphery of your vision. If you can just barely see the objects when they shake, but not at all when they're held still, then you won't be able to tell what color they are (Figure 25.17). Try it and see!

Another distinguishing feature of the rods and cones is the intensity of light to which they respond. The cones require more energy than the rods before they will "fire" an impulse through the nervous system. If the intensity of light is very low, the things we see have no color. We see low intensities with our rods. Dark-adapted vision is almost entirely due to the rods, while vision in bright light is due to the cones. Stars, for example, look white to us. Yet most stars are actually brightly colored. A time exposure of the stars with a camera reveals reds and red-oranges for the "cooler" stars and blues and blue-violets for the "hotter" stars. The starlight is too weak, however, to fire the color-perceiving cones in the retina. So we see the stars with our rods and perceive them as white or, at best, as only faintly colored. Females have a slightly lower threshold of firing for the cones, however, and can see a bit more color than males. So if she says she sees colored stars and he says she doesn't, she is probably right!

We find that the rods "see" better than the cones toward the blue end of the color spectrum. The cones can see a deep red where the rods see no light at all. Red light may as well be black as far as the rods can tell. Thus, if you have two colored objects—say, blue and red—the blue will appear much brighter than the red in dim light, though the red might be much brighter than the blue in bright light. The effect is quite interesting. Try this: In a dark room, find a magazine or something that has colors, and, before you know for sure what the colors are, judge the lighter and darker areas. Then carry the magazine into the light. You should see a remarkable shift between the brightest and dimmest colors.*

The rods and cones in the retina are not connected directly to the optic nerve, but, interestingly enough, are connected to many other cells that are connected to each other. While many of these cells are interconnected, only a few carry information to the optic nerve. Through these interconnections a certain amount of information is combined from several visual receptors and "digested" in the retina. In this way the light signal is "thought about" before it goes to the optic nerve and then to the main body of the brain. So some brain functioning occurs in the eye itself. The eye does some of our "thinking."

This thinking is betrayed by the iris, the colored part of the eye that expands and contracts and regulates the size of the pupil as the iris contracts for admitting more or less light as the intensity of light changes. It so happens that the relative size of this enlargement or contraction is also related to our emotions. If we see, smell, taste, or hear something that is pleasing to us, our pupils automatically increase in size. If we see, smell, taste, or hear something repugnant to us, our pupils automatically contract. Many card players have betrayed the value of a hand by the size of their pupils! (The study of the size of the pupil as a function of attitudes is called *pupilometrics*.)

The brightest light that the human eye can perceive without damage is some 500 million times brighter than the dimmest light that can be perceived. Look at a nearby light bulb. Then turn to look into a dimly lit closet. The difference in light intensity may be more than a million to one. Because of an effect called *lateral inhibition,* we don't perceive the actual differences in brightness. The brightest places in our visual field are prevented from outshining the rest, for whenever a receptor cell on our retina sends a strong brightness signal to our brain, it also signals neighboring cells to dim their responses. In this way, we even out our visual field, which allows us to discern detail in very bright areas and in dark areas as well. (Camera film is not so good at this. A photograph of a scene with strong differences of intensity may be overexposed in one area and underexposed in another.) Lateral inhibition exaggerates the difference in brightness at the edges of places in our visual field. Edges, by definition, separate one thing from another. So we accentuate differences. The gray rectangle on the left in Figure 25.19 appears darker than the gray rectangle on the right when the edge that separates them is in our view. But cover the edge with your pencil or your finger, and they look equally bright. That's because both rectangles *are* equally bright; each rectangle is shaded

She loves you...

She loves you not?

FIGURE 25.18 The size of your pupils depends on your mood.

---

* This phenomenon is called the *Purkinje effect* after the Czech physiologist who discovered it.

FIGURE 25.19 Both rectangles are equally bright. Cover the boundary between them with your pencil and see.

lighter to darker, moving from left to right. Our eye concentrates on the boundary where the dark edge of the left rectangle joins the light edge of the right rectangle, and our eye-brain system assumes that the rest of the rectangle is the same. We pay attention to the boundary and ignore the rest.

Questions to ponder: Is the way the eye picks out edges and makes assumptions about what lies beyond similar to the way we sometimes make judgments about other cultures and other people? Don't we in the same way tend to exaggerate the differences on the surface while ignoring the similarities and subtle differences within?

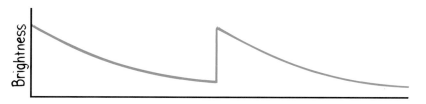

FIGURE 25.20 Graph of brightness levels for the rectangles in Figure 25.19.

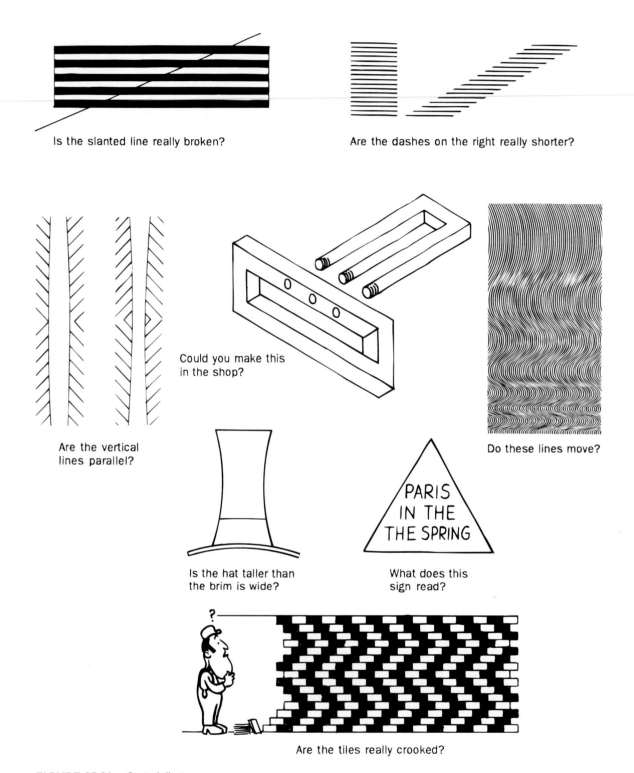

Is the slanted line really broken?

Are the dashes on the right really shorter?

Are the vertical lines parallel?

Could you make this in the shop?

Do these lines move?

Is the hat taller than the brim is wide?

PARIS IN THE THE SPRING

What does this sign read?

?

Are the tiles really crooked?

FIGURE 25.21 Optical illusions.

Summary of Terms

Electromagnetic wave   An energy-carrying wave emitted by vibrating charge (often electrons) that is composed of oscillating electric and magnetic fields that regenerate one another.

Electromagnetic spectrum   The range of electromagnetic waves extending in frequency from radio waves to gamma rays.

Transparent   The term applied to materials through which light can pass in straight lines.

Opaque   The term applied to materials that absorb light without re-emission and thus through which light cannot pass.

Shadow   A shaded region that appears where light rays are blocked by an object.

Umbra   The darker part of a shadow where all the light is blocked.

Penumbra   A partial shadow that appears where some of the light is blocked and other light can fall.

Solar eclipse   The event wherein the moon blocks light from the sun and the moon's shadow falls on part of the earth.

Lunar eclipse   The event wherein the moon passes into the shadow of the earth.

Suggested Reading

Falk, D.S., Brill, D.R., and Stork, D. *Seeing the Light: Optics in Nature.* New York: Harper & Row, 1986.

Review Questions

1. Does light make up a relatively large part or a relatively small part of the electromagnetic spectrum?

Electromagnetic Waves

2. What does a *changing magnetic field* induce?
3. What does a *changing electric field* induce?
4. What are the two components of an electromagnetic wave?

Electromagnetic Wave Velocity

5. Why does an electromagnetic wave in space never slow down?
6. Why does an electromagnetic wave in space never speed up?
7. What do electric and magnetic fields contain and transport?
8. What is light?

The Electromagnetic Spectrum

9. What is the principal difference between a *radio wave* and *light*?
10. What is the principal difference between *light* and an *X ray*?
11. How much of the measured electromagnetic spectrum does light occupy?
12. What color does visible light of the lowest frequencies appear? Of the highest?
13. How does the frequency of a radio wave compare to the frequency of the vibrating electrons that produce it?
14. How is the wavelength of light related to its frequency?
15. What is the wavelength of a wave that has a frequency of 1 Hz and travels at 300,000 km/s?
16. What is the wavelength of a wave that has a frequency of 2 Hz and travels at 300,000 km/s?
17. In what sense do we say that outer space is not really empty?

Transparent Materials

18. The sound coming from one tuning fork can force another to vibrate. What is the analogous effect for light?
19. In what region of the electromagnetic spectrum is the resonant frequency of electrons in glass?
20. What is the fate of the energy in ultraviolet light that is incident upon glass?
21. What is the fate of the energy in visible light that is incident upon glass?
22. Why are your answers for Questions 20 and 21 different?
23. How does the frequency of re-emitted light in a transparent material compare to the frequency of the light that stimulates its re-emission?
24. How does the average speed of light in glass compare with its speed in a vacuum?
25. Why is it that infrared waves often called *heat waves*?

Opaque Materials

26. Why do opaque materials become warmer when light shines on them?
27. Why are metals shiny?
28. Why are there several answers to the question: Is the earth's atmosphere transparent?

Shadows

29. Distinguish between an *umbra* and a *penumbra*.
30. Do the earth and the moon always cast shadows? What do we call the occurrence where one passes within the shadow of the other?

Seeing Light—The Eye

31. Distinguish between the *rods* and *cones* of the eye and between their functions.
32. What are the two unusual features of vision on the periphery of human vision?
33. Why do red-hot and blue-hot stars look white to the eye?
34. Why do we say that the eye does some of our thinking?
35. Why are we able to discern detail in both very bright areas and very dim areas at the same time?

Projects

1. Compare the size of the moon on the horizon with its size higher in the sky. One way to do this is to hold at arm's length various objects that will just barely block out the moon. Experiment until you find something just right, perhaps a thick pencil or pen. You'll find that the object will be less than a centimeter wide, depending on the length of your arms. Is the moon really bigger when near the horizon?
2. Which eye do you use more? To test which you favor, hold a finger up at arm's length. With both eyes open, look past it at a distant object. Now close your right eye. If your finger appears to jump to the right, then you use your right eye more. Check with friends who are both left-handed and right-handed. Is there a correlation between dominant eye and dominant hand?

3. Monitor the change of size of your own eye pupil. Here's how. Punch a small hole in a piece of paper with a small-size paper clip. Hold the paper a few centimeters from your eye and look at overhead lights or the sky through the hole. Don't focus on the pinhole, but relax and stare at the hole—and put it so close that you wouldn't be able to focus on it anyway (remove your glasses if you wear any). You see a nice circle. But that circle is not the hole. It's your pupil! To demonstrate this, cover the other eye with your free hand. Both pupils are locked, so if one changes, the other does as well. Covering the other eye and diminishing light to it make that pupil expand. You'll note they both expand as you see the circle of light expand while you look through the pinhole. Nice?

Exercises

1. Which have the longest wavelengths: light waves, X rays, or radio waves?
2. Which has the shorter wavelengths, ultraviolet or infrared? Which has the higher frequencies?
3. What is it, exactly, that waves in a light wave?
4. We hear people talk of "ultraviolet light" and "infrared light." Why are these terms misleading? Why are we less likely to hear people talk of "radio light" and "X-ray light"?
5. Which requires a physical medium in which to travel: light, sound, or both? Explain.
6. When astronomers observe a supernova explosion in a distant galaxy, they see a sudden, simultaneous rise in visible light and other forms of electromagnetic radiation. Is this evidence to support the idea that the speed of light is independent of frequency? Explain.
7. What is the same about radio waves and light? What is different about them?
8. What evidence can you cite to support the idea that light can travel in a vacuum?
9. Why would you expect the speed of light to be slightly less in the atmosphere than in a vacuum?
10. The intensity of light falls off as the inverse square of the distance from the source. Does this mean that light energy is lost? Explain.
11. If you fire a bullet through a tree, it will slow down inside the tree and emerge at a speed less than the speed at which it entered. Does light, then, similarly slow down when it passes through glass and also emerge at a lower speed? Defend your answer.
12. Pretend a person can walk only at a certain pace—no faster, no slower. If you time her uninterrupted walk across a room of known length, you can calculate her walking speed. If, however, she stops momentarily along the way to greet others in the room, the extra time spent in her brief interactions gives an *average* speed across the room that is less than her walking speed. How is this like light passing through glass? In what way is it not?
13. Is glass transparent to light of frequencies that match its own natural frequencies? Explain.
14. Short wavelengths of visible light interact more frequently with the atoms in glass than do longer wavelengths. Does this interaction time tend to speed up or slow down the average speed of light in glass?
15. What determines whether a material is transparent or opaque?

16. You can get a sunburn on a cloudy day, but you can't get a sunburn even on a sunny day if you are behind glass. Explain.

17. Suppose that sunlight falls on both a pair of reading glasses and a pair of dark sunglasses. Which pair of glasses would you expect to become warmer? Defend your answer.

18. Why does a high-flying plane cast little or no shadow on the ground below, while a low-flying plane casts a sharp shadow?

19. Only some of the people on the daytime side of the earth can witness a solar eclipse when it occurs, whereas all the people on the nighttime side of the earth can witness a lunar eclipse when it occurs. Why is this so?

20. Lunar eclipses are always eclipses of a full moon. That is, the moon is always seen full just before and after the earth's shadow passes over it. Why is this? Why can we never have a lunar eclipse when the moon is in its crescent or half-moon phase?

21. What astronomical event would be seen by observers on the moon (a) at the time the earth was seeing a lunar eclipse? (b) at the time the earth was experiencing a solar eclipse?

22. Light from a location on which you concentrate your attention falls on your fovea, which contains only cones. If you wish to observe a weak source of light, like a faint star, why should you not look *directly* at the source?

23. Why do objects illuminated by moonlight lack color?

24. Why do we not see color at the periphery of our vision?

25. From your experimentation with Figure 25.15, is your blind spot located noseward from your fovea or to the outside of it?

26. When your sweetheart holds you and looks at you with contracted eye pupils and says, "I love you," should you believe it?

27. Can we infer that a person with large pupils is generally happier than a person with small pupils? Why or why not?

28. Ships determine the ocean depth by bouncing sonar waves from the ocean bottom and measuring the round trip time. How does an airplane similarly determine its distance to the ground below?

29. When we look at the sun, we are seeing it as it was 8 minutes ago. So we can only see the sun "in the past." When you look at the back of your own hand, do you see it "now" or "in the past"?

30. Light from a camera flash weakens with distance in accord with the inverse-square law. Comment on an airline passenger who takes a flash photo of a city at night-time from a high-flying plane, or on the practice of taking flash photos of distant players during a nighttime football game.

Problems

1. In about 1675 the Danish astronomer Olaus Roemer found that light from eclipses of one of Jupiter's moons took an extra 16.5 min to travel 300,000,000 km across the diameter of the earth's orbit around the sun. What approximate value for the speed of light did Roemer deduce? How much does it differ from our modern value? (Roemer's measurement, although not accurate by modern standards, was the first demonstration that light travels at a finite, not an infinite, speed.)

2. The sun is 1.50×10^{11} meters from the earth. How long does it take for the sun's light to reach the earth? How long does it take light to cross the diameter of earth's orbit? Compare this time with the time measured by Roemer in the seventeenth century (preceding problem).

3. How long does it take for a pulse of laser light to reach the moon and bounce back to earth?

4. The nearest star beyond the sun is Alpha Centauri, 4.2×10^{16} meters away. If we received a radio message from this star today, how long ago was it sent?

5. Blue-green light has a frequency of about 6×10^{14} Hz. Use the relationship $c = f\lambda$ to find the wavelength of this light in air. How does this wavelength compare with the size of an atom, which is about 10^{-10} m?

6. The wavelength of light changes as light goes from one medium to another, while the frequency remains the same. Is the wavelength longer or shorter in water than in air? Explain in terms of the equation speed = frequency × wavelength. A certain blue-green light has a wavelength of 600 nm (6×10^{-7} m) in air. What is its wavelength in water, where light travels at 75% of its speed in air? In Plexiglas, where light travels at 67% of its speed in air?

7. A certain radar installation used to track airplanes transmits electromagnetic radiation of wavelength 3 cm. (a) What is the frequency of this radiation, measured in billions of hertz (GHz)? (b) What is the time required for a pulse of radar waves to reach an airplane 5 km away and return?

8. A ball with the same diameter as a lightbulb is held halfway between the bulb and a wall, as shown in the sketch. Construct light rays (similar to those in Figure 25.13), and show that the diameter of the umbra on the wall is the same as the diameter of the ball and that the diameter of the penumbra is three times the diameter of the ball.

Lamp   Ball   Wall

26

Color

Red, blue, and green spotlights shine on a white wall. The lights add to make white, but why do shadows appear in brilliant colors?

Roses are red and violets are blue; colors intrigue artists and physics types too. To the physicist, the colors of objects are not in the substances of the objects themselves or even in the light they emit or reflect. Color is a physiological experience and is in the eye of the beholder. So when we say that light from a rose is red, in a stricter sense we mean that it appears red. Many organisms, including people with defective color vision, will not see the rose as red at all.

The colors we see depend on the frequency of the light we see. Lights of different frequencies are perceived as different colors; the lowest-frequency light we detect appears to most people as the color red, and the highest frequency as violet. Between them range the infinite number of hues that make up the color spectrum of the rainbow. By convention these hues are grouped into the seven colors of red, orange, yellow, green, blue, indigo, and violet. These colors together appear white. The white light from the sun is a composite of all the visible frequencies.

Except for light sources such as lamps, lasers, and gas discharge tubes (which we will treat in Chapter 29), most of the objects around us reflect rather than emit light. They reflect only part of the light that is incident upon them, the part that gives them their color.

Selective Reflection

FIGURE 26.1 The colors of things depend on the colors of light that illuminate them.

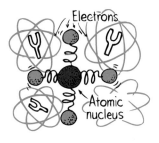

FIGURE 26.2 The outer electrons in an atom vibrate and resonate just as weights on springs would do. As a result, atoms and molecules behave somewhat like optical tuning forks.

FIGURE 26.3 The square on the left *reflects* all the colors illuminating it. In sunlight it is white. When illuminated with blue light, it is blue. The square on the right *absorbs* all the colors illuminating it. In sunlight it is warmer than the white square.

A rose, for example, doesn't emit light; it reflects light (Figure 26.1). If we pass sunlight through a prism and then place a deep-red rose in various parts of the spectrum, the rose will appear brown or black in all parts of the spectrum except in the red. In the red part of the spectrum, the petals also will appear red, but the green stem and leaves will appear black. This shows that the red rose has the ability to reflect red light, but it cannot reflect other kinds of light; the green leaves have the ability to reflect green light and likewise cannot reflect other kinds of light. When the rose is held in white light, the petals appear red and the leaves green, because the petals reflect the red part of the white light and the leaves reflect the green part of the white light. To understand why objects reflect specific colors of light, we must turn our attention to the atom.

Light is reflected from objects in a manner similar to the way sound is "reflected" from a tuning fork when another that is nearby sets it into vibration. A tuning fork can be made to vibrate even when the frequencies are not matched, although at significantly reduced amplitudes. The same is true of atoms and molecules. Using the spring model of the atom we discussed in the previous chapter, we can think of atoms and molecules as three-dimensional tuning forks with electrons that behave as tiny oscillators that vibrate as if attached by invisible springs (Figure 26.2). Electrons can be forced into vibration (oscillation) by the vibrating (oscillating) electric fields of electromagnetic waves.* Once vibrating, these electrons send out their own electromagnetic waves just as vibrating acoustical tuning forks send out sound waves.

Different materials have different natural frequencies for absorbing and emitting radiation. In one material, electrons oscillate readily at certain frequencies; in another material, they oscillate readily at different frequencies.

At the resonant frequencies where the amplitudes of oscillation are large, light is absorbed. But at frequencies below and above the resonant frequencies, light is re-emitted. If the material is transparent, the re-emitted light passes through it. If the material is opaque, the light passes back into the medium from which it came. This is reflection.

Usually a material will absorb light of some frequencies and reflect the rest. If a material absorbs light of most visible frequencies and reflects red, for example, the material appears red. If it reflects light of all the visible frequencies, like the white part of this page, it will be the same color as the light that shines on it. If a material absorbs all the light that shines on it, it reflects none and is black.

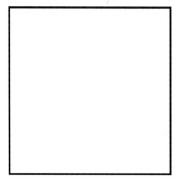

* The words *oscillation* and *vibration* both refer to periodic motion—motion that regularly repeats.

When white light falls on a flower, light of some frequencies is absorbed by the cells in the flower and some is reflected. Cells that contain chlorophyll absorb most of the light and reflect the green part that falls on it. The petals of a red rose, on the other hand, reflect primarily red light, with a lesser amount of blue. Interestingly enough, the petals of most yellow flowers, like daffodils, reflect red and green as well as yellow. Yellow daffodils reflect a broad band of frequencies. The reflected colors of most objects are not pure single-frequency colors, but are composed of a spread of frequencies.

An object can only reflect light with those frequencies that are present in the illuminating light. The appearance of a colored object therefore depends on the kind of light used. A candle flame emits light that is deficient in blue; its light is yellowish. An incandescent lamp emits light that is richer toward the lower frequencies, enhancing the reds. In a fabric with a little bit of red in it, for example, the red will be more apparent under an incandescent lamp than when illuminated with a fluorescent lamp. Fluorescent lamps are richer in the higher frequencies, so blues are enhanced under them. With various kinds of illumination, it is difficult to tell the true color of objects. Colors appear different in daylight from how they appear when illuminated by either kind of lamp (Figure 26.5).

FIGURE 26.4 The bunny's dark fur absorbs all the radiant energy in incident sunlight and therefore is black. Light fur on other parts of the body reflects light of all frequencies and therefore is white.

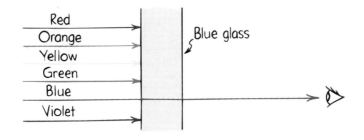

FIGURE 26.5 Color depends on the light source.

Selective Transmission

The color of a transparent object depends on the color of the light it transmits. A red piece of glass appears red because it absorbs all the colors that compose white light, except red, which it *transmits*. Similarly, a blue piece of glass appears blue because it transmits primarily blue light and absorbs light of the other colors that illuminate it. The piece of glass contains dyes or *pigments*—fine particles that selectively absorb light of certain frequencies and selectively transmit others. From an atomic point of view, electrons in the pigment atoms selectively absorb illuminating light of certain frequencies. Light of other frequencies is re-emitted from atom to atom in the glass. The energy of the absorbed light increases the kinetic energy of the molecules and the glass is warmed. Ordinary window glass is colorless because it transmits light of all visible frequencies equally well.

FIGURE 26.6 Only blue light is transmitted; light of all the other frequencies is absorbed and warms the glass.

Questions

1. When red light shines on a red rose, why do the leaves become warmer than the petals?
2. When green light shines on a rose, why do the petals look black?
3. What is the color of the bunny shown in Figure 26.4 when illuminated with only green light?
4. If you hold a match, a candle flame, or any small source of white light between you and a piece of red glass, you'll see two reflections from the glass: one from the front surface and one from the back surface. What color reflections will you see?

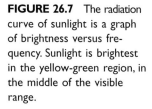

Mixing Colored Light

You can easily demonstrate the fact that white light from the sun is a composite of all the visible frequencies by passing sunlight through a prism and observing the rainbow-colored spectrum. The distribution of solar frequencies is uneven, being most intense in the yellow-green part of the spectrum. It is interesting to note that our eyes have evolved to have maximum sensitivity in this range. That's why more new fire engines are painted yellow-green, particularly at airports where visibility is vital. This also explains why at night we see better under the illumination of yellow sodium-vapor lamps than under common tungsten-filament lamps of the same brightness.

The graphical distribution of brightness versus frequency is called the *radiation curve* of sunlight (Figure 26.7). Most whites produced from reflected sunlight share this frequency distribution.

FIGURE 26.7 The radiation curve of sunlight is a graph of brightness versus frequency. Sunlight is brightest in the yellow-green region, in the middle of the visible range.

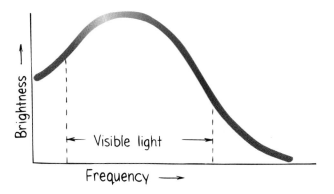

Answers

1. The leaves absorb rather than reflect red light, so the leaves become warmer.
2. The petals absorb rather than reflect the green light. Since green is the only color illuminating the rose, and green contains no red to be reflected, the rose reflects no color at all and appears black.
3. Black and green. Fur that is white in white light reflects all colors of incident light, so is green. The black remains black. Why?
4. Reflection from the front surface is white, because the light doesn't go far enough into the colored glass for absorption of nonred light. The back surface has only red to reach it, so red is what you see reflected.

We have seen that all the visible colors mixed together produce white. Interestingly enough, the perception of white also results from the combination of only red, green, and blue light. We can understand this by simplifying the solar radiation curve to that shown in Figure 26.7, where we divide it into three regions as in Figure 26.8. Light in the lowest third of the spectral distribution stimulates the cones in our eyes that are sensitive to low frequencies and appears red; light in the middle third stimulates the mid-frequency-sensitive cones and appears green; light in the high-frequency third stimulates the higher-frequency-sensitive cones and appears blue.

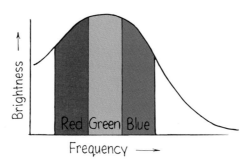

FIGURE 26.8 Radiation curve of sunlight divided into three regions—red, green, and blue. These are the additive primary colors.

If we project these three colors on a screen, overlapping each other, their colors will add to produce white. When the beams of light reflect off a white screen, the light seen is an additive mixture because the lights are added together before the observer sees them. If two of the three colors are added, then another color will be produced (Figure 26.9). By adding various amounts of red, green, and blue, the colors to which each of our three types of cones are sensitive, we can produce any color in the spectrum. For this reason, red, green, and blue are called the **additive primary colors**. A close examination of the picture on most color television tubes will reveal that the picture is an assemblage of tiny spots, each less than a millimeter across. When the screen is lit,

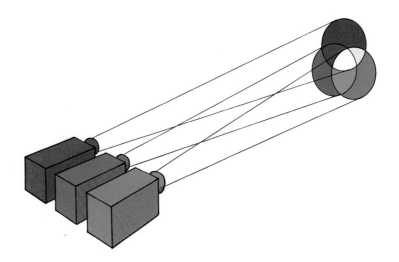

FIGURE 26.9 Color addition by the mixing of colored lights. When three projectors shine red, green, and blue light on a white screen, the overlapping parts produce different colors. White is produced where all three overlap.

FIGURE 26.10 The white golf ball appears white when illuminated with red, green, and blue lights of equal intensities. Why are the shadows of the ball cyan, magenta, and yellow?

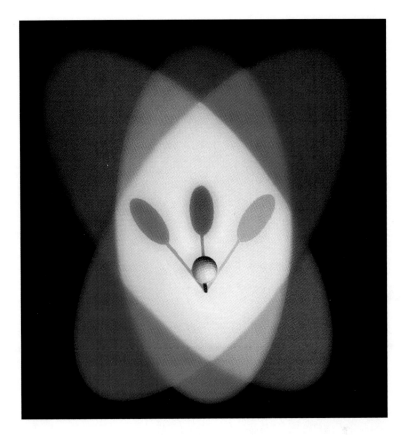

some of the spots are red, some green, some blue; red and green, for example, average to appear yellow; red and blue appear magenta; and green and blue appear cyan. The mixtures of combinations of red, green, and blue at a distance provide a complete range of colors, plus white.*

Mixing Colored Pigments

Every artist knows that if you mix red, green, and blue paint, the result will not be white, but a muddy dark brown. Red and green paint certainly do not combine to form yellow, as is the rule for mixing colored light. The mixing of paints is an entirely different process from the mixing of colored lights. Paint is composed of pigments—tiny solid particles that produce their characteristic colors by the processes of selective absorption or selective transmission of light, depending on their frequencies. When red, green, and blue pigments are combined, every color in the white light that illuminates them is absorbed, and the mixture is black. (In practice, absorption is less than 100%, so the mixture appears brown rather than black.)

* It's interesting to note that the "black" you see on the darkest scenes on a black-and-white TV tube is simply the color of the tube face itself, which is more a light gray than black. Because our eyes are sensitive to the contrast with the illuminated parts of the screen, we see this gray as black.

Any of the three additive primaries can be produced by a combination of any two of cyan, magenta, or yellow pigments. The mixture of absorbing pigments results in a subtraction of colors; the observer sees the light left over after absorption has taken place. For this reason, cyan, magenta, and yellow are called the **subtractive primary colors**. In painting or printing, the primaries are often said to be red, yellow and blue. Here we are loosely speaking of magenta, yellow, and cyan. They can be combined to produce any color in the spectrum in painting or printing.

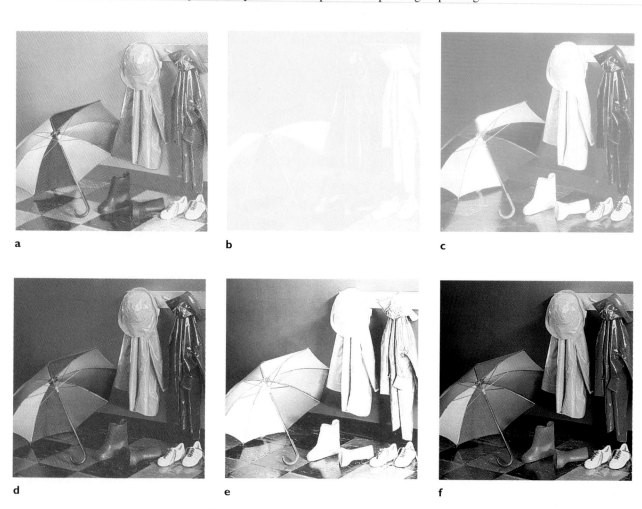

a b c

d e f

FIGURE 26.11 In the printing of color illustrations and photographs, only four colors of ink are used; (a) magenta, (b) yellow, (c) cyan, which when combined produce (d), and (e) black, which when added, produces the finished result, (f).

Color printing is an interesting application of color mixing. Three photographs (color separations) are taken of the illustration to be printed: one through a magenta filter, one through a yellow filter, and one through a cyan filter. This produces three negatives, each with a different pattern of exposed areas that corresponds to the filter used and the color distribution in the original illustration. Light is shone through these negatives onto metal plates specially treated to hold printer's ink only in areas that have been exposed to light. The ink deposits are regulated on different parts of the plate by

FIGURE 26.12 Dyes or pigments, such as those in inks, paints, and photographic materials, like the three transparencies shown, absorb and effectively subtract light of some frequencies and transmit only part of the spectrum. The subtractive primary colors are yellow, magenta, and cyan. When white light passes through overlapping sheets of these colors, light of all frequencies is blocked (subtracted) and we have black. Where only yellow and cyan overlap, light of all frequencies except green is subtracted. Various proportions of yellow, cyan, and magenta dyes will produce nearly any color in the spectrum.

tiny dots. Inkjet printers deposit various combinations of magenta, cyan, yellow, and black inks. Examine the color in any of the figures in this or any book with a magnifying glass and see how the overlapping dots of these colors give the appearance of many colors. Or look at a billboard up close.

Complementary Colors

Here's what happens when two of the three additive primary colors are combined:

$$\text{Red} + \text{Blue} = \text{Magenta}$$

$$\text{Red} + \text{Green} = \text{Yellow}$$

$$\text{Blue} + \text{Green} = \text{Cyan}.$$

We say that magenta is the opposite of green; cyan is opposite red; and yellow is opposite blue. Now, when we add in the opposite color, we get white.

$$\text{Magenta} + \text{Green} = \text{White} (= \text{Red} + \text{Blue} + \text{Green})$$

$$\text{Yellow} + \text{Blue} = \text{White} (= \text{Red} + \text{Green} + \text{Blue})$$

$$\text{Cyan} + \text{Red} = \text{White} (= \text{Blue} + \text{Green} + \text{Red})$$

When two colors are added together to produce white, they are called **complementary colors**. We see that all the rules of color addition and subtraction can be deduced from Figures 26.9, 26.10, and 26.12.

When we look at the colors on a soap bubble or soap film, we see cyan, magenta and yellow predominantly. What does this tell us? It tells us that some primary colors have been subtracted from the original white light! (How this happens will be discussed in Chapter 28.)

FIGURE 26.13 Approximate ranges of frequencies we sense as the additive primary colors and the subtractive primary colors.

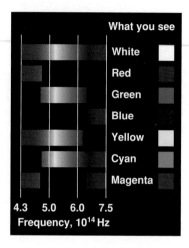

Questions

1. From Figure 26.9, find the complements of cyan, of yellow, and of red.
2. Red + blue = _____ .
3. White − red = _____ .
4. White − blue = _____ .

Why the Sky Is Blue

If a sound beam of a particular frequency is directed to a tuning fork of similar frequency, the tuning fork will be set into vibration and will redirect the beam in multiple directions. The tuning fork *scatters* the sound. A similar process occurs with the scattering of light from molecules and larger specks of matter that are far apart from one another—as in the atmosphere. In this scattering process in the atmosphere, light is absorbed and re-emitted at the same frequency.*

We know that atoms and molecules behave like tiny optical tuning forks and re-emit light waves that shine on them. Very tiny particles do the same. The tinier the

Answers

1. Red, blue, cyan.
2. Magenta.
3. Cyan.
4. Yellow.

---

* This type of scattering is called *Rayleigh scattering* (after Lord Scattering) and occurs whenever the scattering particles are much smaller than the wavelength of incident light and have resonances at frequencies higher than those of the scattered light.

FIGURE 26.14 A beam of light falls on an atom and increases the vibrational motion of electrons in the atom. The vibrating electrons, in turn, re-emit light in various directions. Light is scattered.

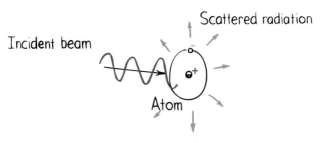

particle, the higher the frequency of light it will scatter. This is similar to the way small bells ring with higher notes than larger bells. The nitrogen and oxygen molecules that make up most of the atmosphere are like tiny bells that "ring" with high frequencies when energized by sunlight. Like sound from the bells, the re-emitted light is sent in all directions. It is scattered.

Most of the ultraviolet light from the sun is absorbed by a thin protective layer of ozone gas in the upper atmosphere. The remaining ultraviolet sunlight that passes through the atmosphere is scattered by atmospheric particles and molecules. Of the visible frequency light, violet is scattered the most, followed by blue, green, yellow, orange, and red, in that order. Red is scattered only a tenth as much as violet. Although violet light is scattered more than blue, our eyes are not very sensitive to violet light. The lesser amount of blue predominates in our vision, so we see a blue sky!

The blue of the sky varies in different places under different conditions. A principal factor is the water-vapor content of the atmosphere. On clear dry days the sky is a much deeper blue than on clear days with high humidity. Places where the upper air is exceptionally dry, such as Italy and Greece, have beautifully blue skies that have inspired painters for centuries. Where there are a lot of particles of dust and other particles larger than oxygen and nitrogen molecules, light of lower frequencies is also scattered strongly. This makes the sky less blue; it takes on a whitish appearance. After a heavy rainstorm when the particles have been washed away, the sky becomes a deeper blue.

FIGURE 26.15 In clean air the scattering of high-frequency light gives us a blue sky. When the air is full of particles larger than molecules, lower-frequency light is also scattered, which adds to give a whitish sky.

The grayish haze in the skies of large cities is the result of particles emitted by car and truck engines and by factories. Even when idling, a typical automobile engine emits more than 100 billion particles per second. Most are invisible but act as tiny centers to which other particles adhere. These are the primary scatterers of lower frequency light. For the larger of these particles, absorption rather than scattering takes place and a brownish haze is produced. Yuk!

Why Sunsets Are Red

Light of lower frequencies is scattered the least by nitrogen and oxygen molecules. Therefore red, orange, and yellow light are transmitted through the atmosphere much more than violet and blue. Red, which is scattered the least, passes through more atmosphere than any other color. Therefore, when white light passes through a thick atmosphere, higher-frequency light is scattered most while light of lower frequencies is transmitted with minimal scattering. And a thickened atmosphere is presented to sunlight at sunset.

At noon the sunlight travels through the minimum amount of atmosphere to reach the earth's surface. Then a relatively small amount of high-frequency light is scattered from sunlight, enough to make the sun somewhat yellow. As the day progresses and the sun is lower in the sky (Figure 26.16), the path through the atmosphere is longer, and more blue is scattered from the sunlight. The removal of blue leaves the transmitted light more reddish in appearance. The sun becomes progressively redder, going from yellow to orange and finally to a red-orange at sunset.*

FIGURE 26.16 A sunbeam must travel through more kilometers of atmosphere at sunset than at noon. As a result, more blue is scattered from the beam at sunset than at noon. By the time a beam of initially white light gets to the ground, only the lower frequencies survive to produce a red sunset.

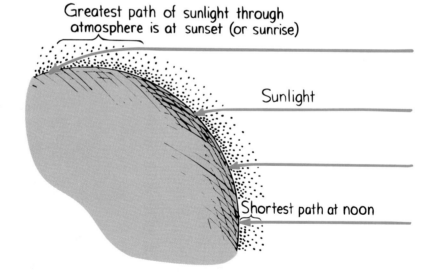

The colors of the sunset are consistent with our rules for color mixing. When blue is subtracted from white light, the complementary color that is left is yellow. When higher-frequency violet is subtracted, the resulting complementary color is orange. When medium-frequency green is subtracted, magenta is left. The combinations of resulting colors vary with atmospheric conditions, which change from day to day and give us a variety of sunsets.

* Sunsets and sunrises would be unusually colorful if particles larger than atmospheric molecules were more abundant in the air. This was the case all over the world for three years following the eruption of the volcano Krakatoa in 1883, when micrometer-sized particles were spewed out in abundance and spread throughout the world's atmosphere. This occurred to a lesser extent following the 1991 eruption of Mount Pinatubo in the Philippines.

Questions

1. If molecules in the sky scattered low-frequency light more than high-frequency light, how would the colors of the sky and sunsets appear?

2. Distant dark mountains are bluish in color. What is the source of this blueness? (*Hint:* Exactly what is between us and the mountains we see?)

3. Distant snow-covered mountains reflect a lot of light and are bright. But they look yellowish, depending on how far away they are. Why do they appear yellowish? (*Hint:* What happens to the reflected white light in traveling from the mountain to us?)

Why Clouds Are White

Water droplets in a variety of sizes—most of them microscopic—make up clouds. The different-size droplets result in a variety of colors of scattered light: low frequencies from larger droplets and high frequencies from tinier droplets of water molecules. The

FIGURE 26.17 A cloud is composed of a variety of droplet sizes. The tiniest scatter blue, slightly larger scatter green, and still larger scatter reds. The overall result is a white cloud.

Answers

1. If low-frequency light were scattered more, red light would be scattered out of the sunlight on its long path through the atmosphere at sunset, and the sunlight that reached your eye would be predominantly blue and violet. So sunsets would appear blue! And the sky at midday would be reddish, not blue.

2. If we look at distant dark mountains, very little light from them reaches us, and the blueness of the atmosphere between us and the mountains predominates. The blueness is of the low-altitude "sky" between us and the mountains. That's why distant mountains look blue!

3. The reason that bright snow-covered mountains appear yellow is because the blue in the white light from the snowy mountains is scattered on its way to us. So by the time the light gets to us, it is weak in the high frequencies and strong in the low frequencies—hence it is yellowish. For greater distances, farther away than mountains are usually seen, they would appear orange for the same reason a sunset appears orange.

    Why do we see the scattered blue when the background is dark, but not when the background is bright? Because the scattered blue is faint. A faint color will show itself against a dark background, but not against a bright background. For example, when we look from the earth's surface at the atmosphere against the darkness of space, the atmosphere is sky blue. But astronauts above who look below through the same atmosphere to the bright surface of the earth do not see the same blueness.

overall result is a white cloud. The electrons in a tiny droplet vibrate together and in step, which results in the scattering of a greater amount of energy than when the same number of electrons vibrate separately. Hence, clouds are bright!

Absorption occurs for larger droplets, and the scattered light intensity is less. The clouds are darker. Further increase in the size of the drops causes them to fall to earth, and we have rain.

The next time you find yourself admiring a crisp blue sky, or delighting in the shapes of bright clouds, or watching a beautiful sunset, think about all those ultra-tiny optical tuning forks vibrating away—you'll appreciate these everyday wonders of nature even more!

Why Water Is Greenish Blue

We often see a beautiful deep blue when we look at the surface of a lake or the ocean. But that isn't the color of water; it's the reflected color of the sky. The color of water itself, as you can see by looking at a piece of white material under water, is a pale greenish blue.

FIGURE 26.18 Water is cyan because it absorbs red. The froth in the waves is white because, like clouds, it is composed of a variety of tiny clusters of water droplets that scatter all the visible frequencies.

Although water is transparent to light of nearly all the visible frequencies, it strongly absorbs infrared waves. This is because water molecules resonate to the frequencies of infrared. The energy of the infrared waves is transformed into internal energy in the water, which is why sunlight warms water. Water molecules somewhat resonate in the visible red, which causes red light to be a little more strongly absorbed than blue light in water. Red light is reduced to a quarter of its initial brightness by 15 meters of water. There is very little red light in the sunlight that penetrates below 30 meters of water. When red is taken away from white light, what color remains? This question can be asked in another way: What is the complementary color of red? The complementary color of red is cyan—a bluish green color. In seawater, the color of everything at these depths looks greenish.

Many crabs and other sea creatures that appear black in deep water are found to be red when they are raised to the surface. At these depths, black and red look the same. Apparently the selection mechanism of evolution could not distinguish between black and red at such depths in the ocean.

So while the sky is blue because blue is strongly scattered by molecules in the atmosphere, water is bluish green because red is absorbed by molecules in the water. We see that the colors of things depend on which colors are scattered or reflected by molecules and also on which colors are absorbed by molecules.*

Interestingly enough, the color we see is not in the world around us—the color is in our heads. The world is filled with a montage of vibrations—electromagnetic waves that stimulate the sensation of color when the vibrations interact with the cone-shaped receiving antennae in the retinas of our eyes. How nice that eye-brain interactions produce the beautiful colors we see.

FIGURE 26.19   There are no blue pigments in the feathers of a blue jay. Instead there are tiny alveolar cells in the barbs of its feathers that scatter light—mainly high-frequency light. So a blue jay is blue for the same reason the sky is blue—scattering.

Summary of Terms

Additive primary colors   The three colors of light—red, blue, and green—that when added in certain proportions will produce any color in the spectrum.

Subtractive primary colors   The three colors of absorbing pigments—magenta, yellow, and cyan—that when mixed in certain proportions will reflect any color in the spectrum.

Complementary colors   Any two colors of light that when added produce white light.

Review Questions

1. What is the relationship between the frequency of light and its color?

Selective Reflection

2. What happens to light when it falls upon a material that has a natural frequency equal to the frequency of the light?
3. What happens to light when it falls upon a material that has a natural frequency above or below the frequency of the light?
4. Distinguish between the white of this page and the black of this ink in terms of what happens to the white light that falls on both.
5. How does the color of an object differ when illuminated by candlelight and by the light from a fluorescent lamp?

Selective Transmission

6. What is a *pigment*?
7. What color light is transmitted through a piece of red glass?

8. Which will warm more quickly in sunlight, a clear or a colored piece of glass? Why?

Mixing Colored Light

9. What is the evidence for the statement that white light is a composite of all the colors of the spectrum?
10. What is the color of the peak frequency of solar radiation?
11. What color light are our eyes most sensitive to?
12. What is a *radiation curve*?
13. What frequency ranges of the radiation curve do red, green, and blue light occupy?
14. Why are red, green, and blue called the *additive primary colors*?

Mixing Colored Pigments

15. Why does a mixture of red and green pigments appear blackish?
16. Why does a mixture of cyan and yellow pigments appear blue?
17. Why are magenta, yellow, and cyan called the *subtractive primary colors*?
18. If you look with a magnifying glass at pictures printed in full color in magazines, you'll notice three colors of ink plus black. What are these colors?

Complementary Colors

19. What is the resulting color of equal intensities of red, blue, and green light combined?
20. What is the resulting color of equal intensities of blue light and green light combined?
21. What is the resulting color of equal intensities of red light and cyan light combined?
22. Why are red and cyan called *complementary colors*?

* Scattering by small, widely spaced particles in the irises of blue eyes, rather than any pigments, accounts for their color. Absorption by pigments accounts for brown eyes.

Why the Sky Is Blue

23. Which interacts more with high-pitched sounds, small bells or large bells?
24. Which interacts more with high-frequency light, small particles or large particles?
25. True or false: The sky is blue because oxygen and nitrogen molecules are blue in color.
26. Why does the sky sometimes appear whitish?
27. Why is the sky a deeper blue after a heavy rainstorm?

Why Sunsets Are Red

28. True or false: The sky at sunset is reddish because at sunset oxygen and nitrogen molecules are red in color.
29. Why does the sun look reddish at sunrise and sunset, but not at noon?
30. Why does the color of a sunset vary from day to day?

Why Clouds Are White

31. What is the evidence for a variety of particle sizes in a cloud?
32. What is the evidence for extra big particles in a rain cloud?

Why Water Is Greenish Blue

33. What part of the electromagnetic spectrum is most absorbed by water?
34. What part of the visible electromagnetic spectrum is most absorbed by water?
35. What color results when red is subtracted from white light?
36. Why does water appear cyan?
37. What color would a red crab appear very deep in the water?

Projects

1. Stare at a piece of colored paper for 45 seconds or so. Then look at a plain white surface. The cones in your retina receptive to the color of the paper become fatigued, so you see an afterimage of the complementary color when you look at a white area. This is because the fatigued cones send a weaker signal to the brain. All the colors produce white, but all the colors minus one produce the complementary to the missing color. Try it and see!
2. Cut a disk a few centimeters or so in diameter from a piece of cardboard; punch two holes a bit off-center, big enough for a piece of string loop as shown in the sketch.

Twirl the disk as shown, so the string winds up like a rubber band on a model airplane. Then tighten the string by pulling outward and the disk will spin. If half the disk is colored yellow and the other half blue, when it is spun the color will be mixed and appear nearly white (how close to white depends on the hues of the colors). Try this for other complementary colors.

3. Fashion a cardboard tube covered at each end with metal foil. Punch a hole in each end with a pencil, one about 3 or so millimeters in diameter and the other twice as big. Put your eye to the small hole and look through the tube at the colors of things against the black background of the tube. You'll see colors that look very different from how they appear against ordinary backgrounds.
4. Simulate a sunset: Add a few drops of milk to a glass of water and look through it at a light bulb. The bulb appears to be red or pale orange, while light scattered to the side appears blue. Try it and see.

Exercises

1. Why will the leaves of a rose be heated more than the petals when illuminated with red light? What does this have to do with people in the hot desert wearing white clothes?
2. In a dress shop with only fluorescent lighting, a customer insists on taking dresses into the daylight at the doorway to check their color. Is she being reasonable? Explain.
3. If the sunlight were somehow green instead of white, what color garment would be most advisable on an uncomfortably hot day? On a very cold day?
4. Why do we not list black and white as colors?
5. Why are the interiors of optical instruments black?
6. Fire engines used to be red. Now many of them are yellow-green. Why the change?
7. What is the color of common tennis balls, and why?
8. The radiation curve of the sun (Figure 26.8) shows that the brightest light from the sun is yellow-green. Why then do we see the sun as whitish instead of yellow-green?
9. What color does red cloth appear when it is illuminated by sunlight? By light from a neon sign? By cyan light?

10. Why does a white piece of paper appear white in white light, red in red light, blue in blue light, and so on for every color?

11. A spotlight is coated so that it won't transmit yellow light from its white-hot filament. What color is the emerging beam of light?

12. How could you use the spotlights at a play to make the yellow clothes of the performers suddenly change to black?

13. Suppose two flashlight beams are shone on a white screen, one beam through a pane of blue glass and the other through a pane of yellow glass. What color appears on the screen where the two beams overlap? Suppose, instead, that the two panes of glass are placed in the beam of a single flashlight. What then?

14. Does a color television work by color addition or by color subtraction? Which dots must be struck by electrons to create the color yellow? Magenta? White?

15. On a TV screen, red, green, and blue spots of fluorescent materials are illuminated at a variety of relative intensities to produce a full spectrum of colors. What dots are activated to produce yellow? Magenta? White?

16. What colors of ink do color ink-jet printers use to produce a full range of colors? Do the colors form by color addition or by color subtraction?

17. On a photo print you see your best friend wearing a blue sweater. What color is the sweater on the negative?

18. What color will be transmitted through overlapping cyan and magenta filters?

19. Look at sunburned red feet when they are under water. Why don't they look as red as when they are above water?

20. Why does blood of injured deep-sea divers look greenish black in underwater photographs taken with natural light, but red when flashbulbs are used?

21. By reference to Figure 26.9, complete the following equations:
    Yellow light + blue light = —————— light.
    Green light + —————— light = white light.
    Magenta + yellow + cyan = —————— light.

22. Check Figure 26.9 to see if the following three statements are accurate. Then fill in the last statement. (All colors are combined by the addition of light.)
    Red + green + blue = white.
    Red + green = yellow = white − blue.
    Red + blue = magenta = white − green.
    Green + blue = cyan = white − —————— .

23. In which of these cases will a ripe banana appear black: When illuminated with red light? With yellow light? With green light? With blue light?

24. When white light is shone on red ink dried on a glass plate, the color that is transmitted is red. But the color that is reflected is not red. What is it?

25. Stare intently at an American flag. Then turn your view to a white area. What colors do you see in the image of the flag that appears on the wall?

26. Why can't we see stars in the daytime?

27. Why is the sky a darker blue when you are at high altitudes? (*Hint:* What color is the "sky" on the moon?)

28. Can stars be seen from the moon in the "daytime" when the sun is shining?

29. Will people in space habitats see the sun against a background of stars, or will they see stars only when they are in the shadow of their habitat? Explain.

30. What does the fact that one can get a sunburn in the shade have to do with the fact that the ultraviolet light not absorbed by the atmosphere is very much scattered by it?

31. Pilots sometimes wear glasses that transmit yellow light and absorb light of most other colors. Why does this help them see more clearly?

32. Does light travel faster through the lower atmosphere or the upper atmosphere?

33. Comment on the statement, "Oh, that beautiful red sunset is just the leftover colors that weren't scattered on their way through the atmosphere."

34. If the sky on a certain planet in the solar system were normally orange, what color would its sunsets be?

35. Tiny particles, like tiny bells, scatter high-frequency waves more than low-frequency waves. Large particles, like large bells, mostly scatter low frequencies. Intermediate-size particles and bells mostly scatter intermediate frequencies. What does this have to do with the whiteness of clouds?

36. Very big particles, like droplets of water, absorb more radiation than they scatter. What does this have to do with the darkness of rain clouds?

37. If the atmosphere of the earth were about fifty times thicker, would ordinary snowfall still seem white or would it be some other color? What color?

38. The atmosphere of Jupiter is more than 1000 km thick. From the surface of this planet, would you expect to see a white sun?

39. The refraction of all the sunsets and sunrises around the world causes the redness of the moon during a lunar eclipse. Explain how this is so.

40. You're explaining to a youngster at the seashore why the water is cyan colored. The youngster points to the whitecaps of overturning waves and asks why they are white. What is your answer?

27

Reflection and Refraction

: **T**he slowing of light in water produces refraction.

Most of the things we see around us do not emit their own light. They are visible because they re-emit light reaching their surface from a primary source, such as the sun or a lamp, or from a secondary source, such as the illuminated sky. When light falls on the surface of a material it is either re-emitted without change in frequency or is absorbed in the material and turned into heat.* Usually both of these processes occur in varying degrees. When the re-emitted light is returned into the medium from which it came, it is *reflected,* and the process is **reflection**. When the re-emitted light bends from its original course and proceeds in straight lines from molecule to molecule within a transparent material, it is *refracted,* and the process is **refraction**.

Reflection

When this page is illuminated with sunlight or lamplight, electrons in its atoms vibrate more energetically in response to the oscillating electric fields of the illuminating light. The energized electrons re-emit the light by which we see the page. When the page is illuminated by white light, it appears white, which reveals the fact that the electrons

* Another less common fate is absorption followed by re-emission at lower frequencies—fluorescence (Chapter 29).

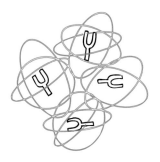

reemit light of all the visible frequencies. Very little absorption occurs. The ink that comprises the lettering that you are now reading is a different story. Except for a bit of reflection, it absorbs light of all the visible frequencies and therefore appears black. And the colored inks are absorbing their complementary colors and reflecting what you see.

FIGURE 27.1   Light interacts with atoms as sound interacts with tuning forks.

Principle of Least Time*

We all know that light ordinarily travels in straight lines. In going from one place to another, light will take the most efficient path and travel in a straight line. This is true if there is nothing to obstruct the passage of light between the places under consideration. If light is reflected from a mirror, the kink in the otherwise straight-line path is described by a simple formula. If light is refracted, as when it goes from air into water, still another formula describes the deviation of light from the straight-line path. Before thinking of light with these formulas, we will first consider an idea that underlies all the formulas that describe light paths. This idea was formulated by the French scientist Pierre de Fermat in about 1650, and it is called **Fermat's principle of least time**. His idea is this: Out of all possible paths that light might take to get from one point to another, it takes the path that requires the *shortest time*.

Law of Reflection

We can understand reflection by the principle of least time. Consider the following situation. In Figure 27.2 we see two points, A and B, and an ordinary plane mirror beneath. How can we get from A to B in the shortest time? The answer is simple enough—go straight from A to B! But if we add the condition that the light must strike the mirror in going from A to B in the shortest time, the answer is not so easy. One way would be to go as quickly as possible to the mirror and then to B, as shown by the solid lines in Figure 27.3. This gives us a short path to the mirror but a very long path from the mirror to B. If we instead consider a point on the mirror a little to the right, we slightly increase the first distance, but we considerably decrease the second distance, and so the total path length shown by the dashed lines—and therefore the travel time—is less. How can we find the exact point on the mirror for which the time is shortest? We can find it very nicely by a geometric trick.

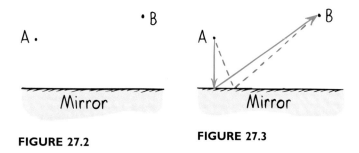

FIGURE 27.2   **FIGURE 27.3**

* This material and many of the examples of least time are adapted from R. P. Feynman, R. B. Leighton, and M. Sands, *The Feynman Lectures on Physics,* Vol. I, Chap. 26, 1963 by California Institute of Technology. Reading, MA: Addison-Wesley.

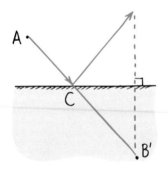

FIGURE 27.4

We construct on the opposite side of the mirror an artificial point, B', which is the same distance "through" and below the mirror as the point B is above the mirror (Figure 27.4). The shortest distance between A and this artificial point B' is simple enough to determine: It's a straight line. Now this straight line intersects the mirror at a point C, the precise point of reflection for the shortest path and hence the path of least time for the passage of light from A to B. Inspection will show that the distance from C to B equals the distance from C to B'. We see that the length of the path from A to B' through C is equal to the length of the path from A to B bouncing off point C along the way.

Further inspection and a little geometrical reasoning will show that the angle of incident light from A to C is equal to the angle of reflection from C to B. This is the **law of reflection**, and it holds for all angles (Figure 27.5).

The angle of incidence equals the angle of reflection.

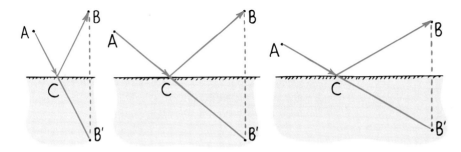

FIGURE 27.5 Reflection.

The law of reflection is illustrated with arrows representing light rays in Figure 27.6. Instead of measuring the angles of incident and reflected rays from the reflecting surface, it is customary to measure them from a line perpendicular to the plane of the reflecting surface. This imaginary line is called the *normal*. The incident ray, the normal, and the reflected ray all lie in the same plane.

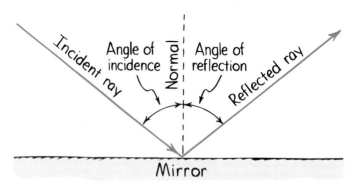

FIGURE 27.6 The law of reflection.

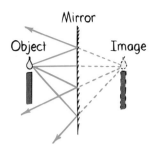

FIGURE 27.7 A virtual image is formed behind the mirror and is located at the position where the extended reflected rays (dashed lines) converge.

Question The constructions of artificial points B′ in Figures 27.4 and 27.5 show how light encounters point C in reflecting from A to B. By similar construction, show that light that originates at B and reflects to A also encounters the same point C.

Plane Mirrors

Suppose a candle flame is placed in front of a plane mirror. Rays of light are sent from the flame in all directions. Figure 27.7 shows only four of the infinite number of rays leaving one of the infinite number of points on the flame. When these rays encounter the mirror, they are reflected at angles equal to their angles of incidence. The rays diverge from the flame and on reflection diverge from the mirror. These divergent rays appear to emanate from a particular point behind the mirror (where the dashed lines intersect). An observer sees an image of the flame at this point. The light rays do not actually come from this point, so the image is called a *virtual image*. The image is as far behind the mirror as the object is in front of the mirror, and image and object have the same size. When you view yourself in a mirror, for example, the size of your image is the same as the size your twin would appear if located as far behind the mirror as you are in front—as long as the mirror is flat.

FIGURE 27.8 Marjorie's image is as far behind the mirror as she is in front. Note that she and her image have the same color of clothing—evidence that light doesn't change frequency upon reflection. Interestingly, her left and right axis is no more reversed than her up and down axis. The axis that *is* reversed, as shown to the right, is front and back. That's why it seems her left hand faces the right hand of her image.

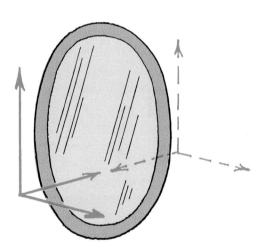

When the mirror is curved, the sizes and distances of object and image are no longer equal. We will not get into curved mirrors in this text, except to say that the law of reflection still holds. A curved mirror behaves as a succession of flat mirrors, each at a slightly different angular orientation from the one next to it. At each point, the angle

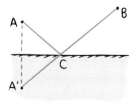

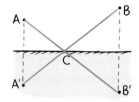

Answer Construct an artificial point A′ as far below the mirror as A is above; then draw a straight line from B to A′ to find C, as shown at the left. Both constructions superimposed, at right, show that C is common to both. We see that light will follow the same path if it goes in the opposite direction. Whenever you see somebody else's eyes in a mirror, be assured that they can also see yours.

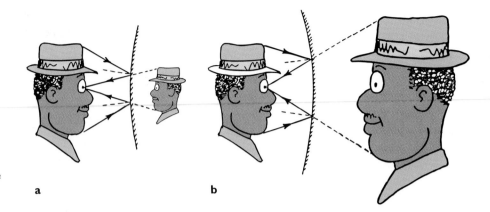

FIGURE 27.9 (a) The virtual image formed by a *convex* mirror (a mirror that curves outward) is smaller and closer to the mirror than the object. (b) When the object is close to a *concave* mirror (a mirror that curves inward like a "cave"), the virtual image is larger and farther away than the object. In either case the law of reflection holds for each ray.

a   b

of incidence is equal to the angle of reflection (Figure 27.9). Note that in a curved mirror, unlike in a plane mirror, the normals (shown by the dashed black lines) at different points on the surface are not parallel to one another.

Whether the mirror is plane or curved, the eye-brain system cannot ordinarily tell the difference between an object and its reflected image. So the illusion that an object exists behind a mirror (or in some cases in front of a concave mirror) is merely due to the fact that the light from the object enters the eye in exactly the same manner, physically, as it would have entered if the object really were at the image location.

Questions

1. What evidence can you cite to support the claim that the frequency of light does not change upon reflection?
2. If you wish to take a picture of your image while standing 5 m in front of a plane mirror, for what distance should you set your camera to provide sharpest focus?

Only part of the light that strikes a surface is reflected. On a surface of clear glass, for example, and for normal incidence (light perpendicular to the surface) only about 4% is reflected from each surface, while on a clean and polished aluminum or silver surface, about 90% of incident light is reflected.

Diffuse Reflection

When light is incident on a rough surface, it is reflected in many directions. This is called **diffuse reflection** (Figure 27.10). If the surface is so smooth that the distances between successive elevations on the surface are less than about one-eighth the wavelength of the light, there is very little diffuse reflection, and the surface is said to be

FIGURE 27.10 Diffuse reflection. Although the reflection of each single ray obeys the law of reflection, the many different surface angles that light rays encounter in striking a rough surface cause reflection in many directions.

Answers

1. Simply stand in front of a mirror and compare the color of your shirt with the color of its image. The fact that the color is the same is evidence that the frequency of light does not change upon reflection.
2. You should set your camera for 10 m; the situation is equivalent to your standing 5 m in front of an open window and viewing your twin standing 5 m in back of the window.

FIGURE 27.11 The open-mesh parabolic dish is a diffuse reflector for short-wavelength light but polished for long-wavelength radio waves.

polished. A surface therefore may be polished for radiation of long wavelength but not polished for light of short wavelength. The wire-mesh "dish" shown in Figure 27.11 is very rough for light waves and is hardly mirrorlike. But for long-wavelength radio waves it is "polished" and is an excellent reflector.

Light reflecting from this page is diffuse. The page may be smooth to a radio wave, but to a light wave it is rough. Rays of light hitting on this page encounter millions of tiny flat surfaces facing in all directions. The incident light therefore is reflected in all directions. This is a desirable circumstance. It enables us to see objects from any direction or position. Diffuse reflection lets us see the road surface illuminated by our headlights on a dry night. But when the road is wet, there is less diffuse reflection and the road is harder to see. Most of our environment is seen by diffuse reflection.

An undesirable circumstance related to diffuse reflection is the ghost image that occurs on a TV set when the TV signal bounces off buildings and other obstructions. For antenna reception, this difference in path lengths for the direct signal and the reflected

FIGURE 27.12 A magnified view of the surface of ordinary paper.

signal produces a slight time delay. The ghost image is normally displaced to the right, the direction of scanning in the TV tube, because the reflected signal arrives at the receiving antenna later than the direct signal. Multiple reflections may produce multiple ghosts.

Refraction

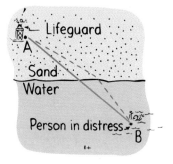

FIGURE 27.13 Refraction.

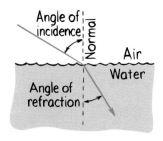

FIGURE 27.14 Refraction.

Recall from Chapter 25 that the average speed of light is lower in glass and other transparent materials than through empty space. Light travels at different speeds in different materials.* It travels at 300,000 kilometers per second in a vacuum, at a slightly lower speed in air, and at about three-fourths that speed in water. In a diamond, light travels at about 40% of its speed in a vacuum. Light bends when it changes speed and passes obliquely from one medium to another. This is **refraction**. It is a common observation that a ray of light bends and takes a longer path when it encounters glass or water at an oblique angle. But the longer path taken is nonetheless the path requiring the least time. A straight-line path would take a longer time. We can illustrate this with the following situation.

Imagine that you are a lifeguard at a beach and you spot a person in distress in the water. We show the relative positions of you, the shoreline, and the person in distress in Figure 27.13. You are at point A, and the person is at point B. You can run faster than you can swim. Should you travel in a straight line to get to B? A little thought will show that a straight-line path would not be the best choice, because if you instead spent a little bit more time traveling farther on land, you would save a lot more time in swimming a lesser distance in the water. The path of shortest time is shown by the dashed-line path, which clearly is not the path of the shortest distance. The amount of bending at the shoreline of course depends on how much faster you can run than swim. The situation is similar for a ray of light incident upon a body of water, as shown in Figure 27.14. The angle of incidence is larger than the angle of refraction by an amount that depends on the relative speeds of light in air and in water.

Question Suppose our lifeguard in the preceding example were a seal instead of a human being. How would its path of least time from A to B differ?

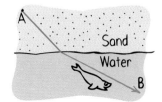

Answer The seal can swim faster than it can run and its path would bend as shown; likewise with light emerging from the bottom of a piece of glass into air.

* Just how much the speed of light differs from its speed in a vacuum is given by the index of refraction, n, of the material; n is the ratio of the speed of light in a vacuum to the speed of light in the material:

$$n = \frac{\text{speed of light in vacuum}}{\text{speed of light in material}}$$

For example, the speed of light in a diamond is $(1/2.4)c$, so $n = 2.4$. For a vacuum, $n = 1$.

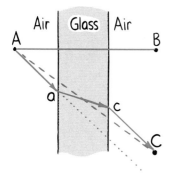

FIGURE 27.15 Refraction through glass. Although dashed line AC is the shortest path, light goes a slightly longer path through the air from A to a, then a shorter path through the glass to c, and then to C. The emerging light is displaced but parallel to the incident light.

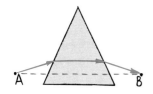

FIGURE 27.16   A prism.

FIGURE 27.17   A curved prism.

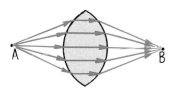

FIGURE 27.18   A converging lens.

Consider the pane of thick window glass in Figure 27.15. When light goes from point A through the glass to point B, it will go in a straight-line path. In this case, light encounters the glass perpendicularly, and we see that the shortest distance through both air and glass corresponds to the shortest time. But what about light that goes from point A to point C? Will it travel in the straight-line path shown by the dashed line? The answer is no, for if it did so it would be spending more time inside the glass, where light travels more slowly than in air. The light will instead take a less-inclined path through the glass. The time saved by taking the resulting shorter path through the glass more than compensates for the added time required to travel the slightly longer path through the air. The overall path is the path of least time. The result is a parallel displacement of the light beam, because the angles in and out are the same. You'll notice this displacement when you look through a thick pane of glass at an angle. The more your angle of viewing differs from perpendicular, the more pronounced the displacement.

Another example of interest is the prism, in which opposite faces of the glass are not parallel (Figure 27.16). Light that goes from point A to point B will not follow the straight-line path shown by the dashed line, because too much time would be spent in the glass. Instead, the light will follow the path shown by the solid line—a path that is a bit farther through the air—and pass through a thinner section of the glass to make its trip to point B. By this reasoning, one might think that the light should take a path closer to the upper vertex of the prism and seek the minimum thickness of glass. But if it did, the extra distance through the air would result in an overall longer time of travel. The path followed is the path of least time.

Interestingly enough, a properly curved prism will provide many paths of equal time from a point A on one side to a point B on the opposite side (Figure 27.17). The curve decreases the thickness of the glass correctly to compensate for the extra distances light travels to points higher on the surface. For appropriate positions of A and B and for the appropriate curve on the surfaces of this modified prism, all light paths are of exactly equal time. In this case, all the light from A that is incident on the glass surface is focused on point B. We see that this shape is simply the upper half of a converging lens (Figure 27.18, and treated in more detail later in this chapter).

Whenever we watch a sunset, we see the sun for several minutes after it has sunk below the horizon. The earth's atmosphere is thin at the top and dense at the bottom. Since light travels faster in thin air than it does in dense air, light from the sun can get to us more quickly if, instead of just going in a straight line, it avoids the denser air by taking a higher and longer path to penetrate the atmosphere at a steeper tilt (Figure 27.19).

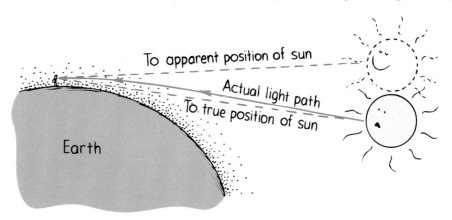

FIGURE 27.19   Because of atmospheric refraction, when the sun is near the horizon it appears higher in the sky.

Since the density of the atmosphere changes gradually, the light path bends gradually to produce a curved path. Interestingly enough, this path of least time provides us with a slightly longer period of daylight each day. Furthermore, when the sun (or moon) is near the horizon, the rays from the lower edge are bent more than the rays from the upper edge. This produces a shortening of the vertical diameter, causing the sun to appear elliptical (Figure 27.20).

FIGURE 27.20   The sun is distorted by differential refraction.

We are all familiar with the mirage we see while driving on a hot road. The sky appears to be reflected from water on the distant road, but when we get there, the road is dry. Why is this so? The air is very hot just above the road surface and cooler above. Light travels faster through the thinner hot air than through the denser cool air above. So light, instead of coming to us from the sky in straight lines, also has least-time paths by which it curves down into the hotter region near the road for a while before reaching our eyes (Figure 27.21). A mirage is not, as many people mistakenly believe, a "trick of the mind." A mirage is formed by real light and can be photographed, as shown in Figure 27.22.

When we look at an object over a hot stove or over hot pavement, we see a wavy, shimmering effect. This is due to the various least-time paths of light as it passes

FIGURE 27.21   A mirage is the result of refraction through the air.

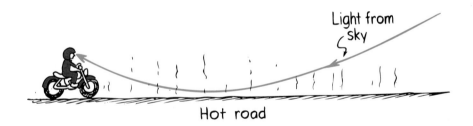

Hot road

FIGURE 27.22 A mirage. The apparent wetness of the road is not reflection of the sky by water but refraction of sky light through the warmer and less-dense air near the road surface.

through varying temperatures and therefore varying densities of air. The twinkling of stars results from similar phenomena in the sky, where light passes through unstable layers in the atmosphere.

Question If the speed of light were the same in air of various temperatures and densities, would there still be slightly longer daytimes, twinkling stars at night, mirages, and slightly squashed suns at sunset?

In the foregoing examples, how does light seemingly "know" what conditions exist and what compensations a least-time path requires? When approaching window glass at an angle, how does light know to travel a bit farther in air to save time in taking a less-inclined angle and therefore a shorter path through the pane of glass? In approaching a prism or a lens, how does light know to travel a greater distance in air to reach a thinner portion of the glass? How does light from the sun know to travel above the atmosphere an extra distance before taking a shortcut through the denser air to save time? How does sky light above know that it can reach us in minimum time if it dips toward a hot road before tilting upward to our eyes? The principle of least time appears to be noncausal. It seems as if light has a mind of its own, that it can "sense" all the possible paths, calculate the times for each, and choose the one that requires the least time. Is this the case? As intriguing as all this may seem, the answer is no—there is no reason to assign foresight to light. It turns out the path of least time has a straightforward physics explanation in terms of the speed at which light moves in different media.

Answer No.

Cause of Refraction

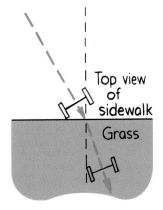

FIGURE 27.23 The direction of the rolling wheels changes when one wheel slows down before the other wheel.

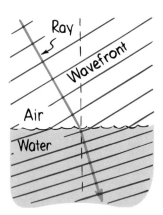

FIGURE 27.24 The direction of the light waves changes when one part of the wave slows down before the other part.

When light bends as it passes obliquely from one medium to another, we call the process *refraction*. The cause of refraction is the changing of the average speed of light in going from one transparent medium to another. We can understand this by considering the action of a pair of toy cart wheels connected to an axle as the wheels roll gently downhill from a smooth sidewalk onto a grass lawn. If the wheels meet the grass at some angle (Figure 27.23), they will be deflected from their straight-line course. The direction of the rolling wheels is shown by the dashed line. Note that on meeting the lawn, where the wheels roll slower owing to interaction with the grass, the wheel on the left in the figure slows down first. This is because it meets the grass while the right wheel is still on the smooth sidewalk. The faster-moving right wheel tends to pivot about the slower-moving left wheel. It travels farther during the same time that the left wheel travels a shorter distance in the grass. This action bends the direction of the rolling wheels toward the "normal," the lightly dashed line perpendicular to the grass-sidewalk border.

A light wave bends in a similar way (Figure 27.24). Note the direction of light shown by the solid arrow (the light ray) and wave fronts at right angles to the ray. (If the light source were close, the wave fronts would appear as segments of circles; but if we assume the distant sun is the source, the waves form practically straight lines.) In any case, we see that the wave fronts are everywhere perpendicular to the light rays. In the figure the wave meets the water surface at an angle, so the left portion of the wave slows down in the water while the part still in the air travels at speed c. The ray or beam of light remains perpendicular to the wave front and bends at the surface, just as the wheels bend to change direction when they roll from the sidewalk into the grass. In both cases the bending is caused by a change in speed.*

The changeable speed of light provides a wave explanation for mirages. Sample wavefronts coming from the top of a tree on a hot day are shown in Figure 27.25. If the temperature of the air were uniform, the average speed of light would be the same in all parts of the air; light traveling toward the ground would meet the ground. But the air is warmer and less dense near the ground and the wavefronts pick up speed as they travel

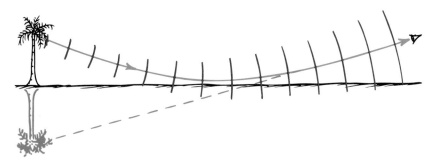

FIGURE 27.25 A wave explanation of a mirage. Wave fronts of light travel faster in the hot air near the ground and bend upward.

* The quantitative law of refraction, called *Snell's law,* was first worked out in 1621 by W. Snell, a Dutch astronomer and mathematician: $n_1 \sin \theta_1 = n_2 \sin \theta_2$, where n_1 and n_2 are the indices of refraction of the media on either side of the surface, and θ_1 and θ_2 are the respective angles of incidence and refraction. If three of these values are known, the fourth can be calculated from this relationship.

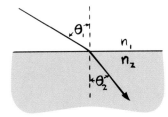

FIGURE 27.26 When light slows down in going from one medium to another, such as from air to water, it bends toward the normal. When it speeds up in traveling from one medium to another, such as from water to air, it bends away from the normal.

downward, making them bend upward. So when the observer looks downward he sees the top of the tree—this is a mirage.

The refraction of light is responsible for many illusions; one of them is the apparent bending of a stick partly immersed in water. The submerged part seems closer to the surface than it really is. Likewise when you view a fish in the water. The fish appears nearer to the surface and closer (Figure 27.27). Because of refraction, submerged objects appear to be magnified. If we look straight down into water, an object submerged 4 meters beneath the surface will appear to be 3 meters deep.

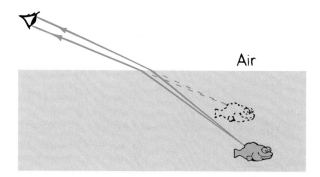

FIGURE 27.27 Because of refraction, a submerged object appears to be nearer the surface than it actually is.

We see that we can interpret the bending of light at the water surface in at least two ways. We can say that the light that leaves the fish and reaches the observer's eye does so in the least time by taking a shorter path upward toward the water surface and a correspondingly longer path through the air. In this view, least time dictates the path taken. Or we can say that the waves of light that happen to be directed upward at an angle toward the water surface are bent off-kilter as they speed up when emerging into the air and that these waves reach the observer's eye. In this view, the change in speed from water to air dictates the path taken, and this path turns out to be a least-time path. Whichever view we choose, the results are the same.

FIGURE 27.28 Because of refraction, the full root beer mug appears to hold more root beer than it actually does.

Question If the speed of light were the same in all media, would refraction still occur when light passes from one medium to another?

Dispersion

We know that the average speed of light is less than c in a transparent medium; how much less depends on the nature of the medium and on the frequency of light. The speed of light in a transparent medium depends on its frequency. Recall from Chapter 25 that light whose frequencies match the natural or resonant frequencies of the electron oscillators in the atoms and molecules of the transparent medium is absorbed, and light with

Answer No.

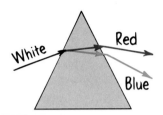

FIGURE 27.29 Dispersion by a prism makes the components of white light visible.

frequencies near the resonant frequencies interacts more often in the absorption/re-emission sequence and therefore travels more slowly. Since the natural or resonant frequency of most transparent materials is in the ultraviolet part of the spectrum, higher-frequency light travels more slowly than lower-frequency light. Violet light travels about 1% slower in ordinary glass than red light. Light waves with colors between red and violet travel at their own intermediate speeds.

So light waves of different frequencies travel at different speeds in transparent materials. Because they travel at different speeds, they refract differently and bend by different amounts. When light is bent twice, as in a prism, the separation of the different colors of light is quite noticeable. This separation of light into colors arranged according to their frequency is called *dispersion* (Figure 27.29). It is what enabled Isaac Newton to form a spectrum when he held a glass prism in sunlight.

Rainbows

A most spectacular illustration of dispersion is the rainbow. For a rainbow to be seen, the sun must be shining in one part of the sky and water drops in a cloud or in falling rain must be present in the opposite part of the sky. When we turn our backs toward the sun, we see the spectrum of colors in a bow. Seen from an airplane near mid-day, the bow forms a complete circle. All rainbows would be completely round if the ground were not in the way.

The beautiful colors of rainbows are dispersed from the sunlight by millions of tiny spherical water droplets that act like prisms. We can better understand this by considering an individual raindrop, as shown in Figure 27.30. Follow the ray of sunlight as it enters the drop near its top surface. Some of the light here is reflected (not shown), and the remainder is refracted into the water. At this first refraction, the light is dispersed into its spectrum colors, violet being deviated the most and red the least. Reaching the opposite side of the drop, each color is partly refracted out into the air (not shown) and partly reflected back into the water. Arriving at the lower surface of the drop, each color is again reflected (not shown) and refracted into the air. This second refraction is similar to that of a prism, where refraction at the second surface increases the dispersion already produced at the first surface.

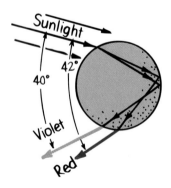

FIGURE 27.30 Dispersion of sunlight by a single raindrop.

Two refractions and a reflection can actually result in the angle between the incoming and outgoing rays being anything between zero and about 42° (0° corresponding to a full 180° reversal of the light). There is a strong concentration of light intensity, however, near the maximum angle of 42°. That is what is shown in Figure 27.30.

Although each drop disperses a full spectrum of colors, an observer is in a position to see the concentrated light of only a single color from any one drop (Figure 27.31). If violet light from a single drop reaches the eye of an observer, red light from the same drop is incident elsewhere toward the feet. To see red light, one must look to a drop higher in the sky. The color red will be seen when the angle between a beam of sunlight and the light sent back by a drop is 42°. The color violet is seen when the angle between the sunbeams and deflected light is 40°.

Why does the light dispersed by the raindrops form a bow? The answer to this involves a little geometric reasoning. First of all, a rainbow is not the flat two-dimensional arc it appears to be. It appears flat for the same reason a spherical burst of

FIGURE 27.31 Sunlight is incident on two sample raindrops as shown and emerges from them as dispersed light. The observer sees the red light from the upper drop and the violet light from the lower drop. Innumerable drops produce the whole spectrum.

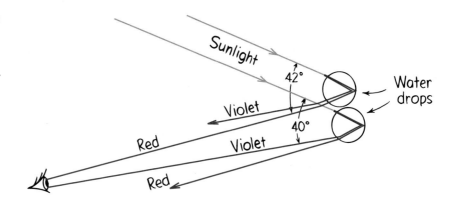

fireworks high in the sky appears as a disc—because of a lack of distance cues. The rainbow you see is actually a three-dimensional cone with the apex at your eye (Figure 27.32). Consider a glass cone, the shape of those paper cones you sometimes see at drinking fountains. If you hold the tip of such a glass cone against your eye, what would you see? You'd see the glass as a circle. Likewise with a rainbow. All the drops that disperse the rainbow's light toward *you* lie in the shape of a cone—a cone of different layers with drops that disperse red to your eye on the outside, orange beneath the red, yellow beneath the orange, and so on all the way to violet on the inner conical surface. The thicker the region containing water drops, the thicker the conical edge you look through, and the more vivid the rainbow.

FIGURE 27.32 When your eye is located between the sun (not shown off to the left) and the water-drop region, the rainbow you see is the edge of a 3-dimensional cone that extends through the water-drop region. Violet is dispersed by drops that form a 40° conical surface; red is seen from drops along a 42° conical surface, with other colors in between. (Innumerable layers of drops form innumerable 2-dimensional arcs like the four suggested here.)

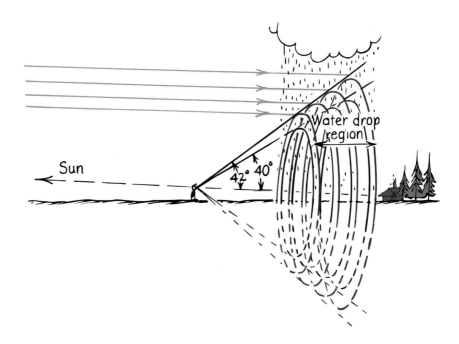

To see this further, consider only the deflection of red light. You see red when the angle between the incident rays of sunlight and dispersed rays make a 42° angle. Of course beams are dispersed 42° from drops all over the sky in all directions—up, down, and sideways. But the only red light *you* see is from drops that lie on a cone with a side-to-axis angle of 42°. Your eye is at the apex of this cone as shown in Figure 27.33. To see violet, you look 40° from the conical axis (so the thickness of glass in the cone of the previous paragraph is tapered—very thin at the tip and thicker with increased distance from the tip).

FIGURE 27.33 Only raindrops along the dashed line disperse red light to the observer at a 42° angle; hence, the light forms a bow.

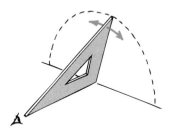

Your cone of vision that intersects the cloud of drops that creates your rainbow is different from that of a person next to you. So when a friend says, "look at the pretty rainbow," you can reply, "Okay, move aside so I can see it too." Everybody sees his or her own personal rainbow.

Another fact about rainbows: A rainbow always faces you squarely, because of the lack of distance cues mentioned earlier. When you move, your rainbow moves with you. So you can never approach the side of a rainbow, or see it end-on as in the exaggerated view of Figure 27.32. You *can't* get to its end. Hence the expression "looking for the pot of gold at the end of the rainbow" means pursuing something you can never reach.

FIGURE 27.34 Light sent back by water droplets at angles less than 40° smears to produce a bright whitish glow inside the rainbow. Dispersed light concentrated between 40° and 42° produces the primary rainbow. The only light sent back at angles greater than 42° is from double reflection in drops, which produces the fainter secondary rainbow.

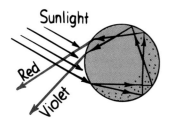

FIGURE 27.35 Double reflection in a drop produces a secondary bow.

Often a larger, secondary bow with colors reversed can be seen arching at a greater angle around the primary bow. We won't treat this secondary bow except to say that it is formed by similar circumstances and is a result of double reflection within the raindrops (Figure 27.35). Because of this extra reflection (and extra refraction loss), the secondary bow is much dimmer and reversed.

Questions

1. If you point to a wall with your arm extended to make about a 42° angle to the wall, then rotate your arm in a full circle while keeping the 42° angle to the wall normal, what shape does your arm describe? What shape on the wall does your finger sweep out?
2. If light traveled at the same speed in raindrops as it does in air, would we still have rainbows?

Total Internal Reflection

FIGURE 27.36 Light emitted in the water is partly refracted and partly reflected at the surface. The length of the arrows indicates the proportions refracted and reflected. Beyond the critical angle the beam is totally internally reflected.

Some Saturday night when you're taking your bath, fill the tub extra deep and bring a waterproof flashlight into the tub with you. Put the bathroom light out. Shine the submerged light straight up and then slowly tip it away from the surface. Note how the intensity of the emerging beam diminishes and how more light is reflected from the water surface to the bottom of the tub. At a certain angle, called the **critical angle**, you'll notice that the beam no longer emerges into the air above the surface. The intensity of the emerging beam reduces to zero where it tends to graze the surface. When the flashlight is tipped beyond the critical angle (48° from the normal for water), you'll notice that all the light is reflected back into the tub. This is *total internal reflection*. The light striking the air-water surface obeys the law of reflection: The angle of incidence is equal to the angle of reflection. The only light emerging from the water surface is that which is diffusely reflected from the bottom of the bathtub. This procedure is shown in Figure 27.36. The proportion of light refracted and light internally reflected is indicated by the relative lengths of the arrows.

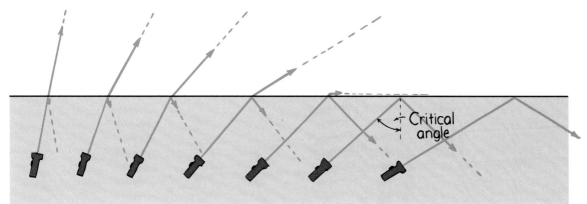

Answers

1. Your arm describes a cone, and your finger sweeps out a circle. Likewise with rainbows.
2. No.

Total internal reflection occurs in materials where the speed of light is less than the speed of light outside. The speed of light is less in water than in air, so all light rays in water that reach the surface at more than an incident angle of 48° are reflected back into the water. So your pet goldfish in the bathtub looks up to see a reflected view of the sides and bottom of the bathtub. Directly above, it sees a compressed view of the outside world (Figure 27.37). The outside 180° view from horizon to opposite horizon is seen through an angle of 96°—twice the critical angle. A lens that similarly compresses a wide view is called a *fisheye lens*.

FIGURE 27.37 An observer underwater sees a circle of light at the still surface. Beyond a cone of 96° (twice the critical angle) an observer sees a reflection of the water interior or bottom.

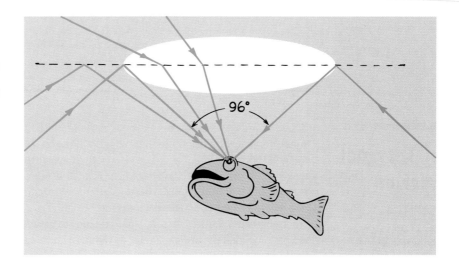

Total internal reflection occurs in glass surrounded by air, for the speed of light in glass is less than in air. The critical angle for glass is about 43°, depending on the type of glass. So light in the glass that is incident at angles greater than 43° to the surface is totally internally reflected. No light escapes beyond this angle; instead, all of it is reflected back into the glass—even if the outside surface is marred by dirt or dust. Hence the usefulness of glass prisms (Figure 27.38). A little light is lost by reflection before it

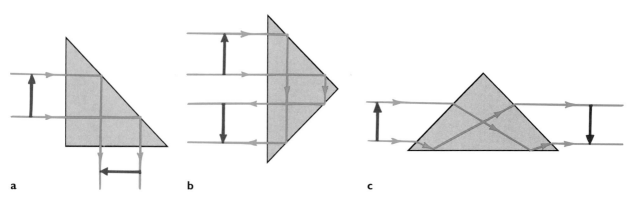

a   b   c

FIGURE 27.38 Total internal reflection in a prism. In (a) the prism changes the direction of the light beam by 90°, in (b) by 180°, and in (c) does not change the direction but instead turns the image upside down.

FIGURE 27.39 Total internal reflection in a pair of prisms.

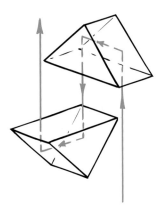

FIGURE 27.40 Prism bincolours.

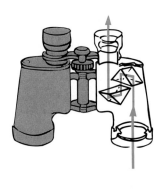

enters the prism, but once inside, reflection from the 45°-slanted face is total—100%. In contrast, silvered or aluminized mirrors reflect only about 90% of incident light. Hence the use of prisms instead of mirrors in many optical instruments.

A pair of prisms each reflecting light through 180° is shown in Figure 27.39. Binoculars use pairs of prisms to lengthen the light path between lenses and thus eliminate the need for long barrels. So a compact set of binoculars is as effective as a longer telescope (Figure 27.40). Another advantage of prisms is that whereas the image of a straight telescope is upside down, reflection by the prisms in binoculars reinverts the image, so things are seen right-side up.

The critical angle for a diamond is about 24.5°, smaller than for any other known substance. The critical angle varies slightly for different colors, because the speed of light varies slightly for different colors. Once light enters a diamond gemstone, most is incident on the sloped backsides at angles greater than 24.5° and is totally internally reflected (Figure 27.41). Because of the great slowdown in speed as light enters a diamond, refraction is pronounced and because of the frequency-dependence of the speed, there is great dispersion. Further dispersion occurs as the light exits through the many facets at its face. Hence we see unexpected flashes of a wide array of colors. Interestingly enough, when these flashes are narrow enough to be seen by only one eye at a time, the diamond "sparkles."

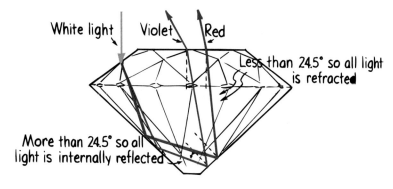

FIGURE 27.41 Paths of light in a diamond. Rays that strike the inner surface of a diamond at angles greater than the critical angle, about 24.5° depending on color, are internally reflected and exit via refraction at the top surface.

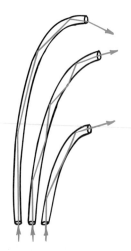

Total internal reflection also underlies the operation of optical fibers, or light pipes (Figure 27.42). An optical fiber "pipes" light from one place to another by a series of total internal reflections, much as a bullet ricochets down a steel pipe. Light rays bounce along the inner walls, following the twists and turns of the fiber. Optical fibers are used in decorative table lamps and to illuminate instrument displays on automobile dashboards from a single bulb. Dentists use them with flashlights to get light where they want it. Bundles of thin flexible glass or plastic fibers are used to see what is going on in inaccessible places, such as the interior of a motor or a patient's stomach. They can be made small enough to snake through blood vessels or through tubes such as the urethra. Light shines down some of the fibers to illuminate the scene and is reflected back along others.

Optical fibers are important in communications. In many places, thin glass fibers have replaced thick, bulky, and expensive copper cables to carry thousands of simultaneous telephone messages between major switching centers. Undersea copper cables are also being replaced by optical fibers. More information can be carried in the high frequencies of visible light than in the lower frequencies of electric current. In many aircraft, control signals are fed from the pilot to the control surfaces by means of optical fibers. Signals are carried in the modulations of laser light. Unlike electricity, light is indifferent to temperature and fluctuations in surrounding magnetic fields, so the signal is clearer. Also, it is much less likely to be tapped by eavesdroppers.

Optical fibers in nature are found on the polar bear—the hairs of its fur! The hairs of a polar bear are actually transparent optical fibers. Its fur appears white because visible light is reflected from the rough inner surface of each hollow hair. However, the hairs trap ultraviolet light. Like light within an optical fiber, the radiant energy is conducted through the hairs to the bear's skin. The skin is very efficient at absorbing all the solar energy it can get, and is black.

FIGURE 27.42 The light is "piped" from below by a succession of total internal reflections until it emerges at the top ends.

FIGURE 27.43 The hairs of the polar bear are transparent light pipes that direct ultraviolet light to its skin—which is guess what color? Black! So what's white and black and warm all over? A polar bear under the Arctic sun.

Lenses

A very practical case of refraction occurs in lenses. We can understand a lens by analyzing equal-time paths as we did earlier, or we can assume that it consists of a set of several matched prisms and blocks of glass arranged in the order shown in Figure 27.44. The prisms and blocks refract incoming parallel light rays so they converge to (or diverge from) a point. The arrangement shown in Figure 27.44a converges the light, and we call such a lens a **converging lens**. Note that it is thicker in the middle.

FIGURE 27.44 A lens may be thought of as a set of prisms.

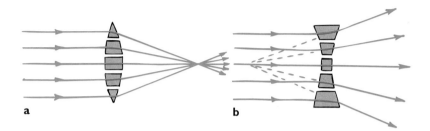

The arrangement in Figure 27.44b is different. The middle is thinner than the edges, and it diverges the light; such a lens is called a **diverging lens**. Note that the prisms diverge the incident rays in a way that makes them appear to come from a single point in front of the lens. In both lenses the greatest deviation of rays occurs at the outermost prisms, for they have the greatest angle between the two refracting surfaces. No deviation occurs exactly in the middle, for in that region the glass faces are parallel to each other. Real lenses are not made of prisms, of course, as is indicated in Figure 27.44; they are made of a solid piece of glass with surfaces ground usually to a circular curve. In Figure 27.45 we see how smooth lenses refract waves.

FIGURE 27.45 Wave fronts travel more slowly in glass than in air. In (a), the waves are retarded more through the center of the lens, and convergence results. In (b), the waves are retarded more at the edges, and divergence results.

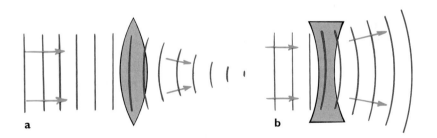

Some key features in lens description are shown for a converging lens in Figure 27.46. The *principal axis* of a lens is the line joining the centers of curvatures of its surfaces. The *focal point* is the point to which a beam of parallel light, parallel to the principal axis, converges. Incident parallel beams that are not parallel to the principal axis focus at points above or below the focal point. All such possible points make up a *focal plane*. Since a lens has two surfaces, it has two focal points and two focal planes.

FIGURE 27.46 Key features of a converging lens.

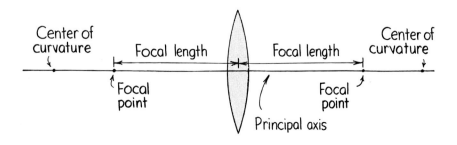

When the lens of a camera is set for distant objects, the film is in the focal plane behind the lens in the camera.

In the diverging lens, an incident beam of light parallel to the principal axis is not converged to a point, but is diverged—so the light appears to come from a point in front of the lens. The *focal length* of a lens, whether converging or diverging, is the distance between the center of the lens and its focal point. For a thin lens, the focal lengths on either side are equal, even when the curvatures on the two sides are different.

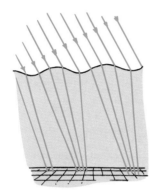 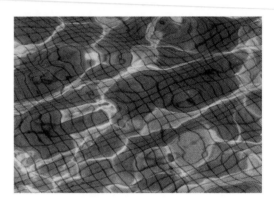

FIGURE 27.47 The moving patterns of bright and dark areas at the bottom of the pool result from the uneven surface of water, which behaves like a blanket of undulating lenses. A fish looking upward at the sun would see the sun shimmering in intensity. Because of similar irregularities in the atmosphere, we see the stars twinkle.

Image Formation by a Lens

At this moment, light is reflecting from your face onto the page of this book. Light that reflects from your forehead, for example, strikes every part of the page. Likewise for the light that reflects from your chin. Every part of the page is illuminated with reflected light from your forehead, your nose, your chin, and every part of your face. You don't see an image of your face on the page because there is too much overlapping of light. But put a barrier with a pinhole in it between your face and the page, and the light that reaches the page from your forehead will not overlap the light from your chin. Likewise for the rest of your face. Without this overlapping, there will be an image of your face on the page. It will be very dim, for very little light reflected from your face gets through the pinhole. To be seen, you'd have to shield the page from other light sources.

FIGURE 27.48 Image formation. (a) No image appears on the wall because rays from all parts of the object overlap all parts of the wall. (b) A single small opening in a barrier prevents overlapping rays from reaching the wall; a dim upside-down image is formed. (c) A lens converges the rays upon the wall without overlapping; more light makes a brighter image.

The first cameras had no lenses and admitted light through a small pinhole. You can see why the image is upside down by the sample rays in Figure 27.48b. Long exposure times were required because of the small amount of light admitted by the pinhole. A somewhat larger hole would admit more light, but overlapping rays would produce a blurry image. Too large a hole would allow too much overlapping and no image would be discernible. That's where a converging lens comes in (Figure 27.48c). The lens converges light onto the screen without the unwanted overlapping of rays. Whereas the first

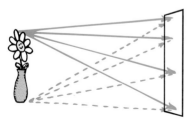

a

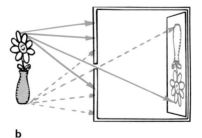

b

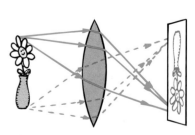

c

pinhole cameras were useful only for still objects because of the long exposure time required, moving objects can be taken with the lens camera because of the short exposure time. Now you know why photographs taken with lens cameras came to be called *snapshots*.

The simplest use of a converging lens is a magnifying glass. Magnification occurs when an image is observed through a wider angle with the use of a lens than without the lens. With unaided vision, an object far away is seen through a relatively small angle of view, while the same object when closer is seen through a larger angle of view. This wider angle permits the perception of greater detail (Figure 27.49). A magnifying glass is simply a converging lens that can increase the angle of view.

FIGURE 27.49

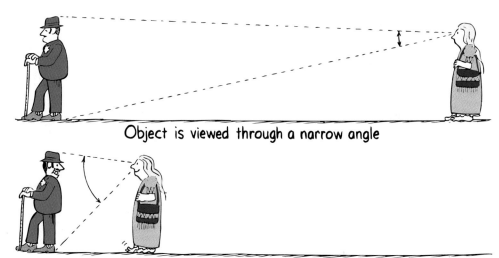

Object is viewed through a narrow angle

Object is viewed through a wide angle

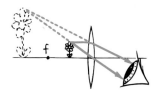

FIGURE 27.50 When an object is near a converging lens (inside its focal point *f*), the lens acts as a magnifying glass to produce a virtual image. The image appears larger and farther from the lens than the object.

When we use a magnifying glass, we hold it close to the object we wish to examine. This is because a converging lens provides a larger right-side-up image only when the object is inside the focal point (Figure 27.50) The larger image will be farther from the lens than the object. If a screen is placed at the image distance, no image will appear on it. This is because no light is actually directed to the image position. The rays that reach our eye, however, behave as *if* they came from the image position, so we call this a **virtual image**.

When the object is far away enough to be outside the focal point of a converging lens, instead of a virtual image a **real image** is formed. Figure 27.51 shows a case in

FIGURE 27.51 When an object is far from a converging lens (beyond its focal point), a real upside-down image is formed.

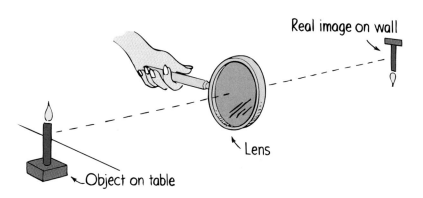

Real image on wall

Lens

Object on table

which a converging lens forms a real image on a screen. A real image is inverted. A similar arrangement is used for projecting slides and motion pictures on a screen and for projecting a real image on the film of a camera. Real images with a single lens will always be inverted.

A diverging lens, when used alone, produces only a diminished virtual image. It makes no difference how far or how near the object is. When a diverging lens is used alone the image is always virtual, erect, and smaller than the object. A diverging lens is often used as a "finder" on a camera. When you look at the object to be photographed through such a lens, you see a virtual image that approximates the same proportions as the photograph.

FIGURE 27.52 A diverging lens forms a virtual, right-side-up image of Jamie and his cat.

Question Why is the greater part of the photograph in Figure 27.52 out of focus?

Answer Both Jamie and his cat and the virtual image of Jamie and his cat are "objects" for the lens of the camera that took this photograph. Since the objects are at different distances from the camera lens, their respective images are at different distances with respect to the film in the camera. So only one can be brought into focus. The same is true of your eyes. You cannot focus on near and far objects at the same time.

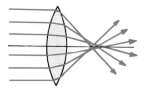

FIGURE 27.53 Spherical aberration.

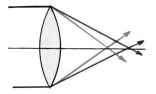

FIGURE 27.54 Chromatic aberration.

Lens Defects

No lens provides a perfect image. A distortion in an image is called an **aberration**. By combining lenses in certain ways, aberrations can be minimized. For this reason, most optical instruments use compound lenses, each consisting of several simple lenses, instead of single lenses.

Spherical aberration results from light that passes through the edges of a lens focusing at a slightly different place from where light passing through the center of the lens focuses (Figure 27.53). This can be remedied by covering the edges of a lens, as with a diaphragm in a camera. Spherical aberration is corrected in good optical instruments by a combination of lenses.

Chromatic aberration is the result of light of different colors having different speeds and hence different refractions in the lens (Figure 27.54). In a simple lens (as in a prism), red light and blue light do not come to focus in the same place. *Achromatic lenses,* which combine simple lenses of different kinds of glass, correct this defect.

The pupil of the eye regulates the amount of light that enters the eye by changing its size. Vision is sharpest when the pupil is smallest because light then passes through only the center of the eye's lens, where spherical and chromatic aberrations are minimal. Also, the eye then acts more like a pinhole camera, so minimum focusing is required for a sharp image. You see better in bright light because your pupils are smaller.*

Astigmatism of the eye is a defect that results when the cornea is curved more in one direction than the other, somewhat like the side of a barrel. Because of this defect, the eye does not form sharp images. The remedy is eyeglasses with cylindrical lenses that have more curvature in one direction than in another.

Questions

1. If light traveled at the same speed in glass as in air, would glass lenses still alter the direction of light rays?
2. Why is there chromatic aberration in light that passes through a lens, but no chromatic aberration in light that reflects from a mirror?

Answers

1. No.
2. Light of different frequencies travels at different speeds in a transparent medium and therefore refracts at different angles, which produces chromatic aberration. But the angles at which light reflects has nothing to do with the frequency of light. Light of one color reflects the same as any other color. Mirrors are therefore preferable to lenses in telescopes because with reflection there is no chromatic aberration.

---

* If you wear glasses and happen to misplace them, or if you find it difficult to read small print as in a telephone book, squint, or even better, hold a pinhole (in a piece of paper or whatever) in front of your eye, close to the page. You'll see the print clearly, and because you're close, it will seem magnified. Try it and see!

An option for those with poor sight today is wearing eyeglasses. The advent of eyeglasses probably occurred in Italy in the late 1200s. (Curiously enough, the telescope wasn't invented until some 300 years later. If, in the meantime, anybody viewed objects through a pair of lenses separated along their axes, such as fixed at the ends of a tube, there is no record of it.) In recent times a present-day option to wearing eyeglasses is contact lenses. Now there is an alternative to both eyeglasses and contacts for people with poor eyesight. Experimental and controversial techniques now allow eye surgeons to shave the cornea of the eye to a proper shape for normal vision. In tomorrow's world, the wearing of eyeglasses and contact lenses may be a thing of the past. We really do live in a rapidly changing world. And that can be nice.

Summary of Terms

Reflection   The return of light rays from a surface in such a way that the angle at which a given ray is returned is equal to the angle at which it strikes the surface. When the reflecting surface is irregular, light is returned in irregular directions; this is *diffuse reflection*.

Refraction   The bending of an oblique ray of light when it passes from one transparent medium to another. This is caused by a difference in the speed of light in the transparent media. When the change in medium is abrupt (say, from air to water), the bending is abrupt; when the change in medium is gradual (say, from cool air to warm air), the bending is gradual, which accounts for mirages.

Fermat's principle of least time   Light takes the path that requires the least time when it goes from one place to another.

Law of reflection   The angle of incidence equals the angle of reflection. The incident and reflected rays lie in a plane that is normal to the reflecting surface.

Critical angle   The minimum angle of incidence inside a medium at which a light ray is totally reflected.

Total internal reflection   The total reflection of light when it strikes a boundary with a less dense medium at an angle greater than the critical angle.

Converging lens   A lens that is thicker in the middle than at the edges and refracts parallel rays of light passing through it to a focus.

Diverging lens   A lens that is thinner in the middle than at the edges, causing parallel rays of light passing through it to diverge as if from a point.

Virtual image   An image formed by light rays that do not converge at the location of the image. Mirrors, converging lenses used as magnifying glasses, and diverging lenses all produce virtual images.

Real image   An image formed by light rays that converge at the location of the image. A real image, unlike a virtual image, can be displayed on a screen.

Aberration   Distortion in an image produced by a lens, which to some degree is present in all optical systems.

Suggested Reading

Greenler, R. *Rainbows, Halos, and Glories*. New York: Cambridge University Press, 1980.

Review Questions

1. Distinguish between *reflection* and *refraction*.

Reflection

2. What does incident light do to the electron clouds that surround the atomic nucleus?
3. What do electron clouds do when they are made to oscillate with greater energy?

Principle of Least Time

4. What is Fermat's principle of least time?

Law of Reflection

5. What is the law of reflection?

Plane Mirrors

6. Relative to the distance of an object in front of a plane mirror, how far behind the mirror is the image?
7. Does the law of reflection hold for curved mirrors? Explain.
8. What fraction of the light shining straight at a piece of clear glass is reflected from the first surface?

Diffuse Reflection

9. Does the law of reflection hold for diffuse reflection? Explain.
10. How can a surface be polished for some waves and not others?

Refraction

11. Light bends when it passes obliquely from one medium to another. If it didn't bend, it would take more time to get from a point in one medium to a point in the other medium. Explain.

12. How does the angle at which light strikes a pane of window glass compare to the angle at which it passes out the other side?

13. How does the path at which a ray of light strikes a prism compare to the path after emergence from the other side?

14. Does light travel faster in thin air or in dense air? What does this have to do with the duration of daylight?

15. What is a mirage?

16. Why do stars twinkle?

17. Does light "know" which path takes the least time from one place to another, or does the path it follows have some other explanation?

Cause of Refraction

18. What is the cause of refraction?

19. When a cart wheel rolls from a smooth sidewalk onto a plot of grass, the interaction of the wheel with the blades of grass slows the wheel. What slows light when it passes from air into glass or water?

20. When light passes from one material into another, it may bend toward the normal to the surface. Or it may bend away from the normal. When does it do which?

21. Does refraction make a swimming pool seem deeper or shallower?

Dispersion

22. What happens to light of a certain frequency when it is incident upon a material whose natural frequency is the same as the frequency of the light?

23. Which travels more slowly in glass, red light or violet light?

24. If light of different frequencies has different speeds in a material, does it also refract at different angles as it enters the same material? Explain.

Rainbows

25. What is it that prevents all rainbows from being complete circles?

26. Does a single raindrop illuminated by sunlight deflect light of a single color or does it disperse a spectrum of colors?

27. Does a viewer see a single color or a spectrum of colors coming from a single faraway drop?

28. Why is a secondary rainbow dimmer than a primary bow?

Total Internal Reflection

29. What is meant by *critical angle*?

30. When is light totally reflected in glass?

31. When is light totally reflected in a diamond?

32. What are two advantages of using prisms in binoculars?

33. Light normally travels in straight lines, but it "bends" in an optical fiber. Explain.

Lenses

34. Distinguish between a *converging lens* and a *diverging lens*.

35. What is the *focal length* of a lens?

Image Formation by a Lens

36. Distinguish between a *virtual image* and a *real image*.

37. What kind of lens can be used to produce a real image? A virtual image?

Lens Defects

38. Distinguish between *spherical aberration* and *chromatic aberration*.

39. Why is vision sharpest when the pupils of the eye are very small?

40. What is astigmatism, and how can it be corrected?

Projects

1. Make a pinhole camera, as illustrated below. Cut out one end of a small cardboard box, and cover the end with tissue or wax paper. Make a clean-cut pinhole at the other end. (If the cardboard is thick, make it through a piece of tinfoil placed over an opening in the cardboard.) Aim the camera at a bright object in a darkened room, and you will see an upside-down image on the tissue paper. If in a dark room you replace the tissue paper with unexposed photographic film, cover the back so it is light tight, and cover the pinhole with a removable flap, you are ready to take a picture. Exposure times differ depending principally on the kind of film and amount of light. Try different exposure times, starting with about 3 seconds. Also try boxes of various lengths. You'll find everything in focus in your photographs, but the pictures will not have clear-cut, sharp outlines. A great difference between your pinhole camera and a commercial one is the glass lens, which is larger than the pinhole and therefore admits more light in less time. It is because a lens camera is so fast that the pictures it takes are called *snapshots*.

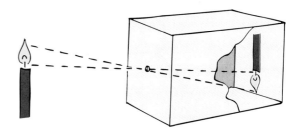

2. If you don't have a prism, you can produce a spectrum by placing a tray of water in bright sunlight. Lean a pocket mirror against the inside edge of the pan and adjust it until a spectrum appears on the wall or ceiling.

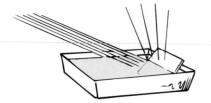

3. Stand a pair of mirrors on edge with the faces parallel to each other. Place an object such as a coin between the mirrors and look at the reflections in each mirror. Impressive?

4. Set up two pocket mirrors at right angles and place a coin between them. You'll see four coins. Change the angle of the mirrors and see how many images of the coin you can see. With the mirrors at right angles, look at your face. Then wink. What do you see? You now see yourself as others see you. Hold a printed page up to the double mirrors and contrast its appearance with the reflection of a single mirror.

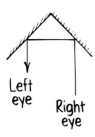

Left eye   Right eye

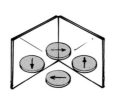

5. Determine the magnifying power of a lens by focusing on the lines of a ruled piece of paper. Count the spaces between the lines that fit into one magnified space, and you have the magnifying power of the lens. You can do the same with binoculars and a distant brick wall. Hold

Magnified space

3 spaces fit into one magnified space

the binoculars so that only one eye looks at the bricks through the eyepiece while the other eye looks directly at the bricks. The number of bricks seen with the unaided eye that will fit into one magnified brick gives the magnification of the instrument.

6. Look at the reflections of overhead lights from the two surfaces of eyeglasses, and you will see two fascinatingly different images. Why are they different?

Exercises

1. Fermat's principle is of least time rather than of least distance. Would least distance apply for reflection? For refraction? Why are your answers different?

2. Her eye at point P looks into the mirror. Which of the numbered cards can she see reflected in the mirror?

Mirror

3. Cowboy Joe wishes to shoot his assailant by ricocheting a bullet off a mirrored metal plate. To do so, should he simply aim at the mirrored image of his assailant? Explain.

4. Trucks often have signs on the back that say, "If you can't see my mirrors, I can't see you." Explain the physics here.

5. Why is the lettering on the front of some vehicles "backward"?

AMBULANCE

6. Is it possible to design a rearview mirror for the inside of an automobile that would show a correct view of front license plates of cars behind? Would such a mirror be practical or impractical?

7. How is the image you see of yourself in a mirror different from what someone looking straight at you from in front sees? Defend your answer.

8. When you look at yourself in the mirror and wave your right hand, your beautiful image waves the left hand. Then why don't the feet of your image wiggle when you shake your head?

9. Car mirrors are uncoated on the front surface and silvered on the back surface. When the mirror is properly adjusted, light from behind reflects from the silvered surface into the driver's eyes. Good. But this is not so good at night with the glare of headlights from behind. This problem is solved by the wedge shape of the mirror (see sketch). When the mirror is tilted slightly upward to the "nighttime" position, glare is directed upward toward the ceiling, away from the driver's eyes. Yet the driver can still see cars behind in the mirror. Explain.

10. To reduce the glare of the surroundings, the windows of some department stores slant inward at the bottom, rather than being vertical. How does this reduce glare?

11. A person in a dark room looking through a window can clearly see a person outside in the daylight, whereas the person outside cannot see the person inside. Explain.

12. Which kind of road surface is easier to see when driving at night, a pebbled uneven surface or a mirror-smooth surface? Explain.

13. Why is it difficult to see the roadway in front of you when driving on a rainy night?

14. What must be the minimum length of a plane mirror in order for you to see a full view of yourself?

15. What effect does your distance from the plane mirror have in the above answer? (Try it and see!)

16. Hold a pocket mirror at almost arm's length from your face and note the amount of your face you can see. To see more of your face, should you hold the mirror closer or farther, or would you have to have a larger mirror? (Try it and see!)

17. The diagram shows a person and her twin at equal distances on opposite sides of a thin wall. Suppose a window is to be cut in the wall so each twin can see a complete view of the other. Show the size and location of the smallest window that can be cut in the wall to do the job. (*Hint*: Draw rays from the top of each twin's head to the other twin's eyes. Do the same from the feet of each to the eyes of the other.)

18. Does the reflection of a scene in calm water look the same as the scene itself only upside down? (*Hint*: Can you ever see the reflection of the top of a stone that extends above water?)

19. Why does reflected light from the sun or moon appear as a column in the body of water as shown? How would it appear if the water surface were perfectly smooth?

20. What is wrong with the cartoon of the man looking at himself in the mirror? (Have a friend face a mirror as shown, and you'll see.)

21. Show with a simple diagram that when a mirror with a fixed beam incident on it is rotated through a certain angle, the reflected beam is rotated through an angle twice as large. (This doubling of displacement makes irregularities in ordinary window glass more evident.)

22. A pair of toy-cart wheels are rolled obliquely from a smooth surface onto two plots of grass, a rectangular plot as shown in *a* and a triangular plot as shown in *b*. The ground is on a slight incline so that after slowing down in the grass, the wheels will speed up again when emerging on the smooth surface. Finish each sketch by showing some positions of the wheels inside the plots and on the other sides, thereby indicating the direction of travel.

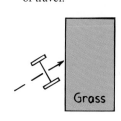
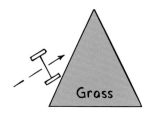

23. A pulse of red light and a pulse of blue light enter a glass block normal to its surface at the same time. After passing through the block, which pulse exits first?

24. Place a glass test tube in water and you can see the tube. Place it in transparent liquid benzene, and you can't see it. What does this tell you about the speed of light in the benzene and in the glass?

25. A beam of light bends as shown in a, while the edges of the immersed square bend as shown in b. Do these pictures contradict each other? Explain.

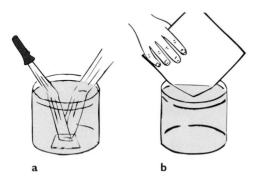

a   b

26. If you were spearing a fish, would you aim above, below, or directly at the observed fish to make a direct hit? If instead you zapped the fish with a laser, would you aim above, below, or directly at the observed fish? Defend your answers.

27. If the fish in the previous exercise were small and blue, and your laser light were red, what corrections should you make? Explain.

28. When a fish looks upward at an angle of 45°, does it see the sky or the reflection of the bottom? Defend your answer.

29. If you were to send a beam of laser light to a space station above the atmosphere and just above the horizon, would you aim the laser above, below, or at the visible space station? Defend your answer.

30. What accounts for the large shadows cast by the ends of the thin legs of the water strider? What accounts for the ring of bright light around the shadows on the bottom?

31. Transparent plastic swimming-pool covers called *solar heat sheets* have thousands of small lenses made of air-filled bubbles. The lenses in these sheets are advertised to focus heat from the sun into the water and raise its temperature. Do you think the lenses of such sheets do in fact direct more solar energy into the water? Defend your answer.

32. Would the average intensity of sunlight measured by a light meter at the bottom of the pool in Figure 27.47 be different if the water were still?

33. When you stand with your back to the sun, you see a rainbow as a circular arc. Could you move off to one side and then see the rainbow as the segment of an ellipse rather than the segment of a circle (such as Figure 27.32 suggests)? Defend your answer.

34. Two observers standing apart from one another do not see the "same" rainbow. Explain.

35. A rainbow viewed from an airplane may form a complete circle. Where will the shadow of the airplane appear? Explain.

36. How is a rainbow similar to the halo sometimes seen around the moon on a frosty night? How are rainbows and halos different?

37. What is responsible for the rainbow-colored fringe commonly seen at the edges of a spot of white light from the beam of a lantern or slide projector?

38. Rays of light in water that shine up to the water-air boundary at angles of more than 48° to the normal are totally reflected. No rays beyond 48° refract outside. How about the other way around? Is there an angle at which light rays in air meeting the air-water boundary will totally reflect? Or will some light be refracted at all angles?

39. Why will goggles allow a swimmer under water to focus more clearly on what he is looking at?

40. If a fish wore goggles above the water surface, why would vision be better for the fish if the goggles were filled with water? Explain.

41. Cover the top half of a camera lens. What effect does this have on the pictures taken?

42. Would refracting telescopes and microscopes magnify if light had the same speed in glass as it does in air? Explain.

43. Consider a simple magnifying glass under water. Will it magnify more or less? Explain why.

44. What would be the effect of a pinhole camera (Figure 27.48b and Project 1) that has two pinholes instead of one? Multiple holes?

45. What condition must exist for a converging lens to produce a virtual image? Can a diverging lens ever produce a real image?

46. Can you take a photograph of your image in a plane mirror and focus the camera on both your image and the mirror frame? Explain.

47. In terms of focal length, how far behind the camera lens is the film located when very distant objects are being photographed?

48. Why do you have to put slides into a slide projector upside down?

49. Maps of the moon are upside down. Why?

50. Why do older people who do not wear glasses read books farther away from their eyes than do younger people?

Problems

1. If you take a photograph of your image in a plane mirror, how many meters away should you set your focus if you are 2 m in front of the mirror?

2. Suppose you walk toward a mirror at 2 m/s. How fast do you and your image approach each other? (The answer is *not* 2 m/s.)

3. When light strikes glass perpendicularly, about 4% is reflected at each surface. How much light is transmitted through a pane of window glass?

4. No glass is perfectly transparent. Mainly because of reflections, about 92% of light passes through an average sheet of clear windowpane. The 8% loss is not noticed through a single sheet, but through several sheets it is apparent. How much light is transmitted by two sheets?

5. The diameter of the sun makes an angle of 0.53° from earth. How many minutes does it take for the sun to move one solar diameter in an overhead sky? (Remember that it takes 24 hours, or 1440 minutes, for the sun to move through 360°.) How does your answer compare with the time it takes for the sun to disappear once its lower edge meets the horizon at sunset? (Does refraction affect your answer?)

6. Consider two ways that light might hypothetically get from its starting point S to its final point F by way of a mirror: by reflecting from point A or by reflecting from point B. Since light travels at a fixed speed in air, the path of least time will also be the path of least distance. Show by calculation that the path SBF is shorter than the path SAF. How does this result tend to support the principle of least time?

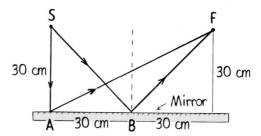

28

Light Waves

Interference colors produced by doubly-
reflected light from the top and bottom
surfaces of a thin film of gasoline on water.

Throw a rock in a quiet pool, and waves appear along the surface of the water. Strike a tuning fork, and waves of sound spread in all directions. Light a match, and waves of light similarly expand in all directions—at the enormously high speed of 300,000 kilometers per second. In this chapter we'll study the wave nature of light. In the next chapter we'll see that light has a particle nature also. Here we'll investigate some of the wave properties of light: diffraction, interference, and polarization.

FIGURE 28.1 Water waves.

Huygens' Principle

Although Galileo is credited as being the first to design a pendulum device to operate a system of toothed wheels, it was a Dutchman, Christiaan Huygens, who made the first pendulum clock. Huygens is most remembered, however, for his idea about waves.* The wave crests shown in Figure 28.1 form concentric circles—called *wave fronts*. Huygens proposed that wave fronts of light waves spreading out from a point source can be regarded as overlapped crests of tiny secondary waves (Figure 28.2). In short, wave fronts are made up of tinier wave fronts. This idea is called **Huygens' principle**.

* In 1665, 20 years before Huygens made his hypothesis about wave fronts, the English physicist Robert Hooke first proposed a wave theory of light.

516

FIGURE 28.2 These drawings are from Huygens' book *Treatise on Light*. Light from A expands in wave fronts, every point of which behaves as if it were a new source of waves. Secondary wavelets starting at *b,b,b,b*, form a new wave front (*d,d,d,d*); secondary wavelets starting at *d,d,d,d* form still another new wave front (DCEF).

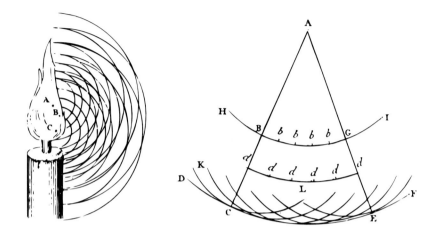

Consider the spherical wave front in Figure 28.3. We can see that if all points along the wave front AA′ are sources of new wavelets, a short time later the new over-lapping wavelets will form a new surface, BB′, which can be regarded as the envelope of all the wavelets. In the figure we show only a few of the infinite number of wavelets

FIGURE 28.3 Huygens' principle applied to a spherical wave front.

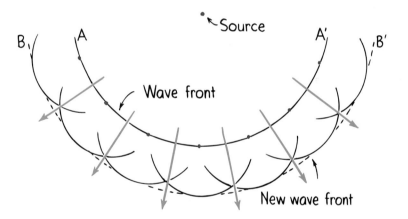

from a few secondary point sources along AA′ that combine to produce the smooth envelope BB′. As the wave spreads, a segment appears less curved. Very far from the original source, the waves nearly form a plane—as do waves from the sun, for example. A Huygens wavelet construction for plane wave fronts is shown in Figure 28.4. We see the laws of reflection and refraction illustrated via Huygens' principle in Figure 28.5.

FIGURE 28.4 Huygens' principle applied to a plane wave front.

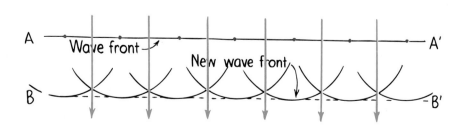

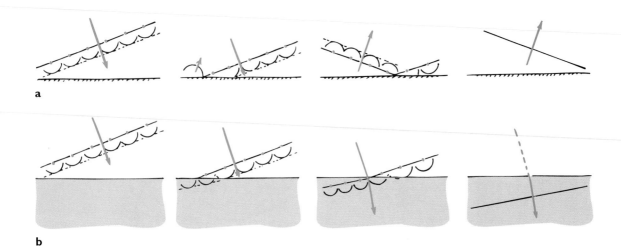

a

b

FIGURE 28.5 Huygens' principle applied to (a) reflection and (b) refraction.

Plane waves can be generated in water by successively dipping a horizontally held straightedge such as a meter stick into the surface (Figure 28.6). The photographs in Figure 28.7 are top views of a ripple tank, where plane waves are produced in water by an oscillating straightedge. The straightedge is not shown, but the plane waves produced are shown incident upon openings of various sizes. In Figure 28.7a, where the opening is wide, we see the plane waves continue through the opening without change—except at the corners, where the waves are bent into the shadow region, as predicted by Huygens' principle. As the width of the opening is narrowed as in Figure 28.7b, less and less of the incident wave is transmitted, and the spreading of waves into the shadow region becomes more pronounced. When the opening is small compared with the wavelength of the incident wave as in Figure 28.7c, the truth of Huygens' idea that every part of a wave front can be regarded as a source of new wavelets becomes quite apparent. As the waves are incident upon the narrow opening, the water sloshing up and down in the opening is easily seen to act as a "point" source of the new waves that fan out on the other side of the barrier. We say the waves are *diffracted* as they spread into the shadow region.

FIGURE 28.6 The oscillating meter stick makes plane waves in the tank of water. Water oscillating in the opening acts as a source of waves that fan out on the other side of the barrier.

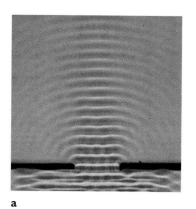
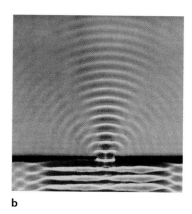
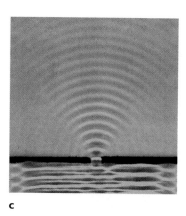

a   b   c

FIGURE 28.7   Plane waves passing through openings of various sizes. The smaller the opening, the greater the bending of the waves at the edges.

Diffraction

FIGURE 28.8   (a) Light casts a sharp shadow with some fuzziness at its edges when the opening is large compared with the wavelength of the light. (b) When the opening is very narrow, diffraction is more apparent and the shadow is fuzzier.

In the previous chapter we saw that light can be bent from its ordinary straight-line path by reflection and by refraction, and now we see another way light bends. Any bending of light by means other than reflection and refraction is called **diffraction**. The diffraction of plane water waves shown in Figure 28.7 occurs for all kinds of waves, including light waves.

When light passes through an opening that is large compared with the wavelength of light, it casts a shadow such as that shown in Figure 28.8a. We see a rather sharp boundary between the light and dark area of the shadow. But if we pass light through a thin razor slit in a piece of opaque cardboard, we see that the light diffracts (Figure 28.8b). The sharp boundary between the light and dark area disappears, and the light spreads out like a fan to produce a bright area that fades into darkness without sharp edges. The light is diffracted.

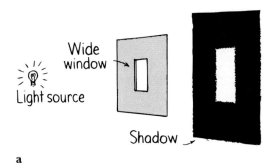
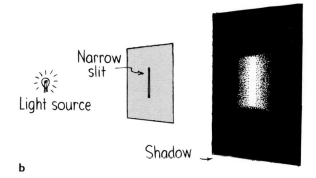

a   b

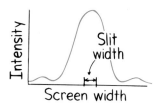

FIGURE 28.9   Graphic interpretation of diffracted light through a single thin slit.

A graph of the intensity distribution for diffracted light through a single thin slit appears in Figure 28.9. Because of diffraction, there is a gradual increase in light intensity rather than an abrupt change from dark to light. A photodetector sweeping across the screen would sense a gradual change from no light to maximum light. (Actually, there are slight fringes of intensity to either side of the main pattern; we will see shortly that these are evidence of interference that is more pronounced with a double slit or multiple slits.)

FIGURE 28.10 Diffraction fringes are evident in the shadows of laser light, which is of a single frequency. These fringes would be filled in by multitudes of other fringes if the source were white light.

Diffraction is not confined to narrow slits or to openings in general but can be seen for all shadows. On close examination, even the sharpest shadow is blurred slightly at the edge. When the light is of a single color (monochromatic), diffraction can produce *diffraction fringes* at the edge of the shadow, as in Figure 28.10. In white light, the fringes merge together to create a fuzzy blur at the edge of a shadow.

The amount of diffraction depends on the wavelength of the wave compared with the size of the obstruction that casts the shadow. Long waves are better at filling in shadows, which is why foghorns emit low-frequency sound waves—to fill in any "blind spots." Likewise for radio waves of the standard AM broadcast band, which are very long compared with the size of most objects in their path. The wavelength of AM radio waves ranges from 180 to 550 meters, and the waves readily bend around buildings and other objects that might otherwise obstruct them. A long-wavelength radio wave doesn't "see" a relatively small building in its path—but a short-wavelength radio wave does. The short radio waves of the FM band range from 2.8 to 3.4 meters and don't bend very well around buildings. This is one of the reasons that FM reception is often poor in localities where AM comes in loud and clear. In the case of radio reception, we don't wish to "see" objects in the path of radio waves, so diffraction is nice.

Diffraction is not so nice when we wish to see a very small object with a microscope. If the object is appreciably larger than the wavelengths of light illuminating it, then the effects of diffraction will be relatively small (the effect is similar to that shown in Figure 28.11c). As the size of the object under view approaches the size of the wavelength of light illuminating it, the details in the object become more and more difficult to see because diffraction blurs the edges (similar to the effect shown in Figure 28.11a). If the object is as small as or smaller than the illuminating wavelengths of light, no detail can be seen at all. No amount of magnification or perfection of microscope design can overcome this fundamental diffraction limit.

FIGURE 28.11 (a) Waves tend to spread into the shadow region. (b) When the wavelength is about the size of the object, the shadow is soon filled in. (c) When the wavelength is short compared with the width of the object, a sharper shadow is cast.

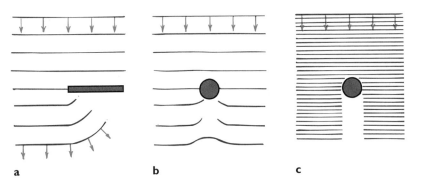

a b c

To circumvent this problem, microscopists illuminate tiny objects with electron beams rather than light. They use electron microscopes, which take advantage of the fact that all matter has wave properties: A beam of electrons has a wavelength smaller than visible light. (We discussed this briefly back in Chapter 10, and will look at this in greater detail in Chapter 30.) So more detail can be seen with an electron microscope than with an optical microscope that is restricted to the longer waves of visible light. In an electron microscope, electric and magnetic fields, rather than optical lenses, are used to focus and magnify images.

The fact that smaller details can be better seen with smaller wavelength waves is cleverly employed by the dolphin in scanning its environment with ultrasound. The echoes of long-wavelength sound give the dolphin an overall image of objects in its surroundings. To examine more detail, the dolphin emits sound of shorter wavelengths. Recall from Chapter 19 that the multiple reflections of very short waves provide the dolphin an acoustical image of the insides of the object scanned. The dolphin has always done naturally what the physician has only recently been able to do with an ultrasonic imaging device.

> **Question** Why does a microscopist use blue light to illuminate the objects viewed?

Interference

Spectacular illustrations of diffraction are shown in Figure 28.12. Physicist Chuck Manka made these by placing pieces of photographic film in the shadow of a screw illuminated with laser light. The fringes appear in both figures. These fringes are produced by **interference**, which we first discussed in Chapter 18. Constructive and

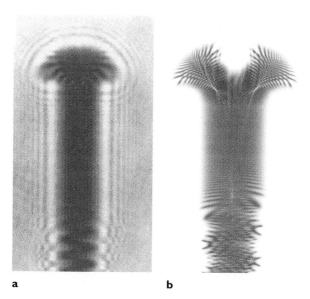

a   b

FIGURE 28.12 (a) The shadow of a screw in laser light shows fringes produced by destructive interference of the diffracted light. (b) A longer exposure shows fringes within the shadow produced by constructive and destructive interference.

> **Answer** For the objects viewed, less diffraction results from the short waves of blue light relative to other, longer waves of other colors of light. So the microscopist sees more detail with short-wave blue light, just as a dolphin beautifully investigates fine detail in its environment by the echoes of ultra-short wavelengths of sound.

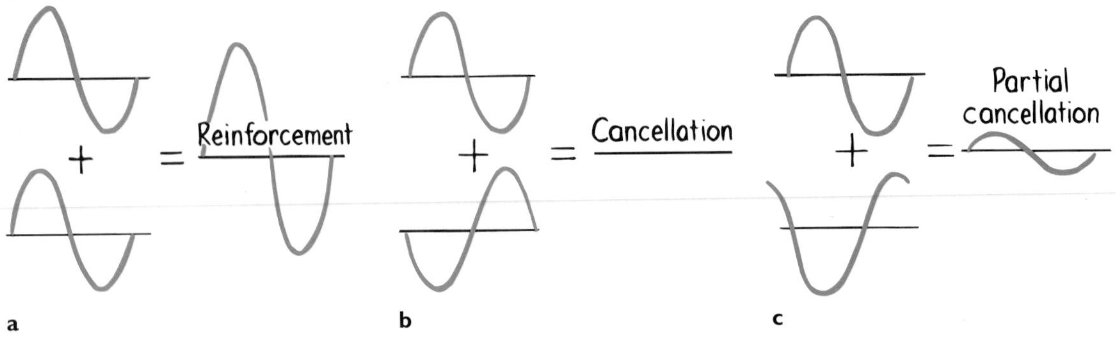

a   Reinforcement

b   Cancellation

c   Partial cancellation

FIGURE 28.13   Wave interference.

destructive interference are reviewed in Figure 28.13. We see that the adding, or superposition, of a pair of identical waves in phase with each other produces a wave of the same frequency but with twice the amplitude. If the waves are exactly one-half wavelength out of phase, their superposition results in complete cancellation. If they are out of phase by other amounts, partial cancellation occurs.

The interference of water waves is a common sight, as shown in Figure 28.14. In some places crests overlap crests, while in other places crests overlap troughs of other waves.

FIGURE 28.14   Interference of water waves.

Under more carefully controlled conditions, interesting patterns are produced when two sources of waves are placed side by side (Figure 28.15). Drops of water are allowed to fall at a controlled frequency into shallow tanks of water (ripple tanks) while their patterns are photographed from above. Note that areas of constructive and destructive interference extend as far as the right-side edges of the ripple tanks, where the number of these regions and their size depend on the distance between the wave

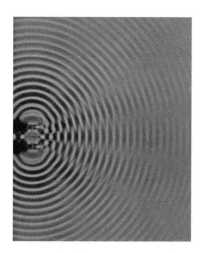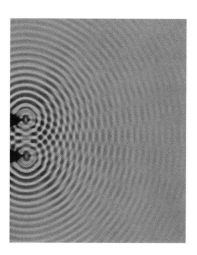

FIGURE 28.15   Interference patterns of overlapping waves from two vibrating sources.

sources and on the wavelength (or frequency) of the waves. Interference is not restricted to easily seen water waves, but is a property of all waves.

In 1801 the wave nature of light was convincingly demonstrated when the British physicist and physician Thomas Young performed his now famous interference experiment.* Young found that light directed through two closely spaced pinholes recombined to produce fringes of brightness and darkness on a screen behind. The bright fringes of light resulted from light waves from the two holes arriving crest to crest, while the dark areas resulted from light waves arriving trough to crest. Figure 28.16 shows Young's

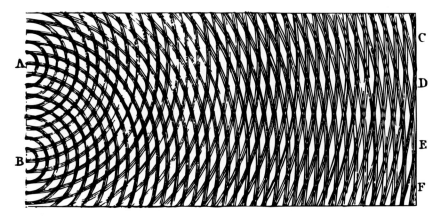

FIGURE 28.16   Thomas Young's original drawing of a two-source interference pattern. The dark circles represent wave crests; the white spaces between the crests represent troughs. Constructive interference occurs where crests overlap crests or troughs overlap troughs. Letters C, D, E, and F mark regions of destructive interference.

* Thomas Young read fluently at the age of 2; by 4, he had read the Bible twice; by 14, he knew eight languages. In his adult life he was a physician and scientist, contributing to an understanding of fluids, work and energy, and the elastic properties of materials. He was the first person to make progress in deciphering Egyptian hieroglyphics. No doubt about it—Thomas Young was a bright guy!

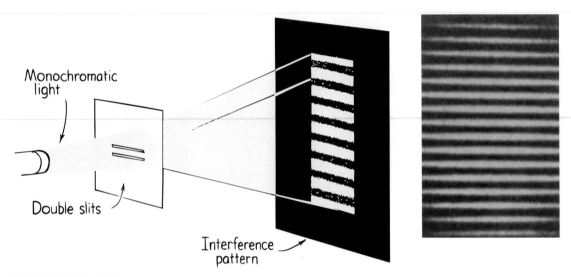

FIGURE 28.17 When monochromatic light passes through two closely spaced slits, a striped interference pattern is produced.

drawing of the pattern of superimposed waves from the two sources. His experiment is now done with two closely spaced slits instead of pinholes, so the fringe patterns are straight lines (Figure 28.17).

FIGURE 28.18 Bright fringes occur when waves from both slits arrive in phase; dark areas result from the overlapping of waves that are out of phase.

We see in Figure 28.19 how the series of bright and dark lines results from the different path lengths from the slits to the screen. For the central bright fringe, the paths from the two slits are the same length, and the waves arrive in phase and reinforce each other. The dark fringes on either side of the central fringe result from one path being longer (or shorter) by one-half wavelength, so the waves arrive half a wavelength out of phase (180° out of phase). The other sets of dark fringes occur where the paths differ by odd multiples of one-half wavelength: $\frac{3}{2}$, $\frac{5}{2}$, and so on.

FIGURE 28.19 Light from O passes through slits M and N and produces an interference pattern on the screen S.

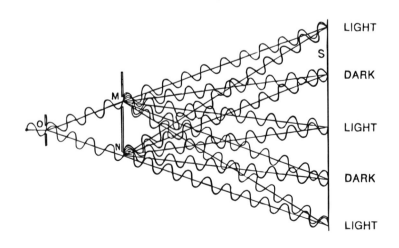

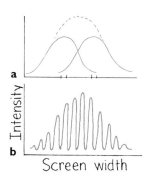

FIGURE 28.20 The light that diffracts through each of the double slits does not form a superposition of intensities as suggested in (a). The intensity pattern, because of interference, is as shown in (b).

In performing this double-slit experiment, suppose we cover one of the slits so that light passes through only a single slit. Then light will fan out and illuminate the screen to form a simple diffraction pattern, as discussed earlier (Figure 28.8b and 28.9). If we cover the other slit and allow light to pass only through the slit just uncovered, we get the same illumination on the screen, only displaced somewhat because of the difference in slit location. If we didn't know better, we might expect that with both slits open, the pattern would simply be the sum of the single-slit diffraction patterns, as suggested in Figure 28.20a. But this doesn't happen. Instead, the pattern formed is one of alternating light and dark bands, as shown in Figure 28.20b. We have an interference pattern. Interference of light waves does not, by the way, create or destroy light energy; it merely redistributes it.

Interference patterns are not limited to single and double slits. A multitude of closely spaced slits makes up a *diffraction grating*. These devices, like prisms, disperse white light into colors. Whereas a prism separates the colors of light by refraction, a diffraction grating separates colors by interference. These are used in devices called *spectroscopes*, which we will discuss in the next chapter, and more commonly in items such as costume jewelry and automobile bumper stickers. These materials are ruled with tiny grooves that diffract light into a brilliant spectrum of colors. This is also seen in the colors dispersed by the feathers of some birds, and in the beautiful colors dispersed by the pits on the reflective surface of audio and video laser discs.

FIGURE 28.21 A diffraction grating disperses light into colors by interference among light beams diffracted by many slits or grooves. It may be used in place of a prism in a spectrometer.

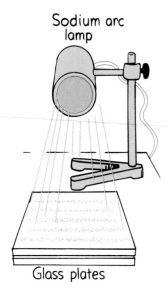

Sodium arc lamp

Glass plates

FIGURE 28.22   Interference fringes produced when monochromatic light is reflected from two plates of glass with an air wedge between them.

Single-Color Thin Film Interference

Another way interference fringes can be produced is by the reflection of light from the top and bottom surfaces of a thin film. A simple demonstration can be set up with a monochromatic light source and a couple of pieces of glass. A sodium-vapor lamp provides a good source of monochromatic light. The two pieces of glass are placed one atop the other, as shown in Figure 28.22. A very thin piece of paper is placed between the plates at one edge. This leaves a very thin wedge-shaped film of air between the plates. If the eye is in a position to see the reflected image of the lamp, the image will not be continuous but will be made up of dark and bright bands.

The cause of these bands is the interference between the waves reflected from the glass on the top and bottom surfaces of the air wedge, as shown in the exaggerated view in Figure 28.23. The light reflecting from point P comes to the eye by two different paths. In one of these paths the light is reflected from the top of the air wedge; in the other path it is reflected from the lower side. If the eye is focused on point P, both rays reach the same place on the retina of the eye. But these rays have traveled different distances and may meet in phase or out of phase, depending on the thickness of the air wedge—that is, on how much farther one ray has traveled than the other. When we look over the whole surface of the glass, we see alternate dark and bright regions—the dark portions where the air thickness is just right to produce destructive interference and the bright portions where the air wedge is just the proper amount thinner or thicker to result in the reinforcement of light. So the dark and bright bands are caused by the interference of light waves reflected from the two sides of the thin film.*

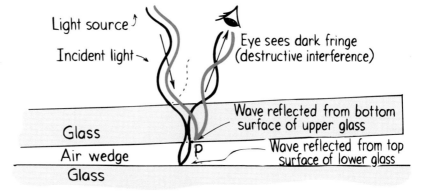

Light source

Incident light

Eye sees dark fringe (destructive interference)

Glass

Wave reflected from bottom surface of upper glass

Air wedge

Wave reflected from top surface of lower glass

Glass

P

FIGURE 28.23   Reflection from the upper and lower surfaces of a "thin film of air."

If the surfaces of the glass plates used are perfectly flat, the bands are uniform. But if the surfaces are not perfectly flat, the bands are distorted. The interference of light provides an extremely sensitive method for testing the flatness of surfaces. Surfaces that produce uniform fringes are said to be optically flat—this means that surface irregularities are small compared with the wavelength of visible light (Figure 28.24).

* Phase shifts at some reflecting surfaces also contribute to interference. For simplicity and brevity, our concern with this topic will be limited to this footnote: In short, when light in a medium is reflected at the surface of a second medium in which the speed of transmitted light is less (when there is a greater index of refraction), there is a 180° phase shift (that is, half a wavelength). However, no phase shift occurs when the second medium is one that transmits light at a higher speed (and has a lower index of refraction). In our air-wedge example, no phase shift occurs for reflection at the upper glass-air surface, and a 180° shift does occur at the lower air-glass surface. So at the apex of the air wedge where the thickness approaches zero, the phase shift produces cancellation, and the wedge is dark. Likewise with a soap film so thin that its thickness is appreciably smaller than the wavelength of light.

FIGURE 28.24 Optical flats used for testing the flatness of surfaces.

When a plano-convex lens is placed on an optically flat plate of glass and illuminated from above with monochromatic light, a series of light and dark rings is produced. This pattern is known as *Newton's rings* (Figure 28.25). These light and dark rings are the same kinds of fringes observed with plane surfaces. This is a useful testing technique in polishing precision lenses.

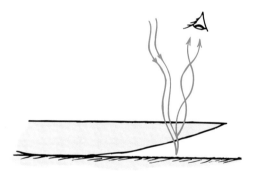

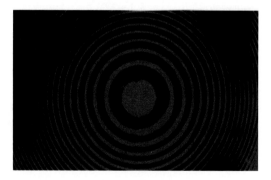

FIGURE 28.25 Newton's rings.

Question How would the spacings between Newton's rings differ when illuminated by red and by blue light?

Interference Colors by Reflection from Thin Films

We have all noticed the beautiful spectrum of colors reflected from a soap bubble or from gasoline on a wet street. We have seen the beautiful colors reflected from some bird feathers that seem to change in hue as the bird moves. These colors are produced by the *interference* of light waves. This phenomenon is often called *iridescence* and is observed in thin films.

Answer The rings would be more widely spaced for longer-wavelength red light than for the shorter waves of blue light. Do you see the geometrical reason for this?

A soap bubble appears iridescent in white light when the thickness of the soap film is about the same as the wavelength of light. Light waves reflected from the outer and inner surfaces of the film travel different distances. When illuminated by white light, the film may be just the right thickness at one place to cause the destructive interference of, say, yellow light. When yellow light is subtracted from white light, the mixture left will appear as the complementary color of yellow, which is blue. At another place where the film is thinner, a different color may be canceled by interference, and the light seen will be its complementary color. The same thing happens to gasoline on a wet street (Figure 28.26). Light reflects from both the upper gasoline surface and the lower gasoline-water surface. If the thickness of the gasoline is such that it cancels blue, as the figure suggests, then the gasoline surface appears yellow to the eye. This is because the blue is subtracted from the white, leaving the complementary color, yellow. The different colors, then, correspond to different thicknesses of the thin film, providing a vivid "contour map" of microscopic differences in surface "elevations."

FIGURE 28.26 The thin film of gasoline is just the right thickness so that blue light reflected from the top surface of the gasoline is canceled by light of the same wavelength reflected from the water.

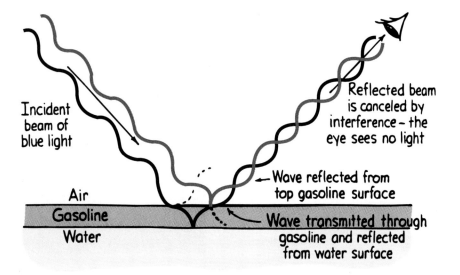

Over a wider field of view, different colors can be seen even if the thickness of the gasoline film is uniform. This has to do with the apparent thickness of the film: Light reaching the eye from different parts of the surface is reflected at different angles and traverses different thicknesses. If the light is incident at a grazing angle, for example, the ray transmitted to the gasoline's lower surface travels a longer distance. Longer waves are canceled in this case, and different colors appear.

Dishes washed in soapy water and poorly rinsed have a thin film of soap on them. Hold such a dish up to a light source so that interference colors can be seen. Then turn the dish to a new position, keeping your eye on the same part of the dish, and the color will change. Light reflecting from the bottom surface of the transparent soap film is canceling light reflecting from the top surface of the soap film. Light waves of different wavelengths are canceled for different angles. You'll notice that the interference colors are predominantly cyan, magenta and yellow, due to the subtraction of primary red, green, and blue.

Interference provides a way to measure the wavelength of light and other electromagnetic radiation. It also makes it possible to measure extremely small distances with great accuracy. Instruments called *interferometers*, which use the principle of interference, are the most accurate instruments known for measuring small distances.

Questions

1. What color will appear to be reflected from a soap bubble in sunlight when its thickness is such that green light is canceled?
2. In the left column are the colors of certain objects. In the right column are various ways in which colors are produced. Match the right column to the left.

(a) yellow daffodil	(1) interference
(b) blue sky	(2) diffraction
(c) rainbow	(3) selective reflection
(d) peacock feathers	(4) refraction
(e) soap bubble	(5) scattering

Polarization

FIGURE 28.27 A vertically polarized plane wave and a horizontally polarized plane wave.

Interference and diffraction provide the best evidence that light is wavelike in nature. Wave motion in general can be either longitudinal or transverse. Sound in air travels in longitudinal waves, where the vibratory motion is *along* the direction of wave propagation. But when we shake a taut rope, the vibratory motion that propagates along the rope is perpendicular, or *transverse*, to the rope. Both longitudinal and transverse waves exhibit interference and diffraction effects. Are light waves, then, longitudinal or transverse? The fact that light waves can be **polarized** demonstrates that they are transverse.

If we shake a taut rope up and down as in Figure 28.27, a transverse wave travels along the rope in a plane. We say that such a wave is *plane-polarized.** If the rope is shaken up and down vertically, the wave is vertically plane-polarized; that is, the waves traveling along the rope are confined to a vertical plane. If we shake the rope from side to side, we produce a horizontally plane-polarized wave.

A single vibrating electron can emit an electromagnetic wave that is plane-polarized. The plane of polarization will match the vibrational direction of the electron.

Answers

1. The composite of all the visible wavelengths except green results in the complementary color, magenta. See Figure 26.9.
2. a-3; b-5; c-4; d-2; e-1.

---

* Light may also be circularly polarized and elliptically polarized, which are also transverse polarizations. But we will not study these cases.

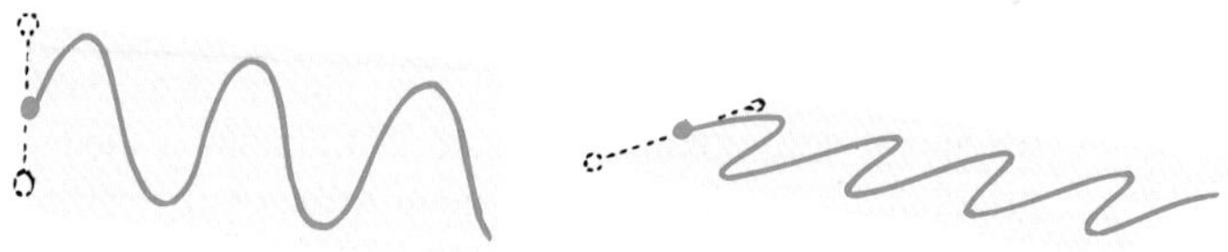

FIGURE 28.28 (a) A vertically polarized plane wave from a charge vibrating vertically. (b) A horizontally polarized plane wave from a charge vibrating horizontally.

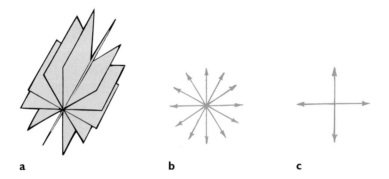

a   b

A vertically accelerating electron, then, emits light that is vertically polarized, while a horizontally accelerating electron emits light that is horizontally polarized (Figure 28.28).

A common light source such as an incandescent lamp, a fluorescent lamp, a candle flame, or an arc lamp emits light that is nonpolarized. This is because there is no preferred vibrational direction for the accelerating electrons producing the light. The number of planes of vibration might be as numerous as the accelerating electrons producing them. A few planes are represented in Figure 28.29a. We can represent all these planes by radial lines (Figure 28.29b) or, more simply, by vectors in two mutually perpendicular directions (Figure 28.29c), as if we had resolved all the vectors of Figure 28.29b into horizontal and vertical components. This simpler schematic representation is a useful shorthand. In this case, the diagram represents nonpolarized light. Polarized light would be represented by a single vector.

FIGURE 28.29
Representations of plane-polarized waves.

a   b   c

All transparent crystals of a noncubic natural shape have the property of transmitting light of one polarization differently from light of another polarization. Certain crystals* not only divide unpolarized light into two internal beams polarized at right angles to each other but also strongly absorb one beam while transmitting the other (Figure 28.30). Tourmaline is one such common crystal, but unfortunately the transmitted light is colored. Herapathite, however, does the job without discoloration. Microscopic crystals of herapathite are embedded between cellulose sheets in uniform alignment and are used in making Polaroid filters. Some Polaroid sheets consist of certain aligned molecules rather than tiny crystals.†

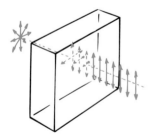

FIGURE 28.30 One component of the incident nonpolarized light is absorbed, resulting in emerging polarized light.

If you look at nonpolarized light through a Polaroid filter, you can rotate the filter in any direction, and the light will appear unchanged. But if this light is polarized, then as you rotate the filter, you can progressively cut off more and more of the light until it is blocked out. An ideal Polaroid will transmit 50 percent of incident nonpolarized light. That 50 percent is, of course, polarized. When two Polaroids are arranged so that their polarization axes are aligned, light will be transmitted through both (Figure 28.31). If

* Called *dichroic*.

† The molecules are polymeric iodine in a sheet of polyvinyl alcohol or polyvinylene.

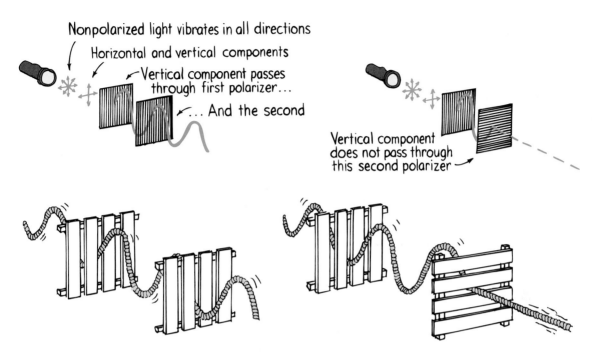

FIGURE 28.31 A rope analogy illustrates the effect of crossed Polaroids.

their axes are at right angles to each other, no light will penetrate the pair. Actually, some light of the shorter wavelengths does get through, but not to any significant degree. When Polaroids are used in pairs like this, the first one is called the *polarizer* and the second one the *analyzer*.

Much of the light reflected from nonmetallic surfaces is polarized. The glare from glass or water is a good example. Except for normal incidence, the reflected ray contains more vibrations parallel to the reflecting surface, while the transmitted beam contains more vibrations at right angles to the surface (Figure 28.33). Skipping flat rocks off the

FIGURE 28.32 Polaroid sunglasses block out horizontally vibrating light. When the lenses overlap at right angles, no light gets through.

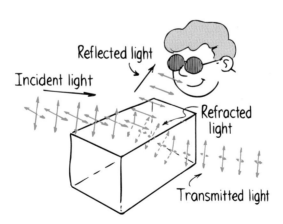

FIGURE 28.33 Most glare from nonmetallic surfaces is polarized. Here we see that the components of incident light that are parallel to the surface are reflected, while the components that are perpendicular to the surface are refracted into the medium. Since most of the glare we encounter is from horizontal surfaces, the polarization axes of Polaroid sunglasses are vertical.

a   b   c

FIGURE 28.34   Light is transmitted when the axes of the Polaroids are aligned (a), but absorbed when Ludmila rotates one so that the axes are at right angles to each other (b). When she inserts a third Polaroid at an angle between the crossed Polaroids, light is again transmitted (c). Why? (For the answer, after you have given this some thought see Appendix D, "More About Vectors.")

surface of a pond is analogous. When the rocks hit parallel to the surface, they easily reflect; but if they hit with their faces at right angles to the surface, they "refract" into the water. The glare from reflecting surfaces can be appreciably diminished with the use of Polaroid sunglasses. The polarization axes of the lenses are vertical, as most glare reflects from horizontal surfaces. Properly aligned Polaroid eyeglasses enable us to see the projection of stereoscopic movies or slides on a flat screen in three dimensions.

Question   Which pair of glasses is best suited for automobile drivers? (The polarization axes are shown by the straight lines.)

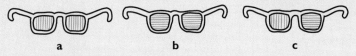

a   b   c

Three-Dimensional Viewing

Vision in three dimensions depends primarily on the fact that both eyes give their impressions simultaneously (or nearly so), each eye viewing the scene from a slightly

Answer   Glasses a are best suited because the vertical axis blocks horizontally polarized light, which composes much of the glare from horizontal surfaces. Glasses c are suited for viewing 3-D movies.

FIGURE 28.35 The crystal structure of ice in stereo. You'll see depth when your brain combines the views of your left eye looking at the left figure and your right eye looking at the right figure. To accomplish this, focus your eyes for distant viewing before looking at this page. Without changing your focus, look at the page, and each figure will appear double. Then adjust your focus so that the two inside images overlap to form a central composite image. Practice makes perfect. (If you instead *cross* your eyes to overlap the figures, near and far are reversed!)

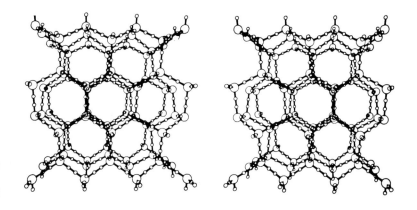

different angle. To convince yourself that each eye sees a different view, hold an upright finger at arm's length and see how it appears to shift position from left to right in front of the background as you alternately close each eye. The drawings of Figure 28.35 illustrate a stereo view of the crystal structure of ice.

FIGURE 28.36 A stereo view of snowflakes. View these in the same way as Figure 28.35.

FIGURE 28.37 With your eyes focused for distant viewing, the second and fourth lines appear to be farther away; if you cross your eyes, the second and fourth lines appear closer.

The test of all knowledge is experiment.
Experiment is the *sole judge* of scientific "truth."
                    Richard P. Feynman

The test of all knowledge is experiment.
Experiment is the *sole judge* of scientific "truth."
                    Richard P. Feynman

FIGURE 28.38 A stereoscopic viewer.

The familiar handheld stereoscopic viewer (Figure 28.38) simulates the effect of depth. In this device, there are two photographic transparencies (or slides) taken from slightly different positions. When they are viewed at the same time, the arrangement is such that the left eye sees the scene as photographed from the left, and the right eye sees it as photographed from the right. As a result, the objects in the scene sink into relief in correct perspective, giving apparent depth to the picture. The device is constructed so that each eye sees only the proper view. There is no chance for one eye to see both views. If you remove the slides from the hand viewer and project each view on a screen by slide projector (so that the views are superimposed), a blurry picture results. This is because each eye sees both views simultaneously. This is where Polaroid filters come in. If you place the Polaroids in front of the projectors so that one is horizontal and the other vertical, and you view the polarized image with polarized glasses of the same orientation, each eye will see the proper view as with the stereoscopic viewer (Figure 28.39). You then will see an image in three dimensions.

FIGURE 28.39 A 3-D slide show using Polaroids. The left eye sees only polarized light from the left projector, the right eye sees only polarized light from the right projector, and both views merge in the brain to produce depth.

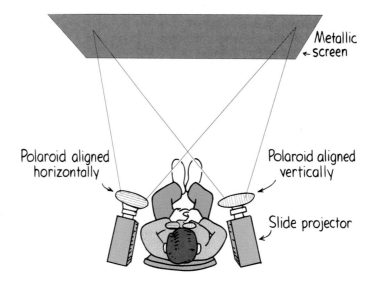

Depth is also seen in computer-generated stereograms, as in Figure 28.40. Here the slightly different patterns are not obvious in a casual view. Use the procedure for viewing the previous stereo figures. Once you've mastered the viewing technique, head for the local mall and check the variety of stereograms in posters and books.

FIGURE 28.40 A computer-generated stereogram.

Holography

Perhaps the most exciting illustration of interference is the **hologram**, a two-dimensional photographic plate illuminated with laser light that allows you to see a faithful reproduction of a scene in three dimensions. The hologram was invented and named by Dennis Gabor in 1947, 10 years before lasers were invented. *Holo* in Greek means "whole," and *gram* in Greek means "message" or "information." A hologram contains the whole message or entire picture. When illuminated with laser light, the image is so realistic that you can actually look around the corners of objects in the image and see the sides.

In ordinary photography, a lens is used to form an image of an object on photographic film. Light reflected from each point on the object is directed by the lens only to a corresponding point on the film. And all the light that reaches the film comes from the object being photographed. In the case of holography, however, no image-forming lens is used. Instead, each point of the object being "photographed" reflects light to the *entire* photographic plate, so every part of the plate is exposed with light reflected from every part of the object. Most importantly, the light used to make a hologram must be of a single frequency and all parts exactly in phase: It must be *coherent*. If, for example, white light were used, the diffraction fringes for one frequency would be washed out by other frequencies. Only a laser can produce such light (we will treat lasers in detail in the next chapter). Holograms are made with laser light.

A conventional photograph is a recording of an image, but a hologram is a recording of the interference pattern resulting from the combination of two sets of wave fronts. One set of wave fronts is from light reflected by the object, and the other set is from a *reference beam*, which is diverted from the illuminating beam and sent directly to the photographic plate (Figure 28.41). The developed photograph has no recognizable image. The hologram is simply a hodgepodge of whirly lines, tiny fringed areas of varying density—dark where wave fronts from the object and reference beam arrived in phase and light where the wave fronts were out of phase. The hologram is a photographic pattern of microscopic interference fringes.

FIGURE 28.41 A simplified arrangement for making a hologram. The laser light that exposes the photographic plate consists of two parts: the reference beam reflected from the mirror and light reflected from the object. The wave fronts of these two parts interfere to produce microscopic fringes on the photographic plate. The exposed and developed plate is then a hologram.

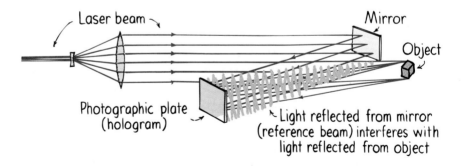

When a hologram is placed in a beam of coherent light, the light is diffracted by the microscopic fringes to produce wave fronts identical in form to the original wave fronts reflected by the object. When viewed with the eye or any other optical instrument, the diffracted wave fronts produce the same effect as the original reflected wave fronts. You look through the hologram and see a full, realistic three-dimensional image as though you were viewing the original object through a window. Parallax is

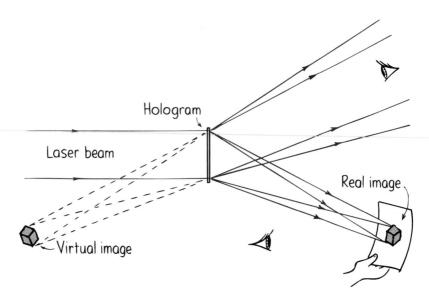

FIGURE 28.42 When laser light is transmitted through the hologram, the divergence of diffracted light produces a three-dimensional image that can be seen when looking *through* the hologram, like looking through a window. This is a *virtual image,* for it appears to be only in back of the hologram, like your virtual image in a mirror. By refocusing your eyes, you can see all parts of the virtual image, near and far, in sharp focus. Converging diffracted light produces a *real image* in front of the hologram that can be projected on a screen. Since the image is three-dimensional, you cannot see the entire image in sharp focus for any single position on a flat screen.

evident when you move your head and see down the sides of the object or when you lower your head and look underneath the object. Holographic pictures are extremely realistic.

Interestingly enough, if the hologram is made on film, you can cut it in half and still see the entire image on each half. And you can cut one of the pieces in half again, and again. This is because every part of the hologram has received and recorded light from the entire object. Similarly, light outside an open window fills the entire window, so you can see an outside view from every part of this open window. Because vast amounts of information are recorded in a tiny area, the film used for holograms must have a resolving power much greater than ordinary fine-grain photographic film. The optical storage of information via holograms is finding wide application in computers.

Even more interesting is holographic magnification. If holograms are made using a short-wavelength light and viewed with a longer-wavelength light, the resulting image is magnified in the same proportion as the wavelengths. Holograms made with X rays would be magnified thousands of times when viewed with visible light and appropriate geometric viewing arrangements. Because holograms require no lenses, the possibilities of an X-ray microscope are particularly attractive.

And as for television, today's two-dimensional screens may appear as quaint to your children as your grandparents' early radio receivers appear to you now.

Light is fascinating—especially when it is diffracted by the interference fringes of a hologram.

Summary of Terms

Huygens' principle   Every point on a wave front can be regarded as a new source of wavelets, which combine to produce the next wave front, whose points are sources of further wavelets, and so on.

Diffraction   The bending of light that passes around an obstacle or through a narrow slit, causing the light to spread and to produce light and dark fringes.

Interference   The result of superposing different waves, usually of the same wavelength. Constructive interference results from crest-to-crest reinforcement; destructive interference results from crest-to-trough cancellation. The interference of selected wavelengths of light produces colors known as *interference colors*.

Polarization   The alignment of the transverse electric vectors that make up electromagnetic radiation. Such waves of aligned vibrations are said to be *polarized*.

Hologram   A two-dimensional microscopic interference pattern that shows three-dimensional optical images.

Suggested Reading

Boleman, J. *Physics—A Window on Our World, Third Edition*. Prentice Hall, Englewood Cliffs, NJ, 1995. This delightful text has many figures in stereo, with optional glasses provided.

Falk, D. S., Brill, D. R., and Stork, D. *Seeing the Light: Optics in Nature*. New York: Harper & Row, 1985.

Review Questions

Huygens' Principle

1. According to Huygens, how does every point on a wave front behave?
2. Will plane waves incident upon a small opening in a barrier fan out on the other side or continue as plane waves?

Diffraction

3. Is diffraction more pronounced through a small opening or through a large opening?
4. For an opening of a given size, is diffraction more pronounced for a larger wavelength or a smaller wavelength?
5. Which is more easily diffracted around buildings, AM or FM radio waves? Why?
6. What are some of the assets and liabilities of diffraction?

Interference

7. Is interference restricted to only some types of waves or does it occur for all types of waves?
8. What exactly did Thomas Young demonstrate in his famous experiment with light?
9. Does the energy of light disappear when light destructively interferes?

Single-Color Thin Film Interference

10. What is monochromatic light?
11. What accounts for the light and dark bands when monochromatic light reflects from a glass pane atop another glass pane?
12. What is the cause of Newton's rings ?
13. Newton's rings show up nicely when viewed with monochromatic light—long wavelength light shows wide-spaced rings and short wavelength light shows close-spaced rings. Why are these rings not seen when viewed with white light?

Interference Colors by Reflection from Thin Films

14. What produces iridescence?
15. What causes the spectrum of colors seen in gasoline splotches on a wet street? Why are these not seen on a dry street?
16. What accounts for the different colors in either a soap bubble or a layer of gasoline on water?
17. If you look at a soap bubble from different angles so you're viewing different apparent thicknesses of soap film, will you see different colors? Explain.

Polarization

18. What phenomenon distinguishes longitudinal waves from transverse waves?
19. Is polarization characteristic of all types of waves?
20. How does the direction of polarization of light compare to the direction of vibration of the electron that produced it?
21. Why will light pass through a pair of Polaroids when the axes are aligned but not when the axes are at right angles to each other?
22. How much ordinary light will an ideal Polaroid transmit?
23. When *ordinary* light is incident at an oblique angle upon water, what can you say about the *reflected* light?
24. What is the advantage of Polaroid sunglasses over regular sunglasses?

Three-Dimensional Viewing

25. Why would depth not be perceived if you viewed duplicates of ordinary slides in a stereo viewer (Figure 28.38), rather than the pairs of slides taken with a stereo camera?

26. What role do polarization filters play in a 3-D slide show?

Holography

27. How does a hologram differ from a conventional photograph?
28. How is *coherent* light different from ordinary light?
29. When coherent light is diffracted onto photographic film, interference fringes are produced. What would happen to these fringes if light of other frequencies were mixed in with the coherent light?
30. Why is coherent light necessary to make a hologram?

Projects

1. With a razor blade, cut a slit in a card and look at a light source through it. You can vary the size of the opening by bending the card slightly. See the interference fringes? Try it with two closely spaced slits.
2. Next time you're in the bathtub, froth up the soapsuds and notice the colors of highlights from the illuminating light overhead on each tiny bubble. Notice that different bubbles reflect different colors, due to the different thicknesses of soap film. If a friend is bathing with you, compare the different colors that you each see reflected from the same bubbles. You'll see that they're different—for what you see depends on your point of view!
3. Do this one at your kitchen sink. Dip a dark-colored coffee cup (dark colors make the best background for viewing interference colors) in dishwashing detergent, and remove it so that a soap film remains across the top. Then hold it sideways and look at the reflected light from the soap film that covers its mouth. Swirling colors appear as the soap runs down to form a wedge that grows thicker at the bottom with time. The top becomes thinner, so thin that it appears black. This tells us that its thickness is less than one-fourth the thickness of the shortest waves of visible light. Whatever its wavelength, light reflecting from the inner surface reverses phase, re-joins light reflecting from the outer surface, and cancels. The film soon becomes so thin it pops.

4. When you're wearing Polaroid sunglasses, view the glare from nonmetallic surfaces such as a road or body of water. Then tip your head from side to side, and see how the glare intensity changes as you vary the magnitude of the electric vector component aligned with the polarization axis of the glasses. Also notice the polarization of different parts of the sky when you hold the sunglasses in your hand and rotate them.
5. Place a bottle of light corn syrup between two sheets of Polaroid. Place a white light source behind the syrup and Polaroids. Then look through the Polaroids and syrup and view spectacular colors as you rotate one of the Polaroids.

6. See spectacular interference colors with a polarized-light microscope. Any microscope, including an inexpensive toy microscope, can be converted into a polarized-light microscope by fitting a piece of Polaroid inside the eyepiece and taping another onto the stage of the microscope. Mix drops of naphthalene and benzene on a slide and watch the growth of crystals. Rotate the eyepiece and change the colors. Or simply observe the interference colors of pieces of cellophane.

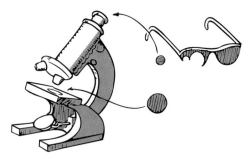

7. Make some slides for a slide projector by sticking some crumpled cellophane onto pieces of slide-sized Polaroid. (Also try strips of cellophane tape overlapped at different angles and experiment with different brands of transparent tape.) Project them onto a large screen or white wall and rotate a second, slightly larger piece of Polaroid in front of the projector lens in rhythm with your favorite music. You'll have your own light show.

Exercises

1. Why can the sunlight that illuminates the earth be approximated by plane waves, whereas the light from a nearby lamp cannot?

2. In our everyday environment, diffraction is much more evident for sound waves than for light waves. Why is this so?

3. Why do *radio waves* diffract around buildings, while *light waves* do not?

4. Why are TV broadcasts in the VHF range easier to receive in areas of marginal reception than broadcasts in the UHF range? (*Hint*: UHF has higher frequencies than VHF.)

5. Can you think of a reason why TV channels of lower numbers might give better pictures in regions of poor TV reception? (*Hint*: Lower channel numbers represent lower carrier frequencies.)

6. Two loudspeakers a meter or so apart emit pure tones of the same frequency and loudness. When a listener walks past in a path parallel to the line that joins the loudspeakers, the sound is heard to alternate from loud to soft. What is going on?

7. In the preceding exercise, suggest a path along which the listener could walk so as not to hear alternate loud and soft sounds.

8. How are interference fringes of light analogous to the varying intensity that you hear as you walk past a pair of speakers giving out the same sound?

9. Light illuminates two closely spaced thin slits and produces an interference pattern on a screen behind. How will the distance between the fringes of the pattern differ for red light and blue light?

10. A prism bends blue light more than red light. Is the same true of a diffraction grating?

11. When white light diffracts upon passing through a thin slit, as in Figure 28.8b, different color components diffract by different amounts, so that a rainbow of colors appears. Which color is diffracted through the greatest angle? Which color is diffracted through the smallest angle?

12. Suppose you place a diffraction grating in front of a camera lens and take a picture of illuminated streetlights. What will you expect to see in your photograph?

13. What happens to the distance between interference fringes if the two slits in Exercise 9 are moved farther apart?

14. Why is Thomas Young's experiment more effective with slits than with the pinholes he first used?

15. In performing Young's interference experiment, you should pass monochromatic light through a single narrow opening before it reaches the double openings. Explain why this makes the fringes clear. (*Hint*: What would be the result if light reaching the double openings came from several different directions?)

16. Light is incident normal to a grating consisting of hundreds of thin, parallel, equally spaced slits. Explain carefully why a maximum of intensity on the screen occurs when the distances from two adjacent slits to a distant observation point differ by exactly one wavelength. Why do the other slits not cause destructive interference, but instead actually contribute to the constructive interference providing the screen is far enough away?

17. A pattern of fringes is produced when monochromatic light passes through a pair of thin slits. Would such a pattern be produced by three parallel thin slits? By thousands of such slits? Give an example to support your answer.

18. For complete cancellation of light reflected out of phase from the two surfaces of a thin film, what else besides frequency and wavelength must be the same for both parts of the recombined wave?

19. The colors of peacocks and hummingbirds are the result not of pigments, but of ridges in the surface layers of their feathers. By what physical principle do these ridges produce colors?

20. The colored wings of many butterflies are due to pigmentation, but others, such as the Morpho butterfly, show colors that do not result from any pigmentation. When the wing is viewed from different angles, the colors change. How are these colors produced?

21. Why do the iridescent colors seen in some seashells (such as abalone shells) change as the shells are viewed from different positions?

22. When dishes are not properly rinsed after washing, different colors are reflected from their surfaces. Explain.

23. Why are interference colors more apparent for thin films than for thick films?

24. Will the light from two very close stars produce an interference pattern? Explain.

25. If you notice the interference patterns of a thin film of oil or gasoline on water, you'll note that the colors form complete rings. How are these rings similar to the lines of equal elevation on a contour map?

26. Polarized light is a part of nature, but polarized sound is not. Why?

27. Why will an ideal Polaroid filter transmit 50% of incident nonpolarized light?

28. Why may an ideal Polaroid filter transmit anything from zero to 100% of incident polarized light?

29. What percentage of light is transmitted by two ideal Polaroids, one on top of the other with their polarization axes aligned? With their axes at right angles to each other?

30. How can you determine the polarization axis for a single sheet of Polaroid?

31. Why do Polaroid sunglasses reduce glare, whereas nonpolarized sunglasses simply cut down on the total amount of light reaching the eyes?

32. Most of the glare from nonmetallic surfaces is polarized, the axis of polarization being parallel to that of the reflecting surface. Would you expect the polarization axis of Polaroid sunglasses to be horizontal or vertical? Why?

33. How can a single sheet of Polaroid film be used to show that the sky is partially polarized? (Interestingly enough, unlike humans, bees and many insects can discern polarized light and use this ability for navigation.)

34. Light will not pass through a pair of Polaroid sheets aligned perpendicularly. But if a third Polaroid is sandwiched between the two, with its alignment half way between the alignments of the other two (that is, with its axis making a 45° angle with each of the other two alignment axes), some light does get through. Why?

35. Why does a painting look less flat when viewed with one eye? (If you're not on to this, look at paintings with one eye and see the difference.)

36. Why will viewing Figures 28.35 to 28.37 and Figure 28.40 with only one eye produce no sensation of depth?

37. Why did practical holography have to await the advent of the laser?

38. How is magnification accomplished with holograms?

39. If you are viewing a hologram and you close one eye, will you still perceive depth? Explain.

40. Make up a multiple-choice question that would check a classmate's understanding of diffraction, interference, or polarization.

29

Light Emission

: **D**ifferent colors of light are emitted by different kinds of atoms.

If energy is pumped into a metal antenna in such a way that it causes free electrons to vibrate to and fro a few hundred thousand times per second, a radio wave is emitted. If the free electrons could be made to vibrate to and fro on the order of a million billion times per second, a visible light wave would be emitted. But light is not produced from metallic antennae, nor is it exclusively produced by atomic antennae via oscillations of electrons in atoms, as discussed in previous chapters. We now distinguish between light reflected, refracted, scattered, and diffracted by objects and light emitted by objects. In this chapter we discuss the physics of light sources—of light *emission*.

The details of light emission from atoms involve the transitions of electrons from higher to lower energy states within the atom. This emission process can be understood in terms of the familiar planetary model of the atom that we discussed in Chapter 10. Just as each element is characterized by the number of electrons that occupy the shells surrounding its atomic nucleus, so also each element possesses its own characteristic pattern of electron shells, or energy states. These states are found only at certain radii and energies. Because these states can only

FIGURE 29.1 Simplified view of electrons orbiting in discrete shells about the nucleus of an atom.

have certain energies, we say they are *discrete*. We call these discrete states *quantum states,* and we'll return to them in detail in the next two chapters. For now we'll concern ourselves only with their role in light emission.

Excitation

An electron farther from the nucleus has a greater electric potential energy with respect to the nucleus than an electron nearer the nucleus. We say that the more distant electron is in a higher energy state, or, equivalently, a higher energy level. In a sense, this is similar to the energy of a spring door or a pile driver. The wider the door is pulled open, the greater its spring potential energy; the higher the ram of a pile driver is raised, the greater its gravitational potential energy.

FIGURE 29.2 The different orbits in an atom are like steps. When an electron is raised to a higher orbit, the atom is excited. When the electron falls to a lower step, it releases energy in the form of light.

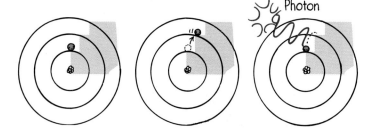

When an electron is in any way raised to a higher energy level, the atom is said to be *excited*. The electron's higher position is only momentary, for like the pushed-open spring door, it soon returns to its lowest-energy state. The atom loses its temporarily acquired energy when the electron returns to a lower level and emits radiant energy. The atom has undergone the process of **excitation** and *de-excitation.*

Just as each electrically neutral element has its own number of electrons, each element also has its own characteristic set of energy levels. Electrons dropping from higher to lower energy levels in an excited atom emit with each jump a throbbing pulse of electromagnetic radiation called a *photon,* the frequency of which is related to the energy transition of the jump. We think of this photon as a localized corpuscle of pure energy— a "particle" of light—which is ejected from the atom. The frequency of the photon is directly proportional to its energy. In shorthand notation,

$$E \sim f$$

A photon in a beam of red light, for example, carries an amount of energy that corresponds to its frequency. Another photon of twice the frequency has twice as much energy and is found in the ultraviolet part of the spectrum. If many atoms in a material are excited, many photons with many frequencies are emitted that correspond to the many different levels excited. These frequencies correspond to characteristic colors of light from each chemical element.

The light emitted in the glass tubes in an advertising sign is a familiar consequence of excitation. The different colors in the sign correspond to the excitation of different gases. Although it is common to refer to any of these as "neon," only the red light is that of neon. At the ends of the glass tube that contains the neon gas are electrodes.

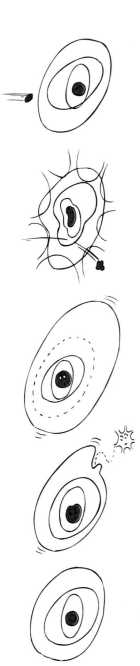

Electrons are boiled off these electrodes and are jostled back and forth at high speeds by a high ac voltage. Millions of high-speed electrons vibrate back and forth inside the glass tube and smash into millions of target atoms; each smash often boosts orbital electrons into higher energy levels by an amount of energy equal to the decrease in kinetic energy of the bombarding electron, and this energy is radiated as the characteristic red light of neon when the electrons fall back to their stable orbits. The process occurs and recurs many times, as neon atoms continually undergo a cycle of excitation and de-excitation. The overall result of this process is the transformation of electrical energy into radiant energy.

The colors of various flames are due to excitation. Different atoms in the flame emit colors characteristic of their energy-level spacings. Common table salt placed in a flame, for example, produces the characteristic yellow of sodium. Every element, excited in a flame or otherwise, emits its own characteristic color—a composite of the many colors that make up its characteristic spectrum.

Street lamps provide another example. City streets are no longer illuminated by incandescent lamps but are now illuminated with the light emitted by gases such as mercury vapor. Not only is the light brighter, it is less expensive. Whereas most of the energy in an incandescent lamp is converted to heat, most of the energy put into a mercury-vapor lamp is converted to light. The light from these lamps is rich in blues and violets and therefore is a different "white" than the light from an incandescent lamp. See if your instructor has a spare prism or diffraction grating you can borrow. Look through the prism or grating at the light from the street lamp and see the discreteness of the colors, which indicates the discreteness of the atomic levels. Also note that the colors from different mercury-vapor lamps are identical, showing that the atoms of mercury are identical.

Excitation is illustrated in the aurora borealis, as featured on the cover of this book. High-speed electrons that originate in the solar wind strike atoms and molecules in the upper atmosphere. They emit light exactly as occurs in a neon tube. The different colors in the aurora correspond to the excitation of different gases—oxygen atoms produce a greenish-white color, nitrogen molecules produce red-violet, and nitrogen ions produce a blue-violet color. Auroral emissions are not restricted to visible light, and also include infrared, ultraviolet, and X-ray radiation.

The excitation/de-excitation process can be accurately described only by quantum mechanics. An attempt to view the process in terms of classical physics runs into contradictions. Classically, an accelerating electric charge produces electromagnetic radiation. Does this explain light emission by excited atoms? An electron does accelerate in a transition from a higher to a lower energy level. Just as the innermost planets of the solar system have greater orbital speeds than those in the outermost orbits, the electrons in the innermost orbits of the atom have greater speeds. An electron gains speed in dropping to lower energy levels. Fine—the accelerating electron radiates a photon! But not so fine—the electron is continually undergoing acceleration (centripetal acceleration) in any orbit, whether or not it changes energy levels. According to classical physics, it should continually radiate energy. But it doesn't. All attempts to explain the emission of light by an excited atom in terms of a classical model have been unsuccessful. We shall simply say that light is emitted when electrons in an atom make a "quantum jump" from a higher to a lower energy level and that the energy and frequency of the emitted photon are described by the relationship $E \sim f$.

FIGURE 29.3 Excitation and de-excitation.

Question Suppose a friend suggests that for a first-rate operation, the gaseous neon atoms in a neon tube should be periodically replaced with fresh atoms because the energy of the atoms tends to be used up with continued excitation, producing dimmer and dimmer light. What do you say to this?

Emission Spectra

Every element has its own characteristic pattern of electron energy levels and therefore emits light with its own characteristic pattern of frequencies, its **emission spectrum**, when excited. This pattern can be seen when light is passed through a prism, or better, when it is first passed through a thin slit and then focused through a prism onto a viewing screen behind. Such an arrangement of slit, focusing lenses, and prism (or diffraction grating) is called a **spectroscope**, one of the most useful instruments of modern science (Figure 29.4).

FIGURE 29.4 A simple spectroscope.

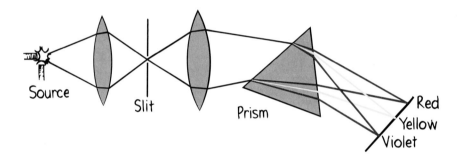

Each component color is focused at a definite position according to its frequency and forms an image of the slit on the screen, photographic film, or appropriate detector. The different-colored images of the slit are called *spectral lines*. Some typical spectral patterns labeled by wavelengths are shown in Figure 29.5. It is customary to refer to colors in terms of their wavelengths rather than their frequencies. A given frequency corresponds to a definite wavelength.*

If the light given off by a sodium-vapor lamp is analyzed in a spectroscope, a single yellow line predominates—a single image of the slit. If we narrow the width of the slit, we find that this line is really composed of two very close lines. These lines correspond

Answer The neon atoms don't give out any energy that is not imparted to them by the electric current in the tube and therefore don't get "used up." Any single atom may be excited and re-excited without limit. If the light is, in fact, becoming dimmer and dimmer, it is probably because a leak exists. Otherwise there is no advantage whatsoever to changing the gas in the tube, for a "fresh" atom is indistinguishable from a "used" one. Both are ageless and older than the solar system.

* Recall from Chapter 18 that $v = f\lambda$, where v is the wave speed, f is the wave frequency, and λ (lambda) is the wavelength. For light, v is the constant c, so we see from $c = f\lambda$ the relationship between frequency and wavelength; namely, $f = c/\lambda$ and $\lambda = c/f$.

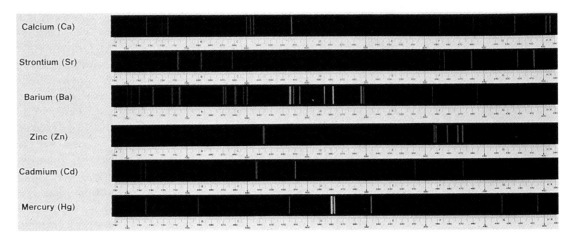

Calcium (Ca)

Strontium (Sr)

Barium (Ba)

Zinc (Zn)

Cadmium (Cd)

Mercury (Hg)

FIGURE 29.5   Some typical spectral patterns.

to the two predominant frequencies of light emitted by excited sodium atoms. The rest of the spectrum looks dark. (Actually, there are many other lines, often too dim to be seen with the naked eye.)

The same happens with all glowing vapors. The light from a mercury-vapor lamp shows a pair of bright yellow lines close together (but in different positions from those of sodium), a very intense green line, and several blue and violet lines. A neon tube produces a more complicated pattern of lines. The color of any excited gas, like the red of neon, is characteristic of the gas—no other gas glows with exactly the same color. And whatever the color, it is a composite of many spectral colors. So we find that the light emitted by each element in the vapor phase produces its own characteristic pattern of lines. These lines correspond to the electron transitions between atomic energy levels and are as characteristic of each element as are fingerprints of people. The spectroscope therefore is widely used in chemical analysis.

The next time you see evidence of atomic excitation, perhaps the green flame produced when a piece of copper is placed in a fire, squint your eyes and see if you can imagine electrons jumping from one energy level to another in a pattern characteristic of the atom being excited—a pattern that gives off a color unique to that atom. That's what's happening!

Question Spectral patterns are not shapeless smears of light but, instead, consist of fine and distinct straight lines. Why is this so?

Answer The spectral lines are simply images of the slit, which is itself a thin, straight opening through which light is admitted before being spread by the prism (or diffraction grating). When the slit is adjusted to make its most narrow opening, closely spaced lines can be resolved (distinguished from one another). A wider slit admits more light, which permits easier detection of dimmer radiant energy. But width is at the expense of resolution when closely spaced lines blur together.

Incandescence

FIGURE 29.6 The sound of an isolated bell rings with a clear and distinct frequency, whereas the sound emanating from a box of bells crowded together is discordant. Likewise with the difference between the light emitted from atoms in the gaseous phase and that from atoms in the solid phase.

Light that is produced as a result of high temperature has the property of **incandescence** (from a Latin word meaning "to grow hot"). It can have a reddish tint, as from the heating element of a toaster, or a bluish tint, as from a particularly hot star. Or it can be white, as from the familiar incandescent lamp. What sets incandescent light apart from the light of a neon tube or mercury-vapor lamp is that it contains an infinite number of frequencies, spread smoothly across the spectrum. Does this mean that an infinite number of energy levels characterizes the tungsten atoms making up the filament of the incandescent lamp? The answer is no; if the filament were vaporized and then excited, the tungsten gas would emit light with a finite number of frequencies and produce an overall bluish color. Light emitted by atoms far from one another in the gaseous phase is quite different from the light emitted by the same atoms closely packed in the solid phase. This is analogous to the differences in sound from an isolated ringing bell and from a box crammed with ringing bells (Figure 29.6). In a gas the atoms are far apart. Electrons undergo transitions between energy levels within the atom quite unaffected by the presence of neighboring atoms. But when the atoms are closely packed, as in a solid, electrons of the outer orbits make transitions not only within the energy levels of their "parent" atoms but also between the levels of neighboring atoms. They bounce around over dimensions larger than a single atom, resulting in an infinite variety of transitions—hence the infinite number of radiant energy frequencies.

As might be expected, incandescent light depends on temperature, for it is a form of thermal radiation. A plot of radiated energy over a wide range of frequencies for two different temperatures is shown in Figure 29.7. (Recall that we treated the radiation curve for sunlight back in Chapter 26 and discussed blackbody radiation in Chapter 15.) As the solid is heated further, more high-energy transitions occur, and higher frequency radiation is emitted. In the brightest part of the spectrum the predominant frequency of emitted radiation, the *peak frequency*, is directly proportional to the absolute temperature of the emitter:

$$\bar{f} \sim T$$

We use the bar above the f to indicate the peak frequency, for radiations of many frequencies are emitted from the incandescent source. If the temperature of an object (in kelvins) is doubled, the peak frequency of emitted radiation is doubled. The electromagnetic waves of violet light have nearly twice the frequency of red light waves. A violet-hot star therefore has nearly twice the surface temperature of a red-hot star.* The

FIGURE 29.7 Radiation curves for an incandescent solid.

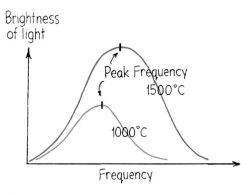

* If you study this topic further, you will find that the time rate at which an object radiates energy (the radiated power) is proportional to the fourth power of its Kelvin temperature. So a doubling of temperature corresponds to a doubling of the frequency of radiant energy but a sixteen-fold increase in the rate of emission of radiant energy.

temperature of incandescent bodies, whether they be stars or blast-furnace interiors, can be determined by measuring the peak frequency (or color) of radiant energy they emit.

Question From the radiation curves shown in Figure 29.7, which emits the higher average frequency of radiant energy—the 1000°C source or the 1500°C source? Which emits more radiant energy?

Absorption Spectra

When we view white light from an incandescent source with a spectroscope, we see a continuous rainbow-colored spectrum. If a gas is placed between the source and the spectroscope, however, careful inspection will show that the spectrum is not quite continuous. This is an **absorption spectrum**, and there are dark lines distributed throughout it; these dark lines against a rainbow-colored background are like emission lines in reverse. These are *absorption lines*.

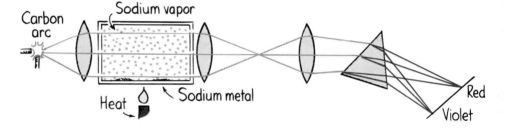

FIGURE 29.8 Experimental arrangement for demonstrating the absorption spectrum of a gas.

Atoms absorb light as well as emit light. An atom will most strongly absorb light having the frequencies to which it is tuned—the same frequencies it emits. When a beam of white light passes through a gas, the atoms of the gas absorb light of selected frequencies from the beam. This absorbed light is re-radiated, but in *all* directions instead of only in the direction of the incident beam. When the light remaining in the beam spreads out into a spectrum, the frequencies that were absorbed show up as dark lines in the otherwise continuous spectrum. The positions of these dark lines correspond exactly to the positions of lines in an emission spectrum of the same gas (Figure 29.9).

FIGURE 29.9 Emission and absorption spectra.

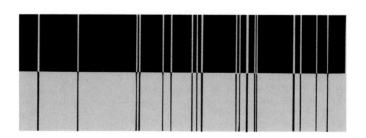

Answer The 1500°C radiating source emits the higher average frequencies, as noted by the extension of the curve to the right. The 1500°C source is the brighter and also emits more radiant energy, as noted by its greater vertical displacement.

Although the sun is a source of incandescent light, the spectrum it produces, upon close examination, is not continuous. There are many absorption lines, called *Fraunhofer lines* in honor of the Bavarian optician J. D. Lines, who first observed and mapped them accurately. Similar lines are found in the spectra produced by the stars. These lines indicate that the sun and stars are each surrounded by an atmosphere of cooler gases that absorb some of the frequencies of light coming from the main body. Analysis of these lines reveals the chemical composition of the atmospheres of such sources. We find from these analyses that the stellar elements are the same elements that exist on earth. An interesting sidelight is that in 1868, spectroscopic analysis of sunlight showed some spectral lines different from any known on earth. These lines identified a new element, which was named *helium*, after Helios, the sun. Helium was discovered in the sun before it was discovered on earth. How about that!

We can determine the speed of stars by studying the spectra they emit. Just as a moving sound source produces a Doppler shift in its pitch (Chapter 18), a moving light source produces a Doppler shift in its light frequency. The frequency (not the speed!) of light emitted by an approaching source increases, while the frequency of light from a receding source decreases. The corresponding spectral lines are displaced toward the red end of the spectrum for receding sources. Since the universe is expanding, almost all the galaxies show a red shift in their spectra.

We shall see in Chapter 31 how the spectra of elements enable us to determine atomic structure.

Fluorescence

So we see that thermal agitation and bombarding by particles such as high-speed electrons are not the only means of imparting excitation energy to an atom. An atom may be excited by absorbing a photon of light. From the relationship $E \sim f$, we see that high-frequency light, such as ultraviolet, which lies beyond the visible spectrum, delivers more energy per photon than lower-frequency light. Many substances undergo excitation when illuminated with ultraviolet light.

Many materials that are excited by ultraviolet light emit visible light upon de-excitation. The action of these materials is called **fluorescence**. In these materials a photon of ultraviolet light excites the atom, boosting an electron to a higher energy state. In this upward quantum jump, the atom is likely to leapfrog over several intermediate energy states. So when the atom de-excites, it may make smaller jumps, emitting photons with less energy.

FIGURE 29.10 In fluorescence the energy of the absorbed ultraviolet photon boosts the electron in an atom to a higher energy state. When the electron then returns to an intermediate state, the photon emitted is less energetic and therefore of a lower frequency than the ultraviolet photon.

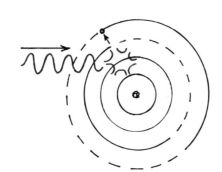
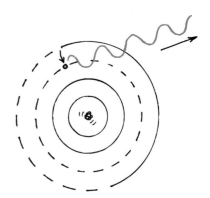

This excitation and de-excitation process is like leaping up a small staircase in a single bound, and then coming down one or two steps at a time rather than leaping all the way down in a single bound. Since the photon energy released at each step is less than the total energy originally in the ultraviolet photon, lower-frequency photons are emitted. Hence, ultraviolet light shining on the material causes it to glow an overall red, yellow, or whatever color is characteristic of the material. Fluorescent dyes are used in paints and fabrics to make them glow when bombarded with ultraviolet photons in sunlight. These are the Day-Glo colors, which are spectacular when illuminated with an ultraviolet lamp.

FIGURE 29.11 An excited atom may de-excite in several combinations of jumps.

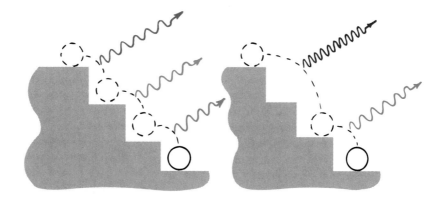

Question Why would it be impossible for a fluorescent material to emit ultraviolet light when illuminated by infrared light?

Detergents that make the claim of cleaning your clothes "whiter than white" use the principle of fluorescence. Such detergents contain a fluorescent dye that converts the ultraviolet light in sunlight into blue visible light, so clothes dyed in this way appear to reflect more blue light than they otherwise would. This makes the clothes appear whiter.*

The next time you visit a natural science museum, go to the geology section and take in the exhibit of minerals illuminated with ultraviolet light. You'll notice that different minerals radiate different colors. This is to be expected because different minerals are composed of different elements, which in turn have different sets of electron energy levels. Seeing the radiating minerals is a beautiful visual experience, which is even more fascinating when integrated with your knowledge of nature's submicroscopic happenings. High-energy ultraviolet photons strike the minerals, causing the excitation of

Answer Photon energy output would be greater than photon energy input, which would violate the law of conservation of energy.

---

* Interestingly enough, the same detergents marketed in some countries are adjusted for a rosier, warmer effect.

atoms in the mineral structure. The frequencies of light that you see correspond to the tiny energy-level spacings as the energy cascades down. Every excited atom emits its characteristic frequencies, with no two different minerals giving off exactly the same color light. Beauty is in both the eye and the mind of the beholder.

Fluorescent Lamps

The common fluorescent lamp consists of a cylindrical glass tube with electrodes at each end (Figure 29.12). In the lamp, as in the neon sign tube, electrons are boiled from one of the electrodes and forced to vibrate to and fro at high speeds within the tube by the ac voltage. The tube is filled with very low-pressure mercury vapor, which is excited by the impact of the high-speed electrons. Much of the emitted light is in the ultraviolet region. This is the primary excitation process. The secondary process occurs when the ultraviolet light strikes *phosphors*, a powdery material on the inner surface of the tube. The phosphors are excited by the absorption of the ultraviolet photons and fluoresce, giving off a multitude of lower frequency photons that combine to produce white light. Different phosphors can be used to produce different colors of light.

FIGURE 29.12 A fluorescent tube. Ultraviolet (UV) light is emitted by gas in the tube excited by an alternating electric current. The UV light, in turn, excites phosphors on the inner surface of the glass tube, which emit white light.

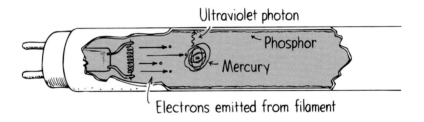

Phosphorescence

When excited, certain crystals as well as some large organic molecules remain in a state of excitement for a prolonged period of time. Their electrons are boosted into higher orbits and become "stuck." As a result, there is a time delay between the processes of excitation and de-excitation. Materials that exhibit this peculiar property are said to have **phosphorescence**. The element phosphorus, which is used in luminous clock dials and in other objects that are made to glow in the dark, is a good example. Atoms or molecules in these materials are excited by incident visible light. Rather than de-exciting immediately, as fluorescent materials do, many of the atoms remain in a state of excitement, sometimes for as long as several hours—although most undergo de-excitation rather quickly. If the source of excitation is removed—for example, if the lights are put out—an afterglow occurs while millions of atoms spontaneously undergo gradual de-excitation.

A TV screen is slightly phosphorescent, the glow decaying rather quickly, but just slowly enough so that successive scans of the picture blend into one another. The afterglow of some phosphorescent light switches in the home may last more than an hour. Likewise for luminous clock dials, excited by visible light. Some older clock dials glow indefinitely in the dark, not because of a long time delay between excitation and de-excitation, but because they contain radium or some other radioactive material that continuously supplies energy to keep the excitation process going. Such dials are no longer common because of the potential harm of the radioactive material to the user, especially if in a wristwatch or pocket watch.

Many living creatures—from bacteria to fireflies and larger animals like jelly-fish—chemically excite molecules in their bodies that give off light. We say that such living things are *bioluminescent*. Under some conditions certain fish become luminescent when they swim, but remain dark when still. Schools of these fish hang motionless and are not seen, but when alarmed streak the depths with sudden light, creating a sort of deep-sea fireworks. The mechanism of bioluminescence is not well understood and is currently being researched.

Lasers

The phenomena of excitation, fluorescence, and phosphorescence underlie the operation of a most intriguing instrument, the **laser** (**l**ight **a**mplification by **s**timulated **e**mission of **r**adiation).* Stimulated emission is a relatively new technique, although Einstein predicted it in 1917. To understand how a laser operates, we must first discuss coherent light.

Light emitted by a common lamp is incoherent; that is, photons of many frequencies and in many phases of vibration are emitted. The light is as incoherent as the footsteps on an auditorium floor when a mob of people are chaotically rushing about. Incoherent light is chaotic. A beam of incoherent light spreads out after a short distance, becoming wider and wider and less intense with increased distance.

FIGURE 29.13 Incoherent white light contains waves of many frequencies (and wavelengths) that are out of phase with one another.

Even if the beam is filtered so that it is formed of single frequency waves (monochromatic), it is still incoherent, for the waves are out of phase with one another. The slightest differences in their directions result in a spreading of the beam with increased distance.

FIGURE 29.14 Light of a single frequency and wavelength still contains a mixture of phases.

A beam of photons having the same frequency, phase, and direction—that is, a beam of photons that are identical copies of one another—is said to be *coherent*. Only a beam of coherent light will not noticeably spread and diffuse.

FIGURE 29.15 Coherent light: All the waves are identical and in phase.

---

* A word constructed from the initials of a phrase is called an *acronym.*

A laser is an instrument that produces a beam of coherent light. One model, the earliest, consists of a ruby crystal rod a few centimeters long that has been "doped" with impurities—atoms of chromium. The ruby rod is surrounded by a photoflash tube that sends high-intensity green light into the ruby (Figure 29.16). Excitation of the chromium atoms occur as outer electrons are boosted to higher energy levels. The electrons fall back to an intermediate level and then, more slowly, to their lower level. At this last stage they emit photons of red light; the ruby fluoresces. The key to the laser's operation is that, for at least a brief time, there are more atoms in an upper energy level than in a lower one, the reverse of normal conditions.

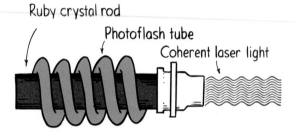

FIGURE 29.16   A ruby crystal laser.

Photons of red light within the crystal trigger the de-excitation of neighboring atoms that are still in the excited state. Since there are more excited atoms than unexcited ones, the light is more likely to stimulate emission than to be absorbed. Instead of a change from one photon to no photons (absorption), there is a change from one photon to two photons (stimulated emission). The photons whose emission is stimulated by incident photons will be identical in frequency, phase, and direction to the incident photons. These photons then may further stimulate the radiation of other excited atoms, a cascade that produces a beam of coherent light. Most of this light initially escapes through the sides of the crystal in random directions. Light traveling along the crystal axis, however, is reflected from mirrors that have been coated to selectively reflect light of the desired wavelength. One mirror is totally reflecting, while the other is partially reflecting. The reflected waves reinforce each other after each round-trip reflection between the mirrors, thereby setting up a resonance condition in which the light builds up to an appreciable intensity. The light that escapes in a pulse through the more transparent-mirrored end makes up the laser beam. The process is repeated every time the photoflash tube is fired.

Shortly after the development of the ruby laser, physicists made a gas laser that produced a continuous beam. The first such laser used a mixture of helium and neon gases. When a high voltage is applied to electrodes at the ends of a tube of the mixed gases, the electrical discharge excites helium atoms to a high-energy state. In collisions, the helium atoms transfer their excitation to a prolonged excited state in the neon atoms. This establishes more neon atoms in an excited state than in certain lower-energy states, the condition required for "lasing." In addition to crystal lasers and gas lasers, other new types have joined the laser family: glass lasers, chemical and liquid lasers, and semiconductor lasers. Present models produce beams ranging from infrared through ultraviolet. Some models can be tuned to various frequency ranges. Most exciting is the prospect of an X-ray laser beam.

The laser is not a source of energy. It is simply a converter of energy that takes advantage of the process of stimulated emission to concentrate a certain fraction of its

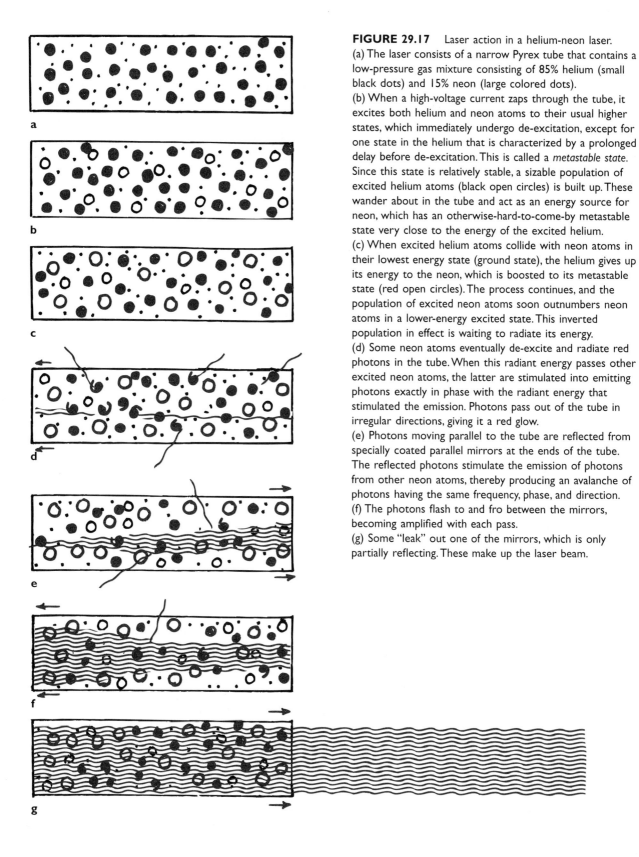

FIGURE 29.17 Laser action in a helium-neon laser.

(a) The laser consists of a narrow Pyrex tube that contains a low-pressure gas mixture consisting of 85% helium (small black dots) and 15% neon (large colored dots).

(b) When a high-voltage current zaps through the tube, it excites both helium and neon atoms to their usual higher states, which immediately undergo de-excitation, except for one state in the helium that is characterized by a prolonged delay before de-excitation. This is called a *metastable state*. Since this state is relatively stable, a sizable population of excited helium atoms (black open circles) is built up. These wander about in the tube and act as an energy source for neon, which has an otherwise-hard-to-come-by metastable state very close to the energy of the excited helium.

(c) When excited helium atoms collide with neon atoms in their lowest energy state (ground state), the helium gives up its energy to the neon, which is boosted to its metastable state (red open circles). The process continues, and the population of excited neon atoms soon outnumbers neon atoms in a lower-energy excited state. This inverted population in effect is waiting to radiate its energy.

(d) Some neon atoms eventually de-excite and radiate red photons in the tube. When this radiant energy passes other excited neon atoms, the latter are stimulated into emitting photons exactly in phase with the radiant energy that stimulated the emission. Photons pass out of the tube in irregular directions, giving it a red glow.

(e) Photons moving parallel to the tube are reflected from specially coated parallel mirrors at the ends of the tube. The reflected photons stimulate the emission of photons from other neon atoms, thereby producing an avalanche of photons having the same frequency, phase, and direction.

(f) The photons flash to and fro between the mirrors, becoming amplified with each pass.

(g) Some "leak" out one of the mirrors, which is only partially reflecting. These make up the laser beam.

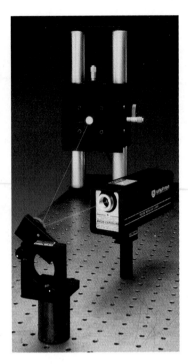

FIGURE 29.18 (left) A helium-neon laser. (right) Lasers are common tools in most school laboratories.

ISBN 0-321-00971-1

FIGURE 29.19 This text-book's unique identification is contained in the barcode that appears on the back cover.

energy (commonly 1%) into radiant energy of a single frequency moving in a single direction. Like all devices, a laser can put out no more energy than is put in.

Some lasers are so intense (and concentrated) that eye surgeons use them to "weld" detached retinas back into place without making an incision. The light is simply brought to focus in the region where the welding is to take place.

Whereas radio wavelengths span hundreds of meters and television waves span many centimeters, wavelengths of laser light are measured in millionths of a centimeter. As a result, an enormous number of messages bunched into a very narrow band of frequencies can be carried on these ultrashort waves. Communications can be carried in a laser beam directed through space, through the atmosphere, or through optical fibers (light pipes) that can be bent like cables.

The laser is at work in supermarket checkout counters, where code-reading machines scan the universal product code (UPC) symbol printed on packages and on the back cover of this book (Figure 29.19). Laser light is reflected from the bars and spaces and converted to an electric signal as the symbol is scanned. The signal rises to a high value when reflected from a bright space and falls to a low value when reflected from a dark bar. The high and low values are converted to 1's and 0's, respectively, and the information is in the binary code ready for computer processing.

Lasers measure and detect pollutants in exhaust gases. Different gases absorb light at characteristic wavelengths and leave their "fingerprints" on a reflected beam of laser light. The specific wavelength and amount of light absorbed are analyzed by a computer, which produces an immediate tabulation of the pollutants.

Just after its invention in 1958, the laser was touted as being a solution looking for a problem. It didn't have to look very far or wait very long. It has ushered in a whole new technology—the promise of which we have only begun to tap. The future for laser applications seems unlimited.

FIGURE 29.20 The Nova laser at Lawrence Livermore National Laboratory, used to study inertial confinement fusion. One of the world's largest and most powerful lasers, the Nova, with its ten beams, is capable of producing up to 100 trillion watts of power in a burst of light lasting for one-billionth of a second.

Summary of Terms

Excitation The process of boosting one or more electrons in an atom or molecule from a lower to a higher energy level. An atom in an excited state will usually decay (de-excite) rapidly to a lower state by the emission of a photon. The energy of the photon is proportional to its frequency: $E \sim f$.

Emission spectrum The distribution of wavelengths in the light from a luminous source.

Spectroscope An optical instrument that separates light into its constituent frequencies in the form of spectral lines.

Incandescence The state of glowing while at a high temperature, caused by electrons bouncing around over dimensions larger than the size of an atom, emitting radiant energy in the process. The peak frequency of radiant energy is proportional to the absolute temperature of a heated substance:

$$\bar{f} \sim T$$

Absorption spectrum A continuous spectrum, like that of white light, interrupted by dark lines or bands that result from the absorption of light of certain frequencies by a substance through which the radiant energy passes.

Fluorescence The property of certain substances to absorb radiation of one frequency and to re-emit radiation of lower frequency. It occurs when an atom is boosted up to a higher excited state and loses its energy in two or more downward jumps to lower energy states.

Phosphorescence A type of light emission that is the same as fluorescence except for a delay between excitation and de-excitation, which provides an afterglow. The delay is caused by atoms being excited to energy levels that do not decay rapidly. The afterglow may last from fractions of a second to hours, or even days, depending on the type of material, temperature, and other factors.

Laser (light amplification by stimulated emission of radiation) An optical instrument that produces a beam of coherent monochromatic light.

Review Questions

1. If electrons are made to vibrate to and fro at a few hundred thousand hertz, radio waves are emitted. What class of waves would be emitted if electrons were made to vibrate to and fro at a few million billion hertz?

2. What does it mean to say an energy state is *discrete*?

Excitation

3. Which has the greatest potential energy with respect to the atomic nucleus—electrons in inner electron shells or electrons in outer electron shells?

4. What does it mean to say an atom is *excited*?

5. How is a ball rolling down a stairway analogous to an electron undergoing de-excitation?

6. What is the name given to a single throbbing pulse of electromagnetic radiation?

7. What is the relationship between the *difference in energy* between energy levels and the *energy of the photon* that is emitted by a transition between those levels?

8. How is the *energy* of a photon related to its vibrational frequency?

9. Which has the higher *frequency*, red or blue light?

10. Which has the greater *energy* per photon, red or blue light?

11. An electron loses some of its kinetic energy when it bombards a neon atom in a glass tube. What becomes of this energy?

12. Can a neon atom in a glass tube be excited more than once?

13. What do the variety of colors displayed in the flame of a burning log represent?

14. Which converts the greater percentage of its energy into heat, an incandescent lamp or a mercury-vapor lamp?

15. If you look through a prism or diffraction grating at the light emitted by a mercury-vapor lamp, you'll see a variety of colors. What do these colors represent?

Emission Spectra

16. What is it that every element has that makes it emit its own characteristic colors of light?

17. What is a *spectroscope*, and what does it do?

18. Why do the colors of light in a spectroscope appear as lines?

Incandescence

19. What is the "color" of all the visible light frequencies mixed together equally?

20. When a gas glows, discrete colors are emitted. When a solid glows, the colors are smudged. Why?

21. How is the peak frequency of emitted light related to the temperature of its incandescent source?

Absorption Spectra

22. Distinguish between *emission spectra* and *absorption spectra*.

23. Since an absorbing gas re-emits the light it absorbs, why are there dark lines in an absorption spectrum? That is, why doesn't the re-emitted light simply fill in the dark places?

24. How was helium discovered?

25. How can astrophysicists tell whether a star is receding or approaching earth?

Fluorescence

26. Name three ways in which atoms can be excited.

27. Why is ultraviolet light rather than infrared light effective in making certain materials fluoresce?

28. If atoms of a substance absorb ultraviolet light and emit red light, what becomes of the "missing" energy?

Fluorescent Lamps

29. Distinguish between the primary and secondary excitation processes that occur in a fluorescent lamp.

30. What enables a fluorescent lamp to give off white light?

Phosphorescence

31. Distinguish between *fluorescence* and *phosphorescence*.

32. What is responsible for the afterglow of phosphorescent materials?

Lasers

33. Distinguish between *monochromatic light* and *coherent light*.

34. Does the red light emitted by a helium-neon laser come from helium or neon?

35. What is a *metastable state*?

36. In the operation of a helium-neon laser, why is it important that the metastable state of helium be relatively stable? (What would be the effect of this state de-exciting too rapidly?) (Refer to Figure 29.17.)

37. In the operation of a helium-neon laser, why is it important that the metastable state in the helium atom closely match the energy level of a more-difficult-to-come-by metastable state in neon?

38. How is the inverted population of excited neon atoms in a helium-neon laser achieved?

39. How does the avalanche of photons in a laser beam differ from the hordes of photons emitted by an incandescent lamp?

40. A friend speculates that scientists in a certain country have developed a laser that puts out far more energy than is put into it. Your friend asks for your response to this speculation. What is your response?

Project

Borrow a diffraction grating from your physics instructor. The common kind looks like a photographic slide, and light passing through it or reflecting from it is diffracted into its component colors by thousands of finely ruled lines. Look through the grating at the light from a sodium-vapor street lamp. If it's a low-pressure lamp, you'll see the nice yellow spectral "line" that dominates sodium light (actually, it's two closely spaced lines). If the street lamp is round, you'll see circles instead of lines; look through a slit cut in cardboard or whatever, and you'll see lines. What happens with the now-common high-pressure sodium lamps is more interesting. Because of the collisions of excited atoms, you'll see a smeared-out spectrum that is nearly continuous, almost like that of an incandescent lamp. Right at the yellow location where you'd expect to see the sodium line is a dark area. This is the sodium absorption band. It is due to the cooler sodium, which surrounds the high-pressure emission region. You should view this a block or so away so that the line, or circle, is small enough to allow the resolution to be maintained. Try this. It is very easy to see!

Exercises

1. Have you ever watched a fire and noticed that the burning of different materials often produces flames of different colors? Why is this so?

2. Green light is emitted when electrons in a substance make a particular energy-level transition. If blue light were instead emitted from the same substance, would it correspond to a greater or lesser change of energy in the atom?

3. Ultraviolet light causes sunburns, whereas visible light, even of greater intensity, does not. Why is this so?

4. If we double the frequency of light, we double the energy of each of its photons. If we instead double the wavelength of light, what happens to the photon energy?

5. Why doesn't a neon sign finally "run out" of excited atoms and produce dimmer and dimmer light?

6. If light were passed through a round hole instead of a thin slit in a spectroscope, how would the spectral "lines" appear? What is the drawback of a hole relative to a slit?

7. If we use a prism or a diffraction grating to compare the red light from a common neon tube and the red light from a helium neon laser, what striking difference do we see?

8. What is the evidence for the claim that iron exists in a relatively cool outer layer of the sun?

9. How might the Fraunhofer lines in the spectrum of sunlight that are due to absorption in the sun's atmosphere be distinguished from those due to absorption by gases in the earth's atmosphere?

10. How does the light that astronomers see from distant stars and galaxies tell them that the same atoms with the same properties exist throughout the universe? Why are spectral lines often referred to as "atomic fingerprints"?

11. What difference does an astronomer see between the emission spectrum of an element in a receding star and a spectrum of the same element in the lab? (*Hint*: This relates to information in Chapter 18.)

12. A blue-hot star is about twice as hot as a red-hot star. But the temperatures of the gases in advertising signs are about the same, whether they emit red or blue light. What is your explanation?

13. Which has the greatest energy, a photon of infrared light, of visible light, or of ultraviolet light?

14. Does atomic excitation occur in solids as well as in gases? How does the radiant energy from an incandescent solid differ from the radiant energy emitted by an excited gas?

15. A lamp filament is made of tungsten. Why do we get a continuous spectrum rather than a tungsten line spectrum when light from an incandescent lamp is viewed with a spectroscope?

16. How can a hydrogen atom, which has only one electron, have so many spectral lines?

17. (a) Light from an incandescent source is passed through sodium vapor and then examined with a spectroscope. What is the appearance of the spectrum? (b) The incandescent source is switched off and the sodium is heated until it glows. How does the spectrum of the glowing sodium compare with the previously observed spectrum?

18. Your friend reasons that if ultraviolet light can activate the process of *fluorescence*, infrared light ought to also. Your friend looks to you for approval or disapproval of this idea. What is your position?

19. When ultraviolet light falls on certain dyes, visible light is emitted. Why does this not happen when infrared light falls on these dyes?

20. Why are fabrics that fluoresce when exposed to ultraviolet light so bright in sunlight?

21. Why do different fluorescent minerals emit different colors when illuminated with ultraviolet light?

22. In a certain material, visible light causes electrons to jump from lower to higher energy states in the atoms, while ultraviolet light ionizes the atoms by ejecting electrons from them. Why do the two kinds of light behave differently?

23. The forerunner to the laser involved microwaves rather than visible light. What does *maser* mean?

24. A ruby laser is activated by a photoflash tube that emits green light. Why would a laser composed of a green crystal and a photoflash tube that emits red light not work?

25. How do the avalanches of photons in a laser beam differ from the hordes of photons emitted by an incandescent lamp?

26. A friend speculates that scientists in a certain country have developed a laser that puts out far more energy than is put into it. Your friend asks for your response to this speculation. What is your response?

27. A laser cannot put out more energy than is put into it. A laser can, however, produce pulses of light with more power output than the power input required to run the laser. Explain.

28. In the equation $\bar{f} \sim T$, what is $\bar{f}$? What is T ?

29. We know that a lamp filament at 2500K emits white light. Does the lamp filament also emit radiant energy when it is at room temperature?

30. We know that the sun emits radiant energy. Does the earth similarly emit radiant energy? If so, what is the difference in radiation emitted by each?

31. Since every object has some temperature, every object radiates energy. Why, then, can't we see objects in the dark?

32. If we continue heating a piece of initially room-temperature metal in a dark room, it will begin to glow visibly. What will be its first visible color, and why?

33. We can heat a piece of metal to red hot and to white hot. Can we heat it until the metal glows blue hot?

34. How do the surface temperatures of reddish, bluish, and whitish stars compare?

35. If you see a red-hot star, you can be sure that its peak intensity is in the infrared region. Why is this? And if you see a "violet-hot" star, you can be sure its peak intensity is in the ultraviolet. Why is this?

36. We see a "green-hot" star not green, but white. Why? (*Hint*: Consider the radiation curve back in Figure 26.7.)

37. Part a in the sketch below shows a radiation curve of an incandescent solid and its spectral pattern as produced with a spectroscope. Part b shows the "radiation curve" of an excited gas and its emission spectral pattern. Part c shows the curve produced when a cool gas is between an incandescent source and the viewer; the corresponding spectral pattern is left as an exercise for you to construct. Part d shows the spectral pattern of an incandescent source as seen through a piece of green glass; you are to sketch in the corresponding radiation curve.

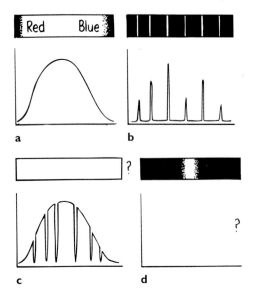

38. Consider just four of the energy levels in a certain atom, as shown in the diagram. How many spectral lines will result from all possible transitions among these levels? Which transition corresponds to the highest frequency light emitted? To the lowest frequency?

n = 4 ————————
n = 3 ————————
n = 2 ————————

n = 1 ————————

39. An electron de-excites from the fourth quantum level in the diagram above to the third and then directly to the ground state. Two photons are emitted. How does the sum of their frequencies compare to the frequency of the single photon that would be emitted by de-excitation from the fourth level directly to the ground state?

40. Suppose the four energy levels in Exercise 38 were somehow evenly spaced. How many spectral lines would result?

Problem

In the diagram, the energy difference between states A and B is twice the energy difference between states B and C. In a transition (quantum jump) from C to B, an electron emits a photon of wavelength 600 nm. (a) What is the wavelength emitted when the photon jumps from B to A? (b) When it jumps from C to A?

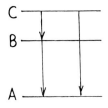

30

· · · · · · · · · · · · ·

Light Quanta

The blue beam of the argon laser is diffracted into smaller beams.

The classical physics that we have so far studied deals with two categories of phenomena: particles and waves. According to our everyday experience, "particles" are tiny objects like bullets. They have mass and obey Newton's laws—they *travel* through space in straight lines unless a force acts upon them. Likewise, according to our everyday experience, "waves," like waves in the ocean, are phenomena that *extend* in space. When a wave travels through an opening or around a barrier, the wave diffracts and different parts of the wave interfere. Therefore, particles and waves are easy to distinguish from each other. In fact, they have properties that are mutually exclusive. Nonetheless, the question of how to classify light was a mystery for centuries.

One of the early theories about the nature of light is that of Plato, who thought that light consisted of streamers emitted by the eye. Euclid later also held this view. The Pythagoreans, on the other hand, believed that light emanated from luminous bodies in the form of very fine particles, while Empedocles, a predecessor of Plato, taught that light is composed of high-speed waves of some sort. For more than 2000 years, the questions remained unanswered. Does light consist of waves or particles?

In 1704 Isaac Newton described light as a stream of particles or corpuscles. He held this view despite his knowledge of what we now call polarization and despite his experiment with light reflecting from glass plates, in which he noticed fringes of brightness and darkness (Newton's rings). He knew that his particles of light had to have certain wave properties too. Christiaan Huygens, a contemporary of Newton, advocated a wave theory of light.

With all this history as background, Thomas Young in 1801 performed the "double-slit experiment," which seemed to prove, finally, that light is a wave phenomenon. This view was reinforced in 1862 by Maxwell's prediction that light carries energy in oscillating electric and magnetic fields. Twenty-five years later, Hertz used sparking electric circuits to demonstrate the reality of electromagnetic waves (of radio frequency). In 1905, however, Albert Einstein published a Nobel Prize-winning paper that challenged the wave theory of light by arguing that light interacts with matter, not in continuous waves as Maxwell envisioned, but in tiny packets of energy that we now call *photons*. This discovery didn't wipe out light waves. It revealed, instead, that light is both wave *and* particle. In this chapter we shall enter the world of the very small and look into some of the strange and exciting aspects of quantum reality.

Birth of the Quantum Theory

As the twentieth century arrived, new technologies enabled scientists to design experiments to explore the behavior of very small particles. With the discovery of the electron in 1897 and the investigation of radioactivity at about the same time, experimenters began to probe the atomic structure of matter. In 1900 the German theoretical physicist Max Planck hypothesized that the vibrating electrons in incandescent lights could only have energies restricted to certain values. As a result, radiation was emitted in discrete bundles of energy, which he called **quanta** (*quanta* is the plural form of *quantum*, just as *momenta* is the plural form of *momentum*). Planck's hypothesis began a revolution of ideas that has completely changed the way we think about the physical world. We will see that the rules we apply to the everyday macroworld, the Newtonian laws that work so well for large objects like baseballs and planets, simply don't apply to events in the microworld of the atom. In the macroworld the study of motion is called *mechanics;* in the microworld, where new rules hold sway, the study of motion is **quantum mechanics**. More broadly, the body of laws developed from 1900 to the late 1920s that describe all quantum phenomena of the microworld are known as *quantum physics*.

Quantization and Planck's Constant

Quantization, the idea that the natural world is granular rather than smoothly continuous, is certainly not a new idea to physics. Matter is quantized; the mass of a brick of gold, for example, is equal to some whole-number multiple of the mass of a single gold atom. Electricity is quantized, as electric charge is always some whole-number multiple of the charge of a single electron.

Max Planck

Quantum physics states that in the microworld of the atom, the amount of energy in any system is quantized—not all values of energy are possible. This is analogous to saying a campfire can only be so hot. It might burn at 450°C or it might burn at 451°C, but in no way can it burn at 450.5°C. Believe it? Well, you shouldn't, for as far as our macroscopic thermometers can measure, a campfire can burn at any temperature as long as it's above that required for combustion. But the energy of the campfire, interestingly enough, is the composite energy of a great number and great variety of elemental units of energy. A simpler example is the energy in a beam of laser light, which is a whole-number multiple of a single lowest value of energy—one quantum. The quanta of light, and electromagnetic radiation in general, are the photons.

Recall from the previous chapter that the energy of a photon E is proportional to its frequency f: $E \sim f$. When the energy of a photon is divided by its frequency, the single number that results is the proportionality constant, called **Planck's constant**, h.* We shall see that Planck's constant is a fundamental constant of nature that serves to set a lower limit on the smallness of things. It ranks with the velocity of light as a basic constant of nature and appears again and again in quantum physics. We can insert this constant in the above proportion and express it as an exact equation:

$$E = hf$$

This equation gives the smallest amount of energy that can be converted to light with frequency f. The radiation of light is not emitted continuously but is emitted as a stream of photons, with each photon throbbing at a frequency f and carrying an energy hf.

Question How much total energy is in a monochromatic beam composed of n photons of frequency f?

The "new" physics tells us that the physical world is a coarse, grainy place rather than the smooth, continuous place with which we are familiar. The "commonsense" world described by classical physics seems smooth and continuous because quantum graininess is on a very small scale compared with the sizes of things in the familiar world. Planck's constant is small in terms of familiar units. But you don't have to go all the way to the quantum world to encounter graininess underlying apparent smoothness. For example, the blending areas of black, white, and gray in the photograph of Max Planck and other photographs in this book do not look smooth at all when viewed through a magnifying glass. With magnification you can see that a printed photograph consists of many tiny dots. In a similar way, we live in a world that is a blurred image of the grainy world of atoms.

Answer The energy in a beam of light containing n quanta is $E = nhf$.

---

* Planck's constant, h, has the numerical value 6.6×10^{-34} J·s.

Question What does the term *quantum* mean?

Physicists were reluctant to adopt Planck's revolutionary quantum notion. Before it could be taken seriously, the quantum idea would have to be verified by something besides emission from an incandescent source. A verification was supplied five years later by Einstein, who extended Planck's ideas to explain the photoelectric effect in the Nobel Prize-winning paper we mentioned earlier. (Even then, scientists were slow to accept so revolutionary an idea. Only after Niels Bohr's work on atomic structure in 1913—covered in the next chapter—was the quantum generally accepted. Einstein's Nobel Prize was delayed until 1921.)

Photoelectric Effect

In the latter part of the nineteenth century, several investigators noticed that light was capable of ejecting electrons from various metal surfaces. This is the **photoelectric effect**, now used in electric eyes, in the photographer's light meter, and in picking up sound from the sound track of motion pictures.

An arrangement for observing the photoelectric effect is shown in Figure 30.1. Light shining on the negatively charged photosensitive metal surface liberates electrons. The liberated electrons are attracted to the positive plate and produce a measurable current. If we instead charge this plate with just enough negative charge that it repels electrons, the current can be stopped. We can then calculate the energies of the ejected electrons from the easily measured potential difference between the plates.

FIGURE 30.1 An apparatus used for observing the photoelectric effect. Reversing the polarity and stopping the electron flow provides a way to measure the energy of the electrons.

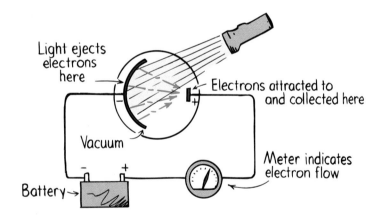

Answer A *quantum* is the smallest elemental unit of a quantity. Radiant energy, for example, is composed of many quanta, each of which is called a *photon*. So the more photons in a beam of light, the more energy in that beam.

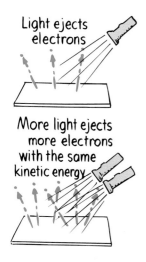

Light ejects electrons

More light ejects more electrons with the same kinetic energy

FIGURE 30.2 The photoelectric effect depends on intensity.

The photoelectric effect was not particularly surprising to early investigators. The ejection of electrons could be accounted for by classical physics, which pictures the incident light waves building an electron's vibration up to greater and greater amplitudes until it finally breaks loose from the metal surface, just as water drops boil off the surface of hot water. It should take considerable time to heat the surface of a weak source sufficiently to boil off electrons. Instead it was found that electrons are ejected as soon as the light is turned on—but not as many are ejected as with a strong source. Careful examination of the photoelectric effect led to several observations that were quite contrary to the classical wave picture:

1. The time lag between turning on the light and the ejection of the first electrons was not affected by the brightness or frequency of the light.
2. The effect was easy to observe with violet or ultraviolet light but not with red light.
3. The rate at which electrons were ejected was proportional to the brightness of the light.
4. The maximum energy of the ejected electrons was not affected by the brightness of the light. However, there were indications that the electrons' energy did depend on the frequency of the light.

The lack of any appreciable time lag was especially difficult to understand in terms of the wave picture. According to the wave theory, an electron in dim light should, after some delay, build up enough vibrational energy to eject, while an electron in bright light should be ejected almost immediately. However, this didn't happen. It was not unusual to observe an electron being ejected immediately, even under the dimmest light. The observation that the brightness of light in no way affected the energies of ejected electrons was also perplexing. The stronger electric fields of brighter light did not cause electrons to be ejected at greater speeds. More electrons were ejected in brighter light, but not at greater speeds. A weak beam of ultraviolet light, on the other hand, produced a smaller number of ejected electrons but at much higher speeds. This was most puzzling.

Einstein produced the answer in 1905, the same year he explained Brownian motion and set forth his theory of special relativity. His clue was Planck's quantum theory of radiation. Planck had assumed that the emission of light in quanta was due to restrictions on the vibrating atoms that produced the light. That is, he assumed that energy in *matter* is quantized, but that radiant energy is continuous. Einstein, on the other hand, attributed quantum properties to light itself and viewed radiation as a hail of particles. To emphasize this particle aspect, we speak of photons (by analogy with electrons, protons, and neutrons) whenever we are thinking of the particle nature of light. One photon is completely absorbed by each electron ejected from the metal. The absorption is an all-or-nothing process and is immediate, so there is no delay as "wave energies" build up.

A light wave has a broad front, and its energy is spread out along this front. For the light wave to eject a single electron from a metal surface, all its energy would somehow have to be concentrated on that one electron. But this is as improbable as an ocean wave hurling a boulder far inland with an energy equal to that of the whole wave. Therefore, instead of thinking of light encountering a surface as a continuous train of waves,

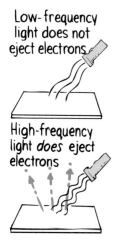

Low-frequency light does not eject electrons

High-frequency light *does* eject electrons

FIGURE 30.3 The photoelectric effect depends on frequency.

the photoelectric effect suggests we conceive of light encountering a surface or any detector as a succession of corpuscles, or photons. The number of photons in a light beam controls the brightness of the *whole* beam, whereas the frequency of the light controls the energy of each *individual* photon.

Experimental verification of Einstein's explanation of the photoelectric effect was made 11 years later by the American physicist Robert Millikan. Every aspect of Einstein's interpretation was confirmed, including the direct proportionality of photon energy to frequency. It was for this (and not for his theory of relativity) that Einstein received the Nobel prize.

The photoelectric effect proves conclusively that light has particle properties. We cannot conceive of the photoelectric effect on the basis of waves. On the other hand, we have seen that the phenomenon of interference demonstrates convincingly that light has wave properties. We cannot conceive of interference in terms of particles. In classical physics, this appears to be and is contradictory. From the point of view of quantum physics, light has properties resembling both. It is "just like a wave" or "just like a particle," depending on the particular experiment. So we think of light as both, as a wavepacket. How about "wavicle"? Quantum physics calls for a new way of thinking.

Questions

1. Will brighter light eject more electrons from a photosensitive surface than dimmer light of the same frequency?
2. Will high-frequency light eject a greater number of electrons than low-frequency light?

Wave-Particle Duality

The wave and particle nature of light is evident in the formation of optical images. We understand the photographic image produced by a camera in terms of light waves, which spread from each point of the object, refract as they pass through the lens system, and converge to focus on the photographic film. The path of light from the object through the lens system and to the focal plane can be calculated using methods developed from the wave theory of light.

But now consider carefully the way in which the photographic image is formed. The photographic film consists of an emulsion that contains grains of silver halide crystal, each grain containing about 10^{10} silver atoms. Each photon that is absorbed gives up its energy hf to a single grain in the emulsion. This energy activates surrounding crystals in the entire grain and is used in development to complete the photochemical process. Many photons activating many grains produce the usual photographic exposure. When a photograph is taken with exceedingly feeble light, we find that the image is built up by individual photons that arrive independently and are seemingly random in their distribution. We see this strikingly illustrated in Figure 30.4, which shows how an exposure progresses photon by photon.

Answers

1. Yes. The number of ejected electrons depends on the number of incident photons.
2. Not necessarily. The energy (not the number) of ejected electrons depends on the frequency of the illuminating photons. A bright source of blue light, for example, may eject more electrons at lower energy than a dim violet source.

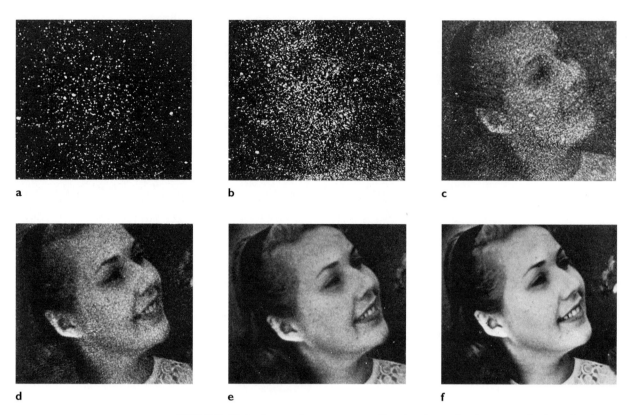

FIGURE 30.4 Stages of exposure revealing the photon-by-photon production of a photograph. The approximate numbers of photons at each stage are (a) 3×10^3, (b) 1.2×10^4, (c) 9.3×10^4, (d) 7.6×10^5, (e) 3.6×10^6, and (f) 2.8×10^7.

Double-Slit Experiment

Let's return to Thomas Young's double-slit experiment, which we discussed in terms of waves in Chapter 28. Recall that when we pass monochromatic light through a pair of closely spaced thin slits, we produce an interference pattern (Figure 30.5). Now let's consider the experiment in terms of photons. Suppose we dim our light source so that in

FIGURE 30.5 (a) Arrangement for double-slit experiment. (b) Photograph of interference pattern. (c) Graphic representation of pattern.

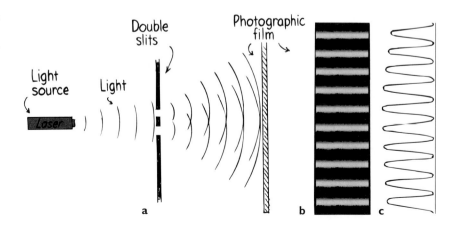

a   b   c

FIGURE 30.6 Stages of two-slit interference pattern. The pattern of individually exposed grains progresses from (a) 21 photons to (b) 1000 photons to (c) 10,000 photons. As more photons hit the screen, a pattern of interference fringes appears.

effect only one photon at a time reaches the barrier with the thin slits. If film behind the barrier is exposed to the light for a very short time, the film gets exposed as simulated in Figure 30.6. Each spot represents the place where the film has been exposed to a photon. If the light is allowed to expose the film for a longer time, a pattern of fringes begins to emerge as in Figure 30.6b and c. This is quite amazing. Spots on the film are seen to progress photon by photon to form the same interference pattern characterized by waves!

FIGURE 30.7 Single-slit diffraction pattern.

If we cover one slit so that photons hitting the photographic film can only pass through a single slit, the tiny spots on the film accumulate to form a single-slit diffraction pattern (Figure 30.7). We find that photons hit the film at places they would not hit if both slits were open! If we think about this classically, we are perplexed and may ask how photons passing through the single slit "know" that the other slit is covered and therefore fan out to produce the wide single-slit diffraction pattern. Or, if both slits are open, how do photons traveling through one slit "know" that the other slit is open and avoid certain regions, proceeding only to areas that will ultimately fill to form the fringed double-slit interference pattern?* The modern answer is that the wave nature of light is not some average property that shows up only when many photons act together. Each single photon has wave as well as particle properties. But the photon displays different aspects at different times. *A photon behaves as a particle when it is being emitted by an atom or absorbed by photographic film or other detectors and behaves as a wave in traveling from a source to the place where it is detected.* So the photon strikes the film as a particle but travels to its position as a wave that interferes constructively. The fact that light exhibits both wave and particle behavior was one of the interesting surprises of the early twentieth century. Even more surprising was the discovery that objects with mass also exhibit a dual wave-particle behavior.

---

* From a pre-quantum point of view, this wave-particle duality is indeed mysterious. This leads some people to believe that quanta have some sort of consciousness, with each photon or electron having "a mind of its own." The mystery, however, is like beauty. It is in the mind of the beholder rather than in nature itself. We conjure models to understand nature, and when inconsistencies arise, we sharpen or change our models. The wave-particle duality of light doesn't fit a model built on classical ideas. An alternate model is that quanta have minds of their own. Another model is quantum physics. In this book we subscribe to the latter.

Particles as Waves: Electron Diffraction

Louis de Broglie

If a photon of light has both wave and particle properties, why can't a material particle (one with mass) also have both wave and particle properties? This question was posed by the French physicist Louis de Broglie while he was still a graduate student in 1924. His answer constituted his doctoral thesis in physics and later won him the Nobel Prize in physics. According to de Broglie, every particle of matter is somehow endowed with a wave to guide it as it travels. Under the proper conditions, then, every particle will produce an interference or diffraction pattern. All bodies—electrons, protons, atoms, mice, you, planets, suns—have a wavelength that is related to the body's momentum by

$$\text{Wavelength} = \frac{h}{\text{momentum}}$$

where h is Planck's constant. A body of large mass and ordinary speed has such a small wavelength that interference and diffraction are negligible; rifle bullets fly straight and do not pepper their targets far and wide with detectable interference patches.* But for smaller particles such as electrons, diffraction can be appreciable.

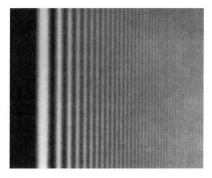

a

b

FIGURE 30.8 Fringes produced by the diffraction of (a) light and (b) an electron beam.

A beam of electrons can be diffracted in the same way a beam of photons can be diffracted, as is evident in Figure 30.8 above. Beams of electrons directed through double slits exhibit interference patterns, just as photons do. The double-slit experiment discussed in the last section can be performed with electrons as well as with photons. For electrons the apparatus is more complex, but the procedure is essentially the same. The

FIGURE 30.9 An electron microscope makes practical use of the wave nature of electrons. The wavelength of electron beams is typically thousands of times shorter than the wavelength of visible light, so the electron microscope is able to distinguish detail not visible with optical microscopes.

* A bullet of mass 0.02 kg traveling at 330 m/s, for example, has a de Broglie wavelength of $h/mv = \dfrac{6.6 \times 10^{-34}\,\text{J·s}}{(0.02\,\text{kg})(330\,\text{m/s})} = 10^{-34}$ m, an incredibly small size a million million million millionth the size of a hydrogen atom. An electron traveling at 2% of the speed of light, on the other hand, has a wavelength 10^{-10} m, which is equal to the diameter of the hydrogen atom. Diffraction effects for electrons are measurable, whereas diffraction effects for bullets are not.

intensity of the source can be reduced to direct electrons one at a time through a double-slit arrangement, producing the same remarkable results as with photons. Like photons, electrons strike the screen as particles, but the *pattern* of arrival is wavelike. The angular deflection of electrons to form the interference pattern agrees perfectly with calculations using de Broglie's equation for the wavelength of an electron.

This wave-particle duality is not restricted to photons and electrons. In Figure 30.11 we see the results of a similar procedure that uses a standard electron microscope. The electron beam of very low current density is directed through an electrostatic biprism that diffracts the beam. A pattern of fringes produced by individual electrons builds up step by step and is displayed on a TV monitor. The image is gradually filled by electrons to produce the interference pattern customarily associated with waves.

Neutrons, protons, whole atoms, and to an immeasurable degree even high-speed rifle bullets exhibit a duality of particle and wave behavior.

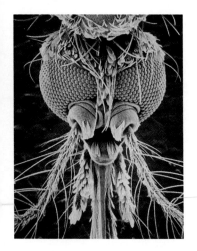

FIGURE 30.10 Detail of a female mosquito head as seen with a scanning electron microscope at a "low" magnification of 200 times.

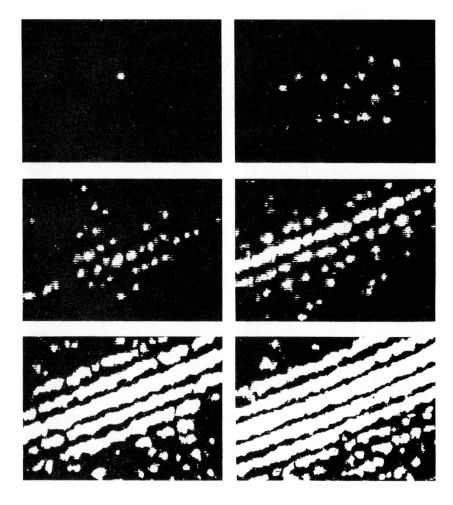

FIGURE 30.11 Electron interference patterns filmed from a TV monitor, showing the diffraction of a very low intensity electron-microscope beam through an electrostatic biprism.

Questions

1. If electrons behaved only like particles, what pattern would you expect on the screen after the electron passed through the double slits?
2. We don't notice the de Broglie wavelength for a pitched baseball. Is this because the wavelength is very large or because it is very small?
3. If an electron and a proton have the same de Broglie wavelength, which particle has the greater speed?

Uncertainty Principle

The wave-particle duality of quanta has inspired interesting discussions about the limits of our ability to accurately measure the properties of small objects. The discussions center on the idea that the act of measuring something affects the quantity being measured.

We know, for example, that if we place a cool thermometer in a cup of hot coffee, the temperature of the coffee is altered as it gives heat to the thermometer. The measuring device alters the quantity being measured. But we can correct for these errors in measurement if we know the initial temperature of the thermometer, the masses and specific heats involved, and so forth. Such corrections fall well within the domain of classical physics—these are *not* the uncertainties of quantum physics. Quantum uncertainties belong to the microworld of the atom, and stem from the wave nature of matter. A wave by its very nature occupies some space and lasts for some time. It cannot be squeezed to a point in space or limited to a single instant of time, for then it would not be a wave. This inherent "fuzziness" of a wave gives a fuzziness to measurement at the quantum level. Innumerable experiments have shown that any measurement that in any way probes a system necessarily disturbs the system by at least one quantum of action, *h*—Planck's constant. So any measurement that involves interaction between the measurer and what is being measured is subject to this minimum inaccuracy.

Answers

1. If electrons behaved only like particles, they would form two bands, as indicated in a. Because of their wave nature, they actually produce the pattern shown in b.

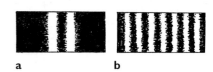

    a   b

2. We don't notice the wavelength of a pitched baseball because it is extremely small—on the order of 10^{-20} times smaller than the atomic nucleus.
3. The same wavelength means that the two particles have the same momentum. This means that the less massive electron must travel faster than the heavier proton.

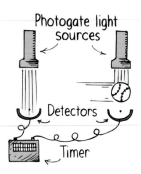

Photogate light sources

Detectors

Timer

FIGURE 30.12 The ball's speed is measured by dividing the distance between the photogates by the time difference in crossing light paths. Photons hitting the ball alter its motion much less than the motion of an oil supertanker is altered when a few fleas bump into it.

Werner Heisenberg

We distinguish between passive observing and probing. If we look at a cup of hot coffee across the room and judge that it is hot by observing the steam rising from it, this act of "measuring" neither adds nor subtracts energy from the coffee—it involves no *probing*. Placing a thermometer in it is a different story. We physically interact with the coffee and thereby subject it to alteration. The quantum contribution to this alteration, however, is completely dwarfed by classical uncertainties and is negligible. Quantum uncertainties are significant only in the atomic and subatomic realm.

Compare the acts of making measurements of a pitched baseball and an electron. We can measure the speed of a pitched baseball by having it fly through a pair of photogates that are a known distance apart (Figure 30.12). The ball is timed as it interrupts beams of light in the gates. The accuracy of the ball's measured speed has to do with uncertainties in the measured distance between the gates and in the timing mechanisms. Interactions between the macroscopic ball and the photons it encounters are insignificant. But not so in the case of measuring submicroscopic things like electrons. Even a single photon bouncing off an electron appreciably alters the motion of the electron—and in an unpredictable way. If we wished to observe an electron and determine its whereabouts with light, the wavelength of the light would have to be very short. We fall into a dilemma. A short wavelength that can better "see" the tiny electron corresponds to a large quantum of energy, which greatly alters the electron's state of motion. If, on the other hand, we use a long wavelength that corresponds to a smaller quantum of energy, the change we induce to the electron's state of motion will be smaller, but the determination of its position by the coarser wave will be less accurate. The act of observing something as tiny as an electron probes the electron and in so doing produces a considerable uncertainty in either its position or its motion. Although this uncertainty is completely negligible for measurements of position and motion regarding everyday (macroscopic) objects, it is a predominant fact of life in the atomic domain.

The uncertainty of measurement in the atomic domain was first stated mathematically by the German physicist Werner Heisenberg* and is called the **uncertainty principle**. It is a fundamental principle in quantum mechanics. Heisenberg found that when the uncertainties in the measurement of momentum and position for a particle are multiplied together, the product must be equal to or greater than Planck's constant, h, divided by 2π, which is represented as $\hbar$ (called *h-bar*). We can state the uncertainty principle in a simple formula:

$$\Delta p \ \Delta x \geq \hbar$$

The Δ here means "uncertainty of": Δp is the uncertainty of momentum (the symbol for momentum is conventionally p), and Δx is the uncertainty of position. The product of these two uncertainties must be equal to or greater than ($\geq$) the size of $\hbar$. For minimum uncertainties, the product will equal $\hbar$; the product of larger uncertainties will be greater than $\hbar$. But in no case can the product of the uncertainties be less than $\hbar$. The significance of the uncertainty principle is that even in the best of conditions, the lower limit of uncertainty is $\hbar$. This means that if we wish to know the momentum of an electron

---

* As a teenager, Heisenberg was fascinated with Plato's basic tenet that one can never understand the universe until the smallest components of matter are known. Young Heisenberg decided to devote his life to studying the smallest particles of matter.

with great accuracy (small Δp), the corresponding uncertainty in position will be large. Or if we wish to know the position with great accuracy (small Δx), the corresponding uncertainty in momentum will be large. The sharper one of these quantities is, the less sharp is the other.*

The uncertainty principle operates similarly with energy and time. We cannot measure a particle's energy with complete precision in an infinitesimally short span of time. The uncertainty in our knowledge of energy, ΔE, and the duration taken to measure the energy, Δt, are related by the expression[†]

$$\Delta E \, \Delta t \geq \hbar$$

The greatest accuracy we can ever hope to attain is that case in which the product of the energy and time uncertainties equals $\hbar$. The more accurately we determine the energy of a photon, an electron, or a particle of whatever kind, the more uncertain we will be of the time during which it has that energy.

The uncertainty principle is relevant only to quantum phenomena. The inaccuracies in measuring the position and momentum of a baseball due to the interactions of observation, for example, are completely negligible. But the inaccuracies in measuring the position and momentum of an electron are far from negligible. This is because the uncertainties in the measurements of these subatomic quantities are comparable to the magnitudes of the quantities themselves.[‡]

There is a danger in applying the uncertainty principle to areas outside of quantum mechanics. Some people conclude from statements about the interaction between the observer and the observed that the universe does not exist "out there," independent of all acts of observation, and that reality is created by the observer. Others interpret the uncertainty principle as nature's shield of forbidden secrets. Some critics of science use the uncertainty principle as evidence that science itself is uncertain. The reality of the universe whether observed or not, nature's secrets, and the uncertainties of science have very little to do with Heisenberg's uncertainty principle. The profundity of the uncertainty principle has to do with the unavoidable interaction between nature at the atomic level and the means by which we probe it.

---

* Only in the classical limit where $\hbar$ becomes zero could the uncertainties of both position and momentum be arbitrarily small. Planck's constant is greater than zero, and we cannot in principle simultaneously know both quantities with absolute certainty.

† We can see that this is consistent with the uncertainty in momentum and position. Recall that Δ momentum = force $\times$ Δ time and that Δ energy = force $\times$ Δ distance. Then

$$\hbar = \Delta \text{ momentum } \Delta \text{ distance}$$

$$= (\text{force} \times \Delta \text{ distance}) \, \Delta \text{ time}$$

$$= \Delta \text{ energy } \Delta \text{ time}$$

‡ The uncertainties in measurements of momentum, position, energy, or time that are related to the uncertainty principle for a pitched baseball are only 1 part in about 10 million billion billion billion (10^{-34}). Quantum effects are negligible even for the swiftest bacterium, where the uncertainties are about 1 part in a billion (10^{-9}). Quantum effects become evident for atoms, where the uncertainties can be as large as 100%. For electrons moving in an atom, quantum uncertainties dominate, and we are in the full-scale quantum realm.

Questions

1. Is Heisenberg's uncertainty principle applicable to the practical case of using a thermometer to measure the temperature of a glass of water?

2. A Geiger counter measures radioactive decay by registering the electrical pulses produced in a gas tube when high-energy particles pass through it. The particles emanate from a radioactive source—say, radium. Does the act of measuring the decay rate of radium alter the radium or its decay rate?

3. Can the quantum principle that we cannot observe something without changing it be reasonably extrapolated to support the claim that you can make a stranger turn around and look at you by staring intently at his back?

Complemen-tarity

The realm of quantum physics seems confusing. Light waves that interfere and diffract deliver their energy in particle packages of quanta. Electrons that move through space in straight lines and experience collisions as if they were particles distribute themselves spatially in interference patterns as if they were waves. In this confusion there is an underlying order. The behavior of light and electrons is confusing in the same way! Light and electrons both exhibit wave and particle characteristics.

The Danish physicist Niels Bohr, one of the founders of quantum physics, formulated an explicit expression of the wholeness inherent in this dualism. He called his expression of this wholeness **complementarity**. As Bohr expressed it, quantum phenomena exhibit complementary (mutually exclusive) properties—appearing either as particles or as waves—depending on the type of experiment conducted. Experiments

Answers

1. No. Although we probably subject the temperature of water to a change by the act of probing it with a thermometer, especially one appreciably colder or hotter than the water, the uncertainties that relate to the precision of the thermometer are quite within the domain of classical physics. The role of uncertainties at the subatomic level is inapplicable here.

2. Not at all, because the interaction involved is between the Geiger counter and the particles, not between the Geiger counter and the radium. It is the behavior of the particles that is altered by measurement, not the radium from which they emanate. See how this ties into the next question.

3. No. Here we must be careful in defining what we mean by *observing*. If our observation involves probing (giving or extracting energy), we indeed change to some degree that which we observe. For example, if we shine a light source onto the person's back, our observation consists of probing, which, however slight, physically alters the configuration of atoms on his back. If he senses this, he may turn around. But simply staring intently at his back is observing in the passive sense. The light you receive or block by blinking, for example, has already left his back. So whether you stare, squint, or close your eyes completely, you in no physical way alter the atomic configuration on his back. Shining a light or otherwise probing something is not the same thing as passively looking at something. A failure to make the simple distinction between *probing* and *passive observation* is at the root of much nonsense that is said to be supported by quantum physics. Better support for the above claim would be positive results from a simple and practical test, rather than the assertion that it rides on the hard-earned reputation of quantum theory.

Predictability and Chaos

We can make predictions about an orderly system when we know the initial conditions. For example, we can state precisely where a launched rocket will land, where a given planet will be at a particular time, or when an eclipse will occur. These are examples of events in the Newtonian macroworld. Similarly, in the quantum microworld we can predict where an electron is likely to be in an atom, and the probability that a radioactive particle will decay in a given time interval. Predictability in orderly systems, both Newtonian and quantum, depends on knowledge of initial conditions.

Some systems however, whether Newtonian or quantum, are not orderly—they are inherently unpredictable. These are called "chaotic systems." Turbulent water flow is an example. No matter how precisely we know the initial conditions of a piece of floating wood as it flows downstream, we cannot predict its location later downstream. A feature of chaotic systems is that slight differences in initial conditions result in wildly different outcomes later. Two identical pieces of wood just slightly apart at one time can be vastly far apart soon thereafter.

Weather is chaotic. Small changes in one day's weather can produce big (and largely unpredictable) changes a week later. Meteorologists try their best, but they are bucking the hard fact of chaos in nature. This barrier to good prediction first led the scientist Edward Lorenz to ask, "Does the flap of a butterfly's wings in Brazil set off a tornado in Texas?" We now talk about the *butterfly effect* when we are dealing with situations where very small effects can amplify into very big effects.

Interestingly, chaos is not all hopeless unpredictability. Even in a chaotic system there can be patterns of regularity. There is *order in chaos*. Scientists have learned how to treat chaos mathematically and how to find the parts of it that are orderly.

designed to examine individual exchanges of energy and momentum bring out particlelike properties, while experiments designed to examine spatial distribution of energy bring out wavelike properties. The wavelike properties of light and particlelike properties of light complement one another—both are necessary for the understanding of "light." Which part is emphasized depends on what question one puts to nature.

The idea that opposites are components of a wholeness is not new. Ancient Eastern cultures incorporated it as an integral part of their world view. This is demonstrated

in the yin-yang diagram of T'ai Chi Tu (Figure 30.13). One side of the circle is called *yin*, and the other side is called *yang*. Where there is yin, there is yang. Only the union of yin and yang forms a whole. Where there is low, there is also high. Where there is night, there is also day. Where there is birth, there is also death. A whole person integrates yin (emotion, intuition, feminine traits, right brain) with yang (reason, logic, masculine traits, left brain). Each has aspects of the other. For Niels Bohr, the yin-yang diagram symbolized the principle of complementarity. In later life Bohr wrote broadly on the implications of complementarity. In 1947, when he was knighted for his contributions to physics, he chose for his coat of arms the yin-yang symbol.

FIGURE 30.13 Opposites are seen to complement one another in the yin-yang symbol of Eastern cultures.

Summary of Terms

Quantum theory The theory that describes the microworld, where many quantities are granular (in units called *quanta*), not continuous, and where particles of light (*photons*), and particles of matter, such as electrons, exhibit wave as well as particle properties.

Planck's constant A fundamental constant, h, which relates the energy of light quanta to their frequency:

$$h = 6.6 \times 10^{-34} \text{ joule-second}$$

Photoelectric effect The emission of electrons from a metal surface when light shines on it.

Uncertainty principle The principle formulated by Heisenberg, stating that Planck's constant, h, sets a limit on the accuracy of measurement. According to the uncertainty principle, it is not possible to measure exactly both the position and the momentum of a particle at the same time, nor the energy and the time during which the particle has that energy.

Complementarity The principle enunciated by Niels Bohr stating that the wave and particle aspects of both matter and radiation are necessary, complementary parts of the whole. Which part is emphasized depends on what experiment is conducted (i.e., on what question one puts to nature).

Suggested Reading

Cole, K. C. *Sympathetic Vibrations: Reflections on Physics as a Way of Life*. New York: Morrow, 1984. A delightful unraveling of the theories of Bohr, Einstein, and other developers of quantum physics, with emphasis on the human side of physics.

Gleick, James. *Chaos: Making a New Science*. New York: Viking, 1987. See Physics Potpourri on p. 64.

March, R. H. *Physics for Poets*, 3rd ed. New York: McGraw-Hill, 1992. This textbook presents an interesting account of contemporary physics and its historical roots.

Review Questions

1. Did the findings of Young, Maxwell, and Hertz support the wave theory or the particle theory of light?
2. Did Einstein's photon explanation of the photoelectric effect support the wave theory or the particle theory of light?

Birth of the Quantum Theory

3. What exactly did Max Planck consider quantized, the energy of vibrating atoms or the energy of light itself?
4. Distinguish between the study of *mechanics* and the study of *quantum mechanics*.

Quantization and Planck's Constant

5. What does it mean to say that a certain quantity is *quantized*?
6. Why is the energy of a campfire not a whole-number multiple of a single quantum, although the energy of a laser beam is?
7. What is a quantum of light called?
8. In the previous chapter we learned the formula $E \sim f$. In this chapter we learned the formula $E = hf$. Explain the difference between these two formulas. What is h?
9. In the formula $E = hf$, does f stand for wave frequency, as defined in Chapter 18?
10. Which has the lower energy quanta, red light or blue light? Radio waves or X rays?

Photoelectric Effect

11. What evidence can you cite for the wave nature of light? For the particle nature of light?
12. Which are more successful in dislodging electrons from a metal surface, photons of violet light or photons of red light? Why?

13. Why won't a very bright beam of red light impart more energy to an ejected electron than a feeble beam of violet light?

14. In considering the interaction of light with matter, how did Einstein extend the quantum idea of Planck?

15. Does the brightness of a beam of light primarily depend on the frequency of photons or on the number of photons?

16. Einstein proposed his explanation of the photoelectric effect in 1905. When were his views on this confirmed?

Wave-Particle Duality

17. Why do photographs in a book or magazine look grainy when magnified?

18. Does light behave primarily as a wave or as a particle when it interacts with the crystals of matter in photographic film?

Double-Slit Experiment

19. Does light travel from one place to another in a wavelike way or a particlelike way?

20. Does light interact with a detector in a wavelike way or a particlelike way?

21. When does light behave as a wave? When does it behave as a particle?

Particles as Waves: Electron Diffraction

22. What evidence can you cite for the wave nature of particles?

23. When electrons are diffracted through a double slit, do they arrive at the screen in a wavelike way or a particlelike way? Is the pattern of hits wavelike or particlelike?

Uncertainty Principle

24. In which of the following are quantum uncertainties significant: measuring simultaneously the speed and location of a baseball; of a spitball; of an electron.

25. What is the uncertainty principle with respect to motion and position?

26. If measurements show a precise position for an electron, can those measurements show precise momentum also? Explain.

27. If measurement shows a precise value for the energy radiated by an electron, can that measurement show a precise time for this event as well? Explain.

28. Is there a distinction between passively observing something and actively probing something? In which case does one affect the behavior of that something?

Complementarity

29. What is the principle of complementarity?

30. According to the principle of complementarity, how are the opposite wavelike and particlelike properties of light reconciled?

Exercises

1. Distinguish between *classical physics* and *quantum physics*.

2. What does it mean to say that something is quantized?

3. How can a photon's energy be given by the formula $E = hf$ when f in the formula is a wave frequency?

4. The frequency of violet light is about twice that of red light. How does the energy of a violet photon compare with the energy of a red photon?

5. We speak of photons of red light and photons of green light. Can we speak of photons of white light? Why or why not?

6. A beam of red light and a beam of blue light have exactly the same energy. Which beam contains the greater number of photons?

7. One of the technical challenges facing the original developers of color television was the design of an image tube (camera) for the red portion of the image. Why was finding a material that would respond to red light more difficult than finding materials to respond to green and blue light?

8. Silver bromide (AgBr) is a light-sensitive substance used in some types of photographic film. To expose the film, it must be illuminated with light which has sufficient energy to break apart the AgBr molecules. Why do you suppose this film may be handled without exposure in a darkroom illuminated with red light? How about blue light? How about very bright red light relative to very dim blue light?

9. Suntanning produces cell damage in the skin. Why is ultraviolet radiation capable of producing this damage, while visible radiation, even if more intense, is not?

10. In the photoelectric effect, does brightness or frequency determine the kinetic energy of the ejected electrons? The number of the ejected electrons?

11. A very bright source of red light has much more energy than a dim source of blue light, but the red light has no effect in ejecting electrons from a certain photosensitive surface. Why is this so?

12. Why does light striking a metal surface eject only electrons, not protons?

13. Explain how the photoelectric effect is used to open automatic doors when someone approaches.

14. If you shine an ultraviolet light on the metal ball of a negatively charged electroscope (shown in Exercise 8 in Chapter 21), it will discharge. But if the electroscope is positively charged, it won't discharge. Can you venture an explanation?

15. Discuss how the reading of the meter in Figure 30.1 will vary as (a) the photosensitive plate is illuminated by light of various colors at a given intensity and (b) illuminated at various intensities for a given color.

16. Explain briefly how the photoelectric effect is used in the operation of at least two of the following: an electric eye, a photographer's light meter, the sound track of a motion picture.

17. Does the photoelectric effect *prove* that light is made of particles? Do interference experiments *prove* that light is composed of waves? (Is there a distinction between what something *is* and how it *behaves*?)

18. Does Einstein's explanation of the photoelectric effect invalidate Young's explanation of the double-slit experiment? Explain.

19. The camera that took the photograph of the woman's face (Figure 30.4) used ordinary lenses that are known to refract waves. Yet the step-by-step formation of the image is evidence of photons. How can this be? What is your explanation?

20. What evidence can you cite for the wave nature of light? For the particle nature of light?

21. What laboratory device utilizes the wave nature of electrons?

22. How might an atom get enough energy to become ionized?

23. When a photon hits an electron and gives it energy, what happens to the frequency of the photon after bouncing from the electron? (This occurs, and is called the Compton Effect.)

24. If a proton and an electron have identical speeds, which has the longer wavelength?

25. One electron travels twice as fast as another. Which has the longer wavelength?

26. Does the de Broglie wavelength of a proton become longer or shorter as its velocity increases?

27. If a particle could gain mass while maintaining constant speed, what would happen to its de Broglie wavelength?

28. If a cannonball and a BB have the same speed, which has the longer wavelength?

29. We don't notice the wavelength of moving matter in our ordinary experience. Is this because the wavelength is extraordinarily large or extraordinarily small?

30. What principal advantage does an electron microscope have over an optical microscope?

31. Would a beam of protons in a "proton microscope" exhibit greater or less diffraction than electrons of the same speed in an electron microscope? Defend your answer.

32. Suppose nature were entirely different so that an infinite number of photons were needed to make up even the tiniest amount of radiant energy, the wavelength of material particles were zero, light had no particle properties, and matter had no wave properties. This would be the classical world described by the mechanics of Newton and the electricity and magnetism of Maxwell. What would be the value of Planck's constant for such a world with no quantum effects?

33. If Planck's constant, *h*, were equal to zero rather than a tiny number, how would the uncertainty principle differ?

34. Suppose you lived in a hypothetical world where you'd be knocked down by a single photon, where matter would be so wavelike that it would be fuzzy and hard to grasp, and where the uncertainty principle would impinge on simple measurements of position and speed in a laboratory, making results irreproducible. In such a world, how would Planck's constant compare to what it is in this world?

35. Comment on the idea that the theory one accepts determines the meaning of one's observations and not vice versa.

36. A friend says, "If an electron is not a particle, then it must be a wave." What is your response? (Do you hear "either-or" statements like this often?)

37. Consider one of the many electrons on the tip of your nose. If somebody looks at it, will its motion be altered? How about if it is looked at with one eye closed? With two eyes, but crossed? Does Heisenberg's uncertainty principle apply here?

38. Do we alter that which we attempt to measure in a public opinion survey? Does Heisenberg's uncertainty principle apply here?

39. If the past behavior of a system is determined exactly and is understood, does it follow that the future behavior of that system can be exactly predicted? (Is there a distinction between a *determined* system and a *predictable* system?)

40. To measure the exact age of Old Methuselah, the oldest living tree in the world, a Nevada professor of dendrology, aided by an employee of the U.S. Bureau of Land Management, in 1965 cut the tree down and counted its rings. Is this an example of the uncertainty principle in its extreme or an example of arrogance and criminal stupidity?

Problems

1. A typical wavelength of infrared radiation emitted by your body is 25 μm (2.5×10^{-5} m). What is the energy per photon of such radiation?

2. What is the de Broglie wavelength of an electron that strikes the back of the face of a TV screen at $\frac{1}{10}$ the speed of light?

3. You decide to roll a 0.1-kg ball across the floor so slowly that it will have a small momentum and a large de Broglie wavelength. If you roll it at 0.001 m/s, what will its wavelength be? How does this compare with the de Broglie wavelength of the high-speed electron in the previous problem?

ATOMIC AND NUCLEAR PHYSICS

"Know nukes!" The natural heat of the earth that warms this natural hot spring, or that powers a geyser or powers a volcano, comes from nuclear power –– the radioactivity of minerals in the earth's core. Power from the atomic nucleus has been with us since the earth was formed and is not restricted to today's nuclear reactors, or "nukes," as they are called. How about that!

31

The Atom and the Quantum

: **A** scanning tunneling microscope image of 48 iron atoms on a copper surface.

We discussed the atom as a building block of matter in Chapter 10 and as an emitter of light in the preceding chapters. We know that the atom is composed of a central nucleus surrounded by a complex arrangement of electrons; the study of this atomic structure is called *atomic physics*. In this chapter we will outline some of the developments that led to our present understanding of the atom. We will trace developments in atomic physics from classical to quantum physics. In the two following chapters, we will learn about nuclear physics—the study of the structure of the atomic nucleus. This knowledge of the atom and its implications are having a profound impact on human society.

We begin our study of atomic and nuclear physics with a brief look at some events at the turn of the century that led to our current understanding of the atom.

Discovery of the Atomic Nucleus

Half a dozen years after Einstein announced the photoelectric effect, the British physicist Ernest Rutherford performed his now-famous gold-foil experiment. This 1911 experiment showed that the atom was mostly empty space, with most of its mass packed into the central region—the *nucleus*. Rutherford and his team directed a beam of positively charged particles (alpha particles) from a radioactive source through a very thin

FIGURE 31.1 The occasional large-angle scattering of alpha particles from gold atoms led Rutherford to the discovery of the small, very massive nuclei at their centers.

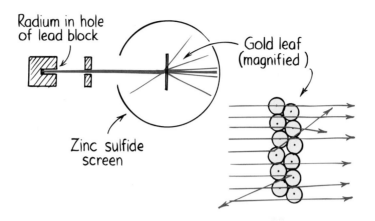

Ernest Rutherford

gold leaf. They measured the angles at which the particles were deflected from their straight-line path as they emerged from the gold foil. Rutherford determined the angles of particle deflection by noting spots of light on a zinc sulfide screen around the gold leaf (Figure 31.1). Most particles continued, as expected, in a basically straight-line path after going through the foil. But, surprisingly, some particles were scattered back along their incident paths. As Rutherford later put it, "It was as though you had fired a 15-inch shell at a piece of tissue paper and it came back and hit you." Rutherford reasoned that the particles that were only slightly deflected traveled through mostly empty space in the gold foil, while any particle deflected through a large angle must have experienced a very strong force by coming close to a concentration of positive charge at the center of an atom. Rutherford had discovered the atomic nucleus.

Atomic Spectra: Clues to Atomic Structure

During the period of Rutherford's experiments, chemists were using the spectroscope (discussed in Chapter 29) for chemical analysis, while physicists were busy trying to find order in the confusing arrays of spectral lines. It had long been known that the lightest element, hydrogen, has a far more orderly spectrum than the other elements (Figure 31.2). An important sequence of lines in the hydrogen spectrum starts with a line in the red region, followed by one in the blue, then by several lines in the violet, and many in the ultraviolet. Spacing between successive lines becomes smaller and smaller from the first in the red to the last in the ultraviolet, until the lines become so close they seem to merge. A Swiss schoolteacher, J. J. Balmer, first expressed the wavelengths of these lines in a single mathematical formula in 1884. Balmer, however, could give no reason why his formula worked so successfully. His guess that series for other elements might follow a similar formula proved correct, leading to the prediction of lines that had not yet been measured.

FIGURE 31.2 A portion of the hydrogen spectrum (higher frequency is to the right).

Another regularity in atomic spectra was found by J. Rydberg. He noticed that the sum of the frequencies of two lines in the spectrum of hydrogen sometimes equals the frequency of a third line. This relationship was later advanced as a general principle by W. Ritz and is called the **Ritz combination principle**. It states that the spectral lines of any element include frequencies that are either the sum or the difference of the frequencies of two other lines. Like Balmer, Ritz was unable to offer an explanation for this regularity. These regularities were the clues Danish physicist Niels Bohr used to understand the structure of the atom itself.

Bohr Model of the Atom

Niels Bohr

In 1913 Bohr applied the quantum theory of Planck and Einstein to the nuclear atom of Rutherford and formulated the well-known planetary model of the atom.* Bohr reasoned that Planck's quantized electron energy states (discussed in the previous chapter) corresponded to electrons orbiting at different distances from the nucleus. He reasoned that light is emitted when electrons make a transition from a larger to a smaller orbit (a higher to a lower energy). Further, Bohr realized that the frequency of emitted radiation is given by $E = hf$, where E is the difference in the atom's energy when the electron is in the different orbits. So the frequencies of the emitted spectral lines told Bohr what the energy *differences* in the atom are; from there he could take the next step and figure out the energies of the individual orbits.

Bohr's planetary model of the atom begged a major question. Accelerated electrons, according to Maxwell's theory, radiate energy in the form of electromagnetic waves. So an electron accelerating around a nucleus should radiate energy continuously. This radiating away of energy should cause the electron to spiral into the nucleus (Figure 31.4). Bohr boldly broke with classical physics by stating that the electron doesn't radiate light while it accelerates around the nucleus in a single orbit, but that radiation of light takes place only when the electron jumps orbit from a higher energy level to a lower energy level. The energy of the emitted photon is equal to the *difference* in energy between the two energy levels, $E = hf$. Color depends on the jump. So the quantization of light energy neatly corresponds to the quantization of electron energy.

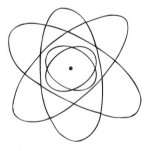

FIGURE 31.3 The Bohr model of the atom. Although this model is very over-simplified, it is still useful in understanding light emission.

FIGURE 31.4 According to classical theory, an electron accelerating around its orbit should continuously emit radiation. This loss of energy should cause it to spiral rapidly into the nucleus. But this does not happen.

---

* This model, like most models, has major defects, because the electrons do not revolve in planes as planets do. Later, the model was revised; "orbits" became "shells" and "clouds." We use *orbit* because it was, and still is, commonly used. Electrons are not just bodies, like planets, but rather are like waves concentrated in certain parts of the atom.

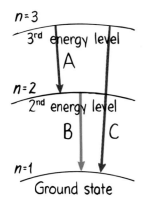

FIGURE 31.5 Three of the many energy levels in an atom. The total energy radiated in jumps A + B equals the energy radiated in jump C. The sum of the frequencies of light emitted from jump A and jump B therefore equals the frequency of light emitted from jump C.

Bohr's views, as outlandish as they seemed at the time, explained the regularities being found in atomic spectra. Bohr's explanation of the Ritz combination principle is shown in Figure 31.5. If an electron is raised to the third energy level, it can return to its initial level by a single jump from the third to the first level—or by a double jump, first to the second level and then to the first level. These two return paths will produce three spectral lines. Note that the sum of the energy jumps along paths A and B is equal to the energy jump C. Since frequency is proportional to energy, the frequencies of light emitted along path A and path B when added equal the frequency of light emitted when the transition is along path C. Now we see why the sum of two frequencies in the spectrum is equal to a third frequency in the spectrum.

Bohr was able to account for X rays in heavier elements, showing that they are emitted when electrons jump from outer to innermost orbits. He predicted X-ray frequencies that were later experimentally confirmed. Bohr was also able to calculate the "ionization energy" of a hydrogen atom—the energy needed to knock the electron out of the atom completely. This also was verified by experiment.

Using measured frequencies of X rays as well as visible, infrared, and ultraviolet light, scientists could map energy levels of all the atomic elements. Bohr's model had electrons orbiting in neat circles (or ellipses) arranged in groups or shells. This model of the atom accounted for the general chemical properties of the elements. It also predicted a missing element, which led to the discovery of hafnium.

Bohr solved the mystery of atomic spectra while providing an extremely useful model of the atom. He was quick to point out that his model was to be interpreted as a crude beginning, and the picture of electrons whirling about the nucleus like planets about the sun was not to be taken literally (to which popularizers of science paid no heed). His sharply defined orbits were conceptual representations of an atom whose later description involved a wave description—quantum mechanics. Still, his planetary model of the atom, with electrons occupying discrete energy levels, underlies the more complex models of the atom today.

Questions

1. What is the maximum number of paths for de-excitation available to a hydrogen atom excited to level number 3 in changing to the ground state?

2. Two predominant spectral lines in the hydrogen spectrum, an infrared one and a red one, have frequencies 2.7×10^{14} Hz and 4.6×10^{14} Hz respectively. Can you predict a higher-frequency line in the hydrogen spectrum?

Answers

1. Three, as shown in Figure 31.5.

2. The sum of the frequencies is $2.7 \times 10^{14} + 4.6 \times 10^{14} = 7.3 \times 10^{14}$ Hz, which happens to be the frequency of a violet line in the hydrogen spectrum. Can you see that if the infrared line is produced by a transition similar to path A in Figure 31.5 and the red line corresponds to path B, then the violet line corresponds to path C?

Relative Sizes of Atoms

The diameters of the electron orbits in the Bohr model of the atom are determined by the amount of electrical charge in the nucleus. For example, the positive proton in the hydrogen atom holds one electron in an orbit at a certain radius. If we double the positive charge in the nucleus, the orbiting electron will be pulled into a tighter orbit with half its former radius since the electrical attraction is doubled. This doesn't quite happen, however, because the double charge in the nucleus can attract and hold a second orbital electron, which diminishes the effect of the positive nucleus. This added electron makes the atom electrically neutral. The atom is no longer hydrogen. It is helium.

The two orbital electrons assume an orbital configuration characteristic of helium. An additional proton in the nucleus pulls the two electrons into an even closer orbit and, furthermore, holds a third electron in a second orbit. This is the lithium atom, atomic number 3. We can continue with this process, increasing the positive charge of the nucleus and adding successively more electrons and more orbits all the way up to atomic numbers above 100, to the "synthetic" radioactive elements.*

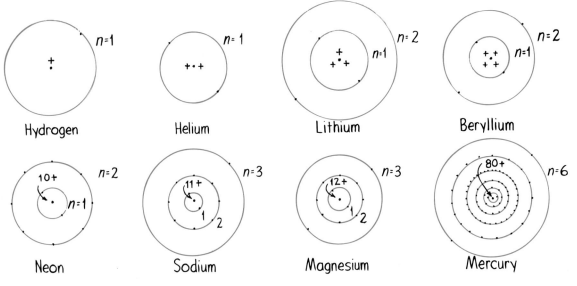

FIGURE 31.6   Orbital models for some light and heavy atoms drawn to approximate scale. Note that the heavier atoms are not much larger than the lighter atoms. (Adapted from *Descriptive College Physics* 3/e by Harvey E. White. ©1971 Litton Educational Publishing Co. Reprinted by permission of Wadsworth, Inc.)

We find that as the nuclear charge increases and additional electrons are added in outer orbits, the inner orbits shrink in size because of the stronger nuclear attraction. This means that the heavier elements are not much larger in diameter than the lighter elements. The diameter of the uranium atom, for example, is only about three hydrogen diameters, even though it is 238 times as massive. The schematic diagrams in Figure 31.6 are drawn approximately to the same scale.

We find that each element has an arrangement of electron orbits unique to that element. For example, the radii of the orbits for the sodium atom are the same for all sodium atoms, but different from the radii of the orbits for other kinds of atoms. When we consider the 92 naturally occurring elements, we find that there are 92 distinct patterns or electron orbital configurations—a different pattern for each element.

* Each orbit will hold only so many electrons. A rule of quantum mechanics is that an orbit is filled when it contains a number of electrons given by $2n^2$ where n is 1 for the first orbit, 2 for the second orbit, 3 for the third orbit, and so on. For $n = 1$, there are 2 electrons; for $n = 2$, there are $2(2^2)$ or 8 electrons; for $n = 3$, there are a maximum of $2(3^2)$ or 18 electrons, etc. The number n is called the principal quantum number. Because of complexities that arise in heavy atoms, the $2n^2$ rule is strictly valid only for the lighter atoms.

Explanation of Quantized Energy Levels: Electron Waves

So we see that a photon is emitted when an electron makes a transition from a higher to a lower energy level, and the frequency of the photon is equal to the energy-level difference divided by Planck's constant, h. If an electron jumps down a large energy-level difference, the emitted photon has a high frequency—perhaps ultraviolet. If an electron jumps through a lesser energy difference, the emitted photon is lower in frequency—perhaps it is a photon of red light. Each element has its own characteristic energy levels; thus, transitions of electrons between these levels result in each element emitting its own characteristic colors. Each of the elements emits its own unique pattern of spectral lines.

The idea that electrons may occupy only certain levels was very perplexing to early investigators and to Bohr himself. It was perplexing because the electron was considered to be a particle, a tiny BB whirling around the nucleus like a planet whirling around the sun. Just as a satellite can orbit at any distance from the sun, it would seem that an electron should be able to orbit around the nucleus at any radial distance—depending, of course, like the satellite, on its speed. Moving among all orbits would enable the electrons to emit all energies of light. But this doesn't happen. It can't. Why the electron occupies only discrete levels is understood by considering the electron to be, not a particle, but a *wave*.

Louis de Broglie introduced the concept of matter waves in 1924. He hypothesized that a wave is associated with every particle and that the wavelength of a matter wave is inversely related to a particle's momentum. These de Broglie **matter waves** behave just like other waves; they can be reflected, refracted, diffracted, and caused to interfere. Using the idea of interference, de Broglie showed that the discrete values of radii of Bohr's orbits are a natural consequence of standing electron waves. A Bohr orbit exists where an electron wave closes on itself constructively. The electron wave becomes a standing wave, like a wave on a music string. In this view the electron is thought of not as a particle located at some point in the atom but as though its mass and charge were spread out into a standing wave surrounding the atomic nucleus, with an integral number of wavelengths fitting evenly into the circumferences of the orbits (Figure 31.7). The circumference of the innermost orbit, according to this picture, is equal

FIGURE 31.7 (a) Orbital electrons form standing waves only when the circumference of the orbit is equal to an integral number of wavelengths. In (b) the wave does not close in on itself in phase, and therefore it undergoes destructive interference.

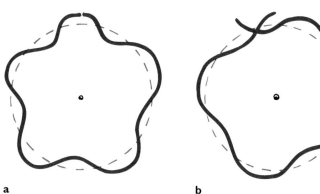

a   b

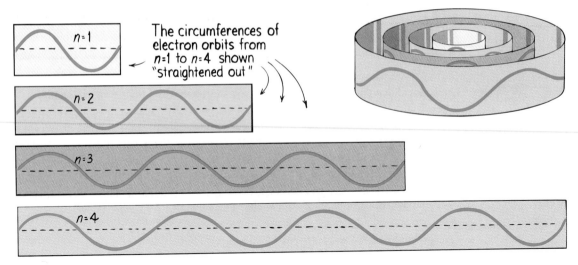

FIGURE 31.8 The electron orbits in an atom have discrete radii because the circumferences of the orbits are integral multiples of the electron wavelengths. The wavelengths differ for different elements (and also for different orbits within the elements), resulting in discrete orbital radii, or energy levels, which are characteristic of each element. The figure is greatly oversimplified, as the standing waves make up spherical and ellipsoidal shells rather than flat, circular ones.

to one wavelength. The second orbit has a circumference of two electron wavelengths, the third three, and so forth (Figure 31.8). This is similar to a chain necklace made of paper clips. No matter what size necklace is made, its circumference is equal to some multiple of the length of a single paper clip.* Since the circumferences of electron orbits are discrete, it follows that the radii of these orbits, and hence the energy levels, are also discrete.

This model explains why electrons don't spiral closer and closer to the nucleus, causing atoms to shrink to tiny dots. If each electron orbit is described by a standing wave, the circumference of the smallest orbit can be no smaller than one wavelength—no fraction of a wavelength is possible in a circular (or elliptical) standing wave.

In the still more modern wave model of the atom, electron waves move not only around the nucleus, but also in and out, toward and away from the nucleus. The electron wave is spread out in three dimensions. This leads to the picture of an electron "cloud," as we shall see.

Quantum Mechanics

The mid-1920s saw many changes in physics. Not only was the particle nature of light established experimentally, but material particles were found to have wave properties. Starting with de Broglie's matter waves, the Austrian-German physicist Erwin Schrödinger formulated an equation that describes how matter waves change under the influence of external forces. Schrödinger's equation plays the same role in **quantum mechanics** that Newton's equation (acceleration = force/mass) plays in classical

---

* For each orbit the electron has a unique speed, which determines its wavelength. Electron speeds decrease and wavelengths increase as the radii of orbits increase. So to make our analogy accurate, we would have to use not only more paper clips to make longer necklaces, but larger paper clips for each larger orbit as well.

Erwin Schrödinger

physics.* The matter waves in Schrödinger's equation are mathematical entities that are not directly observable, so the equation provides us with a purely mathematical rather than a visual model of the atom—which puts it beyond the scope of this book. So our discussion of it will be brief.†

In **Schrödinger's wave equation**, the thing that "waves" is the nonmaterial *matter wave amplitude*—a mathematical entity called a *wave function,* represented by the symbol ψ (the Greek letter psi). The wave function given by Schrödinger's equation represents the possibilities that can occur for a system. For example, the location of the electron in a hydrogen atom may be anywhere from the center of the nucleus to a radial distance approaching infinity. An electron's possible position and its probable position at a particular time are not the same. A physicist can calculate its probable position by multiplying the wave function by itself ($|\psi|^2$). This produces a second mathematical entity called a *probability density function,* which tells us at a given time the probability per unit volume for each of the possibilities represented by ψ. To a chemist the "orbital" is in fact a 3-dimensional graphical picture of $|\psi|^2$.

FIGURE 31.9 Probability distribution of an electron cloud.

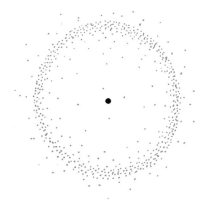

Experimentally, there is a finite probability (chance) of finding an electron in a particular region at any instant. The mathematical value of this probability lies between the limits 0 and 1, where 0 indicates never and 1 indicates always. For example, if the probability is 0.4 for finding the electron within a certain radius, this signifies a 40% chance that the electron will be there. So the Schrödinger equation cannot tell a physicist where an electron can be found in an atom at any moment, but only the *likelihood* of finding it there—or, for a large number of measurements, what fraction of measurements will find the electron in each region. When an electron's position in its Bohr energy level (state) is repeatedly measured and each of its locations is plotted as a dot, the resulting pattern resembles a sort of electron cloud (Figure 31.9). An individual electron may at various times be detected anywhere in this probability cloud; it even has an

---

* Schrödinger's wave equation, strictly for math types, is

$$\left(-\frac{\hbar^2}{2m}\nabla^2 + V\right)\psi = i\hbar\frac{\partial\psi}{\partial t}$$

† Our short treatment of this complex subject is hardly conducive to any real understanding of quantum mechanics. At best it serves as a brief overview and possible introduction to further study. The reading suggested at the end of the chapter may be quite useful.

FIGURE 31.10 From the Bohr model of the atom to the modified model with de Broglie waves to a wave model with the electrons distributed in a "cloud" throughout the atomic volume.

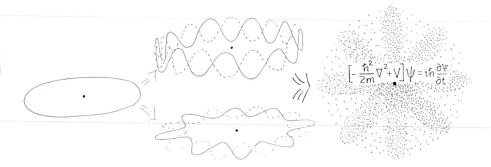

$$\left[-\frac{\hbar^2}{2m}\nabla^2 + V\right]\psi = i\hbar\frac{\partial\psi}{\partial t}$$

extremely small but finite probability of momentarily existing inside the nucleus. It is detected most of the time, however, close to an average distance from the nucleus, which fits the orbital radius described by Niels Bohr.

Questions

1. Consider 100 photons diffracting through a thin slit to form a diffraction pattern. If we detect five photons in a certain region in the pattern, what is the probability (between 0 and 1) of detecting a photon in this region?

2. Open a second identical slit and the diffraction pattern is one of bright and dark bands. Suppose the region where 5 photons hit before now has none. A wave theory says waves that hit before are now canceled by waves from the other slit—that crests and troughs combine to 0. But our measurement is of photons that either make a hit or don't. How does quantum mechanics reconcile this?

Most physicists, but not all, view quantum mechanics as a fundamental theory of nature. Interestingly enough, Albert Einstein, one of the founders of quantum physics, never accepted it as fundamental; he considered the probabilistic nature of quantum phenomena as the outcome of a deeper, as yet undiscovered, physics. He stated, "Quantum mechanics is certainly imposing. But an inner voice tells me it is not yet the real thing. The theory says a lot, but does not really bring us closer to the secret of 'the Old One'."*

Answers

1. We have approximately a 0.05 probability of detecting a photon at this location. In quantum mechanics we say $|\psi|^2 \approx 0.05$. The true probability could be somewhat more or less than 0.05. Put the other way around, if the true probability is 0.05, the number detected could be somewhat more or less than 5.

2. Quantum mechanics says that photons propagate as waves and are absorbed as particles, with the probability of absorption governed by the maxima and minima of wave interference. Where the combined wave from the two slits has zero amplitude, the probability of a particle being absorbed is zero.

---

* Although Einstein practiced no religion, he often invoked God as the "Old One" in his statements about the mysteries of nature.

Correspondence Principle

If a new theory is valid, it must account for the verified results of the old theory. This is the **correspondence principle**, first articulated by Bohr. New theory and old must correspond; that is, they must overlap and agree in the region where the results of the old theory have been fully verified.*

When the techniques of quantum mechanics are applied to macroscopic systems rather than atomic systems, the results are essentially identical to those of classical mechanics. For a large system such as the solar system, where classical physics is successful, the Schrödinger equation leads to results that differ from classical theory only by infinitesimal amounts. The two domains blend when the de Broglie wavelength is small compared with the dimensions of the system or of the pieces of matter in the system. It is, in fact, impractical to use quantum mechanics in the domains where classical physics is successful. But at the atomic level, quantum physics reigns and is the only theory that gives results consistent with what is observed.

Summary of Terms

Ritz combination principle The statement that the frequencies of some spectral lines of the elements are either the sums or the differences of the frequencies of two other lines.

Quantum mechanics The theory of the microworld based on wave functions and probabilities, developed especially by Werner Heisenberg (1925) and Erwin Schrödinger (1926).

Schrödinger's wave equation A fundamental equation of quantum mechanics, which relates probability wave amplitudes to the forces acting on a system. It is as basic to quantum mechanics as Newton's laws of motion are to classical mechanics.

Correspondence principle The rule that a new theory must give the same results as the old theory where the old theory is known to be valid.

Suggested Reading

Cline, Barbara L. *The Questioners: Physicists and the Quantum Theory.* New York: Crowell, 1973. A fascinating account of the development of quantum physics with emphasis on the participating physicists.

Gamow, George. *Thirty Years That Shook Physics.* New York: Dover, 1985. A historical tracing of quantum theory by someone who was part of it.

Hey, A. J., & Walters, P. *The Quantum Universe.* New York: Cambridge University Press, 1987. A broad view of modern physics with many illustrations.

Pagels, H. R. *The Cosmic Code: Quantum Physics as the Language of Nature.* New York: Simon & Schuster, 1982. An elegant and highly recommended book for the general reader.

Trefil, J. *Atoms to Quarks.* New York: Scribner's, 1980. A nice development of quantum theory in the early chapters, which leads on to particle physics.

Review Questions

1. Distinguish between *atomic physics* and *nuclear physics*.

Discovery of the Atomic Nucleus

2. Why are alpha particles repelled by the atomic nucleus?
3. Why do most alpha particles fired through a piece of gold foil emerge almost undeflected?
4. Why do a few alpha particles fired at a piece of gold foil bounce backward?

Atomic Spectra: Clues to Atomic Structure

5. What did J. J. Balmer discover about the spectrum of hydrogen?
6. What did J. Rydberg and W. Ritz discover about atomic spectra?

Bohr Model of the Atom

7. What relationship between electron orbits and light emission did Bohr postulate?

---

* This is a general rule not only for good science, but for all good theory—even in areas as far removed from science as government, religion, and ethics.

8. According to Bohr, can a single electron in one excited state give off more than one photon when it jumps to a lower-energy state?

9. What is the relationship between the energy differences of orbits in an atom and the light emitted by the atom?

10. In what way, according to Bohr, is an atom like a tiny solar system? In what way is it different?

Relative Sizes of Atoms

11. Why is a helium atom smaller than a hydrogen atom?

12. Why are heavy atoms not much larger than the hydrogen atom?

Explanation of Quantized Energy Levels: Electron Waves

13. How does treating the electron as a wave rather than as a particle solve the riddle of why electron orbits are discrete?

14. According to the simple de Broglie model, how many wavelengths are there in an electron wave in the first orbit? In the second orbit? In the nth orbit?

15. How can we explain why electrons don't spiral into the attracting nucleus?

Quantum Mechanics

16. What does the wave function ψ represent?

17. Distinguish between a *wave function* and a *probability density function*.

18. How does the probability cloud of the electron in a hydrogen atom relate to the orbit described by Niels Bohr?

Correspondence Principle

19. Why is the correspondence principle a test for the validity of a new idea?

20. Would it be correct to say that an electron is a wave or a particle? Or would it be correct to say that an electron, whatever it is, *behaves* as if it were a wave and a particle?

Exercises

1. How does Rutherford's model of the atom account for the back-scattering of alpha particles directed at the gold leaf?

2. At the time of Rutherford's gold leaf experiment, scientists knew that negatively charged electrons exist within the atom, but they did not know where the positive charge resides. What information about the positive charge was provided by Rutherford's experiment?

3. Uranium is 238 times more massive than hydrogen. Why, then, isn't the diameter of the uranium atom 238 times that of the hydrogen atom?

4. Why does classical physics predict that atoms should collapse?

5. If the electron in a hydrogen atom obeyed classical mechanics instead of quantum mechanics, would it emit a continuous spectrum or a line spectrum? Explain.

6. Why are spectral lines often referred to as "atomic fingerprints"?

7. Figure 31.5 shows three transitions among three energy levels that would produce three spectral lines in a spectroscope. If the energy spacing between the levels were equal, would this affect the number of spectral lines?

8. How can elements with low atomic numbers have so many spectral lines?

9. How does the wave model of electrons orbiting the nucleus account for discrete energy values rather than arbitrary energy values?

10. Why do atoms that have the same number of electron shells decrease in size with increasing atomic number?

11. Why do helium and lithium differ by only one electron yet exhibit very different chemical behavior?

12. The Ritz combination principle can be considered to be a statement of energy conservation. Explain.

13. Why does no stable electron orbit exist in an atom for a circumference of 2.5 de Broglie wavelengths?

14. Can a particle be diffracted? Can it exhibit interference?

15. What does the amplitude of a matter wave have to do with probability?

16. If Planck's constant, h, were larger, would atoms be larger also? Defend your answer.

17. If the world of the atom is so uncertain and subject to the laws of probabilities, how can we accurately measure such things as light intensity, electric current, and temperature?

18. Why do we say that light has wave properties? Why do we say that light has particle properties?

19. Why do we say that electrons have particle properties? Why do we say that electrons have wave properties? Is there a contradiction here? Explain.

20. When only a few photons are observed, classical physics fails. When many are observed, classical physics is valid. Which of these two facts is consistent with the correspondence principle?

21. What does Bohr's correspondence principle say about quantum mechanics versus classical mechanics?

22. Does the correspondence principle have application to macroscopic events in the everyday macroworld?

23. Richard Feynman in his book *The Character of Physical Law* states: "A philosopher once said, 'It is necessary for the very existence of science that the same conditions always produce the same results.' Well, they don't!" Who was speaking of classical physics, and who was speaking of quantum physics?

24. What does the wave nature of matter have to do with the fact that we can't walk through solid walls, as Hollywood often shows in special effects films?

25. Largeness or smallness has meaning only relative to something else. Why do we usually call the speed of light "large" and Planck's constant "small"?

26. Make up a multiple-choice question that would check a classmate's understanding of the different domains of classical mechanics and quantum mechanics.

Problems

1. The higher the energy level occupied by an electron in the hydrogen atom, the larger the atom. The diameter of the atom is proportional to n^2, where $n = 1$ labels the lowest, or "ground" state, $n = 2$ is the second state, $n = 3$ is the third state, and so on. If the atom's diameter is 1×10^{-10} m in its lowest energy state, what is its diameter in state number 50? How many unexcited atoms could be fit within this one giant atom?

2. We can define zero energy in the hydrogen atom to be the energy of the lowest, or "ground," state. Energies of successive excited states above ground state are proportional to $100 - (100/n^2)$, for the quantum numbers $n = 1, 2, 3$, etc. So on this scale the energy of level $n = 2$ is $[100 - (\frac{100}{4})] = 75.0$, for $n = 3$, $[100 - (\frac{100}{9})] = 88.9$, and for $n = 4$, $[100 - (\frac{100}{16})] = 93.8$, and so on. (a) Sketch, approximately to scale, a diagram that includes the ground state and the lowest four excited states ($n = 1$ to 5). (b) The most prominent red line in the spectrum of hydrogen is caused by an electron transition from state 3 to state 2. Will the transition from states 4 to 3 produce a higher frequency or lower frequency spectral line? (c) What about the 2-to-1 transition?

32

The Atomic Nucleus and Radioactivity

The tracks of elementary particles are atomic versions of the ice-crystal trails left in the sky by jet planes.

In the previous chapter we were concerned with the study of the clouds of electrons that make up the atom. That was *atomic physics*. In this chapter we will burrow beneath the electrons and go deeper into the atom—to the atomic nucleus. The study of the nucleus was made possible by the chance discovery of radioactivity in 1896, which in turn was based on the discovery of X rays two months earlier. So to begin our study of *nuclear physics*, we'll consider X rays.

X Rays and Radioactivity

Before the turn of the twentieth century the German physicist Wilhelm Roentgen discovered a "new kind of ray" produced by a beam of "cathode rays" (later found to be electrons) striking the glass surface of a gas-discharge tube. He named these **X rays**—rays of an unknown nature. Roentgen found that X rays could pass through solid materials, could ionize the air, showed no refraction in glass, and were undeflected by magnetic fields. Today we know that X rays are high-frequency electromagnetic waves, usually emitted by the de-excitation of the innermost orbital electrons of atoms. Whereas the electron current in a fluorescent lamp excites the outer electrons of atoms and produces ultraviolet and visible photons, a more energetic beam of electrons excites the innermost electrons and produces higher-frequency photons of X radiation.

FIGURE 32.1 X rays emitted by excited metallic atoms in the electrode pass more readily through flesh than through bone and produce an image on the film.

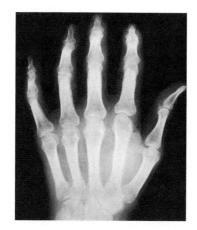

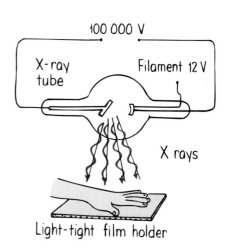

Marie Curie

X-ray photons have high energy and can pass through many layers of atoms before being absorbed or scattered. X rays do this when they pass through your soft tissues to produce images of the bones inside your body (Figure 32.1) In a modern X-ray tube, the target of the electron beam is a metal plate rather than the glass wall of the tube.

Two months after Roentgen announced his discovery of X rays, the French physicist Antoine Henri Becquerel tried to find out whether any elements spontaneously emit X rays. To do this, he wrapped a photographic plate in black paper to keep out the light and then put pieces of various elements against the wrapped plate. From Roentgen's work, Becquerel knew that if these materials emitted X rays, the rays would go through the paper and blacken the plate. He found that although most elements produced no effect, uranium did give out rays. It was soon discovered that similar rays are emitted by other elements, such as thorium, actinium, and two new elements discovered by Marie and Pierre Curie—polonium and radium. The emission of these rays was evidence of much more drastic changes in the atom than atomic excitation. These rays were the result not of changes in the electron energy states of the atom, but of changes occurring within the central atomic core—the nucleus. These rays were the result of a spontaneous chipping apart of the atomic nucleus—*radioactivity.*

Alpha, Beta, and Gamma Rays

The radioactive elements emit three distinct types of rays, called by the first three letters of the Greek alphabet, α, β, γ—*alpha, beta,* and *gamma,* respectively. **Alpha rays** have a positive electrical charge, **beta rays** have a negative charge, and **gamma rays** have no charge at all. The three rays can be separated by putting a magnetic field across their

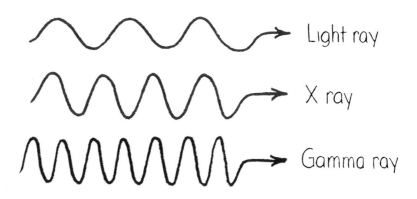

FIGURE 32.2 A gamma ray is simply electromagnetic radiation, much higher in frequency and energy than light and X rays.

FIGURE 32.3 In a magnetic field, alpha rays bend one way, beta rays bend the other way, and gamma rays don't bend at all. The combined beam comes from a radioactive source placed at the bottom of a hole drilled in a lead block.

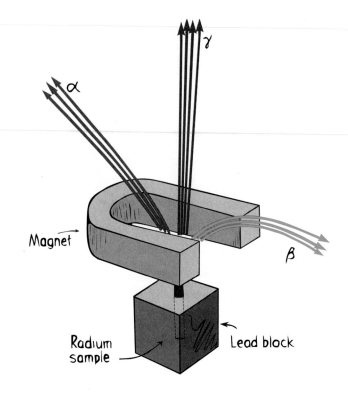

paths (Figure 32.3). Further investigation has shown that an alpha ray is a stream of helium nuclei, and a beta ray is a stream of electrons. Hence, we often call these **alpha particles** and **beta particles**. A gamma ray is electromagnetic radiation (a stream of photons) whose frequency is even higher than that of X rays. Whereas X rays originate in the electron cloud outside the atomic nucleus, gamma rays originate in the nucleus. Gamma photons provide information about nuclear structure, much as visible and X-ray photons give information about atomic electron structure.

FIGURE 32.4 Alpha particles are the least penetrating and can be stopped by a few sheets of paper. Beta particles will readily pass through paper, but not through a sheet of aluminum. Gamma rays penetrate several centimeters into solid lead.

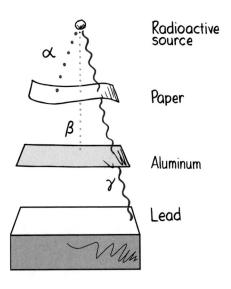

The Nucleus

As described in earlier chapters, the atomic nucleus occupies only a few quadrillionths of the volume of the atom. Thus, the atom is mostly empty space. The nucleus is composed of **nucleons**, which when electrically charged are protons and when electrically neutral are neutrons. The positive charge of the proton is the same in magnitude as the negative charge of the electron. Nucleons have nearly 2000 times the mass of electrons, so the mass of an atom is practically equal to the mass of its nucleus. The neutron's mass is slightly greater than the proton's. We'll see that when an electron is ejected from a neutron (beta emission), the neutron becomes a proton.

Nuclear radii range from about 10^{-15} meter for hydrogen to about seven times larger for uranium. Some nuclei are spherical but most deviate from that shape in the "football" way and a few in the "doorknob" way. Protons and neutrons within the nucleus move relatively freely, yet provide a "skin" that gives the nucleus some properties of a liquid drop. Just as there are energy levels for the orbital electrons of an atom, there are energy levels within the nucleus. Whereas electrons making transitions to lower orbits emit photons of light, similar changes of energy states within the nucleus result in the emission of gamma-ray photons.

The emission of alpha particles is a quantum phenomenon that can be understood in terms of waves and probability. Just as orbital electrons form a probability cloud about the nucleus, inside the radioactive nucleus there is a similar probability cloud for the clustering of the two protons and two neutrons that comprise an alpha particle. A tiny part of the alpha particle's probability wave extends outside the nucleus, meaning that there is a small chance that the alpha particle will be outside. Once outside, it is hurled violently away by electric repulsion. The electron emitted in beta decay, on the other hand, is not "there" before it is emitted. It is created at the moment of radioactive decay when a neutron is transformed into a proton.

In addition to alpha, beta, and gamma rays, more than 200 various other particles have been detected coming from the nucleus when it is clobbered by energetic particles. We do not think of these so-called elementary particles as being buried within the nucleus and then popping out, just as we do not think of a spark as being buried in a match before it is struck. These particles, like the electrons in beta decay, come into being when the nucleus is disrupted. There are regularities in the masses of these particles as well as the particular characteristics of their creation. Because of these similarities, they are categorized together as a fairly compact family of six subnuclear particles—the **quarks**.

Two of the six quarks are the fundamental building blocks of all nucleons. An unusual property of quarks is that they carry fractional electrical charges. One kind, the *up* quark, carries $+\frac{2}{3}$ the proton charge, and another kind, the *down* quark, has $-\frac{1}{3}$ the proton charge. (The name *quark*, inspired by a quotation from *Finnegans Wake* by James Joyce, was chosen in 1963 by Murray Gell-Mann, who first proposed their existence.) Each quark has an antiquark with opposite electric charge. The proton consists of the combination *up up down*, and the neutron of *up down down*. The other four quarks bear the whimsical names *strange, charm, top,* and *bottom*. All of the several hundred particles that feel the strong nuclear force appear to be composed of some combination of the six quarks. As with magnetic poles, no quarks have been isolated and experimentally observed—most investigators think quarks by their nature cannot be isolated.

Elementary particles that are not involved with the nuclear forces belong to a class of six particles called *leptons,* which are not composed of quarks. The six leptons are the electron and its associated neutrino, the muon and its neutrino, and a heavier particle called *tau* and its neutrino. Leptons feel electromagnetic forces and a weak nuclear force, but are unaffected by the strong nuclear force. At the present time, the six quarks and six leptons (and their antiparticles) are thought to be the truly *elementary particles,* particles not composed of more basic entities. They are the building blocks of which all matter is composed. Investigation of elementary particles is at the frontier of our present knowledge and the area of much current excitement and research.

Isotopes

The nucleus of a hydrogen atom contains a single proton. Helium has two protons, lithium has three, and so forth. Every succeeding element in the list of elements has one more proton than the preceding element. In neutral atoms, there are as many protons in the nucleus as there are electrons outside the nucleus. The number of protons in the nucleus is the same as the atomic number.

The number of neutrons in the nucleus of a given element may vary somewhat. For example, every atom of chlorine has 17 protons and hence 17 orbital electrons that determine what chemical compounds it can make. But the number of accompanying neutrons varies. Atoms that have like numbers of protons but unlike numbers of neutrons are called **isotopes**. The two most common chlorine isotopes have 35 and 37 times the mass of a single nucleon. We denote these by $^{35}_{17}Cl$ and $^{37}_{17}Cl$, where the bottom number (17 in this case) refers to the **atomic number** and the top number (35 or 37 in this case) refers to the **atomic mass number** (approximately but not exactly equal to the *atomic mass*).

The mass number corresponds to the total number of nucleons in the nucleus. Since for both isotopes of chlorine, 17 of these are protons, the lighter isotope has $35 - 17 = 18$ neutrons, and the heavier isotope has $37 - 17 = 20$ neutrons. In nature the isotopes of chlorine are found mixed, with three times as many $^{35}_{17}Cl$ atoms as there are $^{37}_{17}Cl$ atoms, so the average atomic mass of naturally occurring chlorine is about 35.5. Chlorine is only one of many elements found in nature that consist of two or more stable isotopes. Even hydrogen, the lightest element, has three known isotopes (Figure 32.5). The double-weight hydrogen isotope $^{2}_{1}H$ is called *deuterium.* "Heavy water" is the name usually given to H_2O in which one or both of the H's are deuterium atoms. In all hydrogen compounds in nature such as methane and water there is 1 atom of deuterium to about 6000 atoms of "ordinary" hydrogen. The triple-weight hydrogen isotope $^{3}_{1}H$, which is not stable but lives long enough to be a known constituent of atmospheric water, is called *tritium.* Tritium is present only in extremely minute amounts—less than 1 per 10^{17} atoms of ordinary hydrogen.

FIGURE 32.5 Three isotopes of hydrogen. Each nucleus has a single proton, which holds a single orbital electron, which in turn determines the chemical properties of the atom. The different number of neutrons changes the mass of the atom but not its chemical properties.

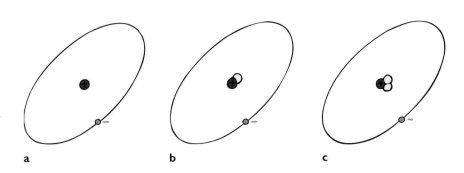

a   b   c

There are three isotopes of uranium that naturally occur in the earth's crust; the most common is $^{238}_{92}$U. In briefer notation we can drop the atomic number and simply say U-238. (We will see in the next chapter that it is the isotope U-235 that undergoes nuclear fission.) Of the 83 elements present on earth in significant amounts, 20 have a single stable isotope. The others have from 2 to 10 stable isotopes. Taking radioactive isotopes into account, over 2000 distinct isotopes are known.

Question State the numbers of protons and neutrons in each of the following nuclei: $^1_1$H, $^{14}_6$C, $^{235}_{92}$U.

Why Atoms Are Radioactive

FIGURE 32.6 The nuclear strong interaction is a very short-range force. For nucleons very close or in contact, it is very strong. But a few nucleon diameters away it is nearly zero.

The positively charged and closely spaced protons in a nucleus have huge electrical forces of repulsion between them. Why don't they fly apart because of this huge repulsive force? Because there is an even more formidable force within the nucleus—the nuclear force. Both neutrons and protons are bound to each other by this attractive force. The nuclear force is much more complicated than the electrical force and is only now becoming understood. The principal part of the nuclear force, the part that holds the nucleus together, is called the *strong interaction.** The strong interaction is an attractive force that acts between protons, neutrons, and particles called *mesons,* all of which are called *hadrons.* This force acts over only a very short distance (Figure 32.6). It is very strong between nucleons about 10^{-15} meter apart, but close to zero at greater separations. So the strong nuclear interaction is a short-range force. Electrical interaction, on the other hand, weakens as the inverse square of separation distance and is a relatively long-range force. So as long as protons are close together, as in small nuclei, the nuclear force easily overcomes the electrical force of repulsion. But for distant protons, like those on opposite edges of a large nucleus, the attractive nuclear force may be small in comparison to the repulsive electrical force. Hence, a larger nucleus is not as stable as a smaller nucleus.

The presence of the neutrons also plays a large role in nuclear stability. It turns out that a proton and a neutron can be bound together a little more tightly, on average, than two protons or two neutrons. As a result, many of the first 20 or so elements have equal numbers of neutrons and protons.

For heavier elements, it is a different story, because protons repel each other electrically and neutrons do not. If you have a nucleus with 28 protons and 28 neutrons, for example, it can be made more stable by replacing two of the protons with neutrons,

---

Answer 1 proton and no neutron in $^1_1$H; 6 protons and 8 neutrons in $^{14}_6$C; 92 protons and 143 neutrons in $^{235}_{92}$U.

---

* Fundamental to the strong interaction is the *color force* (which has nothing to do with visible color). This color force interacts between quarks and holds them together by the exchange of "gluons." Read more about this in H. R. Pagels, *The Cosmic Code: Quantum Physics as the Language of Nature* (New York: Simon & Schuster, 1982).

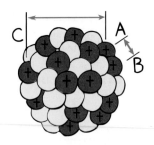

FIGURE 32.7 Proton A both attracts (nuclear force) and repels (electrical force) proton B, but mainly repels proton C (because the nuclear attraction is weaker at greater distances). The greater the distance between A and C, the more unstable the nucleus because more pairs of protons repel than attract. (The yellow particles are neutrons.)

resulting in Fe-56, the iron isotope with 26 protons and 30 neutrons. The inequality of neutron and proton numbers becomes more pronounced for still heavier elements. For example, in U-238, which has 92 protons, there are 146 (238 − 92) neutrons. If the uranium nucleus were to have equal numbers of protons and neutrons, 92 protons and 92 neutrons, it would fly apart at once because of the electrical repulsion forces. The extra 54 neutrons are needed for relative stability. Even so, the U-238 nucleus is still unstable because of the electrical forces.

To put the matter in another way: There is an electrical repulsion between *every pair* of protons in the nucleus, but there is not a substantial nuclear attractive force between every pair (Figure 32.7). Every proton in the uranium nucleus exerts a repulsion on each of the other 91 protons—those near and those far. However, each proton (and neutron) exerts an appreciable nuclear attraction only on those nucleons that happen to be near it.

We find that all nuclei having more than 82 protons are unstable. In this unstable environment, alpha and beta emissions take place. The force responsible for beta emission is called the *weak interaction:* it acts on leptons as well as nucleons. When an electron is created in beta decay, another lighter particle called an *antineutrino* is also created and also shoots out of the nucleus.

Half-Life

Radioactive elements decay to other elements when they emit subatomic particles. The radioactive decay rate of an element is measured in terms of a characteristic time, the **half-life**. The half-life of a radioactive isotope is the time needed for half of any given quantity of that isotope to decay. Radium-226, for example, has a half-life of 1620 years. This means that half of any given specimen of radium-226 will be converted into other elements by the end of 1620 years. In the next 1620 years, half of the remaining radium will decay, leaving only one-fourth the original number of radium atoms. By a succession of disintegrations the radium ultimately becomes lead. After 20 half-lives, an initial quantity of radioactive atoms will be diminished 1 millionfold. The isotopes of some elements have a half-life of less than a millionth of a second, while uranium-238,

FIGURE 32.8 Every 1620 years the amount of radium decreases by half.

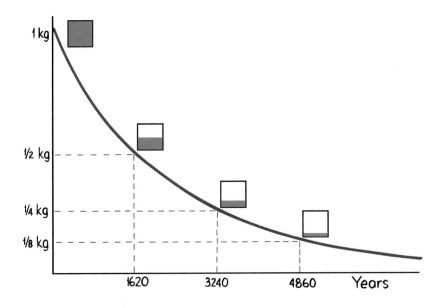

for example, has a half-life of 4.5 billion years. Every isotope of every radioactive element has its own characteristic half-life.

The neutron can be stable inside the atomic nucleus, but by itself it is unstable and has a half-life of 12 minutes. Many elementary particles have even shorter half-lives. The muon, a close relative of the electron that is produced by the bombardment of cosmic rays against the upper atmosphere, has a half-life of 2 millionths of a second (2×10^{-6} s). The shortest half-lives of elementary particles are on the order of 10^{-23} second, the time light takes to travel a distance equal to the diameter of a nucleus.

The half-lives of radioactive elements and elementary particles appear to be absolutely constant, unaffected by any external conditions, however drastic. Wide temperature and pressure extremes, strong electric and magnetic fields, and even violent chemical reactions have no detectable effect on the rate of decay of a given element. Any of these stresses, although severe by ordinary standards, is far too mild to affect the nucleus in the deep interior of the atom.

How do we measure half-lives? Observing a specimen and waiting until half of it is decayed won't work if the half-life is long. The half-life of a radioactive substance is related to its rate of disintegration. The shorter the half-life of an element, the faster is the rate of disintegration and the more active is the element. The half-life can be determined from the rate of disintegration, which can be measured in the laboratory.

Radiation Detectors

Ordinary thermal motions of atoms bumping one another in a gas or liquid are not energetic enough to dislodge electrons, and the atoms remain neutral. But when an energetic particle such as an alpha or a beta particle shoots through matter, electrons one after another are knocked from the atoms in the particle's path. The result is a trail of freed electrons and positively charged ions. This ionization process is responsible for the harmful effects of high-energy radiation in living cells. Ionization also makes it relatively easy to trace the paths of high-energy particles. We will briefly discuss four radiation detection devices.

1. A *Geiger counter* consists of a central wire in a hollow metal cylinder filled with gas. An electrical voltage is applied across the cylinder and wire so that the wire is more positive than the cylinder. If radiation enters the tube and ionizes an atom in the gas, the freed electron is attracted to the positively charged central wire. As this electron is accelerated toward the wire, it collides with other atoms and knocks out more electrons, which in turn produce more electrons, and so on, resulting in a cascade of electrons moving toward the wire. This makes a short pulse of electric current, which activates a counting device connected to the tube. Amplified, this pulse of current makes the familiar clicking sound we associate with radiation detectors.

FIGURE 32.9 Radiation detectors. (a) A Geiger counter detects incoming radiation by its ionizing effect on enclosed gas in the tube. (b) A scintillation counter detects incoming radiation by flashes of light that are produced when charged particles or gamma rays pass through it.

a

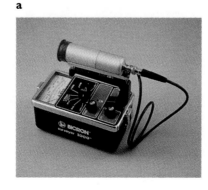

b

2. A *scintillation counter* uses the fact that certain substances are easily excited and emit light when charged particles or gamma rays pass through them. Tiny flashes of light, or scintillations, are converted into electric signals by special photo-multiplier tubes. A scintillation counter is much more sensitive to gamma rays than a Geiger counter and, in addition, can measure the energy of charged particles or gamma rays absorbed in the detector.

3. A *cloud chamber* shows a visible path of ionizing radiation in the form of fog trails. It consists of a cylindrical glass chamber closed at the upper end by a glass window and at the lower end by a movable piston. Water vapor or alcohol vapor in the chamber can be saturated by adjusting the piston.

   The radioactive sample is placed inside the chamber, as shown in Figure 32.10, or outside the thin glass window. When radiation passes through the chamber, ions are produced along its path. If the saturated air in the chamber is then suddenly cooled by motion of the piston, tiny droplets of moisture condense about these ions and form vapor trails showing the paths of the radiation. These are the atomic versions of the ice-crystal trails left in the sky by jet planes.

   Even simpler is the continuous cloud chamber. This has a steady supersaturated vapor, because it sits on a slab of dry ice so there is a temperature gradient from near room temperature at the top to very low temperature at the bottom. In either version the fog tracks that form are illuminated with a lamp and may be seen or photographed through the glass top. The chamber may be placed in a strong electric or magnetic field, which will bend the paths in a manner that provides information about the charge, mass, and momentum of the radiation particles. Positively and negatively charged particles will bend in opposite directions.

4. The particle trails seen in a *bubble chamber* are minute bubbles of gas in liquid hydrogen (Figure 32.11). The liquid hydrogen is heated under pressure in a glass and stainless steel chamber to a point just short of boiling. If the pressure in the chamber is suddenly released at the moment an ion-producing particle enters, a thin trail of bubbles is left along the particle's path. All the liquid erupts to a boil, but in the few thousandths of a second before this happens, photographs are taken

Radioactive sample · Vapor trails · Piston

FIGURE 32.10   The cloud chamber.

FIGURE 32.11   Tracks of elementary particles in a bubble chamber. Two particles have been destroyed at the points from which the spirals emanate, and four others are created in the collision.

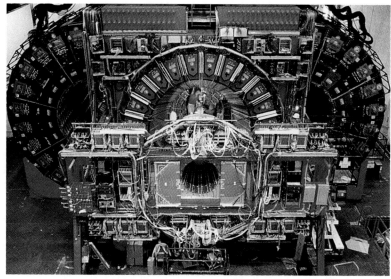

a

b

FIGURE 32.12 (a) Installation of the Big European Bubble Chamber (BEBC) at CERN, near Geneva, typical of the large bubble chambers used in the 1970s to study particles produced by high-energy accelerators. The 3.7-m cylinder contained liquid hydrogen at −173°C. (b) The collider detector at Fermilab, which detects and records myriads of events when particle beams collide. The detector stands two stories tall, weighs 4500 tons, and was built by a collaboration of more than 170 physicists from the United States, Japan, and Italy.

of the particle's brief-lived trail. As with the cloud chamber, a magnetic field in the bubble chamber reveals the charge and relative mass of the particles being studied. Bubble chambers have been widely used by researchers in recent decades, but presently there is greater interest in spark chambers.

The principles underlying the four radiation detection devices cited are at the heart of more complex and larger particle detectors, such as the Collider Detector at Fermilab (Figure 32.12b). In this enormous detector, particles produced by proton-antiproton collisions pass through gas-filled chambers and ionize the gas along their paths in cloud-chamber fashion. In Geiger-counter fashion, freed electrons from the gas are collected by networks of high-voltage wires in the chamber, the locations of which reveal the paths of particles. Scintillation counters detect the amount of light given off, which enables a measure of the energy of the particles.

Questions

1. If a sample of radioactive isotopes has a half-life of 1 day, how much of the original sample will be left at the end of the second day? The third day?

2. Which will give a higher counting rate on a radiation detector, radioactive material that has a short half-life or the same amount of radioactive material that has a long half-life?

Answers

1. One-fourth of the original sample will be left—the three-fourths that underwent decay is now a different element (or perhaps several other different elements) altogether. At the end of 3 days, $\frac{1}{8}$ of the original sample will remain.

2. The material with the shorter half-life is more active and will give a higher counting rate on a radiation detector.

Natural Transmutation of Elements

When a nucleus emits an alpha or a beta particle, a different element is formed. The changing of one chemical element to another is called **transmutation**. Consider uranium, for example. Uranium-238 has 92 protons and 146 neutrons. When an alpha particle is ejected, the nucleus is reduced by two protons and two neutrons (an alpha particle is a helium nucleus consisting of two protons and two neutrons). The 90 protons and 144 neutrons left behind are then the nucleus of a new element. This element is *thorium*. We can express this reaction as

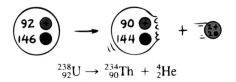

$$^{238}_{92}U \rightarrow\ ^{234}_{90}Th\ +\ ^{4}_{2}He$$

An arrow is used here to show that the $^{238}_{92}U$ changes into the other elements. When this happens, energy is released, partly in the form of gamma radiation, partly in the kinetic energy of the alpha particle ($^{4}_{2}He$), and partly in the kinetic energy of the thorium atom. In this and all such equations, the mass numbers at the top balance (238 = 234 + 4) and the atomic numbers at the bottom also balance (92 = 90 + 2).

Thorium-234, the product of this reaction, is also radioactive. When it decays, it emits a beta particle. Recall that a beta particle is an electron—not an orbital electron, but one from the nucleus—from one of the neutrons. You may find it useful to think of a neutron as a combined proton and electron (although it's not really the case) because when the neutron emits an electron, it becomes a proton.* So in the case of thorium, which has 90 protons, beta emission leaves it with one less neutron and one more proton. The new nucleus then has 91 protons and is no longer that of thorium, but the nucleus of the element *protactinium*. Although the atomic number has increased by 1 in this process, the mass number (protons + neutrons) remains the same.

The nuclear equation is

$$^{234}_{90}Th \rightarrow\ ^{234}_{91}Pa\ +\ ^{0}_{-1}e$$

We write an electron as $^{0}_{-1}e$. The 0 indicates that its mass is insignificant when compared to that of the protons and neutrons that alone contribute to the mass number. The −1 is the charge of the electron. Remember that this electron is a beta particle from the nucleus and not an electron from the surrounding electron cloud.

We can see that when an element ejects an alpha particle from its nucleus, the mass number of the resulting atom decreases by 4, and its atomic number *decreases* by

---

* Beta emission is always accompanied by the emission of a neutrino (actually an antineutrino), a neutral particle that travels at about the speed of light. The neutrino ("little neutral one") was postulated by Wolfgang Pauli in 1930 and detected in 1956. At this writing, whether or not the neutrino has mass is not known. Neutrinos are so numerous in the universe that if they do have mass, they might comprise more than half the mass of the universe—perhaps enough to halt the present expansion and ultimately close the cycle from the Big Bang to the Big Crunch. Neutrinos may be the "glue" that holds the universe together.

2. The resulting atom belongs to an element two spaces back in the periodic table. When an element ejects a beta particle (electron) from its nucleus, the mass of the atom is practically unaffected so there is no change in mass number, but its atomic number *increases* by 1. The resulting atom belongs to an element one place forward in the periodic table. Gamma emission results in no change in either the mass number or the atomic number. So we see that the emission of an alpha or beta particle by an atom produces a different atom in the periodic table. Alpha emission lowers the atomic number and beta emission increases it. Radioactive elements can decay backward or forward in the periodic table.*

The radioactive decay of $^{238}_{92}U$ to $^{206}_{82}Pb$, an isotope of lead, is shown in Figure 32.13. The steps in the decay process are shown in the diagram, where each nucleus that plays a part in the series is shown by a burst. The vertical column containing the burst shows its atomic number, and the horizontal row shows its mass number. Each

FIGURE 32.13 U-238 decays to Pb-206 through a series of alpha and beta decays.

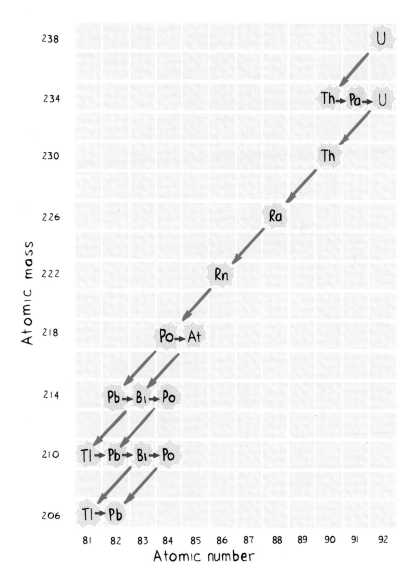

* Sometimes a nucleus emits a positron, which is the "antiparticle" of an electron. In this case, a proton becomes a neutron, and the atomic number is decreased.

green arrow shows an alpha decay, and each red arrow shows a beta decay. Notice that some of the nuclei in the series can decay in both ways. This is one of several similar radioactive series that occur in nature.

Questions

1. Complete the following nuclear reactions.
   a. $^{228}_{88}Ra \rightarrow \, ^{?}_{?}? + \, ^{0}_{-1}e$
   b. $^{209}_{84}Po \rightarrow \, ^{205}_{82}Pb + \, ^{?}_{?}?$
2. What finally becomes of all the uranium-238 that undergoes radioactive decay?

Artificial Transmutation of Elements

The alchemists of old tried vainly for over 2000 years to cause the transmutation of one element to another. Enormous efforts were expended and elaborate rituals performed in the quest to change lead into gold. They never succeeded. Lead in fact can be changed to gold, but not by the chemical means employed by the alchemists. Chemical reactions involve alterations of the outermost shells of the electron clouds of atoms and molecules. To change an element from one kind to another, one must go deep within the electron clouds to the central nucleus, which is immune to the most violent chemical reactions. To change lead to gold, three positive charges must be extracted from the nucleus. Ironically enough, transmutations of atomic nuclei were constantly going on all around the alchemists, as they are around us today. Radioactive decay of minerals in rocks has been occurring since their formation. But this was unknown to the alchemists, and even if they had discovered these radiations, the atoms so transmuted would most likely have escaped their notice.

Ernest Rutherford, in 1919, was the first of many investigators to succeed in transmuting a chemical element. He bombarded nitrogen nuclei with alpha particles and succeeded in transmuting nitrogen into oxygen:

$$^{14}_{7}N + \, ^{4}_{2}He \rightarrow \, ^{17}_{8}O + \, ^{1}_{1}H$$

Rutherford's source of alpha particles was a radioactive piece of ore. From a quarter-of-a-million cloud-chamber tracks photographed on movie film, he showed seven examples of atomic transmutation. Analysis of tracks bent by a strong external magnetic field showed that when an alpha particle collided with a nitrogen atom, a proton bounced out and the heavy atom recoiled a short distance. The alpha particle disappeared. The alpha particle was absorbed in the process, transforming nitrogen to oxygen.

Answers

1. a. $^{228}_{88}Ra \rightarrow \, ^{228}_{89}Ac + \, ^{0}_{-1}e$
   b. $^{209}_{84}Po \rightarrow \, ^{205}_{82}Pb + \, ^{4}_{2}He$
2. All uranium-238 will ultimately become lead-206. On the way to becoming lead, it will exist as various isotopes of various elements, as indicated in Figure 32.13.

Since Rutherford's announcement in 1919, we have created many such nuclear reactions, first with natural bombarding projectiles from radioactive ores and then with still more energetic projectiles, protons and electrons hurled by giant particle accelerators. Artificial transmutation has been used to produce the hitherto unknown elements from atomic number 93 to 112, the first eleven of which are named *neptunium, plutonium, americium, curium, berkelium, californium, einsteinium, fermium, mendelevium, nobelium,* and *lawrencium.* The United States scientists who claim discovery of elements 104 and 105 have recommended the names *rutherfordium* for element 104 and *hahnium* for 105. Russian researchers, who also claim the discovery of these elements, have proposed different names. Generally accepted names for elements 106 to 109 are *seaborgium, nielsbohrium, hassium,* and *meitnerium.* Elements 110 and 111, discovered in late 1994, and element 112, discovered in early 1996, are known as *ununnilium, unununium,* and *ununbium.* All these artificially made elements have short half-lives. Whatever transuranic elements existed naturally when the earth was formed have long since decayed.

Radioactive Isotopes

All elements have isotopes. Some of the isotopes of a given element may be stable, while most are radioactive. Radioactive isotopes of all the elements have been made by bombardment with neutrons and other particles. As such isotopes have become available and inexpensive, their use in scientific research and industry has expanded tremendously.

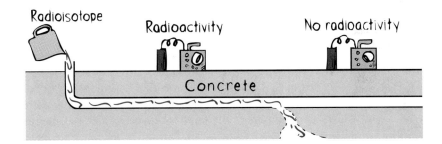
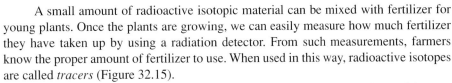

FIGURE 32.14 Tracking pipe leaks with radioactive isotopes.

A small amount of radioactive isotopic material can be mixed with fertilizer for young plants. Once the plants are growing, we can easily measure how much fertilizer they have taken up by using a radiation detector. From such measurements, farmers know the proper amount of fertilizer to use. When used in this way, radioactive isotopes are called *tracers* (Figure 32.15).

Tracers are widely used in medicine to study the process of digestion and the way chemical substances move about in the body. Food containing a small amount of radioactive material is fed to a patient and traced through the body with a radiation detector. The same method can be used to study the circulation of the blood. Large amounts of radioactive iodine taken into the body are used to combat cancer of the thyroid gland. Since iodine tends to collect in the thyroid gland, the radioactive isotope lodges where it can destroy the malignant cells. Radioactive isotopes are an increasingly powerful tool for the modern biologist and physician.

Engineers can study how the parts of an automobile engine wear away during use by making the cylinder walls radioactive. While the engine is running, the piston rings

FIGURE 32.15 Radioisotopes are used to check the action of fertilizers in plants and the progress of food in digestion.

rub against the cylinder walls. The tiny particles of radioactive metal that are worn away fall into the lubricating oil, where they can be measured with a radiation detector. By repeating this test with different oils, the engineer can determine which oil gives the least wear and longest life to the engine.

Tire manufacturers also employ radioactive isotopes. If a known fraction of the carbon atoms used in an automobile tire is radioactive, the amount of rubber left on the road when the car is braked can be estimated through a count of the radioactive atoms.

There are hundreds more examples of the use of radioactive isotopes. The important thing is that this technique provides a way to detect and count atoms in samples of materials too small to be seen with a microscope.

Carbon Dating

A constant bombardment of cosmic rays on the earth's atmosphere results in the transmutation of many atoms in the upper atmosphere. Protons, neutrons, and a host of other particles are sprayed throughout the atmosphere. Most of the protons quickly capture stray electrons and become hydrogen atoms in the upper atmosphere, but the neutrons keep going for long distances because they have no charge and therefore do not interact electrically with matter. Sooner or later many of them collide with the nuclei of atoms in the lower atmosphere. If they interact with the nucleus of a nitrogen atom, the following reaction takes place:

$$^{14}_{7}N + ^{1}_{0}n \rightarrow ^{14}_{6}C + ^{1}_{1}H$$

So some of the nitrogen in the atmosphere is converted to carbon and hydrogen. Most carbon that exists is the stable isotope C-12, $^{12}_{6}C$. Because of cosmic-ray bombardment, about one-millionth of 1 percent of the carbon in the atmosphere is C-14. Both of these atoms join with oxygen and become carbon dioxide, which is taken in by plants. Hence, all plants have a tiny bit of radioactive C-14 in them. All animals eat plants (or eat plant-eating animals), and therefore have a little C-14 in them. All living things contain some C-14.

Carbon-14 is a beta emitter and decays back into nitrogen by the following reaction:

$$^{14}_{6}C \rightarrow ^{14}_{7}N + ^{0}_{-1}e$$

So the C-14 in living things changes into stable nitrogen. But because living things breathe, this decay is accompanied by a replenishment of C-14, and a radioactive equilibrium is reached where there is a fixed ratio of C-14 to C-12. When a plant or an animal dies, however, replenishment stops. The percentage of C-14 steadily decreases—at a known rate. The longer an organism is dead, the less C-14 remains.*

---

* A 1-g sample of contemporary carbon contains about 5×10^{22} atoms, 6.5×10^{10} of which are C-14 atoms, and has a beta disintegration rate of about 13.5 decays per minute.

20,920 BC   15,190 BC   9460 BC   3730 BC   2000 AD

FIGURE 32.16 The radioactive carbon isotopes in the skeleton diminish by one-half every 5730 years.

The half-life of C-14 is 5730 years, which means that half the C-14 atoms present in a body decay in that time. Half the remaining C-14 atoms decay in the following 5730 years, and so forth. The radioactivity of once-living things therefore gradually decreases at a predictable rate (Figure 32.16).

We can measure the radioactivity of plants and animals today and compare this with the radioactivity of ancient organic matter. If we extract a small, but precise, quantity of carbon from an ancient wooden ax handle, for example, and find it has one-half as much radioactivity as an equal quantity of carbon extracted from a living tree, then the old wood must have come from a tree that was cut down or made from a log that died 5730 years ago. In this way, we can probe as much as 50,000 years into the past to find out such things as the age of ancient civilizations.

Carbon dating would be an extremely simple and accurate dating method if the amount of radioactive carbon in the atmosphere had been constant over the ages. But it hasn't been. Fluctuations in the sun's magnetic field as well as changes in the strength of the earth's magnetic field affect cosmic-ray intensities in the earth's atmosphere, which in turn produce fluctuations in the production of C-14. In addition, changes in the earth's climate affect the amount of carbon dioxide in the atmosphere. The oceans are great reservoirs of carbon dioxide. When the oceans are cold, they release less carbon dioxide into the atmosphere than when they are warm. These complications require complex corrective procedures for the accurate dating of ancient organic objects. Nevertheless, recalibration and refined techniques make carbon dating the most versatile chronometric method we have. Present laser-enrichment techniques using samples containing only a few milligrams of carbon are soon expected to double the current 50,000-year range. Similar results are now possible with a technique that bypasses a radioactive measure altogether. It involves an improved atomic acceleration-mass spectrometer device that effectively makes a $\frac{C\text{-}14}{C\text{-}12}$ count directly. There are also techniques involving other, longer lived radioactive isotopes for assigning dates to more ancient relics.

Question Suppose an archaeologist extracts a gram of carbon from an ancient ax handle and finds it one-fourth as radioactive as a gram of carbon extracted from a freshly cut tree branch. About how old is the ax handle?

Answer Assuming the ratio of $\frac{C\text{-}14}{C\text{-}12}$ was the same when the ax was made, the ax handle is two half-lives of C-14 or about 11,460 years old.

Uranium Dating

The dating of older, but nonliving, things is accomplished with radioactive minerals, such as uranium. The naturally occurring isotopes U-238 and U-235 decay very slowly and ultimately become isotopes of lead—but not the common lead isotope Pb-208. For example, U-238 decays through several stages to finally become Pb-206, whereas U-235 finally becomes the isotope Pb-207. Most of the lead isotopes 206 and 207 that now exist were at one time uranium. The older the uranium-bearing rock, the higher the percentage of these remnant isotopes. We know the half-lives of the uranium isotopes and we can determine the percentage of each lead isotope in a rock sample, so we can calculate the time of decay for uranium in the rock. In this way the ages of rocks have been found to be as much as 3.7 billion years. Samples from the moon, where there has been less obliteration of the early rocks than on earth, have been dated at 4.2 billion years, which is "only" 400 million years short of the well-established 4.6-billion-year age of the earth and solar system.

Effects of Radiation on Humans

A common misconception is that radioactivity is something new in the environment. But radioactivity has been around far longer than the human race. It is as much a part of our environment as the sun and the rain. It is what warms the interior of the earth and makes it molten. In fact, radioactive decay inside the earth is what heats the water that spurts from a geyser or that wells up from a natural hot spring. Even the helium in a child's balloon is the offspring of radioactivity. Its nuclei are nothing more than the alpha particles that were once shot out of radioactive nuclei.

As Figure 32.17 shows, most of the radiation we encounter originates in our natural surroundings. It is in the ground we stand on and in the bricks and stones of surrounding buildings. This natural background radiation was present before humans emerged in the world. If our bodies couldn't tolerate it, we wouldn't be here. Even the cleanest air we breathe is somewhat radioactive due to cosmic ray bombardment. At sea level the protective blanket of the atmosphere reduces cosmic-ray intensity, while at higher altitudes radiation is more intense. In Denver, the "mile-high city," a person receives more than twice as much radiation from cosmic rays as at sea level. A couple of round-trip flights between distant places such as New York and San Francisco exposes us to as much radiation as we receive in a normal chest X ray. (Is the air time of airline personnel limited because of this extra radiation?)

FIGURE 32.17 Origins of radiation exposure for an average individual in the United States.

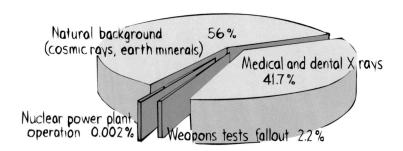

Natural background 56%
(cosmic rays, earth minerals)

Medical and dental X rays 41.7%

Nuclear power plant operation 0.002%

Weapons tests fallout 2.2%

Exposure to radiation greater than normal background should be avoided because of the damage it can do. The cells of living tissue are composed of intricately structured molecules in a watery, ion-rich brine. When X radiation or nuclear radiation encounters this highly ordered soup, it produces chaos on the atomic scale. A beta particle, for example, passing through living matter collides with a small percentage of the molecules and leaves a randomly dotted trail of altered or broken molecules along with newly formed, chemically active ions and free radicals. Free radicals are unbonded, electrically neutral, very chemically active atoms or molecular fragments. The ions and free radicals may break even more molecular bonds or they may quickly form strong new bonds, creating molecules that may be useless or harmful to the cell. Gamma radiation produces a similar effect. As a high-energy gamma-ray photon moves through matter, it may rebound from an electron and give the electron a high kinetic energy. The electron then may career through the tissue, creating havoc in the ways described above. All types of high-energy radiation break or alter the structure of some molecules and create conditions in which other molecules will be formed that may be harmful to life processes.

We are bombarded most by what harms us least—neutrinos. Neutrinos are the most weakly interacting particles. They have near-zero mass, no charge, and are produced frequently in radioactive decays. They are the most common high-speed particles known, zapping the universe, and passing unhindered through our bodies by the billions every second. They pass completely through the earth with only occasional encounters. It would take a "piece" of lead 6 light-years in thickness to absorb half the neutrinos incident upon it. About once per year on the average, a neutrino triggers a nuclear reaction in your body. We don't hear much about neutrinos because they ignore us.

Of the radiations we have focused upon in this chapter, gamma radiation is the most penetrating and therefore the hardest to shield against. This, combined with its ability to interact with the matter in our bodies, makes it potentially the most dangerous radiation. It emanates from radioactive materials and makes up a substantial part of the normal background radiation. Exposure to it should be minimized.

The cells are able to repair most kinds of molecular damage if the radiation is not too intense. A cell can survive an otherwise lethal dose of radiation if the dose is spread over a long period of time to allow intervals for healing. When radiation is sufficient to kill a cell, the dead cells can be replaced by new ones. Important exceptions to this are most nerve cells, which are irreplaceable. Sometimes a cell will survive with a damaged DNA molecule. Defective genetic information will be transmitted to its daughter cell when it reproduces, and a cell mutation will occur. This will sometimes be insignificant. But if significant, it will probably result in cells that do not function as well as the original one. In rare cases it will be an improvement. A genetic change of this type could also be part of the cause of a cancer that will develop in the tissue at a much later time.

The concentration of disorder produced along the trajectory of a particle depends upon its energy, charge, and mass. Gamma-ray photons and very energetic beta particles spread their damage out over a long track. They penetrate deeply with widely separated interactions, like a very fast BB fired through a hailstorm. Slow, massive, highly charged particles such as low-energy alpha particles do their damage in the shortest distance. They have collisions that are close together, more like a bull charging through a flock of sleepy sheep. They do not penetrate deeply because their energy is absorbed by many closely spaced collisions. Particles that produce especially concentrated damage in living things are the assorted nuclei (called *heavy primaries*) flung outward by the sun in solar flares. These include all the elements found on earth. Heavy primaries are also found in a small percentage of cosmic radiation. Some of this radiation is captured in the earth's magnetic field and some is stopped by collisions in the atmosphere, so practically none reaches the earth. We are shielded from most of these dangerous particles by the very property that makes them a threat—their tendency to have many collisions close together.

FIGURE 32.18 The internationally used symbol to indicate an area where radioactive material is being handled or produced.

Astronauts do not have this protection, and they absorb large doses of radiation during the time they spend in space. Every few decades there is an exceptionally powerful solar flare that would almost certainly kill any conventionally protected astronaut who is unprotected by the earth's atmosphere and magnetic field.

Generally, radiation dosage is measured in *rads* (short for radiation), a unit of absorbed energy of ionizing radiation. The number of rads indicates the amount of radiation energy absorbed per gram of exposed material. However, when concerned with the potential ability of radiation to affect human beings, we measure the radiation in *rems* (*r*oentgen *e*quivalent *m*an). In calculating the dosage in rems, we multiply the number of rads by a factor that allows for the different health effects of different types of radiation. For example, 1 rad of slow alpha particles has the same biological effect as 10 rads of fast electrons. We call both of these dosages 10 rems.

The average person in the United States is exposed to about 0.2 rem a year. This comes from within the body itself, from the ground, buildings, cosmic rays, diagnostic X rays, television, and so on. It varies widely from place to place on the planet, but it is strongest near the poles where the earth's magnetic field does not act as a shield. It is also stronger at higher altitudes where the atmosphere provides less protection.

The lethal dose of radiation is on the order of 500 rems; that is, a person has about a 50% chance of surviving a dose of this magnitude if it is received over a short period of time. Under radiotherapy—the use of radiation to kill cancer cells—a patient may receive localized doses in excess of 200 rems each day for a period of weeks. A typical diagnostic chest X ray exposes a person to from 5 to 30 millirems, less than one ten-thousandth of the lethal dose. However, even small doses of radiation can produce long-term effects due to mutations within the body's tissues. And, because a small fraction of any X-ray dose reaches the gonads, some mutations occasionally occur that are passed on to the next generation. Although medical X rays for diagnosis and therapy have a far larger effect on the human genetic heritage than any other artificial source of radiation, the effect from all artificial sources of radiation is quite small when compared with the effect from natural sources on earth.

Taking all causes into account, most of us will receive a lifetime exposure of less than 20 rems, distributed over several decades. This makes us a little more susceptible to cancer and other disorders. But more significant is the fact that all living beings have always absorbed natural radiation and that the radiation received in the reproductive cells has produced genetic changes in all species for generation after generation. Small mutations selected by nature for their contributions to survival over billions of years can gradually come up with some interesting organisms—*us,* for example!

Summary of Terms

X ray   Electromagnetic radiation of higher frequencies than ultraviolet emitted by electrons jumping to the lowest energy state in atoms.

Alpha particle   The nucleus of a helium atom, which consists of two neutrons and two protons, ejected by certain radioactive elements.

Beta particle   An electron (or positron) emitted during the radioactive decay of certain nuclei.

Alpha ray   A stream of helium nuclei ejected by certain radioactive nuclei.

Beta ray   A stream of beta particles ejected by certain radioactive nuclei.

Gamma ray   High-frequency electromagnetic radiation emitted by the nuclei of radioactive atoms.

Nucleon   A nuclear proton or neutron; the collective name for either or both.

Quarks   The elementary constituent particles or building blocks of nuclear matter.

Isotopes   Atoms whose nuclei have the same number of protons but different numbers of neutrons.

Atomic number   A number associated with an atom, equal to the number of protons in the nucleus or, equivalently, to the number of electrons in the electron cloud of a neutral atom.

Atomic mass number   A number associated with an atom, equal to the number of nucleons (protons plus neutrons) in the nucleus.

Half-life   The time required for half the atoms in a sample of a radioactive isotope to decay.

Transmutation   The conversion of an atomic nucleus of one element into an atomic nucleus of another element through a loss or gain in the number of protons.

Suggested Reading

Cobb, Charles E. "Living with Radiation," *National Geographic* 175, no. 4 (April 1989): 403–437. An account of the effects of radiation on people, from those exposed to radon gas in their homes to survivors of nuclear blasts.

Waldrop, M. Mitchell, "The Shroud of Turin: An Answer Is at Hand." *Science,* 30 Sept. 1988, pp. 1750–1751. Describes the events leading up to the final dating of the Shroud.

Review Questions

X Rays and Radioactivity

1. What is the principal difference between a beam of X rays and a beam of light?
2. What did the physicist Roentgen discover coming from the spot where a cathode-ray beam struck a glass surface?
3. What did the physicist Becquerel discover about uranium?
4. What two elements did Pierre and Marie Curie discover?

Alpha, Beta, and Gamma Rays

5. How do the electric charges of alpha, beta, and gamma rays differ?
6. How does the source differ for a beam of gamma rays and a beam of X rays?

The Nucleus

7. Give two examples of a nucleon.

8. How does the mass of a nucleon compare with the mass of an electron?
9. When beta emission occurs, what change takes place in an atomic nucleus?
10. In what way is the emission of gamma rays from a nucleus similar to the emission of light from an atom?
11. Why does an alpha particle leave at high speed once it gets outside an atomic nucleus?
12. What are *quarks?*

Isotopes

13. Distinguish between an *isotope* and an *ion.*
14. Distinguish between *atomic number* and *atomic mass number.*
15. Distinguish between *deuterium* and *tritium.*

Why Atoms Are Radioactive

16. Why does the repulsive electric force of protons in the atomic nucleus not cause the protons to fly apart?
17. Why do protons in a very large nucleus have a greater chance of flying apart by electrical repulsion?
18. Why is a smaller nucleus generally more stable than a larger nucleus?
19. Distinguish between a *hadron* and a *lepton.*

Half-Life

20. What is meant by *radioactive half-life?*
21. What is the half-life of Ra-226? Of a muon?

Radiation Detectors

22. What kind of trail is left when an energetic particle shoots through matter?
23. What radiation detectors operate primarily by sensing the trails left by energetic particles that shoot through matter?

Natural Transmutation of Elements

24. What exactly is *transmutation?*
25. When thorium, atomic number 90, decays by emitting an alpha particle, what is the atomic number of the resulting nucleus?
26. When thorium decays by emitting a beta particle, what is the atomic number of the resulting nucleus?
27. How does the atomic mass change for each of the above two reactions?
28. What is the effect on the makeup of a nucleus when it emits an alpha particle? A beta particle? A gamma ray?
29. What is the long-range fate of all the uranium that exists in the world?

Artificial Transmutation of Elements

30. The alchemists of old believed that elements could be changed to other elements. Were they correct? Were they effective? Why or why not?

31. When did the first successful intentional transmutation of an element occur?

Radioactive Isotopes

32. How are radioactive isotopes produced?

33. What is a radioactive *tracer?*

Carbon Dating

34. Which is radioactive, C-12 or C-14?

35. Why is there more C-14 in new bones than in old bones of the same mass?

36. Why would the carbon dating method be useless for dating old coins but not old pieces of cloth?

Uranium Dating

37. Why is lead found in all deposits of uranium ores?

Effects of Radiation on Humans

38. From where does most of the radiation you encounter originate?

39. Which is worse: having cells in your body *damaged* by radiation or *killed* by radiation?

40. Is radioactivity in the world something relatively new? Defend your answer.

Project

Some watches and clocks have luminous hands that continuously glow. Some of these have traces of radium bromide mixed with zinc sulfide. (Safer clock faces use light rather than radioactive disintegration as a means of excitation and become progressively dimmer in the dark.) If you have a luminous watch or clock available, take it into a completely dark room and, after your eyes have become adjusted to the dark, examine the luminous hands with a very strong magnifying glass or the eyepiece of a microscope or telescope. You should be able to see individual tiny flashes, which together seem to be a steady source of light to the unaided eye. Each flash occurs when an alpha particle ejected by a radium nucleus strikes a molecule of zinc sulfide.

Exercises

1. X rays are most similar to which of the following: alpha, beta, or gamma rays?

2. Why is a sample of radioactive material always a little warmer than its surroundings?

3. Some people say that all things are possible. Is it at all possible for a hydrogen nucleus to emit an alpha particle? Defend your answer.

4. Why are alpha and beta rays deflected in opposite directions in a magnetic field? Why are gamma rays undeflected?

5. The alpha particle has twice the electric charge of the beta particle but, for the same kinetic energy, deflects less than the beta in a magnetic field. Why is this so?

6. How do the paths of alpha, beta, and gamma radiations compare in an electric field?

7. Which type of radiation—alpha, beta, or gamma— results in the greatest change in mass number? Atomic number?

8. Which type of radiation—alpha, beta, or gamma— results in the least change in mass number? Atomic number?

9. In bombarding atomic nuclei with proton "bullets," why must the protons be accelerated to high energies if they are to make contact with the target nuclei?

10. Just after an alpha particle leaves the nucleus, would you expect it to speed up? Defend your answer.

11. Why would you expect alpha particles, with their greater charge, to be less able to penetrate into materials than beta particles of the same energy?

12. Which interaction tends to hold an atomic nucleus together and which interaction tends to push it apart?

13. What evidence supports the contention that the strong nuclear interaction can dominate over the electrical interaction at short distances within the nucleus?

14. Exactly what is a positively charged hydrogen atom?

15. Why do different isotopes of the same element have the same chemical properties?

16. If a sample of a radioactive isotope has a half-life of 1 year, how much of the original sample will be left at the end of the second year? Third year? Fourth year?

17. A sample of a particular radioisotope is placed near a Geiger counter, which is observed to register 160 counts per minute. Eight hours later the detector counts at a rate of 10 counts per minute. What is the half-life of the material?

18. The isotope Cesium-137, which has a half life of 30 years, is a product of nuclear power plants. How long will it take for this isotope to decay to about one-sixteenth its original amount?

19. If you keep track of 1000 people born in the year 2000 and find that half of them are still living in 2060, does this mean that one-quarter of them will still be living in 2120 and one-eighth of them in 2180? What is different about the death rates of people and the "death rates" of radioactive atoms?

20. Radiation from a point source obeys the inverse-square law. If a Geiger counter 1 m from a small sample reads 360 counts per minute, what will be its counting rate 2 m from the source? 3 m?

21. Why do the charged particles in bubble chambers move in spiral paths rather than the circular or helical paths they would ideally follow?

22. From Figure 32.13, how many alpha and how many beta particles are emitted in the radioactive decay of a U-238 nucleus to a Pb-206 nucleus?

23. If an atom has 104 electrons, 157 neutrons, and 104 protons, what is its approximate atomic mass? What is the name of this element?

24. When $^{226}_{88}$Ra decays by emitting an alpha particle, what is the atomic number of the resulting nucleus? What is the resulting atomic mass?

25. When $^{218}_{84}$Po emits a beta particle, it transforms into a new element. What are the atomic number and atomic mass of this new element? What are they if the polonium instead emits an alpha particle?

26. State the number of neutrons and protons in each of the following nuclei: $^2_1$H, $^{12}_6$C, $^{56}_{26}$Fe, $^{197}_{79}$Au, $^{90}_{38}$Sr, and $^{238}_{92}$U.

27. How is it possible for an element to decay "forward in the periodic table"—that is, to decay to an element of higher atomic number?

28. When radioactive phosphorus (P) decays, it emits a positron. Will the resulting nucleus be another isotope of phosphorus? If not, what?

29. How could a physicist test the following statement? "Strontium-90 is a pure beta source."

30. People who work around radioactivity wear film badges to monitor their radiation exposure. These badges are small pieces of photographic film enclosed in a light-proof wrapper. What kind of radiation do these devices monitor?

31. Elements above uranium in the periodic table do not exist in any appreciable amounts in nature because they have short half-lives. Yet there are several elements below uranium in atomic number with equally short half-lives that do exist in appreciable amounts in nature. How can you account for this?

32. Your friend says that the helium used to inflate balloons is a product of radioactive decay. Another friend says no way. With whom do you agree?

33. Another friend, fretful about living near a fission power plant, wishes to get away from radiation by traveling to the high mountains and sleeping out at night on granite outcroppings. What comment do you have about this?

34. Still another friend has journeyed to the mountain foothills to escape the effects of radioactivity altogether. While bathing in the warmth of a natural hot spring she wonders aloud how the spring gets its heat. What do you tell her?

35. Coal contains minute quantities of radioactive materials, yet because of the vast amount of coal burned there is more radiation emitted by a coal-fired power plant than a fission power plant. What does this indicate about the shielding that typically surrounds these power plants?

36. A friend uses a Geiger counter to check the local normal background radiation. It clicks. Another friend, whose tendency is to fear most that which is least understood, makes an effort to keep away from the Geiger counter and looks to you for advice. What do you say?

37. If you hold a Geiger counter and find that it is clicking regularly exactly once each second, does that suggest that (a) you are detecting cosmic rays, (b) you are near a weak radioactive source, or (c) the Geiger counter is malfunctioning?

38. Is carbon dating advisable for measuring the age of materials (a) a few years old? (b) a few thousand years old? (c) a few million years old?

39. The age of the Dead Sea Scrolls was found by carbon dating. Could this technique have worked if they were carved in stone tablets? Explain.

40. Make up two multiple-choice questions that would check a classmate's understanding of radioactive dating.

Problems

1. A certain radioactive substance has a half-life of one hour. If you start with 1 g of the material at noon, how much will be left at 3:00 PM? at 6:00 PM? at 10:00 PM?

2. Suppose that you measure the intensity of radiation from carbon-14 in an ancient piece of wood to be 6% of what it would be in a freshly cut piece of wood. How old is this artifact?

33

. . . .

Nuclear Fission and Fusion

Test firing of the Particle Beam Fusion Accelerator II at Sandia
National Laboratories.

The greater force
is nuclear

Critical deformation

The greater force
is electrical

In December 1938, two German scientists, Otto Hahn and Fritz Strassmann, made
an accidental discovery that was to change the world. While bombarding a sample
of uranium with neutrons in the hope of creating new heavier elements, they were
astonished to find chemical evidence for the production of barium, an element
about half the mass of uranium. They were reluctant to believe their own results.
Hahn sent news of this discovery to his former colleague Lise Meitner, a refugee
from Nazism working in Sweden. Over the Christmas holidays, she discussed it with
her nephew Otto Frisch, also a German refugee, visiting her from Denmark where
he worked with Niels Bohr. Together, they came up with the explanation. The
uranium nucleus, activated by neutron bombardment, had split in half. Frisch and
Meitner named the process *fission,* after the similar process of cell division in
biology.

FIGURE 33.1 Nuclear deformation may result in repulsive electrical forces overcoming
attractive nuclear forces, in which case fission occurs.

Nuclear Fission

Nuclear fission involves a delicate balance within the nucleus between nuclear attraction and the electrical repulsion between protons. In all known nuclei the nuclear forces dominate. In uranium, however, this domination is tenuous. If the uranium nucleus is stretched into an elongated shape (Figure 33.1), the electrical forces may push it into an even more elongated shape. If the elongation passes a critical point, nuclear forces yield to electrical ones, and the nucleus separates. This is fission. The absorption of a neutron by a uranium nucleus supplies enough energy to cause such an elongation. The resultant fission process may produce many different combinations of smaller nuclei. A typical example is

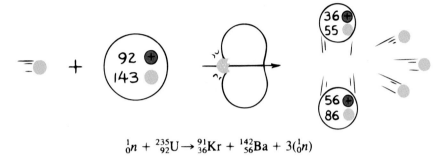

$$^1_0n + {}^{235}_{92}U \rightarrow {}^{91}_{36}Kr + {}^{142}_{56}Ba + 3(^1_0n)$$

This reaction releases energy of about 200,000,000 electron volts.* (By comparison, the explosion of the TNT molecule releases 30 electron volts.) The combined mass of the fission fragments and neutrons produced in fission is less than the mass of the original uranium atom. The tiny amount of missing mass converted to this awesome amount of energy is in accord with Einstein's relation $E = mc^2$. The energy of fission is mainly in the form of kinetic energy of the fission fragments that fly apart from one another, with some kinetic energy given to ejected neutrons and the rest to gamma radiation.

The scientific world was jolted by the news of nuclear fission—not only because of the enormous energy release but also because of the extra neutrons liberated in the process. A typical fission reaction releases an average of about two or three neutrons. These new neutrons can in turn cause the fissioning of two or three other atomic nuclei, releasing more energy and a total of from four to nine more neutrons. If each of these splits just one nucleus, the next step in the reaction will produce between 8 and 27 neutrons, and so on. Thus, a **whole chain reaction** can proceed at an ever-accelerating rate (Figure 33.2).

Why doesn't a chain reaction start in naturally occurring uranium deposits?[†] Chain reactions don't ordinarily happen because fission occurs mainly for the rare isotope U-235, which makes up only 0.7 percent of the uranium in pure uranium metal. The prevalent isotope U-238 absorbs neutrons but does not ordinarily undergo fission, so a chain reaction can be quickly snuffed by the neutron-absorbing U-238 nuclei. With rare exceptions, naturally occurring uranium is too "impure" to undergo a chain reaction spontaneously.

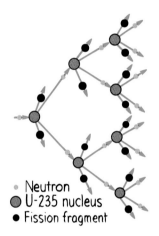

- Neutron
- U-235 nucleus
- Fission fragment

FIGURE 33.2 A chain reaction.

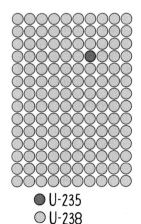

- U-235
- U-238

FIGURE 33.3 Only 1 part in 140 of naturally occurring uranium is U-235.

---

* The electron volt (eV) is defined as the amount of energy an electron acquires in accelerating through a potential difference of 1 V.

[†] There is evidence that at least one did occur—almost two billion years ago when isotopic abundances were different. It occurred in unusually rich concentrations of ore under very unusual circumstances. See *Scientific American,* July 1976.

If a chain reaction occurred in a chunk of pure U-235 the size of a baseball, an enormous explosion would likely result. If the chain reaction were started in a smaller chunk of pure U-235, however, no explosion would occur. This is because a neutron ejected by a fission event travels a certain average distance through the material before it encounters another uranium nucleus and triggers another fission event. If the piece of uranium is too small, a neutron is likely to escape through the surface before it "finds" another nucleus. On the average, fewer than one neutron per fission will be available to trigger more fission, and the chain reaction will die out. In a bigger piece, a neutron can move farther through the material before reaching a surface. Then more than one neutron from each fission event, on the average, will be available to trigger more fission. The chain reaction will build up to enormous energy. We can also understand this geometrically. Recall the concept of scaling in Chapter 11. Small pieces of material have more surface relative to volume than large pieces (there is more skin on a kilogram of small potatoes than on a single one-kilogram large potato). The larger the piece of fission fuel, the less surface area it has relative to its volume.

The **critical mass** is the amount of mass for which each fission event produces, on the average, one additional fission event. It is just enough to "hold even." A *subcritical* mass is one in which the chain reaction dies out. A *supercritical* mass is one in which the chain reaction builds up explosively.

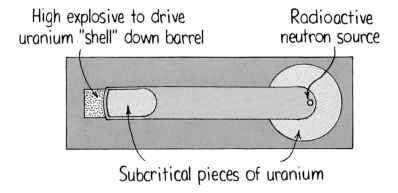

High explosive to drive uranium "shell" down barrel

Radioactive neutron source

Subcritical pieces of uranium

FIGURE 33.4 Simplified diagram of an idealized uranium fission bomb. (In an actual "gun-type" weapon, only one of the two pieces of uranium is fired toward the other one, which is the "target.")

In Figure 33.4 there are two pieces of pure U-235, each of them subcritical. Neutrons readily reach a surface and escape before a sizable chain reaction builds up. But if one of these pieces is suddenly driven into the other, the average distance a neutron can move within the material is increased and fewer neutrons escape through the surface. If the timing is right and the combined mass is greater than the critical mass, we have a nuclear fission bomb.

A chunk of U-235, believed to be about the size of a bowling ball (after being assembled), was used in the historic Hiroshima blast in 1945. Separating this much fissionable material from natural uranium was one of the principal and most difficult tasks of the secret Manhattan Project during World War II. Project scientists used two methods of isotope separation. One method depended on the fact that lighter U-235 moves at a slightly faster average speed than U-238 at the same temperature. Combined with fluorine to make the gas uranium hexafluoride, the faster isotope has a higher rate of diffusion through a thin membrane or small opening, resulting in a slightly enriched gas containing U-235 on the other side (Figure 33.5). Diffusion through thousands of chambers ultimately produced a sufficiently enriched sample of U-235. The other method, used only for partial enrichment, consisted of shooting uranium ions into a magnetic field. The smaller-mass U-235 ions were deflected more by the magnetic field than the

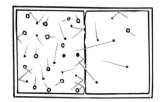

FIGURE 33.5 Lighter molecules move faster than heavier ones at the same temperature and diffuse more readily through a thin membrane.

U-238 ions and were collected atom by atom through a slit positioned to catch them. After a couple of years, the two methods together netted a few tens of kilograms of U-235.

> **Question** Why will molecules of gas in uranium hexafluoride composed with U-235 move slightly faster than molecules in the same gas composed with U-238 at the same temperature?

Uranium-isotope separation today is more easily accomplished with a gas centrifuge. Uranium hexafluoride gas is whirled in a drum at tremendously high rim speeds (on the order of 1500 kilometers per hour). Under centrifugal force, the heavier U-238 gravitates to the outside like milk in a dairy separator, and gas rich in the lighter U-235 is extracted from the center. Engineering difficulties, only recently overcome, prevented the use of this method in the Manhattan Project.

Nuclear Fission Reactors

A chain reaction cannot ordinarily take place in *pure* natural uranium, since it is mostly U-238. The neutrons released by fissioning U-235 atoms are fast neutrons, readily captured by U-238 atoms, which do not fission. A crucial experimental fact is that *slow* neutrons are far more likely to be captured by U-235 than by U-238.* If neutrons can be slowed down, there is an increased chance that a neutron released by fission will cause fission in another U-235 atom, even amid the more plentiful and otherwise neutron-absorbing U-238 atoms. This increase may be enough to allow a chain reaction to take place.

Within less than a year after the discovery of fission, scientists realized that a chain reaction with ordinary uranium metal might be possible if the uranium were broken up into small lumps and separated by a material that would slow down neutrons. Enrico Fermi, who came to America from Italy at the beginning of 1939, led the construction of the first nuclear reactor—or *atomic pile,* as it was called—in a squash court underneath the grandstands of the University of Chicago's Stagg Field. His group achieved the first self-sustaining controlled release of nuclear energy on December 2, 1942.

Three fates are possible for a neutron in ordinary uranium metal. It may (1) cause fission of a U-235 atom, (2) escape from the metal into nonfissionable surroundings, or (3) be absorbed by U-238 without causing fission. To make the first fate more probable, the uranium was divided into discrete parcels and buried at regular intervals in nearly 400 tons of graphite, a familiar form of carbon. A simple analogy clarifies the function

Enrico Fermi. It was said jokingly that when Fermi left Stockholm to return to his native Italy after receiving the Nobel Prize in December 1938, he got lost and ended up in New York. In fact, he and his Jewish wife Laura carefully planned this escape from Fascist Italy. Fermi became an American citizen in 1945.

> **Answer** At the same temperature, the molecules of both compounds have the same kinetic energy ($\frac{1}{2}mv^2$). So the molecule composed with the less massive U-235 must have a correspondingly higher speed.

---

* This is similar to the selective absorption of different frequencies of light. Different isotopes of the same element are chemically almost identical, but can have quite different nuclear properties, and absorb neutrons differently. Similarly, atoms of different elements absorb light differently.

FIGURE 33.6 A detailed view of an early atomic reactor. A controlled chain reaction occurs in the uranium slugs embedded in the graphite.

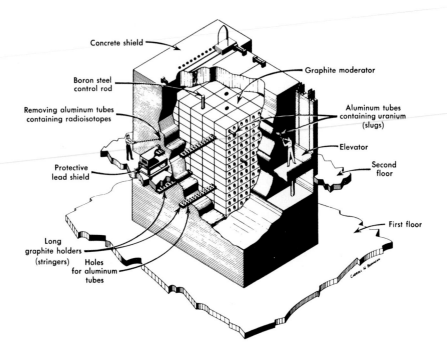

of the graphite: If a golf ball rebounds from a massive wall, it loses hardly any speed; but if it rebounds from a baseball, it loses considerable speed. The case of the neutron is similar. If a neutron rebounds from a heavy nucleus, it loses hardly any speed; but if it rebounds from a lighter carbon nucleus, it loses considerable speed. The graphite was said to "moderate" the neutrons.* The whole apparatus is called a *reactor* (Figure 33.6).

Now an uncontrolled reactor would not be a reactor, but a technical accident about to happen. A runaway chain reaction is prevented by control rods made of cadmium or boron metal, which easily absorb neutrons. The control rods can be moved into and out of the reactor to control the flux of neutrons within. When fully inserted into the reactor, the control rods snuff the chain reaction. If the rods are fully removed, the reaction can build to a level capable of melting the reactor. Because "bomb-grade" isotopes are so highly diluted with U-238, an explosion like that of a nuclear bomb is not possible.

FIGURE 33.7 The bronze plaque at Chicago's Stagg Field commemorates Enrico Fermi's historic fission chain reaction.

---

* *Heavy water,* which contains the heavy hydrogen isotope deuterium, is an even more effective moderator. This is because the nucleus of deuterium, being lighter than a carbon nucleus, moderates the neutrons even better, without absorbing neutrons.

FIGURE 33.8 An artist's depiction of the setting beneath the stands at the University of Chicago's Stagg Field where Enrico Fermi and his colleagues constructed the first nuclear reactor.

Questions

1. What is the function of a *moderator* in a nuclear reactor?
2. What is the function of *control rods* in a nuclear reactor?

Plutonium

When a U-238 nucleus absorbs a neutron, no fission occurs. The nucleus that is created, U-239, is radioactive. With a half-life of 24 minutes, it emits a beta particle and becomes an isotope of the first synthetic element beyond uranium—the transuranic element called *neptunium* (named after the first planet discovered via Newton's law of gravitation). This isotope of Neptunium is also radioactive, with a half-life of 2.3 days, so it soon emits a beta particle and becomes an isotope of *plutonium* (named after Pluto, the second planet to be discovered via Newton's law of gravitation). The half-life of this isotope, Pu-239, is about 24,000 years. Like U-235, Pu-239 will undergo fission when it captures a neutron.

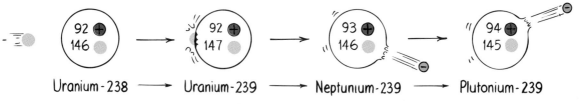

FIGURE 33.9 Within about half an hour after absorbing a neutron, U-238 emits a beta particle, resulting in a neutron becoming a proton. The atom is no longer uranium, but neptunium. After the neptunium atom, in turn, emits a beta particle, it becomes plutonium. (In both events, an antineutrino, not shown, is also emitted.)

Answers

1. A moderator slows down neutrons that are normally too fast to be easily absorbed by fissionable isotopes.
2. Control rods absorb more neutrons when they are pushed into the reactor and fewer neutrons when they are pulled out of the reactor. They thereby control the number of neutrons that participate in the chain reaction.

Even before Fermi's pile went critical, physicists realized that reactors could be used to make plutonium, and they set about designing much larger ones. The reactors constructed at Hanford, Washington, for plutonium production during World War II were 200 million times more powerful than Fermi's pile. By mid-1945, they had manufactured kilograms of this element not found in nature and unknown a few years earlier. Since plutonium is an element distinct from uranium, it can be separated from uranium in the reactor by ordinary chemical methods. Consequently, the reactor provides a process for making pure fissionable material more easily than by separating U-235 from natural uranium. The atomic bomb detonated over Nagasaki was a plutonium bomb.

Although the process of separating plutonium from uranium is simple in theory, it is very difficult in practice. This is because large quantities of radioactive fission products are formed in addition to plutonium. All chemical processing must be done by remote control to protect the staff against radiation. Also, the element plutonium is chemically toxic in the same sense as are lead and arsenic. It attacks the nervous system and can cause paralysis; death can follow if the dose is sufficiently large. Fortunately, plutonium doesn't remain in its elemental form for long but rapidly combines with oxygen to form three compounds, PuO, PuO_2, and Pu_2O_3, all of which are chemically inert. They will not dissolve in water or in biological systems. These plutonium compounds do not attack the nervous system and have been found to be biologically harmless.

Plutonium in any form, however, is radioactively toxic. It is more toxic than uranium and less toxic than radium. Pu-239 emits alpha particles that have a very short range and high energy transfer, which kill rather than mutate the cells they encounter (unlike the beta particles emitted by radium). Since damaged cells rather than dead cells contribute to cancer, plutonium ranks low as a cancer-producing substance. The greatest danger that plutonium presents to humans is its use in nuclear fission bombs. Its greatest potential benefit is in breeder reactors.

Breeder Reactors

When U-238 is mixed with fissionable isotopes in a reactor, neutrons liberated by fission convert the abundant U-238 into fissionable Pu-239. Or when Th-232 is mixed with fissionable isotopes, it similarly is converted to fissionable U-233. So in addition to the production of energy, fission fuel is bred from normally nonfissionable isotopes in the process. Such a reactor is a *breeder reactor.*

A breeder reactor effectively breeds more fuel than it consumes.* For about every two fissionable isotopes put into the reactor, three new fissionable isotopes are pro-

FIGURE 33.10 Pu-239 or U-233, like U-235, undergoes fission when it captures a neutron.

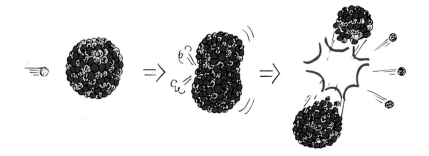

* Neutron capture is sufficient in a typical "nonbreeder" U.S. power reactor of the light-water type to convert almost 1 percent of the predominantly U-238 supply to plutonium over the 3-year period the fuel resides in the core. As the supply of U-238 decreases, a supply of Pu-239 increases and actually predominates over uranium in energy output near the end of the fuel cycle. To various extents, all reactors breed fissionable fuel.

duced. This is like nearly filling your gas tank with water, adding some gasoline, then driving your car and having more gasoline after your trip than you started with, at the expense of common water. After the initial cost of building a reactor, this is a very economical method of producing vast amounts of energy. With a breeder reactor, a power utility can, after a few years of operation, breed twice as much fuel as its original fuel.

The principal use of nuclear reactors in the world today, whether conventional or breeder, is the production of electric power. It is interesting to note that fission reactors are simply nuclear furnaces, which, like fossil-fuel furnaces, do nothing more elegant than boil water to produce steam for a turbine (Figure 33.11). The benefits of fission power are plentiful electricity; the conservation of the many billions of tons of coal, oil, and natural gas that every year are literally turned to heat and smoke and that in the long run may be far more precious as sources of organic molecules than as sources of heat; and the elimination of the megatons of sulfur oxides and other poisons that are put into the air each year by the burning of these fuels.

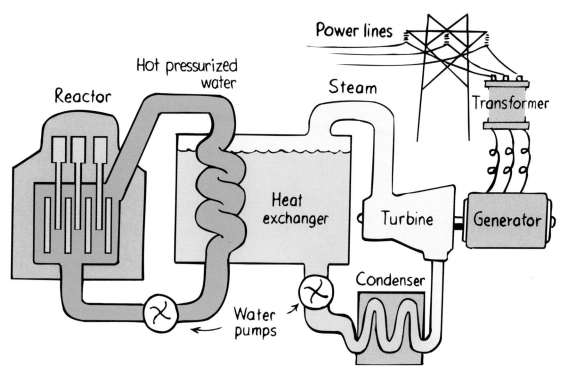

FIGURE 33.11 Diagram of a nuclear fission power plant.

The drawbacks include the problems of storing radioactive wastes, the production of plutonium and the danger of nuclear weapons proliferation, the low-level release of radioactive materials into the air and groundwater, and, most importantly, the risk of an accidental release of large amounts of radioactivity.

Reasoned judgment requires us not only to examine the benefits and drawbacks of fission power but also to compare its benefits and drawbacks with those of alternate power sources. For a variety of reasons, public opinion in the United States and much of Europe is now against fission power plants. Nukes are on the decline and fossil-fuel power plants are on the upswing.

Mass-Energy Equivalence

FIGURE 33.12 Work is required to pull a nucleon from an atomic nucleus. This work goes into mass energy.

From Einstein's mass-energy equivalence, $E = mc^2$, we can think of mass as congealed energy (more about this in Chapter 34). Mass is like a super storage battery. It stores energy—vast quantities of energy—which can be released if and when the mass decreases. If you stacked up 238 bricks, the mass of the stack would be equal to the sum of the masses of the bricks. At the nuclear level, things are different. The mass of a nucleus is not simply the masses of individual nucleons that compose it. Consider the work that would be required to separate nucleons from an atomic nucleus.

Recall that work, which is transferred energy, is equal to the product of force and distance. Imagine that you can reach into a U-238 nucleus and, pulling with a force even greater than the attractive nuclear force, remove one nucleon. That would require a lot of work. Then keep repeating the process until you end up with 238 nucleons, stationary and well separated. What happened to all the work that you have done? You started with a stationary nucleus containing 238 particles and ended with 238 stationary particles. The work that you did has to show up somewhere as additional energy. It shows up as *mass energy*. The separated nucleons have a total mass greater than the mass of the original nucleus, and the extra mass, multiplied by the square of the speed of light, is exactly equal to your energy input: $\Delta E = \Delta mc^2$.

One way to interpret this mass change is to say that an average nucleon inside a nucleus has less mass than a nucleon outside the nucleus. How much less depends on which nucleus. The mass difference is related to the "binding energy" of the nucleus.

FIGURE 33.13 The mass spectrometer. Ions of a fixed speed are directed into the semicircular "drum," where they are swept into semicircular paths by a strong magnetic field. Because of different inertia, heavier ions are swept into curves of larger radii and lighter ions are swept into curves of smaller radii. The radius of the curve is directly proportional to the mass of the ion. Using C-12 as a standard, the masses of all the elements and their isotopes are easily determined.

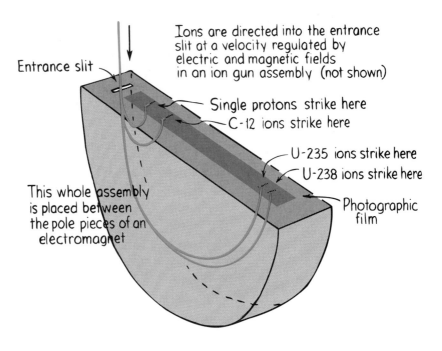

Ions are directed into the entrance slit at a velocity regulated by electric and magnetic fields in an ion gun assembly (not shown)

Entrance slit

Single protons strike here
C-12 ions strike here

U-235 ions strike here
U-238 ions strike here

Photographic film

This whole assembly is placed between the pole pieces of an electromagnet

For uranium, the mass difference is about 0.7%, or 7 parts in a thousand. The 0.7% reduced nucleon mass in uranium indicates the binding energy of the nucleus, or how much work it would take to disassemble the nucleus.

The experimental verification of this conclusion is one of the triumphs of modern physics. The masses of nucleons and the isotopes of the various elements can be measured with an accuracy of 1 part per million or better. One means of doing this is with the *mass spectrometer* (Figure 33.13).

In the mass spectrometer, charged ions are directed into a magnetic field, where they deflect into circular arcs. The greater the inertia of the ion, the more it resists deflection and the greater the radius of its curved path. All the ions entering this device have the same speed. The magnetic force sweeps the heavier ions into larger arcs and the lighter ions into smaller arcs. The ions pass through exit slits where they may be collected, or they strike photographic film where they may be simply detected. An isotope is chosen as a standard, and its position on the film of the mass spectrometer is used as a reference point. The standard is the common isotope of carbon, C-12. The mass of the C-12 nucleus is assigned the value of 12 atomic mass units. The atomic mass unit (amu) is defined to be precisely one-twelfth the mass of the common carbon-12 atom. With this reference, the amu's of the other atomic nuclei are measured. The masses of the proton and neutron are greater alone than when in a nucleus. They are 1.00728 and 1.00867 amu, respectively.

A plot of nuclear masses for the elements from hydrogen through uranium is shown in Figure 33.14. The curve slopes upward with increasing atomic number as expected: elements are more massive as atomic number increases. The curvature is attributed to the fact that neutron number grows faster than atomic number.

A more interesting curve results if we plot the nuclear mass *per nucleon* from hydrogen through uranium. To obtain the nuclear mass per nucleon, we divide the nuclear mass by the number of nucleons in the particular nucleus. The resulting graph, Figure 33.15, indicates the different average effective masses of nucleons in atomic nuclei, and is the crux of energy release in nuclear reactions. The greatest mass per nucleon occurs for the proton alone, because it isn't bound to anything, so has no binding energy to pull its mass down. Progressing beyond hydrogen, the masses of nucleons in heavier

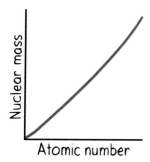

FIGURE 33.14 The plot shows how nuclear mass increases with increasing atomic number.

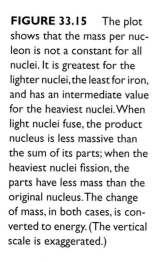

FIGURE 33.15 The plot shows that the mass per nucleon is not a constant for all nuclei. It is greatest for the lighter nuclei, the least for iron, and has an intermediate value for the heaviest nuclei. When light nuclei fuse, the product nucleus is less massive than the sum of its parts; when the heaviest nuclei fission, the parts have less mass than the original nucleus. The change of mass, in both cases, is converted to energy. (The vertical scale is exaggerated.)

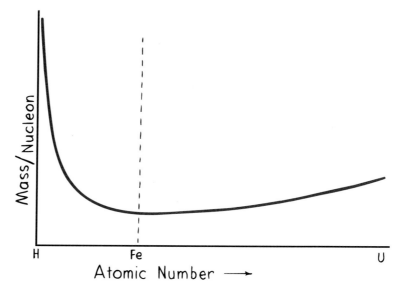

Table 33.1 Relative masses and masses/nucleon of some isotopes

Isotope	Symbol	Mass (amu)	Mass/Nucleon (amu)
Neutron	n	1.008665	1.008665
Hydrogen	$^1_1$H	1.007825	1.007825
Deuterium	$^2_1$H	2.01410	1.00705
Tritium	$^3_1$H	3.01605	1.00535
Helium-4	$^4_2$He	4.00260	1.00065
Carbon-12	$^{12}_6$C	12.00000	1.000000
Iron-58	$^{58}_{26}$Fe	57.93328	0.99885
Copper-63	$^{63}_{29}$Cu	62.92960	0.99888
Krypton-90	$^{90}_{36}$Kr	89.91959	0.99911
Barium-143	$^{143}_{56}$Ba	142.92054	0.99944
Uranium-235	$^{235}_{92}$U	235.04395	1.00019

nuclei are effectively smaller. Pulling apart an iron nucleus, represented at the lowest point on the curve, would take more work per nucleon than pulling apart any other nucleus. Iron holds its nucleons more tightly than any other nucleus does. Beyond iron, the average effective mass of nucleons increases. For elements lighter than iron and heavier than iron, the binding energy per nucleon is less than it is in iron.

From the graph we can see why energy is released when a uranium nucleus is split into nuclei of lower atomic number. The masses of the fission fragments lie about halfway between uranium and hydrogen on the horizontal scale of the graph. Most important, note that the mass per nucleon in the fission fragments is *less than* the mass per nucleon when the same set of nucleons are combined in the uranium nucleus. When this decrease in mass is multiplied by the speed of light squared, it equals 200,000,000 electron volts, the energy yielded by each uranium nucleus that undergoes fission.

FIGURE 33.16 The mass of a nucleus is *not* equal to the sum of the masses of its parts. (a) The fission fragments of a heavy nucleus like uranium are less massive than the uranium nucleus. (b) Two protons and two neutrons are more massive in their free states than when combined to form a helium nucleus.

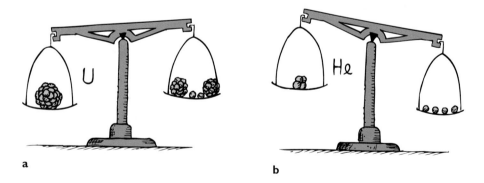

a

b

We can think of the mass-per-nucleon curve as an energy valley that starts at the highest point (hydrogen), slopes steeply down to the lowest point (iron), and then slopes more gradually up to uranium. Iron is at the bottom of the energy valley, which is the place with the greatest binding energy per nucleon. Any nuclear transformation that moves lighter nuclei toward iron by combining them or moves heavier nuclei toward iron by dividing them releases energy.

Table 33.2 Energy gain from fission of uranium

Reaction:	$^{235}U + n \rightarrow\ ^{143}Ba + ^{90}Kr + 3n + \Delta m$
Mass balance:	$235.04395 + 1.008665 = 142.92054 + 89.91959 + 3(1.008665) + \Delta m$
Mass defect:	$\Delta m = 0.186$ amu
Energy gain:	$\Delta E = \Delta m c^2 = 0.186 \times 931$ MeV $= 173.6$ MeV
Energy gain/nucleon:	$\Delta E/236 = 173.6$ MeV$/236 = 0.74$ MeV$/$nucleon

(When m is expressed in amu, c^2 is equivalent to 931 MeV)

So the decrease in mass is detectable in the form of energy—much energy—when heavy nuclei undergo fission. A drawback to this process involves the fission fragments. They are radioactive isotopes because of the greater-than-normal number of neutrons for elements of these atomic numbers. A more promising long-range source of energy is to be found on the left side of the energy valley.

Nuclear Fusion

Inspection of the graph in Figure 33.15 will show that the steepest part of the energy hill is from hydrogen to iron. Energy is gained as light nuclei *fuse*, or combine, rather than split apart. This process is **nuclear fusion**, the opposite of nuclear fission. Whereas energy is released when heavy nuclei split apart in the fission process, energy is released when light nuclei fuse together. After fusion, the total mass of the nuclei formed in the fusion process is less than the total mass of the nuclei that fused.

FIGURE 33.17 Fictitious example: the "hydrogen magnets" weigh more when apart than when together. (From *The New College Physics: A Spiral Approach* by A. V. Baez. Copyright © 1967 by W. H. Freeman and Company. Reprinted with permission.)

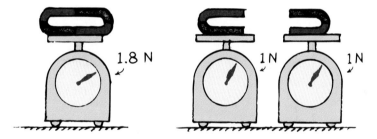

Atomic nuclei are positively charged. For fusion to occur, they normally must collide at very high speed in order to overcome electrical repulsion. The required speeds correspond to the extremely high temperatures found in the center of the sun and other stars. Fusion brought about by high temperatures is called **thermonuclear fusion**—that is, the welding together of atomic nuclei by high temperature. In the hot central part of the sun, approximately 657 million tons of hydrogen are converted into 653 million tons of helium each second. The missing 4 million tons of mass is discharged as radiant energy. Such reactions are, quite literally, nuclear burning. Most of the energy of nuclear fusion is in the kinetic energy of fragments, mainly neutrons. When the neutrons are stopped and captured, the energy of fusion is turned into heat. In fusion reactions of the future, part of this heat will be transformed into electricity.

Thermonuclear fusion is analogous to ordinary chemical combustion. In both chemical and nuclear burning, a high temperature starts the reaction; the release of energy by the reaction maintains a high enough temperature to spread the fire. The net result of the chemical reaction is a combination of atoms into more tightly bound molecules. In nuclear reactions, the net result is more tightly bound nuclei. The difference between chemical and nuclear burning is essentially one of scale.

Questions

1. First it was stated that nuclear energy is released when atoms split apart. Now it is stated that nuclear energy is released when atoms combine. Is this a contradiction? How can energy be released by opposite processes?

2. To get energy from the element iron, should iron be fissioned or fused?

In fission reactions, the amount of matter that is converted to energy is about 0.1%; in fusion, it can be as much as 0.7%. These numbers hold true whether the process takes place in bombs or reactors or stars.

Some typical fusion reactions are shown in Figure 33.18. Note that all reactions produce at least a pair of particles. For example, a pair of deuterium ions fusing produce a helium-3 nucleus and a neutron rather than a lone helium nucleus. Either reaction is okay as far as adding the nucleons and charges is concerned, but the lone-nucleus case is not okay with conservation of momentum and energy. If a lone helium nucleus flies away after the reaction, it adds momentum that wasn't there initially. Or if it remains motionless, there's no mechanism for energy release. So because a single product particle can't move and it can't sit still, it isn't formed. Fusion normally requires the creation of at least two particles to share the released energy.*

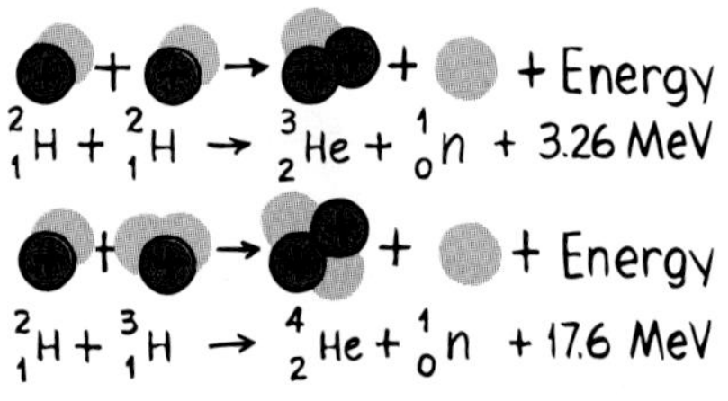

FIGURE 33.18 Two of many fusion reactions.

$$^2_1H + {}^2_1H \rightarrow {}^3_2He + {}^1_0n + 3.26\ MeV$$

$$^2_1H + {}^3_1H \rightarrow {}^4_2He + {}^1_0n + 17.6\ MeV$$

Answers

1. Energy is released only in a nuclear reaction in which the mass per nucleon decreases. Light nuclei, such as hydrogen, lose mass when they combine (fuse) to form heavier nuclei. They release energy in this reaction. Heavy nuclei, such as uranium, lose mass when they split to become lighter nuclei. For energy release, "Lose Mass" is the name of the game—any game.

2. Iron will release no energy at all, because it is at the very bottom of the energy valley. If fused with something else, it climbs the right side of the hill and gains mass. If fissioned, it climbs the left side of the hill and gains mass. In gaining mass, it absorbs energy instead of releasing energy.

* One of the reactions in the sun's proton-proton fusion cycle does have a one-particle final state. It is proton + deuteron goes to He-3. This happens because the density in the center of the sun is great enough that "spectator" particles share in the energy release. So even in this case, the energy released goes to two or more particles. Fusion in the sun involves more complicated (and slower!) reactions where a small part of the energy also appears in the form of gamma rays and neutrinos. The neutrinos escape unhindered from the center of the sun and bathe the solar system.

Table 33.3   Energy gain from fusion of hydrogen

Reaction:	$^2H + {}^3H \rightarrow {}^4_2H + n + \Delta m$
Mass balance:	$2.01410 + 3.01605 = 4.00260 + 1.008665 + \Delta m$
Mass defect:	$\Delta m = 0.01888$ amu
Energy gain:	$\Delta E = \Delta mc^2 = 0.18888 \times 931$ MeV $= 17.6$ MeV
Energy gain/nucleon:	$\Delta E/5 = 17.6$ MeV$/5 = 3.5$ MeV/nucleon

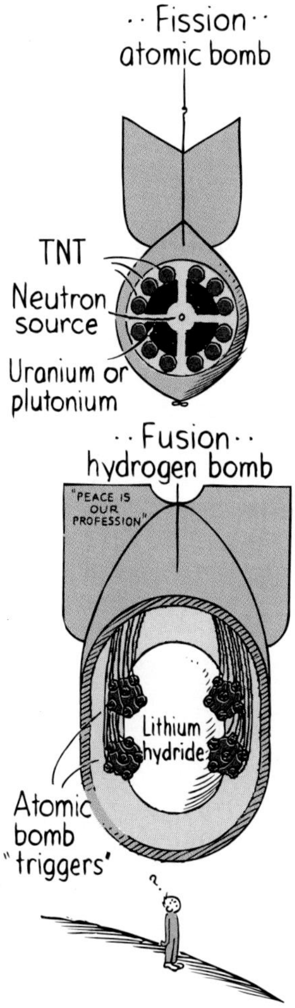

·· Fission ··
atomic bomb

TNT
Neutron
source
Uranium or
plutonium

·· Fusion ··
hydrogen bomb

"PEACE IS
OUR
PROFESSION"

Lithium
hydride

Atomic
bomb
"triggers"

FIGURE 33.19   Fission and fusion bombs.

Table 33.3 shows the energy gain from the fusion of hydrogen isotopes deuterium and tritium. This is the reaction slated to power tomorrow's fusion power plants. The high-energy neutrons will escape from the plasma in the reactor vessel and heat a surrounding blanket of material to provide useful energy. The helium nuclei remaining behind will help to keep the plasma hot.

Although the fusion energy per reaction of individual hydrogen atoms is less than the energy given up by the fissioning of individual uranium atoms, gram for gram fusion is several times more energy-producing than the fission process. This is because there are many more hydrogen atoms in a gram of hydrogen than there are heavier uranium atoms in a gram of uranium.

Elements heavier than hydrogen and lighter than iron give off energy when fused. But they give off much less energy/fusion reaction than hydrogen. The fusion of these heavier elements occurs in the advanced stages of a star's evolution. The energy released per gram during the various fusion stages from helium to iron amounts only to about one-fifth of the energy released in the fusion of hydrogen to helium. Hydrogen, most notably in the form of deuterium, is the choicest fuel for fusion.

Prior to the development of the atomic bomb, the temperatures required to initiate nuclear fusion on earth were unattainable. When it was found that the temperatures inside an exploding atomic bomb are four to five times the temperature at the center of the sun, the thermonuclear bomb was but a step away. This first hydrogen bomb was detonated in 1952. Whereas the critical mass of fissionable material limits the size of a fission bomb (atomic bomb), no such limit is imposed on a fusion bomb (thermonuclear, or hydrogen, bomb). Just as there is no limit to the size of an oil-storage depot, there is no theoretical limit to the size of a fusion bomb. Like the oil in a storage depot, any amount of fusion fuel can be stored with safety until it is ignited. Although a mere match can ignite an oil depot, nothing less energetic than a fission bomb can ignite a thermonuclear bomb. We can see that there is no such thing as a "baby" hydrogen bomb. It cannot be less energetic than its fuse, which is an atomic bomb.

The hydrogen bomb is another example of a discovery applied to destructive rather than constructive purposes. The potential constructive side of the picture is the controlled release of vast amounts of clean energy.

Controlling Fusion

Carrying out fusion reactions under controlled conditions ordinarily requires temperatures of hundreds of millions of degrees. Producing and sustaining such high temperatures along with reasonable densities is the goal of much current research. High temperatures are obtained in electric arcs, where injected gases are ionized and become plasmas. Further heating is accomplished by electromagnetic resonance techniques and magnetic compression.

It might seem that finding a wall to contain the super-hot plasma is impossible, given that all known materials melt and vaporize below 4000°C, while fusion generally requires temperatures exceeding 100 million degrees. It turns out that walling in the plasma is not the most challenging problem of thermonuclear power. A magnetic field is nonmaterial, can exist at any temperature, and can exert powerful forces on charged particles in motion. "Magnetic walls" of sufficient strength have been designed to contain plasmas in a kind of magnetic straitjacket. The magnetic fields can be regulated to compress the plasma and produce fusion temperatures. The plasma will not melt the walls of the electromagnets even if its temperature is a billion degrees, because its total heat content is very small due to its very low density. To have a controlled reactor rather than a bomb, plasma densities are kept low, about 10,000 times lower than atmospheric density. The plasma must also be kept off the walls of the container, not because it will melt it but because contact will slow the ions and therefore cool the plasma.

A few ions at 1 million degrees are moving fast enough to overcome electrical repulsion and slam together, but the energy output of these fusion reactions is much smaller than the energy used to heat the plasma. Even at 100 million degrees, more energy must be put into the plasma than is given off by fusion. At about 350 million degrees, the fusion reactions can produce as much energy as is put into the plasma. This is *break-even*. Between break-even in the laboratory and practical power generation, there is still a long, hard road to travel. Power must be sustained far beyond its present fraction of a second, and ways must be found to harness the energy output.

Although net energy production has been achieved in several fusion devices, instabilities in the plasma have thus far prevented a sustained reaction. One of the most difficult problems has been to devise a field system that will hold the plasma in a stable position long enough for an ample number of ions to fuse. A variety of magnetic-confinement devices are the subject of much present-day research.

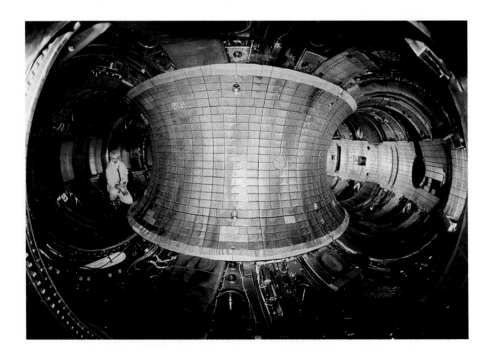

FIGURE 33.20 Interior view of the Tokamak Fusion Test Reactor (TFTR) vacuum vessel at the Princeton Plasma Physics Laboratory. Magnetic fields confine plasma that fuses to produce energy.

Another approach uses high-energy lasers to bypass magnetic confinement altogether. One technique is to aim an array of laser beams at a common point and drop solid pellets of hydrogen isotopes through the synchronous cross fire (Figure 33.21). The energy of the multiple beams crushes the deuterium-tritium fuel to a density more than 20 times that of lead. A fusion "burn" consumes most of the hydrogen to produce helium nuclei and neutrons that fly apart with kinetic energies equivalent to the mass loss of the fusion reaction. This outward propagation occurs in about a tenth of a billionth of a second and produces several hundred times more energy than is delivered by the laser beams that compress and ignite the pellets. Like the succession of small fuel/air explosions in an automobile engine's cylinders that convert into a smooth flow of mechanical power, the successive ignition of dropping pellets in a fusion power plant may similarly produce a steady stream of electrical power.* The success of this technique requires precise timing, for the necessary compression must occur before a shock wave causes the pellet to disperse. Success also requires the development of more efficient lasers. Break-even has not yet been achieved with laser fusion.

FIGURE 33.21 Fusion with multiple laser beams. Pellets of frozen deuterium are rhythmically dropped into synchronized laser cross fire. The resulting heat is carried off by molten lithium to produce steam.

FIGURE 33.22 Pellet chamber at Lawrence Livermore National Laboratory. The laser source is Nova, one of the most powerful lasers in the world, which directs 10 beams into the target region.

* The rate of pellet fusion is 5 per second on the projected Cascade power plant, now on the drawing boards at Lawrence Livermore National Laboratory. (For comparison, approximately 20 explosions per second occur in each automobile engine cylinder in a car that travels at highway speed.) Such a plant could produce 1000 million W of electric power, enough to supply a city of about 600,000 people. Five fusion burns per second will provide about the same power as 60 L of fuel oil or 70 kg of coal per second from conventional power plants.

Still other approaches involve the bombardment of fuel pellets—not by laser light but by beams of electrons and ions. Whatever the method, we are still looking forward to the great day in the twenty-first century when fusion power becomes a reality.

Fusion power is nearly ideal. Fusion reactors cannot become "supercritical" and get out of control because fusion requires no critical mass. Furthermore, there is no air pollution because the only product of the thermonuclear combustion is helium (good for children's balloons). Except for some radioactivity in the inner chamber of the fusion device because of high-energy neutrons, the byproducts of fusion are not radioactive. Disposal of radioactive waste is not a major problem. Furthermore, there is no air pollution because there is no combustion. The problem of thermal pollution, characteristic of conventional steam-turbine plants, can be avoided by direct generation of electricity with MHD generators or by similar techniques using charge-particle fuel cycles that employ a direct energy conversion.

The fuel for nuclear fusion is hydrogen, the most plentiful element in the universe. The simplest reaction is the fusion of the hydrogen isotopes deuterium (^2_1H) and tritium (^3_1H). Deuterium is found in ordinary water and tritium can be produced by the fusion reactor. For example, 30 liters of seawater contain 1 gram of deuterium, which when fused releases as much energy as 10,000 liters of gasoline or 80 tons of TNT. Natural tritium is much scarcer, but given enough to get started, a controlled thermonuclear reactor will breed it from deuterium in ample quantities. Because of the abundance of fusion fuel, the amount of energy that can be released in a controlled manner is virtually unlimited.

The development of fusion power has been slow and difficult, already extending over nearly fifty years. It is one of the biggest scientific and engineering challenges that we face. Yet there is every reason to believe that it will be achieved and will be a primary energy source for future generations.*

Questions

1. Fission and fusion are opposite processes, yet each releases energy. Isn't this contradictory?
2. Would you expect the temperature of the core of a star to increase or decrease as a result of the fusion of intermediate elements to manufacture heavy elements?

Answers

1. No, no, no! This is contradictory only if the same element is said to release energy by both the processes of fission and fusion. Only the fusion of light elements and the fission of heavy elements result in a decrease in nucleon mass and a release of energy.
2. Energy is absorbed and the star core tends to cool at this late stage of its evolution. This, however, allows the star to collapse, which produces an even greater temperature.

* Public outcry against electricity was common in the previous century. Those with the loudest voices likely knew the least about electricity. Today there is public outcry against nuclear power—"No Nukes!" The position of this book, in contrast, is "Know Nukes!"—first know something about the promise as well as the drawbacks of nuclear power, before saying yes or no.

Fusion Torch and Recycling

A fascinating application for the abundant energy that fusion of whatever kind may provide is the *fusion torch,* a star-hot flame or high-temperature plasma into which all waste materials—whether liquid sewage or solid industrial refuse—could be dumped. In the high-temperature region the materials would be reduced to their constituent atoms and separated by a mass-spectrometer-type device into various bins ranging from hydrogen to uranium. In this way, a single fusion plant could, in principle, not only dispose of thousands of tons of solid wastes per day but also provide a continuous supply of fresh raw material—thereby closing the cycle from use to reuse.

This would be a major turning point in materials economy (Figure 33.23). Our present concern for recycling materials will reach a grand fruition with this or a comparable achievement, for it would be recycling with a capital *R*! Rather than gut our planet further for raw materials, we'd be able to recycle our existing stock over and over again, adding new material only to replace the relatively small amounts that are lost. Fusion power can produce abundant electrical power, can desalinate water, can help to cleanse our world of pollution and wastes, can recycle our materials, and

in so doing can provide the setting for a better world—not in the far-off future, but perhaps in the twenty-first century. If and when fusion power plants become a reality, they are likely to have an even more profound impact upon almost every aspect of human society than did the harnessing of electromagnetic energy in the last century.

When we think of our continuing evolution, we can see that the universe is well suited to those who will live in the future. If people are one day to dart about the universe in the same way we are able to jet about the world today, their supply of fuel is assured. The fuel for fusion is found in every part of the universe, not only in the stars but also in the space between them. About 91% of the atoms in the universe are estimated to be hydrogen. For people of the future, the supply of raw materials is also assured; all the elements result from fusing more and more hydrogen nuclei together. Simply put, if you fuse 8 deuterium nuclei, you have oxygen; 26, you have iron; and so forth. Future humans might synthesize their own elements and produce energy in the process, just as the stars do. Humans may one day travel to the stars in ships fueled by the same energy that makes the stars shine.

FIGURE 33.23 A closed materials economy could be achieved with the aid of the fusion torch. In contrast to present systems, (a) which are based on inherently wasteful linear material economies, a stationary-state system (b) would be able to recycle the limited supply of material resources, thus alleviating most of the environmental pollution associated with present methods of energy utilization. (Redrawn from "The Prospects of Fusion Power" by William C. Gough and Bernard J. Eastlund. Copyright © February 1971 by Scientific American, Inc. All rights reserved.)

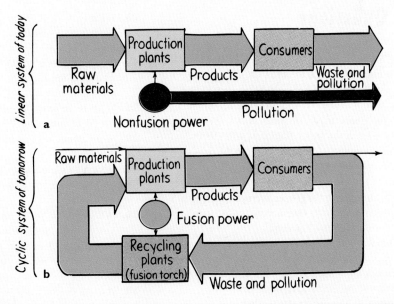

Summary of Terms

Nuclear fission   The splitting of the nucleus of a heavy atom, such as uranium-235, into two main parts, accompanied by the release of much energy.

Chain reaction   A self-sustaining reaction in which the products of one reaction event stimulate further reaction events.

Critical mass   The minimum mass of fissionable material in a reactor or nuclear bomb that will sustain a chain reaction.

Mass-energy equivalence   The relationship between mass and energy as given by the equation $E = mc^2$.

Nuclear fusion   The combination of the nuclei of light atoms to form heavier nuclei, with the release of much energy.

Thermonuclear fusion   Nuclear fusion produced by high temperature.

Suggested Reading

Sessler, Andrew M., Stix, Thomas H., and Rosenbluth, Marshall N. "Build the International Thermonuclear Experimental Reactor?" *Physics Today,* June 1996.

White, Peter T. "A Comeback for Nuclear Power?" *National Geographic* 180. No. 2 ("Aug 1991"): 60–89. A look at the prospects for commercial nuclear power and other sources of electric energy.

Review Questions

Nuclear Fission

1. What is the role of electrical forces in nuclear fission?
2. How does the energy release of a single uranium fission compare to that of a molecule of TNT when it explodes?
3. Which has the greater mass, a uranium nucleus before it undergoes fission or the fission fragments after fission?
4. When a nucleus undergoes fission, what can the neutrons that are ejected do?
5. Why does a chain reaction not occur in uranium mines?
6. Where do neutrons have the greater average path length inside the material—in two separated pieces of uranium or in the same pieces stuck together?
7. Which has more total surface area, a whole apple or an apple cut in two pieces?
8. Which will leak more neutrons, two separate pieces of uranium or the same pieces stuck together?
9. Is a chain reaction more likely to occur in two separate pieces of U-235 or in the same pieces stuck together?

10. What were the two methods used to separate U-235 from U-238 in the Manhattan Project during World War II?

Nuclear Fission Reactors

11. What was the function of graphite in the first atomic reactor?
12. How is the chain reaction in a nuclear reactor controlled?

Plutonium

13. What isotope is produced when U-238 absorbs a neutron?
14. What isotope is produced when U-239 emits a beta particle?
15. What isotope is produced when Np-239 emits a beta particle?
16. What do U-235 and Pu-239 have in common?
17. When is plutonium chemically toxic and when is it not?
18. Is plutonium radioactively more toxic or less toxic than uranium?

Breeder Reactors

19. What is the effect of putting small amounts of Pu-239 with large amounts of U-238?
20. Name three isotopes that undergo nuclear fission.
21. How does a breeder reactor breed nuclear fuel?

Mass-Energy Equivalence

22. Is work required to pull a nucleon out of an atomic nucleus? Does the nucleon, once outside, then have more mass than the average mass per nucleon of the original nucleus?
23. Is the mass per nucleon of a nucleus greater than, less than, or the same as the mass of a nucleon outside a nucleus?
24. Does the mass of a nucleon after it has been pulled from an atomic nucleus depend on which nucleus it was extracted from?
25. What is the basic difference between the graphs of Figure 33.14 and Figure 33.15?
26. What atomic nucleus has the greatest mass per nucleon? The least?
27. How does the mass per nucleon in uranium compare with the mass per nucleon in the fission fragments of uranium?
28. What becomes of the missing mass when a uranium nucleus undergoes fission?
29. If an iron nucleus splits in two, would its fission fragments have more mass per nucleon or less mass per nucleon?
30. If a pair of iron nuclei were fused, would the product nucleus have more mass per nucleon or less mass per nucleon?

Nuclear Fusion

31. Which releases more energy, the fissioning of a uranium nucleus or the fusing of a pair of hydrogen nuclei? Which releases more energy, the fissioning of nuclei in a gram of uranium or the fusing of nuclei in a gram of hydrogen? Why are your answers different?

32. When a pair of hydrogen isotopes are fused, is the mass of the product nucleus more or less than the total mass of the hydrogen nuclei?

33. For helium to release energy, should it be fissioned or fused?

34. What exactly is *thermonuclear* fusion?

35. How is nuclear burning similar to chemical burning?

Controlling Fusion

36. What kind of containers are used to contain multi-million-degree plasmas?

37. How is fusion accomplished with lasers?

38. How do the product particles of fusion reactions differ from the product particles of fission reactions?

39. In what form is energy initially released in nuclear fusion?

Fusion Torch and Recycling

40. If a star-hot flame is positioned between a pair of large electrically charged plates, one positive and the other negative, and materials dumped into the flame are dissociated into bare nuclei and electrons, in which direction will the nuclei move? In which direction will the electrons move?

41. Suppose the negative plate has a hole in it so that atomic nuclei that move toward and pass through it end up as a beam. Further suppose the beam is then directed between the pole pieces of a powerful electromagnet. Will the beam of charged nuclei continue in a straight line or will the beam be deflected?

42. Supposing the beam is deflected, will all nuclei, both light and heavy, deflect by the same amount? (Can you see that this device is a version of the mass spectrometer and that it effectively separates atomic nuclei?)

43. In a bucketful of seawater are minute amounts of gold. You can't separate them from the water with an ordinary magnet, but if the bucket is dumped into the fusion torch we are describing, a magnet can separate them. If hydrogen atoms are collected in bin #1 and uranium atoms are collected in bin #92, what will be the bin number for gold?

44. What would be the outcome of dumping sewage and refuse in a fusion torch instead of into the ocean or underground?

45. What is the fuel for nuclear fusion? What is the most abundant element in the universe? Why will there be no shortage of energy and material in the "hydrogen age"?

Exercises

1. Why is it that uranium ore doesn't spontaneously undergo a chain reaction?

2. Why will nuclear fission probably not be used directly for powering automobiles? How could it be used indirectly?

3. Why does a neutron make a better nuclear bullet than a proton or an electron?

4. Why will the escape of neutrons be proportionally less in a large piece of fissionable material than in a smaller piece?

5. Which shape is likely to need more material for a critical mass, a cube or a sphere? Explain.

6. Does the average distance that a neutron travels through fissionable material before escaping increase or decrease when two pieces of fissionable material are assembled into one piece? Does this assembly increase or decrease the probability of an explosion?

7. U-235 releases an average of 2.5 neutrons per fission, while Pu-239 releases an average of 2.7 neutrons per fission. Which of these elements might you therefore expect to have the smaller critical mass?

8. Why does plutonium not occur in appreciable amounts in natural ore deposits?

9. Why, after a uranium fuel rod reaches the end of its fuel cycle (typically 3 years) does most of its energy come from the fissioning of plutonium?

10. If a nucleus of $^{232}_{90}$Th absorbs a neutron and the resulting nucleus undergoes two successive beta decays (emitting electrons), what nucleus results?

11. The water that passes through a reactor core does not pass into the turbine. Instead, heat is transferred to a separate water cycle that is entirely outside the reactor. Why it this done?

12. Why is carbon better than lead as a moderator in nuclear reactors?

13. Is the mass of an atomic nucleus greater or less than the sum of the masses of the nucleons composing it? Why don't the nucleon masses add up to the total nuclear mass?

14. The energy release of nuclear fission is tied to the fact that the heaviest nuclei have about 0.1% more mass per nucleon than nuclei near the middle of the periodic table of elements. What would be the effect on energy release if the 0.1% figure were instead 1%?

15. In what way are fission and fusion reactions similar? What are the main differences in these reactions?

16. How is chemical burning similar to nuclear fusion?

17. To predict the approximate energy release of either a fission or a fusion reaction, explain how a physicist makes use of the curve in Figure 33.15 or a table of nuclear masses, and the equation $E = mc^2$.

18. Which process would release energy from gold, fission or fusion? From carbon? From iron?

19. If a uranium nucleus were to fission into three nuclei of equal size instead of two, would more energy or less energy be released? Defend your answer in terms of Figure 33.15.

20. Suppose the curve of Figure 33.15 for mass per nucleon versus atomic number took the shape of the curve shown in Figure 33.14. Then would nuclear fission reactions produce energy? Would nuclear fusion reactions produce energy? Defend your answers.

21. The "hydrogen magnets" in Figure 33.17 weigh more when apart than when combined. What would be the basic difference if the fictitious example instead consisted of "nuclear magnets" half as heavy as uranium?

22. Which produces more energy, the fissioning of a single uranium atom or the fusing of a pair of deuterium atoms? The fissioning of a gram of uranium or the fusing of a gram of deuterium? (Why are your answers different?)

23. Why is there, unlike fission fuel, no limit to the amount of fusion fuel that can be safely stored in one locality?

24. If a fusion reaction produces no appreciable radioactive isotopes, why does a hydrogen bomb produce significant radioactive fallout?

25. List at least two major advantages to power production by fusion rather than by fission.

26. Nuclear fusion is a present hope for abundant future energy. Yet the energy that has always sustained us has been the energy of nuclear fusion. Explain.

27. Explain how radioactive decay has always warmed the earth from the inside, and nuclear fusion has always warmed the earth from the outside.

28. What effect on the mining industry can you foresee in the disposal of urban waste by a fusion torch coupled with a mass spectrometer?

29. The world has never been the same since the discovery of electromagnetic induction and its applications to electric motors and generators. Speculate and list some of the worldwide changes that are likely to follow the advent of successful fusion reactors.

30. Compare pollution generated by conventional fossil-fuel power plants with pollution caused by nuclear-fission power plants. Consider thermal pollution, chemical pollution, and radioactive pollution. Discuss your findings.

Problems

1. The kiloton, which is used to measure the energy released in an atomic explosion, is equal to 4.2×10^{12} J (approximately the energy released in the explosion of 1000 tons of TNT). Recalling that 1 kilocalorie of energy raises the temperature of 1 kg of water by $1°C$ and that 4184 joules is equal to 1 kilocalorie, calculate how many kilograms of water can be heated through $50°C$ by a 20-kiloton atomic bomb.

2. The isotope of lithium used in a hydrogen bomb is Li-6, whose nucleus contains 3 protons and 3 neutrons. When a Li-6 nucleus absorbs a neutron, a nucleus of the heaviest hydrogen isotope, tritium, is produced. What is the other product of this reaction? Which of these two products fuels the explosive reaction?

3. An important fusion reaction in both hydrogen bombs and controlled fusion reactors is the "DT reaction," in which a deuteron and a triton (nuclei of heavy hydrogen isotopes) combine to form an alpha particle and a neutron with the release of much energy. Use momentum conservation to explain why the neutron resulting from this reaction gets about 80% of the energy, with the alpha particle getting only about 20%.

P a r t

RELATIVITY

Before the advent of Special Relativity, people thought the stars were beyond human reach. But distance is relative -- it depends on motion. In a frame of reference moving almost as fast as light, distance contracts and time stretches enough to allow future astronauts access to the stars and beyond! We *are* like Sarah's chickie on opening page 1, at the verge of a whole new beginning. Newton's physics got us to the moon; Einstein's physics points us to the stars. We live at an exciting time!

34

Special Theory of Relativity

: Time—nature's way of seeing that everything doesn't all happen at once.

As an eager young student of physics in the 1890s, Albert Einstein was troubled by a difference between Newton's laws of mechanics and Maxwell's laws of electromagnetism. Newton's laws were independent of the state of motion of an observer; Maxwell's laws were not—or so it seemed. Someone at rest and someone in motion would find that the *same* laws of mechanics apply to a moving object being studied, but they would find that *different* laws of electricity and magnetism apply to a moving charge being studied. Newton's laws suggest that there is no such thing as absolute motion; only relative motion matters. But Maxwell's laws seemed to suggest that motion is absolute.

In a celebrated 1905 paper titled "On the Electrodynamics of Moving Bodies," written when he was 26, Einstein showed that Maxwell's laws can, after all, like Newton's laws, be interpreted as being independent of the state of motion of an observer—but at a cost! The cost of achieving this unified view of nature's laws is a total revolution in how we understand space and time.

Einstein showed that as the forces between electric charges are affected by motion, the very measurements of space and time are also affected by motion. All measurements of space and time depend on relative motion.

For example, the length of a rocket ship poised on its launching pad and the ticks of clocks within are found to change when the ship is set into motion at high speed. It has always been common sense that we change our position in space when we move, but Einstein flouted common sense and stated that in moving we also change our rate of proceeding into the future—time itself is altered. Einstein went on to show that a consequence of the interrelationship between space and time is an interrelationship between mass and energy, given by the famous equation $E = mc^2$.

These are the ideas that make up this chapter—the ideas of special relativity—ideas so remote from your everyday experience that understanding them requires stretching your mind. It will be enough to become acquainted with these ideas, so be patient with yourself if you don't understand them right away. Perhaps in some future era when high-speed interstellar space travel is commonplace, your descendants will find that relativity makes common sense.

Motion Is Relative

Recall from Chapter 3 that whenever we talk about motion, we must always specify the position from which motion is being observed and measured. For example, a person who walks along the aisle of a moving train may be walking at a speed of 1 kilometer per hour relative to his seat, but 60 kilometers per hour relative to the railroad station. We call the place from which motion is observed and measured a **frame of reference**. An object may have different velocities relative to different frames of reference.

To measure the speed of an object, we first choose a frame of reference and pretend that the frame of reference is standing still. Then we measure the speed with which the object moves relative to the frame of reference. In the foregoing example, if we pretend the train is standing still, then the speed of the walking person is 1 kilometer per hour. If we pretend that the ground is standing still, then the speed of the walking person is 60 kilometers per hour. But the ground is not really still, for the earth spins like a top about its polar axis. Depending on how near the train is to the equator, the speed of the walking person may be as much as 1600 kilometers per hour relative to a frame of reference at the center of the earth. And the center of the earth is moving relative to the sun. If we place our frame of reference on the sun, the speed of the person walking in the train, which is on the orbiting earth, is nearly 110,000 kilometers per hour. And the sun is not at rest, for it orbits the center of our galaxy, which moves with respect to other galaxies.

Michelson-Morley Experiment

Isn't there some reference frame that is still? Isn't space itself still, and can't measurements be made relative to still space? In 1887 the American physicists A. A. Michelson and E. W. Morley attempted to answer these questions by performing an experiment that was designed to measure the motion of the earth through space.

FIGURE 34.1 The Michelson-Morley interferometer, which splits a light beam into two parts and then recombines them to form an interference pattern after they have traveled different paths. Rotation was accomplished in their experiment by floating a massive sandstone slab in mercury. This schematic diagram shows how the half-silvered mirror splits the beam into two rays. The clear glass assured that both rays traverse the same amount of glass. Four mirrors at each end were used to lengthen the paths.

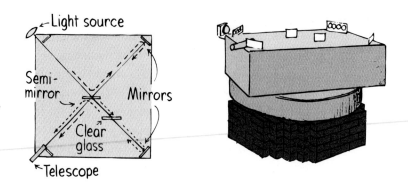

Because light travels in waves, it was then assumed that something in space vibrates—a mysterious something called *ether*, thought to fill all space and serve as a frame of reference attached to space itself. These physicists used a very sensitive apparatus called an *interferometer* to make their observations (Figure 34.1). In this instrument, a beam of light from a monochromatic source was separated into two beams with paths at right angles to each other; these were reflected and recombined to show whether there was any difference in average speed over the two back-and-forth paths. The interferometer was set with one path parallel to the motion of the earth in its orbit; then either Michelson or Morley carefully watched for any changes in average speed as the apparatus was rotated to put the other path parallel to the motion of the earth. The interferometer was sensitive enough to measure the difference in the round-trip times of light going with and against the earth's orbital velocity of 30 kilometers per second and going back and forth across the earth's path through space. But no changes were observed. None. Something was wrong with the sensible idea that the speed of light measured by a moving receiver should be its usual speed in a vacuum, c, plus or minus the contribution from the motion of the source or receiver. Many repetitions and variations of the Michelson and Morley experiment by many investigators showed the same null result. This was one of the puzzling facts of physics when the twentieth century opened.

One interpretation of the bewildering result was suggested by the Irish physicist G. F. FitzGerald, who proposed that the length of the experimental apparatus shrank in the direction in which it was moving by just the amount required to counteract the presumed variation in the speed of light. The factor that would account for the shrinkage, $\sqrt{1 - v^2/c^2}$, was worked out by the Dutch physicist Hendrik A. Lorentz. This arithmetical factor accounted for the discrepancy, but neither FitzGerald nor Lorentz had a suitable theory for why this was so. Interestingly, the same factor was derived by Einstein in his 1905 paper, where he showed it to indeed have a theoretical basis.

How much the Michelson and Morley experiment influenced Einstein, if at all, is unclear. In any event, Einstein advanced the idea that the speed of light in free space is the same in all reference frames, an idea that was contrary to the classical ideas of space and time. Speed is a ratio of distance through space to a corresponding interval of time. For the speed of light to be a constant, the classical idea that space and time are independent of each other had to be rejected. Einstein saw that space and time are linked, and with simple postulates, he developed a profound relationship between space and time.

Postulates of the Special Theory of Relativity

FIGURE 34.2 The speed of light is measured to be the same in all frames of reference.

FIGURE 34.3 The speed of a light flash emitted by the space station is measured to be *c* by observers on both the space station and the rocket ship.

Einstein saw no need for the ether. Gone with the stationary ether was the notion of an absolute frame of reference. All motion is relative, not to any stationary hitching post in the universe, but to arbitrary frames of reference. A rocket ship cannot measure its speed with respect to empty space, but only with respect to other objects. If, for example, rocket ship A drifts past rocket ship B in empty space, spaceman A and spacewoman B will each observe the relative motion, and from this observation each will be unable to determine who is moving and who is at rest, if either.

This is a familiar experience to a passenger on a train who looks out his window and sees the train on the next track moving by his window. He is aware only of the relative motion between his train and the other train and cannot tell which train is moving. He may be at rest relative to the ground and the other train may be moving, or he may be moving relative to the ground and the other train may be at rest, or they both may be moving relative to the ground. The important point here is that if you were in a train with no windows, there would be no way to determine whether the train was moving with uniform velocity or was at rest. This is the first of Einstein's postulates of the special theory of relativity:

All laws of nature are the same in all uniformly moving frames of reference.

On a jet airplane going 700 kilometers per hour, for example, coffee pours as it does when the plane is at rest; we swing a pendulum and it swings as it would if the plane were at rest on the runway. There is no physical experiment we can perform, even with light, to determine our state of uniform motion. The laws of physics within the uniformly moving cabin are the same as those in a stationary laboratory.

Any number of experiments can be devised to detect accelerated motion, but none can be devised, according to Einstein, to detect a state of uniform motion. Therefore absolute motion has no meaning.

It would be very peculiar if the laws of mechanics varied for observers moving at different speeds. It would mean, for example, that a pool player on a smooth-moving ocean liner would have to adjust her style of play to the speed of the ship, or even to the season as the earth varies in its orbital speed about the sun. It is our common experience that no such adjustment is necessary. And, according to Einstein, this same insensitivity to motion extends to electromagnetism. No experiment, mechanical or electrical or optical has ever revealed absolute motion. That is what the first postulate of relativity means.

One of the questions that Einstein as a youth asked his schoolteacher was, "What would a light beam look like if you traveled along beside it?" According to classical physics, the beam would be at rest to such an observer. The more Einstein thought about this, the more convinced he became that one could not move with a light beam. He finally came to the conclusion that no matter how close a person comes to the speed of light, he would still measure the light leaving him at 300,000 kilometers per second. This was the second postulate in his special theory of relativity:

The speed of light in free space has the same measured value regardless of the motion of the source or the motion of the observer; that is, the speed of light is invariant.

To illustrate this statement, consider a rocket ship departing from the space station shown in Figure 34.3. A flash of light traveling at 300,000 kilometers per second, or *c*, is emitted from the station. Regardless of the velocity of the rocket, an observer in the rocket sees the flash of light pass her at the same speed c. If a flash is sent to the station from the moving rocket, observers on the station will measure the speed of the flash to be c. The speed of light is measured to be the same regardless of the speed of the source or receiver. *All* observers who measure the speed of light will find it has the same value c. The more you think about this, the more you think it doesn't make sense. We will see that the explanation has to do with the relationship between space and time.

Simultaneity

An interesting consequence of Einstein's second postulate occurs with the concept of **simultaneity**. We say that two events are simultaneous if they occur at the same time. Consider, for example, a light source in the exact center of the compartment of a rocket ship (Figure 34.4). When the light source is switched on, light spreads out in all directions at speed c. Because the light source is equidistant from the front and back ends of the compartment, an observer inside the compartment finds that light reaches the front end at the same instant it reaches the back end. This occurs whether the ship is at rest or moving at constant velocity. The events of hitting the back end and hitting the front end occur *simultaneously*.

FIGURE 34.4 From the point of view of the observer who travels with the compartment, light from the source travels equal distances to both ends of the compartment and therefore strikes both ends simultaneously.

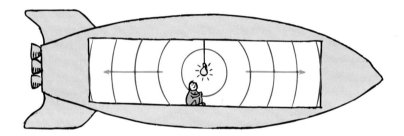

But time in one frame of reference may be different from time in another frame. To an outside observer who views the same two events in another frame of reference, say from a planet not moving with the ship, these same two events are *not* simultaneous. As light travels out from the source, this observer sees the ship move forward, so the back of the compartment moves toward the beam while the front moves away from it. The beam going to the back of the compartment therefore has a shorter distance to travel than the beam going forward (Figure 34.5). Since the speed of light is the same in both

FIGURE 34.5 The events of light striking the front and back of the compartment are not simultaneous from the point of view of an observer in a different frame of reference. Because of the ship's motion, light that strikes the back of the compartment doesn't have as far to go and strikes sooner than light that strikes the front of the compartment.

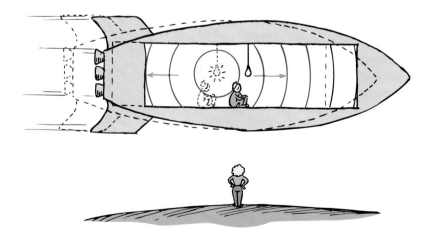

directions, this observer sees the event of light hitting the back of the compartment *before* seeing the event of light hitting the front of the compartment. (Of course, we are making the assumption that the observer can discern these slight differences.) A little thought will show that an observer in another rocket ship that passes the ship in the opposite direction would report that the light reaches the front of the compartment first.

> **Two events that are simultaneous in one frame of reference need not be simultaneous in a frame moving relative to the first frame.**

This nonsimultaneity of events in one frame that are simultaneous in another is a purely relativistic result—a consequence of light always having the same speed for all observers.

Questions

1. How is the nonsimultaneity of hearing thunder *after* seeing lightning similar to relativistic nonsimultaneity?
2. Suppose that the planet-bound observer in Figure 34.5 sees a pair of lightning bolts simultaneously strike the front and rear ends of the high-speed rocket ship compartment. Will the lightning strikes be simultaneous to an observer in the middle of the rocket ship compartment? (We assume here that an observer can detect any slight differences in time for light to travel from the ends to the middle of the compartment.)

Spacetime

When we look up at the stars, we realize that we are actually looking backward in time. The stars we see farthest away are the stars we are seeing longest ago. The more we think about this, the more apparent it becomes that space and time must be intimately tied together.

The space we live in is three-dimensional; that is, we can specify the position of any location in space with three dimensions. For example, these dimensions could be north-south, east-west, and up-down. If we are at the corner of a rectangular room and wish to specify the position of any point in the room, we can do this with three numbers. The first would be the number of meters the point is along a line joining the side wall and the floor; the second would be the number of meters the point is along a line joining the adjacent back wall and the floor; and the third would be the number of meters the point lies above the floor or along the vertical line joining the walls at the corner.

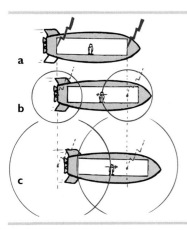

Answers

1. It isn't! The duration between hearing thunder and seeing lightning has nothing to do with moving observers or relativity. In such a case you simply correct for the time the signals (sound and light) take to reach you. The relativity of simultaneity is a genuine discrepancy between observations made by observers in relative motion, and not simply a disparity between different travel times for different signals.
2. No; an observer in the middle of the compartment will see the lightning that hits the front end of the compartment before seeing the lightning that hits the rear end. This is shown in positions a, b, and c to the left. In a, we see both lightning bolts striking the ends of the compartment simultaneously. In position b, light from the front lightning bolt reaches the observer. Slightly later, in c, light from the rear lightning bolt reaches the observer.

Clock Watching on a Trolley Car Ride

Pretend you are Einstein at the turn of the twentieth century in a trolley car that provided the high speed travel back then. Suppose the trolley car is moving in a direction away from a huge clock in a village square. The clock reads 12 noon. To say it reads 12 noon is to say that light that carries the information "12 noon" left the clock and traveled for some time and some distance before bringing that information to you. You wonder about this delay. If the trolley car traveled as fast as the light, then it would keep up with the information "12 noon." Traveling at the speed of light, you would always see the clock reading 12 noon. From your point of view, it would seem that time at the village square is frozen! So if the trolley car is not moving, you see the village-square clock move into the future at the rate of 60 minutes per hour; if you move at the speed of light, you see minutes on the clock taking infinite time. These are the two extremes. What's in between? How would the advance of the clock's hands be viewed at speeds less than the speed of light? A little thought will show that you will receive the "1 o'clock" message anywhere from 60 minutes to an infinity of time after you receive the "12 noon" message, depending on your speed between zero and the speed of light. From your high-speed (but less than c) frame

of reference, the clock and all events in the reference frame of the clock will be seen in slow motion. If you reverse direction and travel at high speed back toward the clock, earthly events will be seen speeded up. When you return, will the effects of going and coming compensate? Amazingly, no! Time will be stretched. Your clock and the village clock will disagree. This is time dilation. We'll see in the following pages that you'll be in a completely different realm of time than those that stayed behind—as evidenced by their advanced ages.

Physicists speak of these three lines as the *coordinate axes* of a reference frame (Figure 34.6). Three numbers—the distances along the x-axis, the y-axis, and the z-axis—will specify the position of a point in space.

We specify the size of objects with three dimensions. A box, for example, is described by its length, width, and height. But the three dimensions do not give a complete picture. There is a fourth dimension—time. The box was not always a box of given length, width, and height. It began as a box only at a certain point in time, on the day it was made. Nor will it always be a box. At any moment it may be crushed, burned, or otherwise destroyed. So the three dimensions of space are a valid description of the box only during a certain specified period of time. We cannot speak meaningfully about space without implying time. Things exist in **spacetime**. Each object, each person, each planet, each star, each galaxy exists in what physicists call "the spacetime continuum."

Two side-by-side observers at rest relative to one another share the same reference frame. Both would agree on measurements of space and time intervals between given events, so we say they share the same realm of spacetime. If there is relative motion

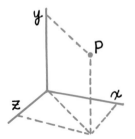

FIGURE 34.6 Point P can be specified with three numbers: the distance along the x-axis, the y-axis, and the z-axis.

between them, however, the observers will not agree on these measurements of space and time. At ordinary speeds, differences in their measurements are imperceptible, but at speeds near the speed of light—so-called relativistic speeds—the differences are appreciable. Each observer is in a different realm of spacetime, and her measurements of space and time differ from the measurements of an observer in some other realm of spacetime. The measurements differ not haphazardly but in such a way that each observer will always measure the same ratio of space and time for light; the greater the measured distance in space, the greater the measured interval of time. This constant ratio of space and time for light, *c,* is the unifying factor between different realms of spacetime and is the essence of Einstein's second postulate.

FIGURE 34.7 All space and time measurements of light are unified by *c.*

Time Dilation

Time is not an absolute but is relative, depending on the motion of an observer relative to what is being observed. Imagine, for example, that we are somehow able to observe a flash of light bouncing back and forth between a pair of parallel mirrors. If the distance between the mirrors is fixed, then the arrangement constitutes a sort of "light clock," because the back and forth trips of the light flash take equal intervals of time (Fig-

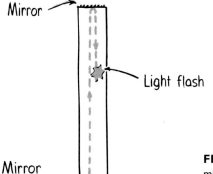

FIGURE 34.8 A light clock. Light will bounce up and down between parallel mirrors and "tick off" equal intervals of time.

ure 34.8). Suppose our light clock is inside a transparent high-speed spaceship. If we travel along with the ship and watch the light clock (Figure 34.9a), we will see the flash of light reflecting straight up and down between the two mirrors, just as it would if the spaceship were at rest. Our observations will show no relativistic effects because there is no relative motion between us and our light clock; we share the same reference frame in spacetime.

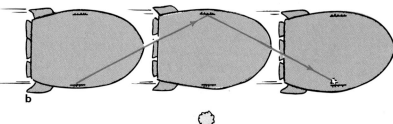

FIGURE 34.9   (a) An observer moving with the spaceship observes the light flash moving vertically between the mirrors of the light clock. (b) An observer who is passed by the moving ship observes the flash moving along a diagonal path.

FIGURE 34.10 The longer distance taken by the light flash in following the diagonal path must be divided by a correspondingly longer time interval to yield an unvarying value for the speed of light.

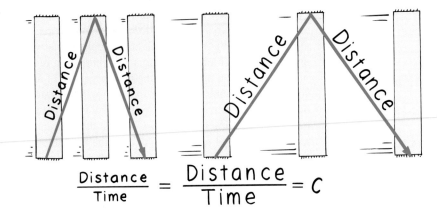

$$\frac{Distance}{Time} = \frac{Distance}{Time} = c$$

If, instead, we stand at some relative rest position and observe the spaceship whizzing by us at an appreciable speed—say, half the speed of light—things are quite different. We no longer share the same reference frame, for in this case there is relative motion between the observer and the observed. We will not see the path of the light in simple up-and-down motion as before. Because the light flash keeps up with the horizontally moving light clock, we will see the flash following a diagonal path (Figure 34.9b). Notice that the flash travels a *longer distance* as it moves between the mirrors in our position of spacetime than it does in the reference frame of an observer riding with the ship. Since the speed of light is the same in all reference frames (Einstein's second postulate), the flash must travel for a longer time between the mirrors in our frame than in the reference frame of an observer on board. This follows from the definition of speed, simply stated, as a ratio of distance to time. *The longer diagonal distance must be divided by a correspondingly longer time interval to yield an unvarying value for the speed of light.* Thus, from our relative position of rest, we measure a longer time interval between ticks when a clock is in motion than when it is at rest (Figure 34.10). We have considered a light clock in our example, but the same is true for any kind of clock. Moving clocks appear to run slow. This stretching out of time is called **time dilation**, which has nothing to do with the mechanics of clocks, but instead arises from the nature of time itself. Time dilation applies to *all* properly functioning timepieces.

FIGURE 34.11 When we see the rocket at rest, we see it traveling at the maximum rate in time: 24 hours per day. If we could see the rocket traveling at the maximum rate through space (the speed of light), we'd see its time standing still.

The exact relationship of time dilation for different frames of reference in spacetime can be derived from Figure 34.10 with simple geometry and algebra.* The relationship between the time t_0 in the observer's own frame of reference and the relative time t measured in another frame of reference is

$$t = \frac{t_0}{\sqrt{1 - \dfrac{v^2}{c^2}}}$$

where v represents the relative velocity between the observer and the observed and c is the speed of light. The quantity $\sqrt{1 - (v^2/c^2)}$ is the same factor used by Lorentz to explain length contraction. We call the inverse of this quantity the *Lorentz factor* γ (gamma). That is

$$\gamma = \frac{1}{\sqrt{1 - \dfrac{v^2}{c^2}}}$$

Then we can express the time dilation equation more simply as

$$t = \gamma \, t_0$$

Let's look at the terms in γ. Some mental tinkering will show that γ is always greater than 1 for any speed v greater than zero. Note that since speed v is always less than c, the ratio v/c is always less than 1; likewise for v^2/c^2. Can you see it follows that γ is greater than 1? Now consider the case where $v = 0$. This ratio v^2/c^2 is zero, and for

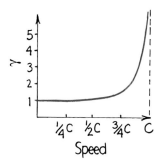

FIGURE 34.12 A plot of the Lorentz factor γ as a function of speed.

---

* The light clock is shown in three successive positions in the figure below. The diagonal lines represent the path of the light flash as it starts from the lower mirror at position 1, moves to the upper mirror at position 2, and then back to the lower mirror at position 3. Distances on the diagram are marked ct, vt, and ct_0, which follows from the fact that the distance traveled by a uniformly moving object is equal to its speed multiplied by the time.

The symbol t_0 represents the time it takes for the flash to move between the mirrors as measured from a frame of reference fixed to the light clock. This is the time for straight up or down motion. The speed of light is c, and the path of light is seen to move a vertical distance ct_0. This distance between mirrors is at right angles to the motion of the light clock and is the same in both reference frames.

The symbol t represents the time it takes the flash to move from one mirror to the other as measured from a frame of reference in which the light clock moves with speed v. Since the speed of the flash is c and the time to go from position 1 to position 2 is t, the diagonal distance traveled is ct. During this time t, the clock (which travels horizontally at speed v) moves a horizontal distance vt from position 1 to position 2.

These three distances make up a right triangle in the figure, in which ct is the hypotenuse and ct_0 and vt are legs. A well-known theorem of geometry (the Pythagorean theorem) states that the square of the hypotenuse is equal to the sum of the squares of the other two sides. If we apply this to the figure, we obtain:

$$c^2 t^2 = c^2 t_0^2 + v^2 t^2$$
$$c^2 t^2 - v^2 t^2 = c^2 t_0^2$$
$$t^2 [1 - (v^2/c^2)] = t_0^2$$
$$t^2 = \frac{t_0^2}{1 - (v^2/c^2)}$$
$$t = \frac{t_0}{\sqrt{1 - (v^2/c^2)}}$$

Path of light as seen from a position of rest

ct ct_0 vt

Mirrors at position 1   Mirrors at position 2   Mirrors at position 3

everyday speeds where v is negligibly small compared to c, it's practically zero. Then $1 - (v^2/c^2)$ has a value of 1, as has $\sqrt{1 - (v^2/c^2)}$, which makes $\gamma = 1$. Then we find $t = t_0$—time intervals appear the same in both reference frames. For higher speeds, v/c is between zero and 1, and $1 - (v^2/c^2)$ is less than 1; likewise, $\sqrt{1 - (v^2/c^2)}$. This makes γ greater than 1, so t_0 multiplied by a factor greater than 1 produces a value greater than t_0—an elongation—a dilation of time.

To consider some numerical values, assume that v is 50% the speed of light. Then we substitute $0.5c$ for v in the time-dilation equation and after some arithmetic find that $\gamma = 1.15$; so $t = 1.15\, t_0$. This means that if we viewed a clock on a spaceship traveling at half the speed of light, we would see the second hand take 1.15 minutes to make a revolution, whereas if the spaceship were at rest, we would see it take 1 minute. If the spaceship passes us at 87% the speed of light, $\gamma = 2$; and $t = 2\, t_0$. We would measure time events on the spaceship taking twice the usual intervals, for the hands of a clock on the ship would turn only half as fast as those on our own clock. Events on the ship would seem to take place in slow motion. At 99.5% the speed of light, $\gamma = 10$ and $t = 10\, t_0$; we would see the second hand of the spaceship's clock take 10 minutes to sweep through a revolution requiring 1 minute on our clock.

To put these figures another way, at $0.995\, c$, the moving clock would appear to run at a tenth of our rate; it would tick only 6 seconds while our clock ticks 60 seconds. At $0.87\, c$, the moving clock ticks at half rate and shows 30 seconds to our 60 seconds; at $0.50c$, the moving clock ticks $\frac{1}{1.15}$ as fast and ticks 52 seconds to our 60 seconds. We see that moving clocks run slow.

Nothing is unusual about a moving clock itself; it is simply ticking to the rhythm of a different time. The faster a clock moves, the slower it appears to run as viewed by an observer not moving with the clock. If it were possible to make a clock fly by us at the speed of light, the clock would not appear to be running at all. We would measure the interval between ticks to be infinite. The clock would be ageless! If we could move with such an imaginary clock, however, the clock would not show any slowing down of time. To us the clock would be operating normally. This is because there would be no motion of the clock relative to us. The v in γ would then be zero, and $t = t_0$; we and the clock would share the same frame in spacetime.

If a person whizzing past us checked a clock in our reference frame, he would find our clock to be running as slowly as we find his to be. We each see each other's clock running slow. There is really no contradiction here, for it is physically impossible for two observers in relative motion to refer to one and the same realm of spacetime. The measurements made in one realm of spacetime need not agree with the measurements made in another realm of spacetime. The measurement that all observers always agree on, however, is the speed of light.

Time dilation has been confirmed in the laboratory innumerable times with particle accelerators. The lifetimes of fast-moving radioactive particles increase as the speed goes up, and the amount of increase is just what Einstein's equation predicts.

Time dilation has been confirmed also for not-so-fast motion. In 1971 four cesium-beam atomic clocks were twice flown on regularly scheduled commercial jet flights around the world, once eastward and once westward, to test Einstein's theory of relativity with macroscopic clocks. The clocks indicated different times after their round trips. Relative to the atomic time scale of the U.S. Naval Observatory, the observed time differences, in billionths of a second, were in accord with relativistic prediction. Now, with atomic clocks orbiting the earth as part of the global positioning system, adjustments for the effects of time dilation are essential in order to use signals from the clocks to pinpoint locations on earth.

This all seems very strange to us only because it is not our common experience to deal with measurements made at relativistic speeds or atomic-clock-type measurements at ordinary speeds. The theory of relativity does not make common sense. But common sense, according to Einstein, is that layer of prejudices laid down in the mind prior to the age of 18. If we spent our youth zapping through the universe in high-speed spaceships, we would probably be quite comfortable with the results of relativity.

Questions

1. If you are moving in a spaceship at a high speed relative to the earth, would you notice a difference in your pulse rate? In the pulse rate of the people back on earth?
2. Will observers A and B agree on measurements of time if A moves at half the speed of light relative to B? If both A and B move together at half the speed of light relative to the earth?
3. Does time dilation mean that time really passes more slowly in moving systems or that it only seems to pass more slowly?

The Twin Trip

A dramatic illustration of time dilation is provided by identical twins, one an astronaut who takes a high-speed round-trip journey in the galaxy while the other stays home on earth. When the traveling twin returns, he is younger than the stay-at-home twin. How much younger depends on the relative speeds involved. If the traveling twin maintains a speed of 50% the speed of light for 1 year (according to clocks aboard the spaceship),

Answers

1. There would be no relative speed between you and your own pulse, which share the same frame of reference, so you would notice no relativistic effects in your own pulse. There would, however, be a relativistic effect between you and people back on earth. You would find their pulse rate slower than normal (and they would find your pulse rate slower than normal). Relativity effects are always attributed to the other guy.
2. When A and B move relative to each other, each will observe a slowing of time in the frame of reference of the other. So they will not agree on measurements of time. When they are moving in unison, however, they share the same frame of reference and will agree on measurements of time. They will see each other's time as passing normally, and they will each see events on earth in the same slow motion.
3. The slowing of time in moving systems is not merely an illusion resulting from motion. Time really does pass more slowly in a moving system compared to one at relative rest, as we shall see in the next section. Read on!

FIGURE 34.13 The traveling twin does not age as fast as the stay-at-home twin.

1.15 years will have elapsed on earth. If the traveling twin maintains a speed of 87% the speed of light for a year, then 2 years will have elapsed on earth. At 99.5% the speed of light, 10 earth years would pass in one spaceship year. At this speed the traveling twin would age a single year while the stay-at-home twin ages 10 years.

One question often arises: Since motion is relative, why doesn't the effect work equally well the other way around? Why wouldn't the traveling twin return to find his stay-at-home twin younger than himself? We will show that from the frames of reference of both the earthbound twin and traveling twin, it is the earthbound twin who ages more.

First, consider a spaceship hovering at rest relative to a distant planet. Suppose the spaceship sends regularly spaced brief flashes of light to the planet (Figure 34.14). Some time will elapse before the flashes get to the planet, just as 8 minutes elapse before sunlight gets to the earth. The light flashes will encounter the receiver on the planet at speed c. Since there is no relative motion between the sender and receiver, successive flashes will be received as frequently as they are sent. For example, if a flash is sent from the ship every 6 minutes, then after some initial delay, the receiver will receive a flash every 6 minutes. With no motion involved, there is nothing unusual about this.

FIGURE 34.14 When no motion is involved, the light flashes are received as frequently as the spaceship sends them.

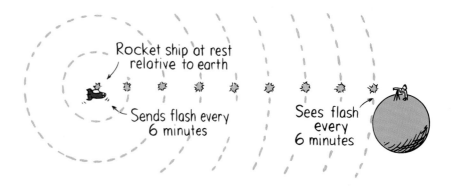

When motion is involved, the situation is quite different. It is important to note that the speed of the flashes will still be *c*, no matter how the ship or receiver may move. How frequently the flashes are seen, however, very much depends on the relative motion involved. When the ship travels toward the receiver, the receiver sees the flashes more frequently. This happens not only because time is altered due to motion, but mainly because each succeeding flash has less distance to travel as the ship gets closer to the receiver. If the spaceship emits a flash every 6 minutes, the flashes will be seen at intervals of less than 6 minutes. Suppose the ship is traveling fast enough for the flashes to be seen twice as frequently. Then they are seen at intervals of 3 minutes (Figure 34.15).

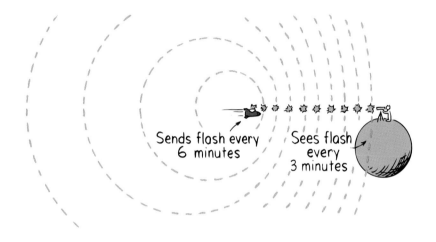

FIGURE 34.15 When the sender moves toward the receiver, the flashes are seen more frequently.

If the ship recedes from the receiver at the same speed and still emits flashes at 6-minute intervals, these flashes will be seen half as frequently by the receiver, that is, at 12-minute intervals (Figure 34.16). This is mainly because each succeeding flash has a longer distance to travel as the ship gets farther away from the receiver.

The effect of moving away is just the opposite of moving closer to the receiver. So if the flashes are received twice as frequently when the spaceship is approaching

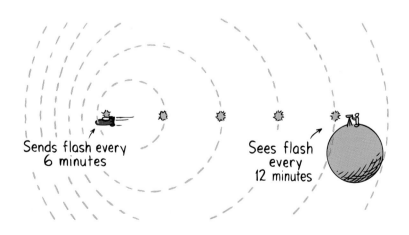

FIGURE 34.16 When the sender moves away from the receiver, the flashes are spaced farther apart and are seen less frequently.

(6-minute flash intervals are seen every 3 minutes), they are received half as frequently when it is receding (6-minute flash intervals are seen every 12 minutes).*

The light flashes make up a light clock. In the frame of reference of the receiver, events that take 6 minutes in the spaceship are seen to take 12 minutes when the spaceship recedes and only 3 minutes when the ship is approaching.

Questions

1. If the spaceship travels for 1 h and emits a flash every 6 min, how many flashes will be emitted?
2. The ship sends equally spaced 6-min flashes while approaching the receiver at constant speed. Will these flashes be equally spaced when they encounter the receiver?
3. If the receiver sees these flashes at 3-min intervals, how much time will occur between the first and the last flash (in the frame of reference of the receiver)?

Let's apply this doubling and halving of flash intervals to the twins. Suppose the traveling twin recedes from the earthbound twin at the same high speed for 1 hour and then quickly turns around and returns in 1 hour. Follow this line of reasoning with the help of Figure 34.17. The traveling twin takes a round trip of 2 hours, according to all clocks aboard the spaceship. This trip will not be seen to take 2 hours from the earth frame of reference, however. We can see this with the help of the flashes from the ship's light clock.

As the ship recedes from the earth, it emits a flash of light every 6 minutes. These flashes are received on earth every 12 minutes. During the hour of going away from the earth, a total of ten flashes are emitted. If the ship departs from the earth at noon, clocks aboard the ship read 1 PM when the tenth flash is emitted. What time will it be on earth when this tenth flash reaches the earth? The answer is 2 PM. Why? Because the time it

Answers

1. The ship will emit a total of ten flashes in 1 h, since (60 min)/(6 min) = 10.
2. Yes; as long as the ship moves at constant speed, the equally spaced flashes will be seen equally spaced but more frequently. (If the ship accelerated while sending flashes, then they would not be seen at equally spaced intervals.)
3. All ten flashes will be seen in 30 min, since 10 × (3 min) = 30 min.

---

* This reciprocal relationship (halving and doubling of frequencies) is a consequence of the constancy of the speed of light and can be illustrated with the following example: Suppose a sender on earth emits flashes 3 min apart to a distant observer on a planet that is at rest relative to the earth. The observer, then, sees a flash every 3 min. Now suppose a second observer travels in a spaceship between the earth and the planet at a speed great enough to allow him to see the flashes half as frequently—6 min apart. This halving of frequency occurs for a speed of recession of $0.6c$. We can see that the frequency will double for a $0.6c$-speed of approach by supposing that the spaceship emits its own flash every time it sees an earth flash, that is, every 6 min. How does the observer on the distant planet see these flashes? Since the earth flashes and the spaceship flashes travel together at the same speed c, the observer will see not only the earth flashes every 3 min but the spaceship flashes every 3 min as well. So although a person on the spaceship emits flashes every 6 min, the observer sees them every 3 min at twice the emitting frequency. So for a speed of recession where frequency appears halved, frequency appears doubled for the same speed of approach. If the ship were traveling faster so that the frequency of recession were $1/3$ or $1/4$ as much, then the frequency of approach would be three- or fourfold, respectively. This reciprocal relationship does not hold for waves that require a medium. In the case of sound waves, for example, a speed that results in a doubling of emitting frequency for approach produces $\frac{2}{3}$ (not $\frac{1}{2}$) the emitting frequency for recession.

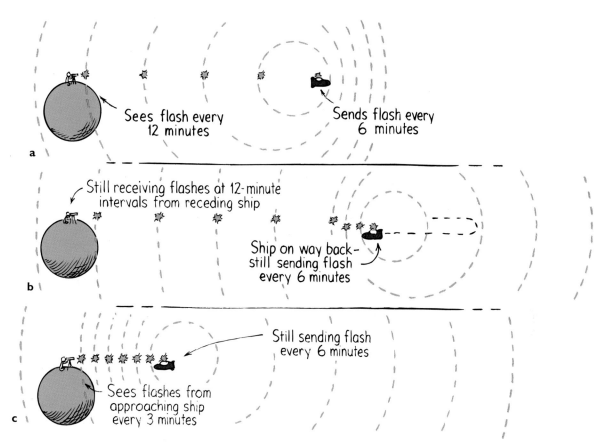

a Sees flash every 12 minutes — Sends flash every 6 minutes

b Still receiving flashes at 12-minute intervals from receding ship — Ship on way back—still sending flash every 6 minutes

c Still sending flash every 6 minutes — Sees flashes from approaching ship every 3 minutes

FIGURE 34.17   The spaceship emits flashes each 6 min during a 2-hour trip. During the first hour, it recedes from the earth. During the second hour, it approaches the earth.

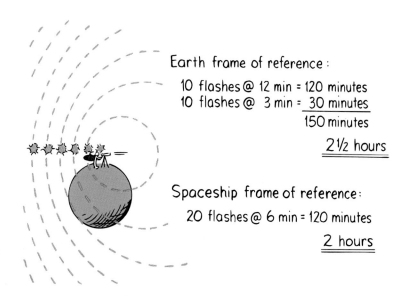

Earth frame of reference :
10 flashes @ 12 min = 120 minutes
10 flashes @ 3 min = <u>30 minutes</u>
150 minutes

<u>2½ hours</u>

Spaceship frame of reference :
20 flashes @ 6 min = 120 minutes

<u>2 hours</u>

FIGURE 34.18   The trip that takes 2 h in the frame of reference of the spaceship takes $2\frac{1}{2}$ h in the earth's frame of reference.

takes the earth to receive 10 flashes at 12-minute intervals is 10 × (12 minutes), or 120 minutes (= 2 hours).

Suppose the spaceship is somehow able to turn around suddenly in a negligibly short time and return at the same high speed. During the hour of return it emits ten more flashes at 6-minute intervals. These flashes are received every 3 minutes on earth, so all ten flashes come in 30 minutes. A clock on earth will read 2:30 PM when the spaceship completes its 2-hour trip. We see that the earthbound twin has aged half an hour more than the twin aboard the spaceship!

The result is the same from either frame of reference. Consider the same trip again, only this time with flashes emitted from the earth at regularly spaced 6-minute intervals in earth time. From the frame of reference of the receding spaceship, these flashes are received at 12-minute intervals (Figure 34.19a). This means that five flashes are seen by the spaceship during the hour of receding from earth. During the spaceship's hour of approaching, the light flashes are seen at 3-minute intervals (Figure 34.19b), so twenty flashes will be seen.

So we see that the spaceship receives a total of twenty-five flashes during its 2-hour trip. According to clocks on the earth, however, the time it took to emit the twenty-five flashes at 6-minute intervals was 25 × (6 minutes), or 150 minutes (= 2.5 hours). This is shown in Figure 34.20.

So both twins agree on the same results, with no dispute as to who ages more. While the stay-at-home twin remains in a single reference frame, the traveling twin has experienced two different frames of reference, separated by the acceleration of the spaceship in turning around. The spaceship has in effect experienced two different realms of time, while the earth has experienced a still different but single realm of time. The twins can meet again at the same place in space only at the expense of time.

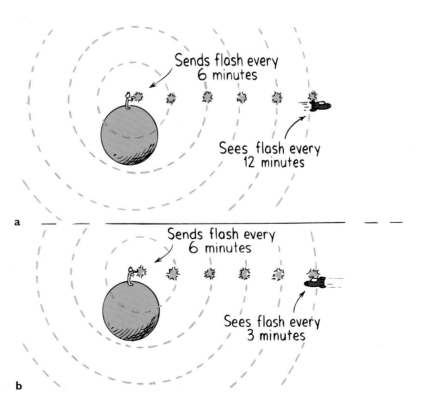

FIGURE 34.19 Flashes sent from earth at 6-min intervals are seen at 12-min intervals by the ship when it recedes and at 3-min intervals when it approaches.

FIGURE 34.20 A time interval of $2\frac{1}{2}$ h on earth is seen to take 2 h in the spaceship's frame of reference.

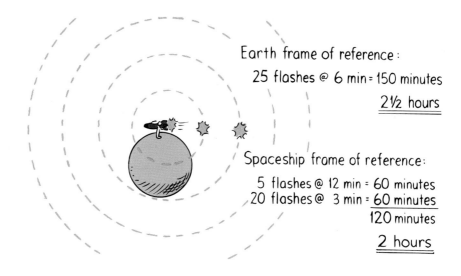

Earth frame of reference:
25 flashes @ 6 min = 150 minutes
2½ hours

Spaceship frame of reference:
5 flashes @ 12 min = 60 minutes
20 flashes @ 3 min = 60 minutes
120 minutes
2 hours

Question Since motion is relative, can't we say as well that the rocket ship is at rest and the earth moves, in which case the twin on the rocket ship ages more?

Addition of Velocities

Most people know that if you walk at 1 km/h along the aisle of a train that moves at 60 km/h, your speed relative to the ground is 61 km/h if you walk in the same direction as the moving train, and 59 km/h if you walk in the opposite direction. What most people know is *almost* correct. Taking special relativity into account, one's speeds are *very nearly* 61 km/h and 59 km/h respectively.

For everyday objects in uniform motion we ordinarily combine velocities by the simple rule:

$$V = v_1 + v_2$$

But this rule can't apply for light, which always has the same velocity c. Strictly speaking, the above rule is an approximation of the relativistic rule for adding velocities. We'll not treat the long derivation, and simply state the rule. In general,

$$V = \frac{v_1 + v_2}{1 + \dfrac{v_1 v_2}{c^2}}$$

The numerator of this formula makes common sense. But this simple sum of two velocities is altered by the second term in the denominator, which is significant only when both v_1 and v_2 are nearly c.

Answer No, not unless the earth then undergoes the turnaround and returns, as our rocket ship did in the twin-trip example. The situation is not symmetrical, for one twin remains in a single reference frame in spacetime during the trip while the other makes a distinct change of reference frame, as evidenced by the acceleration in turning around.

As an example, consider a spaceship moving away from you at a velocity of $0.5c$. It fires a rocket that thrusts in the same direction also away from you at a speed of $0.5c$ relative to the spaceship. How fast does the rocket move relative to you? The non-relativistic rule would say the rocket moves at the speed of light in your reference frame. But in fact,

$$V = \frac{0.5c + 0.5c}{1 + \dfrac{0.25c^2}{c^2}} = \frac{c}{1.25} = 0.8c$$

which illustrates another consequence of relativity: No material object can travel as fast or faster than light.

Suppose the spaceship instead fires a pulse of laser light in the same direction. How fast does the pulse move in your frame of reference?

$$V = \frac{0.5c + c}{1 + \dfrac{0.5c^2}{c^2}} = \frac{1.5c}{1.5} = c$$

No matter what the relative velocities between two frames, light moving at c in one frame will be seen to be moving at c in any other frame. Try chasing light and you can never catch it.

Space Travel

One of the old arguments advanced against the possibility of human interstellar travel is that our life span is too short—at least for the distant stars. It was argued, for example, that the nearest star (after the sun), Alpha Centauri, is 4 light-years away, and a round trip even at the speed of light would require 8 years.* The center of our galaxy is some 30,000 light-years away, so it has been reasoned that a person traveling even at the speed of light would require a lifetime of 30,000 years to make such a voyage. But these arguments fail to take into account time dilation. Time for a person on earth and time for a person in a high-speed rocket ship are not the same.

FIGURE 34.21 From the earth frame of reference, light takes 30,000 years to travel from the center of our galaxy to our solar system. From the frame of reference of a high-speed spaceship, the trip takes less time. From the frame of reference of light itself, the trip takes no time. There is no time in a speed-of-light frame of reference.

* A light-year is the distance light travels in 1 year, 9.46×10^{12} km.

A person's heart beats to the rhythm of the realm of time it is in; and one realm of time seems the same as any other realm of time to the person, but not to an observer who stands outside the person's frame of reference—for she can see the difference. For example, astronauts traveling at 99% the speed of light could go to the star Procyon (10.4 light-years distant) and back in 21 years. It would take light itself 20.8 years to make the same round trip. Because of time dilation, it would seem that only 3 years had gone by to the astronauts. This is what all their clocks would tell them—and biologically they would be only 3 years older. It would be the space officials greeting them on their return who would be 21 years older!

At higher speeds the results are even more impressive. At a rocket speed of 99.99% the speed of light, travelers could travel slightly more than 70 light-years in a single year of their own time; at 99.999% the speed of light, this distance would be pushed appreciably farther than 200 light-years. A 5-year trip would take them farther than light travels in 1000 earth-time years!

Present technology does not permit such journeys. The problems of getting enough propulsive energy and shielding against radiation are presently both prohibitive. Spaceships traveling at relativistic speeds would require billions of times the energy used to put a space shuttle into orbit. Even some kind of interstellar ramjet that scooped up interstellar hydrogen gas for burning in a fusion reactor would have to overcome the enormous retarding effect of scooping up the hydrogen at high speeds. And the space travelers would encounter interstellar particles just as if they had a large particle accelerator pointed at them. No way of shielding such intense particle bombardment for prolonged periods of time is presently known.

If and when these problems are overcome and space travel becomes a routine experience, people will have the option of taking a trip and returning to any future century of their choosing. For example, one might depart from earth in a high-speed ship in the year 2050, travel for 5 years or so, and return in the year 2500. One could live among the earthlings of that period for a while and depart again to try out the year 3000 for style. People could keep jumping into the future with some expense of their own time, but they could not trip into the past. They could never return to the same era on earth that they bid farewell to. Time, as we know it, travels one way—forward. Here on earth we move constantly into the future at the steady rate of 24 hours per day. A deep-space astronaut leaving on a deep-space voyage must live with the fact that, upon her return, much more time will have elapsed on earth than she has subjectively experienced on her voyage. The credo of all star travelers, whatever their physiological condition, will be permanent farewell.

We can see into the past, but we cannot go into the past. For example, we experience the past when we look at the night skies. The starlight impinging on our eyes left those stars dozens, hundreds, even millions of years ago. What we see is the stars as they were long ago. We are thus eyewitnesses to ancient history—and can only speculate about what may have happened to the stars in the interim.

If we are looking at light that left a star, say, 100 years ago, then it follows that any sighted beings in that solar system are seeing us by light that left *here* 100 years ago and that, further, if they possessed supertelescopes, they might very well be able to eyewitness earthly events of a century ago—the aftermath of the American Civil War, for instance. They would see our past, but they would still see events in a forward direction; they would see our clocks running clockwise.

We can speculate about the possibility that time might just as well move counter-clockwise into the past as clockwise into the future. Why is it, we might ask, that in space we can move forward or back, left or right, up or down, but we can move only in one direction through time? Quite interestingly, the mathematics of elementary-particle interactions permits "time reversal," although there are some particle interactions that favor only one direction in time. Hypothetical particles that move backward in time are called *tachyons*. In any case, for the complex organism called a human being, time has only one direction.*

This conclusion is blithely ignored in a limerick that is a favorite with scientist types:

> There was a young lady named Bright
> Who traveled much faster than light.
> She departed one day
> In a relative way
> And returned on the previous night.

Even with our heads fairly well into relativity, we may still unconsciously cling to the idea that there is an absolute time and compare all these relativistic effects to it— recognizing that time changes this way and that way for this speed and that speed, yet feeling that there still is some basic or absolute time. We may tend to think that the time we experience on earth is fundamental and that other times differ from it. This is understandable; we're earthlings. But the idea is confining. From the point of view of observers elsewhere in the universe, we may be moving at relativistic speeds; they see us moving in slow motion. They may see us living lifetimes a hundred times as long as theirs, just as with supertelescopes we would see them living lifetimes a hundredfold ours. There is no universally standard time. None.

We think of time and then we think of the universe. We think of the universe and we wonder about what went on before the universe began. We wonder about what will happen if the universe ceases to exist in time. But the concept of time applies to events and entities within the universe—not to the universe as a whole. Time is "in" the universe; the universe is not "in" time. Without the universe, there is no time; no before, no after. Likewise, space is "in" the universe; the universe is not "in" a region of space. There is no space "outside" the universe. Spacetime exists within the universe. Think about that!

FIGURE 34.22 The Lorentz contraction. The meter stick is measured to be half as long when traveling at 87% the speed of light relative to the observer.

Length Contraction

As objects move through spacetime, space as well as time undergoes changes in measurement. The lengths of objects appear to be contracted when they move by us at relativistic speeds. This **length contraction** is really a space contraction, quite different from the contraction of matter suggested by FitzGerald and worked out by Lorentz. Nevertheless, because Einstein's formula is the same as Lorentz's, we call the effect the Lorentz contraction. It is expressed mathematically as

$$L = L_0 \sqrt{1 - \frac{v^2}{c^2}}$$

where v is the relative velocity between the observed object and the observer, c is the speed of light, L is the measured length of the moving object, and L_0 is the measured

* It has been speculated that if we moved backward through time, we wouldn't know it, for then we would remember our future and would think it was our past!

FIGURE 34.23 As speed increases, length in the direction of motion decreases. Lengths in the perpendicular direction do not change.

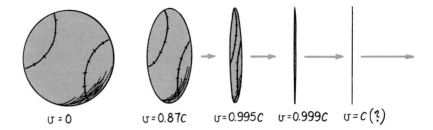

$v = 0$ $v = 0.87c$ $v = 0.995c$ $v = 0.999c$ $v = c(?)$

length of the object at rest.* Suppose that an object is at rest and $v = 0$. Upon substitution of $v = 0$ in the equation, we find $L = L_0$, as we would expect. At 87% the speed of light, an object would appear to be contracted to half its original length. At 99.5% the speed of light, it would seem to contract to one-tenth its original length. If the object moved at c, its length would be zero. This is one of the reasons we say that the speed of light is the upper limit for the speed of any moving object. Another ditty popular with the science heads is this one:

> There was a young fencer named Fisk,
> Whose thrust was exceedingly brisk.
> So fast was his action
> The Lorentz contraction
> Reduced his rapier to a disk.

As Figure 34.23 indicates, contraction takes place only in the direction of motion. If an object is moving horizontally, no contraction takes place vertically.

FIGURE 34.24 In the frame of reference of our meterstick, its length is 1 meter. Observers in a moving frame see *our* metersticks contracted, while we see *their* metersticks contracted. The effects of relativity are always attributed to "the other guy."

* We can express this as $L = \frac{1}{\gamma}L_0$, where $\frac{1}{\gamma}$ is always 1 or less (because γ is always 1 or greater).

Note that we do not explain how the length contraction equation (or other equations) comes about. We simply state equations as "guides to thinking" about the ideas of special relativity.

Do objects really contract at relativistic speeds? Well, if we attempt to check this out (in principle) and travel alongside the moving object with a meter stick, we notice nothing at all unusual about the length of the object. An observer at rest watching this experiment would report that the reason we don't measure the contraction, which is obvious to him, is that our meter stick is shortened as well. But because we are moving with the object, it appears to us that there is no contraction. This is because our relative velocity with respect to the object being measured is zero. The v in the Lorentz equation refers to the relative velocity between the observed and the observer. And that is zero, so $L = L_0$. It is important to stress this point. The object doesn't contract. A measure of the object from another reference frame contracts. We are simply measuring the distortion of space when we measure such a contraction, just as we measure the distortion of time itself when we find clocks running slow. The distortions are of space and time between different spacetimes, not of objects and events within individual realms of spacetime.

FIGURE 34.25 The Stanford Linear Accelerator is 3.2 km (2 miles) long. But to electrons moving through it at 0.9999995 c, the accelerator is only 3.2 meters long. The electrons start their journey in the foreground. They smash into targets or are otherwise studied in the experimental areas beyond the freeway.

Question A rectangular billboard in space has the dimensions 10 m × 20 m. How fast and in what direction with respect to the billboard would a space traveler have to pass for the billboard to appear square?

Answer The space traveler would have to travel at 0.87c in a direction parallel to the longer side of the board.

Relativistic Momentum

Recall our study of momentum in Chapter 5. We learned that the change of momentum mv of an object is equal to the impulse Ft applied to it: $Ft = \Delta\ mv$, or, $Ft = \Delta p$, where $p = mv$. Apply more impulse to an object that is free to move and the object acquires more momentum. Double the impulse and the momentum doubles. Apply ten times the impulse and the object gains ten times as much momentum. Does this mean that momentum can increase without any limit? The answer is yes. Does this mean that speed can also increase without any limit? The answer is no! Nature's speed limit for material objects is c.

To Newton, infinite momentum would mean infinite mass or infinite speed. But not so in relativity. Einstein showed that a new definition of momentum is required. It is

$$p = \gamma mv$$

where γ is the Lorentz factor (recall that γ is always 1 or greater). This generalized definition of momentum is valid in all uniformly moving reference frames. *Relativistic momentum* is larger than mv by a factor of γ. For everyday speeds much less than c, γ is nearly equal to 1, so p is nearly equal to mv. Newton's definition of momentum is valid at low speed.

At higher speeds γ grows dramatically, and so does relativistic momentum. As speed approaches c, γ approaches infinity! No matter how close to c an object is pushed, it would still require infinite impulse to give it the last bit of speed needed to reach *c*—clearly impossible. Hence we see that no body with mass can be pushed to the speed of light, much less beyond it.

Subatomic particles are routinely pushed to nearly the speed of light. The momenta of such particles may be thousands of times more than the Newton expression *mv* predicts. Classically, the particles behave as if their masses increase with speed. Einstein initially favored this interpretation, and later changed his mind to keep mass a constant, a property of matter that is the same in all frames of reference. So it is γ that changes with speed, not mass. The increased momentum of a high-speed particle is evident in the increased "stiffness" of its trajectory. The more momentum it has, the "stiffer" is its trajectory and the harder it is to deflect.

We see this when a beam of electrons is directed into a magnetic field. Charged particles moving in a magnetic field experience a force that deflects them from their normal paths. For small momentum, the path curves sharply. For large momentum, there is greater stiffness and the path curves only a little (Figure 34.26). Even though one particle may be moving only a little faster than another one—say 99.9% of the speed of light instead of 99% of the speed of light—its momentum will be considerably greater and it will follow a straighter path in the magnetic field. This stiffness must be compensated for in circular accelerators like cyclotrons and synchrotrons, where momentum dictates the radius of curvature. In the linear accelerator shown in Figure 34.25, the particle beam travels in a straight-line path and momentum changes don't produce deviations from a straight-line path. Deviations occur when the beam of electrons is bent at the exit port by magnets (Figure 34.26). Whatever the type of particle accelerator, physicists working with subatomic particles every day verify the correctness of the relativistic definition of momentum and the speed limit imposed by nature.

FIGURE 34.26 If the momentum of the electrons were equal to the Newtonian value *mv*, the beam would follow the dashed line. But because the relativistic momentum γ *mv* is greater, the beam follows the "stiffer" trajectory shown by the solid line.

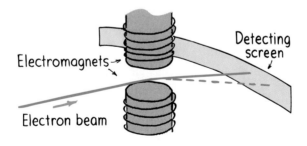

To summarize, we see that as the speed of an object approaches the speed of light, its momentum approaches infinity—which means there is no way that the speed of light can be reached. There is, however, at least one thing that reaches the speed of light—light itself! But light is massless, and the equations that apply to it are different. Light travels always at the same speed. So interestingly, a material particle can never be brought to the speed of light; and light can never be brought to rest.

Mass, Energy, and $E = mc^2$

Einstein not only linked space and time, he also linked mass and energy. A piece of matter, even if at rest and even if not interacting with anything else, has "energy of being." This is called its **rest energy**. Einstein concluded that it takes energy to make mass and that energy is released if mass disappears. Mass is, in effect, a kind of potential energy. Mass stores energy, just as a boulder rolled to the top of a hill stores energy. If the mass of something decreases, as it can do in nuclear reactions, energy is released, just as the boulder rolling to the bottom of the hill releases energy.

The amount of rest energy E is related to the mass m by the most celebrated equation of the twentieth century

$$E = mc^2$$

where c is again the speed of light. This equation gives the total energy content of a piece of stationary matter of mass m.*

Recall from Chapter 33 that tiny decreases of nuclear mass in both nuclear fission and nuclear fusion produced enormous releases of energy, all in accord with $E = mc^2$. To the general public, $E = mc^2$ is synonymous with nuclear energy. If we were to weigh a fully-fueled nuclear power plant, then weigh it again a week later, we'd find it weighs slightly less. Part of the fuel's mass, about 1 part in a thousand, has been converted to energy. Now, interestingly enough, weigh a coal-burning power plant and all the coal and oxygen it consumes in a week, and then weigh it again with all the carbon dioxide and other combustion products that come out during the week, we'll also find it all weighs slightly less. Again, mass has been converted to energy. About 1 part in a billion has been converted. Get this: If both plants produce the same amount of energy, the mass change will be the same for both—whether energy is released by nuclear or chemical mass conversion makes no difference. The chief difference lies in the amount of energy released in each individual reaction and the amount of mass involved. Fissioning of a single uranium nucleus releases 10 million times as much energy as the combustion of carbon to produce a single carbon dioxide molecule. Hence a few truckloads of uranium fuel will power a fission plant while a coal-burning plant consumes many hundred-car trainloads of coal.

FIGURE 34.27 Saying that a power plant delivers 90 million megajoules of energy to its consumers is equivalent to saying that it delivers 1 gram of energy to its consumers, because mass and energy are equivalent.

* When c is in meters per second and m is in kilograms, then E will be in joules. If the equivalence of mass and energy had been understood long ago when physics concepts were first being formulated, there would probably be no separate units for mass and energy. Furthermore, with a redefinition of space and time units, c could equal 1, and $E = mc^2$ would simply be $E = m$.

FIGURE 34.28 In 1 second, 4.5 million tons of mass are converted to radiant energy in the sun. The sun is so massive, however, that in 1 million years only one ten-millionth of the sun's mass will have been converted to radiant energy.

When we strike a match, phosphorus atoms in the match head rearrange themselves and combine with oxygen in the air to form new molecules. The resulting molecules have very slightly less mass than the separate phosphorus and oxygen molecules. From a mass standpoint, the whole is slightly less than the sum of its parts, by amounts that escape our notice. For all chemical reactions that give off energy, there is a corresponding decrease in mass of about one part in a billion.

In contrast, the greater mass decrease of nuclear reactions, one part in a thousand, can be directly measured by a variety of devices. This decrease of mass in the sun by the process of thermonuclear fusion bathes the solar system with radiant energy and nourishes life. The present stage of thermonuclear fusion in the sun has been going on for the past 5 billion years, and there is sufficient hydrogen fuel for fusion to last another 5 billion years. It is nice to have such a big sun!

The equation $E = mc^2$ is not restricted to chemical and nuclear reactions. A change in energy of any object at rest is accompanied by a change in its mass. The filament of a light bulb energized with electricity has more mass than when it is turned off. A hot cup of tea has more mass than the same cup of tea when cold. A wound-up spring clock has more mass than the same clock when unwound. But these examples involve incredibly small changes in mass—too small to be measured. Even the much larger changes of mass in radioactive change were not measured until after Einstein predicted the mass-energy equivalence. Now, however, mass-to-energy and energy-to-mass conversions are measured routinely.

Consider a coin with a mass of 1 g. You'd expect two of the same coins to have a mass of 2 g, ten coins to have a mass of 10 g, and 1000 coins piled in a box to have a mass of 1 kg. Not so if the coins attract or repel each other. Suppose, for example, that each coin carries a negative electric charge, so that each coin repels all other coins. Then forcing them together in the box takes work. This work adds to the mass of the collection. So a box containing 1000 negatively charged coins has more than 1 kg of mass. If, on the other hand, the coins all attracted one another (as nucleons in the nucleus attract one another), it takes work to separate them; then a box of 1000 coins would have a mass less than 1 kg. So the mass of an object is not necessarily equal to the sum of the masses of its parts, as we know from measuring the masses of nuclei. The effect would be dramatically enormous if we could deal with bare charged particles. If we could force together a number of electrons whose masses add separately to 1 g into a 10-cm diameter sphere, the collection would have a mass of 10 trillion kilograms! The equivalence of mass and energy is indeed profound.

Some physicists speculate that the mass of a single electron is merely the energy equivalent of the work required to compress its charge, assuming that if its charge were spread out, it would have no mass at all.*

In ordinary units of measurement, the speed of light c is a large quantity and its square is even larger—hence a small amount of mass stores a large amount of energy. The quantity c^2 is a "conversion factor." It converts the measurement of mass to the measurement of equivalent energy. Or it is the ratio of rest energy to mass: $E/m = c^2$. Its appearance in either form of this equation has nothing to do with light and nothing to do with motion. The magnitude of c^2 is 90 quadrillion (9×10^{16}) joules per kilogram.

---

* San Francisco Sidewalk Astronomer John Dobson speculates that just as a clock becomes more massive when we do work on it by winding it against the resistance of its spring, the mass of the entire universe is nothing more than the energy that has gone into winding it up against mutual gravitation. In this view the mass of the universe is equivalent to the work done in spreading it out. So perhaps each electron has mass because its charge is confined, and the atoms that make up the universe have mass because they are dispersed.

One kilogram of matter has an "energy of being" equal to 90 quadrillion joules. Even a speck of matter with a mass of only 1 milligram has a rest energy of 90 billion joules.

The equation $E = mc^2$ is more than a formula for the conversion of mass into other kinds of energy, or vice versa. It states even more, that energy and mass are the *same thing*. Mass is congealed energy. If you want to know how much energy is in a system, measure its mass. For an object at rest, its energy *is* its mass. Energy, like mass, exhibits inertia. Shake a massive object back and forth; it is energy itself that is hard to shake

Question Can we look at the equation $E = mc^2$ in another way and say that matter transforms into pure energy when it is traveling at the speed of light squared?

The first evidence for the conversion of radiant energy to mass was provided in 1932 by the American physicist Carl Anderson. He and a colleague at Caltech discovered the *positron* by the track it left in a cloud chamber. The positron is the *antiparticle* of the electron, equal in mass and spin to the electron but opposite in charge. When a high-energy photon comes close to an atomic nucleus, it can create an electron and a positron together as a pair, thus creating mass. The created particles fly apart. The positron is not part of normal matter because it lives such a short time in the presence of matter. As soon as it encounters an electron, the pair is annihilated, sending out two gamma rays in the process. Then mass is converted back to radiant energy.

The Correspondence Principle

We introduced the correspondence principle in Chapter 31. Recall that it states that any new theory or any new description of nature must agree with the old where the old gives correct results. If the equations of special relativity are valid, they must correspond to those of classical mechanics when speeds much less than the speed of light are considered.

The relativity equations for time, length, and momentum are:

$$t = \frac{t_0}{\sqrt{1 - \frac{v^2}{c^2}}} = \gamma t_0$$

$$L = L_0 \sqrt{1 - \frac{v^2}{c^2}} = \frac{L_0}{\gamma}$$

$$p = \frac{mv}{\sqrt{1 - \frac{v^2}{c^2}}} = \gamma mv$$

Answer No, no, no! Matter cannot be made to move at the speed of light, let alone the speed of light squared (which is not a speed!). The equation $E = mc^2$ simply means that energy and mass are "two sides of the same coin."

Albert Einstein (1879–1955)

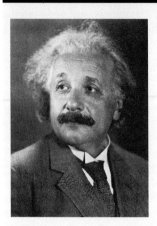

Albert Einstein was born in Ulm, Germany, on March 14, 1879. According to popular legend, he was a slow child and learned to speak at a much later age than average; his parents feared for a while that he might be mentally retarded. Yet his elementary school records show that he was remarkably gifted in mathematics, physics, and playing the violin. He rebelled, however, at the practice of education by regimentation and rote, and was expelled just as he was preparing to drop out at the age of 15. Largely because of business reasons, his family moved to Italy. Young Einstein renounced his German citizenship and went to live with family friends in Switzerland. There he was allowed to take the entrance examinations for the renowned Swiss Federal Institute of Technology in Zurich, at two years younger than normal age. But because of difficulties with the French language, he did not pass the examination. He spent a year at a Swiss preparatory school in Aarau, where he was "promoted with protest in French." He tried the entrance exam again at Zurich and passed. As a student he cut many lectures, preferring to study on his own, and in 1900 he succeeded in passing his examinations by cramming with the help of a friend's meticulous notes. He said later of this, ". . . after I had passed the final examination, I

found the consideration of any scientific problem distasteful to me for an entire year." During this year he became a citizen of Switzerland; he accepted a temporary summer teaching position and tutored two young high school students. He advised their father, a high school teacher himself, to remove the boys from school, where, he maintained, their natural curiosity was being destroyed. Einstein's job as a tutor was shortlived.

It was not until two years after graduation that he got a steady job, as a patent examiner at the Swiss Patent Office in Bern. Einstein held this position for over seven years. He found the work rather interesting, sometimes stimulating his scientific imagination, but mainly freeing him of financial worries while providing time to ponder the problems in physics that puzzled him.

With no academic connections whatsoever, and with essentially no contact with other physicists, he laid out the main lines along which twentieth-century theoretical physics has developed. In 1905, at the age of 26, he earned his Ph.D. in physics and published three major papers. The first was on the quantum theory of light, including an explanation of the photoelectric effect, for which he won the 1921 Nobel Prize in

Note that these equations each reduce to Newtonian values for speeds that are very small compared with c. Then the ratio v^2/c^2 is very small, and for everyday speeds may be taken to be zero. The relativity equations become

$$t = \frac{t_0}{\sqrt{1 - 0}} = t_0$$

$$L = L_0 \sqrt{1 - 0} = L_0$$

$$p = \frac{mv}{\sqrt{1 - 0}} = mv$$

So for everyday speeds, the momentum, length, and time of moving objects are essentially unchanged. The equations of special relativity hold for all speeds, although they differ appreciably from classical equations only for speeds near the speed of light.

Einstein's theory of relativity has raised many philosophical questions. What, exactly, is time? Can we say that it is nature's way of seeing to it that everything does not happen all at once? And why does time seem to move in one direction? Has it always moved *forward*? Are there other parts of the universe where time moves *backward*? Is it

physics. The second paper was on the statistical aspects of molecular theory and Brownian motion, a proof for the existence of atoms. His third and most famous paper was on special relativity. A fourth paper, published in 1915 on the theory of general relativity, presented a new theory of gravitation that included Newton's theory as a special case. These trailblazing papers have greatly affected the course of modern physics.

Einstein's concerns were not limited to physics. He lived in Berlin during World War I and denounced the German militarism of his time. He publicly expressed his deeply felt conviction that warfare should be abolished and an international organization founded to govern disputes between nations. In 1933, while Einstein was visiting the United States, Hitler came to power. Einstein spoke out against Hitler's racial and political policies and resigned his position at the University of Berlin. No longer safe in Germany, Einstein accepted a research position at the Institute for Advanced Study in Princeton, New Jersey.

In 1939, one year before Einstein became an American citizen, and after German scientists fissioned the uranium atom, he was urged by several prominent American scientists to write the famous letter to President Roosevelt pointing out the scientific possibilities of a nuclear bomb. Einstein was a pacifist, but the thought of Hitler developing such a bomb prompted his action. The outcome was the development of the first nuclear bomb, which, ironically, was detonated on Japan after the fall of Germany.

Einstein believed that the universe is indifferent to the human condition, and stated that if humanity were to continue, it must create a moral order. He intensely advocated world peace through nuclear disarmament. Nuclear bombs, Einstein remarked, had changed everything but our way of thinking.

C. P. Snow, who was acquainted with Einstein, in a review of *The Born-Einstein Letters, 1916–1955*, says this of him: "Einstein was the most powerful mind of the twentieth century, and one of the most powerful that ever lived. He was more than that. He was a man of enormous weight of personality, and perhaps most of all, of normal stature . . . I have met a number of people whom the world calls great; of these, he was by far, by an order of magnitude, the most impressive. He was—despite the warmth, the humanity, the touch of the comedian—the most different from other men."

likely that our three-dimensional perception of a four-dimensional world is only a beginning? Could there be a fifth dimension? A sixth dimension? A seventh dimension? And if so, what would the nature of these dimensions be? Perhaps these unanswered questions will be answered by the physicists of tomorrow. How exciting!

Summary of Terms

Frame of reference A vantage point (usually a set of coordinate axes) with respect to which position and motion may be described.

Postulates of the special theory of relativity (1) All laws of nature are the same in all uniformly moving frames of reference. (2) The speed of light in free space has the same measured value regardless of the motion of the source or the motion of the observer; that is, the speed of light is invariant.

Simultaneity Occurring at the same time. Two events that are simultaneous in one frame of reference need not be simultaneous in a frame moving relative to the first frame.

Spacetime The four-dimensional continuum in which all events take place and all things exist: Three dimensions are the coordinates of space and the fourth is of time.

Time dilation The slowing down of time in a frame of reference moving at relativistic speeds.

Length contraction The shrinkage of space, and therefore of matter, in a frame of reference moving at relativistic speeds.

Rest energy The energy of mass itself, as given by the equation $E = mc^2$.

Suggested Reading

Einstein, Albert. *The Meaning of Relativity.* Princeton, N.J.: Princeton University Press, 1950. Written for the average person by Einstein himself.

Epstein, Lewis C. *Relativity Visualized.* San Francisco: Insight Press, 1983.

Gamow, George. *Mr. Tompkins in Wonderland.* New York: Macmillan, 1940. An excellent and very interesting little book.

Gardner, Martin. *The Relativity Explosion.* New York: Vintage Books, 1976.

Holton, Gerald. *Einstein, History, and Other Passions.* Reading, MA: Addison-Wesley, 1996. The rebellion against science at the end of the twentieth century.

Taylor, Edwin F., and Wheeler, John A. *Spacetime Physics.* San Francisco: W. H. Freeman, 1966.

Review Questions

Motion Is Relative

1. If you walk at 1 km/h down the aisle toward the front of a bus that speeds along the road at 60 km/h, what is your speed relative to the ground?

2. In the previous question, what is your approximate speed relative to the sun as you walk down the aisle of the bus?

Michelson–Morley Experiment

3. What does it mean to say that the experiment of Michelson and Morley gave a null result?

4. What hypothesis did G. F. FitzGerald make to explain the findings of Michelson and Morley?

5. What classical idea about space and time was rejected by Einstein?

Postulates of the Special Theory of Relativity

6. Cite at least three examples of Einstein's first postulate.

7. Cite at least three examples of Einstein's second postulate.

Simultaneity

8. Inside the moving compartment of Figure 34.4, light travels a certain distance to the front end and a certain distance to the back end of the compartment. How do these distances compare as seen in the frame of reference of the moving rocket?

9. How do these distances in Question 8 compare as seen in the frame of reference of an observer on a stationary planet?

10. Why are events that are simultaneous in one frame of reference not simultaneous in a frame moving relative to the first frame?

Spacetime

11. How many coordinate axes are usually used to describe three-dimensional space?

12. What does the fourth dimension measure?

13. Under what condition will you and a friend share the same realm of spacetime?

14. Under what condition will you and a friend not share the same realm of spacetime?

15. What is special about the ratio of space traveled and time taken for light?

Time Dilation

16. Suppose the parallel mirrors of a light clock were 150,000 km apart. In the frame of reference of the light clock, how much time would be required for a pulse of light to make a round trip between the mirrors?

17. Suppose the parallel mirrors of a light clock were 150 km apart. In the frame of reference of the light clock, how much time would be required for a pulse of light to make a round trip between the mirrors?

18. Would your answers to the previous two questions be different if your measurements were made from a frame of reference that is moving relative to the light clock? Explain.

19. Time is required for light to travel along a path from one point to another. If this path is seen to be longer because of motion, what happens to the time it takes for light to travel this longer path?

20. What do we call the "stretching out of time"?

21. How do measurements of time intervals differ for events in a frame of reference that moves at 50% the speed of light relative to us?

22. How do measurements of time intervals differ for events in a frame of reference that moves at 99.5% the speed of light relative to us?

23. Suppose a clock accurately shows time passing half as fast in a particular frame of reference as in our own. What is the velocity of this frame of reference relative to us?

24. If we see somebody's clock running slow due to relative motion, will they see our clocks running slow also? Or will they see our clocks running fast? Explain.

25. What is the evidence for time dilation?

26. What does Einstein say about common sense?

The Twin Trip

27. When a flashing light approaches you, each flash that reaches you has a shorter distance to travel. What effect does this have on how frequently you receive the flashes?

28. When a flashing light source approaches you, does the speed of light or the frequency of light—or both—increase?

29. If a flashing light source moves toward you fast enough so the duration between flashes seems half as long, how will the duration between flashes seem if the source is moving away from you at the same speed?

30. How many frames of reference does the stay-at-home twin experience in the twin trip? How many frames of reference does the traveling twin experience?

Addition of Velocities

31. When two velocities v_1 and v_2 are each much less than the speed of light, is the value v_1v_2/c^2 large or small?

32. What is the maximum value of v_1v_2/c^2 in an extreme situation? The smallest?

33. Is the simple rule $V = v_1 + v_2$ consistent with the fact that light can have only one speed in all uniformly moving reference frames?

34. Is the relativistic rule

$$V = \frac{v_1 + v_2}{1 + \dfrac{v_1v_2}{c^2}}$$

consistent with the fact that light can have only one speed in all uniformly moving reference frames?

Space Travel

35. What two main obstacles prevent us from traveling today throughout the galaxy at relativistic speeds?

36. In this century, humans can enjoy jet hopping, traveling from one country to another by way of jet aircraft. In a future century, humans may enjoy "century hopping." How would this be accomplished?

Length Contraction

37. How long would a meter stick appear if it were traveling like a properly thrown spear at 99.5% the speed of light?

38. How long would the meter stick in the previous question appear if it were traveling with its length perpendicular to its direction of motion? (Why is your answer different from the previous question?)

39. If you were traveling in a high-speed rocket ship, would meter sticks on board appear to you to be contracted? Defend your answer.

Relativistic Momentum

40. What would be the momentum of an object if it could be pushed to the speed of light?

41. When a beam of charged particles moves through a magnetic field, what is the evidence that the momentum of the particles is greater than the value mv?

Mass, Energy, and $E = mc^2$

42. Does the equation $E = mc^2$ apply only to reactions that involve the atomic nucleus?

43. A power utility gets its energy from the mass of fuel, whether coal or atomic nuclei. If 1 gram of mass converts to energy in a coal plant, and 1 gram of mass converts to energy in a nuclear plant, why is the energy output at both plants the same?

44. In reference to the preceding question, why then, does it take so much more coal than uranium to produce the same energy?

The Correspondence Principle

45. How does the correspondence principle relate to special relativity?

46. Do the relativity equations for time, length, and momentum hold true for everyday speeds? Explain.

Exercises

1. The idea that force causes acceleration doesn't seem strange. Such ideas of Newtonian mechanics are consistent with our everyday experience. But the ideas of relativity do seem strange. They are hard to grasp. Why is this?

2. If you were in a smooth-riding train with no windows, could you sense the difference between uniform motion and rest? Between accelerated motion and rest? Explain how you could do this with a bowl filled with water.

3. A person riding on the roof of a freight-train car fires a gun pointed forward. (a) Relative to the ground, is the bullet moving faster or slower when the train is moving than when it is standing still? (b) Relative to the freight car, is the bullet moving faster or slower when the train is moving than when the train is standing still?

4. Suppose instead that the person riding on top of the freight car shines a searchlight beam in the direction in which the train is traveling. Compare the speed of the light beam relative to the ground when the train is at rest and when it is moving. How does the behavior of the light beam differ from the behavior of the bullet in Exercise 3?

5. Why did Michelson and Morley at first consider their experiment a failure? (Have you ever encountered other examples where failure has to do not with the lack of ability but with the impossibility of the task?)

6. When you drive down the highway you are moving through space. What else are you moving through?

7. In Chapter 27 we learned that light travels more slowly in glass than in air. Does this contradict the theory of relativity?

8. If two lightning bolts hit exactly the same place at exactly the same time in one frame of reference, is it possible that observers in other frames will see the bolts hitting at different times or at different places?

9. Event A occurs before event B in a certain frame of reference. How could event B occur before event A in some other frame of reference?

10. Suppose that the lightbulb in the rocket ship in Figures 34.4 and 34.5 is closer to the front than to the rear of the compartment, so that the observer in the ship sees the light reaching the front before it reaches the back. Is it still possible that the outside observer will see the light reaching the back first?

11. The speed of light is a speed limit in the universe—at least for the four-dimensional universe we comprehend. No material particle can attain or surpass this limit even when a continuous, unyielding force is exerted on it. Why is this so?

12. Can an electron beam sweep across the face of a cathode-ray tube at a speed greater than the speed of light? Explain.

13. Consider the speed of the point where scissors blades meet when scissors are closed. The closer the blades are to being closed, the faster the point moves. The point could, in principle, move faster than light. Likewise for the speed of the point where an ax meets wood when the ax blade meets the wood almost horizontally. The contact point travels faster than the ax. Similarly, a pair of laser beams that are crossed and moved toward being parallel produce a point of intersection that can move faster than light. Why do these examples not contradict special relativity?

14. Since there is an upper limit on the speed of a particle, does it follow that there is also an upper limit on its momentum? On its kinetic energy? Explain.

15. Light travels a certain distance in, say, 20,000 years. How is it possible that an astronaut, traveling slower than light, could go as far in 20 years of her life as light travels in 20,000 years?

16. Could a human being who has a life expectancy of 70 years possibly make a round-trip journey to a part of the universe thousands of light-years distant? Explain.

17. A twin who makes a long trip at relativistic speeds returns younger than his stay-at-home twin sister. Could he return before his twin sister was born? Defend your answer.

18. Is it possible for a son or daughter to be biologically older than his or her parents? Explain.

19. If you were in a rocket ship traveling away from the earth at a speed close to the speed of light, what changes would you note in your pulse? In your volume? Explain.

20. If you were on earth monitoring a person in a rocket ship traveling away from the earth at a speed close to the speed of light, what changes would you note in his pulse? In his volume? Explain.

21. If you lived in a world where people regularly traveled at speeds near the speed of light, why would it not be safe to make a dental appointment for 10:00 AM next Thursday?

22. How do the measured densities of a body compare at rest and in motion?

23. If stationary observers measure the shape of a passing object to be exactly circular, what is the shape of the object according to observers traveling with it?

24. The formula relating speed, frequency, and wavelength of electromagnetic waves, $c = f\lambda$, was known before relativity was developed. Relativity has not changed this equation but it has added a new feature to it. What is it?

25. Light is reflected from a moving mirror. How is the reflected light different from the incident light and how is it the same?

26. As a meter stick moves past you at nearly the speed of light, your measurements show its momentum to be twice its classical momentum and its length to be 1 m. In what direction is the stick pointing?

27. In the preceding exercise, if the stick is moving in a direction along its length (like a properly thrown spear), how long will you measure its length to be?

28. If a high-speed spaceship appears shrunken to half its normal length, how does its momentum compare with the classical formula $p = mv$?

29. The two-mile linear accelerator at Stanford University in California "appears" to be less than a meter long to the electrons that travel in it. Explain.

30. Electrons end their trip down the Stanford accelerator with an energy thousands of times more than the rest energy they had when they started. In theory, if you could travel with them, would you notice an increase in their energy? Their momentum? In your moving frame of reference, what would be the approximate speed of the target they are about to hit?

31. Two safety pins, identical except that one is latched and one is unlatched, are placed in identical acid baths. After the pins are dissolved, what, if anything, is the difference in the two acid baths?

32. A chunk of radioactive material encased in an idealized perfectly insulating blanket gets warmer as its nuclei decay and release energy. Does the mass of the radioactive material and the blanket change? If so, does it increase or decrease?

33. The electrons that illuminate the screen in a typical television picture tube travel at nearly one-fourth the speed

of light and have an energy nearly 3% greater than the energy of hypothetical non-relativistic electrons traveling at the same speed. Does this relativistic effect tend to increase or decrease your electric bill?

34. How might the idea of the correspondence principle be applied outside physical science?

35. What does the equation $E = mc^2$ mean?

36. Muons are elementary particles that are formed high in the atmosphere by the interactions of cosmic rays with atomic nuclei. Muons are radioactive and have average lifetimes of about two-millionths of a second. Even though they travel at almost the speed of light, they have so far to travel through the atmosphere that very few should be detected at sea level—at least according to classical physics. Laboratory measurements, however, show that muons in great number do reach the earth's surface. What is the explanation?

37. When we look out into the universe, we see into the past. John Dobson, founder of the San Francisco Sidewalk Astronomers, says that we cannot even see the backs of our own hands now—in fact, we can't see anything *now*. Do you agree? Explain.

38. One of the fads of the future might be "century hopping," where occupants of high-speed spaceships would depart from the earth for several years and return centuries later. What are the present-day obstacles to such a practice?

39. Is the statement by the philosopher Kierkegaard that "Life can only be understood backwards; but it must be lived forwards" consistent with the theory of special relativity?

40. Make up four multiple-choice questions, one each that would check a classmate's understanding of (a) time dilation, (b) length contraction, (c) relativistic momentum, and (d) $E = mc^2$.

Problems

Recall from this chapter that the factor gamma (γ) governs both time dilation and length contraction, where

$$\gamma = \frac{1}{\sqrt{1 - (v^2/c^2)}}.$$

When you multiply the time in a moving frame by γ, you get the longer (dilated) time in your fixed frame. When you divide the length in a moving frame by γ, you get the shorter (contracted) length in your fixed frame.

1. Consider a high-speed rocket equipped with a flashing light source. If the frequency of flashes when approaching is seen to increase by a factor of 2, by how much is the period (time interval between flashes) changed? Is this period constant for a constant relative speed? For accelerated motion? Defend your answer.

2. The starship *Enterprise,* passing the earth at 80% of the speed of light, sends a drone ship forward at half the speed of light relative to the *Enterprise.* What is the drone's speed relative to the earth?

3. Pretend that the starship *Enterprise* in the previous problem is somehow traveling at c with respect to the earth, and it fires a drone forward at speed c with respect to itself. Use the equation for the relativistic addition of velocities to show that the speed of the drone with respect to the earth is still c.

4. A passenger on an interplanetary express bus traveling at $v = 0.99c$ takes a five-minute catnap by his watch. How long does the nap last from your vantage point on a fixed planet?

5. According to Newtonian mechanics, the momentum of the bus in the preceding problem is $p = mv$. According to relativity, it is $p = \gamma mv$. How does the actual momentum of the bus moving at $0.99c$ compare with the momentum it would have if classical mechanics were valid? How does the momentum of an electron traveling at $0.99c$ compare with its classical momentum?

6. The bus in the previous problem is 70 feet long according to its passengers and driver. What is its length from your vantage point on a fixed planet?

7. If the bus in Problem 4 slowed to a "mere" 10% of the speed of light, how long would you measure the passenger's catnap to be?

8. If the bus driver in Problem 5 decided to drive at 99.99% of the speed of light in order to make up some time, what would you measure the length of the bus to be?

9. Assume that rocket taxis of the future move about the solar system at half the speed of light. For a one-hour trip as measured by a clock in the taxi, a driver is paid 10 stellars. The taxi driver's union demands that pay be based on earth time instead of taxi time. If their demand is met, what will be the new pay for the same trip?

10. The fractional change of mass to energy in a fission reactor is about 0.1%, or 1 part in a thousand. For each kilogram of uranium that undergoes fission, how much energy is released? If energy costs three cents per megajoule, how much is this energy worth in dollars?

35

General Theory of Relativity

: **S**pace, time, and gravity are beautifully interconnected.

The special theory of relativity is "special" in the sense that it deals with uniformly moving reference frames—that is, reference frames that are not accelerated. The general theory incorporates accelerating reference frames. Underlying it is the idea that the effects of gravitation and acceleration cannot be distinguished from one another. Einstein presented a new theory of gravitation.

Recall that Einstein postulated in 1905 that no observation made inside an enclosed chamber could determine whether the chamber was at rest or moving with constant velocity; that is, no mechanical, electrical, optical, or any other physical measurement that one could perform inside a closed compartment in a smoothly riding train traveling along a straight track (or in an airplane flying through still air with the window curtains drawn) could possibly give any information as to whether the train was moving or at rest (or the plane airborne or at rest on the runway). But if the track were not smooth and straight (or if the air were turbulent), the situation would be entirely different: Uniform motion would give way to accelerated motion, which would be easily noticed. Einstein's conviction that the laws of nature should be expressed in the same form in every frame of reference, accelerated as well as nonaccelerated, was the primary motivation that led him to the general theory of relativity.

Principle of Equivalence

Long before there were real spaceships, Einstein could imagine himself in a vehicle far away from gravitational influences. In such a spaceship at rest or in uniform motion relative to the distant stars, he and everything within the ship would float freely; there would be no "up" and no "down." But when the rocket motors were turned on and the ship accelerated, things would be different; phenomena similar to gravity would be observed. The wall adjacent to the rocket motors would push up against any occupants and become the floor, while the opposite wall would become the ceiling. Occupants in the ship would be able to stand on the floor and even jump up and down. If the acceleration of the spaceship were equal to g, the occupants could well be convinced the ship was not accelerating, but was at rest on the surface of the earth.

FIGURE 35.1 Everything is weightless on the inside of a nonaccelerating spaceship far away from gravitational influences.

FIGURE 35.2 When the spaceship accelerates, an occupant inside feels "gravity."

To examine this new "gravity" in the accelerating spaceship, let's consider the consequence of dropping two balls, say one of wood and the other of lead. When the balls are released, they continue to move upward side by side with the velocity of the ship at the moment of release. If the ship were moving at *constant velocity* (zero acceleration), the balls would remain suspended in the same place since both the ship and the balls would move the same amount. But since the spaceship is accelerating, the floor moves upward faster than the balls, which will soon be intercepted by the floor

(Figure 35.3). Both balls, regardless of their masses, would meet the floor at the same time. Remembering Galileo's demonstration at the Leaning Tower of Pisa, occupants of the ship might be prone to attribute their observations to the force of gravitation.

FIGURE 35.3   To an observer inside the accelerating ship, a lead ball and a wooden ball fall together when released.

The two interpretations of the falling balls are equally valid, and Einstein incorporated this equivalence, or impossibility of distinguishing between gravitation and acceleration, in the foundation of his general theory of relativity. The **principle of equivalence** states that observations made in an accelerated reference frame are indistinguishable from observations made in a Newtonian gravitational field. This equivalence would be interesting but not revolutionary if it applied only to mechanical phenomena, but Einstein went further and stated that the principle holds for all natural phenomena; it holds for optical and all electromagnetic phenomena as well.

Bending of Light by Gravity

A ball thrown sideways in a stationary spaceship in a gravity-free region will follow a straight-line path relative both to an observer inside the ship and to a stationary observer outside the spaceship. But if the ship is accelerating, the floor overtakes the ball just as in our previous example. An observer outside the ship still sees a straight-line path, but to an observer in the accelerating ship, the path is curved; it is a parabola (Figure 35.4). The same holds true for a beam of light.

Imagine that a light ray enters the spaceship horizontally through a side window, passes through a sheet of glass in the middle of the cabin, leaving a visible trace, and then reaches the opposite wall, all in a very short time. The outside observer sees that the light ray enters the window and moves horizontally along a straight line with constant velocity toward the opposite wall. But the spaceship is accelerating upward.

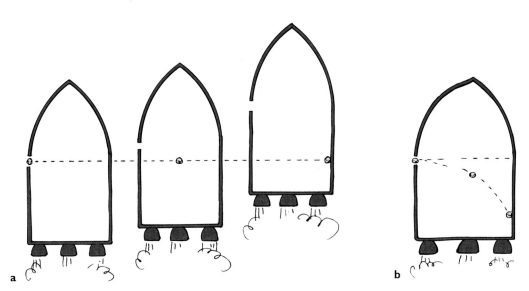

FIGURE 35.4 (a) An outside observer sees a horizontally thrown ball travel in a straight line, and since the ship is moving upward while the ball travels horizontally, the ball strikes the wall somewhat below a point opposite the window. (b) To an inside observer, the ball bends as if in a gravitational field.

During the time it takes for the light to reach the glass sheet, the spaceship moves up some distance, and during the equal time for the light to continue to the far wall, the spaceship moves up a greater distance. So, to observers in the spaceship, the light has followed a downward curving path (Figure 35.5). In this accelerating frame of reference, the light ray is deflected downward toward the floor just as the thrown ball in Figure 35.4 is deflected. The curvature of the slow-moving ball is very pronounced; but if the ball were somehow thrown horizontally across the spaceship cabin at a velocity equal to that of light, its curvature would match the light ray's curvature.

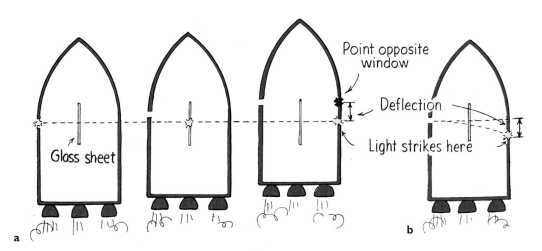

FIGURE 35.5 (a) An outside observer sees light travel horizontally in a straight line, but like the ball in Figure 35.4, it strikes the wall slightly below a point opposite the window. (b) To an inside observer, the light bends as if responding to a gravitational field.

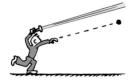

FIGURE 35.6 The trajectory of a flashlight beam is identical to the trajectory that a baseball would have if it could be "thrown" at the speed of light. Both paths curve equally in a uniform gravitational field.

FIGURE 35.7 Starlight bends as it grazes the sun. Point A shows the apparent position; point B shows the true position.

An observer inside the ship feels "gravity" because of the ship's acceleration. The observer is not surprised by the deflection of the thrown ball, but might be quite surprised by the deflection of light. According to the principle of equivalence, if light is deflected by acceleration, it must be deflected by gravity. Yet how can gravity bend light? According to Newton's physics, gravitation is an interaction between masses; a moving ball curves because of the interaction between its mass and the mass of the earth. But what of light, which is pure energy and is massless? Einstein's answer was that light may be massless, but it's not "energy-less." Gravity pulls on the energy of light because energy is equivalent to mass.

This was Einstein's first answer, before he fully developed the general theory of relativity. Later he gave a deeper explanation—that light bends when it travels in a spacetime geometry that is bent. We shall see later in this chapter that the presence of mass results in the bending or warping of spacetime. The mass of the earth is too small to appreciably warp the surrounding spacetime, which is practically flat, so any such bending of light in our immediate environment is not ordinarily detected. Close to bodies of mass much greater than the earth's, however, the bending of light is large enough to detect.

Einstein predicted that starlight passing close to the sun would be deflected by an angle of 1.75 seconds of arc—large enough to be measured. Although stars are not visible when the sun is in the sky, the deflection of starlight can be observed during an eclipse of the sun. (Measuring this deflection has become a standard practice at every total eclipse since the first measurements were made during the total eclipse of 1919.) A photograph taken of the darkened sky around the eclipsed sun reveals the presence of the nearby bright stars. The positions of the stars are compared with those in other photographs of the same area taken at other times in the night with the same telescope. In every instance, the deflection of starlight has supported Einstein's prediction (Figure 35.7).

Light bends in the earth's gravitational field also—but not as much. We don't notice it because the effect is so tiny. For example, in a constant gravitational field of 1 *g*, a beam of horizontally directed light will "fall" a vertical distance of 4.9 meters in 1 second (just as a baseball would), but will travel a horizontal distance of 300,000

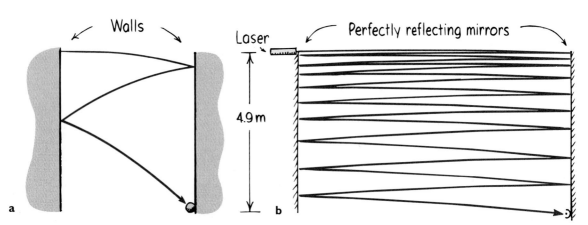

FIGURE 35.8 (a) If a ball is horizontally projected between a vertical pair of parallel walls, it will bounce back and forth and fall a vertical distance of 4.9 m in 1 s. (b) If a horizontal beam of light is directed between a vertical pair of perfectly parallel ideal mirrors, it will reflect back and forth and fall a vertical distance of 4.9 m in 1 s. The number of back and forth reflections is overly simplified in the diagram; if the mirrors were 300 km apart, for example, 1000 reflections would occur in 1 s.

kilometers in that time. Its curve would hardly be noticeable when you're this far from the beginning point. But if the light traveled 300,000 kilometers in multiple reflections between idealized parallel mirrors, the effect would be quite noticeable (Figure 35.8). (Doing this would make a dandy home project for extra credit—like earning credit for a Ph.D.)

Question Why do we not notice the bending of light in our everyday environment?

Gravity and Time: Gravitational Red Shift

According to Einstein's general theory of relativity, gravitation causes time to slow down. If you move in the direction that the gravitational force acts—from the top of a skyscraper to the ground floor, for instance, or from the surface of the earth to the bottom of a well—time will run slower at the point you reach than at the point you left behind. We can understand the slowing of clocks by gravity by applying the principle of equivalence and time dilation to an accelerating frame of reference.

Imagine our accelerating reference frame to be a large rotating disk. Suppose we measure time with three identical clocks, one placed on the disk at its center, a second placed on the rim of the disk, and the third at rest on the ground nearby (Figure 35.9). From the laws of special relativity we know that the clock attached to the center, since it is not moving with respect to the ground, should run at the same rate as the clock on the ground—but not at the same rate as the clock attached to the rim of the disk. The clock at the rim is in motion with respect to the ground and should therefore be observed to be running more slowly than the ground clock—and therefore more slowly than the clock at the center of the disk. Although the clocks on the disk are attached to the same frame of reference, they do not run synchronously; the outer clock runs slower than the inner clock.

FIGURE 35.9 Clocks 1 and 2 are on an accelerating disk, and clock 3 is at rest in an inertial frame. Clocks 1 and 3 run at the same rate, while clock 2 runs slower. From the point of view of an observer at clock 3, clock 2 runs slow because it is moving. From the point of view of an observer at clock 1, clock 2 runs slow because it is in a stronger centrifugal force field.

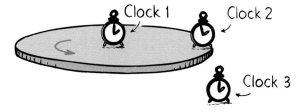

An observer on the rotating disk and an observer at rest on the ground both see the same difference in clock rates between themselves and the clock on the rim. Interpretations of the difference for the two observers are not the same, however. To the observer on the ground, the slower rate of the clock on the rim is due to its motion. But to an observer on the rotating disk, the disk clocks are not in motion with respect to each other; instead, a centrifugal force acts on the clock at the rim, while no such force acts on the clock at the center. The observer on the disk is likely to conclude that the

---

Answer Only because light travels so fast; just as over a short distance we do not notice the curved path of a high-speed bullet, we do not notice the curving of a light beam.

centrifugal force has something to do with the slowing of time. He notices that as he moves in the direction of the centrifugal force, outward from the center to the edge of the disk, time is slowed. By applying the principle of equivalence, which says that any effect of acceleration can be duplicated by gravity, we must conclude that as we move in the direction that a gravitational force acts, time will also be slowed.

This slowing down will apply to all "clocks," whether physical, chemical, or biological. An executive working on the ground floor of a tall city skyscraper will age more slowly than her twin sister working on the top floor. The difference is very small, only a few millionths of a second per decade, because by cosmic standards, the distance is small and the gravitation weak. For larger differences in gravitation, like between the surface of the sun and the surface of the earth, the differences in time are larger (although still tiny). A clock at the surface of the sun should run measurably slower than a clock at the surface of the earth. Years before he completed his general relativity theory, Einstein suggested a way to measure this when he formulated the principle of equivalence in 1907.

All atoms emit light at specific frequencies characteristic of the vibrational rate of electrons within the atom. Every atom is therefore a "clock," and a slowing down of atomic vibration indicates the slowing down of such clocks. An atom on the sun should emit light of a lower frequency (slower vibration) than light emitted by the same kind of atom on the earth. Since red light is at the low-frequency end of the visible spectrum, a lowering of frequency shifts the color toward the red. This effect is called the **gravitational red shift**. The gravitational red shift is observed in light from the sun, but various disturbing influences prevent accurate measurements of this tiny effect. It wasn't until 1960 that an entirely new technique, using gamma rays from radioactive atoms, permitted incredibly precise and confirming measurements of the gravitational slowing of time between the top and bottom floors of a laboratory building at Harvard University.*

So measurements of time depend not only on relative motion, as we learned in the last chapter, but also on gravity. In special relativity, time dilation depends on the *speed* of one frame of reference relative to another one. In general relativity, the gravitational red shift depends on the *location* of one point in a gravitational field relative to another one. As viewed from earth, a clock will be measured to tick more slowly on the surface of a star than on earth. If the star shrinks, its surface moves inward to ever-stronger gravity, which causes time on its surface to slow down more and more. We would measure longer intervals between the ticks of the star clock. But if we made our measurements of the star clock from the star itself, we would notice nothing unusual about the clock's ticking.

Suppose, for example, that an indestructible volunteer stands on the surface of a giant star that begins collapsing. We, as outside observers, will note a progressive slowing of time on the clock of our volunteer as the star surface recedes to regions of stronger gravity. The volunteer himself, however, does not notice any differences in his own time. He is viewing events within his own frame of reference, and he notices nothing unusual. As the collapsing star proceeds toward becoming a black hole and time proceeds normally from the viewpoint of the volunteer, we on the outside perceive time

FIGURE 35.10 If you move from a distant point down to the surface of the earth, you move in the direction that the gravitational force acts—toward a location where clocks run more slowly. A clock at the surface of the earth runs slower than a clock farther away.

---

* In the late 1950s, shortly after Einstein's death, the German physicist Rudolph Mossbauer discovered an important effect in nuclear physics that provides an extremely accurate method of using atomic nuclei as atomic clocks. The *Mossbauer effect*, for which its discoverer was awarded the Nobel Prize, has many practical applications. In late 1959 Robert Pound and Glen Rebka at Harvard University conceived an application that was a test for general relativity and performed the confirming experiment.

for the volunteer as approaching a complete stop; we see him frozen in time with an infinite duration between the ticks of his clock or the beats of his heart. From our view, his time stops completely. The gravitational red shift, instead of being a tiny effect, is dominating.

We can understand the gravitational red shift from another point of view—in terms of the gravitational force acting on photons. As a photon flies from the surface of a star, it is "retarded" by the star's gravity. It loses energy (but not speed). Since a photon's frequency is proportional to its energy, its frequency decreases as its energy decreases. When we observe the photon, we see that it has lower frequency than if it had been emitted by a less massive source. Its time has been slowed, just like the ticking of a clock is slowed. In the case of a black hole, a photon is unable to escape at all. It loses all its energy and all its frequency in the attempt. Its frequency is gravitationally red-shifted to zero, consistent with our observation that the rate at which time passes on a collapsing star approaches zero.

It is important to note the relativistic nature of time in both special relativity and general relativity. In both theories, there is no way that you can extend the duration of your own existence. Others moving at different speeds or in different gravitational fields may attribute a greater longevity to you, but your longevity is seen from *their* frame of reference—never your own. Changes in time are always attributed to "the other guy."

Question Will a person at the top of a skyscraper age more than or less than a person at ground level?

Gravity and Space: Motion of Mercury

FIGURE 35.11 A precessing elliptical orbit.

From the special theory of relativity, we know that measurements of space as well as of time undergo transformations when motion is involved. Likewise with the general theory: Measurements of space differ in different gravitational fields—for example, close to and far away from the sun.

Planets orbit the sun and stars in elliptical orbits and move periodically into regions farther from the sun and closer to the sun. Einstein directed his attention to the varying gravitational fields experienced by the planets orbiting the sun and found that the elliptical orbits of the planets should *precess* (Figure 35.11)—independently of the Newtonian influence of other planets. Near the sun, where the effect of gravity on time is the greatest, the rate of precession should be the greatest; and far from the sun, where time is less affected, any deviations from Newtonian mechanics should be virtually unnoticeable.

Mercury is the planet nearest the sun. If the orbit of any planet exhibits a measurable precession, it should be Mercury, and the fact that the orbit of Mercury does precess—above and beyond effects attributable to the other planets—had been a mystery to astronomers since the early 1800s! Careful measurements showed that Mercury's orbit precesses about 574 seconds of arc per century. Perturbations by the other planets were found to account for all but 43 seconds of arc per century. Even after all known corrections due to possible perturbations by other planets had been applied, the

Answer More—going from the top of the skyscraper to the ground is going in the direction of the gravitational force, so it is going to a place where time runs more slowly.

calculations of physicists and astronomers failed to account for the extra 43 seconds of arc. Either Venus was extra massive or a never-discovered other planet (called Vulcan) was pulling on Mercury. And then came the explanation of Einstein, whose general relativity field equations applied to Mercury's orbit predict an extra 43 seconds of arc per century!

The mystery of Mercury's orbit was solved, and a new theory of gravity was recognized. Newton's law of gravitation, which had stood as an unshakable pillar of science for more than two centuries, was found to be a special limiting case of Einstein's more general theory. If the gravitational fields are comparatively weak, Newton's law turns out to be a good approximation of the new law—enough so that Newton's law, which is easier to work with mathematically, is the law that today's scientists use most of the time, except for cases involving enormous gravitational fields.

Gravity, Space, and a New Geometry

We can begin to understand that measurements of space are altered in a gravitational field by again considering the accelerated frame of reference of our rotating disk. Suppose we measure the circumference of the outer rim with a measuring stick. Recall the Lorentz contraction from special relativity: The measuring stick will appear contracted to any observer not moving along with the stick, while an identical measuring stick moving much more slowly near the center will be nearly unaffected (Figure 35.12). All distance measurements along a *radius* of the rotating disk should be completely unaffected by motion, because motion is perpendicular to the radius. Since only distance measurements parallel to and around the circumference are affected, the ratio of circumference to diameter when the disk is rotating is no longer the fixed constant π (3.14159 . . .), but is a variable depending on angular speed and the diameter of the disk. According to the principle of equivalence, the rotating disk is equivalent to a stationary disk with a strong gravitational field near its edge and a progressively weaker gravitational field toward its center. Measurements of distance, then, will depend on the strength of gravitational field (or more exactly, for relativity buffs, on gravitational potential), even if no relative motion is involved. Gravity causes space to be non-Euclidean; the laws of Euclidean geometry taught in high school are no longer valid when applied to objects in the presence of strong gravitational fields.

FIGURE 35.12 A measuring stick along the edge of the rotating disk appears contracted, while a measuring stick farther in and moving more slowly is not contracted as much. A measuring stick along a radius is not contracted at all. When the disk is not rotating, $C/D = \pi$; but when the disk is rotating, C/D is not equal to π and Euclidean geometry is no longer valid. Likewise in a gravitational field.

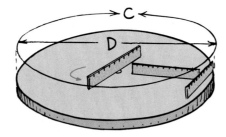

The familiar rules of Euclidean geometry pertain to various figures you can draw on a flat surface. The ratio of the circumference of a circle to its diameter is equal to π; all the angles in a triangle add up to 180°; the shortest distance between two points is a straight line. The rules of Euclidean geometry are valid in flat space, but if you draw these figures on a curved surface like a sphere or a saddle-shaped object, the Euclidean rules no longer hold (Figure 35.13). If you measure the sum of the angles for a triangle in space, you call the space flat if the sum is equal to 180°, spherelike or positively curved if the sum is larger than 180°, and saddlelike or negatively curved if it is less than 180°.

FIGURE 35.13 The sum of the angles for a triangle drawn (a) on a plane surface = 180°, (b) on a spherical surface is greater than 180°, and (c) on a saddle-shaped surface is less than 180°.

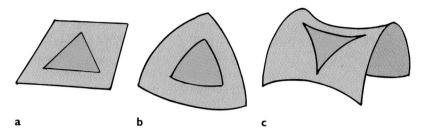

a   b   c

Of course the lines forming the triangles in Figure 35.13 are not all "straight" from the three-dimensional view, but they are the "straightest" or *shortest* distances between two points if we are confined to the curved surface. These lines of shortest distance are called *geodesic lines* or simply **geodesics**.

The path of a light beam follows a geodesic. Suppose three experimenters on Earth, Venus, and Mars measure the angles of a triangle formed by light beams traveling between these three planets. The light beams bend when passing the sun, resulting in the sum of the three angles being larger than 180° (Figure 35.14). So the space around the sun is positively curved. The planets that orbit the sun travel along four-dimensional geodesics in this positively curved spacetime. Freely falling objects, satellites, and light rays all travel along geodesics in four-dimensional spacetime.

FIGURE 35.14 The light rays joining the three planets form a triangle. Since light passing near the sun bends, the sum of the angles of the resulting triangle is greater than 180°.

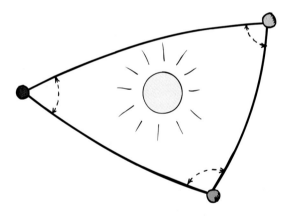

The whole universe may have an overall curvature. If it is negatively curved, it is open-ended and extends without limit; if it is positively curved, it closes in on itself. The surface of the earth, for example, forms a closed curvature, so if you travel along a geodesic, you come back to your starting point. Similarly, if the universe were positively curved, it would be closed, so if you could look infinitely into space through an ideal telescope, you would see the back of your own head! (This is assuming that you waited a long enough time or that light traveled infinitely fast.)

FIGURE 35.15 The geometry of the curved surface of the earth differs from the Euclidean geometry of flat space. Note that the sum of the angles for an equilateral triangle where the sides equal $\frac{1}{4}$ the earth's circumference is clearly greater than 180°, and the circumference is only twice its diameter instead of 3.14 times its diameter. Euclidean geometry is also invalid in curved space.

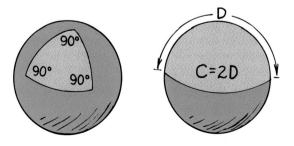

General relativity, then, calls for a *new geometry*: a geometry not only of curved space but of curved time as well—a geometry of curved four-dimensional spacetime. The mathematics of this geometry is too formidable to present here. The essence, however, is that gravity is a manifestation of spacetime geometry; a gravitational field is a geometrical warping of spacetime. The presence of mass results in the curvature or warping of spacetime; by the same token, a curvature of spacetime reveals itself as mass. Instead of visualizing gravitational forces between masses, we abandon altogether the notion of force and instead think of masses responding in their motion to the curvature or warping of the spacetime they inhabit. It is the bumps, depressions, and warpings of geometrical spacetime that *are* the phenomena of gravity.

We cannot visualize the four-dimensional bumps and depressions in spacetime because we are three-dimensional beings. We can get a glimpse of this warping by considering a simplified analogy in two dimensions: a heavy ball resting on the middle of a waterbed. The more massive the ball, the greater it dents or warps the two-dimensional surface. A marble rolled across the bed, but away from the ball, will roll in a relatively straight-line path, whereas a marble rolled near the ball will curve as it rolls across the indented surface. If the curve closes upon itself, its shape resembles an ellipse. The planets that orbit the sun similarly travel along four-dimensional geodesics in the warped spacetime about the sun.

FIGURE 35.16 A two-dimensional analogy of four-dimensional warped spacetime. Spacetime near a star is curved in a way similar to the surface of a waterbed when a heavy ball rests on it.

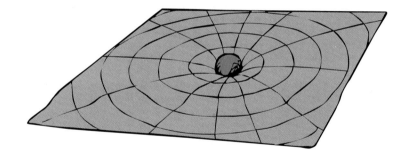

Gravitational Waves

Every object has mass and therefore makes a bump or depression in the surrounding spacetime. When an object moves, the surrounding warp of space and time moves to readjust to the new position. These readjustments produce ripples in the overall geometry of spacetime. This is similar to moving the ball that rests on the surface of the waterbed. A disturbance ripples across the waterbed surface in waves; if we move a more massive ball, then we get a greater disturbance and the production of even stronger waves. The ripples travel outward from the gravitational sources at the speed of light and are called **gravitational waves**.

Any accelerating object produces a gravitational wave. In general, the more massive the moving object and the greater its acceleration, the stronger the resulting gravitational wave. But even the strongest waves produced by ordinary astronomical events are extremely weak—the weakest known in nature. For example, the gravitational waves emitted by a vibrating electric charge are a trillion-trillion-trillion times weaker than the electromagnetic waves emitted by the same charge.

Shake your hand back and forth: You have just produced a gravitational wave. It is not very strong, but it exists.

Newtonian and Einsteinian Gravitation

When Einstein formulated his new theory of gravitation, he realized that if his theory is valid, his field equations must reduce to Newtonian equations for gravitation in the weak-field limit. He showed that Newton's law of gravitation is a special case of the broader theory of relativity. Newton's law of gravitation is still an accurate description of most of the interactions between bodies in the solar system and beyond. From Newton's law, one can calculate the orbits of comets and asteroids and even predict the existence of undiscovered planets. Even today, when computing the trajectories of space probes to the moon and planets, only ordinary Newtonian theory is used. This is because the gravitational field of these bodies is very weak, and from the viewpoint of general relativity, the surrounding spacetime is essentially flat. But for regions of more intense gravitation, where spacetime is more appreciably curved, Newtonian theory cannot adequately account for various phenomena—like the precession of Mercury's orbit close to the sun and, in the case of stronger fields, the gravitational red shift and other apparent distortions in measurements of space and time. These distortions reach their limit in the case of a star that collapses to a black hole, where spacetime completely folds over on itself. Only Einsteinian gravitation reaches into this domain.

We saw in Chapter 31 that Newtonian physics is linked at one end with quantum theory, whose domain is the very light and very small—tiny particles and atoms. And now we have seen that Newtonian physics is linked at the other end with relativity theory, whose domain is the very massive and very large.

We do not see the world the way the ancient Egyptians, Greeks, or Chinese did. It is unlikely that people in the future will see the universe as we do. Our view of the universe may be quite limited, and perhaps filled with misconceptions, but it is most likely clearer than the views of others before us. Our view today stems from the findings of Copernicus, Galileo, Newton, and, more recently, Einstein—findings that were often opposed on the grounds that they diminished the importance of humans in the universe. In the past, being important meant having risen above nature—being apart from nature. We have expanded our vision since then by enormous effort, painstaking observation, and an unrelenting desire to comprehend our surroundings. Seen from today's understanding of the universe, we find our importance lies in being very much a part of nature, not apart from it. We are the part of nature that is becoming more and more conscious of itself.

Summary of Terms

Principle of equivalence Observations made in an accelerated frame of reference are indistinguishable from observations made in a gravitational field.

Gravitational red shift The shift in wavelength toward the red end of the spectrum experienced by light leaving the surface of a massive object, as predicted by the general theory of relativity.

Geodesic The shortest path between points on any surface.

Gravitational wave A gravitational disturbance made by a moving mass that propagates through spacetime.

Suggested Reading

Einstein, Albert. *Relativity: The Special and General Theory.* New York: Crown, 1961. (Orig. pub. 1916.)

Hawking, Stephen W. *A Brief History of Time: From the Big Bang to Black Holes.* New York: Bantam Books, 1988.

Kaufmann, William J. *The Cosmic Frontiers of General Relativity.* Boston: Little, Brown, 1977.

Thorne, Kip S. *Black Holes and Time Warps, Einstein's Outrageous Legacy.* New York: Norton, 1994. An expert's readable account of black holes, neutron stars, gravitational waves, time machines, and more.

Review Questions

1. What is the principal difference between *special relativity* and *general relativity?*

Principle of Equivalence

2. In a spaceship accelerating at *g*, far from earth gravity, you do your maximum number of push-ups. How does this number compare to the number you can normally do here on earth?

3. Exactly what is *equivalent* in the principle of equivalence?

Bending of Light by Gravity

4. Compare the bending of the paths of baseballs and photons by a gravitational field.

5. Why must the sun be eclipsed to measure the deflection of starlight passing near the sun?

Gravity and Time: Gravitational Red Shift

6. Which runs faster, a clock at the top of the Sears Tower in Chicago or a clock on the shore of Lake Michigan?

7. How does the frequency of a particular spectral line observed in sunlight compare with the frequency of that line observed from a source on earth?

8. How does the rate of ticking of a hypothetical clock on the surface of the massive planet Jupiter compare with the rate of ticking of an identical clock on earth?

Gravity and Space: Motion of Mercury

9. Of all the planets, why is Mercury the best candidate for our finding evidence for the effect of gravitation on space?

Gravity, Space, and a New Geometry

10. A measuring stick placed along the circumference of a rotating disk will appear contracted, but if it is oriented along a radius, it will not. Explain.

11. The ratio of circumference to diameter for measured circles on a disk equals π when the disk is at rest, but not when the disk is rotating. Explain.

12. Is the two-dimensional surface of the earth positively or negatively curved? Why?

Gravitational Waves

13. A star 10 light-years away explodes and produces gravitational waves. How long will it take these waves to reach the earth?

14. Why are gravitational waves so difficult to detect?

Newtonian and Einsteinian Gravitation

15. Does Einstein's theory of gravitation invalidate Newton's theory of gravitation? Explain.

Exercises

1. An astronaut awakes in her closed capsule, which actually sits on the moon. Can she tell whether her weight is the result of gravitation or of accelerated motion? Explain.

2. You wake up at night in your berth on a train to find yourself being "pulled" to one side of the train. You naturally assume that the train is rounding a curve but you are puzzled that you don't hear any sounds of motion. Offer another possible explanation that involves only gravity, not acceleration of your frame of reference.

3. Since gravity can duplicate the effects of acceleration, it can also balance the effects of acceleration. Cite how and when an astronaut can experience no net force (as measured by a scale) because of the canceling effects of gravity and acceleration.

4. An astronaut is provided a "gravity" when the ship's engines are activated to accelerate the ship. This requires the use of fuel. Is there a way to accelerate and provide "gravity" without the sustained use of fuel? Explain.

5. In his famous novel *Journey to the Moon,* Jules Verne stated that occupants in a spaceship would shift their orientation from up to down when the ship crossed the point where the moon's gravitation became greater than the earth's. Is this correct? Defend your answer.

6. What happens to the separation distance between two people if they both walk north at the same rate from two different places on the earth's equator? And just for fun, where in the world is a step in every direction a step south?

7. We readily note the bending of light by reflection and refraction, but why is it we do not ordinarily notice the bending of light by gravity?

8. Why do we say that light travels in straight lines? Is it strictly accurate to say that a laser beam provides a perfectly straight line for purposes of surveying? Explain.

9. At the end of 1 s, a horizontally fired bullet has dropped a vertical distance of 4.9 m from its otherwise straight-line path in a gravitational field of 1 *g*. By what distance would a beam of light drop from its otherwise straight-line path if it traveled in a uniform field of 1 *g* for 1 s? For 2 s?

10. Light changes its energy when it "falls" in a gravitational field. This change in energy is not evidenced by a change in speed, however. What is the evidence for this change in energy?

11. Would we notice a slowing down or speeding up of a clock if we carried it to the bottom of a very deep well?

12. If we witness events taking place on the moon, where gravitation is weaker than on earth, would we expect to see a gravitational red shift or a gravitational blue shift? Explain.

13. Armed with highly sensitive detection equipment, you are in the front of a railroad car that is accelerating forward. Your friend at the rear of the car shines a green light toward you. Do you find the light to be red-shifted (lowered in frequency), blue-shifted (increased in frequency), or neither? Explain. (*Hint:* Think in terms of the principle of equivalence. What is your accelerating railroad car equivalent to?)

14. Why will the gravitational field intensity increase on the surface of a shrinking star?

15. Will a clock at the equator run slightly faster or slightly slower than an identical clock at one of the earth's poles?

16. Splitting hairs, should a person who worries about growing old live at the top or at the bottom of a tall apartment building?

17. Prudence and Charity are twins raised at the center of a rotating kingdom. Charity goes to live at the edge of the kingdom for a time and then returns home. Which twin is older when they rejoin? (Ignore any time-dilation effects associated with travel to and from the edge.)

18. Splitting hairs, if you shine a beam of colored light to a friend above in a high tower, will the color of light your friend receives be the same color you send? Explain.

19. Is the color of light emitted from the surface of a massive star red-shifted or blue-shifted by gravity?

20. From our frame of reference on earth, objects slow to a stop as they approach black holes in space because time gets infinitely stretched by the strong gravity near the black hole. If astronauts accidentally falling into a black hole tried to signal back to earth by flashing a light, what kind of "telescope" would we need to see the signals?

21. Would an astronaut falling into a black hole see the surrounding universe red-shifted or blue-shifted?

22. How can we "observe" a black hole if neither matter nor radiation can escape from it?

23. Should it be possible in principle for a photon to circle a star?

24. Why does the gravitational attraction between the sun and Mercury vary? Would it vary if the orbit of Mercury were perfectly circular?

25. In the astronomical triangle shown in Figure 35.14, with sides defined by light paths, the sum of the interior angles is more than 180°. Is there any astronomical triangle with interior angles that sum to less than 180°?

26. Do binary stars (double-star systems that orbit about a common center of mass) radiate gravitational waves? Why or why not?

27. Based on what you know about the emission and absorption of electromagnetic waves, suggest how gravitational waves are emitted and how they are absorbed. (Scientists seeking to detect gravitational waves must arrange for them to be absorbed.)

28. Comparing Einstein's and Newton's theories of gravitation, can the correspondence principle be applied?

29. Make up a multiple-choice question to check a classmate's understanding of the principle of equivalence.

30. Make up a multiple-choice question to check a classmate's understanding of the effect of gravity on time.

Epilogue

I hope that you've enjoyed *Conceptual Physics* and that you value your knowledge of physics as a worthwhile component of your general education. I particularly hope that viewing physics as a study of nature's rules contributes to your sense of wonder and enhances the way you see the physical world, just as you better appreciate a game when you know its rules. I also hope you appreciate knowing that everything in nature is connected—how seemingly diverse phenomena often follow the same basic rules—that, for example, the rules that govern a falling apple also govern the space shuttle orbiting the earth; that the rules discovered by Faraday and Maxwell show that electricity and magnetism connect to become light; that a sky's redness at sunset is connected to its blueness at midday. And I hope that in discovering these things, you understand more than you perhaps expected; you feel the exhilaration that comes in knowing a deep thing well. Knowledge of physics does that for many people.

Physics is generally highly regarded because of its products. But the legacy of physics and science is not fast cars, CD players, and electric toothbrushes. The greatest value of science lies in its methods, treated in the beginning of this book—namely that **hypotheses** are framed so they are capable of being disproved, and ***experiments*** are designed to be reproducible. Being able to distinguish between scientific hypotheses and speculation, and between scientific experiments and unsupported claims, is particularly important in our present time when so much information and hype are publicly perceived to be scientific when they aren't. The influence of those who wish to topple a scientific way of viewing ourselves and the world thrives when skeptical thinking is eroded.

Skeptical thinking, in addition to sharpening common sense, is an essential ingredient in formulating a hypothesis that requires a test for wrongness. *If I'm wrong, how would I know?* Shouldn't this be a key question to accompany any important idea—scientific or otherwise? Applied to social, political, and religious positions, you reduce your chances of being deluded. Socially you see other's points of view more clearly. Politically you see all social movements as experiments. Religiously you see science can be an awesome source of spirituality.

The success of a scientific hypothesis rests not on its beauty, not upon the sincerity with which it is made, not upon the hopes or promises it implies—but on whether or not it stands up to *experiment*—reproducible by investigators in any culture and in any part of the world. Science, like mathematics, has no national language—it is transnational. Like a tree that grows ring by ring, scientific knowledge gained by experiment builds on prior knowledge. The built-in error-correcting methods of science produce a body of knowledge that helps us to better understand the world, to better comprehend ourselves, and to better steer a safe course to our future.

Experimentation requires a patient attitude, for the hypotheses of most scientists fail the test of experiment. Benjamin Franklin said of experimentation, "In going on with these experiments, how many pretty systems do we build, which we soon find ourselves obliged to destroy?" He saw that the experience of experimenting suffices "to make a vain man humble." So whether your activities are scientific or not, when something doesn't work, don't spend your life pretending it does, or worse, trying to convince others of it.

Science provides answers to many previously unanswered questions—particularly questions that ask *how*. Still unanswered are questions that ask *why*—questions about our place in the universe that have stirred the souls of humans for centuries. A variety of proposed answers abound, which gives richness to human diversity. An important message of science, however, is that not knowing is okay—especially in this awesome period of exploding knowledge about who and where we are—when an open mind is better prepared to discover *why* we are. And it's okay to doubt the answers of others. In fact, it's essential to doubt those who claim to be privy to *all* the answers—those who claim attainment of absolute certainty. The tragic part of our past is replete with the power these people wielded. The bright part of our future will be devoid of their influence. And if we find that one of nature's rules is that absolute certainty will always elude us, let it be. *It's okay not to know.* Acknowledging and accepting your limitations is the beginning of wisdom.

Appendix A

Systems of Measurement

Two major systems of measurement prevail in the world today: the *United States Customary System* (USCS, formerly called the British system of units), used in the United States of America and in Burma, and the *Système International* (SI) (known also as the international system and as the metric system), used everywhere else. Each system has its own standards of length, mass, and time. The units of length, mass, and time and sometimes called the *fundamental units* because, once they are selected, other quantities can be measured in terms of them.

United States Customary System

Based on the British Imperial System, the USCS is familiar to everyone in the United States. It uses the foot as the unit of length, the pound as the unit of weight or force, and the second as the unit of time. The USCS is presently being replaced by the international system—rapidly in science and technology (all 1988 Department of Defense contracts) and some sports (track and swimming), but so slowly in other areas and in some specialties it seems the change may never come. For example, we will continue to buy seats on the 50-yard line. Camera film is in millimeters but computer disks are in inches.

For measuring time there is no difference between the two systems except that in pure SI the only unit is the *second* (s, not sec) with prefixes; but, in general, minute, hour, day, year, and so on, with two or more lettered abbreviations (h, not hr), are accepted in the USCS.

Système International

During the 1960 International Conference on Weights and Measures held in Paris, the SI units were defined and given status. Table A.1 shows SI units and their symbols. SI is based on the *metric system,* originated by French scientists after the French revolution in 1791. The orderliness of this system makes it useful for scientific work, and it is used by scientists all over the world. The metric system branches into two systems of units. In one of these the unit of length is the meter, the unit of mass is the kilogram, and the

Table A.1   SI units

Quantity	Unit	Symbol
Length	meter	m
Mass	kilogram	kg
Time	second	s
Force	newton	N
Energy	joule	J
Current	ampere	A
Temperature	kelvin	K

unit of time is the second. This is called the *meter-kilogram-second* (mks) system and is preferred in physics. The other branch is the *centimeter-gram-second* (cgs) system, which because of its smaller values is favored in chemistry. The cgs and mks units are related to each other as follows: 100 centimeters equal 1 meter; 1000 grams equal 1 kilogram. Table A.2 shows several units of length related to each other.

Table A.2   Table conversions between different units of length

Unit of length	Kilometer	Meter	Centimeter	Inch	Foot	Mile
1 kilometer	= 1	1000	100,000	39,370	3280.84	0.62140
1 meter	= 0.00100	1	100	39.370	3.28084	6.21×10^{-4}
1 centimeter	= 1.0×10^{-5}	0.0100	1	0.39370	0.032808	6.21×10^{-6}
1 inch	= 2.54×10^{-5}	0.02540	2.5400	1	0.08333	1.58×10^{-5}
1 foot	= 3.05×10^{-4}	0.30480	30.480	12	1	1.89×10^{-4}
1 mile	= 1.60934	1609.34	160,934	63,360	5280	1

One major advantage of a metric system is that it uses the decimal system, where all units are related to smaller or larger units by dividing or multiplying by 10. The prefixes shown in Table A.3 are commonly used to show the relationship among units.

Table A.3   Some prefixes

Prefix	Definition
micro-	One-millionth: a microsecond is one-millionth of a second
milli-	One-thousandth: a milligram is one-thousandth of a gram
centi-	One-hundredth: a centimeter is one-hundredth of a meter
kilo-	One thousand: a kilogram is 1000 grams
mega-	One million: a megahertz is 1 million hertz

Meter

The standard of length of the metric system orginally was defined in terms of the distance from the north pole to the equator. This distance was thought at the time to be close to 10,000 kilometers. One ten-millionth of this, the meter, was carefully determined and marked off by means of scratches on a bar of platinum-iridium alloy. This bar is kept at the International Bureau of Weights and Measures in France. The standard meter in France has since been calibrated in terms of the wavelength of light—it is 1,650,763.73 times the wavelength of orange light emitted by the atoms of the gas krypton-86. The meter is now defined as being the length of the path traveled by light in a vacuum during a time interval of 1/299,792,458 of a second.

Kilogram

The standard unit of mass, the kilogram, is a block of platinum, also preserved at the International Bureau of Weights and Measures located in France (Figure A.1). The kilogram equals 1000 grams. A gram is the mass of 1 cubic centimeter (cc) of water at a temperature of 4° Celsius. (The standard pound is defined in terms of the standard kilogram; the mass of an object that weighs 1 pound is equal to 0.4536 kilogram.)

Second

The official unit of time for both the USCS and the SI is the second. Until 1956 it was defined in terms of the mean solar day, which was divided into 24 hours. Each hour was divided into 60 minutes and each minute into 60 seconds. Thus, there were 86,400 seconds per day, and the second was defined as 1/86,400 of the mean solar day. This proved unsatisfactory because the rate of rotation of the earth is gradually becoming slower. In 1956 the mean solar day of the year 1900 was chosen as the standard on which to base the second. In 1964, the second was officially defined as the duration of 9,192,631,770 periods of radiation corresponding to the transition between the two hyperfine levels of the ground state of the cesium-133 atom.

Newton

One newton is the force required to accelerate 1 kilogram at 1 meter per second per second. This unit is named after Sir Isaac Newton.

Joule

One joule is equal to the amount of work done by a force of 1 newton acting over a distance of 1 meter. In 1948 the joule was adopted as the unit of energy by the International Conference on Weights and Measures. Therefore, the specific heat of water at 15°C is now given as 4185.5 joules per kilogram Celsius degree. This figure is always associated with the mechanical equivalent of heat—4.1855 joules per calorie.

Ampere

The ampere is defined as the intensity of the constant electric current that, when maintained in two parallel conductors of infinite length and negligible cross section and placed 1 meter apart in a vacuum, would produce between them a force equal to 2×10^{-7} newton per meter length. In our treatment of electric current in this text, we have used the not-so-official but easier-to-comprehend definition of the ampere as being the rate of flow of 1 coulomb of charge per second, where 1 coulomb is the charge of 6.25×10^{18} electrons.

Kelvin

The fundamental unit of temperature is named after the scientist William Thomson, Lord Kelvin. The kelvin is defined to be 1/273.15 the thermodynamic temperature of the triple point of water (the fixed point at which ice, liquid water, and water vapor coexist in equilibrium). This definition was adopted in 1968 when it was decided to change the

FIGURE A.1 The standard kilogram.

name *degree Kelvin* (°K) to *kelvin* (K). The temperature of melting ice at atmospheric pressure is 273.15 K. The temperature at which the vapor pressure of pure water is equal to standard atmospheric pressure is 373.15 K (the temperature of boiling water at standard atmospheric pressure).

Area

The unit of area is a square that has a standard unit of length as a side. In the USCS it is a square with sides that are each 1 foot in length, called 1 square foot and written 1 ft^2. In the international system it is a square with sides that are 1 meter in length, which makes a unit of area of 1 m^2. In the cgs system it is 1 cm^2. The area of a given surface is specified by the number of square feet, square meters, or square centimeters that would fit into it. The area of a rectangle equals the base times the height. The area of a circle is equal to πr^2, where $\pi = 3.14$ and r is the radius of the circle. Formulas for the surface areas of other objects can be found in geometry textbooks.

FIGURE A.2   Unit square.

Volume

The volume of an object refers to the space it occupies. The unit of volume is the space taken up by a cube that has a standard unit of length for its edge. In the USCS one unit of volume is the space occupied by a cube 1 foot on an edge and is called 1 cubic foot, written 1 ft^3. In the metric system it is the space occupied by a cube with sides of 1 meter (SI) or 1 centimeter (cgs). It is written 1 m^3 or 1 cm^3 (or cc). The volume of a given space is specified by the number of cubic feet, cubic meters, or cubic centimeters that will fill it.

In the USCS volumes can also be measured in fluid ounces, liquid pints, dry pints, liquid quarts, dry quarts, gallons, pecks, bushels, and cubic inches as well as in cubic feet. There are 1728 ($12 \times 12 \times 12$) cubic inches in 1 ft^3. A U.S. gallon is a volume of 231 in^3. Four quarts equal 1 gallon. In the SI volumes are also measured in liters. A liter is equal to 1000 cm^3.

FIGURE A.3   Unit volume.

Scientific Notation

It is convenient to use a mathematical abbreviation for large and small numbers. The number 50,000,000 can be obtained by multiplying 5 by 10, and again by 10, and again by 10, and so on until 10 has been used as a multiplier seven times. The short way of showing this is to write the number 5×10^7. The number 0.0005 can be obtained from 5 by using 10 as a divisor four times. The short way of showing this is to write 5×10^{-4} for 0.0005. Thus, 3×10^5 means $3 \times 10 \times 10 \times 10 \times 10 \times 10$, or 300,000; and 6×10^{-3} means $6/(10 \times 10 \times 10)$, or 0.006. Numbers expressed in this shorthand manner are said to be in scientific notation.

Appendix B

More About Motion

When we describe the motion of something, we say how it moves relative to something else (Chapter 3). In other words, motion requires a reference frame (an observer, origin, and axes). We are free to choose this frame's location and to have it moving relative to another frame. When our frame of motion has zero acceleration, it is called an *inertial frame*. In an inertial frame, force causes an object to accelerate in accord with Newton's laws. When our frame of reference is accelerated, we observe fictitious forces and motions (Chapter 7). Observations from a carousel, for example, are different when it is rotating and when it is at rest. Our description of motion and force depends on our "point of view."

We distinguish between *speed* and *velocity* (Chapters 2 and 3). Speed is how fast something moves, or the time rate of change of position (excluding direction): a *scalar* quantity. Velocity includes direction of motion: a *vector* quantity whose magnitude is speed. Objects moving at constant velocity move the same distance in the same time in the same direction.

Another distinction between speed and velocity has to do with the difference between distance and net distance, or *displacement*. Speed is *distance per duration* while velocity is *displacement per duration*. Displacement differs from distance. For example, a commuter who travels 10 kilometers to work and back travels 20 kilometers, but has "gone" nowhere. The distance traveled is 20 kilometers and the displacement is zero. Although the instantaneous speed and instantaneous velocity have the same value at the same instant, the average speed and average velocity can be very different. The average speed of this commuter's round trip is 20 kilometers divided by the total commute time—a value greater than zero. But the average velocity is zero. In science, displacement is often more important than distance. (To avoid information overload, we have not treated this distinction in the text.)

Acceleration is the rate at which velocity changes. This can be a change in speed only, a change in direction only, or both. Negative acceleration is often called *deceleration*.

In Newtonian space and time, space has three dimensions—length, width, and height—each with two directions. We can go, stop, and return in any of them. Time has one dimension, with two directions—past and future. We cannot stop or return, only go. In Einsteinian space-time, these four dimensions merge (Chapter 34).

Computing Velocity and Distance Traveled on an Inclined Plane

Recall from Chapter 2 Galileo's experiments with inclined planes. On page 27 we considered a plane tilted such that the speed of a rolling ball increases at the rate of 2 meters per second each second—an acceleration of 2 m/s^2. So at the instant it starts moving its velocity is zero, and 1 second later it is rolling at 2 m/s, at the end of the next second 4 m/s, the end of the next second 6 m/s, and so on. The velocity of the ball at any instant is simply

$$\text{Velocity} = \text{acceleration} \times \text{time}$$

Or, in shorthand notation,

$$v = at$$

(It is customary to omit the multiplication sign, ×, when expressing relationships in mathematical form. When two symbols are written together, such as the *at* in this case, it is understood that they are multiplied.)

How fast the ball rolls is one thing; how *far* it rolls is another. To understand the relationship between acceleration and distance traveled, we must first investigate the relationship between instantaneous velocity and *average velocity*. If the ball shown in Figure B.1 starts from rest, it will roll a distance of 1 meter in the first second. Question: What will be its average speed? The answer is 1 m/s (because it covered 1 meter in the interval of 1 second). But we have seen that the *instantaneous velocity* at the end of the first second is 2 m/s. Since the acceleration is uniform, the average in any time interval is found the same way we usually find the average of any two numbers: add them and divide by 2. (Be careful not to do this when acceleration is not uniform!) So if we add the initial speed (zero in this case) and the final speed of 2 m/s and then divide by 2, we get 1 m/s for the average velocity.

In each succeeding second we see the ball roll a longer distance down the same slope in Figure B.2. Note the distance covered in the second time interval is 3 meters. This is because the average speed of the ball in this interval is 3 m/s. In the next 1-second interval the average speed is 5 m/s, so the distance covered is 5 meters. It is interesting to see that successive increments of distance increase as a *sequence of odd numbers*. Nature clearly follows mathematical rules!

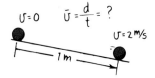

FIGURE B.1 The ball rolls 1 m down the incline in 1 s and reaches a speed of 2 m/s. Its average speed, however, is 1 m/s. Do you see why?

FIGURE B.2 If the ball covers 1 m during its first second, then in each successive second it will cover the odd-numbered sequence of 3, 5, 7, 9 m, and so on. Note that the total distance covered increases as the square of the total time.

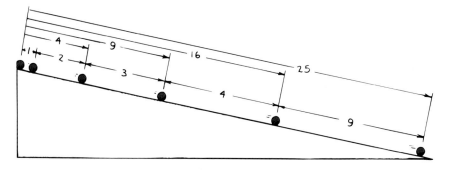

Question During the span of the second time interval, the ball begins at 2 m/s and ends at 4 m/s. What is the *average speed* of the ball during this 1-s interval? What is its *acceleration*?

Answer

$$\text{Average speed} = \frac{\text{beginning} + \text{final speed}}{2} = \frac{2 \text{ m/s} + 4 \text{ m/s}}{2} = \frac{6 \text{ m/s}}{2} = 3 \text{ m/s}$$

$$\text{Acceleration} = \frac{\text{change in velocity}}{\text{time interval}} = \frac{4 \text{ m/s} - 2 \text{ m/s}}{1 \text{ s}} = \frac{2 \text{ m/s}}{1 \text{ s}} = 2 \text{ m/s}^2$$

Investigate Figure B.2 carefully and note the *total* distance covered as the ball accelerates down the plane. The distances go from zero to 1 meter in 1 second, zero to 4 meters in 2 seconds, zero to 9 meters in 3 seconds, zero to 16 meters in 4 seconds, and so on in succeeding seconds. The sequence for *total distances* covered is of the *squares of the time.* We'll investigate the relationship between distance traveled and the square of the time for constant acceleration more closely in the case of free fall.

Computing Distance When Acceleration Is Constant

How far will an object released from rest fall in a given time? To answer this question, let us consider the case in which it falls freely for 3 seconds, starting at rest. Neglecting air resistance, the object will have a constant acceleration of about 10 meters per second each second (actually more like 9.8 m/s^2 but we want to make the numbers easier to follow).

$$\text{Velocity at the } beginning = 0 \text{ m/s}$$

$$\text{Velocity at the } end \text{ of 3 seconds} = (10 \times 3) \text{ m/s}$$

$$Average\ velocity = \tfrac{1}{2} \text{ the sum of these two speeds}$$

$$= \tfrac{1}{2} \times (0 + 10 \times 3) \text{ m/s}$$

$$= \tfrac{1}{2} \times 10 \times 3 = 15 \text{ m/s}$$

$$\text{Distance traveled} = \text{average velocity} \times \text{time}$$

$$= (\tfrac{1}{2} \times 10 \times 3) \times 3$$

$$= \tfrac{1}{2} \times 10 \times 3^2 = 45 \text{ m}$$

We can see from the meanings of these numbers that

$$\text{Distance traveled} = \tfrac{1}{2} \times \text{acceleration} \times \text{square of time}$$

This equation is true for an object falling not only for 3 seconds but for any length of time, as long as the acceleration is constant. If we let *d* stand for the distance traveled, *a* for the acceleration, and *t* for the time, the rule may be written, in shorthand notation,

$$d = \tfrac{1}{2}\,at^2$$

This relationship was first deduced by Galileo. He reasoned that if an object falls for, say, twice the time, it will fall with *twice the average speed.* Since it falls for *twice* the time at *twice* the average speed, it will fall *four* times as far. Similarly, if an object falls for *three* times the time, it will have an average speed *three* times as great and will fall *nine* times as far. Galileo reasoned that the total distance fallen should be proportional to the *square* of the time.

In the case of objects in free fall, it is customary to use the letter *g* to represent the acceleration instead of the letter *a* (*g* because acceleration is due to *gravity*). While the value of *g* varies slightly in different parts of the world, it is approximately equal to 9.8 m/s^2 (32 ft/s^2). If we use *g* for the acceleration of a freely falling object (negligible air resistance), the equations for falling objects starting from a rest position become

$$v = gt$$

$$d = \tfrac{1}{2}\,gt^2$$

Questions

1. An auto starting from rest has a constant acceleration of 4 m/s². How far will it go in 5 s?

2. How far will an object released from rest fall in 1 s? In this case the acceleration is $g = 9.8$ m/s².

3. If it takes 4 s for an object to freely fall to the water when released from the Golden Gate Bridge, how high is the bridge?

Much of the difficulty in learning physics, like learning any discipline, has to do with learning the language—the many terms and definitions. Speed is somewhat different from velocity, and acceleration is vastly different from speed or velocity. Mass and weight are related but are different from each other. Similarly for work, heat, and temperature. Please be patient with yourself as you find learning the similarities and the differences among physics concepts is not an easy task.

Answers

1. Distance $= \frac{1}{2} \times 4 \times 5^2 = 50$ m
2. Distance $= \frac{1}{2} \times 9.8 \times 1^2 = 4.9$ m
3. Distance $= \frac{1}{2} \times 9.8 \times 4^2 = 78.4$ m

   Notice that the units of measurement when multiplied give the proper units of meters for distance:

$$d = \frac{1}{2} \times 9.8 \text{ m/s}^2 \times 16 \text{ s}^2 = 78.4 \text{ m}$$

Graphs—A Way to Express Quantitative Relationships

Graphs, like equations and tables, show how two or more quantities relate to each other. Since investigating relationships between quantities makes up much of the work of physics, equations, tables, and graphs are important physics tools.

Equations are the most concise way to describe quantitative relationships. For example, consider the equation $v = v_0 + gt$. It compactly describes how a freely-falling object's velocity depends on its initial velocity, acceleration due to gravity, and time. Equations are nice shorthand expressions for relationships among quantities.

Tables give values of variables in list form. The dependence of v on t in $v = v_0 + gt$ can be shown by a table that lists various values v for corresponding times t. Table 2.2 on page 28 is an example. Tables are especially useful when the mathematical relationship between quantities is not known, or when numerical values must be given to a high degree of accuracy. Also, tables are handy for recording experimental data.

Graphs *visually* represent relationships between quantities. By looking at the shape of a graph, you can quickly tell a lot about how the variables are related. For this reason, graphs can help clarify the meaning of an equation or table of numbers. And, when the equation is not already known, a graph can help reveal the relationship between variables. Experimental data are often graphed for this reason.

Graphs are helpful in another way. If a graph contains enough plotted points, it can be used to estimate values between the points (interpolation), or following the points (extrapolation).

Cartesian Graphs

The most common and useful graph in science is the *Cartesian* graph. On a Cartesian graph, possible values of one variable are represented on the vertical axis (called the *y-axis*) and possible values of the other variable are plotted on the horizontal axis (*x-axis*).

Figure C.1 shows a graph of two variables, *x* and *y*, that are *directly proportional* to each other. A direct proportionality is a type of *linear* relationship. Linear relationships have straight-line graphs—the easiest kinds of graphs to interpret. On the graph shown in Figure C.1, the continuous straight-line rise from left to right tells you that as

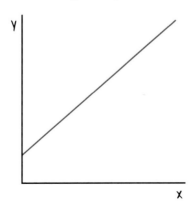

FIGURE C.I

x increases, *y* increases. More specifically, it shows that *y* increases at a constant rate with respect to *x*. As *x* increases, *y* increases. The graph of a direct proportionality often passes through the "origin"— the point at the lower left where *x* = 0 and *y* = 0. In Figure C.1, however, we see the graph begins where *y* has a nonzero value when *x* = 0. The value *y* has a "head start."

Figure C.2 shows a graph of the equation $v = v_0 + gt$. Speed *v* is plotted along the *y*-axis, and time *t* along the *x*-axis. As you can see, there is a linear relationship between *v* and *t*. Note that the initial speed is 10 m/s. If the initial speed were 0, as in dropping an object from rest, then the graph would intercept the origin, where both *v* and *t* are 0. Note that the graph originates at *v* = 10 m/s when *t* = 0, showing a 10-m/s "head start."

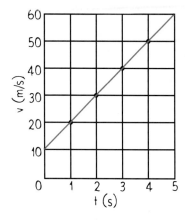

FIGURE C.2

Many physically significant relationships are more complicated than linear relationships, however. If you double the size of a room, the area of the floor increases four times; tripling the size of the room increases the floor area nine times, and so on. This is one example of a *nonlinear* relationship. Figure C.3 shows a graph of another nonlinear relationship: distance versus time in the equation of free fall from rest, $d = \frac{1}{2} gt^2$.

Figure C.4 shows a *radiation curve*. The *curve* (or graph) shows the rather complex nonlinear relationship between intensity *I* and radiation wavelength λ for a glowing object at 2000 K. The graph shows that radiation is most intense when λ equals about 1.4 μm. Which is brighter, radiation at 0.5 μm or radiation at 4.0 μm? The graph can quickly tell you that radiation at 4.0 μm is appreciably more intense.

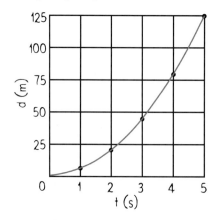

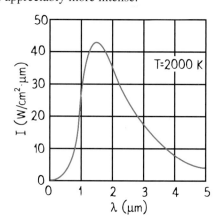

FIGURE C.3   **FIGURE C.4**

Slope and Area Under the Curve

Quantitative information can be obtained from a graph's *slope* and the *area under the curve*. The slope of the graph in Figure C.2 represents the rate at which v increases relative to t. It can be calculated by dividing a segment Δv along the y-axis by a corresponding segment Δt along the x-axis. For example, dividing Δv of 30 m/s by Δt of 3 s gives $\Delta v/\Delta t = 10$ m/s $\cdot$ s $= 10$ m/s^2, the acceleration due to gravity. By contrast, consider the graph in Figure C.5, which is a horizontal straight line. Its slope of zero shows zero acceleration—that is, constant speed. The graph shows that the speed is 30 m/s, acting throughout the entire five-second interval. The rate of change, or slope, of the speed with respect to time is zero—there is no change in speed at all.

The area under the curve is an important feature of a graph because it often has a physical interpretation. For example, consider the area under the graph of v versus t shown in Figure C.6. The shaded region is a rectangle with sides 30 m/s and 5 s. Its area is 30 m/s $\times$ 5 s $= 150$ m. In this example, the area is the distance covered by an object moving at constant speed of 30 m/s for 5 s ($d = vt$).

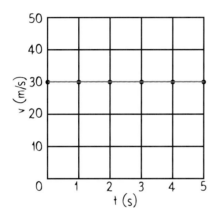

FIGURE C.5

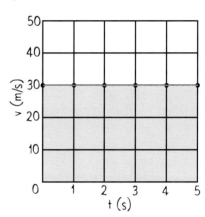

FIGURE C.6

The area need not be rectangular. The area beneath any curve of v versus t represents the distance traveled in a given time interval. Similarly, the area beneath a curve of acceleration versus time gives the change of velocity in a time interval. The area beneath a force-versus-time curve gives the change of momentum. (What does the area beneath a force-versus-distance curve give?) The nonrectangular area under various curves, including rather complicated ones, can be found by way of an important branch of mathematics—*integral calculus*.

Graphing with Conceptual Physics

You may develop basic graphing skills in the laboratory part of this course. The lab *Blind as a Bat* introduces you to graphing concepts. It also gives you a chance to work with a computer and sonic-ranging device. The lab *Trial and Error* will show you the useful technique of converting a nonlinear graph to a linear one to discover a direct proportionality. The area under the curve is the basis of the lab activities *Impact Speed* and *Wrap Your Energy in a Bow*. You will learn about graphing by doing it in other labs as well.

You may also learn in the lab part of your *Conceptual Physics* course that computers can graph data for you. You are not being lazy when you graph your data with a software program. Instead of investing time and energy scaling the axes and plotting points, you spend your time and energy investigating the meaning of the graph, a high level of thinking!

Questions Figure C.7 is a graphical representation of a ball dropped into a mineshaft.

1. How long did the ball take to hit the bottom?
2. What was the ball's speed when it struck bottom?
3. What does the decreasing slope of the graph tell you about the acceleration of the ball with increasing speed?
4. Did the ball reach terminal speed before hitting the bottom of the shaft? If so, about how many seconds did it take to reach its terminal speed?
5. What is the approximate depth of the mineshaft?

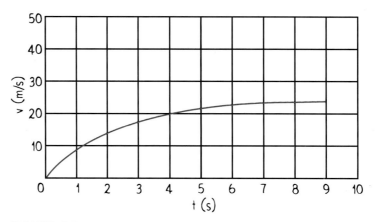

FIGURE C.7

Answers

1. 9 s
2. 25 m/s
3. Acceleration decreases as speed increases (due to air resistance).
4. Yes (since slope curves to zero), about 7 s.
5. Depth is about 170 m. (The area under the curve is about 17 squares, each of which represents 10 m.)

Appendix D

More About Vectors

Vectors and Scalars

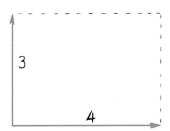

FIGURE D.1

A *vector* quantity is a directed quantity—one that must be specified not only by magnitude (size) but by direction as well. Recall from Chapter 3 that velocity is a vector quantity. Other examples are force, acceleration, and momentum. In contrast, a *scalar* quantity can be specified by magnitude alone. Some examples of scalar quantities are speed, time, temperature, and energy.

Vector quantities may be represented by arrows. The length of the arrow tells you the magnitude of the vector quantity, and the arrowhead tells you the direction of the vector quantity. Such an arrow drawn to scale and pointing appropriately is called a *vector*.

Adding Vectors

Vectors that add together are called *component vectors. The sum of component vectors is called a resultant.*

To add two vectors, make a parallelogram with two component vectors acting as two of the adjacent sides (Figure D.2). (Here our parallelogram is a rectangle.) Then draw a diagonal from the origin of the vector pair; this is the resultant (Figure D.3).

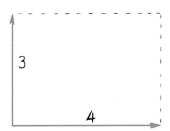

FIGURE D.2

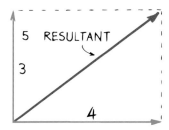

FIGURE D.3

Caution: Do not try to mix vectors! We cannot add apples and oranges, so velocity vector combines only with velocity vector, force vector combines only with force vector, and acceleration vector combines only with acceleration vector—each on its own vector diagram. If you ever show different kinds of vectors on the same diagram, use different colors or some other method of distinguishing the different kinds of vectors.

697

Finding Components of Vectors

Recall from Chapter 3 that to find a pair of perpendicular components for a vector, first draw a dotted line through the tail end of the vector in the direction of one of the desired components. Second, draw another dotted line through the tail end of the vector at right angles to the first dotted line. Third, make a rectangle whose diagonal is the given vector. Draw in the two components. Here we let **F** stand for "total force," **U** stand for "upward force," and **S** stand for "sideways force."

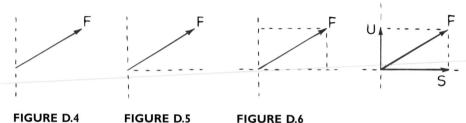

FIGURE D.4   **FIGURE D.5**   **FIGURE D.6**

Examples

1. A man pushing a lawnmower applies a force that pushes the machine forward and also against the ground. In Figure D.7, **F** represents the force applied by the man. We can separate this force into two components. The vector **D** represents the downward component, and **S** is the sideways component, the force that moves the lawnmower forward. If we know the magnitude and direction of the vector **F**, we can estimate the magnitude of the components from the vector diagram.

FIGURE D.7

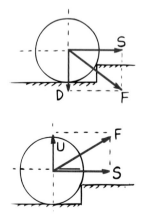

2. Would it be easier to push or pull a wheelbarrow over a step? Figure D.8 shows a vector diagram for each case. When you push a wheelbarrow, part of the force is directed downward, which makes it harder to get over the step. When you pull, however, part of the pulling force is directed upward, which helps to lift the wheel over the step. Note that the vector diagram suggests that pushing the wheelbarrow may not get it over the step at all. Do you see that the height of the step, the radius of the wheel, and the angle of the applied force determine whether the wheelbarrow can be pushed over the step? We see how vectors help us analyze a situation so that we can see just what the problem is!

FIGURE D.8

3. If we consider the components of the weight of an object rolling down an incline, we can see why its speed depends on the angle (Figure D.9). Note that the steeper the incline, the greater the component **S** becomes and the faster the object rolls. When the incline is vertical, **S** becomes equal to the weight, and the object attains maximum acceleration, 9.8 meters per second squared.

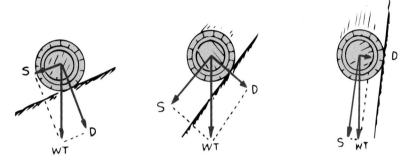

FIGURE D.9

There are two more force vectors that are not shown: the normal force **N**, which is equal and oppositely directed to **D**, and the friction force **f**, acting at the barrel-plane contact.

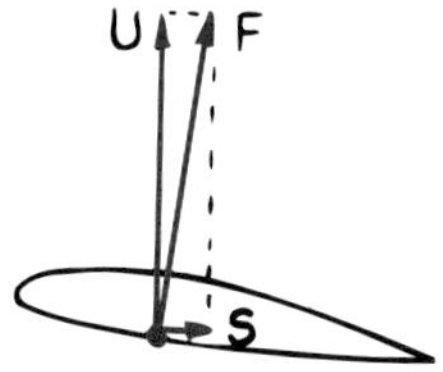

4. When moving air strikes the underside of an airplane wing, the force of air impact against the wing may be represented by a single vector perpendicular to the plane of the wing (Figure D.10). We represent the force vector as acting midway along the lower wing surface, where the dot is, and pointing above the wing to show the direction of the resulting wind impact force. This force can be broken up into two components, one sideways and the other up. The upward component, **U**, is called *lift*. The sideways component, **S**, is called *drag*. If the aircraft is to fly at constant velocity at constant altitude, then lift must equal the weight of the aircraft and the thrust of the plane's engines must equal drag. The magnitude of lift (and drag) can be altered by changing the speed of the airplane or by changing the angle (called *angle of attack*) between the wing and the horizontal.

FIGURE D.10

5. Consider the satellite moving clockwise in Figure D.11. Everywhere in its orbital path, gravitational force *F* pulls it toward the center of the host planet. At position A we see *F* separated into two components: *f*, which is tangent to the path of the projectile, and *f′*, which is perpendicular to the path. The relative magnitudes of these components in comparison to the magnitude of *F* can be seen in the imaginary rectangle they compose; *f* and *f′* are the sides, and *F* is the diagonal. We see that component *f* is along the orbital path but against the direction of motion of the satellite. This force component reduces the speed of the satellite. The other component *f′*, changes the direction of the satellite's motion and pulls it away from its tendency to go in a straight line. So the path of the satellite curves. The satellite loses speed until it reaches position B. At this farthest point from the planet (apogee), the gravitational force is somewhat weaker but perpendicular to the satellite's motion, and component *f* has reduced to zero. Component *f′*, on the other hand, has increased and is now fully merged to become *F*. Speed at this point is not enough for circular orbit, and the satellite begins to fall toward the planet. It picks up speed because the component *f* reappears and is in the direction of motion as shown in position C. The satellite picks up speed until it whips

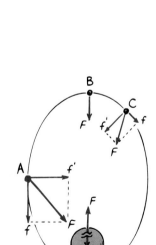

FIGURE D.11

around to position D (perigee), where once again the direction of motion is perpendicular to the gravitational force, f' blends to full F, and f is nonexistent. The speed is in excess of that needed for circular orbit at this distance, and it overshoots to repeat the cycle. Its loss in speed in going from D to B equals its gain in speed from B to D. Kepler discovered that planetary paths are elliptical, but never knew why. Do you?

6. Refer to the Polaroids held by Ludmila back in Chapter 28, in Figure 28.34. In the first picture (a), we see that light is transmitted through the pair of Polaroids because their axes are aligned. The emerging light can be represented as a vector aligned with the polarization axes of the Polaroids. When the Polaroids are crossed (b), no light emerges because light passing through the first Polaroid is perpendicular to the polarization axes of the second Polaroid, with no components along its axis. In the third picture (c), we see that light is transmitted when a third Polaroid is sandwiched at an angle between the crossed Polaroids. The explanation for this is shown in Figure D.12.

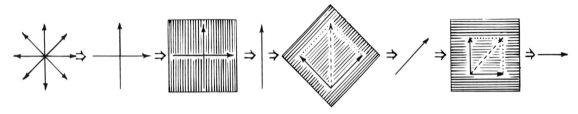

FIGURE D.12

Sailboats

Sailors have always known that a sailboat can sail downwind, in the direction of the wind. Sailors have not always known, however, that a sailboat can sail upwind, against the wind. One reason for this has to do with a feature that is common only to recent sailboats—a finlike keel that extends deep beneath the bottom of the boat to ensure that the boat will knife through the water only in a forward (or backward) direction. Without a keel, a sailboat could be blown sideways.

Figure D.13 shows a sailboat sailing directly downwind. The force of wind impact against the sail accelerates the boat. Even if the drag of the water and all other resistance forces are negligible, the maximum speed of the boat is the wind speed. This is because the wind will not make impact against the sail if the boat is moving as fast as the wind. The wind would have no speed relative to the boat and the sail would simply sag. With no force, there is no acceleration. The force vector in Figure D.13 *decreases* as the boat travels faster. The force vector is maximum when the boat is at rest and the full impact of the wind fills the sail, and is minimum when the boat travels as fast as the wind. If the boat is somehow propelled to a speed faster than the wind (by way of a motor, for example), then air resistance against the front side of the sail will produce an oppositely directed force vector. This will slow the boat down. Hence, the boat when driven only by the wind cannot exceed wind speed.

FIGURE D.13

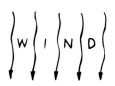

FIGURE D.14

If the sail is oriented at an angle, as shown in Figure D.14, the boat will move forward, but with less acceleration. There are two reasons for this:

1. The force on the sail is less because the sail does not intercept as much wind in this angular position.

2. The direction of the wind impact force on the sail is not in the direction of the boat's motion, but is perpendicular to the surface of the sail. Generally speaking, whenever any fluid (liquid or gas) interacts with a smooth surface, the force of interaction is perpendicular to the smooth surface. * The boat does not move in the same direction as the perpendicular force on the sail, but is constrained to move in a forward (or backward) direction by its keel.

We can better understand the motion of the boat by resolving the force of wind impact, *F*, into perpendicular components. The important component is that which is parallel to the keel, which we label *K*, and the other component is perpendicular to the keel, which we label *T*. It is the component *K*, as shown in Figure D.15 that is responsible for the forward motion of the boat. Component *T* is a useless force that tends to tip the boat over and move it sideways. This component force is offset by the deep keel. Again, maximum speed of the boat can be no greater than wind speed.

Many sailboats sailing in directions other than exactly downwind (Figure D.16) with their sails properly oriented can exceed wind speed. In the case of a sailboat cutting across the wind, the wind may continue to make impact with the sail even after the boat exceeds wind speed. A surfer, in a similar way, exceeds the velocity of the

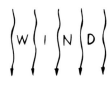

FIGURE D.15   **FIGURE D.16**

---

*You can do a simple exercise to see that this is so. Try bouncing one coin off another on a smooth surface, as shown. Note that the struck coin moves at right angles (perpendicular) to the contact edge. Note also that it makes no difference whether the projected coin moves along path A or path B. See your instructor for a more rigorous explanation, which involves momentum conservation.

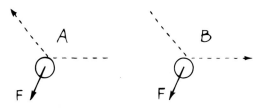

propelling wave by angling his surfboard across the wave. Greater angles to the propelling medium (wind for the boat, water wave for the surfboard) result in greater speeds. A sailcraft can sail faster cutting across the wind than it can sailing downwind.

As strange as it may seem, maximum speed for most sailcraft is attained by cutting into (against) the wind, that is, by angling the sailcraft in a direction upwind! Although a sailboat cannot sail directly upwind, it can reach a destination upwind by angling back and forth in a zigzag fashion. This is called *tacking*. Suppose the boat and sail are as shown in Figure D.17. Component **K** will push the boat along in a forward direction, angling into the wind. In the position shown, the boat can sail faster than the speed of the wind. This is because as the boat travels faster, the impact of wind is increased. This is similar to running in a rain that comes down at an angle. When you run into the direction of the downpour, the drops strike you harder and more frequently; but when you run away from the direction of the downpour, the drops don't strike you as hard or as frequently. In the same way, a boat sailing upwind experiences greater wind impact force, while a boat sailing downwind experiences a decreased wind impact force. In any case the boat reaches its terminal speed when opposing forces cancel the force of wind impact. The opposing forces consist mainly of water resistance against the hull of the boat. The hulls of racing boats are shaped to minimize this resistive force, which is the principal deterrent to high speeds.

Iceboats (sailcraft equipped with runners for traveling on ice) encounter no water resistance and can travel at several times the speed of the wind when they tack upwind. Although ice friction is nearly absent, an iceboat does not accelerate without limits. The terminal velocity of a sailcraft is determined not only by opposing friction forces but also by the change in relative wind direction. When the boat's orientation and speed are such that the wind seems to shift in direction, so the wind moves parallel to the sail rather than into it, forward acceleration ceases—at least in the case of a flat sail. In practice, sails are curved and produce an airfoil that is as important to sailcraft as it is to aircraft. The effects are discussed in Chapter 13.

FIGURE D.17

Appendix E

Exponential Growth and Doubling Time*

Try folding a piece of paper in half, then folding it again upon itself successively for 9 times. You'll soon see that it gets too thick to keep folding. And if you could fold a fine piece of tissue paper upon itself 50 times, it would be more than 20 million kilometers thick! The continual doubling of a quantity builds up astronomically. Give a child a penny on his first birthday, two pennies on his second birthday, four pennies on his third birthday, and so on, doubling the number of pennies every birthday. When this child reaches age 30, he will have accumulated $10,737,418.23! One of the most important things we have trouble perceiving is the process of exponential growth, and why it proliferates out of control.

When a quantity such as money in the bank, population, or the rate of consumption of a resource steadily grows at a fixed percent per year, the growth is said to be *exponential*. Money in the bank may grow at 5 or 6% per year; world population is presently growing at about 2% per year; the electric power generating capacity in the United States grew at about 7% per year for the first three quarters of the twentieth century. The important thing about exponential growth is that the time required for the growing quantity to double in size (increase by 100%) is constant. For example, if the population of a growing city takes 10 years to double from 10,000 to 20,000 people and it continues with exponential growth, in the next 10 years the population will double to 40,000, and in the next 10 years to 80,000, and so on.

There is an important relationship between the percent growth rate and its *doubling time*, the time it takes to double a quantity:[†]

$$\text{doubling time} = \frac{69.2\%}{\text{percent growth per unit time}}$$

$$\approx \frac{70\%}{\text{percent growth rate}}$$

This means that to estimate the doubling time for a steadily growing quantity, we simply divide 70% by the percentage growth rate. For example, when electric power generating capacity in the United States was growing at 7% per year, the capacity doubled every 10 years (since (70%)/(7%/year) = 10 years). If world population grew steadily at 2% per year, the world population would double in 35 years (since (70%)/(2%/year) = 35 years). A city planning commission that accepts what seems like a modest 3.5%-per-year growth rate may not realize that this means that doubling will occur

---

*This appendix is adapted from material written by University of Colorado physics professor Albert A. Bartlett, who asserts, "The greatest shortcoming of the human race is man's inability to understand the exponential function." Look up Professor Bartlett's provocative and still timely article, "Forgotten Fundamentals in the Energy Crisis," in the September 1978 issue of the *American Journal of Physics*, or his revised version in the January 1980 issue of the *Journal of Geological Education*.

[†]For exponential decay we speak about *half-life*, the time for a quantity to reduce to half its value. An example of this case is radioactive decay, treated in Chapter 32.

in 20 years (since (70%)/(3.5%/year) = 20 years). That means double capacity for such things as water supply, sewage-treatment plants, and other municipal services every 20 years.

Steady growth in a steadily expanding environment is one thing, but what happens when steady growth occurs in a finite environment? Consider the growth of bacteria that grow by division, so that one bacterium becomes two, the two divide to become four, the four divide to become eight, and so on. Suppose the division time for a certain kind of bacteria is one minute. This is then steady percentage growth—the number of bacteria grows exponentially with a doubling time of one minute. Further, suppose that one bacterium is put in a bottle at 11:00 A.M. and that growth continues steadily until the bottle becomes full of bacteria at 12 noon.

FIGURE E.1 Graph of a quantity that grows at an exponential rate. Notice that the quantity doubles during each of the successive equal time intervals marked on the horizontal scale. Each of these time intervals represents the doubling time.

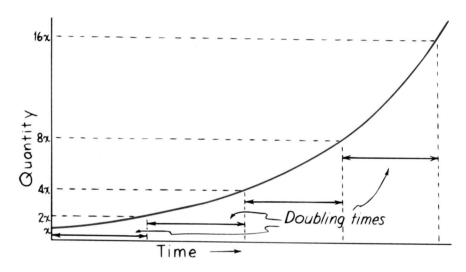

FIGURE E.2

Question When was the bottle half full?

It is startling to note that at 2 minutes before noon the bottle was only $\frac{1}{4}$ full, and at 3 minutes before noon only $\frac{1}{8}$ full. Table E.1 summarizes the amount of space left in the bottle in the last few minutes before noon. If bacteria could think, and if they were concerned about their future, at which time do you think they would sense they were running out of space? Do you think a serious problem would have been evident at, say, 11:55 A.M., when the bottle was only 3% full ($\frac{1}{32}$) and had 97% open space (just yearning for development)? The point here is that there isn't much time between the moment the effects of growth become noticeable and the time when they become overwhelming.

Suppose that at 11:58 A.M. some farsighted bacteria see that they are running out of space and launch a full-scale search for new bottles. And further suppose they consider themselves lucky, for they find three new empty bottles. This is three times as much space as they have ever known. It may seem to the bacteria that their problems are solved—and just in time.

Answer At 11:59 A.M., since the bacteria will double in number every minute!

Table E.1   The Last Minutes in the Bottle

Time	Portion full	Portion empty
11:54 A.M.	$\frac{1}{64}$ (1.5%)	$\frac{63}{64}$ (98.5%)
11:55 A.M.	$\frac{1}{32}$ (3%)	$\frac{31}{32}$ (97%)
11:56 A.M.	$\frac{1}{16}$ (6%)	$\frac{15}{16}$ (94%)
11:57 A.M.	$\frac{1}{8}$ (12%)	$\frac{7}{8}$ (88%)
11:58 A.M.	$\frac{1}{4}$ (25%)	$\frac{3}{4}$ (75%)
11:59 A.M.	$\frac{1}{2}$ (50%)	$\frac{1}{2}$ (50%)
12:00 noon	Full (100%)	None (0%)

Question If the bacteria are able to migrate to the new bottles and their growth continues at the same rate, what time will it be when the three new bottles are filled to capacity?

Table E.2 illustrates that the discovery of the new bottles extends the resource by only two doubling times. In this example the resource is space—such as land area for a growing population. But it could be coal, oil, uranium, or any nonrenewable resource.

Table E.2   Effects of the Discovery of Three New Bottles

Time	Effect
11:58 A.M.	Bottle 1 is $\frac{1}{4}$ full; bacteria divide into four bottles, each $\frac{1}{16}$ full
11:59 A.M.	Bottles 1, 2, 3, and 4 are each $\frac{1}{8}$ full
12:00 noon	Bottles 1, 2, 3, and 4 are each $\frac{1}{4}$ full
12:01 P.M.	Bottles 1, 2, 3, and 4 are each $\frac{1}{2}$ full
12:02 P.M.	Bottles 1, 2, 3, and 4 are each all full

Answer All four bottles will be filled to capacity at 12:02 P.M.!

Question According to a French riddle, a lily pond starts with a single leaf. Each day the number of leaves doubles, until the pond is completely full on the thirtieth day. On what day was the pond half covered? One-quarter covered?

Continued growth and continued doubling lead to enormous numbers. In two doubling times, a quantity will double twice ($2^2 = 4$), or quadruple in size; in three doubling times, its size will increase eightfold ($2^3 = 8$); in four doubling times, it will increase sixteenfold ($2^4 = 16$); and so on. This is best illustrated by the story of the court mathematician in India who years ago invented the game of chess for his king. The king was so pleased with the game that he offered to repay the mathematician, whose request seemed modest enough. The mathematician requested a single grain of wheat on the first square of the chessboard, two grains on the second square, four on the third square, and so on, doubling the number of grains on each succeeding square until all squares had been used. At this rate there would be 2^{63} grains of wheat on the sixty-fourth square alone. The king soon saw that he could not fill this "modest" request, which amounted to more wheat than had been harvested in the entire history of the earth!

FIGURE E.3 A single grain of wheat placed on the first square of the chess board is doubled on the second square, and this number is doubled on the third square, and so on. Note that each square contains one more grain than all the preceding squares combined. Does enough wheat exist in the world to fill all 64 squares in this manner?

As Table E.3 shows, the number of grains on any square is one grain more than the total of all grains on the preceding squares. This is true anywhere on the board. For example, when four gains are placed on the third square, that number of grains is one more than the total of three grains already on the board. The number of grains (eight) on the fourth square is one more than the total of seven grains already on the board. The same pattern occurs everywhere on the board. In any case of exponential growth, a greater quantity is represented in one doubling time than in all the preceding growth.

Answer The pond was half covered on the 29th day, and was one-quarter covered on the 28th day!

Table E.3   Filling the Squares on the Chessboard

Square number	Grains on a square	Total grains thus far
1	1	1
2	2	3
3	$4 = 2^2$	7
4	$8 = 2^3$	15
5	$16 = 2^4$	31
6	$32 = 2^5$	63
7	$64 = 2^6$	127
⋮	⋮	⋮
64	2^{63}	$2^{64} - 1$

This is important enough to be repeated in different words: Whenever steady growth occurs, the numerical count of a quantity that exists after a single doubling time is one greater than the total count of that quantity in the entire history of growth.

Questions

1. In 1996 the population growth rate of the world was 1.4% per year, and the total world population was 5.8 billion. At this rate, how long will it take for the world population to double?

2. What annual percentage increase in world population would be required to double the world population in 100 years?

So if we speak of doubling energy consumption in the next however many years, bear in mind that this means in these years we will consume more energy than has heretofore been consumed during the entire preceding period of steady growth. If power generation continues to use predominantly fossil fuels, then except for some improvements in efficiency, we would burn up in the next doubling time a greater amount of coal, oil, and natural gas than has already been consumed by previous power generation; and except for improvements in pollution control, we can expect to discharge even more toxic wastes into the environment than the millions upon millions of tons already discharged over all the previous years of industrial civilization; and we would expect more human-made calories of heat to be absorbed by the earth's ecosystem than have been absorbed in the entire past! If the 7% growth rate in energy production during the first three-quarters of the twentieth century still persisted, all this would occur in only a

Answers

1. The doubling time is 50 years, since (70%)/(1.4%/year) = 50 years.

2. 0.7%, since (70%)/(0.7%/year) = 100 years. You can rearrange the equation so it reads percent growth rate = (70%)/(doubling time). Using the rearranged equation gives (70%)/(100 years) = 0.7%/year.

single decade. At half this rate, all this would take place in the next 20 years. Clearly this cannot continue.

The consumption of a nonrenewable resource cannot grow exponentially for an indefinite period. The resource is finite and its supply finally expires. This is shown in Figure E.4a, where the rate of consumption, such as barrels of oil per year, is plotted against time, say in years. In such a graph the colored area under the curve represents the supply of the resource. When the supply is exhausted, consumption ceases altogether. This sudden change is rarely the case, for the rate of extracting the supply falls as it becomes more scarce. This is shown in Figure E.4b. Note the area under the curve is equal to the area under the curve in Figure E.4a. Why? Because the total supply is the same in both cases. The difference is the time taken to extract that supply. History shows that the rate of production of a nonrenewableresource rises and falls in a nearly symmetric manner, as shown in Figure E.4c. The time during which production rates rise is approximately equal to the time during which these rates fall to zero or near zero. If we fit the data for U.S. oil production in the contiguous forty-eight states to such a curve, we find that we are to the right of the peak. This suggests that half of the recoverable petroleum that was ever in the ground in the U.S. has already been used and that in the future the domestic petroleum rate of production can only decrease. We'll see. In the meantime we find that we import more oil each year. And each year we consume more oil than in the previous year.

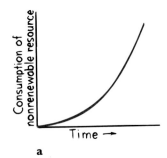
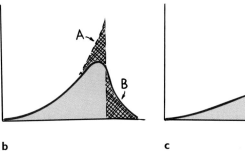
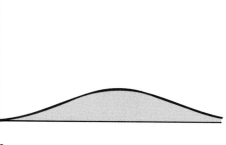

a   b   c

FIGURE E.4   (a) If the exponential rate of consumption for a non-renewable resource continues until it is depleted, consumption falls abruptly to zero. The colored area under this curve represents the total supply of the resource. (b) In practice the rate of consumption levels off and then falls less abruptly to zero. Note that the crosshatched area A is equal to the crosshatched area B. Why? (c) At lower consumption rates, the same resource lasts a longer time.

The consequences of unchecked exponential growth are staggering. It is very important to ask: Is growth really good? In answering this question, bear in mind that human growth is an early phase of life that continues normally through adolescence. Physical growth stops when physical maturity is reached. What do we say of growth that continues in the period of physical maturity? We say that such growth is obesity—or worse, cancer.

FIGURE E.5   A curve showing the rate of consumption of a renewable resource such as agricultural or forest products, where a steady rate of production and consumption can be maintained for a long period, providing this production is not dependent upon the use of a non-renewable resource that is waning in supply.

Questions to Ponder

1. In an economy that has a steady inflation rate of 7% per year, in how many years does a dollar lose half its value?

2. At a steady inflation rate of 7%, what will be the price every 10 years for the next 50 years for a theater ticket that now costs $20? For a coat that now costs $200? For a car that now costs $20,000? For a home that now costs $200,000?

3. If the population of a city with one overloaded sewage treatment plant grows steadily at 5% annually, how many overloaded sewage treatment plants will be necessary 42 years later?

4. In 1996 the population growth rate for the United States was 1.0%, for Mexico 1.9%, and for Rwanda the highest growth rate in the world, 16.5% (taking into account births, deaths, and immigration). If these rates were steady, how long would it take for the population in each of these countries to double?

5. Suppose that the 16.5% growth rate for the population of Rwanda has always been about the same, and will continue at the same rate for another five years. True or false: At the turn of the century there will be more Rwandan teenagers than the total number of Rwandan people who ever lived.

6. If world population doubles in 40 years and world food production also doubles in 40 years, how many people then will be starving each year compared to now?

7. Suppose you get a prospective employer to agree to hire your services for a wage of a single penny for the first day, 2 pennies the second day, and doubling each day thereafter providing the employer keep to the agreement for a month. What will be your total wages for the month?

8. In the previous question, how will your wages for only the 30th day compare to your total wages for the previous 29 days?

9. Oil has been produced in the U.S. for about 120 years. If there remains undiscovered in the country as much oil as all that has been used, what is wrong with the argument that the remaining oil will be sufficient to meet the country's needs for another 120 years?

10. If fusion power were harnessed today, the abundant energy resulting would likely sustain and even further encourage our present appetite for continued growth and in a relatively few doubling times produce an appreciable fraction of the solar power input to the earth. Make an argument that the current delay in harnessing fusion is a blessing for the human race.

Glossary

A (a) Abbreviation for *ampere*. (b) When in lowercase italic *a*, the symbol for *acceleration*.

aberration Distortion in an image produced by a lens or mirror, caused by limitations inherent to some degree in all optical systems. See *spherical aberration* and *chromatic aberration*.

absolute zero Lowest possible temperature that any substance can have; the temperature at which the atoms of a substance have their minimum kinetic energy. The temperature of absolute zero is $-273.15°C$, which is $-459.7°F$ and 0 kelvin.

absorption lines Dark lines that appear in an absorption spectrum. The pattern of lines is unique for each element.

absorption spectrum Continuous spectrum, like that generated by white light, interrupted by dark lines or bands that result from the absorption of light of certain frequencies by a substance through which the light passes.

ac Abbreviation for *alternating current*.

acceleration (a) Rate at which an object's velocity changes with time; the change in velocity may be in magnitude (speed), or direction, or both.

$$\text{acceleration} = \frac{\text{change of velocity}}{\text{time interval}}$$

acceleration due to gravity (g) Acceleration of a freely falling object. Its value near the earth's surface is about 9.8 meters per second each second.

achromatic lenses See *chromatic aberration*.

acoustics Study of the properties of sound, especially its transmission.

action force One of the pair of forces described in Newton's third law. See also *Newton's laws of motion, Law 3*.

additive primary colors Three colors of light—red, blue, and green—that when added in certain proportions will produce any color of the spectrum.

adhesion Molecular attraction between two surfaces making contact.

adiabatic Term applied to expansion or compression of a gas occurring without gain or loss of heat.

adiabatic process Process, often of fast expansion or compression, wherein no heat enters or leaves a system. As a result, a liquid or gas undergoing an expansion will cool, or undergoing a compression will warm.

air resistance Friction, or drag, that acts on something moving through air.

alchemist Practitioner of the early form of chemistry called alchemy, which was associated with magic. The goal of alchemy was to change base metals to gold and to discover a potion that could produce eternal youth.

alloy Solid mixture composed of two or more metals or of a metal and a nonmetal.

alpha particle Nucleus of a helium atom, which consists of two neutrons and two protons, ejected by certain radioactive nuclei.

alpha ray Stream of alpha particles (helium nuclei) ejected by certain radioactive nuclei.

alternating current (ac) Electric current that rapidly reverses in direction. The electric charges vibrate about relatively fixed positions, usually at the rate of 60 hertz.

AM Abbreviation for *amplitude modulation*.

ammeter A device that measures current. See *galvanometer*.

ampere (A) SI unit of electric current. One ampere is a flow of one coulomb of charge per second—6.25×10^{18} electrons (or protons) per second.

amplitude For a wave or vibration, the maximum displacement on either side of the equilibrium (midpoint) position.

amplitude modulation (AM) Type of modulation in which the amplitude of the carrier wave is varied above and below its normal value by an amount proportional to the amplitude of the impressed signal.

amu Abbreviation for *atomic mass unit*.

analog signal Signal based on a continuous variable, as opposed to a digital signal made up of discrete quantities.

aneroid barometer Instrument used to measure atmospheric pressure; based on the movement of the lid of a metal box, rather than on the movement of a liquid.

angle of incidence Angle between an incident ray and the normal to the surface it encounters.

angle of reflection Angle between a reflected ray and the normal to the surface of reflection.

angle of refraction Angle between a refracted ray and the normal to the surface at which it is refracted.

angular momentum Product of a body's rotational inertia and rotational velocity about a particular axis. For an object that is small compared with the radial distance, it is the product of mass, speed, and radial distance of rotation.

$$\text{angular momentum} = mvr$$

antimatter Matter composed of atoms with negative nuclei and positive electrons.

antinode Any part of a standing wave with maximum displacement and maximum energy.

antiparticle Particle having the same mass as a normal particle, but a charge of the opposite sign. The antiparticle of an electron is a positron.

antiproton Antiparticle of a proton; a negatively charged proton.

apogee Point in an elliptical orbit farthest from the focus around which orbiting takes place. See also *perigee*.

Archimedes' principle Relationship between buoyancy and displaced fluid: An immersed object is buoyed up by a force equal to the weight of the fluid it displaces.

armature Part of an electric motor or generator where an electromotive force is produced. Usually the rotating part.

astigmatism Defect of the eye caused when the cornea is curved more in one direction than in another.

atmospheric pressure Pressure exerted against bodies immersed in the atmosphere resulting from the weight of air pressing down from above. At sea level, atmospheric pressure is about 101 kPa.

atom   Smallest particle of an element that has all the element's chemical properties. Consists of protons and neutrons in a nucleus surrounded by electrons.

atomic bonding   Linking together of atoms to form larger structures, such as molecules and solids.

atomic mass number   Number associated with an atom, equal to the number of nucleons (protons plus neutrons) in the nucleus.

atomic mass unit (amu)   Standard unit of atomic mass. It is based on the mass of the common carbon atom, which is arbitrarily given the value of exactly 12. An amu of one is one-twelfth the mass of this common carbon atom.

atomic number   Number associated with an atom, equal to the number of protons in the nucleus, or, equivalently, to the number of electrons in the electron cloud of a neutral atom.

aurora borealis   Glowing of the atmosphere caused by ions from above the atmosphere that dip into the atmosphere; also called northern lights. In the southern hemisphere, they are called aurora australis.

average speed   Path distance divided by time interval.

$$\text{average speed} = \frac{\text{total distance covered}}{\text{time interval}}$$

Avogadro's number 6.02×10^{23} molecules.

Avogadro's principle   Equal volumes of all gases at the same temperature and pressure contain the same number of molecules, 6.02×10^{23} in one mole (a mass in grams equal to the molecular mass of the substance in atomic mass units).

axis   (pl. axes) (a) Straight line about which rotation takes place. (b) Straight lines for reference in a graph, usually the x-axis for measuring horizontal displacement and the y-axis for measuring vertical displacement.

barometer   Device used to measure the pressure of the atmosphere.

beats   Sequence of alternating reinforcement and cancellation of two sets of superimposed waves differing in frequency, heard as a throbbing sound.

bel   Unit of intensity of sound, named after Alexander Graham Bell. The threshold of hearing is 0 bel (10^{-12} watts per square meter). Often measured in decibels (dB, one-tenth of a bel).

Bernoulli's principle   Pressure in a fluid decreases as the speed of the fluid increases.

beta particle   Electron (or positron) emitted during the radioactive decay of certain nuclei.

beta ray   Stream of beta particles (electrons or positrons) emitted by certain radioactive nuclei.

Big Bang   Primordial explosion that is thought to have resulted in the creation of our expanding universe.

bimetallic strip   Two strips of different metals welded or riveted together. Because the two substances expand at different rates when heated or cooled, the strip bends; used in thermostats.

binary code   Code based on the binary number system (which uses a base of 2). In binary code any number can be expressed as a succession of ones and zeros. For example, the number 1 is 1, 2 is 10, 3 is 11, 4 is 100, 5 is 101, 17 is 10001, etc. These ones and zeros can then be interpreted and transmitted electronically as a series of "on" and "off" pulses, the basis for all computers and other digital equipment.

bioluminescence   Light emitted from certain living things that have the ability to chemically excite molecules in their bodies; these excited molecules then give off visible light.

biomagnetism   Magnetic material located in living organisms that may help them navigate, locate food, and affect other behaviors.

black hole   Concentration of mass resulting from gravitational collapse, near which gravity is so intense that not even light can escape.

blind spot   Area of the retina where all the nerves carrying visual information exit the eye and go to the brain; this is a region of no vision.

blue shift   Increase in the measured frequency of light from an approaching source; called the blue shift because the apparent increase is toward the high-frequency, or blue, end of the color spectrum. Also occurs when an observer approaches a source. See also *Doppler effect*.

boiling   Change from liquid to gas occuring beneath the surface of the liquid; rapid vaporization. The liquid loses energy, the gas gains it.

bow wave   V-shaped wave produced by an object moving on a liquid surface faster than the wave speed.

Boyle's law   The product of pressure and volume is a constant for a given mass of confined gas regardless of changes in either pressure or volume individually, as long as temperature remains unchanged.

$$P_1V_1 = P_2V_2$$

breeder reactor   Nuclear fission reactor that not only produces power but produces more nuclear fuel than it consumes by converting a nonfissionable uranium isotope into a fissionable plutonium isotope. See also *nuclear reactor*.

British thermal unit (BTU)   Amount of heat required to change the temperature of 1 pound of water by 1 Fahrenheit degree.

Brownian motion   Haphazard movement of tiny particles suspended in a gas or liquid resulting from bombardment by the fast-moving molecules of the gas or liquid.

BTU   Abbreviation for *British thermal unit*.

buoyancy   Apparent loss of weight of an object immersed or submerged in a fluid.

buoyant force   Net upward force exerted by a fluid on a submerged or immersed object.

butterfly effect   Situation in which a very small change in one place can amplify into a large change somewhere else.

C   Abbreviation for *coulomb*.

cal   Abbreviation for *calorie*.

calorie (cal)   Unit of heat. One calorie is the heat required to raise the temperature of one gram of water 1 Celsius degree. One Calorie (with a capital *C*) is equal to one thousand calories and is the unit used in describing the energy available from food; also called a kilocalorie (kcal).

$$1 \text{ cal} = 4.184 \text{ J} \quad \text{or} \quad 1 \text{ J} = 0.24 \text{ cal}$$

capacitor   Device used to store charge in a circuit.

capillarity   Rise of a liquid in a fine, hollow tube or in a narrow space.

carbon dating   Process of determining the time that has elapsed since death by measuring the radioactivity of the remaining carbon-14 isotopes.

Carnot efficiency   Ideal maximum percentage of input energy that can be converted to work in a heat engine.

carrier wave High-frequency radio wave modified by a lower-frequency wave.

Celsius scale Temperature scale that assigns 0 to the melt-freeze point for water and 100 to the boil-condense point of water at standard pressure (one atmosphere at sea level).

center of gravity (CG) Point at the center of an object's weight distribution, where the force of gravity can be considered to act.

center of mass Point at the center of an object's mass distribution, where all its mass can be considered to be concentrated. For everyday conditions, it is the same as the center of gravity.

centrifugal force Apparent outward force on a rotating or revolving body.

centripetal force Center-directed force that causes an object to follow a curved or circular path.

CG Abbreviation for *center of gravity*.

chain reaction Self-sustaining reaction that, once started, steadily provides the energy and matter necessary to continue the reaction.

charge See *electric charge*.

charging by contact Transfer of electric charge between objects by rubbing or simple touching.

charging by induction Redistribution of electric charges in and on objects caused by the electrical influence of a charged object close by but not in contact.

chemical formula Description that uses numbers and symbols of elements to describe the proportions of elements in a compound or reaction.

chemical reaction Process of rearrangement of atoms that transforms one molecule into another.

chinook Warm, dry wind that blows down from the eastern side of the Rocky Mountains across the Great Plains.

chromatic aberration Distortion of an image caused when light of different colors (and thus different speeds and refractions) focuses at different points when passing through a lens. Achromatic lenses correct this defect by combining simple lenses made of different kinds of glass.

circuit Any complete path along which electric charge can flow. See also *series circuit* and *parallel circuit*.

circuit breaker Device in an electric circuit that breaks the circuit when the current gets high enough to risk causing a fire.

coherent light Light of a single frequency with all photons exactly in phase and moving in the same direction. Lasers produce coherent light. See also *incoherent light* and *laser*.

complementarity Principle enunciated by Niels Bohr stating that the wave and particle aspects of both matter and radiation are necessary, complementary parts of the whole. Which part is emphasized depends on what experiment is conducted (i.e., on what questions one puts to nature.)

complementary colors Any two colors of light that, when added, produce white light.

component Parts into which a vector can be separated and that act in different directions from the vector. See *resultant*.

compound Chemical substance made of atoms of two or more different elements combined in a fixed proportion.

compression (a) In mechanics, the act of squeezing material and reducing its volume. (b) In sound, the region of increased pressure in a longitudinal wave.

concave mirror Mirror that curves inward like a "cave."

condensation Change of phase of a gas into a liquid; the opposite of evaporation.

conduction (a) In heat, energy transfer from particle to particle within certain materials, or from one material to another when the two are in direct contact. (b) In electricity, the flow of electric charge through a conductor.

conduction electrons Electrons in a metal that move freely and carry electric charge.

conductor (a) Material through which heat can be transferred. (b) Material, usually a metal, through which electric charge can flow. Good conductors of heat are generally good electric charge conductors.

cones See *retina*.

conservation of angular momentum When no external torque acts on an object or a system of objects, no change of angular momentum takes place. Hence, the angular momentum before an event involving only internal torques is equal to the angular momentum after the event.

conservation of charge Principle that net electric charge is neither created nor destroyed but is transferable from one material to another.

conservation of energy Principle that energy cannot be created or destroyed. It may be transformed from one form into another, but the total amount of energy never changes.

conservation of energy for machines Work output of any machine cannot exceed the work input.

conservation of momentum In the absence of a net external force, the momentum of an object or system of objects is unchanged.

$$mv_{(\text{before event})} = mv_{(\text{after event})}$$

conserved Term applied to a physical quantity, such as momentum, energy, or electric charge, that remains unchanged during interactions.

constructive interference Combination of waves so that two or more waves overlap to produce a resulting wave of increased amplitude. See also *interference*.

convection Means of heat transfer by movement of the heated substance itself, such as by currents in a fluid.

converging lens Lens that is thicker in the middle than at the edges and refracts parallel rays of light passing through it to a focus. See also *diverging lens*.

convex mirror Mirror that curves outward. The virtual image formed is smaller and closer to the mirror than the object. See also *concave mirror*.

cornea Transparent covering over the eyeball, which helps focus the incoming light.

correspondence principle If a new theory is valid, it must account for the verified results of the old theory in the region where both theories apply.

cosmic ray One of various high-speed particles that travel throughout the universe and originate in violent events in stars.

cosmology Study of the origin and development of the entire universe.

coulomb (C) SI unit of electrical charge. One coulomb is equal to the total charge of 6.25×10^{18} electrons.

Coulomb's law Relationship among electrical force, charges, and distance: The electrical force between two charges varies directly as the product of the charges (q) and inversely as the square of the distance between them. (k is the proportionality constant 9×10^9 N · m^2/C^2) If the charges are alike in sign, the force is repulsive; if the charges are unlike, the force is attractive.

$$F = k\frac{q_1 q_2}{d^2}$$

crest Part of a wave that is highest or the disturbance is greatest. See also *trough*.

critical angle Minimum angle of incidence for which a light ray is totally reflected within a medium.

critical mass Minimum mass of fissionable material in a nuclear reactor or nuclear bomb that will sustain a chain reaction. A subcritical mass is one in which the chain reaction dies out. A supercritical mass is one in which the chain reaction builds up explosively.

crystal Regular geometric shape found in a solid in which the component particles are arranged in an orderly, three-dimensional, repeating pattern.

current See *electric current*.

cyclotron Particle accelerator that imparts high energy to charged particles such as protons, deuterons, and helium ions.

dB Abbreviation for decibel. See *bel*.

dc Abbreviation for *direct current*.

DDT Abbreviation for the chemical pesticide **d**ichloro **d**iphenyl **t**richloroethane.

de-excitation See *excitation*.

de Broglie matter waves All particles have wave properties; in de Broglie's equation, the product of momentum and wavelength equals Planck's constant.

decibel (dB) One-tenth of a *bel*.

density Mass of a substance per unit volume. Weight density is weight per unit volume. In general, any item per space element (e.g., number of dots per area).

$$\text{density} = \frac{\text{mass}}{\text{volume}}$$

$$\text{weight density} = \frac{\text{weight}}{\text{volume}}$$

destructive interference Combination of waves so that crest parts of one wave overlap trough parts of another, resulting in a wave of decreased amplitude. See also *interference*.

deuterium Isotope of hydrogen whose atom has a proton, a neutron, and an electron. The common isotope of hydrogen has only a proton and an electron; therefore, deuterium has more mass.

deuteron Nucleus of a deuterium atom; it has one proton and one neutron.

dichroic crystal Crystal that divides unpolarized light into two internal beams polarized at right angles and strongly absorbs one beam while transmitting the other.

diffraction Bending of light that passes around an obstacle or through a narrow slit, causing the light to spread and to produce light and dark fringes.

diffraction grating Series of closely spaced parallel slits or grooves that are used to separate colors of light by interference.

diffuse reflection Reflection of waves in many directions from a rough surface. See also *polished*.

digital audio Audio reproduction system that uses binary code to record and reproduce sound.

digital signal Signal made up of discrete quantities or signals, as opposed to an analog signal which is based on a continuous signal.

diode Electronic device that restricts current to a single direction in an electric circuit; a device that changes alternating current to direct current.

dipole See *electric dipole*.

direct current (dc) Electric current whose flow of charge is always in one direction.

dispersion Separation of light into colors arranged according to their frequency, for example by interaction with a prism or a diffraction grating.

displaced Term applied to the fluid that is moved out of the way when an object is placed in fluid. A submerged object always displaces a volume of fluid equal to its own volume.

diverging lens Lens that is thinner in the middle than at the edges, causing parallel rays of light passing through it to diverge as if from a point. See also *converging lens*.

Doppler effect Change in frequency of a wave of sound or light due to the motion of the source or the receiver. See also *red shift* and *blue shift*.

echo Reflection of sound.

eddy Changing, curling paths in turbulent flow of a fluid.

efficiency In a machine, the ratio of useful energy output to total energy input, or the percentage of the work input that is converted to work output.

$$\text{Efficiency} = \frac{\text{useful energy output}}{\text{total energy input}}$$

elastic collision Collision in which colliding objects rebound without lasting deformation or heat generation.

elastic limit Distance of stretching or compressing beyond which an elastic material will not return to its original state.

elasticity Property of a solid wherein a change in shape is experienced when a deforming force acts on it, with a return to its original shape when the deforming force is removed.

electric charge Fundamental electrical property to which the mutual attractions or repulsions between electrons or protons is attributed.

electric current Flow of electric charge that transports energy from one place to another. Measured in amperes, where one ampere is the flow of 6.25×10^{18} electrons (or protons) per second.

electric dipole Molecule in which the distribution of charge is uneven, resulting in slightly opposite charges on opposite sides of the molecule.

electric field Force field that fills the space around every electric charge or group of charges. Measured by force per charge (newtons/coulomb).

electric potential Electric potential energy (in joules) per unit of charge (in coulombs) at a location in an electric field; measured in volts and often called voltage.

$$\text{voltage} = \frac{\text{electrical energy}}{\text{charge}} = \frac{\text{joules}}{\text{coulomb}}$$

electric potential energy   Energy a charge has due to its location in an electric field.

electric power   Rate of electrical energy transfer or the rate of doing work, which can be measured by the product of current and voltage.

$$\text{power} = \text{current} \times \text{voltage}$$

electrical force   Force that one charge exerts on another. When the charges are the same sign, they repel; when the charges are opposite, they attract.

electrically polarized   Term applied to an atom or molecule in which the charges are aligned so that one side is slightly more positive or negative than the opposite side.

electrical resistance   Resistance of a material to the flow of electric charge through it; measured in ohms (symbol Ω).

electricity   General term for electrical phenomena, much like gravity has to do with gravitational phenomena, or sociology with social phenomena.

electrode   Terminal, for example of a battery, through which electric current can pass.

electrodynamics   Study of moving electric charge, as opposed to electrostatics.

electromagnet   Magnet whose magnetic properties are produced by electric current.

electromagnetic induction   Phenomenon of inducing a voltage in a conductor by changing the magnetic field near the conductor. If the magnetic field within a closed loop changes in any way, a voltage is induced in the loop. The induction of voltage is actually the result of a more fundamental phenomenon: the induction of an electric field. See also *Faraday's law*.

electromagnetic radiation   Transfer of energy by the rapid oscillations of electromagnetic fields, which travel in the form of waves called electromagnetic waves.

electromagnetic spectrum   Range of frequencies over which electromagnetic radiation can be propagated. The lowest frequencies are associated with radio waves; microwaves have a higher frequency, and then infrared waves, light, ultraviolet radiation, X rays, and gamma rays in sequence.

electromagnetic wave   Energy-carrying wave emitted by vibrating charges (often electrons) that is composed of oscillating electric and magnetic fields that regenerate one another. Radio waves, microwaves, infrared radiation, visible light, ultraviolet radiation, X rays, and gamma rays are all composed of electromagnetic waves.

electromotive force (emf)   Any voltage that gives rise to an electric current. A battery or a generator is a source of emf.

electron   Negative particle in the shell of an atom.

electron volt (eV)   Amount of energy equal to that an electron acquires in accelerating through a potential difference of 1 volt.

electrostatics   Study of electric charges at rest, as opposed to electrodynamics.

element   Substance composed of atoms that all have the same atomic number and, therefore, the same chemical properties.

elementary particles   Subatomic particles. The basic building blocks of all matter, consisting of two classes of particles, the quarks and the leptons.

ellipse   Closed curve of oval shape wherein the sum of the distances from any point on the curve to two internal focal points is a constant.

emf   Abbreviation for *electromotive force*.

emission spectrum   Distribution of wavelengths in the light from a luminous source.

energy   Anything that can change the condition of matter. Commonly defined circularly as the ability to do work; actually only describable (like pornography) by examples.

entropy   Measure of the degree of disorder in a substance or system.

equilibrium   In general, a state of balance. For mechanical equilibrium, the state in which no net forces and no net torques act. In liquids, the state in which evaporation equals condensation. More generally, the state in which no net change of energy occurs.

escape velocity   Velocity that a projectile, space probe, etc., must reach to escape the gravitational influence of the earth or celestial body to which it is attracted.

ether   Hypothetical invisible medium that was formerly thought to be required for the propagation of electromagnetic waves, and thought to fill space throughout the universe.

eV   Abbreviation for *electron volt*.

evaporation   Change of phase from liquid to gas that takes place at the surface of a liquid. The opposite of condensation.

excitation   Process of boosting one or more electrons in an atom or molecule from a lower to a higher energy level. An atom in an excited state will usually decay (de-excite) rapidly to a lower state by the emission of radiation. The frequency and energy of emitted radiation are related by

$$E = hf$$

excited   See *excitation*.

eyepiece   Lens of a telescope closest to the eye; which enlarges the real image formed by the first lens.

fact   Close agreement by competent observers of a series of observations of the same phenomena.

Fahrenheit scale   Temperature scale in common use in the United States. The number 32 is assigned to the melt-freeze point of water, and the number 212 to the boil-condense point of water at standard pressure (one atmosphere, at sea level).

Faraday's law   Induced voltage in a coil is proportional to the product of the number of loops and the rate at which the magnetic field changes within those loops. In general, an electric field is induced in any region of space in which a magnetic field is changing with time. The magnitude of the induced electric field is proportional to the rate at which the magnetic field changes. See also *Maxwell's counterpart to Faraday's law*.

$$\text{voltage induced} \sim \text{number of loops} \times \frac{\text{magnetic field change}}{\text{change in time}}$$

Fermat's principle of least time   Light takes the path that requires the least time when it goes from one place to another.

field   See *force field*.

field lines   See *magnetic field lines*.

flotation   See *principle of flotation*.

fluid   Anything that flows; in particular, any liquid or gas.

fluorescence   Property of certain substances to absorb radiation of one frequency and to re-emit radiation of a lower frequency.

FM   Abbreviation for *frequency modulation*.

focal length   Distance between the center of a lens and either focal point; the distance from a mirror to its focal point.

focal plane   Plane, perpendicular to the principal axis, that passes through a focal point of a lens or mirror. For a converging lens or a concave mirror, any incident parallel rays of light converge to a point somewhere on a focal plane. For a diverging lens or a convex mirror, the rays appear to come from a point on a focal plane.

focal point   For a converging lens or a concave mirror, the point at which rays of light parallel to the principal axis converge. For a diverging lens or a convex mirror, the point from which such rays appear to come.

focus   (pl. foci) (a) For an ellipse, one of the two points for which the sum of the distances to any point on the ellipse is a constant. A satellite orbiting the earth moves in an ellipse that has the earth at one focus. (b) For optics, a focal point.

force   Any influence that tends to accelerate an object; a push or pull; measured in newtons. Force is a vector quantity.

force field   That which exists in the space surrounding a mass, electric charge, or magnet, so that another mass, electric charge, or magnet introduced into this region will experience a force. Examples of force fields are gravitational fields, electric fields, and magnetic fields.

forced vibration   Vibration of an object caused by the vibrations of a nearby object. The sounding board in a musical instrument amplifies the sound through forced vibration.

Fourier analysis   Mathematical method that resolves any periodic wave form into a combination of simple sine waves.

fovea   Area of the retina that is in the center of the field of view; region of most distinct vision.

frame of reference   Vantage point (usually a set of coordinate axes) with respect to which position and motion may be described.

Fraunhofer lines   Dark lines visible in the spectrum of the sun or a star.

free fall   Motion under the influence of gravity only.

free radical   Unbonded, electrically neutral, very chemically active atom or molecular fragment.

freezing   Change in phase from liquid to solid; the opposite of melting.

frequency   For a vibrating body or medium, the number of vibrations per unit time. For a wave, the number of crests that pass a particular point per unit time. Frequency is measured in hertz.

frequency modulation (FM)   Type of modulation in which the frequency of the carrier wave is varied above and below its normal frequency by an amount that is proportional to the amplitude of the impressed signal. In this case, the amplitude of the modulated carrier wave remains constant.

friction   Force that acts to resist the relative motion (or attempted motion) of objects or materials that are in contact.

fulcrum   Pivot point of a lever.

fundamental frequency   See *partial tone*.

fuse   Device in an electric circuit that breaks the circuit when the current gets high enough to risk causing a fire.

g   (a) Abbreviation for *gram*. (b) When in lower-case italic *g*, the symbol for the acceleration due to gravity (at the earth's surface,

9.8 m/s^2). (c) When in lower case bold **g** the gravitational field vector (at the earth's surface, 9.8 N/kg). (d) When in upper-case italic *G*, the symbol for the *universal gravitation constant* (6.67×10^{-11} Nm^2/kg^2).

galvanometer   Instrument used to detect electric current. With the proper combination of resistors, it can be converted to an ammeter or a voltmeter. An ammeter is calibrated to measure electric current. A voltmeter is calibrated to measure electric potential.

gamma ray   High-frequency electromagnetic radiation emitted by atomic nuclei.

gas   Phase of matter beyond the liquid phase, wherein molecules fill whatever space is available to them, taking no definite shape.

general theory of relativity   Einstein's generalization of special relativity, which deals with accelerated motion and features a geometric theory of gravitation.

generator   Machine that produces electric current, usually by rotating a coil within a stationary magnetic field.

geodesic   Shortest path between points on any surface.

geosynchronous orbit   A satellite orbit in which the satellite orbits the earth once each day. When moving westward, the satellite remains at a fixed point (about 42,000 km) above the earth's surface.

global warming   See *greenhouse effect*.

gram (g)   A metric unit of mass. It is one thousandth of a kilogram, and very nearly the mass of one cubic centimeter of water at a temperature of 4° Celsius.

gravitation   Attraction between objects due to mass. See also *law of universal gravitation* and *universal gravitational constant*.

gravitational field   Force field that exists in the space around every mass or group of masses; measured in newtons per kilogram.

gravitational potential energy   Energy that a body possesses because of its position in a gravitational field. On earth, potential energy (PE) equals mass (m) times the acceleration due to gravity (g) times height (h) from a reference level such as the earth's surface.

$$PE = mgh$$

gravitational red shift   Shift in wavelength toward the red end of the spectrum experienced by light leaving the surface of a massive object, as predicted by the general theory of relativity.

gravitational wave   Gravitational disturbance that propagates through spacetime made by a moving mass. (Undetected at this writing.)

graviton   Quantum of gravity, similar in concept to the photon as a quantum of light. (Undetected at this writing.)

greenhouse effect   Warming effect caused by short-wavelength radiant energy from the sun that easily enters the atmosphere and is absorbed by the earth, but when radiated at longer wavelengths cannot easily escape the earth's atmosphere.

grounding   Allowing charges to move freely along a connection from a conductor to the ground.

group   Elements in the same column of the periodic table.

h   (a) Abbreviation for hour (though hr. is often used). (b) When in italic *h*, the symbol for *Planck's constant*.

hadron   Elementary particle that can participate in strong nuclear force interactions.

half-life   Time required for half the atoms of a radioactive isotope of an element to decay. This term is also used to describe decay processes in general.

harmonic   See *partial tone*.

heat capacity   See *specific heat capacity*.

heat engine   Device that changes internal energy to mechanical work.

heat of fusion   Amount of energy that must be added to a kilogram of a solid (already at its melting point) to melt it.

heat of vaporization   Amount of energy that must be added to a kilogram of a liquid (already at its boiling point) to vaporize it.

heat   The energy that flows from one object to another by virtue of a difference in temperature. Measured in *calories* or *joules*.

heat waves   See *infrared waves*.

heavy water   Water (H_2O) that contains the heavy hydrogen isotope deuterium.

hertz (Hz)   SI unit of frequency. One hertz is one vibration per second.

hologram   Two-dimensional microscopic interference pattern that shows three-dimensional optical images.

Hooke's law   Distance of stretch or squeeze (extension or compression) of an elastic material is directly proportional to the applied force. Where Δx is the change in length and k is the spring constant,

$$F = k\Delta x$$

humidity   Measure of the amount of water vapor in the air. Absolute humidity is the mass of water per volume of air. Relative humidity is absolute humidity at that temperature divided by the maximum possible, usually given as a percent.

Huygens' principle   Light waves spreading out from a light source can be regarded as a superposition of tiny secondary wavelets.

hypothesis   Educated guess; a reasonable explanation of an observation or experimental result that is not fully accepted as factual until tested over and over again by experiment.

Hz   Abbreviation for *hertz*.

ideal efficiency   Upper limit of efficiency for all heat engines; it depends on the temperature difference between input and exhaust.

$$\text{ideal efficiency} = \frac{T_{\text{hot}} - T_{\text{cold}}}{T_{\text{hot}}}$$

impulse   Product of force and the time interval during which the force acts. Impulse produces change in momentum.

$$\text{impulse} = Ft = \Delta(mv)$$

incandescence   State of glowing while at a high temperature, caused by electrons bouncing around over dimensions larger than the size of an atom, emitting radiant energy in the process. The peak frequency of radiant energy is proportional to the absolute temperature of the heated substance:

$$\bar{f} \sim T$$

incoherent light   Light containing waves with a jumble of frequencies, phases, and possibly directions. See also *coherent light* and *laser*.

index of refraction (n)   Ratio of the speed of light in a vacuum to the speed of light in another material.

$$\text{n} = \frac{\text{speed of light in vacuum}}{\text{speed of light in material}}$$

induced   (a) Term applied to electric charge that has been redistributed on an object due to the presence of a charged object nearby. (b) Term applied to a voltage, electric field, or magnetic field that is created due to a change in or motion through a magnetic field or electric field.

induction   Charging of an object without direct contact. See also *electromagnetic induction*.

inelastic   Term applied to a material that does not return to its original shape after it has been stretched or compressed.

inelastic collision   Collision in which the colliding objects become distorted and/or generate heat during the collision, and possibly stick together.

inertia   Sluggishness or apparent resistance of an object to change its state of motion. Mass is the measure of inertia.

inertial frame of reference   Unaccelerated vantage point in which Newton's laws hold exactly.

infrared   Electromagnetic waves of frequencies lower than the red of visible light.

infrared waves   Electromagnetic waves that have a lower frequency than visible red light.

infrasonic   Term applied to sound frequencies below 20 hertz, the normal lower limit of human hearing.

in parallel   Term applied to portions of an electric circuit that are connected at two points and provide alternative paths for the current between those two points.

in phase   Term applied to two or more waves whose crests (and troughs) arrive at a place at the same time, so that their effects reinforce each other.

in series   Term applied to portions of an electric circuit that are connected in a row so that the current that goes through one must go through all of them.

instantaneous speed   Speed at any instant.

interaction   Mutual action between objects where each object exerts an equal and opposite force on the other.

insulator   (a) Material that is a poor conductor of heat and that delays the transfer of heat. (b) Material that is a poor conductor of electricity.

intensity   Power per square meter carried by a sound wave, often measured in decibels.

interference   Result of superposing different waves, often of the same wavelength. Constructive interference results from crest-to-crest reinforcement; destructive interference results from crest-to-trough cancellation. The interference of selected wavelengths of light produces colors known as interference colors. See also *constructive interference, destructive interference, interference pattern,* and *standing wave*.

interference pattern   Pattern formed by the overlapping of two or more waves that arrive in a region at the same time.

interferometer   Device that uses the interference of light waves to measure very small distances with high accuracy. Michelson and Morley used an interferometer in their famous experiments with light.

internal energy   The total energy stored in the atoms and molecules within a substance. Changes in internal energy are of principal concern in thermodynamics.

inverse-square law   Law relating the intensity of an effect to the inverse square of the distance from the cause. Gravity, electric, magnetic, light, sound, and radiation phenomena follow the inverse-square law.

$$\text{intensity} \sim \frac{1}{\text{distance}^2}$$

inversely When two values change in opposite directions, so that if one increases and the other decreases by the same amount, they are said to be inversely proportional to each other.

ion Atom (or group of atoms bound together) with a net electric charge, which is due to the loss or gain of electrons. A positive ion has a net positive charge. A negative ion has a net negative charge.

ionization Process of adding or removing electrons to or from the atomic nucleus.

iridescence Phenomenon whereby interference of light waves of mixed frequencies reflected from the top and bottom of thin films produces an assortment of colors.

iris Colored part of the eye that surrounds the black opening through which light passes. The iris regulates the amount of light entering the eye.

isotopes Atoms whose nuclei have the same number of protons but different numbers of neutrons.

J Abbreviation for *joule*.

joule (J) SI unit of work and of all other forms of energy. One joule of work is done when a force of one newton is exerted on an object moved one meter in the direction of the force.

K (a) Abbreviation for *kelvin*. (b) When in lower case k, the abbreviation for the prefix *kilo-*. (c) When in lower case italics *k*, the symbol for the electrical proportionality constant in *Coulomb's law*. It is approximately 9×10^9 N $\cdot$ m^2/C^2. (d) When in lower case italics *k*, the symbol for the spring constant in *Hooke's law*.

kcal Abbreviation for *kilocalorie*.

KE Abbreviation for *kinetic energy*.

Kelvin scale Temperature scale, measured in kelvins K, whose zero (called absolute zero) is the temperature at which it is impossible to extract any more internal energy from a material. 0 K = -273.15°C. There are no negative temperatures on the Kelvin scale.

kelvin SI unit of temperature. A temperature measured in kelvins (symbol K) indicates the number of units above absolute zero. Divisions on the Kelvin scale and Celsius scale are the same size, so a change in temperature of one kelvin equals a change in temperature of one Celsius degree.

Kepler's laws of planetary motion

Law 1: Each planet moves in an elliptical orbit with the sun at one focus.

Law 2: The line from the sun to any planet sweeps out equal areas of space in equal time intervals.

Law 3: The squares of the times of revolution (days, months, or years) of the planets are proportional to the cubes of their average distances from the sun (T$^2 \sim$ R^3 for all planets).

kg Abbreviation for *kilogram*.

kilo- Prefix that means thousand, as in kilowatt or kilogram.

kilocalorie (kcal) Unit of heat. One kilocalorie equals 1000 calories, or the amount of heat required to raise the temperature of one kilogram of water by 1°C. Equal to one food Calorie.

kilogram (kg) Fundamental SI unit of mass. It is equal to 1000 grams. One kilogram is very nearly the amount of mass in one liter of water at 4°C.

kilometer (km) One thousand *meters*.

kilowatt (kW) One thousand *watts*.

kilowatt-hour (kWh) Amount of energy consumed in 1 hour at the rate of 1 kilowatt.

kinetic energy (KE) Energy of motion, equal (nonrelativistically) to half the mass multiplied by the speed squared.

$$KE = \tfrac{1}{2}mv^2$$

km Abbreviation for *kilometer*.

kPa Abbreviation for kilopascal. See *pascal*.

kWh Abbreviation for *kilowatt-hour*.

L Abbreviation for *liter*. (In some textbooks, lowercase l is used.)

laser Optical instrument that produces a beam of coherent light—that is, light with all waves of the same frequency, phase, and direction. The word is an acronym for **l**ight **a**mplification by **s**timulated **e**mission of **r**adiation.

law General hypothesis or statement about the relationship of natural quantities that has been tested over and over again and has not been contradicted. Also known as a *principle*.

law of inertia See *Newton's laws of motion, Law 1*.

law of reflection The angle of incidence for a wave that strikes a surface is equal to the angle of reflection. This is true for both partially and totally reflected waves. See also *angle of incidence* and *angle of reflection*.

law of universal gravitation For any pair of particles (or spherical objects), each particle attracts the other particle with a force that is directly proportional to the product of the masses of the particles, and inversely proportional to the square of the distance between their centers of mass. Where F is the force, m is the mass, d is distance, and G is the gravitation constant:

$$F \sim \frac{m_1 m_2}{d^2} \quad \text{or} \quad F = G\frac{m_1 m_2}{d^2}$$

least time See *Fermat's principle of least time*.

length contraction Shrinkage of space, and therefore of matter, in a frame of reference moving at relativistic speeds.

lens Piece of glass or other transparent material that can bring light to a focus.

lepton Class of elementary particles that are not involved with the nuclear force. It includes the electron and its neutrino, the muon and its neutrino, and the tau and its neutrino.

lever Simple machine made of a bar that turns about a fixed point called the fulcrum.

lever arm Perpendicular distance between an axis and the line of action of a force that tends to produce rotation about that axis.

lift In application of Bernoulli's principle, the net upward force produced by the difference between upward and downward pressures. When lift equals weight, horizontal flight is possible.

light Visible part of the electromagnetic spectrum.

light-year Distance light travels in a vacuum in one year: 9.46×10^{12} km.

line spectrum Pattern of distinct lines of color, corresponding to particular wavelengths, that are seen in a spectroscope when a hot gas is viewed. Each element has a unique pattern of lines.

linear momentum Product of the mass and the velocity of an object. Also called momentum. (This definition applies at speeds much less than the speed of light.)

linear motion Motion along a straight-line path.

linear speed   Path distance moved per unit of time. Also called simply speed.

liquid   Phase of matter between the solid and gaseous phases in which the matter possesses a definite volume but no definite shape: it takes on the shape of its container.

liter (L)   Metric unit of volume. A liter is equal to 1000 cm³.

logarithmic   Exponential.

longitudinal wave   Wave in which the individual particles of a medium vibrate back and forth in the direction in which the wave travels — for example, sound.

Lorentz contraction   See *length contraction*.

loudness   Physiological sensation directly related to sound intensity or volume. Relative loudness, or sound level, is measured in decibels.

lunar eclipse   Event wherein the full moon passes into the shadow of the earth.

m   (a) Abbreviation for *meter*. (b) When in italic *m*, the abbreviation for *mass*.

Mach number   Ratio of the speed of an object to the speed of sound. For example, an aircraft traveling *at* the speed of sound is rated Mach 1.0; traveling at *twice* the speed of sound, Mach 2.0.

machine   Device for increasing (or decreasing) a force or simply changing the direction of a force.

magnet   Any object that has magnetic properties, that is the ability to attract objects made of iron or other magnetic substances. See also *electromagnetism* and *magnetic force*.

magnetic declination   Discrepancy between the orientation of a compass pointing toward magnetic north and the true geographic north.

magnetic domain   Microscopic cluster of atoms with their magnetic fields aligned.

magnetic field   Region of magnetic influence around a magnetic pole or a moving charged particle.

magnetic field lines   Lines showing the shape of a magnetic field. A compass placed on such a line will turn so that the needle is aligned with it.

magnetic force   (a) Between magnets, it is the attraction of unlike magnetic poles for each other and the repulsion between like magnetic poles. (b) Between a magnetic field and a moving charged particle, it is a deflecting force due to the motion of the particle: The deflecting force is perpendicular to the magnetic field lines and the direction of motion. This force is greatest when the charged particle moves perpendicular to the field lines and is smallest (zero) when it moves parallel to the field lines.

magnetic monopole   Hypothetical particle having a single north or a single south magnetic pole, analogous to a positive or negative electric charge.

magnetic pole   One of the regions on a magnet that produces magnetic forces.

magnetic pole reversal   When the magnetic field of an astronomical body reverses its poles, that is, the location where the north magnetic pole existed becomes the south magnetic pole, and the south magnetic pole becomes the north magnetic pole.

magnetism   Property of being able to attract objects made of iron, steel, or magnetite. See also *electromagnetism* and *magnetic force*.

magnetohydrodynamic (MDH) generator   Device generating electric power by interaction of a plasma and a magnetic field.

maser   Instrument that produces a beam of microwaves. The word is an acronym for **m**icrowave **a**mplification by **s**timulated **e**mission of **r**adiation.

mass (*m*)   Quantity of matter in an object; the measurement of the inertia or sluggishness that an object exhibits in response to any effort made to start it, stop it, or change in any way its state of motion; a form of energy.

mass spectrometer   Device that magnetically separates charged ions according to their mass.

mass-energy equivalence   Relationship between mass and energy as given by the equation

$$E = mc^2$$

where *c* is the speed of light.

matter waves   See *de Broglie matter waves*.

Maxwell's counterpart to Faraday's law   A magnetic field is created in any region of space in which an electric field is changing with time. The magnitude of the induced magnetic field is proportional to the rate at which the electric field changes. The direction of the induced magnetic field is at right angles to the changing electric field.

mechanical advantage   Ratio of output force to input force for a machine.

mechanical energy   Energy due to the position or the movement of something; potential or kinetic energy (or a combination of both).

mechanical equilibrium   State of an object or system of objects for which any impressed forces cancel to zero and no acceleration occurs.

mega-   Prefix that means million, as in megahertz or megajoule.

melting   Change in phase from solid to liquid; the opposite of freezing. Melting is a different process from dissolving, in which an added solid mixes with a liquid and the solid dissociates.

meson   Elementary particle with an atomic weight of zero; can participate in the strong interaction.

metastable state   State of excitation of an atom that is characterized by a prolonged delay before de-excitation.

meter (m)   Standard SI unit of length (3.28 feet).

MeV   Abbreviation for million *electron volts*, a unit of energy, or equivalently a unit of mass.

MHD   Abbreviation for *magnetohydrodynamic*.

mi   Abbreviation for mile.

microscope   Optical instrument that forms enlarged images of very small objects.

microwaves   Electromagnetic waves with frequencies greater than radio waves but less than infrared waves.

min   Abbreviation for minute.

mirage   False image that appears in the distance and is due to the refraction of light in the earth's atmosphere.

mixture   Substances mixed together without combining chemically.

MJ   Abbreviation for megajoules, million *joules*.

model   Representation of an idea created to make the idea more understandable.

modulation   Impressing a signal wave system on a higher frequency carrier wave, amplitude modulation (AM) for amplitude signals and frequency modulation (FM) for frequency signals.

molecule   Two or more atoms of the same or different elements bonded to form a larger particle.

momentum   Inertia in motion. The product of the mass and the velocity of an object (provided the speed is much less than the speed of light). Has magnitude and direction and therefore is a vector quantity. Also called linear momentum, and abbreviated p.

$$p = mv$$

monochromatic light   Light made of only one color and therefore waves of only one wavelength and frequency.

muon   Elementary particle in the class of elementary particles called leptons. It is short-lived with a mass that is 207 times that of an electron; may be positively or negatively charged.

music   Scientifically speaking, sound with periodic tones, which appear on an oscilloscope as a regular pattern.

N   Abbreviation for *newton*.

nanometer   Metric unit of length that is 10^{-9} meter (one billionth of a meter).

natural frequency   Frequency at which an elastic object naturally tends to vibrate if it is disturbed and the disturbing force is removed.

neap tide   Tide that occurs when the moon is halfway between a new moon and a full moon, in either direction. The tides due to the sun and the moon partly cancel, so that the high tides are lower than average and the low tides are not as low as average. See also *spring tide*.

net force   Combination of all the forces that act on an object.

neutrino   Elementary particle in the class of elementary particles called leptons. It is uncharged and almost massless; three kinds — electron, muon, and tau neutrinos, are the most common high-speed particle in the universe; more than a billion pass unhindered through each person every second.

neutron   Electrically neutral particle that is one of the two kinds of nucleons that compose an atomic nucleus.

neutron star   Star that has undergone a gravitational collapse in which electrons are compressed into protons to form neutrons.

newton (N)   SI unit of force. One newton is the force applied to a one-kilogram mass that will produce an acceleration of one meter per second per second.

Newton's law of cooling   The rate of cooling of an object—whether by conduction, convection, or radiation—is approximately proportional to the temperature difference between the object and its surroundings.

Newton's laws of motion

Law 1: Every body continues in its state of rest, or of motion in a straight line at constant speed, unless it is compelled to change that state by a net force exerted upon it. Also known as the law of inertia.
Law 2: The acceleration produced by a net force on a body is directly proportional to the magnitude of the net force, is in the same direction as the net force, and is inversely proportional to the mass of the body.
Law 3: Whenever one body exerts a force on a second body, the second body exerts an equal and opposite force on the first.

node   Any part of a standing wave that remains stationary; a region of minimal or zero energy.

noise   Scientifically speaking, sound that corresponds to an irregular vibration of the eardrum produced by some irregular vibration, which appears on an oscilloscope as an irregular pattern.

nonlinear motion   Any motion not along a straight-line path.

normal   At right angles to, or perpendicular to. A normal force acts at right angles to the surface on which it acts. In optics, a normal defines the line perpendicular to a surface about which angles of light rays are measured.

normal force   Component of support force perpendicular to a supporting surface. For an object resting on a horizontal surface, it is the upward force that balances the weight of the object.

northern lights   See *aurora borealis*.

nuclear fission   Splitting of an atomic nucleus, particularly that of a heavy element such as uranium-235, into two lighter elements, accompanied by the release of much energy.

nuclear force   Attractive force within a nucleus that holds neutrons and protons together. Part of the nuclear force is called the strong interaction. The strong interaction is an attractive force that acts between protons, neutrons, and mesons (another nuclear particle); however, it acts only over very short distances (10^{-15} meter). The weak interaction is the nuclear force responsible for beta (electron) emission.

nuclear fusion   Combining of nuclei of light atoms, such as hydrogen, into heavier nuclei, accompanied by the release of much energy. See also *thermonuclear fusion*.

nuclear reactor   Apparatus in which controlled nuclear fission or nuclear fusion reactions take place.

nucleon   Principal building block of the nucleus. A neutron or a proton; the collective name for either or both.

nucleus   (pl. nuclei) Positively charged center of an atom, which contains protons and neutrons and has almost all the mass of the entire atom but only a tiny fraction of the volume.

objective lens   In an optical device using compound lenses, the lens closest to the object observed.

octave   In music, the eighth full tone above or below a given tone. The tone an octave above has twice as many vibrations per second as the original tone; the tone an octave below has half as many vibrations per second.

ohm (Ω)   SI unit of electrical resistance. One ohm is the resistance of a device that draws a current of one ampere when a voltage of one volt is impressed across it.

Ohm's law   Current in a circuit is directly proportional to the voltage impressed across the circuit, and is inversely proportional to the resistance of the circuit.

$$\text{Current} = \frac{\text{voltage}}{\text{resistance}}$$

opaque   Term applied to materials that absorb light without re-emission, and consequently do not allow light through them.

optical fiber   Transparent fiber, usually of glass or plastic, that can transmit light down its length by means of total internal reflection.

oscillation   Same as vibration: a repeating to-and-fro motion about an equilibrium position. Both oscillation and vibration refer to periodic motion, that is, motion that repeats.

oscillatory motion   To-and-fro vibratory motion, such as that of a pendulum.

out of phase   Term applied to two waves for which the crest of one wave arrives coincident with a trough of the second wave. Their effects tend to cancel each other.

overtone   Musical term where the first overtone is the second harmnic. See also *partial tone*.

oxidize   Chemical process in which an element or molecule loses one or more electrons.

ozone Gas, found in a thin layer in the upper atmosphere, composed of molecules of three oxygen atoms. Atmospheric oxygen gas is composed of molecules of two oxygen atoms.

Pa Abbreviation for the SI unit *pascal.*

parabola Curved path followed by a projectile acting under the influence of gravity only.

parallax Apparent displacement of an object when viewed by an observer from two different positions; often used to calculate the distance of stars.

parallel circuit Electric circuit with two or more devices connected in such a way that the same voltage acts across each one and any single one completes the circuit independently of the others. See also *in parallel.*

partial tone One of the many tones that make up one musical sound. Each partial tone (or partial) has only one frequency. The lowest partial of a musical sound is called the fundamental frequency. Any partial whose frequency is a multiple of the fundamental frequency is called a harmonic. The fundamental frequency is also called the first harmonic. The second harmonic has twice the frequency of the fundamental; the third harmonic, three times the frequency, and so on.

pascal (Pa) SI unit of pressure. One pascal of pressure exerts a normal force of one newton per square meter. A kilopascal (kPa) is 1000 pascals.

Pascal's principle Changes in pressure at any point in an enclosed fluid at rest are transmitted undiminished to all points in the fluid and act in all directions.

PE Abbreviation for *potential energy.*

penumbra Partial shadow that appears where some of the light is blocked and other light can fall. See also *umbra.*

percussion In musical instruments, the striking of one object against another.

perigee Point in an elliptical orbit closest to the focus about which orbiting takes place. See also *apogee.*

period In general, the time required to complete a single cycle. (a) For orbital motion, the time required for a complete orbit. (b) For vibrations or waves, the time required for one complete cycle, equal to 1/frequency.

periodic table Chart that lists elements by atomic number and by electron arrangements, so that elements with similar chemical properties are in the same column (group). See foldout following page 202.

perturbation Deviation of an orbiting object (e.g., a planet) from its path around a center of force (e.g., the sun) caused by the action of an additional center of force (e.g., another planet).

phase (a) One of the four main forms of matter: solid, liquid, gas, and plasma. Often called *state.* (b) The part of a cycle that a wave has advanced at any moment. See also *in phase* and *out of phase.*

phosphor Powdery material, such as that used on the inner surface of a fluorescent light tube, that absorbs ultraviolet photons, then gives off visible light.

phosphorescence Type of light emission that is the same as fluorescence except for a delay between excitation and de-excitation, which provides an afterglow. The delay is caused by atoms being excited to energy levels that do not decay rapidly. The afterglow may last from fractions of a second to hours, or even days, depending on such factors as the type of material and temperature.

photoelectric effect Ejection of electrons from certain metals when exposed to certain frequencies of light.

photon Localized corpuscle of electromagnetic radiation whose energy is proportional to its radiation frequency: $E \sim f$, or $E = hf$, where h is Planck's constant.

pigment Fine particles that selectively absorb light of certain frequencies and selectively transmit others.

pitch Term that refers to our subjective impression about the "highness" or "lowness" of a tone, which is related to the frequency of the tone. A high-frequency vibrating source produces a sound of high pitch; a low-frequency vibrating source produces a sound of low pitch.

Planck's constant (h) Fundamental constant of quantum theory that determines the scale of the small-scale world. Planck's constant multiplied by the frequency of radiation gives the energy of a photon of that radiation.

$$E = hf, \quad h = 6.6 \times 10^{-34} \text{ joule-second}$$

plane mirror Flat-surfaced mirror.

plane-polarized wave A wave confined to a single plane.

plasma Fourth phase of matter, in addition to solid, liquid, and gas. In the plasma phase, existing mainly at high temperatures, matter consists of positively charged ions and free electrons.

polarization Aligning of vibrations in a transverse wave, usually by filtering out waves of other directions. See also *plane-polarized wave* and *dichroic crystal.*

polished Describes a surface that is so smooth that the distances between successive elevations of the surface are less than about one-eighth the wavelength of the light or other incident wave of interest. The result is very little diffuse reflection.

positron Antiparticle of an electron; a positively charged electron.

postulates of special relativity

First: All laws of nature are the same in all uniformly moving frames of reference.

Second: The speed of light in free space has the same measured value regardless of the motion of the source or the motion of the observer; that is, the speed of light is invariant.

potential difference Difference in electric potential (voltage) between two points. Free charge flows when there is a difference, and will continue until both points reach a common potential.

potential energy (PE) Energy of position, usually related to the relative position of two things, such as a stone and the earth (gravitational PE), or an electron and a nucleus (electric PE).

power Rate at which work is done or energy is transformed, equal to the work done or energy transformed divided by time; measured in watts.

$$\text{power} = \frac{\text{work}}{\text{time}}$$

precession Wavering of a spinning object, such that its axis of rotation traces out a cone.

pressure Force per surface area where the force is normal (perpendicular) to the surface; measured in pascals. See also *atmospheric pressure.*

$$\text{pressure} = \frac{\text{force}}{\text{area}}$$

primary colors See *additive primary colors* and *subtractive primary colors.*

principal axis Line joining the centers of curvature of the surfaces of a lens. Line joining the center of curvature and the focus of a mirror.

principle General hypothesis or statement about the relationship of natural quantities that has been tested over and over again and has not been contradicted; also known as a law.

principle of equivalence Observations made in an accelerating frame of reference are indistinguishable from observations made in a gravitational field.

principle of flotation A floating object displaces a quantity of fluid of a weight equal to its own weight.

prism Triangular solid of a transparent material such as glass, that separates incident light by refraction into its component colors. These component colors are often called the spectrum.

projectile Any object that moves through the air or through space, acted on only by gravity (and air resistance, if any).

proton Positively charged particle that is one of the two kinds of nucleons in the nucleus of an atom.

pulley Wheel that acts as a lever used to change the direction of a force. A pulley or system of pulleys can also multiply forces.

pupil Opening in the eyeball through which light passes.

quality Characteristic timbre of a musical sound, governed by the number and relative intensities of partial tones.

quantum (pl. quanta) From the Latin word *quantus*, meaning "how much," a quantum is the smallest elemental unit of a quantity, the smallest discrete amount of something. One quantum of electromagnetic energy is called a photon. See also *quantum mechanics* and *quantum theory.*

quantum mechanics Branch of physics concerned with the atomic microworld based on wave functions and probabilities, introduced by Max Planck (1900) and developed by Werner Heisenberg (1925), Erwin Schrödinger (1926), and others.

quantum physics Branch of physics that is the general study of the microworld of photons, atoms, and nuclei.

quantum theory Theory that describes the microworld, where many quantities are granular (in units called quanta), rather than continuous, and where particles of light (photons) and particles of matter (such as electrons) exhibit wave as well as particle properties.

quark One of the two classes of elementary particles. (The other is the lepton.) Two of the six quarks (up and down) are the fundamental building blocks of nucleons (protons and neutrons).

rad Unit used to measure a dose of radiation; the amount of energy (in centijoules) absorbed from ionizing radiation per kilogram of exposed material.

radiant energy Any energy, including heat, light, and X rays, that is transmitted by radiation. It occurs in the form of electromagnetic waves.

radiation (a) Energy transmitted by electromagnetic waves. (b) The particles given off by radioactive atoms such as uranium. Do not confuse radiation with radioactivity.

radiation curve of sunlight See *solar radiation curve.*

radio waves Electromagnetic waves of the longest frequency.

radioactive Term applied to an atom having an unstable nucleus that can spontaneously emit a particle and become the nucleus of another element.

radioactivity Process of the atomic nucleus that results in the emission of energetic particles. See *radiation.*

radiotherapy Use of radiation as a treatment to kill cancer cells.

rarefaction Region of reduced pressure in a longitudinal wave.

rate How fast something happens or how much something changes per unit of time; a change in a quantity divided by the time it takes for the change to occur.

ray Thin beam of light. Also lines drawn to show light paths in optical ray diagrams.

reaction force Force that is equal in strength and opposite in direction to the action force, and one that acts simultaneously on whatever is exerting the action force. See also *Newton's third law.*

real image Image formed by light rays that converge at the location of the image. A real image, unlike a virtual image, can be displayed on a screen.

red shift Decrease in the measured frequency of light (or other radiation) from a receding source; called the *red shift* because the decrease is toward the low-frequency, or red, end of the color spectrum. See also *Doppler effect.*

reflection Return of light rays from a surface in such a way that the angle at which a given ray is returned is equal to the angle at which it strikes the surface. When the reflecting surface is irregular, the light is returned in irregular directions; this is *diffuse reflection.* In general, the bouncing back of a particle or wave that strikes the boundary between two media.

refraction Bending of an oblique ray of light when it passes from one transparent medium to another. This is caused by a difference in the speed of light in the transparent media. In general, the change in direction of a wave as it crosses the boundary between two media in which the wave travels at different speeds.

regelation Process of melting under pressure and the subsequent refreezing when the pressure is removed.

relative Regarded in relation to something else, depending on point of view, or frame of reference. Sometimes referred to as "with respect to."

relative humidity Ratio between how much water vapor is in the air and the maximum amount of water vapor that could be in the air at the same temperature.

relativistic Pertaining to the theory of relativity; or approaching the speed of light.

relativity See *special theory of relativity, postulates of the special theory of relativity,* and *general theory of relativity.*

rem Acronym of **r**oentgen **e**quivalent **m**an, it is a unit used to measure the effect of ionizing radiation on human beings.

resistance See *electrical resistance.*

resistor Device in an electric circuit designed to resist the flow of charge.

rest energy The "energy of being" given by the equation $E = mc^2$.

resolution (a) Method of separating a vector into its component parts. (b) Ability of an optical system to make clear or to separate the components of an object viewed.

resonance Phenomenon that occurs when the frequency of forced vibrations on an object matches the object's natural frequency, producing a dramatic increase in amplitude.

resultant Net result of a combination of two or more vectors.

retina   Layer of light-sensitive tissue at the back of the eye, composed of tiny light-sensitive antennae called rods and cones. Rods sense light and darkness. Cones sense color.

reverberation   Persistence of a sound, as in an echo, due to multiple reflections.

revolution   Motion of an object turning around an axis that lies outside the object.

Ritz combination principle   For an element, the frequencies of some spectral lines are either the sum or the difference of the frequencies of two other lines in that element's spectrum.

rods   See *retina*.

rotation   Spinning motion that occurs when an object rotates about an axis located within the object (usually an axis through its center of mass).

rotational inertia   Reluctance or apparent resistance of an object to change its state of rotation, determined by the distribution of the mass of the object and the location of the axis of rotation or revolution.

rotational speed   Number of rotations or revolutions per unit of time; often measured in rotations or revolutions per second or minute.

rotational velocity   Rotational speed together with a direction for the axis of rotation or revolution.

RPM   Abbreviation for rotations or revolutions per minute.

s   Abbreviation for second.

satellite   Projectile or smaller celestial body that orbits a larger celestial body.

saturated   Term applied to a substance, such as air, that contains the maximum amount of another substance, such as water vapor, at a given temperature and pressure.

scalar quantity   Quantity in physics, such as mass, volume, and time, that can be completely specified by its magnitude, and has no direction.

scale   In music, a succession of notes of frequencies that are in simple ratios to one another.

scaling   Study of how size affects the relationship among weight, strength, and surface area.

scatter   To absorb sound or light and re-emit it in all directions.

scattering   Emission in random directions of light that encounters particles that are small compared to the wavelength of light; more often at short wavelengths (blue) than at long wavelengths (red).

Schrödinger's wave equation   Fundamental equation of quantum mechanics, which interprets the wave nature of material particles in terms of probability wave amplitudes. It is as basic to quantum mechanics as Newton's laws of motion are to classical mechanics.

scientific method   Orderly method for gaining, organizing, and applying new knowledge.

self-induction   Induction of an electric field within a single coil, caused by the interaction of the loops within the same coil. This self-induced voltage is always in a direction opposing the changing voltage that produces it, and is commonly called back electromotive force or back emf.

semiconductor   Device made of material not only with properties that fall between a conductor and an insulator but with resistance that changes abruptly when other conditions change, such as temperature, voltage, and electric or magnetic field.

series circuit   Electric circuit with devices connected in such a way that the electric current through each of them is the same. See also *in series*.

shadow   Shaded region that appears where light rays are blocked by an object.

shell model of the atom   Model in which the electrons of an atom are pictured as grouped in concentric shells around the nucleus.

shock wave   Cone-shaped wave produced by an object moving at supersonic speed through a fluid.

short circuit   Disruption in an electric circuit, caused by the flow of charge along a low-resistance path between two points that should not be directly connected, thus deflecting the current from its proper path; an effective "shortening of the circuit."

SI   Abbreviation for Système International, an international system of units of metric measure accepted and used by scientists throughout the world. See Appendix A for more details.

simple harmonic motion   Vibratory or periodic motion, like that of a pendulum, in which the force acting on the vibrating body is proportional to its displacement from its central equilibrium position and acts toward that position.

simultaneity   Occurring at the same time. Two events that are simultaneous in one frame of reference need not be simultaneous in a frame moving relative to the first frame.

sine curve   Curve whose shape represents the crests and troughs of a wave, as traced out by a pendulum that drops a trail of sand while swinging at right angles to and over a moving conveyor belt.

sine wave   The simplest of waves with only one frequency and the shape of a sine curve.

sliding friction   Contact force produced by the rubbing together of the surface of a moving object with the material over which it slides.

solar constant 1400 J/m^2 received from the sun each second at the top of the earth's atmosphere, expressed in terms of power, 1.4 kW/m^2.

solar eclipse   Event wherein the moon blocks light from the sun and the moon's shadow falls on part of the earth.

solar power   Energy per unit time derived from the sun. See also *solar constant*.

solar radiation curve   Graph of brightness versus frequency (or wavelength) of sunlight.

solid   Phase of matter characterized by definite volume and shape.

solidify   To become solid, as in freezing or the setting of concrete.

sonic boom   Loud sound resulting from the incidence of a shock wave.

sound   Longitudinal wave phenomenon that consists of successive compressions and rarefactions of the medium through which the wave travels.

sound barrier   The pile up of sound waves in front of an aircraft approaching or reaching the speed of sound, believed in the early days of jet aircraft to create a barrier of sound that a plane would have to break through in order to go faster than the speed of sound. The sound barrier does not exist.

spacetime   Four-dimensional continuum in which all events take place and all things exist: Three dimensions are the coordinates of space and the fourth is of time.

special theory of relativity   Comprehensive theory of space and time that replaces Newtonian mechanics when velocities are very large.

Introduced in 1905 by Albert Einstein. See also *postulates of the special theory of relativity.*

specific heat capacity   Quantity of heat required to raise the temperature of a unit mass of a substance by one degree Celsius (or equivalently, by one kelvin). Often simply called specific heat.

spectral lines   Colored lines that form when light is passed through a slit and then through a prism or diffraction grating, usually in a spectroscope. The pattern of lines is unique for each element.

spectrometer   See *spectroscope.*

spectroscope   An optical instrument that separates light into its constituent frequencies or wavelengths in the form of spectral lines. A *spectrometer* is an instrument that can also measure the frequencies or wavelengths.

spectrum   (pl. spectra) For sunlight and other white light, the spread of colors seen when the light is passed through a prism or diffraction grating. The colors of the spectrum, in order from lowest frequency (longest wavelength) to highest frequency (shortest wavelength) are red, orange, yellow, green, blue, indigo, violet. See also *absorption spectrum, electromagnetic spectrum, emission spectrum,* and *prism.*

speed   How fast something moves; the distance an object travels per unit of time; the magnitude of velocity. See also *average speed, linear speed, rotational speed,* and *tangential speed.*

$$\text{speed} = \frac{\text{distance}}{\text{time}}$$

spherical aberration   Distortion of an image caused when the light that passes through the edges of a lens focuses at slightly different points from the point where the light passing through the center of the lens focuses. It also occurs with spherical mirrors.

spring tide   High or low tide that occurs when the sun, earth, and moon are all lined up so that the tides due to the sun and moon coincide, making the high tides higher than average and the low tides lower than average. See also *neap tide.*

stable equilibrium   State of an object balanced so that any small displacement or rotation raises its center of gravity.

standing wave   Stationary wave pattern formed in a medium when two sets of identical waves pass through the medium in opposite directions. The wave appears not to be traveling.

static friction   Force between two objects at relative rest by virtue of contact that tends to oppose sliding.

streamline   Smooth path of a small region of fluid in steady flow.

strong force   Force that attracts nucleons to each other within the nucleus; a force that is very strong at close distances but decreases rapidly as the distance increases. Also called strong interaction. See also *nuclear force.*

strong interaction   See *strong force.*

subcritical mass   See *critical mass.*

sublimation   Direct conversion of a substance from the solid to the vapor phase, or vice versa, without passing through the liquid phase.

subtractive primary colors   The three colors of light-absorbing pigments—magenta, yellow, and cyan—that when mixed in certain proportions will reflect any color in the spectrum.

superconductor   Material that is a perfect conductor with zero resistance to the flow of electric charge.

supercritical mass   See *critical mass.*

superposition principle   In a situation where more than one wave occupies the same space at the same time, the displacements add at every point.

supersonic   Traveling faster than the speed of sound.

support force   Upward force that balances the weight of an object on a surface.

surface tension   Tendency of the surface of a liquid to contract in area and thus behave like a stretched elastic membrane.

tachyon   Hypothetical particle that can travel faster than light and thus move backward in time.

tangent   Line that touches a curve in one place only and is parallel to the curve at that point.

tangential speed   Linear speed along a curved path.

tangential velocity   Component of velocity tangent to the trajectory of a projectile.

tau   The heaviest elementary particle in the class of elementary particles called leptons.

technology   Method and means of solving practical problems by implementing the findings of science.

telescope   Optical instrument that forms images of very distant objects.

temperature   Measure of the average translational kinetic energy per molecule of a substance, measured in degrees Celsius or Fahrenheit, or in kelvins.

temperature inversion   Condition wherein upward convection of air is stopped, sometimes because an upper region of the atmosphere is warmer than the region below it.

terminal speed   Speed attained by an object wherein the resistive forces, often air resistance, counterbalance the driving forces, so motion is without acceleration.

terminal velocity   Terminal speed together with the direction of motion (down for falling objects).

terrestrial radiation   Radiant energy emitted from the earth.

theory   Synthesis of a large body of information that encompasses well-tested and verified hypotheses about aspects of the natural world.

thermal contact   State of two or more objects or substances in contact such that heat can flow from one object or substance to the other.

thermal equilibrium   State of two or more objects or substances in thermal contact when they have reached a common temperature.

thermal pollution   Undesirable heat expelled from a heat engine or other source.

thermodynamics   Study of heat and its transformation to mechanical energy, characterized by two principal laws:

First Law: A restatement of the law of conservation of energy as it applies to systems involving changes in temperature: Whenever heat is added to a system, it transforms to an equal amount of some other form of energy.

Second Law: Heat cannot be transferred from a colder body to a hotter body without work being done by an outside agent.

thermometer   Device used to measure temperature, usually in degrees Celsius, degrees Fahrenheit, or kelvins.

thermonuclear fusion Nuclear fusion brought about by extremely high temperatures; in other words, the welding together of atomic nuclei by high temperature.

thermostat Type of valve or switch that responds to changes in temperature and that is used to control the temperature of something.

time dilation Slowing down of time for an object moving at relativistic speeds.

torque Product of force and lever-arm distance, which tends to produce rotational acceleration.

$$\text{torque} = \text{lever-arm distance} \times \text{force}$$

total internal reflection The 100% reflection (with no transmission) of light that strikes the boundary between two media at an angle greater than the critical angle.

transformer Device for increasing or decreasing voltage or transferring electric power from one coil of wire to another by means of electromagnetic induction.

transistor See *semiconductor*.

transmutation Conversion of an atomic nucleus of one element into an atomic nucleus of another element through a loss or gain in the number of protons.

transparent Term applied to materials that allow light to pass through them in straight lines.

transuranic element Element with an atomic number above 92, which is the atomic number for uranium.

transverse wave Wave with vibration at right angles to the direction the wave is traveling. Light consists of transverse waves.

tritium Unstable, radioactive isotope of hydrogen whose atom has a proton, two neutrons, and an electron.

trough One of the places in a wave where the wave is lowest or the disturbance is greatest in the opposite direction from a crest. See also *crest*.

turbine Paddle wheel driven by steam, water, etc., that is used to do work.

turbogenerator Generator that is powered by a turbine.

ultrasonic Term applied to sound frequencies above 20,000 hertz, the normal upper limit of human hearing.

ultraviolet (UV) Electromagnetic waves of frequencies higher than those of violet light.

umbra Darker part of a shadow where all the light is blocked. See also *penumbra*.

uncertainty principle The principle formulated by Heisenberg, stating that Planck's constant, *h*, sets a limit on the accuracy of measurement at the atomic level. According to the uncertainty principle, it is not possible to measure exactly both the position and the momentum of a particle at the same time, nor the energy and the time associated with a particle simultaneously.

universal gravitational constant The proportionality constant *G* that measures the strength of gravity in the equation for Newton's law of universal gravitation.

$$F = G\frac{m_1 m_2}{d^2}$$

unstable equilibrium State of an object balanced so that any small displacement or rotation lowers its center of gravity.

UV Abbreviation for *ultraviolet*.

V (a) In lower-case italic *v*; the symbol for *speed* or *velocity*. (b) In uppercase V, the abbreviation for *voltage*.

vacuum Absence of matter; void.

Van Allen radiation belts Two donut-shaped belts of radiation that surround the earth.

vaporization The process of a phase change from liquid to vapor; evaporation.

vector Arrow whose length represents the magnitude of a quantity and whose direction represents the direction of the quantity.

vector quantity Quantity in physics that has both magnitude and direction. Examples are force, velocity, acceleration, torque, and electric and magnetic fields.

velocity Speed of an object and its direction of motion; a vector quantity.

vibration Oscillation; a repeating to-and-fro motion about an equilibrium position — a "wiggle in time."

virtual image Image formed by light rays that do not converge at the location of the image. Mirrors, converging lenses used as magnifying glasses, and diverging lenses all produce virtual images. The image can be seen by an observer, but cannot be projected onto a screen.

visible light Part of the electromagnetic spectrum that the human eye can see.

visible spectrum See *electromagnetic spectrum*.

volt (V) SI unit of electric potential. One volt is the electric potential difference across which one coulomb of charge gains or loses one joule of energy. 1 V = 1 J/C

voltage Electrical "pressure" or a measure of electrical potential difference.

$$\text{voltage} = \frac{\text{electric potential energy}}{\text{unit of charge}}$$

voltage source Device, such as a dry cell, battery, or generator, that provides a potential difference.

voltmeter See *galvanometer*.

volume Quantity of space an object occupies.

W (a) Abbreviation for *watt*. (b) When in italic *W*, the abbreviation for *work*.

watt SI unit of power. One watt is expended when one joule of work is done in one second. 1 W = 1 J/s

wave A "wiggle in space and time"; a disturbance that repeats regularly in space and time and that is transmitted progressively from one place to the next with no net transport of matter.

wave front Crest, trough, or any continuous portion of a two-dimensional or three-dimensional wave in which the vibrations are all the same way at the same time.

wave speed Speed with which waves pass a particular point.

$$\text{wave speed} = \text{wavelength} \times \text{frequency}$$

wave velocity Wave speed stated with the direction of travel.

wavelength   Distance between successive crests, troughs, or identical parts of a wave.

weak force   Also called weak interaction. The force within a nucleus that is responsible for beta (electron) emission. See *nuclear force*.

weak interaction   See *nuclear force* and *weak force*.

weight density   See *density*.

weight   Force on a body due to the gravitational attraction of another body (commonly the earth).

weightlessness   Condition of free fall toward or around the earth, in which an object experiences no support force (and exerts no force on a scale).

white light   Light, such as sunlight, that is a combination of all the colors. Under white light, white objects appear white and colored objects appear in their individual colors.

work (W)   Product of the force on an object and the distance through which the object is moved (when force is constant and motion is in a straight line in the direction of the force); measured in joules.

$$\text{work} = \text{force} \times \text{distance}$$

work-energy theorem   Work done on an object is equal to the energy gained by the object.

$$\text{work} = \text{change in energy or } W = \Delta E$$

wormhole   Hypothetical enormous distortion of space and time, similar to a black hole, but opens out again in some other part of the universe.

X ray   Electromagnetic radiation, higher in frequency than ultraviolet, emitted by atoms when the innermost orbital electrons undergo excitation.

zero-point energy   Extremely small amount of kinetic energy that molecules or atoms have even at absolute zero.

Photo Credits

xv   Meidor Hu
1   Charles A. Spiegel
2   Roger Ressmeyer/Starlight-Corbis
8   (l)Jeu de Paume, Paris/Art Resource, NY
      (r)Jay Pasachoff
17   Meidor Hu
18   Felix St. Clair-Renard/The Image Bank
19   The Granger Collection, NY
21   The Granger Collection, NY
32   Al Tielemans/DUOMO
36   Treat Davidson/Photo Researchers, Inc.
42   Richard Megna/Fundamental Photographs
47   Focus on Sports
48   Courtesy NASA
46   Larry Sergean/Comstock, Inc.
58   The Granger Collection, NY
64   R. Mackson/FPG International
69   Fundamental Photographs
75   Paul G. Hewitt
82   Paul G. Hewitt
83   Craig Aurness/Westlight
85   The Harold E. Edgerton 1992 Trust, Courtesy Palm Press, Inc.
86   Glen Allison/Tony Stone Images
87   Terje Rakke/The Image Bank
88   Paul G. Hewitt
92   Paul G. Hewitt
95   D.C. Heath & Company with Educational Development Center, Inc. Newton, MA
100   R. Llewellyn/Superstock, Inc.
101   Hubrich/The Image Bank
102   Courtesy NASA
103   Paul G. Hewitt
104   Paul G. Hewitt
112   Jack Hancock
116   Paul G. Hewitt
118   David Falconer/Tony Stone Images
120   Elliot Erwin/Magnum Photos
121   Meidor Hu
123   Richard Megna/Fundamental Photographs
125   Tony Duffy/Allsport USA
127   Paul G. Hewitt
132   Art by Don Davis/Courtesy NASA

133   Paul G. Hewitt
135   Gerard Lacz/NHPA/Agence Nature
140   Paul G. Hewitt
142   Courtesy NASA/Hubble Space Telescope
143   (l)Erich Lessing/Art Resource, NY
      (c)John R. Freeman & Company
      (r)Scripta Mathematica, Yeshiva University, NY
150   Courtesy NASA
166   Courtesy NASA
171   Diane Schiumo/Fundamental Photographs
175   Courtesy NASA
177   Courtesy NASA
181   Paul G. Hewitt
182   Courtesy IBM
184   (t)Pam Hewitt
      (b)Courtesy University of Chicago
185   Courtesy Lawrence Berkeley Laboratory
187   CNRI/SPL/Photo Researchers
197   Manfred Kage/Peter Arnold, Inc.
201   ©The Harold E. Edgerton 1992 Trust. Courtesy Palm Press, Inc.
204   Meidor Hu
205   (tl)Paul G. Hewitt
      (tr)Raymond V. Schoder, S.J.
      (b)Paul G. Hewitt
206   Joe Sohm/Chromosohm/The Stock Market
210   (l)W. Geirsperger/Camerique/H. Armstrong Roberts
      (r)James P. Rowan
213   Paul G. Hewitt
215   Paul G. Hewitt
216   William Waterfall/The Stock Market
218   Robert Frerck/The Stock Market
224   Milo Patterson
227   Diane Schiumo/Fundamental Photographs
230   Margaret Ellenstein
234   ©Photo Lib Internat, Ltd./Westlight
236   The Granger Collection, NY
239   David E. Hewitt
244   Paul G. Hewitt
245   William Waterfall/The Stock Market
246   The Cousteau Society, Inc.
248   Vito Palmisano/Tony Stone Images
255   Olan Mills
256   B. Brad Lewis
262   (t)Courtesy LU Engineers, Penfield, NY
      (b)Meidor Hu
263   Paul G. Hewitt

270   Tracy Suchocki
271   Cameramann International, Ltd.
272   Nancy Rogers
274   Paul G. Hewitt
277   Paul G. Hewitt
278   David Cavagnaro
279   Paul G. Hewitt
282   (t)Sylvia Means
      (b)Paul G. Hewitt
283   Reprinted with permission of Boeing Company.
286   Paul G. Hewitt
288   Paul McCormick/The Image Bank
289   (t)Paul G. Hewitt
      (c)Paul G. Hewitt
      (b)Ralph A. Reinhold/Animals Animals
290   Paul G. Hewitt
299   (t)Paul G. Hewitt
      (b)Rick Povich Photography
304   Vanscan Thermogram by Daedalus Enterprises, Inc.
308   Paul G. Hewitt
309   Paul G. Hewitt
310   Thomas Ives/The Stock Market
316   Natural Energy Labs of Hawaii Authority
317   R. Krubner/H. Armstrong Roberts
318   Paul G. Hewitt
323   Paul G. Hewitt
324   Paul G. Hewitt
328   Gabe Palmer/Mug Shots/The Stock Market
331   Education Development Center
336   ©The Harold E. Edgerton Trust, Courtesy Palm Press, Inc.
342   Martin Bough/Fundamental Photographs
344   Meidor Hu
346   Terrence McCarthy/San Francisco Symphony
347   (t)Howard Sochurek/The Stock Market
      (b)Laura Pike & Steve Eggen
349   Paul G. Hewitt
350   (l)AP/Wide World
      (c)UPI/Bettmann
      (r)AP/Wide World
352   (t)Paul G. Hewitt
      (b)Norman Synnestvedt
359   David Hewitt
360   The Kobal Collection
361   Paul G. Hewitt
363   ©Ebet Roberts
364   (t)Fotex/R. Dreschler/Shooting Star
      (b)Ronald B. Fitzgerald

Index

Aberration, 509, 510
  astigmatism, 509
  chromatic, 509
  spherical, 509
Absolute zero, 257, 266, 304–5, 319
Absorption lines, 547–48
Absorption spectrum, 547–48, 555
ac. *See* Alternating current
Acceleration, 23, 26, 32, 33, 689
  centripetal, 151, 543
  constant, 691–92
  distinction from velocity, 27
  in free fall, 28–29, 30, 66–67
  on inclined planes, 27
  in nonfree fall, 67
  relationship of, to force and mass,
    60–62, 76
Achromatic lens, 509
Acoustics, 346
Action, 70–76
  at distance, 154
Action force, 70–76
Adams, J. C., 159
Additive primary colors, 474, 477, 483
Adhesion, 228
Adiabatic processes, 308–9, 319
  in atmosphere, 309–11
  in ocean, 311
Age of Enlightenment, 160
Air
  buoyancy of, 273
  as a conductor, 271
  expansion of cool, 274
  sound in, 343–44, 345
Air conditioner, principle behind, 296
Air resistance
  effect of, on projectile motion, 45
  and falling objects, 67–68, 210
Alexander the Great, 19
Alpha Centauri, 652
Alpha particles, 592, 608
Alpha rays, 591–92, 608
Alternating current (ac), 398, 404–5, 412
  converting, to direct current, 405
  and generators, 439–41
Altimeter, 239
Altitude in projectile motion, 44
AM (amplitude modulation), 354, 355, 520
Americium, 603
Ammeter, 426
Amorphous solids, 198–99
Ampere, 398, 399, 687
Ampère, André-Marie, 418
Amplitude, 326, 338
Amplitude modulation (AM), 354, 355, 520

amu (atomic mass unit), 188, 193
Analog signal, 367
Analyzer, 531
Anderson, Carl, 661
Aneroid barometer, 239
Angle of attack, 699
Angular momentum, 133–34, 136
  conservation of, 134–35, 136
Angular speed. *See* Rotational speed
Angular velocity, 133n
Antimatter, 192–93
Antineutrino, 596
Antinodes, 332
Anti-noise technology, 352
Antiparticle, 661
Antiprotons, 192
Antiquarks, 192
Apogee, 173
Arches, 205–6, 213
Archimedes, 219
Archimedes' principle, 219–21, 228, 242,
  249
Area, 687
Aristarchus, 5n, 5–7, 142n
Aristotle, 18–19, 56, 59, 144
Armature, 427
Art, comparison to science, 12
Artificial transmutation, 602–3
Astigmatism, 509
Atmosphere, 234–35
  adiabatic processes in, 309–11
  condensation in, 291–92
  tides in, 153–54
Atmospheric pressure, 236–37, 249
  and barometers, 238–40, 249
  and Boyle's law, 240–41
Atomic bomb, 614, 618, 625
Atomic bonding, 199, 210
Atomic mass number, 594, 609
Atomic mass unit (amu), 188, 193
Atomic nucleus, 593–94
  discovery of, 578–79
Atomic number, 191, 193, 594, 609
Atomic physics, 590
Atomic pile, 615
Atomic spectra, 579–80
Atomic structure, 189–92, 593–94
  characteristics of, 190–91
  classification of, 191–92
Atoms, 182–85, 193, 374
  attraction between, 190
  Bohr model of, 580–81
  electric charge in, 374–75
  excitation of, 542–45, 555
  formation of molecules, 186–87

  mass of, 187–88
  and quantum mechanics, 560, 584–86,
    587
  quarks in, 375n
  radioactivity of, 595–96
  relative sizes of, 582
Aurora australis, 430
Aurora borealis, 430, 543
Avogadro, Amadeo, 188
Avogadro's principle, 188, 193
Axis of rotation, 118–22

Back electromotive force, 445
Bacon, Francis, 9
Balmer, J. J., 579
Barometers, 238–40, 249
Bartlett, Albert A., 703n
Batteries, 400
Beats, 353–55
Becquerel, Antoine Henri, 591
Bell, Alexander Graham, 360
Berkelium, 603
Bernoulli, Daniel, 243, 247
Bernoulli's principle, 243–47, 249
Beta emission, 600n
Beta particles, 592, 608
Beta rays, 591–92, 608
Big Bang theory, 160, 161
Bimetallic strip, 262–63
Binary code, 367
Binoculars, 503
Biology, 14
Bioluminescence, 551
Biomagnetism, 431
Black hole, 157–59, 161
Blind spot, 462
Block and tackle, 108
Blue shift, 334
Bohr, Niels, 562, 572, 580, 587, 612
Bohr model of the atom, 580–81
  and quantum mechanics, 560, 584–87
Boiling, 293, 300
  as cooling process, 294
  and freezing, 294
  and geysers, 293–94
Bomb
  atomic, 614, 618, 625
  hydrogen, 189, 625
Bouncing, and momentum, 87–88
Bow waves, 335–36, 338
Boyle, Robert, 241
Boyle's law, 240–41, 249
Bragg, William Henry, 198
Bragg, William Lawrence, 198
Brahe, Tycho, 143

Breeder reactors, 618–19
British system of units, 59*n*, 62*n*, 684
British thermal unit (BTU), 259*n*
Brown, Robert, 185
Brownian motion, 185, 193
Bubble chamber, 598
Buoyancy, 219
   of air, 242–43, 273
   and displaced liquid, 215–16
Buoyant force, 219, 221, 228
Butterfly effect, 573

Caesar, Julius, 3*n*
Caloric, 306
Calorie, 259–60
Cameras, 506–7
Cantilever beam, 203
Capacitor, 390–91, 393, 405
Capillarity of liquids, 228
Carbon, 401*n*, 604–5, 621
Carbon dating, 604–5
Carnot, Sadi, 313
Carnot efficiency equation, 313–14
Carrier wave, 354, 355
Cartesian graphs, 693–94
Catenary, 205
Cavendish, Henry, 146
Celsius, Anders, 257
Celsius temperature scale, 305
Celsius thermometer, 257
Center of gravity, 123–24, 136
   locating, 124–25
   stability for, 125–27
Center of mass, 71, 123–24, 136
Centigrade thermometer, 257
Centimeter-gram-second system, 685
Centrifugal force, 129, 130–36
Centripetal acceleration, 129, 151, 543
Centripetal force, 128–29, 136
Chain reaction, 613, 630
Change of phase
   boiling, 293–94, 300
   condensation, 290–92, 300
   and energy, 288–89
   evaporation, 288–89, 300
   melting and freezing, 295, 300
Chaotic systems, 573
Charge
   flow of electrical, 398–99
   measurement of, 397
   polarization, 381–83
Charging, 379–81
Chemical reaction, 186, 193
Chemistry, 14
Chinook, 309–10
Chromatic aberration, 509

Circuits, 408–12
   parallel, 410–11, 413
   safety fuses, 412
   series, 408–10, 412
   short, 412
Circular motion, 49–52
Circular orbit, 168–70
Cloud chamber, 598
Clouds, 292, 380
   color of, 481–82
Coherent light, 535, 551
Cohesion, 228
Collider Detector, 599
Collisions
   elastic, 91, 96
   inelastic, 91, 96
   and momentum, 90–93
   more complicated, 94–95
Color(s), 470
   additive primary, 474, 477, 483
   of clouds, 481–82
   complementary, 477–78, 483
   interference, 526–29
   mixing, 473–75
   selective reflection, 471–72
   selective transmission, 472–73
   of sky, 478–80
   subtractive primary, 476, 483
   of sunset, 480
   of water, 482–83
Colored light, mixing, 473–75
Colored pigments, mixing, 475–78
Color force, 595*n*
Compact discs, 367–68
Complementarity, 572–74
Complementary colors, 477–78, 483
Components of vector, 39, 39*n*
Compounds, 193
   versus elements and mixtures, 188–89
Compression, 203–4, 330, 343, 355
Condensation, 290–91, 300
   in the atmosphere, 291–92
   in fog and clouds, 292
Conduction, 270–72, 283, 284
Conduction electrons, 399
Conductors, 377–78, 393
Conservation
   of angular momentum, 134–36
   of electric charge, 374–75, 393
   of energy, 106–10, 108, 113, 172–73
   of momentum, 88–89, 96
Constant velocity, 669
Constructive interference, 331, 521–22
Contact, charging by, 379
Convection, 272, 283, 284
   and buoyancy of warm air, 273

currents, 274–75
   and expansion of cool air, 274
Converging lens, 505, 510
Cooling
   boiling as, 294
   Newton's law of, 279–80, 284
   at night by radiation, 278–79
Coordinate axes, 640
Copernican theory, 3, 20, 142–43
Copernicus, 3, 142, 142*n*, 679
Cornea, 462
Correspondence principle, 587, 661–63
Cosmic rays, 425, 429
Coulomb, 376, 393, 399
Coulomb, Charles, 376, 418
Coulomb's law, 376, 393
   comparison of, to Newton's law of gravitation, 376
   and magnetism, 419
Covalent bonding, 199
Crests of waves, 326
Critical angle, 501, 510
Critical mass, 614, 630
Crystal
   dichroic, 530*n*
   structure of, 198–99
Crystalline solids, 198–99
Curie, Marie, 591
Curie, Pierre, 591
Current. *See* Electric current

Darwin, Charles, 10
da Vinci, Leonardo, 3
dc. *See* Direct current
de Broglie, Louis, 567, 583
de Broglie wavelength, 567*n*
Deceleration, 26, 689
Decibels, 361
De-excitation of atoms, 542
Density, 200–1, 210
   and submerged objects, 221–22
Destructive interference, 331, 522
Deuterium, 594
Diamond, critical angle for, 503
Dichroic crystals, 530*n*
Diffraction, 519–21, 537
   electron, 567–68
Diffraction fringes, 520
Diffraction grating, 525
Diffraction pattern, 198
Diffuse reflection, 490–92
Digital audio, 367
Digital signal, 367
Diode, 405
Direct current (dc), 398, 404–5, 412
   conversion from alternating current, 405

Discrete energy states, 542
Disordered energy, 316–17
Dispersion of light, 497
Displacement, 689
    of water, 219
Distance, 23
    action at, 154
    computing, when acceleration is constant, 691–92
    and gravity, 147–48
    on inclined plane, 689–91
    traveled in free fall, 29
    traveled in projectile motion, 42–44
Diverging lens, 505, 510
Dobson, John, 660n
Doppler, Christian, 334
Doppler effect
    in light, 548
    in sound, 333–34, 338, 347, 355
Double-slit experiment, 523–24, 565
Doubling time, 703–8
Drag, 699
Drift velocity, 406
Dynamic equilibrium, 64

Earth
    magnetic field of, 428–31
    mass of, 146
    measurement of, 3–4
    satellites, 46, 169, 174
    tides on, 153–54
Echo, 345
Eclipse, 153, 155–56, 672
    lunar, 460, 467
    solar, 460, 460n, 467
Eddy, 244
Efficiency, 313n
    and energy conservation, 109–10, 113
Einstein, Albert, 10, 420n, 551, 560, 562, 563, 586, 586n, 613, 620, 662–63, 668, 672, 673, 679
Einsteinian gravitation, 157, 679
Elastic collision, 91, 96
Elasticity, 201–2, 210
Elastic limit, 202
Electrical charge, 373–74
    quantized, 375
Electrical forces, 373
Electrical polarization, 381–83
Electrical resistance, 398, 401, 412
Electric charge
    conservation of, 374–75, 393
    flow of, 398–99
Electric circuits. *See* Circuits
Electric current, 398, 399, 412
    alternating, 404–5

direct, 404–5
    electrical resistance, 401, 412
    electric circuits, 408–12
    electric power, 407–8, 412
    flow of charge, 398–99
    magnetic fields, 423
    Ohm's law of, 401–2
    speed and source of electrons in, 405–7
    voltages sources for, 400
Electric dipoles, 383
Electric energy storage, 390–93
Electric field, 383–85, 393, 447
Electricity, 372, 393
Electric meters, 426
Electric motors, 427–28
Electric potential, 388–90, 393
Electric potential energy, 389, 393
Electric power, measurement of, 407–8, 412
Electric shielding, 386–88
Electric shock, and Ohm's law, 402–4, 412
Electrodes, 442
Electromagnetic induction, 436–38, 448
    and Faraday's law, 438–39, 448
    field induction, 447
    generators and alternating current, 439–41
    power production, 441–42
    power transmission, 446
    self-induction, 445–46
    transformers in, 442–45, 448
Electromagnetic spectrum, 452, 454–55, 467
Electromagnetic waves, 275, 452–55, 467
Electromagnets, 424, 432
Electromotive force, back, 445
Electron(s), 190, 374
    antiparticle of, 661
    charge experiment, 397
    conduction, 399
    diffraction, 567–68
    as electromagnet, 421
    microscope, 520
    speed and source of, 405–7
Electron diffraction, 567–68
Electronic musical instrument, 364
Electron microscope, 520
Electron volt, 613n
Electron waves, 583–84
Electroscope, 395
Electrostatics, 372, 393
    and charge polarization, 381–83
    charging, 379–81
    conductors, 377–78
    and conservation of charge, 374–75, 393
    Coulomb's law, 376, 393
    electrical charges, 373–74
    electrical forces, 373

and electric energy storage, 390–93
    electric field, 383–85, 393
    electric potential, 388–90, 393
    electric shielding, 386–88
    insulators, 377–78
    Van de Graaff generator, 392
Elementary particles, 594
Elements, 183
    artificial transmutation of, 602–3
    versus compounds and mixtures, 188–89
    half-life of, 596–97, 609
    most common, 183n
    natural transmutation of, 600–2
Ellenstein, Marshall, 230
Ellipse, 170, 178
Elliptical orbit, 170–72, 675–76
El Niño, 311, 311n
$E=mc^2$, 658–61
Emerson, Ralph Waldo, 318n
Emission spectrum, 544–45, 555
Empedocles, 559
Energy, 100, 113. *See also* Kinetic energy
    and change of phase, 296–99
    disordered, 316–17
    internal, 306, 319
    kinetic, 105, 110–12, 113, 172–73, 256–58
    for life, 113
    and mass, 658–61
    mass-energy equivalence, 620–23, 630
    mechanical, 102–6
    potential, 103–5, 113, 172–73, 389
    and power, 101–2
    radiant, 275–79
    rest, 658, 663
    sound waves, 348
    storage of electric, 390–92
    and work, 101
    work-energy theorem, 105–6, 113
    zero-point, 257n, 305n
Energy conservation, 106–10, 113, 172–73
    and efficiency, 109–10, 113
    and machines, 108
    and satellite motion, 172–73
Engine, heat, 312–15, 319
Entropy, 318–19
Environment, effect of exponential growth on, 707–8
Equilibrium, 125, 136
    dynamic, 64
    mechanical, 62–63, 76
    static, 62–63
Equivalence
    mass-energy, 620–23, 630
    principle of, 669–70, 679
Eratosthenes, 3n, 3 4

Escape speed, 176n, 176–78, 178, 180
Escape velocity, 157–59, 176
Ether, 636, 637
Euclid, 559
Euclidean geometry, 676
Evaporation, 288–89, 300
Evolution, theory of, 3, 10
Excitation of atoms, 542–45, 555
Expansion, 262–63
    of cool air in convection, 274
    of water, 263–66
Expansion joints, 262
Experiments, 682
Exponential growth, 703–8
Eye
    parts of, 462
    response to light, 464–65
    and seeing color and motion, 463–64
Eyeglasses, 509, 510

Fact, 9, 15
Fahrenheit, Gabriel Daniel, 257n
Fahrenheit temperature scale, 305
Fahrenheit thermometer, 257
Fall
    free, 28–30, 32–33, 66–67, 76
    nonfree, 67
Faraday, Michael, 436, 441
Faraday's law, 438–39, 447, 448
    Maxwell's counterpart to, 447, 448, 453
Fermat, Pierre de, 487
    principle of least time, 487, 510
Fermi, Enrico, 615
Feynman, Richard P., 447n, 589
Field induction, 447
Fields
    electric, 383–85, 393, 447
    force, 154, 383
    gravitational, 383
    magnetic, 420–21, 423, 428–31, 432
Fisheye lens, 502
Fission, nuclear, 612–15, 630
FitzGerald, G. F., 636, 654
Floating mountains, 221
Flotation, principle of, 223–24, 228
Fluids, friction in, 65–66. See also Gases; Liquids
Fluorescence, 548–50, 555
Fluorescent lamps, 550
FM (frequency modulation), 354, 355, 520
Focal length, of lens, 506
Focal plane of lens, 505
Focal point of lens, 505
Foci, 170
Fog, 292

Force, 61–62
    action, 70–76
    back electromotive, 445
    buoyant, 219, 221, 228
    centrifugal, 129–30, 136
    centripetal, 128–29, 136
    color, 595n
    effect of, on acceleration, 60–62, 76
    electrical, 373
    factors, 698–700
    fields, 154, 383
    gravitational, 66, 144–46
    magnetic, 419, 424–28, 432
    net, 61–62, 67, 68, 76
    normal, 63
    reaction, 70–76
    support, 63
    tension, 63
    as vector quantity, 62
    weak, 376n
Forced vibrations, 348, 355
Ford, Kenneth, 352
Fourier, Joseph, 365
Fourier analysis of musical sounds, 364–67, 368
Fovea, 462
Frame of reference, 635, 663
Franklin, Benjamin, 380n, 380–81, 682
Fraunhofer, J. D., 548
Fraunhofer lines, 548
Free fall, 32, 76
    acceleration in, 28–29, 30, 66–67
    distance traveled in, 29, 33
    speed in, 28–29
    state of, 28
Freezing, 295
Freon, 296
Frequency, 327, 338
    fundamental, 362, 368
    natural, 348, 355
    peak, 546
Frequency modulation (FM), 354, 355, 520
Friction, 64–66, 76
    charging by, 379
    fluid, 65–66
Frisch, Otto, 612
Fulcrum, 108, 121–22
Fundamental frequency, 362, 368
Fundamental units, 684
Fuses, safety, 412
Fusion
    laser, 627
    nuclear, 623–29, 630
    thermonuclear, 107, 623, 630
Fusion cycle of the sun, 624n

Fusion power, 628
Fusion torch, 629

Galilei, Galileo, 3, 9, 20–23, 27, 56, 59, 66, 82, 144, 166, 516, 670, 679, 689, 691
Galvani, Luigi, 426n
Galvanometer, 426, 426n
Gamma rays, 275, 591–92, 607, 608
Gases. See also Air; Fluids; Liquids
    in atmosphere, 234–35
    and atmospheric pressure, 236–40, 249
    atomic structure of, 193
    Bernoulli's principle, 243–47, 249
    Boyle's law of, 240–41, 249
    and buoyancy of air, 242–43
    heat conduction in, 270–72
Geiger counter, 597
Gell-Mann, Murray, 593
General theory of relativity, 668–69
    bending of light by gravity, 670–73
    gravitational red shift, 673–75, 679
    gravitational waves, 678, 679
    gravity, space, and new geometry, 676–78
    motion of Mercury, 675–76
    Newtonian and Einsteinian gravitation, 679
    principle of equivalence, 669–70, 679
Generator, 448
    and alternating current, 439–41
    comparison of, to motor, 439
    magnetohydrodynamic (MHD), 441–42
    Van de Graaff, 392
Generator effect, 439
Geodesics, 677, 679
Geysers, 293–94
Gigahertz (GHz), 327
Gilbert, William, 418
Global warming and heat transfer, 280–81
Gluons, 595n
Gold-foil experiment, 579–80
Graphs, 693, 695
    Cartesian, 693–94
    slope and area under curve on, 695
Gravitation
    Einsteinian, 157, 679
    law of, 60
    Newtonian, 144–48, 161, 679
    universal, 159–60
Gravitational fields, 154–57, 161, 383
Gravitational force, 66
Gravitational mass, 145n
Gravitational potential energy, 103
Gravitational red shift, 673–75, 679
Gravitational waves, 678, 679

Gravity
  bending of light by, 670–73
  black holes in, 157–59, 161
  center of, 123–27, 136
  and distance, 147–48
  Einstein's theory of gravitation, 157
  force of, between earth and earth satellite, 180
  gravitational fields, 154–57, 161
  gravitational red shift, 673–75, 679
  motion of Mercury, 675–76
  ocean tides, 150–54
  simulated, 131–33
  space and new geometry, 676–78
  universal gravitation, 159–60
  and weight, 149–50
Greenhouse effect and heat transfer, 280–81, 284
Grounding, 380
Growth, exponential, 703–8
Guitar, electric, 449

Hadrons, 595
Hahn, Otto, 612
Hahnium, 603
Half-life, 596–97, 609, 703n
Halley, Edmund, 145
Hang-time, 32, 47
Harmonic, 362, 368
Harmonic motion, 325n
Heat
  caloric theory of, 306
  capacity for, 260–61
  flow of, 258–59, 266
  measurement of, 259, 266
Heat engine, 312–15, 319
Heat of fusion, 297n
Heat of vaporization, 297n
Heat radiation, 276
Heat transfer
  conduction in, 270–72
  convection in, 272
  and greenhouse effect, 280–81, 284
  Newton's law of cooling and, 279–80, 284
  radiation in, 275–79
  and solar power, 281–83
  and thermos bottle, 283
Heat waves, 457
Heavy primaries, 607
Heavy water, 616n
Heisenberg, Werner, 570
Heliocentric hypothesis, 5n
Helium, 193, 305n, 548
Henry, Joseph, 436, 441
Hertz (Hz), 327, 338

Hertz, Heinrich, 327, 560
Hologram, 535–36, 537
Hooke, Robert, 202, 516n
Hooke's law, 202, 210, 325n
Humidity, relative, 292
Huygens, Christiaan, 516, 560
Huygens' principle, 516–17, 537
Hydrogen, 183, 629
Hydrogen bomb, 189, 625
Hypothesis, 9, 10, 15, 682

Ice, structure of, 264–65
Ideal efficiency, 313–14, 319
Impulse
  definition of, 83, 96
  relationship of, to momentum, 83–87, 96
Incandescence, 546–48, 555
Incidence, angle of, 488
Inclined plane, 21–22
  acceleration on, 27
  computing velocity and distance traveled on, 689–91
Incoherent light, 551
Induction. See also Electromagnetic induction
  charging by, 379–80
  of voltage, 436–38
Inelastic collision, 91, 96
Inelasticity, 201
Inertia, 22, 32, 76
  law of, 56–57, 59–60
  rotational, 118–21, 136
Inertial frame of motion, 689
Inertial mass, 145n
Infrared radiation, 275
Infrasonic sound waves, 343, 355
Instantaneous speed, 24, 32
Instantaneous velocity, 25, 690
Insulators, 271, 377–78, 393
Integral calculus, 695
Intensity of musical sounds, 360–61, 368
Interference, 521–25, 537
  beats caused by, 353–55
  constructive, 331, 521–22
  destructive, 331, 522
  light waves and, 521–29, 537
  in sound, 350–55
  in vibration, 331–33
Interference colors, 526–29, 537
Interference pattern, 331–33, 338
Interferometer, 636
Internal energy, 258, 266
International Bureau of Weights and Measures, 685, 686
International system. See Système International (SI)

Inverse-square law, 144, 147–48, 157, 161
Ionic bonding, 199
Ions, 198, 248, 374–75
Iridescence, 527
Isotopes, 594–95, 609
  radioactive, 603–4

Joule (J), 101, 102, 259, 260, 686
Joule, James, 307
Joyce, James, 593
Jupiter, 159, 177

Kelvin, 687
Kelvin, Lord (Thomson, William), 3, 257, 687
Kelvin temperature scale, 257, 305
Kepler, Johannes, 5n, 143
Kepler's laws of planetary motion, 143–44, 161, 167–68
Kilogram, 59, 76, 686
Kilohertz (kHz), 327
Kilojoule (kJ), 101
Kilopascal, 215n
Kilowatt (kW), 102, 398
Kilowatt-hour (kWh), 407
Kinetic energy, 102, 105, 113, 172–73. See also Energy
  comparison of, to momentum, 110–12
  and temperature, 256–58

Laser fusion, 627
Lasers, 551–55
Lateral inhibition, 464
Law, 9, 15. See also specific
Lawrencium, 603
Length contraction, 654–56, 663
Length in metric system, 685
Lens
  achromatic, 509
  converging, 505, 510
  defects, 509–10
  diverging, 505, 510
  fisheye, 502
  focal length of, 506
  focal plane of, 505
  focal point of, 505
  image formation by, 506–8
  principal axis of, 505
Lenz's law, 445n
Leptons, 593
Lever, 108
Lever arm, 121
Leverrier, Urbain, 159
Life, energy for, 113
Lift, 245, 699

Light
bending of, by gravity, 670–73
coherent, 535, 551
dispersion of, 497
Doppler effect in, 334
electromagnetic waves, 452–55, 467
incoherent, 551
mixing colored, 473–75
and opaque materials, 458–59, 467
refraction of, 492–95
seeing, 462–65
shadows, 459–61, 467
speed of, 453, 492, 496
and transparent materials, 456–57, 467
visible, 275
Light emission, 541
excitation of atoms, 542–45, 555
fluorescence, 548–50, 555
incandescence, 546–48, 555
lasers, 551–55
phosphorescence, 550–51, 555
Lightning, 380–81
Light quanta, 559–60
complementarity, 572–74
double-slit experiment, 565–66
electron diffraction, 567–68
photoelectric effect, 562–64, 574
quantization and Planck's constant, 560–62
quantum theory, 560, 574
uncertainty principle, 569–71, 574
wave-particle duality, 564, 566
Light waves
diffraction, 519–21, 537
holography, 535–36
Huygens' principle, 516–17, 537
interference, 521–29, 537
polarization of, 529–34, 537
Light-year, 652n
Linear motion, 18–32. See also Motion
Linear speed, 49–50, 52. See also Speed
Liquid pressure, 216–18, 228
Liquids. See also Fluids; Gases
Archimedes' principle on, 219–21, 228
atomic structure of, 193
boiling of, 293, 300
and buoyancy, 219, 228
capillarity of, 228
displaced, 215–16
flotation principle of, 223–24
freezing of, 294
friction in, 65–66
Pascal's principle of, 224–26
pressure in, 215–18
surface tension of, 226–27, 228
Lithium, 191

Locke, John, 160
Lodestones, 418
Longitudinal waves, 330–31, 338
Lorentz, Hendrik A., 636, 643, 654
Lorentz contraction, 654
Lorentz factor, 643
Lorenz, Edward, 573
Loudness of musical sounds, 361–62, 368
Lunar eclipse, 460, 467

Machines and energy conservation, 108, 113
Magnetic declination, 428
Magnetic domains, 421–22, 432
Magnetic fields, 420–21, 432
of earth, 428–31
and electric current, 423
of sun, 429
Magnetic force, 419, 432
on current-carrying wires, 425–28
on moving charged particles, 424–25
Magnetic monopoles, 420n
Magnetic poles, 419–20
Magnetic resonance imaging (MRI), 431
Magnetism, 418–31
Magnetohydrodynamic (MHD) generator, 441–42
Magnetohydrodynamic (MHD) power, 249
Magnifying glass, 507
Magplane, 424
Manhattan project, 614
Manka, Chuck, 521
Mass, 57, 59–62, 76
atomic, 187–88
center of, 71, 123–24, 136
critical, 614, 630
definition of, 62n
distinction between volume and, 59
distinction between weight and, 57
of earth, 146
effect of, on acceleration, 60–62, 76
and energy, 658–61
gravitational, 145n
inertial, 145n
molecular, 187–88
as proportional to weight, 66n
standard unit of, 686
subcritical, 614
supercritical, 614
Mass energy, 306
Mass-energy equivalence, 620–23, 630
Mass spectrometer, 621
Mathematics, as language of science, 8
Matter
atomic makeup of, 182–93
change in phase of, 288–99
phases of, 193

Matter wave amplitude, 585
Matter waves, 583
Maximum falling speed, 176n
Maxwell, James Clerk, 447, 453, 454, 560
counterpart to Faraday's law, 447, 448, 453
laws of electromagnetism, 634
Measurement
of distance to sun, 6–7
of earth, 3–4
of heat, 259, 266
of length, 654–56
of moon, 4–6
scientific, 3
scientific notation, 688
of space, 639–41
Système International (SI), 684–87
of time, 641–54
United States Customary System (USCS) of, 684–87
units of, 146n
Mechanical energy, 102–6
kinetic, 105, 110–12
potential energy, 103–5
work-energy theorem, 105–6, 113
Mechanical equilibrium, 62–63, 76
Megahertz (MHz), 327
Megajoule (MJ), 101
Megawatt (MW), 102
Meitner, Lise, 612
Melting, 295
Mercury, orbit of, 675–76
Mesons, 595
Metal in electric shielding, 386–88
Metallic bonding, 199
Meteorology and thermodynamics, 309–11
Meter, 685
Meter-kilogram-second (mks) system, 685
Meters, electric, 426
Metric system. See Système International (SI)
Michell, John, 418
Michelson, A. A., 635
Michelson-Morley experiment, 635–36
Microwave oven, 386
Microwaves, 275
Millikan, Robert
electron charge experiment, 397
photoelectric effect experiment, 564
Mirage, 494, 496–97
Mirrors, plane, 489–90
Mixtures, 193
versus elements and compounds, 188–89

Modulation, 354, 355, 520
Molecules, 186–88, 193
    formation of, 186
    mass of, 187–88
Momentum
    angular, 133–35, 136
    and bouncing, 87–88
    in collisions, 90–95
    compared to kinetic energy, 110–12
    conservation of, 88–89, 96
    definition of, 82, 96
    relationship to impulse, 83–87, 96
    relativistic, 657–58
Monopoles, magnetic, 420n
Moon
    distance to, 6
    measurement of, 4–6
    motion of, 166–68
    and ocean tides, 150–53
    tides on, 154
Morley, E. W., 635
Mossbauer, Rudolph, 674n
Mossbauer effect, 674n
Motion, 689. *See also* Projectile motion;
        Rotational motion; Satellite
        motion
    Aristotle on, 18–19
    Brownian, 185, 193
    circular, 49–52
    Copernicus on, 20
    Galileo on, 20–21
    harmonic, 325n
    inertial frame of, 689
    Kepler's laws of planetary, 143–44, 161,
        167–68
    linear, 18–32
    natural, 18–19
    Newton's laws of, 56–76
    nonlinear, 36–52
    orbital, 675–76
    oscillatory, 325
    projectile, 40–48
    as relative, 36–37, 635–36
    rotational, 121–36
    satellite, 46–48, 166–78
    simple harmonic, 325
    transverse, 329
    violent, 19
    wave, 328
Motor effect, 439
Motors
    comparison to generator, 439
    electric, 427–28
MRI (magnetic resonance imaging), 431
Müller, Erwin, 197
Muon, 596–97

Musical instruments
    electronic, 364
    percussion, 364
    stringed, 363
    wind, 363–64
Musical sound(s). *See also* Sounds
    compact discs, 367–68
    Fourier analysis of, 364–67, 368
    intensity of, 360–61, 368
    loudness of, 361–62, 368
    versus noise, 359–60
    octaves, 364
    phonograph record, 364–65
    pitch, 360, 368
    quality of, 362–63, 368
    scales, 364

Natural frequency, 348, 355
Natural motion, 18–19
Natural philosophy, 13
Natural transmutation, 600–2
Neap tides, 153, 161
Negative charge, 373
Neon lights, 542–43
Neptune, 159
Neptunium, 603, 617
Net force, 61–62, 67, 68, 76
Neutral layer, 203
Neutrinos, 607
Neutrons, 190, 374, 593, 596–97
Newton, 59, 76, 686
Newton, Isaac, 22, 56, 58, 59, 62, 70, 100,
        144–45, 157, 160, 166, 167–68, 560,
        679, 686
    law of cooling, 279–80, 284
    law of universal gravitation, 144–48,
        161
    laws of mechanics, 634
    laws of motion
        first law, 56–57, 59–60
        second law, 60–69, 84, 84n
        third law, 70–76, 89
Newtonian gravitation, 679
Newton-meter (N-m), 101
Newton's rings, 527
Nodes, 332
Noise versus musical sounds, 359–60
Nonfree fall, acceleration in, 67
Nonlinear motion, 36–52
Normal, 488
Normal force, 63
Northern lights, 430
Nuclear fission, 612–15, 630
Nuclear fission reactors, 615–16
Nuclear fusion, 623–25, 630
    controlling, 625–28

fusion torch and recycling, 629
Nuclear magnetic resonance (NMR), 431
Nuclear physics, 590
Nucleons, 190, 593, 608
Nucleus, 189
    atomic, 578–79, 593–94

Ocean
    adiabatic processes in, 311
    tides in, 150–54
Ocean Thermal Energy Conversion (OTEC),
    316
Oersted, Hans Christian, 418, 423, 425, 436
Ohm, 401
Ohm, Georg Simon, 401
Ohm's law, 401–2, 402n, 412
    and electric shock, 402–4
Opaque materials, 458–59, 467
Optical fibers, 504
Optical illusions, 466
Orbit, 374. *See also* Satellite motion
    circular, 168–70
    elliptical, 170–72, 675–76
Order in chaos, 573
Organ of Corti, 361
Oscillation, 471, 471n
Oscillatory motion, 325
Osmium, 200–1
Overloading, of electrical circuits, 411–12

Parabola, 44–45, 52, 205
Parallel circuits, 410, 413
    and overloading, 411
Parallelogram, 38n, 38–39
Parallelogram rule, 62
    of vector addition, 94
Partial tones, 362, 368
Particles
    elementary, 594
    light, 559–74
Pascal, 215n, 225
Pascal, Blaise, 225
Pascal's principle, 224–26, 228
Peak frequency, 546
Pelton, Lester A., 87–88
Pendulum, vibration of, 325, 327
Penumbra, 459–60, 467
Percussion instrument, 364
Perigee, 173
Period, 325, 328, 338
Periodic table, 191–92, 193
Perturbations, 159
Phases, changes in, 288–99
Philosophy, natural, 13
Phosphorescence, 550–51, 555
Phosphors in fluorescent lamps, 550

Photoelectric effect, 562–64, 574
Photograph, production of, 564
Photons, 542, 560, 563
    behavior of, 566
Physics
    as basic science, 13–14
    mathematical structure of, 8
Pigments, 472
    mixing colored, 475–78
Pitch, 343
    of musical sounds, 360, 368
Planck, Max, 560, 563
Planck's constant, 560–62, 561n, 574
Plane, inclined, 21–22, 27
Plane mirrors, 489–90
Plane-polarized wave, 529
Planets
    gravitational field inside, 155–57
    Kepler's laws on motion of, 143–44,
        161
    and universal gravitation, 159–60
Plasma, 193, 249
    atomic structure of, 193
    in everyday world, 248–49
    versus gas, 248
    in nuclear fusion, 625–26
    power of, 249
Plato, 559
Pluto, 159–60
Plutonium, 603, 617–18
Polarization
    electrical, 381–83, 393
    of light waves, 529–34, 537
Polarizer, 531
Poles, magnetic, 419–20
Pollution, thermal, 313
Positive charge, 373
Positrons, 192, 601n, 661
Potential difference, 398–99, 412
Potential energy, 102, 103–5, 113, 172–73,
    388–90, 393
Pound, Robert, 674n
Power, 113, 444
    and energy, 101–2
    fusion, 628
    magnetohydrodynamic, 441–42
    Ocean Thermal Energy Conversion,
        316
    solar, 281–83
    transmission of, 446
    turbogenerator, 441
Precessing elliptical orbit, 675–76
Predictability and chaos, 573
Pressure, 215, 228
    atmospheric, 236–41, 249
    in liquids, 216–18

total, 217n
Primary colors
    additive, 474, 477, 483
    subtractive, 476, 483
Principal axis of lens, 505
Principal quantum number, 582n
Principle, 9. *See also specific*
Probability density function, 585
Procyon, 653
Projectile, 40, 52
Projectile motion. *See also* Satellite motion
    altitude in, 44
    distance traveled in, 42–44
    effect of air resistance on, 45
    and hang-time, 47
    horizontal component of, 40–44
    vertical component of, 41–44
Protactinium, 600
Protons, 190, 374, 593
Pulley, 108
Pupilometrics, 464
Purkinje effect, 464n
Pythagorean theorem, 39n, 643n

Quadrants, 143
Quality of musical sounds, 362–63, 368
Quanta, 375, 560. *See also* Light quanta
Quantization, 560–62
Quantized electrical charge, 375
Quantum effects, 571n
Quantum mechanics, 560, 584–86, 587
Quantum physics, 560
Quantum states, 542
Quantum theory, 560, 574
Quarks, 190, 375n, 593, 609

Radian, 133n
Radiant energy, 275
    absorption and reflection of, 276–77
    emission of, 277–78
Radiation, 275–76, 283, 284
    cooling at night by, 278–79
    effect of, on humans, 606–8, 618
    heat, 276
    infrared, 275
    terrestrial, 280
    ultraviolet, 275
    Van Allen belts, 429
    and X rays, 198, 275, 590–91, 608
Radiation curve, 473
Radiation detector
    bubble chamber, 598
    cloud chamber, 598
    geiger counter, 597
    scintillation counter, 598
Radioactive isotopes, 603–4

Radioactivity
    alpha, beta, and gamma rays, 591–92
    of atoms, 595–96
    half-life, 596–97, 609
Radio broadcasts, 354, 520
Radio waves, 275, 520
Rads, 608
Railroad wheels, 51
Rainbows, 498–501
Rarefaction, 343, 355
    in waves, 330
Rayleigh scattering, 478n
Reaction
    chain, 613, 630
    chemical, 186
Reaction force, 70–76
Reactors
    breeder, 618–19
    nuclear fission, 615–16
Real image, 507, 510
Rebka, Glen, 674n
Record, phonograph, 364–65
Recycling, 629
Red shift, 334
    gravitational, 673–75, 679
Reflection, 486, 510
    diffuse, 490–92
    Fermat's principle of least time, 487, 510
    law of, 487–92, 510
    and mirrors, 489–90
    selective, 471–72
    of sound, 345–46
    total internal, 501–4
Refraction, 486, 510
    cause of, 496–501
    index of, 492n, 526n
    of light, 492–95
    and mirages, 494, 496
    of sound, 346–48
Refrigerator, principle behind, 296
Regelation, 295, 300
Relative humidity, 292
Relativistic momentum, 657–58
Relativity
    general theory of, 145n, 149–50, 668–79
    special theory of, 634–63
Religion, comparison between science and,
    12–13
Rems, 608
Reservoirs, 313
Resistance, electrical, 398, 401, 412
Resonance, 332, 349–50, 355
Rest energy, 658, 663
Resultant, 52, 697
    of vector, 38
Retina, 462

Reverberations, 345
Revolutions per minute (RPM), 49
Ritz, W., 580
Ritz combination principle, 580, 581,
    587
Roemer, Olaus, 469
Roentgen, Wilhelm, 590
Root-mean-square average, 404*n*
Rotational inertia, 118–21, 136
Rotational motion
    angular momentum, 133–35
    center of gravity, 123–27, 136
    center of mass, 123, 136
    centrifugal force, 129–31
    centripetal force, 128–29
    rotational inertia, 118–21
    simulated gravity, 131–33
    torque, 121–22
Rotational speed, 49–50, 52
Rotational velocity, 133*n*
RPM (revolutions per minute), 49
Russell, Bertrand, 16
Rutherford, Ernest, 578, 602
Rutherfordium, 603
Rydberg, J., 580

Safety fuses, 412
Sailboats, 700–2
Satellite, 178
    speed of, 169*n*, 171
Satellite motion, 46–48. *See also* Projectile
        motion
    circular orbit, 168–70
    elliptical orbit, 170–72, 675–76
    and energy conservation, 172–73
    escape speed, 176–78, 178
    gravity in, 48
Saturation, 292
Saturn, 159
Scalar quantity, 25*n*, 37, 52, 697
    speed as, 689
Scalars, and vectors, 697
Scaling, 206–10, 210, 614
Schrödinger, Erwin, 584
    on wave equation, 585, 587
Science
    versus art, 12
    history of, 2
    versus religion, 12–13
    versus technology, 13
Scientific measurement. *See* Measure-
        ment
Scientific method, 9, 15
Scientific notation, 688
Scintillation counter, 598
Second, 686

Selective reflection, 471–72
Selective transmission, 472–73
Self-induction of voltage, 445–46
Semiconductors, 378
Series circuits, 408–10, 412
Shadows, 459–61, 467
Shell, 374
Shock waves, 336–37, 338
Short circuit, 412
SI (Système International), 684–87
Simple harmonic motion, 325
Simulated gravity, 131–33
Simultaneity, 638–39, 663
Sine curve, 326, 338
Sink, 313
Sky, color of, 478–80
Slug, 59*n*, 62*n*
Snell, Willebrord, 496*n*
Snell's law, 496*n*
Snow, C. P., 663
Snow as a conductor, 271
Solar constant, 281–82, 284
Solar eclipse, 460, 460*n*, 467
Solar power, 281–83, 284
Solids
    arches, 205–6
    atomic structure of, 193
    in compression, 203–4
    crystal structure of, 198–99
    density of, 200–1, 210
    elasticity of, 201–2, 210
    scaling, 206–10, 210
    in tension, 203–4
Sonic boom, 336, 338
Sound(s). *See also* Musical sounds
    in air, 343–44, 345
    Doppler effect in, 334
    forced vibrations, 348, 355
    interference in, 350–55
    media transmitting, 344–45
    natural frequency, 348, 355
    origin of, 342
    reflection of, 345–46
    refraction of, 346–48
    resonance, 349–50, 355
    speed of, 345
Sound waves
    energy in, 348
    Fourier analysis of, 364–67, 368
    infrasonic, 343, 355
    ultrasonic, 343, 355
Space
    gravity and new geometry, 676–78
    measurement of, 639–41
    motion of Mercury, 675–76
Spacetime, 639–41, 663, 672

Space travel, 132, 150, 172
    and special relativity, 652–54
    and time, 641–54
Special theory of relativity, 634–35
    addition of velocities, 651–52
    correspondence principle, 661–63
    length contraction, 654–56, 663
    mass, energy, and $E=mc^2$, 658–61
    motion as relative, 635–36
    postulates of, 637–38, 663
    relativistic momentum, 657–58
    simultaneity, 638–39, 663
    spacetime, 639–41, 663
    space travel, 652–54
    time dilation, 641–45, 663
    and twin trip, 645–51
Specific heat capacity, 260–61, 266
Spectral lines, 544
Spectrometer, mass, 621
Spectroscope, 525, 544, 555
Spectrum
    absorption, 547–48, 555
    atomic, 579–80
    electromagnetic, 454–55
    emission, 544–45, 555
Speed, 23, 32, 33
    average, 23–24, 33
    constant, 25
    distinction between velocity and, 25, 37
    of electrons, 405–7
    escape, 176*n*, 176–78, 178
    in free fall, 28–29
    instantaneous, 24, 32
    of light, 453, 457, 457*n*, 492, 496
    linear, 49–50, 52
    maximum falling, 176*n*
    rotational, 49–50, 52
    of satellite, 169*n*, 171
    as scalar quantity, 689
    of sound, 345
    tangential, 49–50, 52
    terminal, 68–69, 76
    vector quantity of, 689
    of waves, 328–29, 338
Spherical aberration, 509
Spring tides, 152–53, 161
Square-cube relationship, 207
Stable equilibrium, 125
Standing wave, 332–33, 338
Static equilibrium, 62–63
Stepped-down voltage, 444
Strassmann, Fritz, 612
Streamlines, 244–47
Stringed musical instruments, 363
Strong interaction, 595
Subcritical mass, 614

Subtractive primary colors, 476, 483
Suchocki, John, 270–71
Sun
  and black holes, 158
  distance to, 6–7
  magnetic field of, 429
  and ocean tides, 150–53
  size of, 7–8
Sunlight, radiation curve of, 473
Sunset, color of, 480
Superconductors, 378, 412, 424
Supercritical mass, 614
Superposition principle, 331
Supersonic craft, 335
Support force, 63
Surface tension of liquids, 226–27, 228
Système International (SI), 684–87

Tachyons, 654
Tacking, 702
T'ai Chi Tu, 574
Tangential speed, 49–50, 52
Tangential velocity, 167–68
Tau, 594
Technology
  anti-noise, 352
  versus science, 13
Television, color on, 474–75, 475n
Temperature, 256–58, 266
  equalization of, 270 –79
  fundamental unit of, 687
  in nuclear fusion, 625–26
Temperature inversion, 310, 310n, 319
Tension, 203–4
Tension force, 63
Terminal speed, 68–69, 76
Terminal velocity, 68
Terrestrial radiation, 280
Tesla, Nikola, 441
Theory, 9, 10, 15. See also specific
Thermal energy, 258
Thermal pollution, 313
Thermodynamics, 304, 319
  absolute zero, 304–5, 319
  adiabatic processes, 308–11, 319
  disordered energy in, 316–17
  entropy, 318–19
  first law of, 306–11, 319
  internal energy, 306, 319
  and meteorology, 309–11
  second law of, 312–16, 319
Thermometers, 257
Thermonuclear fusion, 107, 623, 630
Thermos bottle, and heat transfer, 283
Thermostat, 262–63
Thomson, William (Lord Kelvin), 257, 687

Thorium-232, 618
Thorium-234, 600
Three-dimensional viewing, 532–34
Tides
  on body, 152
  on earth, 153–54
  in the earth and atmosphere, 153–54
  on moon, 154
  neap, 153, 161
  ocean, 150–54
  spring, 152–53, 161
Timbre, 362
Time
  doubling, 703–8
  Fermat's principle of least, 487, 510
  gravitational red shift, 673–75, 679
  in measurement of space, 639–41
  in momentum, 83, 85
  official unit of, 686
  and space travel, 641–54
Time dilation, 641–51, 663
Time rate of change, 23, 26
Torque, 118, 121–22, 133–34, 136
Total internal reflection, 501–4, 510
Total pressure, 217n
Tracers, 603–4
Tracks, railroad, 51
Trajectory, 40–48
Transformers, 442–45, 448
Transistors, 378
Transmission, selective, 472–73
Transmutation of elements, 609
  artificial, 602–3
  natural, 600–2
Transparent materials, 456–57, 467
Transverse motion, 329
Transverse waves, 329–30, 338
Trigonometry, 7
Tritium, 594
Troughs of waves, 326
Turbogenerators, 441
Turns, 443
Twins, and time dilation, 645–51

Ultrasonic sound waves, 343, 355
Ultraviolet radiation, 275
Umbra, 459–60, 467
Uncertainty principle, 569–71, 574
Unified field theory, 376n
United States Customary System (USCS), 684–87
Universal gravitation, 159–60
Universal gravitational constant, 145–46
Universal gravitation law, 144–48, 161
Uranium-233, 618
Uranium-235, 606, 613, 614, 615

Uranium-238, 596, 600, 606, 613, 614, 615, 616, 617, 618
Uranium-239, 617
Uranium dating, 606
Uranus, 159
Useful energy, 110

Vacuum, 236, 239, 240
Vacuum pump, 240
Van Allen, James A., 429n
Van Allen radiation belts, 429
Van de Graaff generator, 392
van der Waals' bonding, 199
Vector(s), 37, 52
  adding, 697
  components of, 39, 39n, 697–700
  resultant of, 38, 697
  and sailboats, 700–2
  and scalars, 697
Vector quantity, 25n, 37–40, 52, 697
  force as, 62
  speed as, 689
Velocity, 23, 32
  acquired in free fall, 28, 33
  angular, 133n
  average, 25, 690
  constant, 25, 669
  distinction between acceleration and, 27
  distinction between speed and, 25, 37
  of electromagnetic waves, 453
  escape, 157–59, 176
  on inclined plane, 689–91
  instantaneous, 25, 690
  rotational, 133n
  and special relativity, 651–52
  tangential, 167–68
  terminal, 68
  as vector quantity, 37
Venus, 676
Vibration, 471, 471n
  definition of, 324
  forced, 348, 355
  interference in, 331–33
  of pendulum, 325, 327
  sympathetic, 349–50
Violent motion, 19
Virtual image, 489, 507, 510
Visible light, 275
Volt, 389, 393, 400
Voltage, 389
  induction of, 436–38
  primary versus secondary, 442–45
  self-induction of, 445–46
  sources of, 400
  stepped-down, 444

stepped-up, 443
Voltmeter, 426
Volume, 76, 687
    determining, 219
    distinction between mass and, 59
von Guericke, Otto, 236
von Jolly, Philipp, 146
von Laue, Max, 198

Water. *See also* Fluids; Liquids
    color of, 482–83
    displacement of, 219
    expansion of, 263–66
    heavy, 616*n*
Watt (W), 102, 398
Watt, James, 102
Wave fronts, 516
Wave function, 585
Wavelength, 326, 328, 338, 567
Wave-particle duality, 564, 566
Waves. *See also* Light waves; Sound
        waves
    bow, 335–36, 338
    carrier, 354, 355

definition of, 324
description of, 325–27
Doppler effect in, 333–34, 338
electromagnetic, 275, 452–55, 467
electron, 583–84
gravitational, 678, 679
heat, 457
interference, 331–33, 521–29, 537
longitudinal, 330–31, 338
matter, 583
motion of, 328
out of phase, 331
plane, 518
plane-polarized, 529
pulse, 343
radio, 275, 520
Schrödinger's equation for, 585, 587
shock, 336–37, 338
speed of, 328–29, 338
standing, 332–33, 338
transverse, 329–30
Weak force, 376*n*
Weak interactions, 596
Web, 204

Weight, 66, 76
    distinction between mass and, 57
    and gravity, 149–50
    as proportional to mass, 66*n*
Weight density, 200*n*, 210
Weightlessness, 149–50, 161
Westinghouse, George, 441
Wheels, railroad, 51
Willey, David, 299
Wind, 274, 309–10
Wind instrument, 363–64
Work, 113, 388*n*
    and energy, 101
Work-energy theorem, 105–6, 113
Wormhole, 158

X rays, 198, 275, 590–91, 608

Yin-yang diagram, 574
Young, Thomas, 523, 523*n*, 560, 565

Zero, absolute, 257, 266, 304–5, 319
Zero acceleration, 669
Zero-point energy, 257*n*, 305*n*

Physical Data

Category	Name	Value
Speeds	Speed of light in a vacuum, c	2.9979×10^8 m/s
	Speed of sound (20°C, 1 atm)	343 m/s
Acceleration	Standard acceleration of gravity, g	9.80 m/s^2
Pressure	Standard atmospheric pressure	1.01×10^5 Pa
Distances	Astronomical unit (A.U.), (average earth-sun distance)	1.50×10^{11} m
	Average earth-moon distance	3.84×10^8 m
	Radius of the sun (average)	6.96×10^8 m
	Radius of earth (equatorial)	6.37×10^6 m
	Radius of earth's orbit	1.50×10^{11} m = 1 AU
	Radius of moon (average)	1.74×10^6 m
	Radius of moon's orbit	3.84×10^8 m
	Radius of Jupiter (equatorial)	7.14×10^7 m
	Radius of hydrogen atom (approx.)	5×10^{-11} m
Masses	Mass of sun	1.99×10^{30} kg
	Mass of earth	5.98×10^{24} kg
	Mass of moon	$7.36 = 10^{22}$ kg
	Mass of Jupiter	1.90×10^{27} kg
	Mass of proton, m_p	$1.6726231 \times 10^{-27}$ kg 938.27231 MeV
	Mass of neutron, m_n	$1.6749286 \times 10^{-27}$ kg 939.56563 MeV
	Mass of electron, m_c	$9.1093897 \times 10^{-31}$ kg 0.51099906 MeV
Charge	Charge of electron, e	1.602×10^{-19} C
Other Constants	Gravitational constant, G	6.67259×10^{-11} m^3/kg·s^2
	Planck's constant, h	$6.6260755 \times 10^{-34}$ J·s $4.1356692 \times 10^{-15}$ eV·s
	Avogadro's number, N_A	6.0221367×10^{23}/mol
	σ	5.67051×10^{-8} W/m^2·K^4

Standard Abbreviations

A	ampere	g	gram	min	minute
amu	atomic mass unit	h	hour	mph	mile per hour
atm	atmosphere	hp	horsepower	N	newton
Btu	British thermal unit	Hz	hertz	Pa	pascal
C	coulomb	in.	inch	psi	pound per square inch
°C	degree Celsius	J	joule	s	second
cal	calorie	K	kelvin	u	unified atomic mass unit
eV	electron volt	kg	kilogram	V	volt
°F	degree Fahrenheit	lb	pound	W	watt
ft	foot	m	meter	Ω	ohm